Introduction to Analysis and Design of Equilibrium Staged Separation Processes

2nd Edition

Introduction to Analysis and Design of Equilibrium Staged Separation Processes

2nd Edition

RAJINDER PAL

Professor of Chemical Engineering
University of Waterloo
Ontario, Canada

CWP
Central West Publishing

This edition has been published by Central West Publishing, Australia

For more information about the books published by Central West Publishing, please visit https://centralwestpublishing.com

Disclaimer

Every effort has been made by the publisher and author while preparing this book; however, no warranties are made regarding the accuracy and completeness of the content. The publisher and author disclaim without any limitation all warranties as well as any implied warranties about sales, along with fitness of the content for a particular purpose. Citation of any website and other information sources does not mean any endorsement from the publisher and author. For ascertaining the suitability of the contents contained herein for a particular lab or commercial use, consultation with the subject expert is needed. In addition, while using the information and methods contained herein, the practitioners and researchers need to be mindful for their own safety, along with the safety of others, including the professional parties and premises for which they have professional responsibility. To the fullest extent of law, the publisher and author are not liable in all circumstances (special, incidental, and consequential) for any injury and/or damage to persons and property, along with any potential loss of profit and other commercial damages due to the use of any methods, products, guidelines, procedures contained in the material herein.

NATIONAL LIBRARY OF AUSTRALIA

A catalogue record for this book is available from the National Library of Australia

ISBN (print): 978-1-922617-31-6

To the memory of my parents,
Smt Karma-Bhari and Shri Khushal Chand

Contents

Preface

Author

Chapter 1 Introduction to Equilibrium Stage Operations 1

1.1 Learning Outcomes 1
1.2 Introduction 1
1.3 Concept of an Equilibrium Stage 4
1.3.1 Distillation 4
1.3.2 Absorption 5
1.3.3 Stripping 6
1.3.4 Liquid-liquid Extraction 7
1.3.5 Leaching 9
1.4 Examples of Industrial Applications of Equilibrium Staged Separation Processes 11
1.4.1 Dehydration of Natural Gas 11
1.4.2 CO_2 Capture and Storage 12
1.4.3 Vegetable Oil Extraction 13
1.4.4 Production of Penicillin Antibiotic 14
1.5 Multi-Stage Contacting of Two Phases 16
1.6 Equilibrium between Phases 18
1.6.1 Definition and Criteria for Phase Equilibrium 18
1.6.2 Relationship between the Compositions of Different Phases at Equilibrium 19
1.6.3 Non-ideal Liquid Solutions 21
1.6.4 Limiting Cases of Non-ideal Solutions 26
1.6.5 More on Henry's Law 28

Notation 29

Tutorial Problems with Solution 30

Short Review Questions 39

Practice Problems 40

Further Reading 44

Chapter 2 Single-Stage Gas Absorption 45

2.1 Learning Outcomes 45
2.2 Introduction 45
2.3 Driving Force for Mass Transfer in Gas Absorption 46
2.4 Analysis of Single-Stage Gas Absorption: Dilute Systems 47
2.5 Analysis of Single-Stage Gas Absorption: Concentrated Systems 51

Notation 55

	Tutorial Problems with Solution	55
	Short Review Questions	67
	Practice Problems	68
Chapter 3	**Single-Stage Liquid Stripping**	**71**
	3.1 Learning Outcomes	71
	3.2 Introduction	71
	3.3 Driving Force for Mass Transfer in Liquid Stripping	72
	3.4 Analysis of Single-Stage Liquid Stripping: Dilute Systems	73
	3.5 Analysis of Single-Stage Liquid Stripping: Concentrated Systems	77
	Notation	79
	Tutorial Problems with Solution	80
	Short Review Questions	85
	Practice Problems	85
Chapter 4	**Multi-Stage Gas Absorption**	**87**
	4.1 Learning Outcomes	87
	4.2 Introduction	87
	4.3 Analysis of Counter-current Multi-Stage Absorption Process	88
	4.4 Design of Multi-Stage Absorption Towers	93
	4.4.1 Graphical Approach: McCabe-Thiele Method	93
	4.4.2 Analytical Approach: Kremser Equations	95
	4.4.3 Minimum L/V Requirement and Pinch Condition	98
	Notation	99
	Tutorial Problems with Solution	100
	Short Review Questions	112
	Practice Problems	113
Chapter 5	**Multi-Stage Liquid Stripping**	**115**
	5.1 Learning Outcomes	115
	5.2 Introduction	115
	5.3 Design of Multi-Stage Liquid Stripping Towers	117
	5.3.1 Graphical Approach: McCabe-Thiele Method	118
	5.3.2 Analytical Approach: Kremser Equations	118
	5.3.3 Minimum V/L (Maximum L/V) Requirement and Pinch Condition	120
	Notation	121

Tutorial Problems with Solution 122

Short Review Questions 131

Practice Problems 132

Chapter 6 Some Definitions and Basic Concepts related to Distillation 135

6.1 Learning Outcomes 135
6.2 Introduction 135
6.3 Vapor Pressure 136
6.3.1 Clapeyron and Clausius-Clapeyron Equations 139
6.3.2 Antoine Equation 140
6.4 Raoult's Law 140
6.5 Volatility and Relative Volatility 141
6.6 Bubble Point and Dew Point 143
6.6.1 Definitions of Bubble and Dew Points 143
6.6.2 Different Classes of Bubble and Dew Points Calculations 146
6.6.3 Bubble and Dew Points Calculations Using Raoult's Law 147
6.6.4 Bubble and Dew Points Calculations Using K_i Values 149
6.6.5 Importance of Bubble and Dew Points Calculations 152
6.7 Construction of P-x-y and T-x-y Phase Diagrams and y-x Equilibrium Curve 153
6.7.1 Construction of Phase Diagrams and Equilibrium Curve Using Raoult's Law 153
6.7.2 Construction of Phase Diagrams and Equilibrium Curve Using K_i Values 154

Notation 155

Tutorial Problems with Solution 155

Short Review Questions 163

Practice Problems 163

Chapter 7 Mass and Energy Balances for Open (Flow) Systems 167

7.1 Learning Outcomes 167
7.2 Introduction 167
7.3 Mass Balance 168
7.4 Energy Balance 170
7.5 Estimation of Enthalpy of Fluids 173
7.5.1 Pure Single-Component Substances 173
7.5.2 Multi-component Gas Mixtures and Liquid Solutions 175

Notation 176

Tutorial Problems with Solution 177

Short Review Questions 180

Practice Problems 181

Chapter 8 Single-Stage Flash Distillation 183

8.1 Learning Outcomes 183
8.2 Introduction 183
8.3 Analysis of Binary Flash Distillation 184
8.3.1 P, ψ - Flash Distillation 184
8.3.2 P, T-Flash Calculation 194
8.4 Analysis of Multi-component Flash Distillation 194

Notation 197

Tutorial Problems with Solution 198

Short Review Questions 205

Practice Problems 205

Chapter 9 Multi-Stage Column Distillation 207

9.1 Learning Outcomes 207
9.2 Introduction 207
9.3 Continuous Single-Stage Distillation Processes 207
9.4 Cascading of Stages 210
9.5 Multi-Stage Distillation Column 215
9.6 Analysis and Design of Multi-Stage Distillation Column 219
9.6.1 External Balances 221
9.6.2 Internal Balances 223
9.6.3 McCabe-Thiele Method of Analysis and Design of Distillation Column 228
9.6.4 Special Applications of Column Distillation 252

Notation 269

Tutorial Problems with Solution 271

Short Review Questions 282

Practice Problems 283

Appendix: The Smoker Equations 286

Chapter 10 Approximate Analytical Method of Designing Distillation Columns 291

10.1 Learning Outcomes 291
10.2 Introduction 291
10.3 Estimation of the Number of Stages 291
10.3.1 Fenske Equation for N_{min} at total reflux 291
10.3.2 Underwood Equation for R_{min} 295
10.3.3 Gilliland Correlation for the Estimation of Number of Theoretical Stages 296

10.3.4 Fenske-Underwood-Gilliland (FUG) Method of Designing Distillation Columns 296
10.4 Estimation of Actual (Non-Equilibrium) Number of Stages 297
10.5 Estimation of Tray/Plate Spacing and Height of Column 300
10.6 Different Types of Trays/Plates 301
10.7 Estimation of Column Diameter 304
10.7.1 Estimation of Flooding Velocity of Vapor 306
10.7.2 Estimation of Column Diameter 309
10.8 Preliminary Mechanical Design of a Distillation Column 310
10.8.1 Selection of Material of Column Construction 310
10.8.2 Shell Thickness of Distillation Column 310

Notation 318

Tutorial Problems with Solution 320

Short Review Questions 325

Practice Problems 326

Chapter 11 Ponchon-Savarit Graphical Method of Designing Distillation Columns 329

11.1 Learning Outcomes 329
11.2 Introduction 329
11.3 Enthalpy-Composition Diagram for Binary Mixtures 330
11.4 Construction of Enthalpy-Composition Diagram for Binary Mixtures 332
11.4.1 Enthalpy of Liquid Solutions at Bubble Point Temperature 333
11.4.2 Enthalpy of Vapor mixtures at Dew Point Temperature 333
11.4.3 Example: Construction of Enthalpy-Composition Diagram for Benzene-Toluene System 334
11.5 Ponchon-Savarit Graphical Method of Analysis and Design of Distillation Columns 337
11.5.1 Enriching Section 337
11.5.2 Stripping Section 341
11.5.3 Feed Line 345
11.5.4 Estimation of Total Number of Stages using Ponchon-Savarit Diagram 346
11.5.5 Limiting Reflux Ratios – Total Reflux and Minimum Reflux 347

Notation 349

Tutorial Problems with Solution 351

Short Review Questions 359

Practice Problems 360

Chapter 12 Batch Distillation – Without and With Reflux 363

12.1 Learning Outcomes 363

12.2	Introduction	363
12.3	Analysis of Batch Distillation	364
12.4	Batch Distillation with Reflux	368
12.5	Batch Steam Distillation	372
Notation		374
Tutorial Problems with Solution		375
Short Review Questions		383
Practice Problems		383
Chapter 13	**Multicomponent Distillation**	**387**
13.1	Learning Outcomes	387
13.2	Introduction	387
13.3	Separation of Multicomponent Mixtures Using a Single Distillation Column	388
13.3.1	Key and Non-Key Components	388
13.3.2	Class 1 and Class 2 Separations	389
13.3.3	Approximate Methods of Designing Multicomponent Distillation Columns	390
13.3.4	Rigorous Calculation Procedure for Rating the Performance of Multicomponent Distillation Columns	397
13.4	Separation of Multicomponent Mixtures into Pure Components Using a Sequence of Distillation Columns	400
13.5	Heuristics Approach for the Development of Separation Sequence	406
Notation		407
Tutorial Problems with Solution		409
Short Review Questions		421
Practice Problems		422
Chapter 14	**Liquid-Liquid Extraction**	**425**
14.1	Learning Outcomes	425
14.2	Introduction	426
14.3	Triangular Phase Diagrams	426
14.3.1	Partially Miscible Solvent-Diluent Pair	426
14.3.2	Completely Immiscible Solvent-Diluent Pair	429
14.4	Single-Stage Liquid-Liquid Extraction	431
14.4.1	Analysis of a Single-Stage LLX Process: Immiscible Solvent-Diluent Pair	431
14.4.2	Analysis of a Single-Stage LLX Process: Partially Miscible Solvent-Diluent Pair	433
14.5	Multi-Stage Liquid-Liquid Extraction	435
14.5.1	Analysis of a Multi-Stage LLX Process: Immiscible Solvent-Diluent Pair	435

14.5.2 Analysis of a Multi-Stage LLX Process: Partially Miscible Solvent-Diluent Pair 439

Notation 448

Tutorial Problems with Solution 450

Short Review Questions 461

Practice Problems 461

Chapter 15 Solid-Liquid Leaching 463

15.1 Learning Outcomes 463
15.2 Introduction 463
15.3 Phase Diagrams for Solid-Liquid Leaching 464
15.4 Analysis of a Single-Stage Solid-Liquid Leaching Process 465
15.5 Multi-Stage Solid-Liquid Leaching 467
15.5.1 Analysis of a Multi-Stage Solid-Liquid Leaching Process 467
15.5.2 McCabe-Thiele Graphical Method of Estimating the Number of Equilibrium Stages for a Multi-Stage Counter-current Leaching Process 474
15.5.3 Application of Kremser Equation 477

Notation 478

Tutorial Problems with Solution 480

Short Review Questions 493

Practice Problems 493

Chapter 16 Analysis and Design of Continuous Contacting Packed Columns 495

16.1 Learning Outcomes 495
16.2 Introduction 495
16.3 Packings and Packing Materials 495
16.3.1 Random Packings 496
16.3.2 Structured Packings 497
16.3.3 Liquid Distribution 497
16.4 Height of Packed Columns 498
16.4.1 HETP Method 498
16.4.2 HTU-NTU Method 500
16.4.3 Relationship between HETP and HTU 506
16.5 Diameter of Packed Columns 507
16.5.1 Flooding in Packed Columns 508
16.5.2 Flooding Curve and its Interpretation 508
16.5.3 Estimation of Column Diameter using the Flooding Curve/Equation 512

Notation 513

Tutorial Problems with Solution 514

Short Review Questions 525

Practice Problems 526

Index **529**

Preface

This book is written with second year chemical engineering undergraduate students in mind. Chemical engineering undergraduate students are generally taught Equilibrium Stage Operations in their second year. This is the first time they are introduced to equilibrium stage-based separation processes. Their background at this stage is quite limited.

A number of excellent textbooks are available on separation processes. They cover the analysis and design of both equilibrium and non-equilibrium (rate-based) separation processes. However, the available textbooks are generally written for more advanced level students. The second year undergraduate students in chemical engineering find the available textbooks to be too difficult to comprehend. The students are overwhelmed with the amount and level of difficulty of information presented in the available textbooks. The present textbook is written to overcome these difficulties. The goal is to present the equilibrium stage concepts and operations in a manner comprehensible to second year chemical engineering students with little or no prior exposure to separation processes.

The book consists of sixteen chapters. It covers single-stage and multi-stage absorption and stripping, flash distillation, multi-stage column distillation, batch distillation with and without reflux, liquid-liquid extraction, and solid-liquid leaching. Although the book is focussed on equilibrium staged separation processes, the final chapter (chapter 16) is devoted to the analysis and design of continuous contacting packed columns as packed columns are becoming increasingly important in practical applications.

Each chapter begins with the student learning outcomes. Near the end of each chapter, a tutorial session is included where a number of problems are solved to illustrate the applications of concepts and theory covered in the chapter. Notation is included separately in each chapter. For the benefit of students, short review questions and practice problems are included at the end of each chapter.

It is hoped that the students will be able to understand the concepts and material presented in this textbook without difficulty. As separation processes are at the core of chemical engineering practice, it is important that students develop a liking for this subject. Understanding and liking of a subject go hand in hand. Students are inspired to pursue a particular field at a more advanced level only if they are able to develop a good understanding and liking for the subject.

Changes Made to the First Edition

- Several chapters have been revised
- The number of practice problems have been doubled
- The number of solved tutorials problems have been increased by 50 percent

I would like to thank my wife Archana and my children Anuva and Arnav for their love and constant support.

Waterloo, January 2022 — Rajinder Pal

Author

Rajinder Pal is a Professor in the Department of Chemical Engineering at the University of Waterloo since 1991, and a Fellow of the Chemical Institute of Canada. He received degrees in Chemical Engineering from I.I.T. Kanpur (B.Tech. 1981) and the University of Waterloo (Ph.D. 1987), followed by a post-doctoral fellowship at the University of Alberta. His research interests include complex fluid rheology, interfacial and transport phenomena, physical properties of composite materials, and non-equilibrium thermodynamics. He has published more than 150 refereed journal articles, 4 books, and 5 book chapters on these and related subjects, including 8 papers in engineering education journals. He has taught over ten different undergraduate and graduate courses related to equilibrium staged separation processes, fluid mechanics, heat and mass transfer, physical chemistry and thermodynamics.

Professor Pal has been the recipient of several awards including the University of Waterloo prestigious Distinguished Teacher Award (2020), Sandford Fleming Foundation Teaching Excellence Award (twice 2001 and 2017), an Outstanding Performance Award from the University of Waterloo (2009), and the Syncrude Canada Innovation Award from the Canadian Society for Chemical Engineering (1998). He is a Fellow of the Chemical Institute of Canada since 1999. He serves on several journal editorial boards and was an Associate Editor of the Canadian Journal of Chemical Engineering for twelve years. He is a registered Professional Engineer in the province of Ontario, Canada.

Chapter 1

Introduction to Equilibrium Stage Operations

1.1 LEARNING OUTCOMES

Upon completion of this chapter, students will be able to:

- Describe the practical importance of separation processes and the relevance of equilibrium stage operations in separation processes
- Explain the concept of equilibrium stage
- Describe different types of equilibrium stage operations (distillation, absorption, stripping, liquid-liquid extraction, and leaching) along with the direction of mass transfer of species in these operations
- Explain the concept of cascading of equilibrium stages in multistage operations
- Explain phase equilibrium and related concepts such as: thermodynamic criterion for phase equilibrium; the relationship between compositions of two phases at equilibrium such as Raoult's law; and the difference between ideal and non-ideal solutions

1.2 INTRODUCTION

Equilibrium stage operations are encountered in separation processes. Separation processes are important from a practical point of view. Many materials in nature exist as mixtures of two or more components. For example, crude oil is a mixture of many components. Most metals found in nature exist as ores, that is, mixtures of metal and solid rock. Vegetable oils are found in oil-bearing seeds (e.g., soybeans and sunflower seeds) as mixtures of oil and oil-free solid material. In many man-made processes, gaseous and liquid streams are generated consisting of mixtures of different components.

From a practical point of view, we would like to separate the mixture into its components or different fractions. There could be different reasons to separate the mixture into its components. In some cases, we are interested in the components that are valuable and therefore, their recovery from the mixture is important for economic reasons. In other cases, the mixture may contain component or components that are toxic or environmentally harmful and therefore they must be removed before the mixture stream is released to the environment.

How can we accomplish separation of a mixture into its components? Consider, for example, a liquid solution consisting of three components (A, B, and C). We would like to separate it into three parts: pure A, pure B, and pure C. How do we achieve it?

Let us assume that the boiling points of the three components at atmospheric pressure are very different from each other, say A boils at 50 °C, B boils at 100 °C, and C boils at 150 °C. We will first heat the mixture contained in a pot at atmospheric pressure to about 50 °C. At this temperature, species "A" will begin to boil (form vapor bubbles) as vapor pressure of species "A" becomes equal to the atmospheric pressure acting on the liquid solution. If we continue to heat the mixture at this temperature, all of "A" will vaporize in the form of bubbles. We will condense the produced vapor of "A" by cooling it and store it as pure liquid "A". So this way we would have separated "A" from the rest of the mixture. The temperature remains nearly constant at 50 °C until all of "A" is boiled off. After all of "A" has vaporized, we will continue to heat the remaining mixture. Now the temperature will rise with heating until we have reached the boiling point of component "B", that is, 100 °C. At this temperature, "B" will begin to come off the liquid solution in the form of vapor bubbles. We will continue to heat until all of "B" has come off (temperature will remain nearly constant at 100 °C until all of "B" has come out). The vapors of "B" are cooled and condensed. Thus we have achieved separation of 3 component mixture into 3 pure components. The temperature-time plot for this process is shown in Figure 1.1.

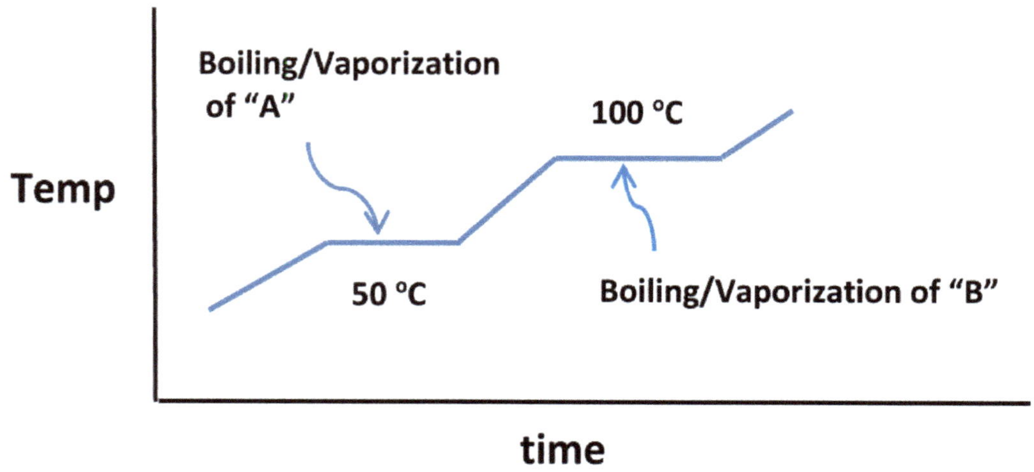

Figure 1.1 Temperature versus time plot of boiling of 3-component liquid solution with different boiling points of components.

It is important to realize that in reality the components of a solution do not act independently of each other and the vapor coming off at any time is a mixture rather than a pure component. The initial vapors will be rich in "more volatile components". With progress in time and continued heating, vapor coming off the liquid solution will become leaner in "more volatile components". Also, the temperature versus time plot is more continuous in nature. This is what generally happens when we are separating a multicomponent solution like crude oil.

In many practical applications, we are not interested in pure components. For example, crude oil has just too many components covering a wide range of molecular weights and hence boiling points. From a practical point of view, we are interested in separating the crude oil into different fractions (see Figure 1.2) covering different ranges of molecular weights: low range of molecular weights (gasoline or petrol), intermediate range of molecular weights (diesel oil), and high range of molecular weights (bitumen).

This process of separation of a liquid solution into pure components or different fractions by vaporization/boiling and subsequent condensation is referred to as "Distillation". Obviously distillation requires heating and the boiling points of different components must be different. If the boiling points of different components of a liquid mixture are the same, that is, we have an azeotropic mixture, then the distillation process will not work. Then the vapor coming off from the liquid solution will have the same composition as that of the liquid solution itself.

Note that the terms vaporization and boiling are often used interchangeably although there are subtle differences. Vaporization can happen from the surface alone as in the case of an evaporation process (evaporation is a surface phenomenon) whereas boiling is a bulk phenomenon that involves formation of vapor bubbles in the liquid solution. Vaporization can occur below the boiling point whereas boiling occurs only at or above the boiling point. Distillation involves boiling.

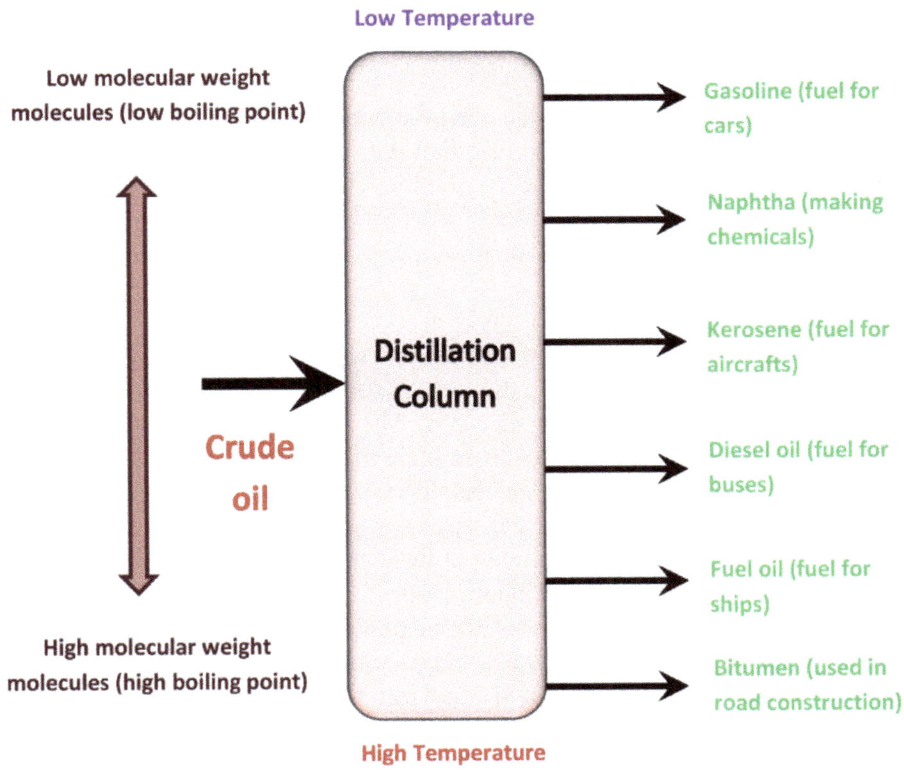

Figure 1.2 Separation of crude oil in a distillation column into different fractions.

Not all separation processes are based on boiling/condensation phenomena. In fact a good number of separation processes do not involve any boiling/condensation at all. Examples of separation processes not based on the difference in boiling points of components are: absorption, stripping, liquid-liquid extraction, and leaching. In these non-distillation separation processes, the raw material or the feed consists of a carrier material plus a solute, that is, Feed = Carrier + Solute. Usually the solute is present in small amounts and it is this solute that we want to separate from the feed or carrier material. We want to separate this solute from the mixture due to reasons such as: it may be a valuable component, or it may be a toxic chemical or pollutant.

How do we remove solute from the feed? By bringing the feed in contact with a solvent and mixing the two thoroughly. The solvent is generally not miscible with the carrier material of the solute so that we have a two-phase mixture. However, the solute has high affinity for the solvent. In fact, one would select a solvent such that the solute has high affinity for it. During the mixing process, the solute is preferentially transferred to the solvent. This way we are able to successfully remove or recover the solute from the feed.

In summary, the equilibrium stage separation processes can be divided into two broad groups: 1) distillation process, and 2) non-distillation processes such as absorption, stripping, liquid-liquid extraction, and leaching. In distillation, the components are separated by exploiting the difference in boiling points whereas in non-distillation separation processes, the solute is separated from the carrier by exploiting the difference in solubilities or affinities of solute between the two phases (carrier versus solvent). It is important to realize that distillation or non-distillation, all separation processes based on equilibrium stage operations involve contact of two phases: distillation involves contact of vapor and liquid phases, absorption and stripping involve contact of gas and liquid phases, liquid-liquid extraction involves contact of two immiscible or partially miscible liquids, and leaching involves contact of solid and liquid phases.

1.3 CONCEPT OF AN EQUILIBRIUM STAGE

An equilibrium stage could be defined as a device where two immiscible streams (phases) enter the device, are intimately mixed, and leave the device at equilibrium with each other. During the mixing and contacting of two phases, the transfer of chemical species (solute, chemical components) takes place from one phase to another.

1.3.1 Distillation

Figure 1.3 represents a distillation type equilibrium stage. Here "A" is assumed to be the more volatile (lower boiling point) component in a binary mixture of "A" and "B". The vapor stream leaving the stage is *in equilibrium* with the liquid stream leaving the stage. The mole fraction of more volatile species y_A in the vapor stream leaving the stage is higher than the mole fraction y_A in the vapor stream entering the stage, that is, enrichment of vapor with more volatile species "A" takes place in the stage. Also, the mole fraction x_A of species "A" in the liquid phase entering the stage is higher than the mole fraction x_A in the liquid phase leaving the stage, that is, depletion of more volatile species "A" occurs in the liquid phase. In other words, species "A" is transferred to the vapor phase from the incoming liquid phase and species "B" is transferred to the liquid phase from the incoming vapor phase. Figure 1.4 shows the directions of mass transfer. In a multi-stage process, the vapor phase could become pure component "A" and the liquid phase may become pure component "B" due to repeated vaporization and condensation. This is how separation of a binary mixture into pure components occurs in a distillation process.

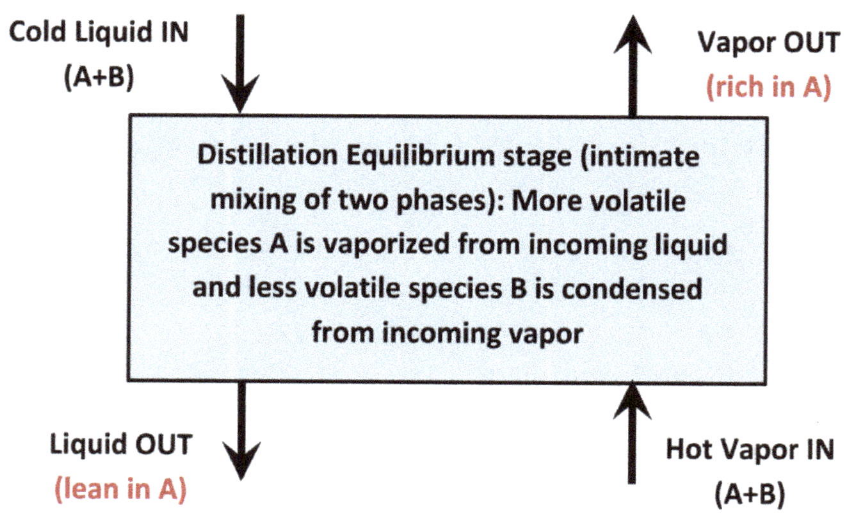

Figure 1.3 Equilibrium stage of a binary distillation process.

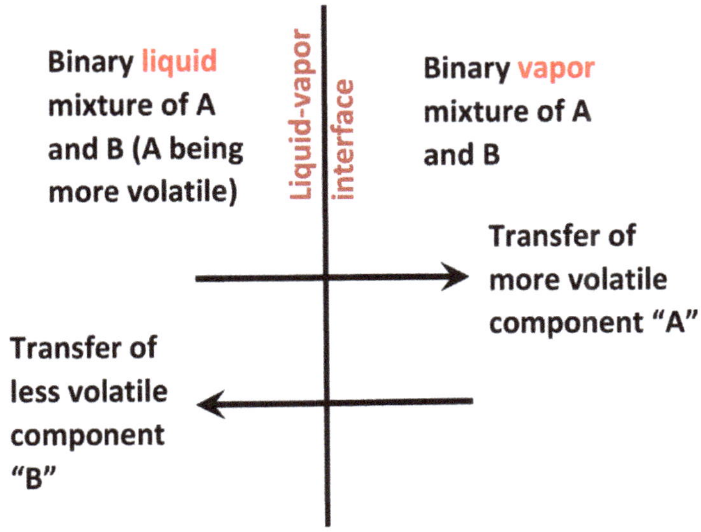

Figure 1.4 Directions of mass transfer of components in a binary distillation process.

1.3.2 Absorption

Figure 1.5 represents an absorption type equilibrium stage. Here "A", the solute present in the gas phase, is to be recovered by contacting with the liquid solvent. The incoming gas and solvent streams are intimately mixed. The gas stream leaving the stage is *in equilibrium* with the liquid stream leaving the stage. Furthermore, y_A in the gas stream leaving the stage is smaller than the y_A in the gas stream entering the stage, that is, depletion of solute "A" from the gas stream takes place during mixing in the stage. Also, x_A in the liquid phase leaving the stage is higher than the x_A in the liquid phase entering the stage. In other words, species "A" is transferred from the gas phase to the liquid phase. Figure 1.6 shows the direction of mass transfer in absorption process.

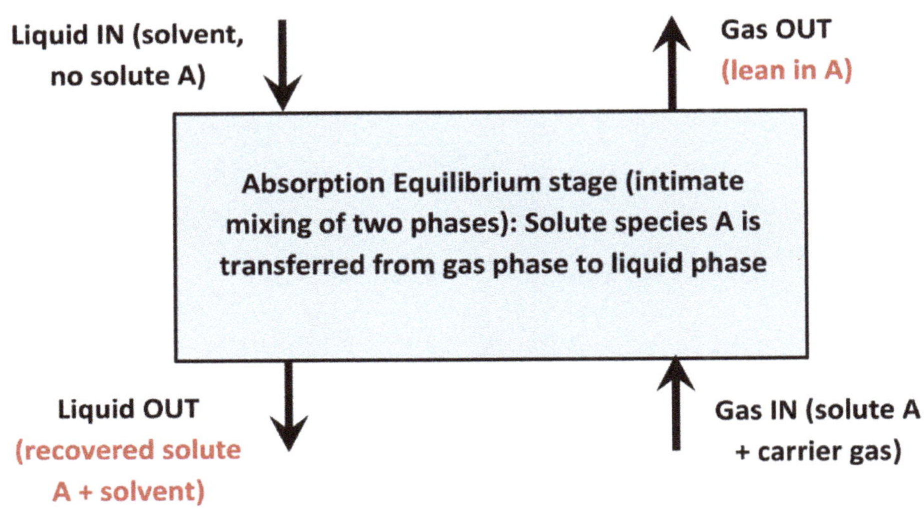

Figure 1.5 Equilibrium stage of an absorption process.

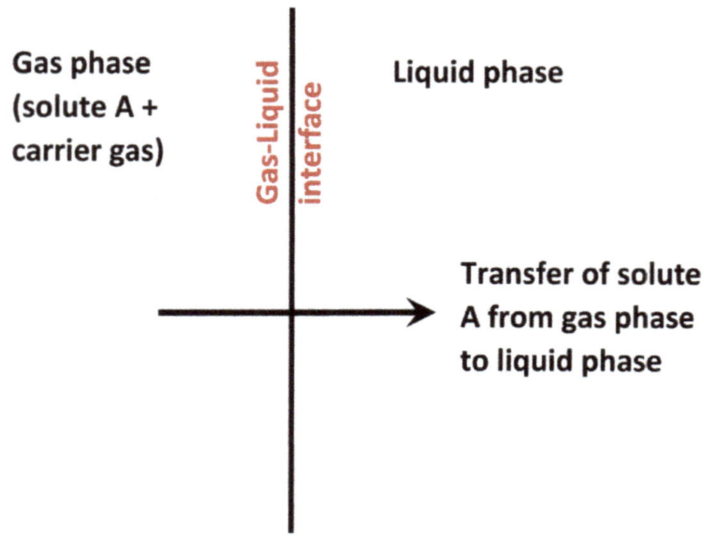

Figure 1.6 Direction of mass transfer of solute in an absorption process.

1.3.3 Stripping

Stripping is a separation process where the solute is transferred from the liquid phase to the gas phase. Now the direction of solute transfer is opposite to that of gas absorption. Figure 1.7 represents a stripping equilibrium stage. The incoming liquid stream containing solute species "A" is intimately mixed with the gas-stream such that the gas stream leaving the stage is *in equilibrium* with the liquid stream leaving the stage. Furthermore, y_A in the gas stream leaving the stage is larger than the y_A in the gas stream entering the stage. Also, x_A in the liquid phase leaving the stage is smaller than the x_A in the liquid phase entering the stage. In other words, solute species "A" is transferred from the liquid phase to the gas phase. Figure 1.8 shows the direction of mass transfer in stripping process.

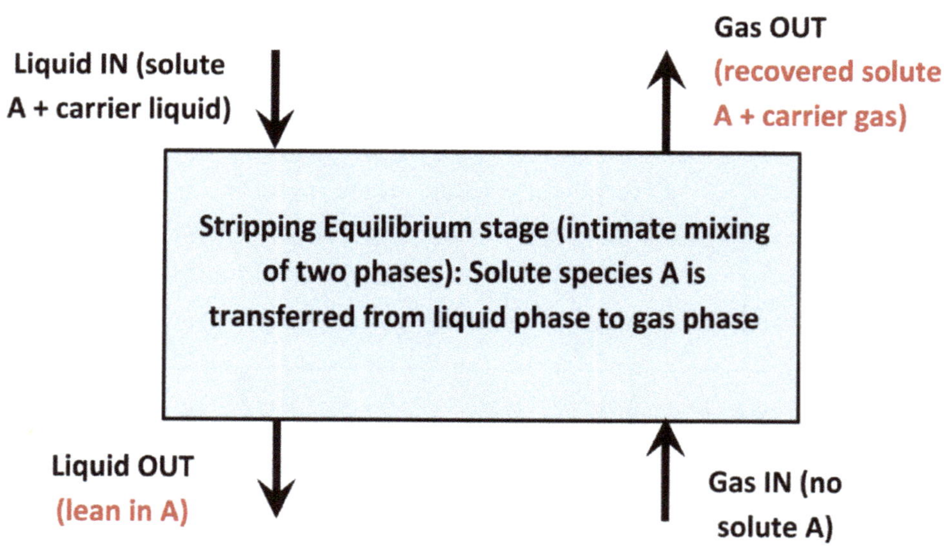

Figure 1.7 Equilibrium stage of a stripping process.

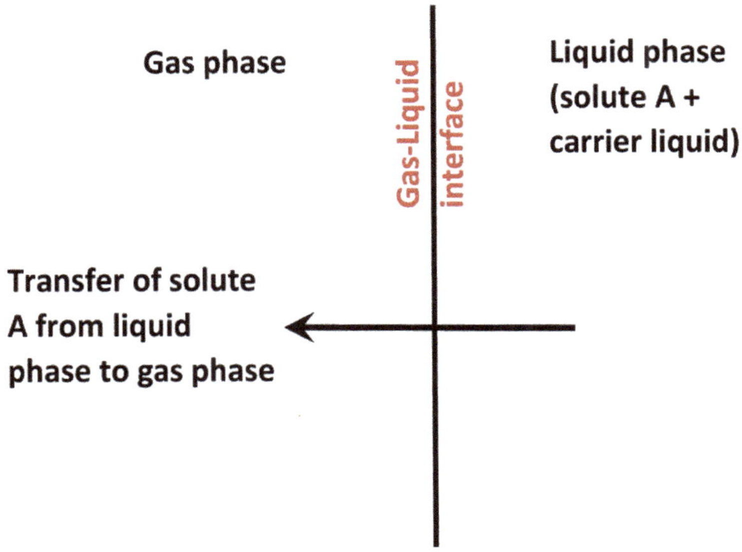

Figure 1.8 Direction of mass transfer of solute in a stripping process.

1.3.4 Liquid-Liquid Extraction

In Liquid-Liquid extraction, the solute (species A) is present in the feed liquid phase (typically aqueous phase) and it is to be recovered in the solvent phase (typically organic solvent). The two phases are mixed intimately resulting in the formation of a dispersion of droplets of one liquid phase (solvent phase) in another (aqueous phase). The two carrier phases of solute are either completely immiscible or only partially miscible. The solvent is selected such that the solute has a high affinity for it as compared with the feed carrier liquid. Figure 1.9 represents a liquid-liquid extraction equilibrium stage. The feed phase (Liquid phase I) containing solute "A" is mixed thoroughly with the solvent phase (Liquid II). The liquid-phase II leaving the stage is *in equilibrium* with the liquid phase I leaving the stage. Note that the solute-depleted liquid phase I leaving the process is referred to as "raffinate" and the solute-rich liquid phase II leaving the process is called "extract". Obviously the concentration of solute y_A in the liquid phase II leaving the stage is larger than the y_A in the liquid phase II entering the stage. Also, x_A in the liquid phase I leaving the stage is less than the x_A in the liquid phase I entering the stage. In other words, species "A" is transferred from the liquid phase I to the liquid phase II. Figure 1.10 shows the direction of mass transfer in the extraction process.

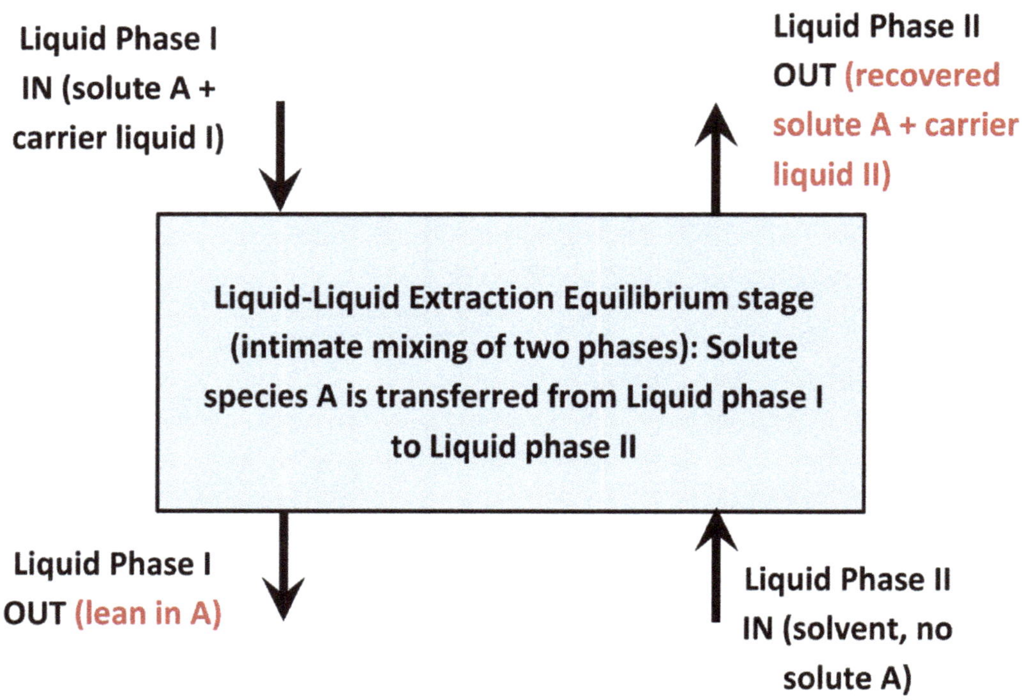

Figure 1.9 Equilibrium stage of a liquid-liquid extraction process.

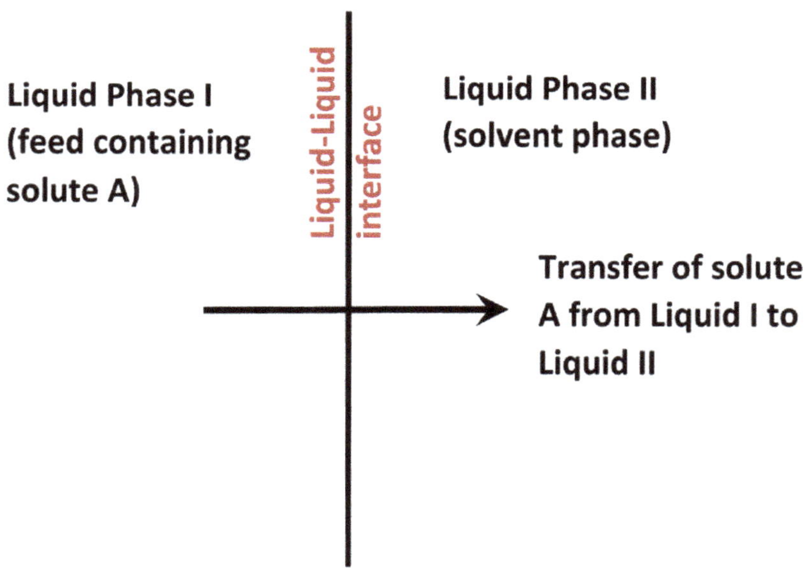

Figure 1.10 Direction of mass transfer of solute in a liquid-liquid extraction process.

Typically two pieces of equipment are involved in a single-stage liquid-liquid extraction process, namely a mixer and a settler, as shown in Figure 1.11. The two liquid streams are mixed in the mixing vessel using a mixer/stirrer and the mixture is passed to a settling vessel for the separation of extract and raffinate phases.

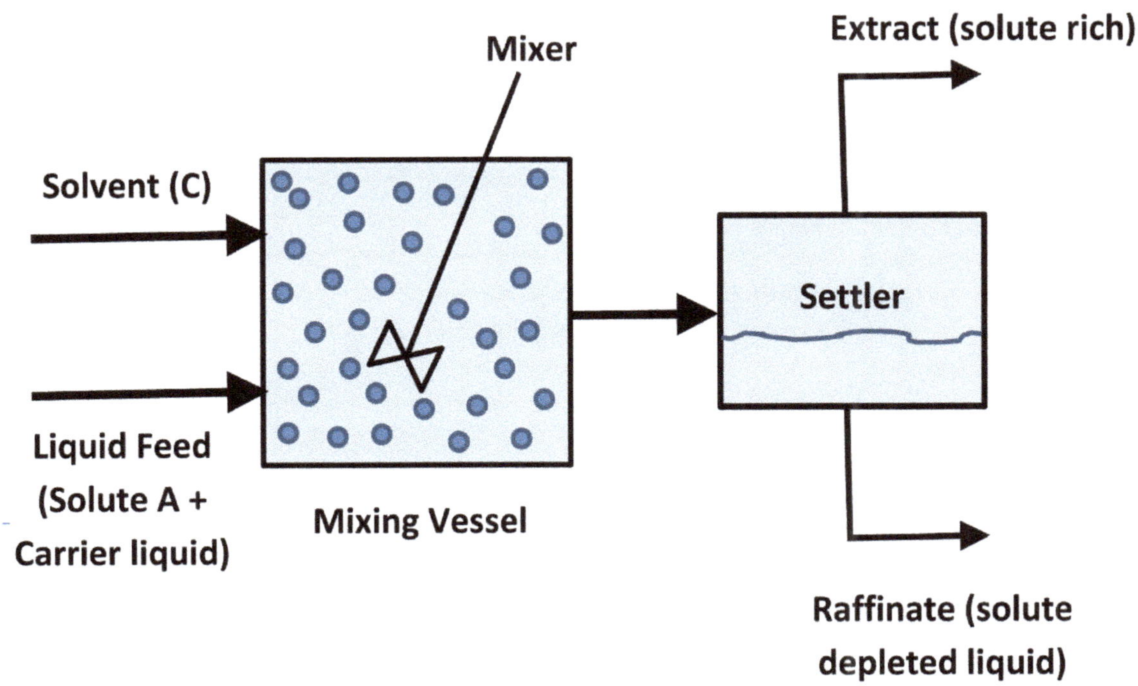

Figure 1.11 Equipment diagram of a single-stage liquid-liquid extraction process.

1.3.5 Leaching

Leaching, also referred to as solid-liquid extraction, is a separation process where the solute present in the solid phase is transferred from the solid phase to the liquid phase. The solid feed in particulate form is contacted with a solvent (liquid) phase. Due to high affinity of solute for the solvent, the solute diffuses from the solid phase to the solvent resulting in separation.

Figure 1.12 represents a leaching equilibrium stage. The incoming liquid (solvent) stream is intimately mixed with the solid particles containing the solute species "A". The depleted solid phase (slurry) leaving the stage is *in equilibrium* with the solvent stream leaving the stage. The concentration of solute species "A" in the solvent stream leaving the stage is larger than the solute concentration in the solvent stream entering the stage. Also, the solute concentration in the slurry phase leaving the stage is smaller than the solute concentration in the solid phase entering the stage. In other words, solute species "A" is transferred from the solid phase to the liquid phase. Figure 1.13 shows the direction of mass transfer in the solid-liquid extraction process.

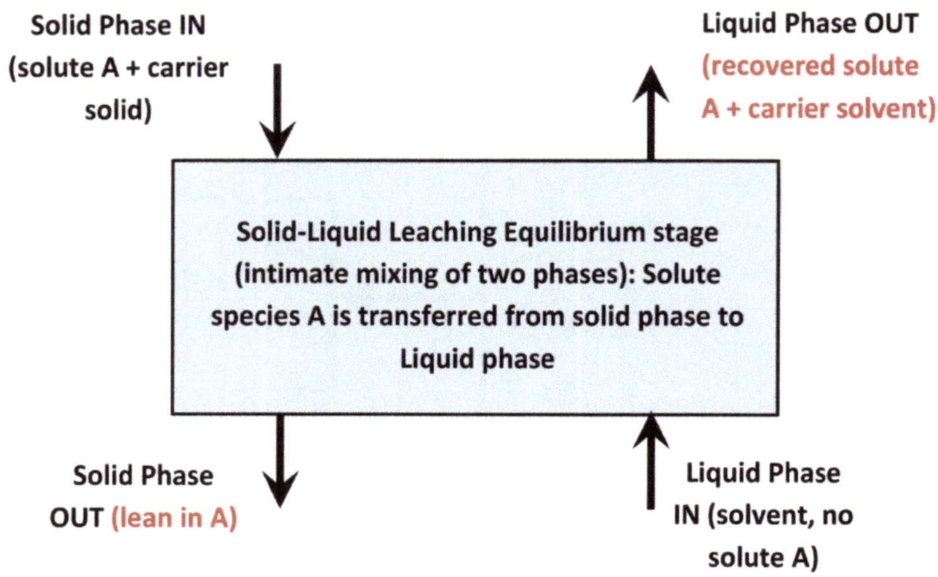

Figure 1.12 Equilibrium stage of a solid-liquid leaching process.

Like liquid-liquid extraction, the single stage of a leaching process involves two pieces of equipment, namely a mixer and a settler, as shown in Figure 1.14. The particulate solids are mixed with the solvent in the mixing vessel using a mixer/stirrer and the mixture is passed to a settling vessel for the separation of solute-rich solvent and solute-depleted solids slurry. The applications of a leaching process are numerous. Some examples include: a) leaching of sugar from sugar beets using hot water as a solvent; b) preparation of instant coffee by passing hot water through ground roasted coffee; c) leaching of pharmaceutical products from roots, stems, and leaves; d) extraction of vegetable oil from oil seeds using organic solvent such as hexane; and e) removal of metals from metal ores; for example, leaching of gold from gold ore (solid) using sodium cyanide solution as a solvent.

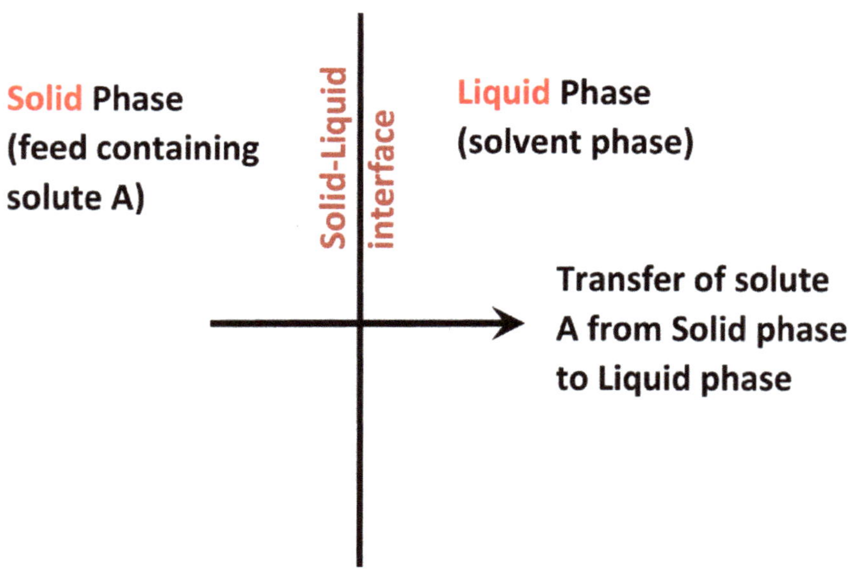

Figure 1.13 Direction of mass transfer of solute in a solid-liquid leaching process.

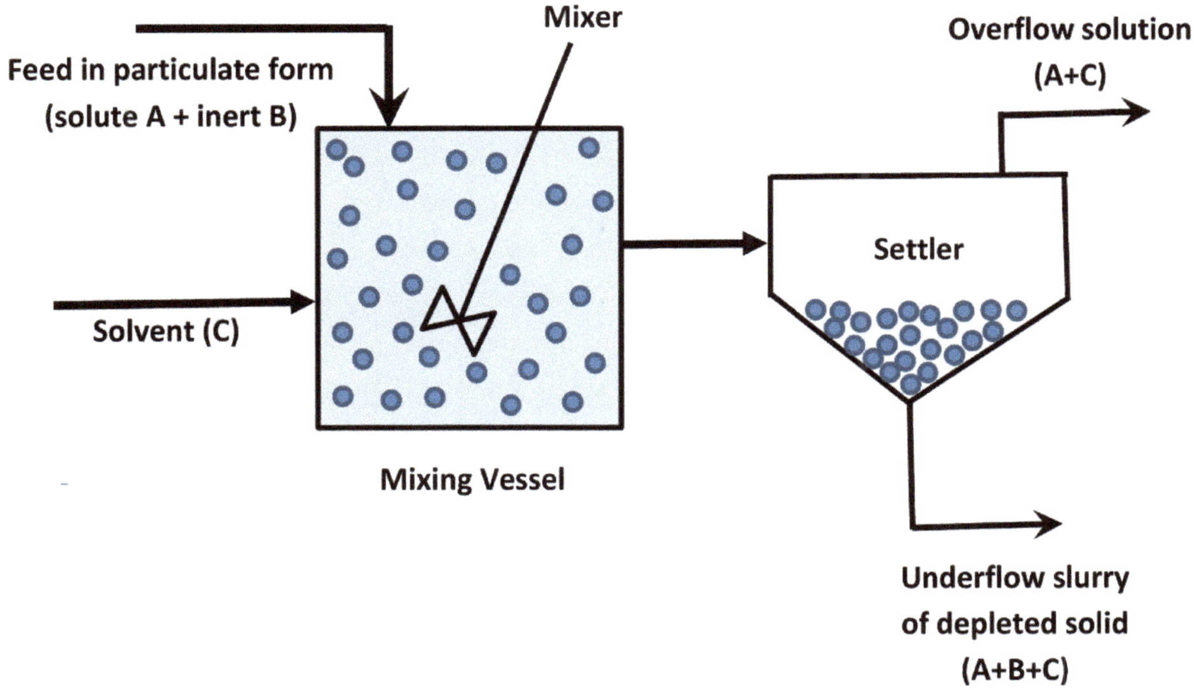

Figure 1.14 Equipment diagram of a single-stage solid-liquid leaching process.

1.4 EXAMPLES OF INDUSTRIAL APPLICATIONS OF EQUILIBRIUM STAGED SEPARATION PROCESSES

1.4.1 Dehydration of Natural Gas

Natural gas is a combustible mixture of hydrocarbon gases consisting of mainly methane (> 85%). Like crude oil, natural gas is trapped under the earth and is extracted from the underground reservoir by drilling a gas well. Natural gas contains water vapor which must be removed in order to avoid gas transmission problems through pipelines. Water vapor causes the formation of solid ice-like methane hydrates which may plug the pipelines and associated equipment. Water vapor also causes corrosion of equipment.

Figure 1.15 shows a simplified process flowsheet for natural gas dehydration. The wet natural gas is passed through an absorption column where dry glycol (triethylene glycol) is used to absorb the water vapor. The water-rich glycol undergoes flashing to knock out any hydrocarbons absorbed in the absorption column. The water-rich glycol is then separated into water vapor and dry glycol using distillation column. The dry glycol is recycled back into the absorption column.

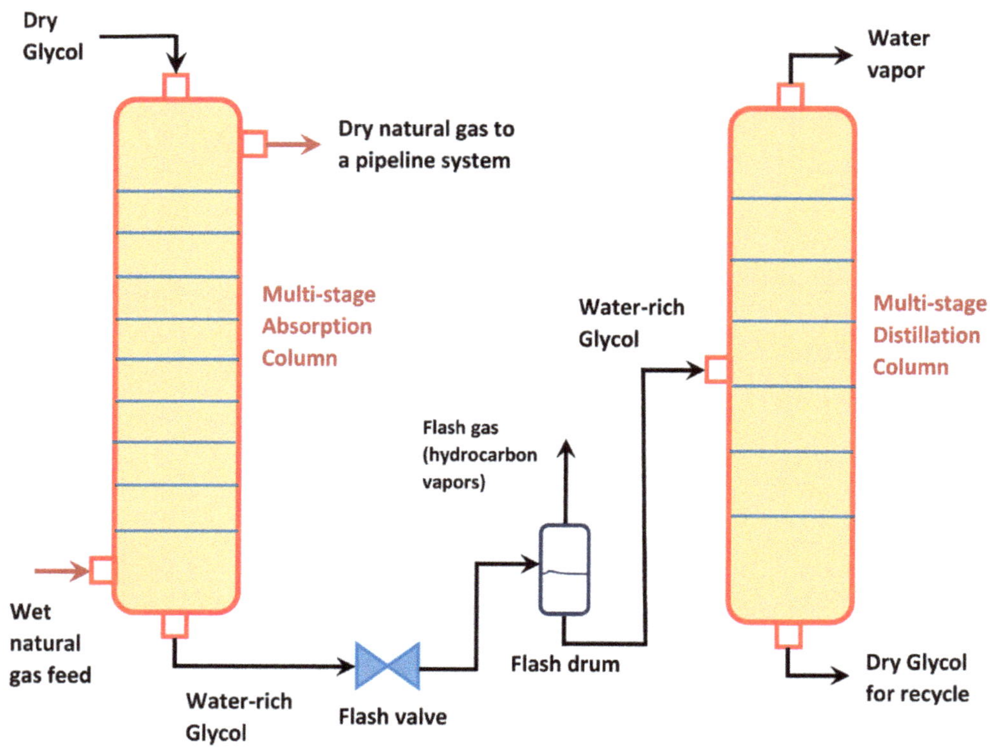

Figure 1.15 Simplified flowsheet of industrial natural gas dehydration plant.

1.4.2 CO_2 Capture and Storage

Carbon dioxide is one of the main "greenhouse gases" responsible for global warming. Thus reducing CO_2 emissions to the environment is becoming increasingly important. As flue gases emitted from fossil fuel fired power plants contain a large amount (> 10%) of CO_2, capturing CO_2 from flue gases is important to mitigate the detrimental effects of CO_2 emission on the environment. Figure 1.16 shows a simplified process flowsheet for CO_2 capturing from flue gas. The flue gas containing CO_2 is passed through a multi-stage absorption column where the flue gas flows in the upward direction and solvent flows in the opposite (downward) direction. Due to intimate mixing of solvent stream and flue gas stream at each stage and high affinity of CO_2 for the solvent phase, the CO_2 is absorbed by the solvent. The cleaned flue gas is discharged to the atmosphere from the top of the absorption column and the CO_2-rich solvent stream is pumped to a multistage stripping column where steam is used to strip CO_2 from the solvent phase to the steam stream. The solvent recovered from the bottom of the stripping column is recycled back to the absorption tower. The CO_2-rich steam emerging from the top of the stripping column is passed through a condenser where steam is condensed into liquid water. The CO_2 is then compressed and stored.

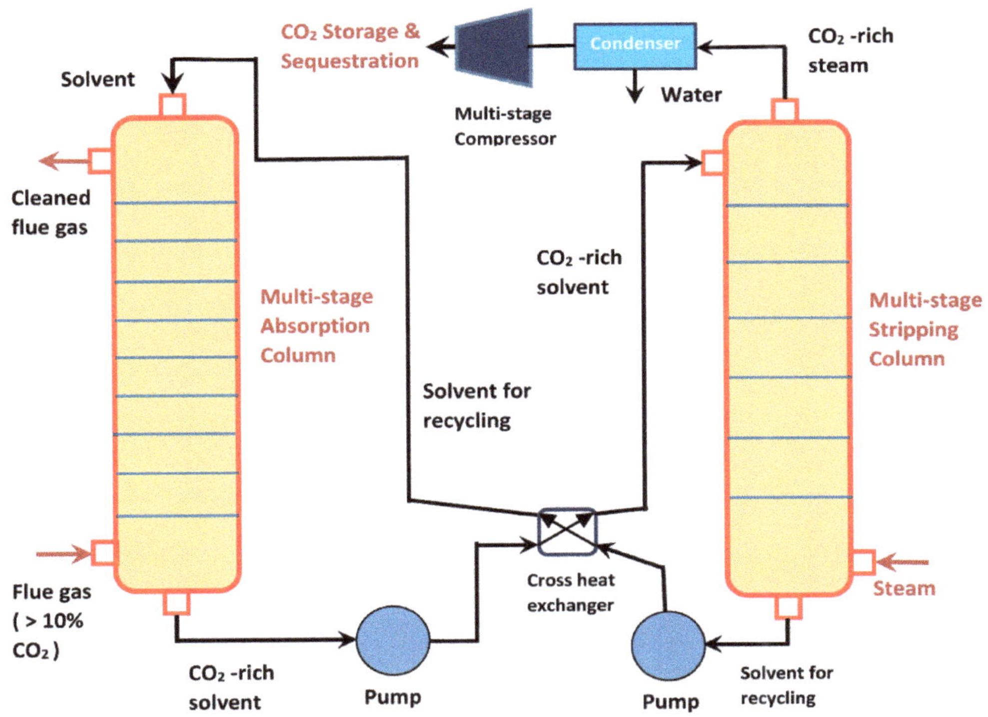

Figure 1.16 Carbon dioxide capture and storage from coal-fired flue gas containing a large amount of CO_2 (> 10%).

1.4.3 Vegetable Oil Extraction

Figure 1.17 shows a simplified process flowsheet of vegetable oil extraction from oil seeds such as soybeans. The seeds are first cleaned to remove any impurities. Then the seeds are crushed to break the seeds and their hulls and to reduce the particle size of the seeds. The next important step is the flaking of the seeds where the seed particles are deformed into flakes resulting in an increase in the surface area of the seeds. Flaking also ruptures the cell walls of the oil cells of the seeds. The increase in the surface area of the seeds and rupturing of the oil cell walls are both beneficial in the next solid-liquid leaching step where the flaked seeds are intimately mixed with the solvent and oil is extracted into the solvent phase. The oil/solvent mixture, also referred to as "miscella", emerging from the leaching process is passed to the distillation column to separate the miscella into solvent and vegetable oil. The solvent stream emerging from the top of the distillation column is recycled back and vegetable oil stream emerging from the bottom of the distillation column is sent for further refining.

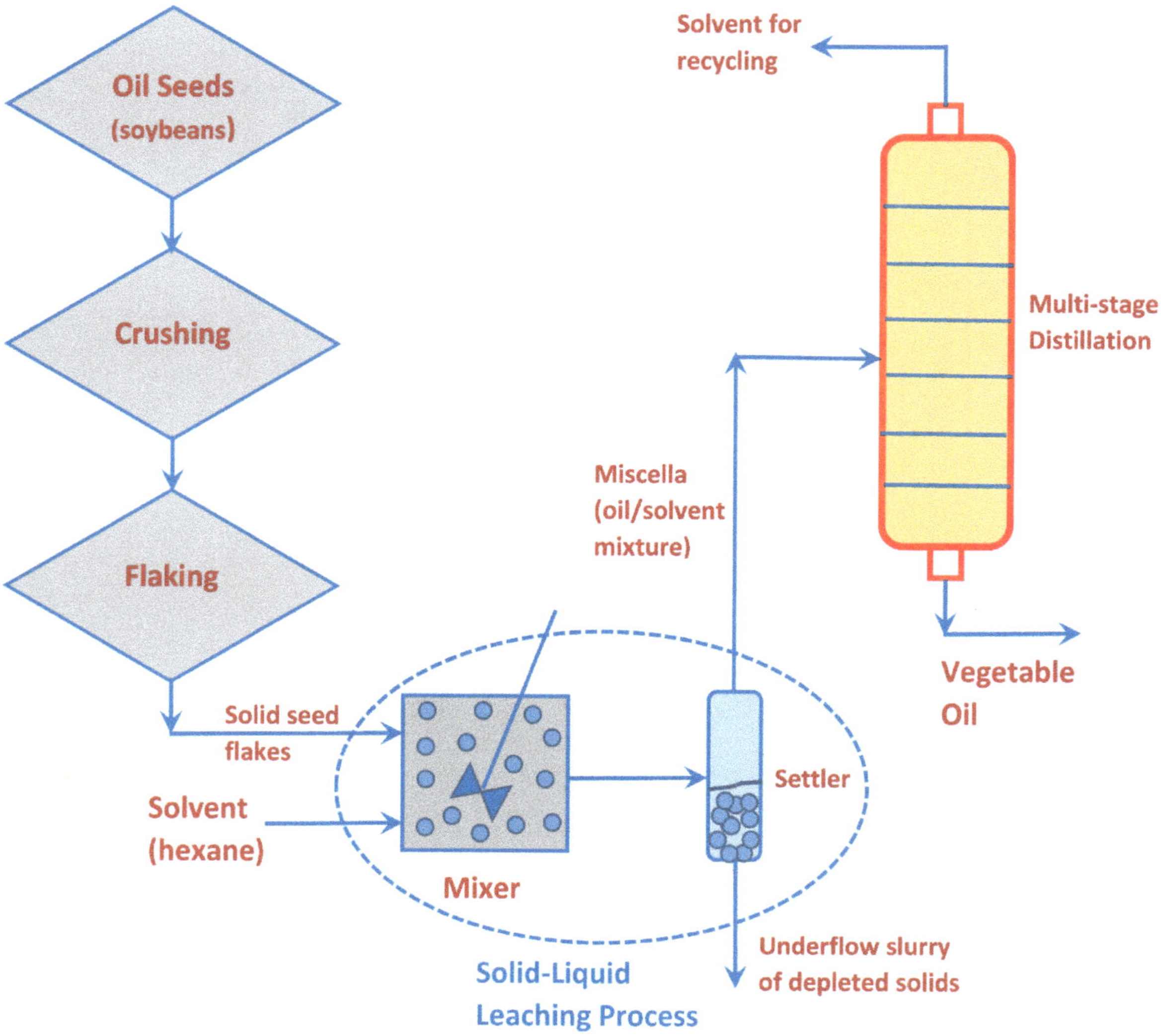

Figure 1.17 Simplified flowsheet of vegetable oil extraction from oil seeds.

1.4.4 Production of Penicillin Antibiotic

Penicillin is the first antibiotic discovered by a Scottish biologist Alexander Fleming in 1928. It inhibits the multiplication and growth of bacteria by interfering with their process of construction of cell walls. In the presence of penicillin, bacteria is unable to synthesize its cell walls and hence it fails to grow and multiply. Penicillin does not affect human cells as they do not have cell walls. Penicillin is produced by filamentous penicillium fungus shown in Figure 1.18 using aerobic fermentation. Figure 1.19 shows a simplified process flowsheet for the production of penicillin. The fermenter is fed with fermentation medium (including nutrients), seed culture (fungus), and air. After fermentation, the filamentous fermentation broth containing penicillin is passed through a filter to remove the biomass (fungus). The resultant fluid consisting of mixture of solute penicillin (species A) and carrier liquid (species B) undergoes liquid-liquid extraction where penicillin is transferred to organic solvent (species C) such as amyl acetate. The penicillin-rich extract phase undergoes further processing where penicillin is crystallized and precipitated out and dried. A sample of dried penicillin is shown in Figure 1.20.

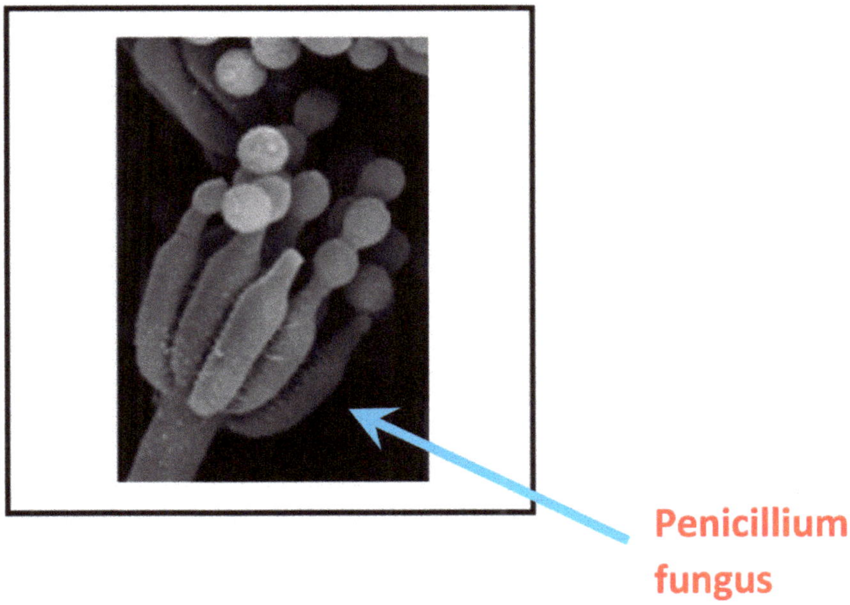

Figure 1.18 Penicillium fungus.

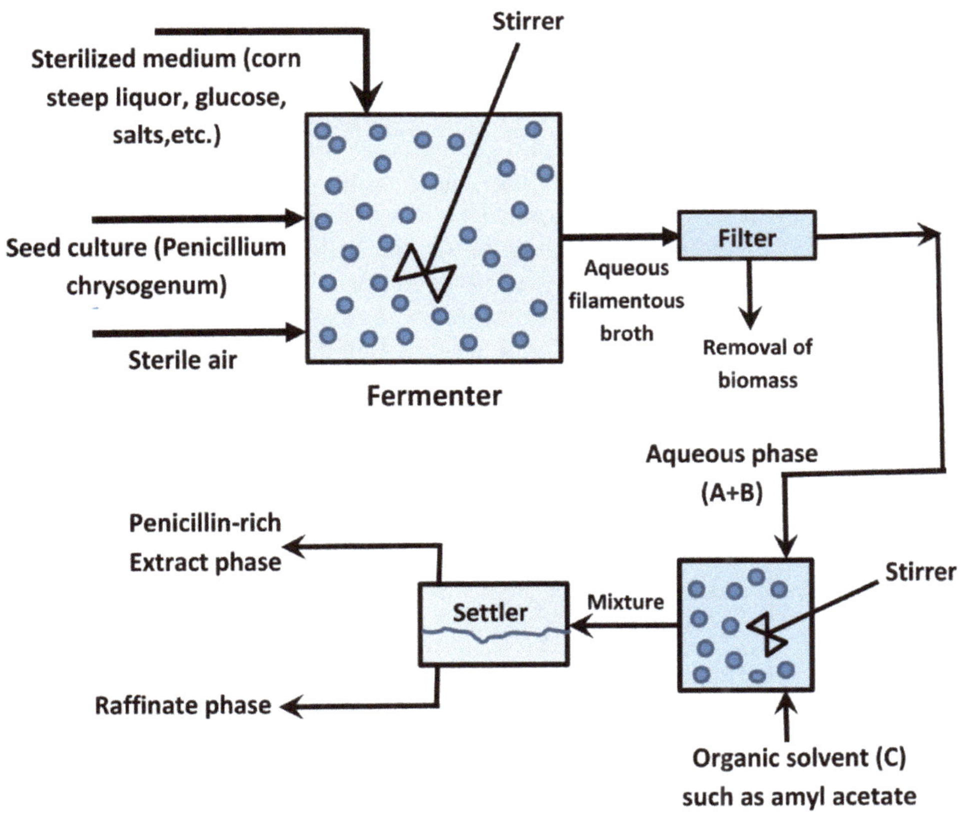

Figure 1.19 Simplified process flowsheet for the production of Penicillin.

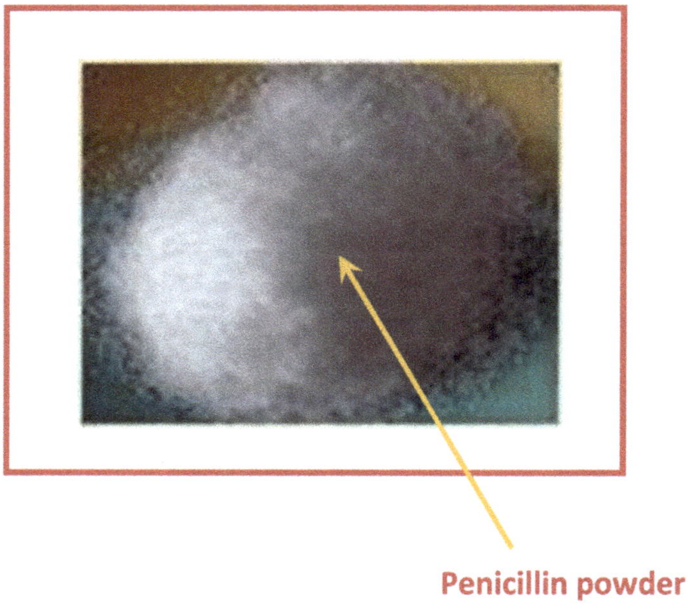

Figure 1.20 White penicillin-sodium salt.

1.5 MULTI-STAGE CONTACTING OF TWO PHASES

The degree of separation of components or the recovery of a solute in a single equilibrium stage is limited. For a high degree of separation of components and greater recoveries of solute, it is necessary to involve multi-stage contacting of two phases. One commonly used multi-stage contacting device in distillation, absorption, and stripping processes is the perforated multi-tray column shown in Figure 1.21. The gas phase or vapor phase enters the column at the bottom and the liquid phase enters the column at the top. The two phases flow in a counter-current manner, that is, in opposite directions. The rising gas or vapor comes in contact with the falling liquid phase on a perforated tray as shown in Figure 1.21. The gas emerges through the small holes on the perforated tray in the form of fine bubbles. The formation of fine bubbles creates a large interfacial area and intense mixing of the two phases on the tray. Both factors, that is, creation of large interfacial area and intense mixing of phases, are favorable for the transfer of species from one phase to the other. The gas phase leaving the perforated tray is assumed to be in equilibrium with the liquid phase leaving the tray through the down-comer. Note that it is the two fluid streams (gas and liquid streams) that are leaving the perforated tray are in equilibrium with each other, if the stage is assumed to be an equilibrium stage. The two incoming streams approaching the tray are not in equilibrium (if the incoming streams are at equilibrium, no transfer of species will take place from one phase to another).

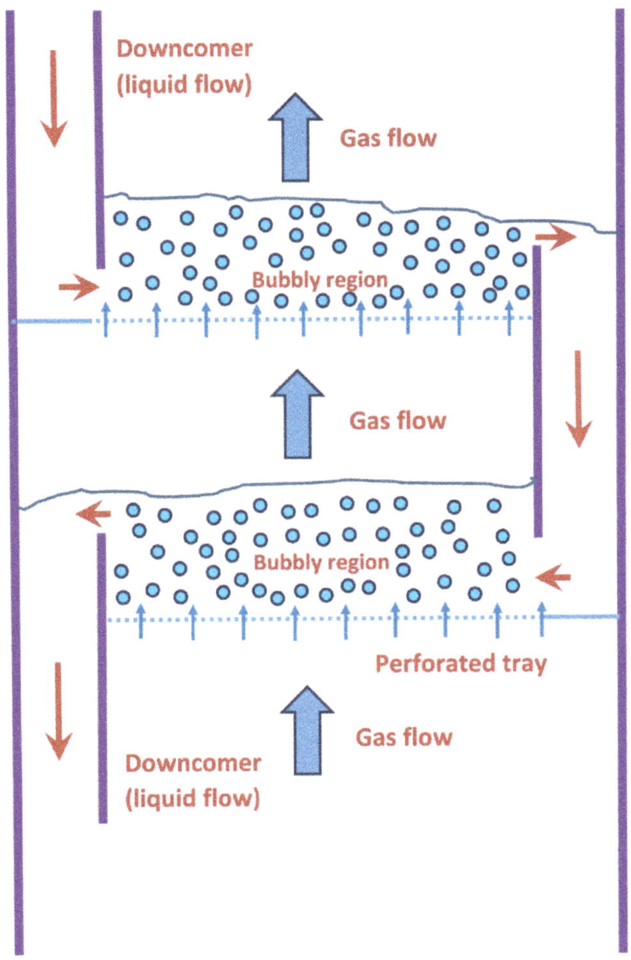

Figure 1.21 Counter-current cascade of two stages (distillation, absorption or stripping).

Figure 1.22 shows counter-current multi-stage contacting of two phases in liquid-liquid extraction and solid-liquid leaching processes. In any given stage, there are two pieces of equipment involved, namely mixer and settler. Feed enters the cascade at the first stage (top stage 1) and pure solvent enters the cascade at the last stage (bottom stage 3 in Figure 1.22). The solvent and feed streams move in opposite directions. The raffinate phase (solute-depleted phase emerging from the bottom of any given settler) becomes feed to the next stage and the extract phase (solute-rich solvent phase emerging from the top of any given settler) becomes solvent stream for the preceding stage. The final extract phase, rich in recovered solute, emerges from the top of the cascade and the final raffinate phase, depleted of solute, emerges from the bottom of the cascade.

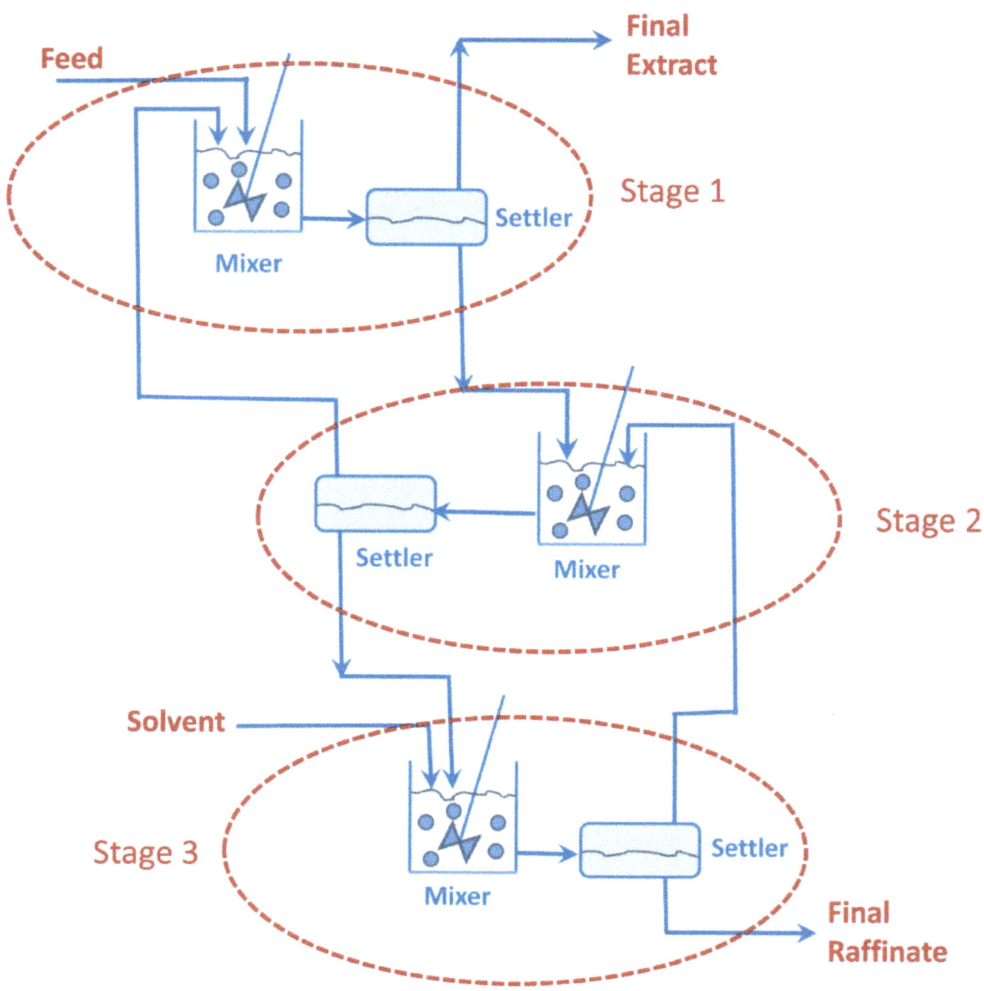

Figure 1.22 Counter-current cascade of three liquid-liquid extraction or solid-liquid leaching stages.

1.6 EQUILIBRIUM BETWEEN PHASES

The two output streams leaving an equilibrium stage are in equilibrium with each other (note that the incoming streams are never at equilibrium otherwise there will be no mass transfer). This is the basic assumption made in the analysis and design of equilibrium stage operations.

1.6.1 Definition and Criteria for Phase Equilibrium

A closed system consisting of two or more phases is said to be in phase equilibrium when the compositions and amounts of the phases present in the system are no longer changing with time, that is, they are time-invariant and there is no further tendency towards any changes in compositions and amounts of various phases. This also means no more separation of any species is possible from one phase to another.

The thermodynamic criterion for phase equilibrium is that the chemical potential of any species is the same in all the phases present. That is the system is free of any chemical potential gradients. Also it is understood that the temperature and pressure are uniform throughout the whole system. Thus:

$$\mu_i\,(phase\ 1) = \mu_i\,(phase\ 2) = \mu_i\,(phase\ 3) = \mu_i\,(phase\ 4) = \cdots \tag{1.1}$$

where μ_i is the chemical potential of species "i". Note that it is the difference in chemical potentials of a species between different phases that drives the movement of species from one phase to another. The species moves in the direction of decreasing chemical potential, just like current flows in the direction of decreasing voltage potential.

1.6.2 Relationship between the Compositions of Different Phases at Equilibrium

When the phases are in equilibrium, the compositions of the phases are related. If the composition of one phase is given, one can calculate the composition of the other phase using the equilibrium relation. One important equilibrium relation relevant to equilibrium staged separation processes is the Raoult law.

Raoult's law is named after the French chemist François-Marie Raoult. Consider a multi-component vapor-liquid system at equilibrium. According to Raoult's law, the partial vapor pressure of any species "i" in the vapor mixture is equal to the vapor pressure of the pure component "i" (at the mixture temperature) multiplied by its mole fraction in the liquid solution. Thus:

$$P_i = x_i P_i^{Sat} \tag{1.2}$$

where P_i is the partial pressure of species "i" in the vapor mixture, x_i is the mole fraction of species "i" in the liquid solution, and P_i^{Sat} is the vapor pressure of pure species "i" at the system temperature. Raoult's law assumes that the vapor mixture is an ideal gas mixture and the liquid solution is an ideal liquid solution. The partial pressure of species "i" in an ideal gas mixture is defined as the pressure that species "i" would exert if it alone occupied the total volume of the gas mixture at the same temperature. For an ideal gas mixture, the partial pressure is given as

$$P_i = y_i P \tag{1.3}$$

where y_i is the mole fraction of species "i" in the vapor mixture and P is the total pressure of the vapor mixture. From Equation (1.3), it follows that:

$$\sum P_i = \sum y_i\, P = P \tag{1.4}$$

This is the Dalton law which states that the sum of the partial pressures of various components of an ideal gas mixture is equal to the total pressure of the gas mixture.

Combining Equations (1.2) and (1.3), we obtain:

$$P_i = y_i P = x_i P_i^{Sat} \tag{1.5}$$

Equation (1.5) could further be re-written as:

$$y_i = K_i x_i \text{ where } K_i = P_i^{Sat}/P \tag{1.6}$$

The K-value (also referred to as distribution coefficient or vapor-liquid equilibrium constant) for any given species is a function of temperature and pressure.

Figure 1.23 shows the plots of partial pressures and total pressure of a binary system of vapor and liquid at equilibrium at a fixed temperature. The data are generated using Raoult's law. Species "A" is more volatile than species "B" (P_A^{Sat} is higher than P_B^{Sat}). The pressures are plotted as a function of mole fraction of more volatile species (species A). As expected, the partial pressure of more volatile species A increases and the partial pressure of less volatile species B decreases with the increase in x_A. The total vapor pressure also increases with the increase in x_A. The plots of partial pressures and total pressure are all linear as expected from Raoult's law. Figure 1.24 shows the same data in the form of P-x-y phase diagram at a fixed

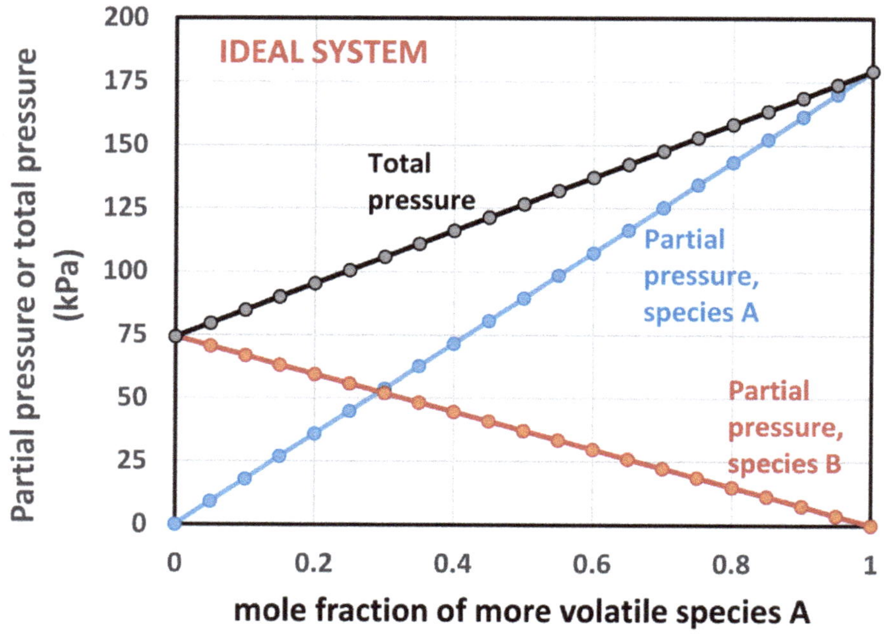

Figure 1.23 Partial pressures and total pressure of a binary vapor-liquid system at equilibrium.

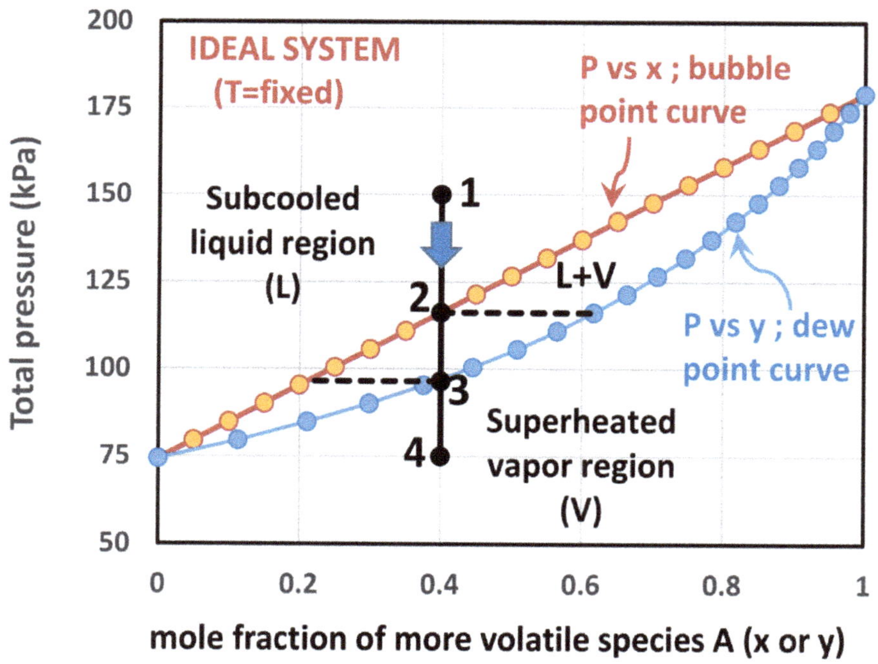

Figure 1.24 P-x-y phase diagram of a binary system at a fixed temperature.

temperature. The P-x plot (referred to as bubble point curve) is linear whereas P-y plot (referred to as dew point curve) is non-linear. Both x and y variables refer to a more volatile species A. Above the bubble point curve, the binary mixture is a single-phase subcooled liquid. Below the dew point curve, the mixture is superheated vapor. The binary mixture exists as a two-phase vapor-liquid system when the pressure falls in between the bubble point and dew point, that is, $P_{bubble-point} < P < P_{dew-point}$. As an example, let us now consider a binary mixture with an overall mole fraction of species A equal to 0.4. We will follow the path indicated by a vertical line in Figure 1.24 starting from point 1. At 1, the binary mixture is subcooled liquid. As the pressure of the system is lowered, the mixture remains as a subcooled liquid until point 2 is reached. At point 2 (bubble point), the first vapor bubble is formed. The vapor is rich in more volatile component A as indicated by a horizontal line. With further lowering of pressure, more vapor is formed until point 3 is reached where the entire mixture exists as a saturated vapor with a few drops of liquid of composition indicated by a horizontal line. Upon lowering the pressure below the dew point, the vapor becomes superheated vapor. Figure 1.25 shows the same data in the form of y-x equilibrium curve. The curve represents liquid – vapor compositions (y versus x) at equilibrium with each other at a fixed temperature.

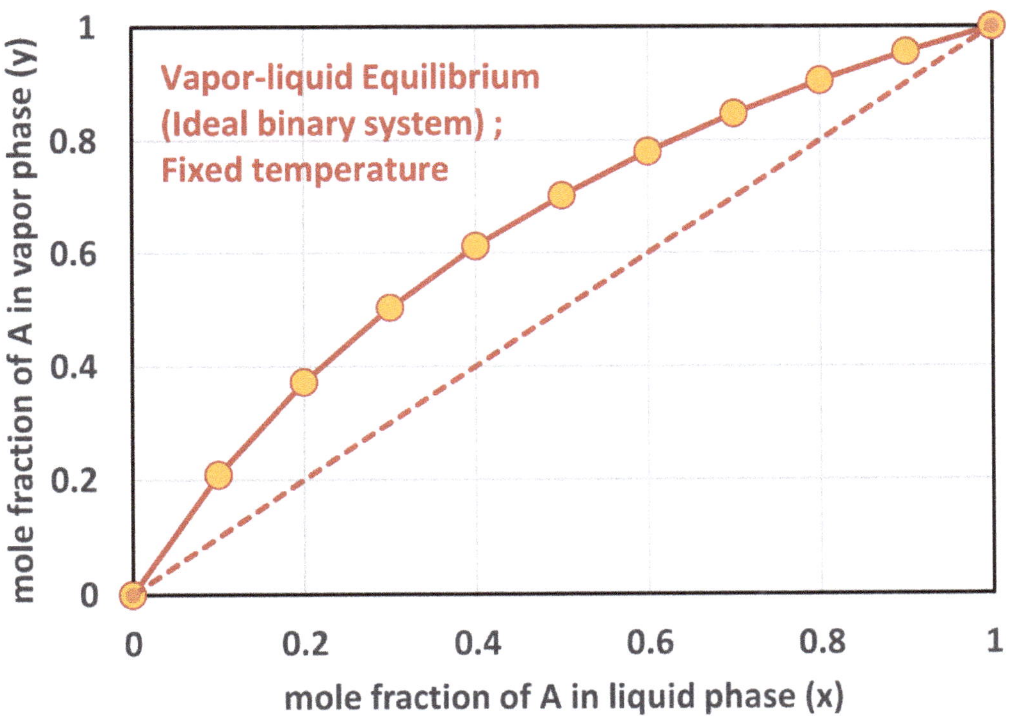

Figure 1.25 Equilibrium y-x curve at constant temperature.

1.6.3 Non-ideal Liquid Solutions

Raoult's law assumes that the liquid solution is ideal and that the vapor mixture is also ideal. More often than not, the liquid solution is a non-ideal solution. The vapor mixture, on the other hand, can often be treated as an ideal gas mixture especially when the total pressure is moderate.

Consider a binary liquid solution of species A and B. The solution would behave like an ideal solution if the interaction between the like molecules (A-A and B-B attractions) is of the same magnitude as the interaction between the unlike molecules (A-B attractions). An example of ideal binary liquid solution is benzene + toluene. However, in the case of non-ideal liquid solutions, the interactions between the unlike molecules are of different magnitude as compared with the interactions between the like molecules. Non-ideal liquid solutions exhibit positive deviation from ideal behavior (that is, Raoult's law) when the attractions between the unlike molecules (A-B attractions) are weaker as compared with the attractions between the like molecules (A-A and B-B attractions). In such solutions, the molecules of species A and species B have a higher tendency to escape to the vapor phase as compared with the ideal solution. Consequently, the partial pressures, and hence the total pressure, are higher as compared with the ideal solution values. An example of a binary solution that exhibits positive deviation from Raoult's law is acetone-carbon tetrachloride system. If the attractions between the unlike molecules (A-B attractions) are stronger as compared with attractions between the like molecules, the solution exhibits negative deviation from Raoult's law as the molecules of species A and species B are not able to escape to the vapor phase as readily as in the ideal solution. Consequently, the partial pressures and hence the total pressure are lower as compared with the ideal solution values. An example of a binary solution that exhibits negative deviation from Raoult's law is acetone-chloroform system.

In the discussion to follow, we will assume that the vapor mixture is ideal but the liquid solution is non-deal. Both positive and negative deviations from ideal behavior will be discussed.

Figure 1.26 shows the plots of partial pressures and total pressure for a binary *non-ideal* vapor-liquid system at equilibrium at a fixed temperature. The partial pressures and the total pressure fall above the values predicted from the Raoult law. Thus the system exhibits *positive deviation* from an ideal behavior. The same data are re-plotted in the form of P-x-y phase diagram in Figure 1.27. The phase diagram now looks quite different from the ideal case (see Figure 1.24). Interestingly, the bubble point curve (P-x curve) and the dew point curve (P-y curve) touch each other at approximately $x_A = y_A = 0.65$. The vapor pressure is also nearly maximum at this composition. The system forms an azeotrope at this composition. Azeotrope could be defined as a liquid mixture of two or more chemical species which boils without any change in chemical composition. In other words, upon boiling an azeotrope gives off vapor that has the same chemical composition as that of the liquid. Consequently azeotropic mixtures cannot be easily separated into components using distillation. Figure 1.28 shows the same data plotted in the form of y-x equilibrium curve. As expected the equilibrium curve crosses the y = x line at azeotropic composition.

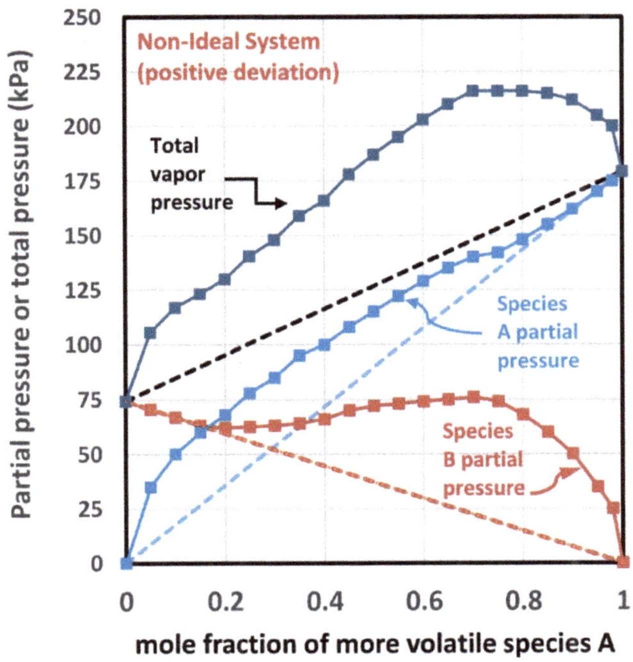

Figure 1.26 Partial pressures and total pressure of a binary non-ideal (positive deviation) vapor-liquid system at equilibrium.

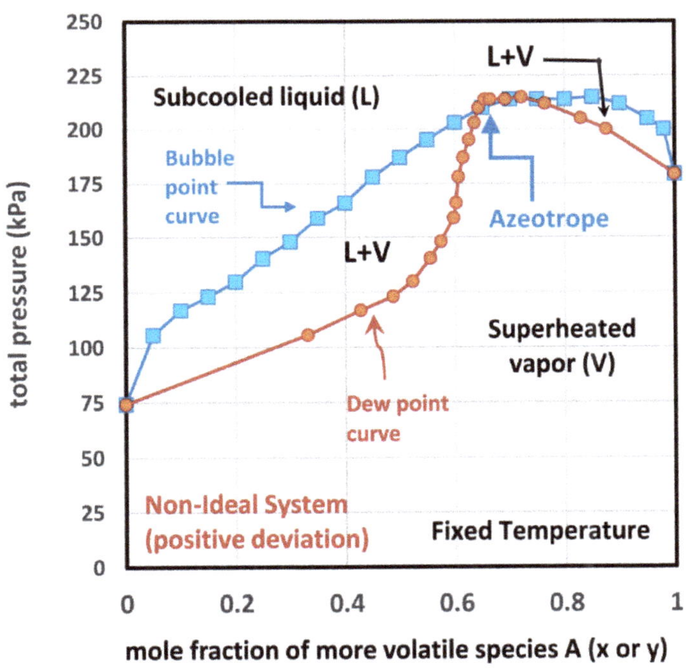

Figure 1.27 P-x-y phase diagram of a binary non-ideal (positive deviation) system at a fixed temperature.

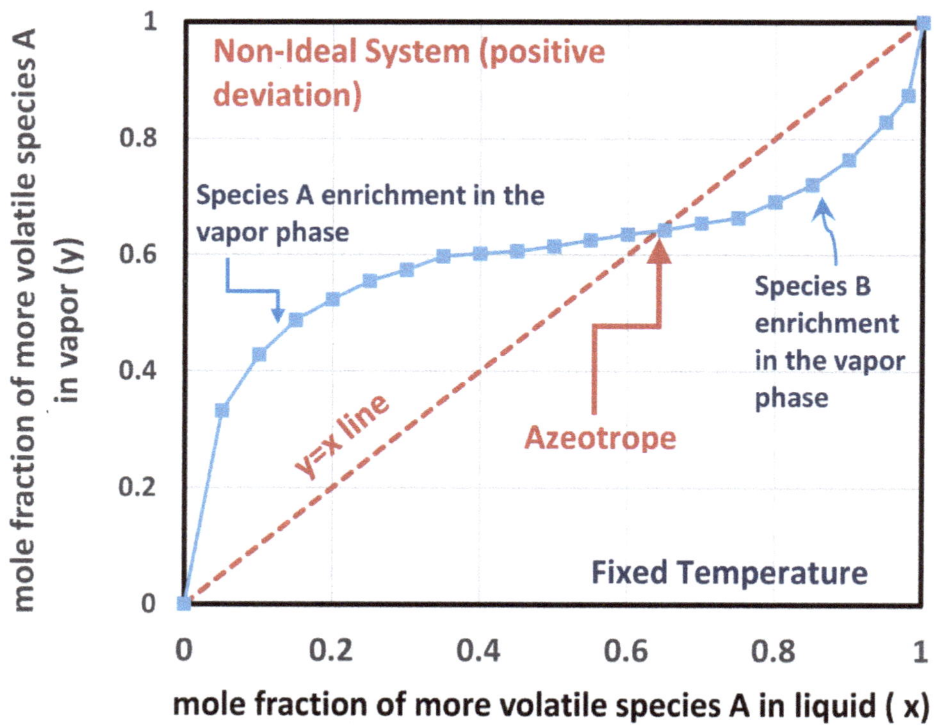

Figure 1.28 Equilibrium y-x curve for non-ideal (positive deviation) vapor-liquid system at constant temperature.

Thus far we have considered a non-ideal binary system that shows a positive deviation from Raoult's law. Figure 1.29 shows the plots of partial pressures and total pressure for a binary *non-ideal* vapor-liquid system exhibiting *negative deviation* from Raoult's law. The partial and total pressures fall below the linear plots predicted by the Raoult law. The same data are re-plotted in Figure 1.30 in the form of the P-x-y phase diagram at a fixed temperature. Now the bubble point and dew point curves exhibit minima where they touch each other. At this composition, approximately $x_A = y_A = 0.35$, the system forms an azeotrope. The formation of an azeotrope at this composition is further confirmed in Figure 1.31 where the same data are plotted in the form of y-x equilibrium curve. The equilibrium curve crosses the y = x line at azeotropic composition.

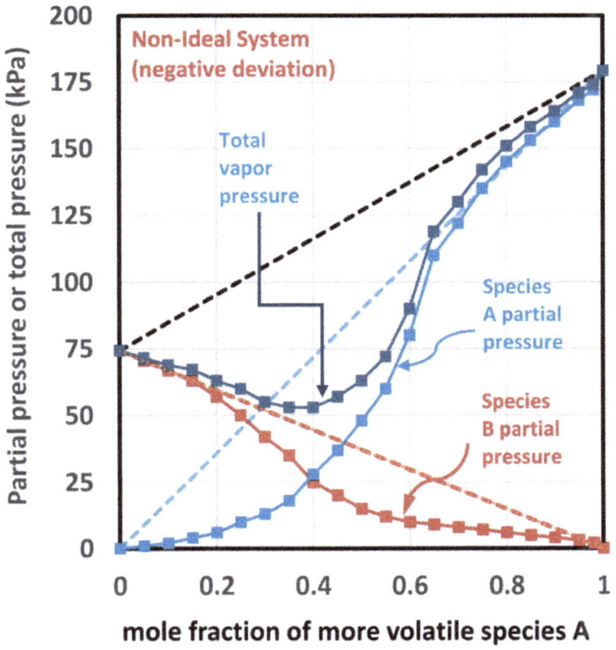

Figure 1.29 Partial pressures and total pressure of a binary non-ideal (negative deviation) vapor-liquid system at equilibrium.

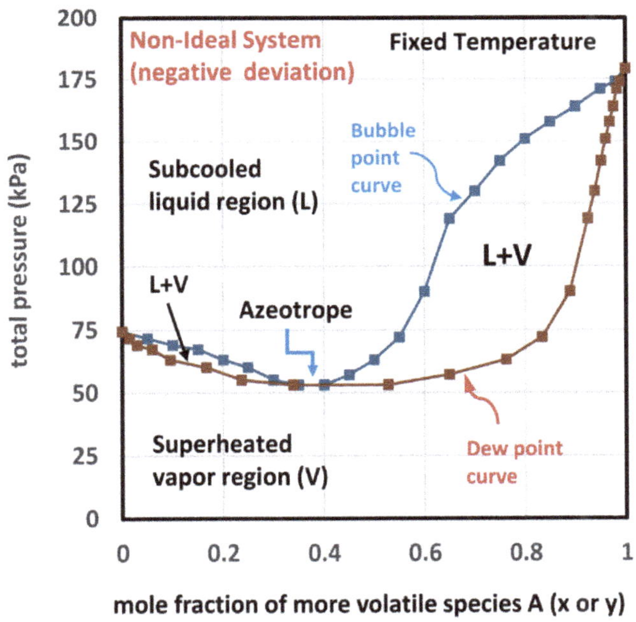

Figure 1.30 P-x-y phase diagram of a binary non-ideal (negative deviation) system at a fixed temperature.

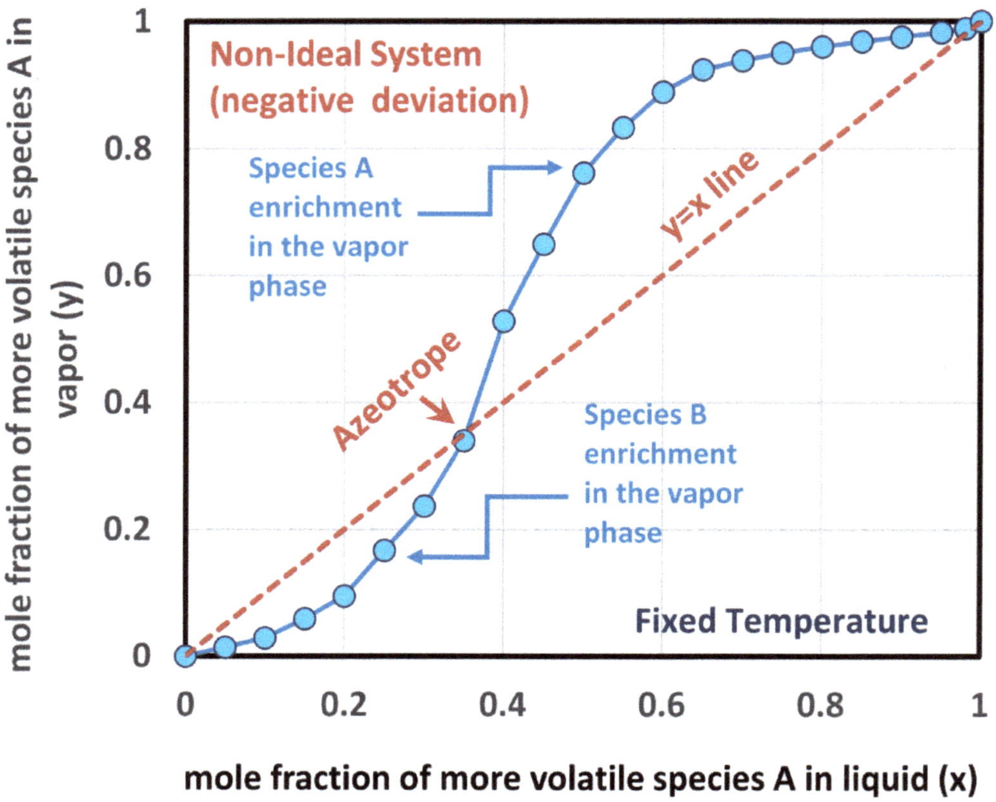

Figure 1.31 Equilibrium y-x curve for non-ideal (negative deviation) vapor-liquid system at constant temperature.

In summary, two types of azeotropes, minimum-boiling azeotrope and maximum-boiling azeotrope, are observed when there occur large deviations from ideal behavior. When the deviation from ideal behavior (Raoult's law) is positive, the total vapor pressure goes through a maximum and hence the boiling point temperature goes through a minimum. Such an azeotrope is referred to as "*minimum boiling azeotrope*". When the deviation from ideal behavior is negative, the total vapor pressure goes through a minimum and hence the boiling point temperature goes through a maximum. The azeotrope formed in this case is called "*maximum boiling azeotrope*".

1.6.4 Limiting Cases of Non-ideal Solutions

Consider a binary mixture of chemical species A and B where species A is more volatile than species B. The vapor mixture is ideal but the liquid solution is non-ideal exhibiting either positive or negative deviation from an ideal behavior. In general, the Raoult law is applicable to such systems in the limits: $x_A \rightarrow 1.0$ and $x_B \rightarrow 1.0$. This is shown clearly in Figure 1.32. When $x_A \rightarrow 1.0$, the Raoult law can be used to calculate the partial pressure of chemical species A. When $x_B \rightarrow 1.0$, the Raoult law can be used to calculate the partial pressure of chemical species B. Thus Raoult's law is applicable to the solvent of a non-ideal solution at small solute concentrations.

Another relation which is as important as the Raoult law is the Henry law. It is applicable in the limits where $x_A \rightarrow 0$ and $x_B \rightarrow 0$. According to the Henry law, the partial pressure of a solute (chemical species present in small concentrations in the non-ideal liquid solution) is linearly related to the mole fraction of the solute in the liquid solution as follows:

$$P_i = K_{H,i} x_i \text{ when } x_i \rightarrow 0 \tag{1.7}$$

where $K_{H,i}$ is the Henry law constant. Figure 1.33 shows the applicability of the Henry law to a binary vapor-liquid system. Note that $K_{H,i}$ is not equal to saturation pressure P_i^{Sat} as observed in the case of ideal solution (Raoult's law). Usually $K_{H,i}$ is very different from the saturation pressure P_i^{Sat} in non-ideal solutions.

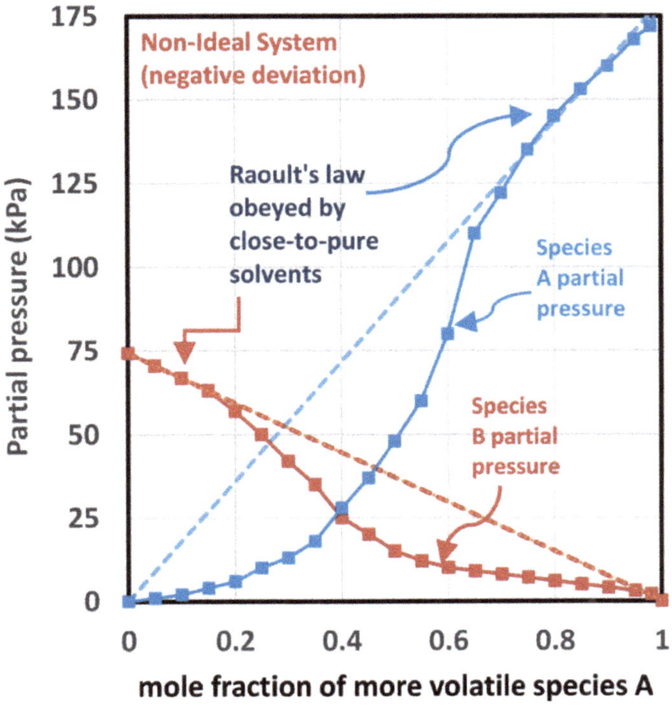

Figure 1.32 Application of Raoult's law to solvents of dilute solutions of solute.

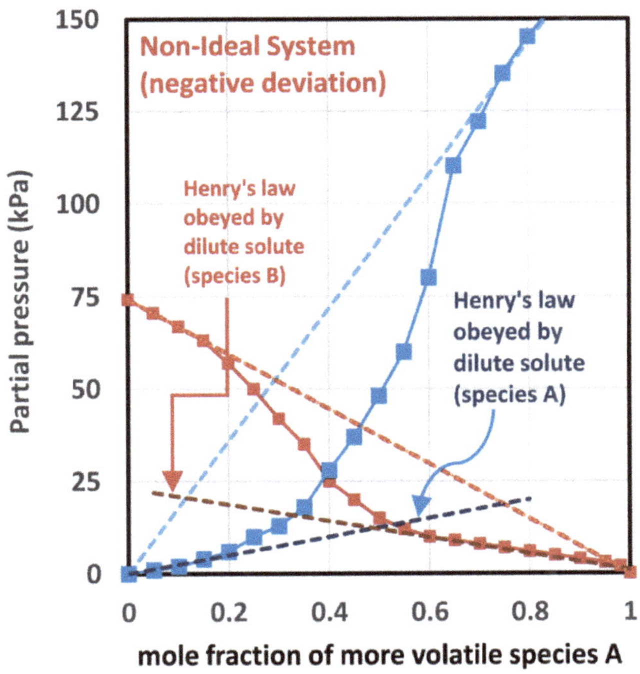

Figure 1.33 Application of Henry's law to solutes of dilute solutions of solute.

In summary, Henry's law is applicable to solute whereas Raoult's law is applicable to solvent when the liquid solution is non-ideal in nature (vapor mixture still assumed to be ideal). Thus, for real solutions at low solute (species A) concentrations:

$$P = P_A + P_B = K_{H,A}x_A + P_B^{Sat}x_B \tag{1.8}$$

In the chapters to follow, we will be using Raoult's law in the analysis and design of distillation processes. The Henry law will be used in the analysis and design of absorption and stripping processes.

1.6.5 More on Henry's Law

In the preceding section, the Henry law was applied to a binary vapor-liquid system at equilibrium where both components (species A and B) were completely miscible with each other in the liquid phase. Both x_A and x_B could be varied from 0 to 1. The Henry law related the partial pressure of the solute in the vapor phase to the mole fraction of the solute in the liquid solution provided that the solute mole fraction in the solution was very small.

Another very important application of the Henry law is to determine the solubility of a sparingly soluble gas species (species A) in a liquid solvent (species B) at equilibrium. As an example, consider a binary (two component) gas-liquid system at equilibrium, shown in Figure 1.34 below.

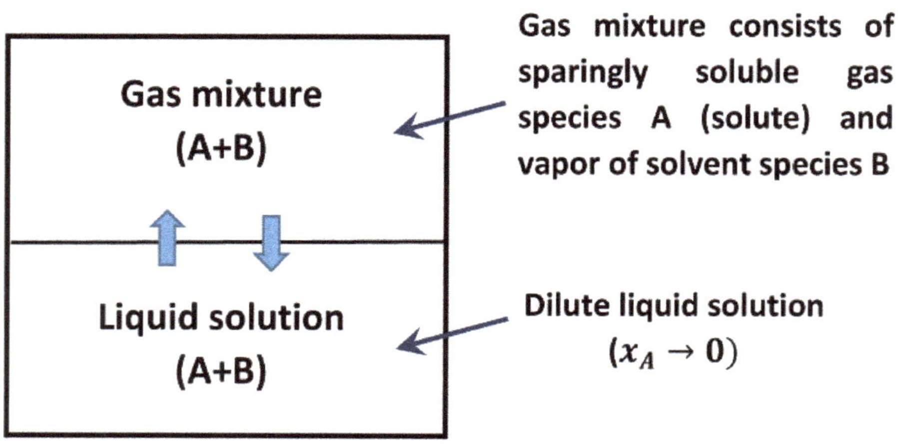

Figure 1.34 Gas mixture in contact with liquid solution at equilibrium

The gaseous species A is only slightly soluble in the liquid species B. At equilibrium, the gas phase consists of gaseous solute species A and vapor of species B. The liquid phase is a dilute solution of solute species A in liquid solvent B. In such applications, Henry's law states that the amount of gas dissolved in the liquid phase is proportional to the partial pressure of the gas above the liquid. If we express the amount of the gas dissolved in the liquid in terms of the mole fraction of the dissolved gaseous species in the liquid, we can express the Henry law as:

$$P_A = K_H x_A \tag{1.9}$$

where K_H is the Henry law constant. Higher the partial pressure P_A, greater is the solubility or mole fraction x_A of the dissolved gaseous species A in the liquid phase. If we compare two different gaseous solute species at the same partial pressures, the one with a higher value of the Henry law constant K_H has lower solubility. Thus, a large value of K_H is indicative of low solubility of solute in liquid and a small value of K_H is indicative of high solubility of solute in liquid. For example, $K_{H,CO_2} = 1.63$ kbars whereas $K_{H,O_2} =$

44.1 kbars in water at 25 °C indicating that carbon dioxide is much more soluble in water as compared with oxygen.

The Henry law is used extensively in gas absorption and stripping processes. For example, in a gas absorption process, the gas phase consisting of gaseous species A (solute) and carrier gas is brought in contact with the liquid solvent (species B), as shown in Figure 1.35 given below. It is assumed that the carrier gas is insoluble in the liquid solvent and that the solvent is non-volatile.

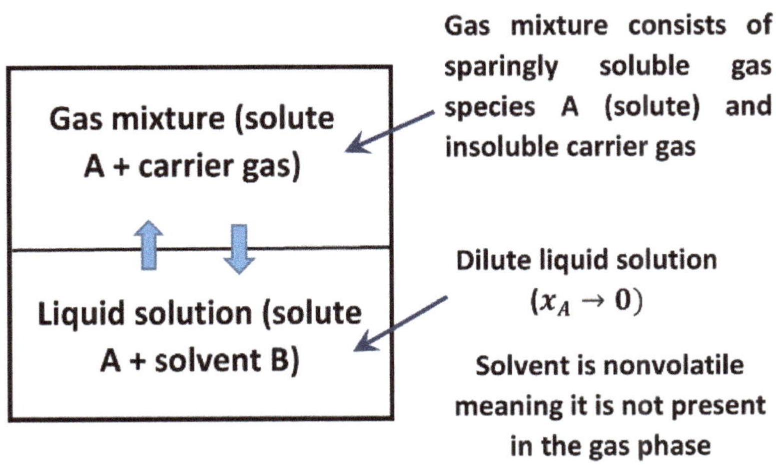

Figure 1.35 Gas mixture in contact with liquid solution at equilibrium.

At equilibrium,

$$P_A = y_A P = K_H x_A \tag{1.10}$$

where y_A is the mole fraction of solute species A in the gas phase and P is the total pressure of the gas. Upon re-arrangement, Equation (1.10) gives:

$$y_A = (K_H/P)x_A = Kx_A \tag{1.11}$$

This is another form of the Henry law where the new Henry law constant is $K = (K_H/P)$. Note that K_H has the units of pressure whereas K is dimensionless.

NOTATION

K	distribution coefficient or Henry law constant
K_i	distribution coefficient of species *i* defined in Equation (1.6)
K_H	Henry law constant
$K_{H,i}$	Henry law constant of species *i* defined in Equation (1.7)
P	total pressure of gas or vapor mixture
P_i	partial pressure of species *i* in gas or vapor mixture
P_i^{Sat}	vapor pressure or saturation pressure of pure species *i*
x_i	mole fraction of species *i* in liquid phase
y_i	mole fraction of species *i* in vapor phase

Greek Letters

μ_i chemical potential of species *i*

TUTORIAL PROBLEMS WITH SOLUTION

Problem 1

A vapor mixture consisting of 3 components (A, B and C) is in equilibrium with a liquid solution having the following composition: $x_A = 0.15, x_B = 0.40,$ and $x_C = 0.45$. The temperature is 80 °C. The vapor pressures of the pure species are given by the following Antoine equations:

$$lnP_A^{Sat} = 14.5463 - \frac{2940.46}{T+237.22}$$

$$lnP_B^{Sat} = 14.2724 - \frac{2945.47}{T+224}$$

$$lnP_C^{Sat} = 14.2043 - \frac{2972.64}{T+209}$$

where T is in °C and the vapor pressures are in kPa. Using Raoult's law, determine P and vapor composition.

Solution

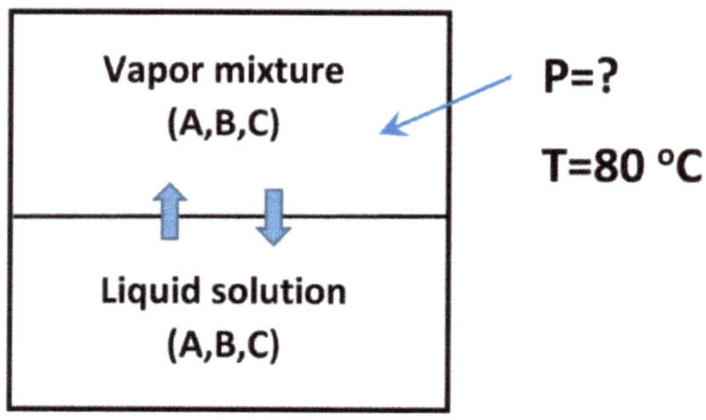

At 80 °C, the vapor pressures of pure species are calculated from the given Antoine equations as:

$$P_A^{Sat} = 195.75\ kPa$$

$$P_B^{Sat} = 97.842\ kPa$$

$$P_C^{Sat} = 50.32\ kPa$$

From Raoult's law:

$$P = x_A P_A^{Sat} + x_B P_B^{Sat} + x_C P_C^{Sat}$$

$$P = 0.15(195.75) + 0.40(97.842) + 0.45(50.32) = 91.143\ kPa$$

Now that P is known, vapor composition can be calculated from the Raoult law as:

$$y_A = \frac{x_A P_A^{Sat}}{P} = \frac{0.15(195.75)}{91.143} = 0.322$$

$$y_B = \frac{x_B P_B^{Sat}}{P} = \frac{0.40(97.842)}{91.143} = 0.429$$

$$y_C = \frac{x_C P_C^{Sat}}{P} = \frac{0.45(50.325)}{91.143} = 0.248$$

Problem 2

A liquid solution consisting of 4 components (species 1,2,3 and 4) is in equilibrium with a vapor mixture containing 25 mole% species 1, 25 mole% species 2, 25 mole% species 3, and 25 mole% species 4. The temperature is 72 °C. For this system, the vapor pressures of the pure species are given by the Antoine equation:

$$lnP_i^{Sat} = A_i - \frac{B_i}{T+C_i}$$

where T is in K and the vapor pressures are in bars. The Antoine coefficients are given in the following table:

Species	**1**	**2**	**3**	**4**
A_i	9.2131	9.1325	9.2164	9.2535
B_i	2477.07	2766.63	2697.55	2911.32
C_i	-39.94	-50.50	-48.78	-56.51

Calculate the liquid composition.

Solution

The vapor composition and temperature are given. At the given temperature of 72 °C, the vapor pressures of pure species are calculated from the given Antoine equations as:

P_1^{Sat} =2.995608 bar ; P_2^{Sat} =0.773332 bar; P_3^{Sat} =1.121237 bar; P_4^{Sat} =0.434814 bar.

From Raoult's law:

$$y_i P = x_i P_i^{Sat}$$

$$\sum x_i = P \sum y_i / P_i^{Sat} = 1$$

Thus,

$$P = \frac{1}{\sum y_i / P_i^{Sat}}$$

For the system under consideration:

$$P = \frac{1}{\left[\frac{y_1}{P_1^{Sat}} + \frac{y_2}{P_2^{Sat}} + \frac{y_3}{P_3^{Sat}} + \frac{y_4}{P_4^{Sat}}\right]} = \frac{1}{\left[\frac{0.25}{2.995608} + \frac{0.25}{0.773332} + \frac{0.25}{1.121237} + \frac{0.25}{0.434814}\right]}$$
$$= 0.830111 \text{ bar}$$

Now that P is known, the liquid composition can be calculated from the Raoult law as:

$$x_1 = \frac{y_1 P}{P_1^{Sat}} = \frac{0.25(0.830111)}{2.995608} = 0.069277$$

$$x_2 = \frac{y_2 P}{P_2^{Sat}} = \frac{0.25(0.830111)}{0.773332} = 0.268355$$

$$x_3 = \frac{y_3 P}{P_3^{Sat}} = \frac{0.25(0.830111)}{1.121237} = 0.185088$$

$$x_4 = \frac{y_4 P}{P_4^{Sat}} = \frac{0.25(0.830111)}{0.434814} = 0.477279$$

Problem 3

A liquid system consisting of 15 moles of benzene (B), 21 moles of toluene (T), and 47 moles of water (W) is at equilibrium with its vapor at 50 °C. Assuming that benzene and toluene mixture follows the Raoult's law, but that neither are miscible with water, calculate the following:

i) the total pressure above the mixture

ii) the composition of the vapor

Given that at 50°C, $P_W^{sat} = 12.3\ kPa$, $P_B^{sat} = 40\ kPa$, $P_T^{sat} = 11.3\ kPa$

Solution

Composition of liquid mixture of benzene and toluene:
$n_B = 15$ moles, $n_T = 21$ moles

$$x_B = \frac{n_B}{n_B + n_T} = \frac{15}{15 + 21} = 0.4167$$

$$x_T = \frac{n_T}{n_B + n_T} = \frac{21}{15 + 21} = 0.5833$$

Water is not soluble with benzene and toluene mixture. Therefore, water acts independently of benzene and toluene mixture and

$$x_w = 1.0$$

i) Total pressure:

$$P = P_w + P_B + P_T = P_W^{Sat} + x_B P_B^{Sat} + x_T P_T^{Sat} = 12.3 + (0.4167 \times 40) + (0.5833 \times 11.3)$$
$$= 35.5593\ kPa$$

ii) Composition of vapor mixture:

$$y_W P = P_W^{Sat} \; ; y_W = P_W^{Sat}/P = 12.3/35.5593 = 0.3459$$

$$y_B P = x_B P_B^{Sat} \; ; y_B = x_B P_B^{Sat}/P = 0.4167 \times 40/35.5593 = 0.4687$$

$$y_T P = x_T P_T^{Sat} \; ; y_T = x_T P_t^{Sat}/P = 0.5833 \times 11.3/35.5593 = 0.1854$$

Problem 4

The equilibrium solubility data of hydrogen gas in liquid water at 25 °C is given as follows:

Mole fraction of hydrogen in water, x_A	Partial pressure of hydrogen, P_A $(bars)$
0	0
0.00002	1.434
0.00004	2.868
0.00005	3.585
0.00007	5.019
0.0001	7.17
0.00015	10.755

Plot the data as P_A $versus$ x_A. From the slope of the plot, determine the Henry law constant, K_H.

Solution

The data are plotted as P_A $versus$ x_A as shown in the figure given below. The data follow a linear relationship passing through the origin, as expected from the Henry law. The Henry law constant (K_H), that is, the slope of P_A $versus$ x_A plot passing through the origin, is 71700 bars or 71.7 kbars.

Note that the slope of a regression line passing through the origin is $\frac{\Sigma x_i y_i}{\Sigma x_i^2}$. In the present problem, y_i represents P_A and x_i represents x_A.

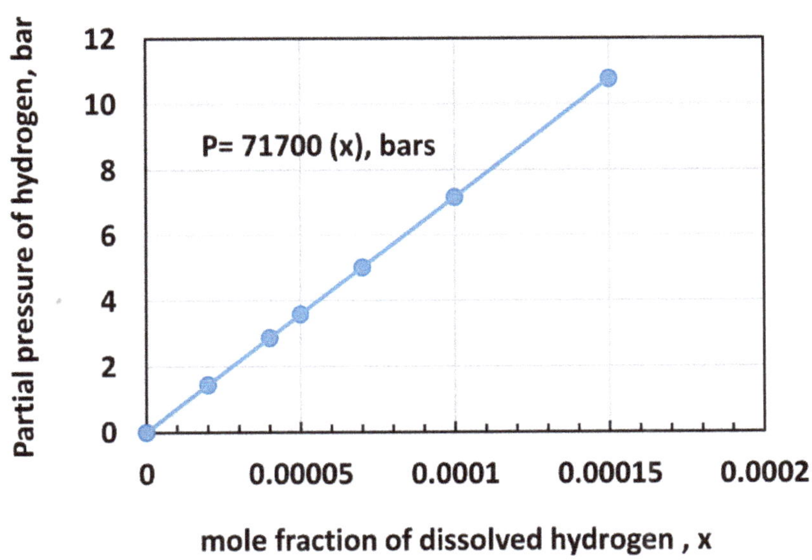

Problem 5

At 298 K, the Henry law constant of carbon dioxide dissolved in water is 1.63×10^3 atm. Estimate the solubility of carbon dioxide in moles per liter in water at 298 K and carbon dioxide partial pressure of 2.5 atm. The molecular weights of carbon dioxide and water are 44 and 18.02 g/mole, respectively. The density of water is 997 g/litre.

Solution

Using Henry's law:

$$P_{CO_2} = (1.63 \times 10^3)x$$

$$x = \frac{P_{CO_2}}{1.63 \times 10^3} = \frac{2.5}{1.63 \times 10^3} = 1.5337 \times 10^{-3}$$

$$x = \frac{n_{CO_2}}{n_{CO_2} + n_W}$$

where n is the number of moles of species in the liquid solution. Upon re-arrangement of this equation, we obtain

$$n_{CO_2} = \frac{x}{1-x} n_W$$

As the mole fraction x is small, we can calculate the number of moles of water per litre of solution as:

$$n_W = \frac{\rho_W}{M_W}$$

where ρ_W is the density of water and M_W is the molecular weight of water. Thus

$$n_W = \frac{\rho_W}{M_W} = \frac{997}{18.02} = 55.33\ moles\ /litre$$

Therefore, the solubility of carbon dioxide in moles per liter is

$$n_{CO_2} = \frac{x}{1-x} n_W = \frac{1.5337 \times 10^{-3}}{1 - 1.5337 \times 10^{-3}} (55.33) = 0.085\ moles/litre$$

Since the molecular weight of carbon dioxide is 44, we can convert this molar concentration to mass concentration as:

$$m_{CO_2} = 0.085 \times 44 = 3.74\ g/litre$$

Problem 6

For a vapor-liquid system consisting of non-ideal liquid solution in equilibrium with the ideal vapor mixture, the following modified Raoult's law is applicable:

$$y_i P = x_i \gamma_i P_i^{Sat}$$

where γ_i is the activity coefficient of species i in the liquid solution. The activity coefficient of a species depends on the temperature and composition of the liquid solution. For ideal liquid solutions $\gamma_i = 1$.

Consider a binary vapor-liquid system of 1-propanol (species 1) and water (species 2) at 70°C. The vapor mixture is ideal whereas the liquid solution is non-ideal. The vapor pressures of the pure components are given by the following Antoine equations:

$$lnP_1^{sat} = 16.0692 - \frac{3448.66}{T - 69.06}$$

$$lnP_2^{sat} = 16.2620 - \frac{3799.89}{T - 46.8}$$

where P is in kPa and T is in K. The activity coefficients of the components in liquid solution at 70°C are given in the following table:

x_1	γ_1	γ_2
0	19.88299	1
0.03	11.09854	1.008265
0.06	7.508266	1.026665
0.09	5.627792	1.050736
0.12	4.491465	1.078766
0.15	3.738355	1.110007
0.18	3.206124	1.144128
0.21	2.811899	1.181008
0.24	2.509292	1.220644
0.27	2.270448	1.26311
0.30	2.077692	1.308536
0.33	1.919302	1.357094
0.36	1.787211	1.408999
0.39	1.675695	1.464502
0.42	1.58059	1.523893
0.45	1.498795	1.587502
0.48	1.427959	1.655705
0.51	1.36627	1.728926
0.54	1.312311	1.807644
0.57	1.264959	1.892401
0.60	1.223319	1.98381
0.63	1.186667	2.082567
0.66	1.154416	2.189461
0.69	1.126087	2.30539
0.72	1.101288	2.431382
0.75	1.079699	2.568612
0.78	1.061058	2.71843
0.81	1.045153	2.882393
0.84	1.031815	3.0623
0.87	1.020912	3.260244
0.90	1.012346	3.47866
0.93	1.006048	3.720402
0.96	1.001979	3.988825
0.98	1.000496	4.184525
1.00	1	4.39518

Construct the P-x-y and y-x diagrams. Does 1-propanol/water system form an azeotrope at 70°C? If yes, what is the composition of the azeotrope?

Solution

The following procedure is used to construct the P-x-y and y-x diagrams:

- Calculate P_1^{sat} and P_2^{sat} at T=70 °C using the given Antoine equations.
- Calculate total pressure as: $P = \sum x_i \gamma_i P_i^{sat} = x_1\gamma_1 P_1^{sat} + x_2\gamma_2 P_2^{sat}$
- Calculate vapor phase composition at each x_1 values as: $y_1 = x_1\gamma_1 P_1^{sat}/P$
- Plot P vs x_1 and P vs y_1. This gives the P-x-y diagram.
- Plot y_1 vs x_1. This gives the y-x diagram.

Note that x_1, y_1 refer to mole fractions of 1-propanol in liquid and vapor phases, respectively. The following table gives the values of x_1, P, and y_1 at 70 °C. The vapor pressures of pure species are calculated from the Antoine equations as: $P_1^{sat} = 32.68716$ kPa and $P_2^{sat} = 31.17746$ kPa.

x_1	P(kPa)	y_1
0	31.17746	0
0.03	41.37549	0.26304
0.06	44.8137	0.328592
0.09	46.36703	0.357066
0.12	47.21478	0.373137
0.15	47.74554	0.383898
0.18	48.11406	0.392065
0.21	48.39019	0.398877
0.24	48.6082	0.404977
0.27	48.78573	0.410733
0.30	48.93194	0.416378
0.33	49.05127	0.42207
0.36	49.14536	0.42793
0.39	49.21402	0.434058
0.42	49.25575	0.440544
0.45	49.26797	0.447473
0.48	49.2472	0.454938
0.51	49.18906	0.463037
0.54	49.08824	0.471878
0.57	48.93843	0.481591
0.60	48.73216	0.492326
0.63	48.46071	0.504263
0.66	48.11384	0.517623
0.69	47.67954	0.53268
0.72	47.14375	0.549777
0.75	46.48991	0.569354
0.78	45.69853	0.591982
0.81	44.74657	0.618418
0.84	43.60673	0.649688
0.87	42.24653	0.687217
0.90	40.62724	0.733047
0.93	38.70241	0.790207
0.96	36.41622	0.8634

0.98	34.65856	0.924715
1.00	32.68716	1

The P-x-y diagram and y-x equilibrium curve for 1-propanol/water system at 70 °C are shown below. The system forms an azeotrope at $x \approx 0.45$.

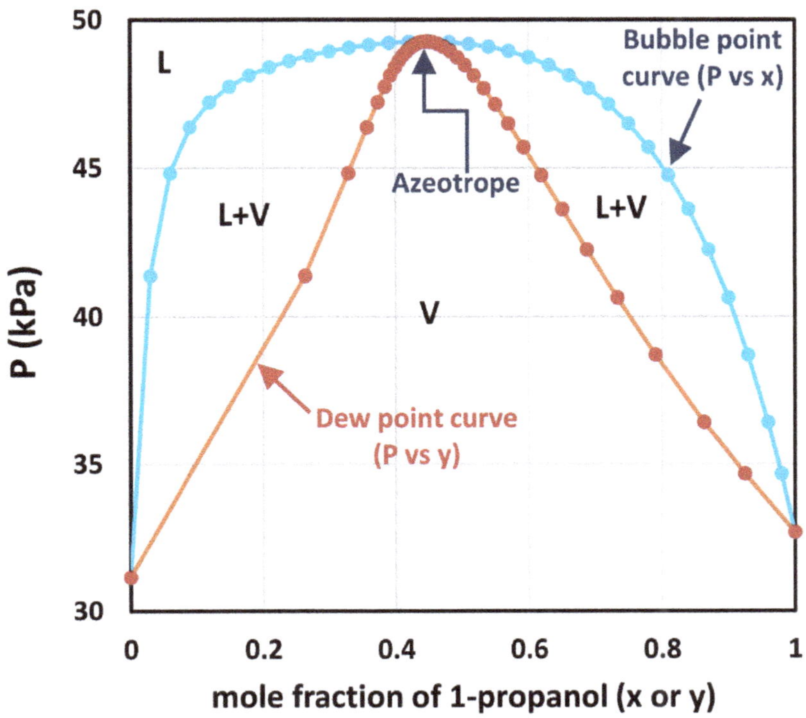

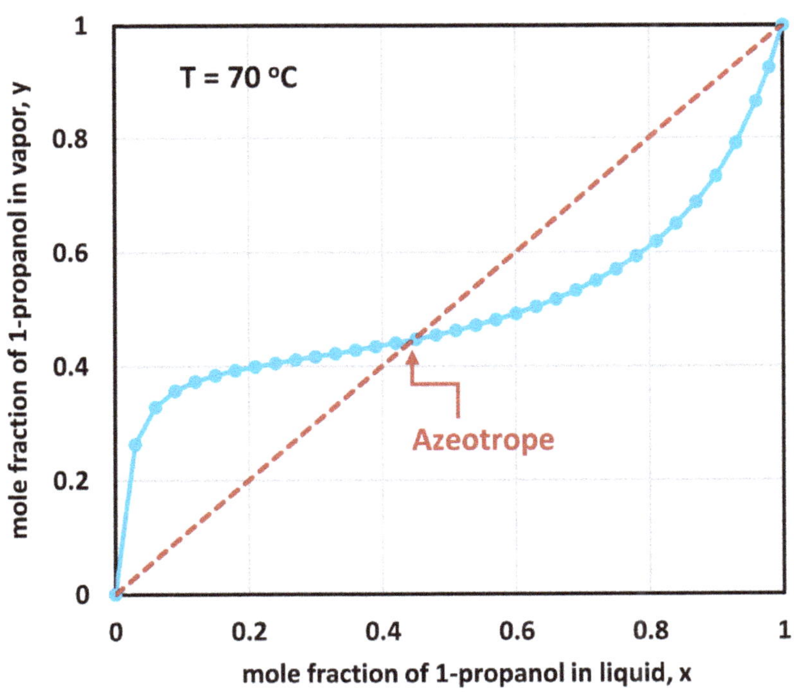

Problem 7

Henry's law constant is strongly dependent on the temperature. It typically increases with the increase in temperature at low temperatures, reaches a maximum value, and then decreases with further increase in temperature. The temperature where a maximum in the Henry law constant occurs is dependent on the critical temperature of the solvent for a given solute.

Harvey (AIChE journal, Vol. 42, 1996, 1491-1494) proposed the following correlation to predict the Henry law constant as a function of temperature.

$$ln(K_H) = ln(P_s^{Sat}) + (A/T^*) + B(1 - T^*)^{0.355}(T^*)^{-1} + C(T^*)^{-0.41}exp(1 - T^*)$$

where K_H is the Henry law constant in MPa, P_s^{Sat} is the vapor pressure of solvent in MPa, T^* is reduced temperature defined as T/T_c (T_c is the critical temperature of the solvent), and A, B, C are constants for a given solute-solvent combination.

The values of constants A, B, C for different solutes in water are given in the table below.

Solute	A	B	C
Hydrogen, H_2	-4.4964	6.0952	5.8390
Helium, He	-3.6674	7.1476	4.5939
Neon, Ne	-3.0496	5.4346	5.2244
Argon, Ar	-7.4316	4.2239	9.6803
Krypton, Kr	-8.6850	3.6077	11.0380
Xenon, Xe	-13.3731	3.8360	14.9540
Nitrogen, N_2	-11.6184	4.9266	13.3445
Oxygen, O_2	-9.4025	4.4923	11.3387
Carbon dioxide, CO_2	-9.4234	4.0087	10.3199
Hydrogen sulfide, H_2S	-5.7131	5.3727	5.4227
Methane, CH_4	-11.0094	4.8362	12.5227
Ethylene, C_2H_4	-11.5733	5.5672	11.9954
Ethane, C_2H_6	-19.7237	4.5051	20.6740

The vapor pressure of solvent (water) can be calculated from the following Antoine equation:

$$lnP_s^{Sat} = 16.2620 - \frac{3799.89}{T + 226.35}$$

where P_s^{Sat} is in kPa and T is in °C.

Using the Harvey correlation, plot K_H (MPa) versus $T(K)$ for hydrogen and carbon dioxide over a temperature range of 250 to 625 K. Which gas has a higher K_H value at any given temperature? The solvent is water which has a critical temperature of T_c = 647.1 K.

Solution

The K_H (MPa) versus $T(K)$ data are generated for hydrogen and carbon dioxide using the Harvey correlation and the Antoine equation. The data are plotted in the figure shown below.

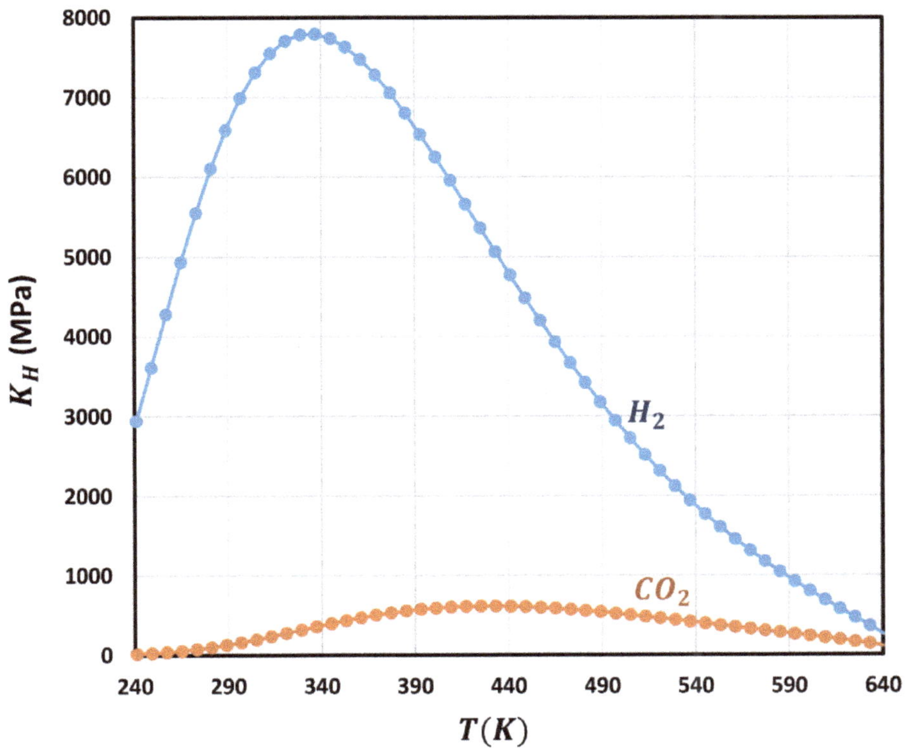

Clearly, the Henry law constant is strongly dependent on the temperature. For a given solute, it increases with the increase in temperature at low temperatures, reaches a maximum value, and then decreases with further increase in temperature. At any given temperature, the K_H value of hydrogen gas is much higher than that of carbon dioxide. A high value of K_H for hydrogen indicates that hydrogen has a very low solubility in water in comparison with carbon dioxide.

Note that the Harvey correlation assumes that the total pressure of the gas is equal to the vapor pressure of the solvent at the given temperature. As the effect of pressure of the gas on Henry law constant is usually negligible, the Harvey correlation can be applied at any pressure as an approximation.

SHORT REVIEW QUESTIONS

1. Give examples of industrial applications of distillation, absorption, stripping, liquid-liquid extraction, and solid-liquid leaching.
2. Define thermodynamic phase equilibrium.
3. What is the criterion of phase equilibrium?
4. Define an equilibrium stage?
5. State the Raoult law.
6. Why does a liquid mixture of two or more different components behave like an ideal solution?
7. Why does a liquid mixture of two or more different components behave like a non-ideal solution with positive deviation from ideal behavior?
8. Why does a liquid mixture of two or more different components behave like a non-ideal solution with negative deviation from ideal behavior?
9. What is an azeotrope? What would be the vapor composition when an azeotropic liquid solution is boiled?
10. State the Henry law

11. Under what conditions are the Raoult and Henry laws applicable to real solutions?

PRACTICE PROBLEMS

Problem 1

Plot the variations of partial pressures and total pressure for a binary vapor-liquid system as functions of mole fraction of a more volatile species A (at a fixed temperature). Both vapor mixture and liquid solution are ideal in nature, that is, the Raoult law is applicable over the entire composition range. The vapor pressures of the pure species at the given temperature are as follows: $P_A^{Sat} = 135.5\ kPa$ and $P_B^{Sat} = 54\ kPa$

Problem 2

Plot the variations of partial pressures and total pressure (on the same graph paper) for a binary vapor-liquid system as functions of mole fraction of a more volatile species A at a fixed temperature. Both vapor mixture and liquid solution are ideal in nature, that is, the Raoult law is applicable over the entire composition range. The vapor pressures of the pure species at the given temperature are as follows: $P_A^{Sat} = 100\ kPa$ and $P_B^{Sat} = 20\ kPa$.

Can we treat the system to be ideal, if the total pressure of the system at the given temperature is 75 kPa when $x_A = x_B = 0.5$? If not, does the system exhibit positive deviation or negative deviation from an ideal behavior?

Problem 3

Consider a 3-component system of species 1, 2 and 3. For this system, the vapor pressures of the pure species are given by the Antoine equations as:

$$lnP_1^{Sat} = 13.7819 - \frac{2726.81}{T+217.572}$$

$$lnP_2^{Sat} = 13.932 - \frac{3056.96}{T+217.625}$$

$$lnP_3^{Sat} = 13.9726 - \frac{3259.93}{T+212.30}$$

where T is in °C and the vapor pressures are in kPa. Species 1 is the most volatile species.

Assuming that the system is ideal in nature and follows the Raoult law, calculate the total pressure and vapor composition at equilibrium, given that $T = 120^oC, x_1 = 0.25, x_2 = 0.35, x_3 = 0.40$.

Problem 4

For the 3-component system of problem 2 given above, calculate the total pressure and liquid composition at equilibrium, given that $T = 110^oC, y_1 = 0.40, y_2 = 0.35, y_3 = 0.25$.

Problem 5

A liquid system consisting of 12 moles of benzene (B), 18 moles of ethylbenzene (EB), and 37 moles of water (W) is at equilibrium with its vapor at 90 °C. Assuming that benzene and ethylbenzene mixture follows the Raoult's law, but that neither are miscible with water, calculate the following:

i) the total pressure above the mixture

ii) the composition of the vapor

For this system, the vapor pressures of the pure species are given by the Antoine equation:

$$lnP_i^{Sat} = A_i - \frac{B_i}{T+C_i}$$

where T is in °C and the vapor pressures are in kPa. The Antoine coefficients are given in the following table:

Antoine Coefficient	Benzene	Ethylbenzene	Water
A_i	13.8594	14.0045	16.2620
B_i	2773.78	3279.47	3799.80
C_i	220.07	213.20	226.35

Problem 6

The solubility data of nitrogen gas in liquid water at 25 °C is given as follows:

Mole fraction of nitrogen in water, x_A	Partial pressure of nitrogen, $P_A\ (bars)$
0	0
0.00002	1.728
0.00004	3.456
0.00005	4.32
0.00007	6.048
0.0001	8.64
0.00015	12.96

Plot the data as P_A $versus$ x_A. From the slope of the plot, determine the Henry law constant K_H. (Use the linear regression formula for the slope of a line passing through the origin).

Problem 7

At 298 K, the Henry law constant of oxygen dissolved in water is $4.259\text{x}10^4$ atm. Estimate the solubility of oxygen in moles per liter in water at 298 K and oxygen partial pressure of 5 atm. Given that the molecular weight of oxygen is 32, determine the solubility of oxygen in water in grams/litre. The density of water is 997 g/litre and its molecular weight is 18.02 g/mole.

Problem 8

At 298 K, the Henry law constant of nitrogen dissolved in water is 86400 bar.

a) Estimate the solubility of nitrogen in grams per liter in water at 298 K and nitrogen partial pressure of 1 bar. The molecular weights of nitrogen and water are 28 and 18.02 g/mole, respectively. The density of water is 997 g/litre.

b) What is the concentration of dissolved nitrogen in moles/litre under atmospheric conditions? The partial pressure of nitrogen gas in atmosphere is 0.78 atm.

Problem 9

The solubility data of some gas in liquid water at 25 °C is given as follows:

Mole fraction of gaseous species in water, x_A	Partial pressure of gas, P_A $(bars)$
0	0
0.00002	0.033
0.00004	0.066
0.00005	0.083
0.00007	0.116
0.0001	0.165
0.00015	0.248

a) Plot the data as P_A $versus$ x_A.

b) From the slope of the plot, determine the Henry law constant K_H in units of bar. (Use the linear regression formula for the slope of a line passing through the origin)

Problem 10

Consider a binary liquid-gas system at equilibrium at 25°C. The liquid solution has the following composition in terms of mole fractions: $x_A = 0.002$ and $x_B = 0.998$. The solute (species A) follows the Henry law and the solvent (species B) follows the Raoult law. The Henry law constant for solute is 2×10^5 kPa. The vapor pressure of pure solvent is given by the following Antoine equation:

$$ln(P^{sat}) = 16.262 - \frac{3799.89}{T + 226.35}$$

where T is in °C and P^{sat} is in kPa.

i) Calculate the partial pressure of species A
ii) Calculate the partial pressure of species B
iii) Calculate the total pressure
iv) Calculate the composition of the gas phase

Problem 11

Starting from the following chemical potential criterion of phase equilibrium, derive the Raoult law.

$$\mu_i^V(T,P) = \mu_i^L(T,P)$$

where $\mu_i^V(T,P)$ is the chemical potential of species i in the vapor mixture and $\mu_i^L(T,P)$ is the chemical potential of species i in the liquid solution.

Hints:

For ideal vapor mixture, $\mu_i^V(T,P) = \mu_{i,pure}^V(T,P) + RTlny_i$

For ideal liquid solution, $\mu_i^L(T,P) = \mu_{i,pure}^L(T,P) + RTlnx_i$

For pure species i in vapor/liquid phase equilibrium, $\mu_{i,pure}^V\left(T,P_i^{Sat}\right) = \mu_{i,pure}^L\left(T,P_i^{Sat}\right)$

For pure liquid species i, $\mu_{i,pure}^L(T,P) \approx \mu_{i,pure}^L\left(T,P_i^{Sat}\right)$

For pure vapor species i, $\mu_{i,pure}^V(T,P) = \mu_{i,pure}^V\left(T,P_i^{Sat}\right) + RTln\left(\frac{P}{P_i^{Sat}}\right)$

Problem 12

Re-do problem 6 of tutorial problems at 60°C. The activity coefficients of the components in liquid solution at 60°C are given in the following table.

x_1	γ_1	γ_2
0	21.2657	1
0.03	11.65824	1.008502
0.06	7.821865	1.027292
0.09	5.834978	1.051778
0.12	4.641893	1.080239
0.15	3.854202	1.111942
0.18	3.298902	1.146573
0.21	2.888255	1.184028
0.24	2.573376	1.224325
0.27	2.32501	1.267557
0.3	2.124645	1.313875
0.33	1.960024	1.36348
0.36	1.82273	1.416611
0.39	1.706797	1.473553
0.42	1.607888	1.534634
0.45	1.522778	1.600229
0.48	1.449025	1.670765
0.51	1.384746	1.746727
0.54	1.32847	1.828668
0.57	1.279034	1.917215
0.6	1.23551	2.013085
0.63	1.197147	2.117095
0.66	1.163341	2.230182
0.69	1.133596	2.353424
0.72	1.107509	2.48806
0.75	1.084752	2.635529
0.78	1.065058	2.797497
0.81	1.048213	2.97591
0.84	1.034048	3.173047
0.87	1.022434	3.391588
0.9	1.013279	3.634703
0.93	1.006523	3.906159
0.96	1.00214	4.210458
0.98	1.000537	4.434199
1	1	4.676792

Problem 13

Re-do problem 7 of the tutorial problems for helium and krypton, that is, plot K_H (MPa) versus $T(K)$ for helium and krypton over a temperature range of 250 to 625 K. Which gas has a higher K_H value at any given temperature? The solvent is water which has a critical temperature of $T_c = 647.1$ K.

Problem 14

The Henry law constants for hydrogen disulfide (H_2S) in water at different temperatures are given in the table below.

$T(K)$	K_H (MPa)	$T(K)$	K_H (MPa)
283.15	37.19	318.15	82.48
288.15	42.86	323.15	89.57
293.15	48.94	333.15	104.36
298.15	55.22	343.15	120.58
303.15	61.71	353.15	136.79
308.15	68.50	363.15	145.91
313.15	75.49	373.15	149.96

Using the Harvey correlation, estimate the Henry law constants of H_2S in water over a temperature range of 250 to 375 K and compare the predicted values with the above experimental data.

FURTHER READING

1. P.C. Wankat, *Separation Process Engineering*, Prentice Hall, New Jersey, 3rd Edition, 2012.
2. C.J. Geankoplis, *Transport Processes and Unit Operations*, Prentice Hall, New Jersey, 3rd Edition, 1993.
3. C.J. Geankoplis, A.A. Hersel, D.H. Lepek, Transport processes and separation process principles, Prentice Hall, New Jersey, 5th Edition, 2018.
4. E.J. Henley, J.D. Seader, *Equilibrium-Stage Separation Operations in Chemical Engineering*, John Wiley & Sons, New York, 1981.
5. J.D. Seader, E.J. Henley, D.K. Roper, *Separation Process Principles Using Process Simulators*, 4th edition, Wiley, New York, 2015.
6. W.L. McCabe, J.C. Smith, P. Harriott, *Unit Operations of Chemical Engineering*, McGraw Hill, 7th Edition, New York, 2005.
7. R.E. Treybal, *Mass Transfer Operations*, McGraw-Hill, New York, 3rd edition, 1980.
8. A.S. Foust, L.A. Wenzel, C.W. Clump, L. Maus, L.B. Anderson, *Principles of Unit Operations*, Wiley & Sons, New York, 1980.
9. F.M. Khoury, *Separation Processes*, CRC Press, 4th edition, Boca Raton, 2015.
10. C.J. King, *Separation Processes*, McGraw-Hill, New York, 2nd Edition, 1980.
11. R.K. Sinnott, *Chemical Engineering Design*, Coulson & Richardson Chemical Engineering Series, Vol. 6, 4th edition, Pergamon Press, Oxford, 2005.
12. J.L. Humphrey, G.E. Keller II, *Separation Process Technology*, McGraw-Hill, New York, 1997.
13. D.W. Green, R.H. Perry, *Perry's Chemical Engineers' Handbook,* McGraw-Hill, New York, 2008.
14. J.M. Smith, H.C. Van Ness, M.M. Abbott, *Introduction to Chemical Engineering Thermodynamics*, 7th edition, McGraw-Hill, New York, 2005.
15. M. Netusil, P. Dtil, Natural gas dehydration, in *Natural Gas*, S. Gupta, Ed., IntechOpen, London, UK, 2012. http://dx.doi.org/10.5772/45802.
16. Y.J. Lin, G.T. Rochelle, Approaching a reversible stripping process for CO_2 capture, *Chemical Engineering Journal*, 283: 1033-1043, 2016.
17. A. Proctor, Soybean oil extraction and processing, in *Soybeans. Chemistry, Technology and Utilization,* Ed. K. Liu, Chapman & Hall, New York, 1997.
18. J.J. Carroll, Henry's law revisited, *Chem. Eng. Progress,* January issue: 49-56, 1999.
19. A.H.Harvey, Semiempirical correlation for Henry's constants over large temperature ranges, AIChE Journal, 42:1491-1494, 1996.

Chapter 2

Single-Stage Gas Absorption

2.1 LEARNING OUTCOMES

Upon completion of this chapter, students will be able to:

- Describe the process of gas absorption or scrubbing and its importance
- Explain the direction and driving force for solute transfer in gas absorption
- Analyze single-stage gas absorption process for dilute and concentrated systems
- Determine the percent recovery of solute from gas phase to liquid phase

2.2 INTRODUCTION

Gas absorption, also known as gas scrubbing, is a separation process in which a solute present in the gas phase is transferred from the gas phase to the liquid phase (see Figure 2.1). The removal of solute from the gas phase to the liquid phase is done either to clean the gas of impurities and pollutants or to recover the valuable solute from the gas phase (the solute could later be recovered from the solute-rich liquid phase by distillation). Thus the gas phase consists of carrier gas + solute and the liquid phase consists of solvent + solute. The carrier gas and solvent are immiscible with each other. However, the solute is soluble in solvent. The solvent is non-volatile and is selected such that the solute has preference for it over the carrier gas. When no chemical reaction is involved between the solute and the solvent, the absorption is referred to as "physical absorption". If the solute to be removed from the gas phase reacts with the solvent, the absorption is referred to as "chemical absorption".

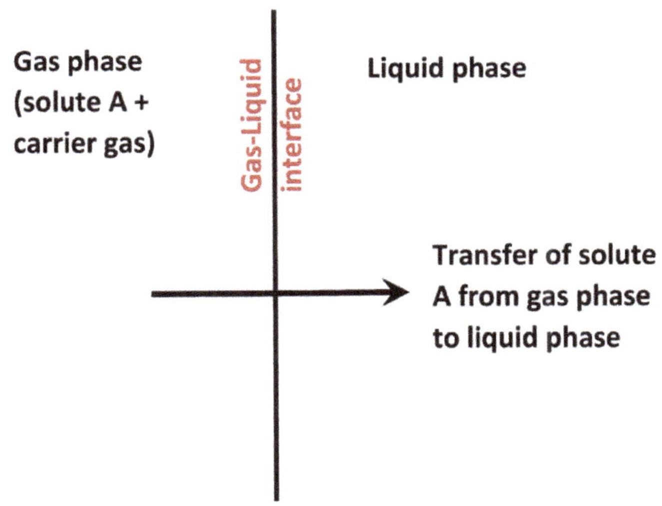

Figure 2.1 Direction of mass transfer of solute in an absorption process.

Here we will restrict ourselves to physical absorption. Examples of commercial applications of physical absorption include: 1) removal of ammonia from ammonia-air mixture using water as solvent; 2) removal of acetone vapor from acetone-air mixture using water as solvent; and 3) removal of sulfur dioxide from sulfur dioxide – air mixture using water as solvent.

Figure 2.2 shows a single-stage contact between gas and liquid phases. The gas containing solute (impurity or valuable material) enters the stage from the bottom and the solvent (liquid) falls in the downward direction from the top. The two streams entering the stage (solute-laden gas and solvent streams) are not at phase equilibrium with each other. Hence the transfer of solute from one phase to another is possible thermodynamically. The two input streams undergo intimate mixing on a perforated (sieve) tray due to formation of bubbles of gas. During the mixing process, the solute is transferred from the gas phase to the solvent stream. The treated gas leaves the stage from the top and the solute-rich liquid comes out of the bottom. Thus we have counter-current flow of gas and liquid streams. In our analysis, it will be assumed that the two streams *leaving the stage* (treated gas stream and solute-rich liquid stream) are in *phase equilibrium* with each other. This is possible if the contact interfacial area between the two phases is large, the mixing is intense, and the residence (contact) time of the two phases is sufficient.

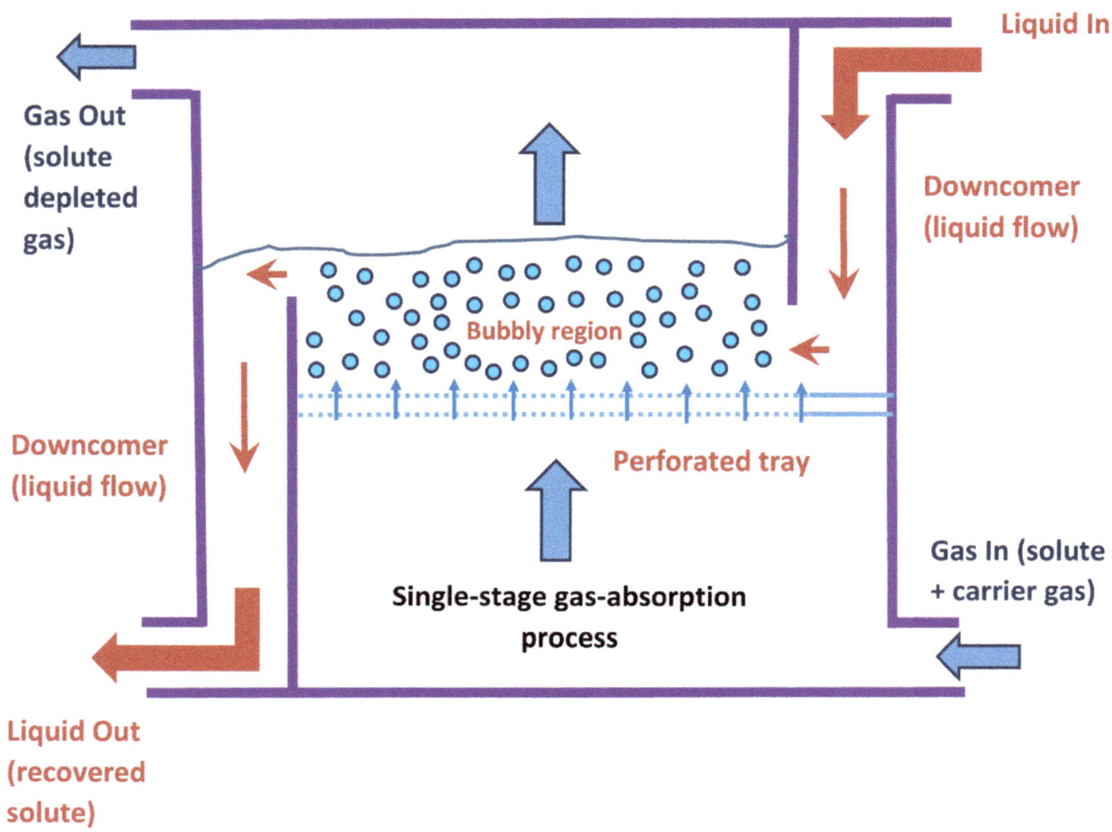

Figure 2.2 Single-stage gas absorption process.

2.3 DRIVING FORCE FOR MASS TRANSFER IN GAS ABSORPTION

Before we proceed with the analysis of a single stage, let us consider the driving force for mass transfer of solute from the gas phase to the liquid phase. As noted in Chapter 1, the driving force for mass transfer is the difference in chemical potentials. The mass flows in the direction of decreasing chemical potential. Thus for solute A to move from the gas phase to the liquid phase, the chemical potential of solute in the gas phase must be higher than that in the liquid phase. Referring to Figure 2.3, the solute (species A) will move from gas phase to liquid phase provided that:

$$\mu_{A,mix}^{G} > \mu_{A,solu}^{L} \tag{2.1}$$

Thus the driving force for mass transfer (transfer of solute A) is:

$$Driving\ force \rightarrow (\mu^G_{A,mix} - \mu^L_{A,solu}) > 0 \tag{2.2}$$

It can be readily shown (proof given in the tutorial of this chapter) that this driving force for mass transfer of solute A can be further expressed as:

$$Driving\ force \rightarrow (y_A - y_A^*) > 0 \tag{2.3}$$

where y_A^* is the mole fraction of solute in the gas phase if the gas phase was in equilibrium with the existing liquid phase. If the existing gas phase is not in equilibrium with the existing liquid phase, y_A^* is a fictitious quantity to be determined from an equilibrium relationship. When the existing gas and liquid phases are in equilibrium, $y_A = y_A^*$ and no further transfer of solute is possible from the gas phase to the liquid phase. When the two phases initially come in contact with each other, the driving force is positive ($y_A > y_A^*$). With continued contact between the two phases, the driving force decreases due to transfer of solute from gas phase to liquid phase. As noted earlier the two phases eventually reach equilibrium if the contact interfacial area between the two phases is large, the mixing is intense, and the residence (contact) time of the two phases in the mixing device is sufficient.

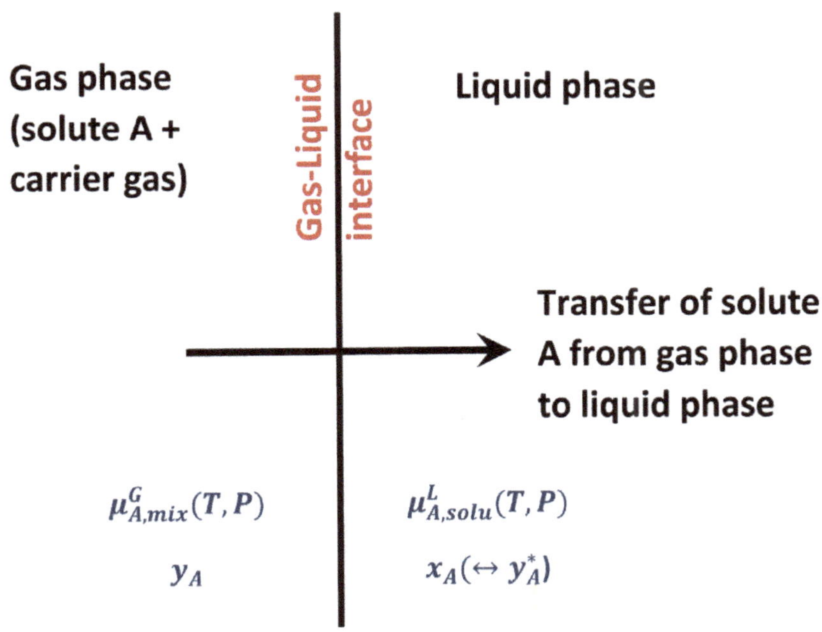

Figure 2.3 Variables governing the driving force for mass transfer of solute from gas phase to liquid phase in absorption process.

2.4 ANALYSIS OF SINGLE-STAGE GAS ABSORPTION: DILUTE SYSTEMS

Figure 2.4 shows a schematic diagram of a single equilibrium stage absorption process. The liquid stream molar flow rates are expressed by L and the solute (species A) concentrations in the liquid stream are expressed in terms of mole fraction x. The molar flow rates of gas stream are given in terms of variable V and the concentrations of solute in the gas stream are given by variable y. The absorption process is of physical type and hence no chemical reaction is involved. Note that it is not necessary to include the subscript "A" in the composition variables of solute (x and y) as we are dealing with only one solute.

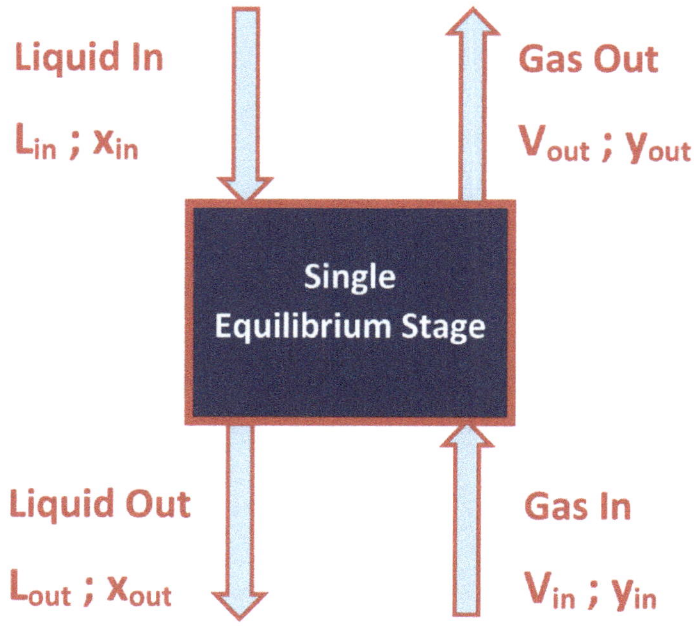

Figure 2.4 Single equilibrium stage absorption.

The overall mass (mole) balance gives:

$$L_{in} + V_{in} = L_{out} + V_{out} \tag{2.4}$$

The solute mole balance gives:

$$x_{in}L_{in} + y_{in}V_{in} = x_{out}L_{out} + y_{out}V_{out} \tag{2.5}$$

Assuming that the system is *dilute* (concentration of solute in any stream is small) and that the mass transfer rate of solute from one stream to another is low, we can regard:

$$L_{in} = L_{out} = L \tag{2.6}$$

$$V_{in} = V_{out} = V \tag{2.7}$$

Consequently, Equation (2.5) reduces to:

$$y_{out} = y_{in} + (-\tfrac{L}{V})(x_{out} - x_{in}) \tag{2.8}$$

This is the *operating line* of a single-stage absorption process. As shown in Figure 2.5, it is a straight line of slope –L/V originating from the point (x_{in}, y_{in}). Given the compositions of the gas and liquid streams entering the stage, the operating line represents a loci of the compositions of the two streams leaving the stage. The operating line is not the solution of the problem, that is, it does not give the values of the compositions of the gas and liquid streams leaving the stage. It simply relates the gas and liquid compositions. The solute concentration increases in the liquid phase and decreases in the gas phase as gas absorption proceeds (indicated by an arrow placed on the operating line). In order to determine the unknown gas and liquid compositions, we need to use the condition of phase equilibrium that the two streams leaving the stage are at equilibrium with each other.

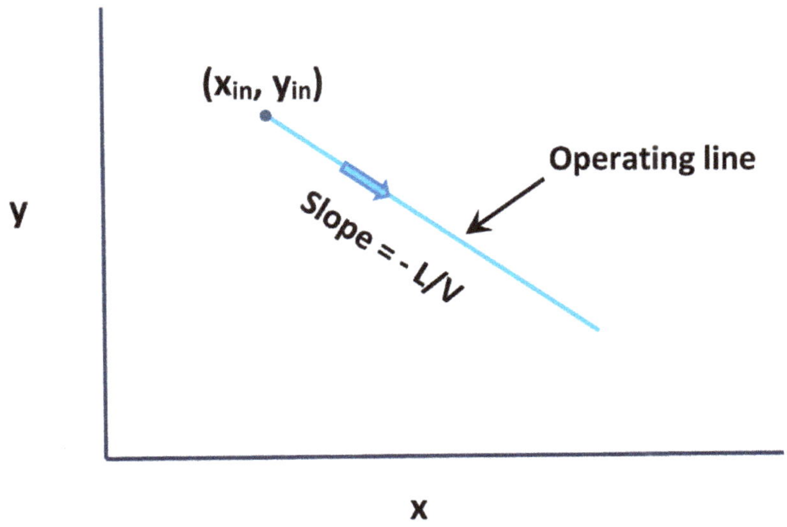

Figure 2.5 Operating line of a single-stage absorption process.

Figure 2.6 shows the graphical solution of a single equilibrium stage absorption process. Given the inlet streams compositions and their flow rates, and the equilibrium relationship (Henry law constant of solute), the intersection of the operating and equilibrium lines gives the unknown compositions (x_{out} and y_{out}).

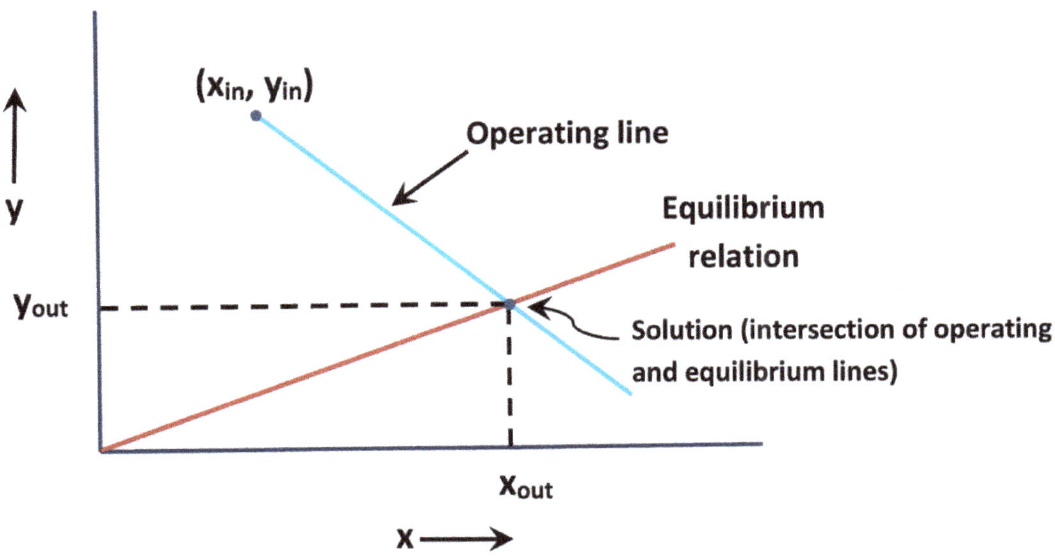

Figure 2.6 Graphical solution of a single equilibrium stage absorption.

The driving force for mass transfer of solute from gas phase to liquid phase is shown in Figure 2.7. When the two phases come in contact with each other, the driving force is initially maximum and as the transfer of solute from gas phase to liquid phase proceeds, the driving force decreases and eventually phase equilibrium is established and consequently, no more transfer of solute from gas to liquid is possible.

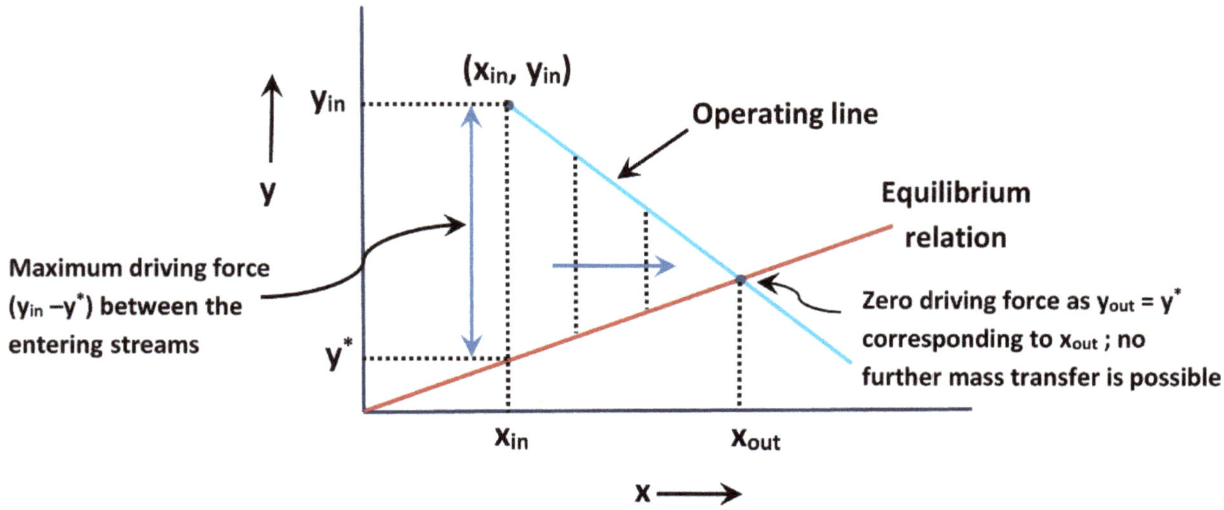

Figure 2.7 The driving force for mass transfer in single-stage absorption.

The percent recovery of solute from a single equilibrium stage absorption process can be calculated once the compositions of the leaving streams are known:

$$\%\,Recovery\ of\ Solute = 100(1 - \frac{y_{out}}{y_{in}}) \tag{2.9}$$

We can readily obtain an analytical solution of unknown compositions of the two discharge streams using the operating line (Equation (2.8)) and the equilibrium relation, that is, Henry law:

$$P_A = K_H x \tag{2.10}$$

$$y = (K_H/P)x = Kx \tag{2.11}$$

where P_A is the partial pressure of the solute in the gas phase, P is the total pressure, K_H is the Henry law constant of solute in units of pressure, and K is the dimensionless Henry law constant or distribution coefficient that depends on temperature and pressure of the system. Note that the Henry law constant K_H for a given system depends mainly on temperature. The effect of pressure on K_H is usually small. A large value of $K_{H,i}$ is indicative of low solubility of solute in liquid and a small value of $K_{H,i}$ is indicative of high solubility of solute in liquid. For example, $K_{H,CO_2} = 1.63$ kbars whereas $K_{H,O_2} = 44.1$ kbars in water at 25 °C indicating that carbon dioxide is much more soluble in water as compared with oxygen.

Since the two leaving streams are at equilibrium with each other, Equation (2.11) can be applied and re-written as:

$$y_{out} = Kx_{out} \tag{2.12}$$

From Equations (2.8) and (2.12), the following expression for y_{out} can easily be derived by eliminating x_{out} :

$$\frac{y_{out}}{y_{in}} = \left(\frac{1}{1+A}\right)\left[1 + \frac{L}{V}\left(\frac{x_{in}}{y_{in}}\right)\right] \tag{2.13}$$

where A is called absorption factor, defined as:

$$A = \frac{|slope\ of\ operating\ line|}{slope\ of\ equilibrium\ line} = \frac{L/V}{K} \tag{2.14}$$

The absorption factor "A" is a measure of degree of separation of solute in absorption process. To see the significance of "A", consider a single equilibrium stage absorption with pure solvent entering the device. In this case, Equation (2.13) could be re-written as:

$$\frac{y_{out}}{y_{in}} = \left(\frac{1}{1+A}\right) \tag{2.15}$$

When $A \to 0, y_{out} = y_{in}$ and no separation of solute occurs. When $A \to \infty, y_{out} = 0$, that is, complete separation of solute occurs.

2.5 ANALYSIS OF SINGLE-STAGE GAS ABSORPTION: CONCENTRATED SYSTEMS

In the preceding section, it was assumed that the system is dilute, that is, the concentration of solute in the gas stream to be treated was small. Consequently, the mass transfer rate of solute from gas stream to liquid stream was small and the flow rates of gas and liquid streams were constant from inlet to outlet. As a result, the operating line was linear with constant slope of $-L/V$.

In this section, we will investigate more *concentrated* systems with large enough mass transfer rates that the flow rates of gas and liquid streams can no longer assumed to be constant from inlet to outlet. When the total flow (carrier + solute) rates of the fluid streams (gas or liquid) are not constant, the operating line is non-linear. The equilibrium relation is also generally non-linear for concentrated systems. For example, Meredith (R.E. Meredith, An extension of Henry's law to higher concentrations, University of California Radiation Laboratory report, UCRL-8426, 1958) proposed the following extension of Henry's law for concentrated liquid solutions:

$$P_A = -K_H \ln(1 - x_A) \tag{2.16}$$

In the limit $x_A \to 0$, $\ln(1 - x_A) = -x_A$ and therefore, Equation (2.16) reduces to the standard Henry law (Equation (2.10)). Note that K_H in Equation (2.16) is the usual Henry law constant obtained for dilute solutions.

Figure 2.8 compares the prediction of Henry law for dilute solutions with the data generated for concentrated solutions by Equation (2.16), assuming Equation (2.16) is valid for concentrated solutions. At high mole fractions of solute in the liquid phase, the data deviate from the Henry law (Equation (2.10)).

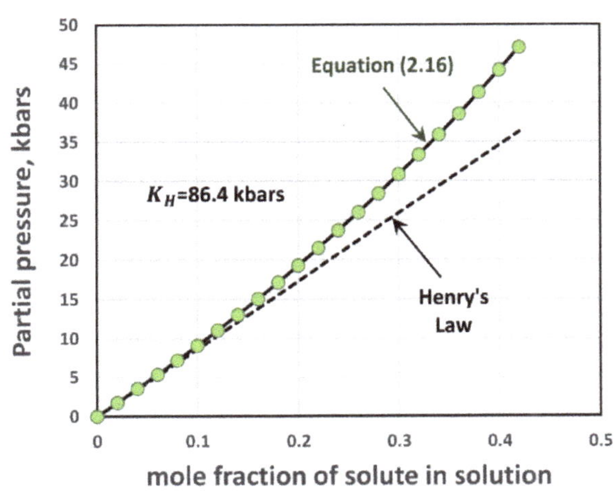

Figure 2.8 Nonlinear behavior of equilibrium relationship in concentrated solutions.

For the system under consideration, the data matches with the Henry law (Equation (2.10)) only when $x \leq 0.05$.

Figure 2.9 shows another example of deviation of data generated for concentrated solutions by Equation (2.16) from the prediction of Henry law for dilute solutions. At high mole fractions of solute in the liquid phase, the data deviate from the Henry law (Equation (2.10)). The data matches with the Henry law (Equation (2.10)) only when x is small.

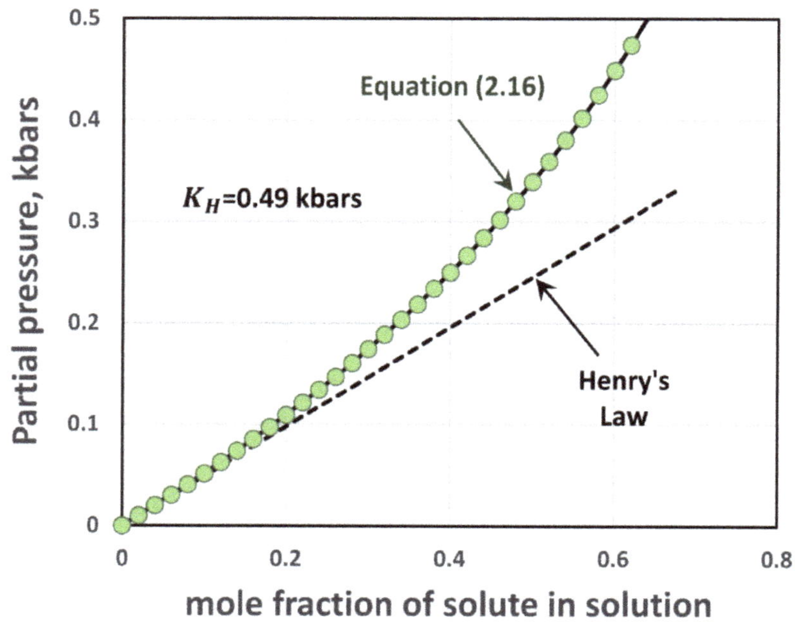

Figure 2.9 Nonlinear behavior of equilibrium relationship in concentrated solutions.

For concentrated systems with large mass transfer rate of solute from gas phase to liquid phase, it is preferred to use mole ratios (instead of mole fractions) as composition variables. The mole ratios are defined as follows:

$$X = \frac{\text{moles of solute in liquid stream}}{\text{moles of carrier (solute–free solvent) in liquid stream}} = \frac{x}{1-x} \tag{2.17}$$

$$Y = \frac{\text{moles of solute in gas stream}}{\text{moles of carrier (solute–free gas) in gas stream}} = \frac{y}{1-y} \tag{2.18}$$

The advantage of using mole ratios is that the operating line for the process turns out to be a linear relation. As already noted, the operating line is non-linear if mole fractions are used as the composition variables in situations where the flow rates of the gas and liquid streams are not constant. However, the Henry law, Equation (2.11), becomes non-linear when mole ratios are used as the composition variables. The Henry law expressed in terms of the mole ratios is linear only at low values of X as shown in Figure 2.10. Thus, for dilute systems, $Y = KX$ where K has the same value as that in $y = Kx$. The equilibrium relation could be extended to higher concentrations in some cases as $Y = K'X$ where K' may be different from K in the Henry law.

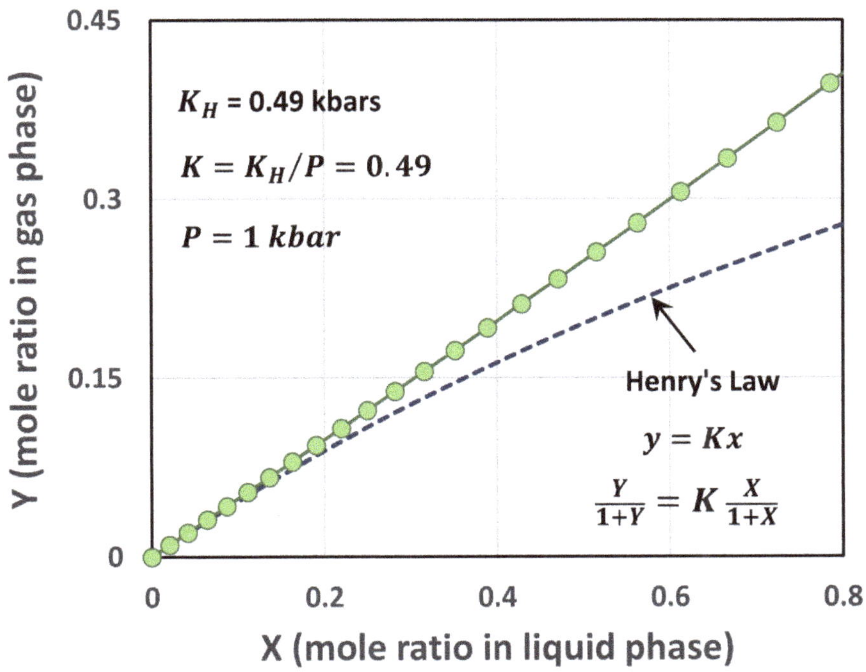

Figure 2.10 Equilibrium relation for concentrated systems plotted using mole ratios (Y vs X) as the composition variables. The data generated for concentrated system using Equation (2.16) follow the Henry law at low values of X.

In order to develop an equation for the operating line in units of mole ratios, we need to consider the solute-free flow rates. Let L' be the carrier (solute-free solvent) flow rate of the liquid stream and V' be the carrier (solute-free gas) flow rate of the gas stream. In any given fluid stream, we generally have both solute and carrier present. However, the bulk of the stream is the carrier fluid. The carrier fluids of the gas and liquid streams are completely immiscible with each other. Only the solute moves from the gas stream to the liquid stream.

From the definitions of X and L' , it follows that the solute flow rate in the liquid stream is XL' and the solute flow rate in the gas stream is YV'. As the carrier fluids are immiscible, their flow rates do not change. Referring to Figure 2.11, it follows that:

$$L'_{in} = L'_{out} = L' \tag{2.19}$$

$$V'_{in} = V'_{out} = V' \tag{2.20}$$

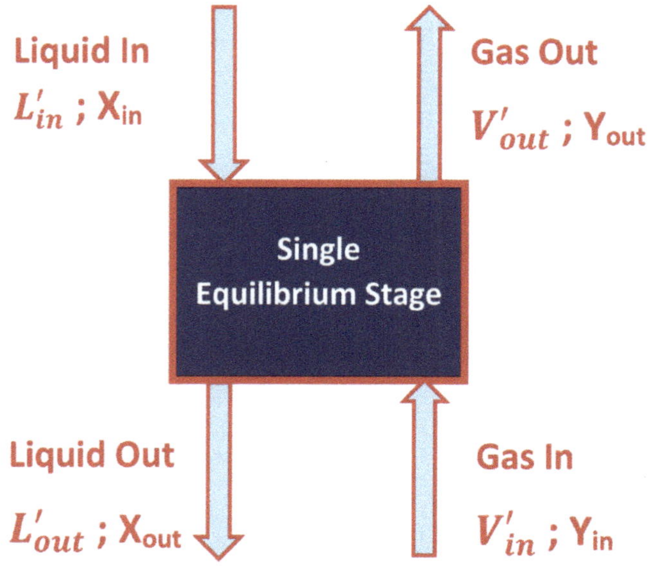

Figure 2.11 Single equilibrium stage absorption.

The solute mole balance gives:

$$X_{in}L'_{in} + Y_{in}V'_{in} = X_{out}L'_{out} + Y_{out}V'_{out} \tag{2.21}$$

From Equations (2.19) - (2.21), it follows that:

$$Y_{out} = Y_{in} + (-\tfrac{L'}{V'})(X_{out} - X_{in}) \tag{2.22}$$

This is the operating line for concentrated systems. It is a linear relation on Y-X scale with a slope of -L'/V'. It represents a loci of the compositions of the two leaving streams.

To solve for Y_{out} and X_{out} , we need the equilibrium relation. In general, the solution is obtained graphically as shown in Figure 2.12. An analytical expression for Y_{out} can be derived only if the equilibrium relation is assumed to be linear, that is:

$$Y_{out} = K'X_{out} \tag{2.23}$$

From Equations (2.22) - (2.23), it follows that:

$$\frac{Y_{out}}{Y_{in}} = \left(\frac{1}{1+A'}\right)\left[1 + \frac{L'}{V'}\left(\frac{X_{in}}{Y_{in}}\right)\right] \tag{2.24}$$

where A' is called absorption factor, defined as:

$$A' = \frac{L'/V'}{K'} \tag{2.25}$$

Once Y_{out} is known, X_{out} can be obtained from the equilibrium relation, Equation (2.23). It should be noted that a mole ratio can exceed a value of unity whereas the mole fraction cannot be larger than unity.

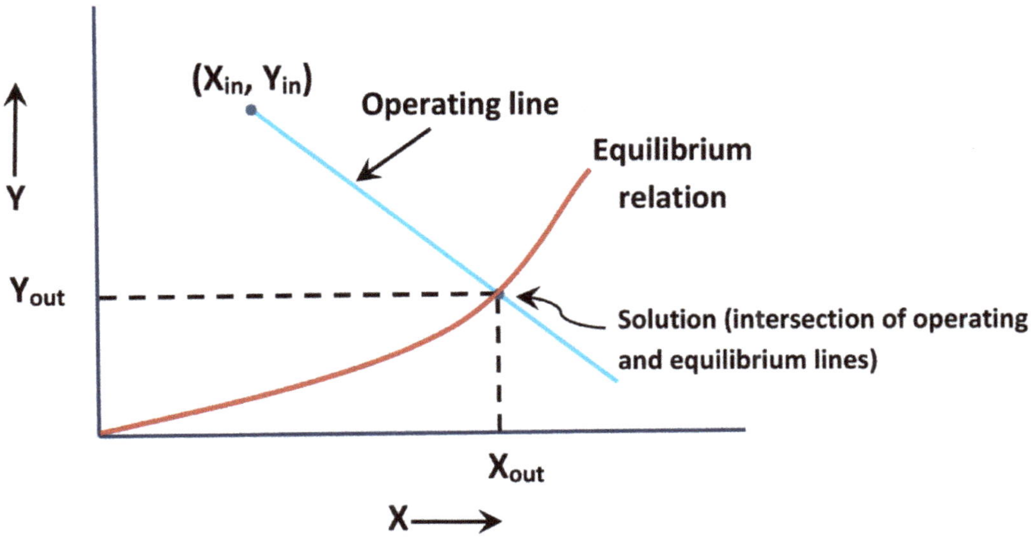

Figure 2.12 Graphical solution of single stage absorption for concentrated systems.

NOTATION

A	absorption factor, defined in Equation (2.14)
A'	absorption factor, defined in Equation (2.25)
K	distribution coefficient of solute, defined in Equation (2.11)
K'	distribution coefficient of solute, defined in Equation (2.23)
K_H	Henry law constant of solute defined in Equation (2.10)
$K_{H,i}$	Henry law constant of species i
L	molar flow rate of liquid stream (solvent + solute)
L'	molar flow rate of solute-free liquid stream, that is, molar flow rate of solvent
P	total pressure of gas
P_A	partial pressure of solute (species A) in the gas phase
V	molar flow rate of gas stream (carrier gas + solute)
V'	molar flow rate of solute-free gas stream, that is, molar flow rate of carrier gas
x, x_A	mole fraction of solute A in liquid phase
X	mole ratio of solute in the liquid phase, defined in Equation (2.17)
y, y_A	mole fraction of solute A in gas phase
y^*	mole fraction of solute A in the gas phase if the gas phase was in equilibrium with the existing liquid phase
Y	mole ratio of solute in the gas phase, defined in Equation (2.18)

Greek Letters

$\mu^G_{A,mix}$	chemical potential of solute (species A) in the gas mixture
$\mu^L_{A,solu}$	chemical potential of solute (species A) in the liquid solution

TUTORIAL PROBLEMS WITH SOLUTION

Problem 1

Starting from the following expression for the driving force for solute mass transfer in absorption,

$$Driving\ force \rightarrow (\mu^G_{A,mix} - \mu^L_{A,solu}) > 0$$

show that this driving force for mass transfer can also be expressed as:

$$Driving\ force \rightarrow (y_A - y^*_A) > 0$$

Solution

Referring to the figure given below and assuming that the gas mixture behaves as an ideal gas mixture and that the liquid solution behaves as an ideal liquid solution, one can write the expressions for the chemical potentials as follows:

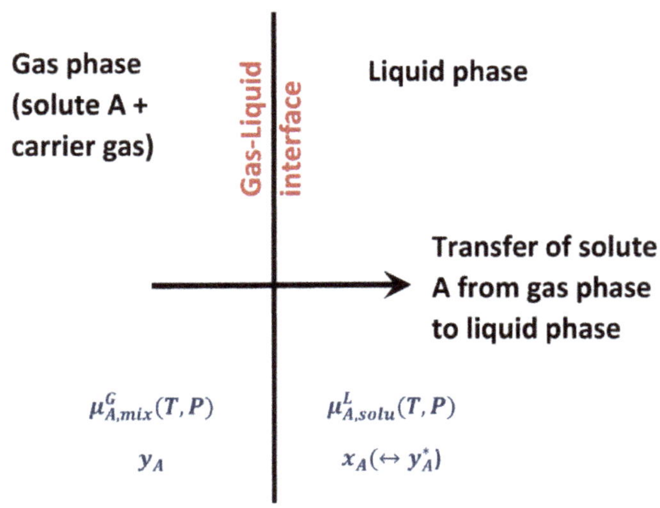

$$\mu^G_{A,mix}(T,P) = \mu^G_{A,pure}(T,P) + RT\ln y_A$$

Taking into account the effect of pressure on the chemical potential of a pure ideal gas, one can write:

$$\mu^G_{A,pure}(T,P) = \mu^G_{A,pure}\left(T,P^{Sat}_A\right) + RT\ln\frac{P}{P^{Sat}_A}$$

Combining the above two equations, we get

$$\mu^G_{A,mix}(T,P) = \mu^G_{A,pure}\left(T,P^{Sat}_A\right) + RT\ln\frac{y_A P}{P^{Sat}_A}$$

This expression gives the chemical potential of species A in an ideal gas mixture. What we need next is an expression of chemical potential of species A in an ideal liquid solution. For an ideal liquid solution:

$$\mu^L_{A,solu}(T,P) = \mu^L_{A,pure}(T,P) + RT\ln x_A$$

Since the effect of pressure on chemical potential of liquid is small, this expression can be re-written as:

$$\mu^L_{A,solu}(T,P) = \mu^L_{A,pure}\left(T,P^{Sat}_A\right) + RT\ln x_A$$

Subtracting the chemical potential of species A in liquid solution from the chemical potential of species A in gas mixture, we obtain:

$$\mu_{A,mix}^{G}(T,P) - \mu_{A,solu}^{L}(T,P) = RT\ln\frac{y_A P}{x_A P_A^{Sat}}$$

Note that $\mu_{A,pure}^{G}\left(T,P_A^{Sat}\right) = \mu_{A,pure}^{L}\left(T,P_A^{Sat}\right)$. This follows from the vapor-liquid equilibrium condition of pure species A. The driving force for mass transfer of species A from gas phase to liquid phase is:

$$\mu_{A,mix}^{G}(T,P) - \mu_{A,solu}^{L}(T,P) > 0$$

This is equivalent to the following condition:

$$RT\ln\frac{y_A P}{x_A P_A^{Sat}} > 0$$

Therefore, the driving force for mass transfer of species A from gas phase to liquid phase is:

$$y_A > \frac{x_A P_A^{Sat}}{P}$$

The mole fraction of species A in the gas phase if the gas phase is in equilibrium with the liquid phase is

$$y_A^* = \frac{x_A P_A^{Sat}}{P}$$

From these two expressions, it follows that the condition for transfer of species A from gas phase to liquid phase is

$$y_A > y_A^*$$

Thus, the driving force for mass transfer of species from gas phase to liquid phase is:

$$Driving\ force \rightarrow (y_A - y_A^*) > 0$$

Problem 2

A gas mixture at 1 bar pressure containing air and solute (some gaseous species) is contacted in a single-stage mixer continuously with pure water at 20 °C. The gas and liquid streams exit at equilibrium with each other. The flow rate of inlet gas stream is 100 kmol/h with a mole fraction of solute of 0.20. The flow rate of liquid water entering the stage is 300 kmol/h. The solute follows the Henry law: $y = 0.142 \times 10^4 x$.

As the mass transfer rate of solute from gas phase to liquid phase is small, you can assume that the molar flow rates of gaseous and liquid streams are constant. For this problem,

a) Determine the compositions of the two discharge streams
b) Determine the percent recovery of solute from the gas phase to the liquid phase
c) Show the process (operating line, equilibrium line, and mole fractions) on the x-y graph

Solution

The absorption factor A can be calculated as:

$$A = \frac{L/V}{K} = \frac{300/100}{0.142\times10^4} = 0.002113$$

As shown in the diagram below, $x_{in} = 0$. Consequently,

$$\frac{y_{out}}{y_{in}} = \left(\frac{1}{1+A}\right) = \frac{1}{1+0.002113} = 0.997891$$

$$y_{out} = 0.997891 \times 0.20 = 0.199578$$

$$x_{out} = y_{out}/K = 0.199578/(0.142 \times 10^4) = 1.40548 \times 10^{-4}$$

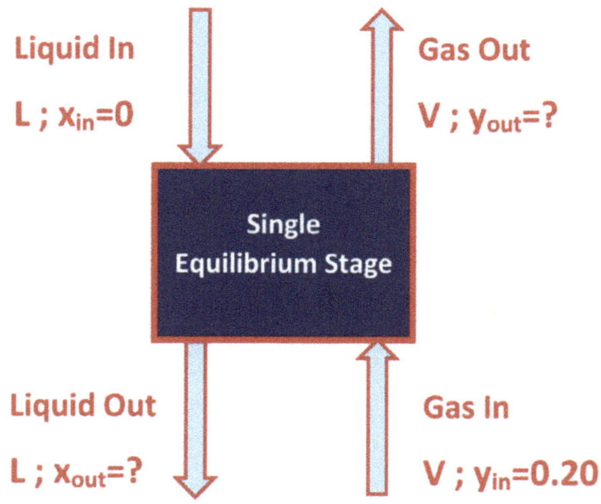

The graphical solution of the problem is shown in the following figure.

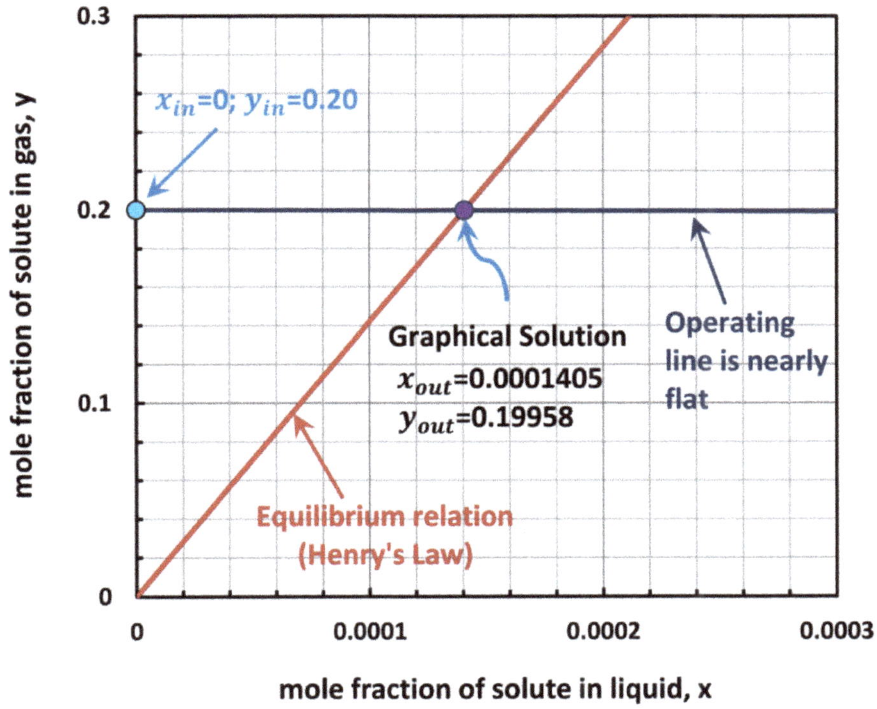

The percent recovery of solute from the gas phase to the liquid phase can be calculated as follows:

$$\%\ Recovery\ of\ Solute = 100\left(1 - \frac{y_{out}}{y_{in}}\right) = 100\left(1 - \frac{0.199578}{0.20}\right) = 0.211$$

The recovery of solute is only 0.211 % in this single-stage absorption process.

Note that the absorption factor A is very small in the present case ($A = 0.002113$) and therefore, the percent recovery of solute is so low.

Problem 3

A gas mixture at 5 bar pressure containing air and solute (some gaseous species) is contacted in a single-stage mixer continuously with pure water at 25 °C. The gas and liquid streams exit at equilibrium with each other. The flow rate of inlet gas stream is 35 kmol/h with a mole fraction of solute of 0.15. The flow rate of liquid water entering the stage is 450 kmol/h. The solute follows the Henry law: $P_{solute} = 10^3 x$, where P_{solute} is in bar. As the mass transfer rate of solute from gas phase to liquid phase is small, you can assume that the molar flow rates of gaseous and liquid streams are constant.

a) Determine the compositions of the two discharge streams
b) Determine the percent recovery of solute from the gas phase to the liquid phase
c) Show the graphical solution (operating line, equilibrium line, and solution) on the y-x graph

Solution: Refer to the diagram given below.

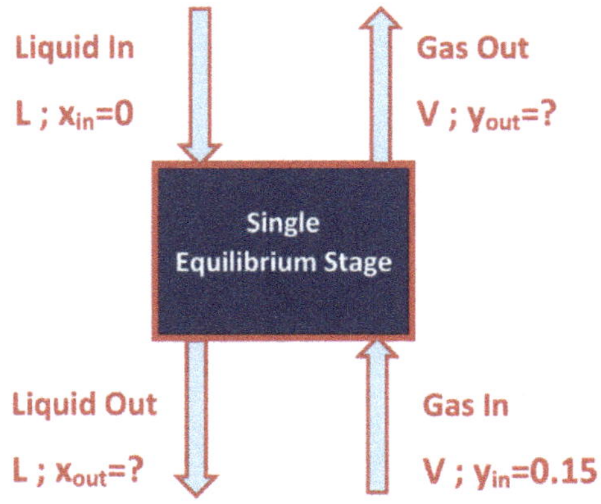

a) The absorption factor A can be calculated as:

$$K = y/x = 10^3/P = 10^3/5 = 200$$

$$A = \frac{L/V}{K} = \frac{450/35}{200} = 0.064286$$

As shown in the diagram above, $x_{in} = 0$. Consequently,

$$\frac{y_{out}}{y_{in}} = \left(\frac{1}{1+A}\right) = \frac{1}{1+0.064286} = 0.9396$$

$$y_{out} = 0.9396 \times 0.15 = 0.14094$$

$$x_{out} = y_{out}/K = 0.14094/(200) = 7.047 \times 10^{-4}$$

b) The percent recovery of solute from the gas phase to the liquid phase can be calculated as follows:

$$\%\ Recovery\ of\ Solute = 100\left(1 - \frac{y_{out}}{y_{in}}\right) = 100\left(1 - \frac{0.14094}{0.15}\right) = 6.04$$

The recovery of solute is only 6.04 % in this single-stage absorption process.

As the absorption factor A is small in the present case ($A = 0.064286$), the percent recovery of solute is low.

c) The graphical solution of the problem is shown in the following figure.

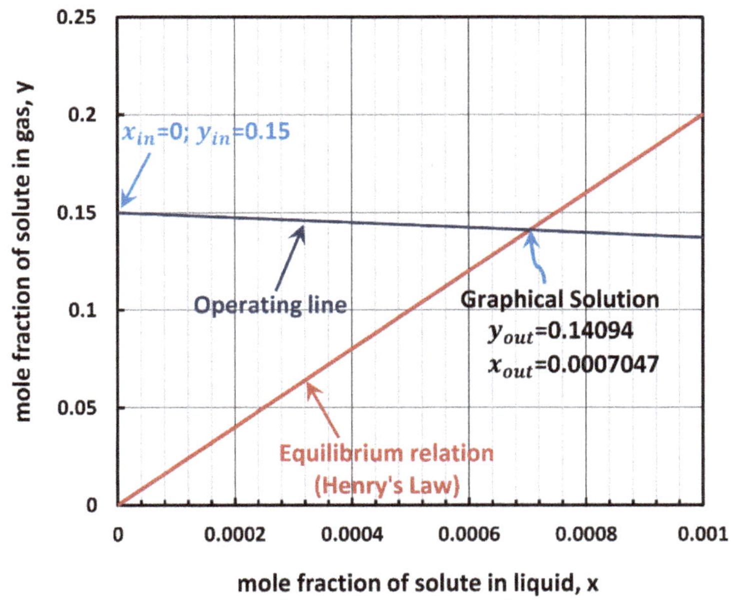

Operating line equation:

$$y_{out} = y_{in} + \left(-\frac{L}{V}\right)(x_{out} - x_{in})$$

$$y_{out} = 0.15 - 12.857x_{out}$$

Equilibrium relation:

$$y = 200x$$

Problem 4

A single-stage absorption is used to recover solute from a gas into a solvent. The mole fraction of solute in the gas entering the absorber is 0.25 and the gas flow rate is 500 kmol/h. Pure solvent enters the absorber at a flow rate of 1500 kmol/h. The output streams leaving the absorber are at equilibrium with each other. The equilibrium relationship (Henry's law) in terms of mole ratios is $Y = 2.8\,X$. The carrier gas is insoluble in the solvent and the solvent is nonvolatile.

Determine the following:

a) the mole ratios and mole fractions of solute in the exit gas and liquid streams
b) the molar flow rates of the exit gas and liquid streams
c) the percent recovery of solute

Solution

Refer to the diagram given below

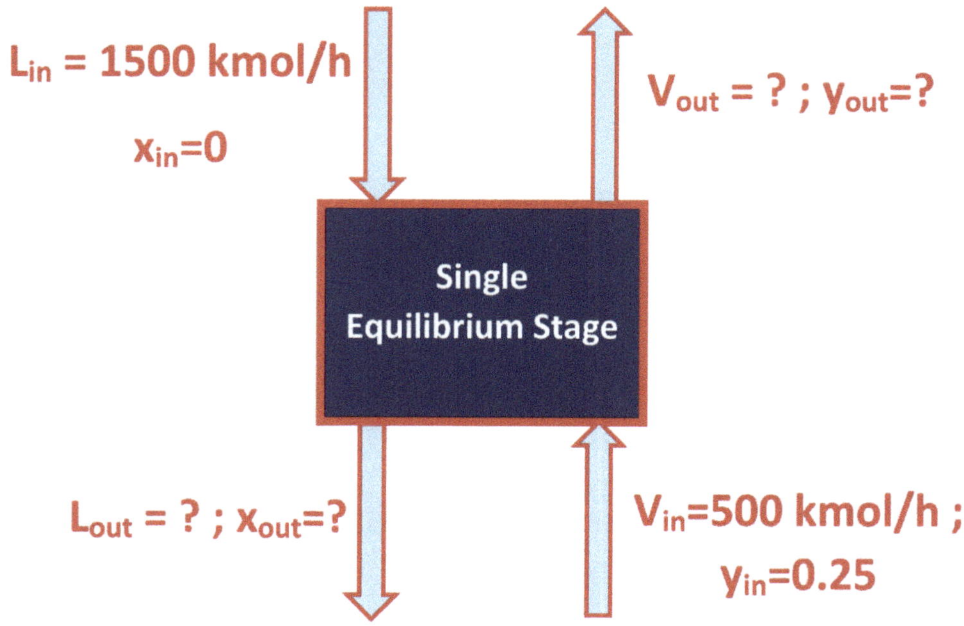

(a) the mole ratios and mole fractions of solute in the exit gas and liquid streams?

$$L' = L_{in}(1 - x_{in}) = 1500(1 - 0) = 1500\ kmol/h$$

$$V' = V_{in}(1 - y_{in}) = 500(1 - 0.25) = 375\ kmol/h$$

$$Y_{in} = \frac{y_{in}}{1 - y_{in}} = \frac{0.25}{1 - 0.25} = 0.3333$$

$$X_{in} = \frac{x_{in}}{1 - x_{in}} = \frac{0}{1 - 0} = 0$$

$$\frac{L'}{V'} = \frac{1500}{375} = 4$$

Equation of the operating line:

$$Y_{out} = Y_{in} - \left(\frac{L'}{V'}\right)(X_{out} - X_{in}) = 0.3333 - 4X_{out}$$

The equilibrium line:

$$Y_{out} = 2.8(X_{out})$$

From the above two equations, it follows that the mole ratios of the output streams are:

$$Y_{out} = 0.13725\ ;\ X_{out} = 0.04902$$

The solution is also shown graphically in the figure given below.

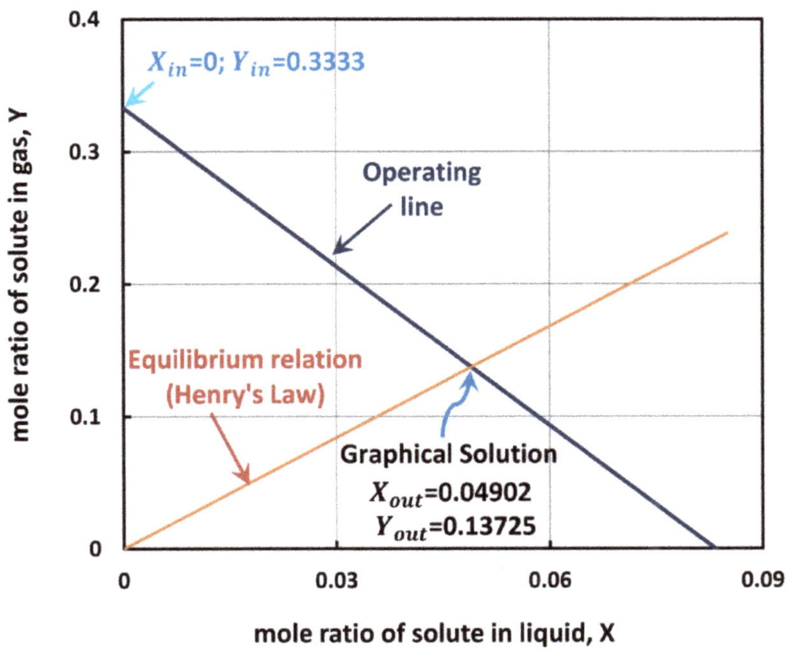

The mole fractions can be calculated from the mole ratios as follows:

$$y_{out} = \frac{Y_{out}}{1 + Y_{out}} = \frac{0.13725}{1 + 0.13725} = 0.12069$$

$$x_{out} = \frac{X_{out}}{1 + X_{out}} = \frac{0.04902}{1 + 0.04902} = 0.04673$$

(b) the molar flow rates of the exit gas and liquid streams?

$$L_{out} = \frac{L'}{1 - x_{out}} = \frac{1500}{1 - 0.04673} = 1573.53\ kmol/h$$

Thus L has increased from 1500 to 1573.53 kmol/h.

$$V_{out} = \frac{V'}{1 - y_{out}} = \frac{375}{1 - 0.12069} = 426.47\ kmol/h$$

The gas flow rate has decreased from 500 to 426.47 kmol/h.

(c) the percent recovery of solute?

$$\%\ Recovery\ of\ Solute = 100\left(1 - \frac{Y_{out}}{Y_{in}}\right) = 100\left(1 - \frac{0.13725}{0.3333}\right) = 58.82$$

Thus the recovery of solute is 58.82% in this single-stage absorption process. This large recovery in a single stage absorption process is not surprising as the solute is highly soluble in solvent as indicated by a small value of Henry's law constant.

The absorption factor $A' = (L'/V')/K'$ in the present case is large (=1.43).

Problem 5

A gas mixture of air and acetone vapor containing 25 mole% acetone is passed through an equilibrium stage so that acetone can be absorbed in water stream. The molar flow rate of gas mixture at the entrance is 250 kmol/h. The pure water stream enters at a molar flow rate of 500 kmol/h. The output streams leaving the absorber are at equilibrium with each other. The equilibrium relationship is $y = -2.5\ln(1-x)$. You can assume that the carrier gas is insoluble in the solvent and the solvent is nonvolatile. The process operates isothermally at a pressure of 1 bar.

Determine the following:

a) the mole ratios and mole fractions of acetone in the exit gas and liquid streams
b) the molar flow rates of the exit gas and liquid streams
c) the percent recovery of acetone

Solution

Refer to the diagram given below

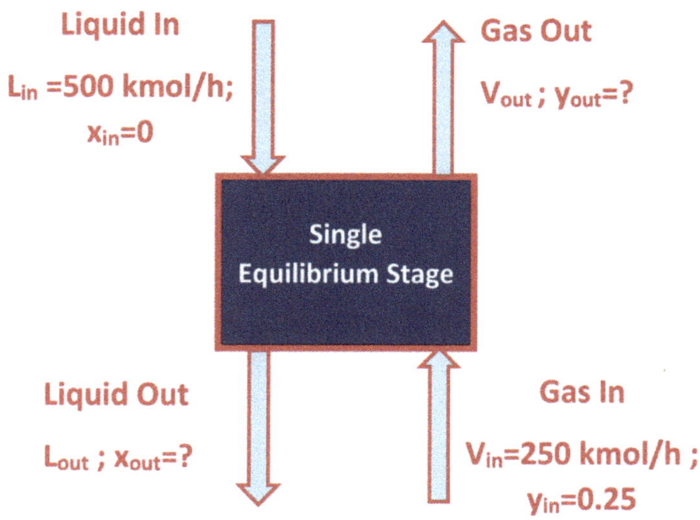

(a) the mole ratios and mole fractions of solute in the exit gas and liquid streams?

$$L' = L_{in}(1 - x_{in}) = 500(1 - 0) = 500\ kmol/h$$

$$V' = V_{in}(1 - y_{in}) = 250(1 - 0.25) = 187.5\ kmol/h$$

$$Y_{in} = \frac{y_{in}}{1 - y_{in}} = \frac{0.25}{1 - 0.25} = 0.3333$$

$$X_{in} = \frac{x_{in}}{1 - x_{in}} = \frac{0}{1 - 0} = 0$$

$$\frac{L'}{V'} = \frac{500}{187.5} = 2.6666$$

Equation of the operating line:

$$Y_{out} = Y_{in} - \left(\frac{L'}{V'}\right)(X_{out} - X_{in}) = 0.3333 - 2.6666X_{out}$$

The equilibrium relation:

$$\frac{Y_{out}}{1 + Y_{out}} = -2.5ln\left[1 - \frac{X_{out}}{1 + X_{out}}\right]$$

The graphical solution of the above two equations gives the following values of the mole ratios of the output streams:

$$Y_{out} = 0.174\,;\; X_{out} = 0.0605$$

The solution is shown graphically in the figure given below.

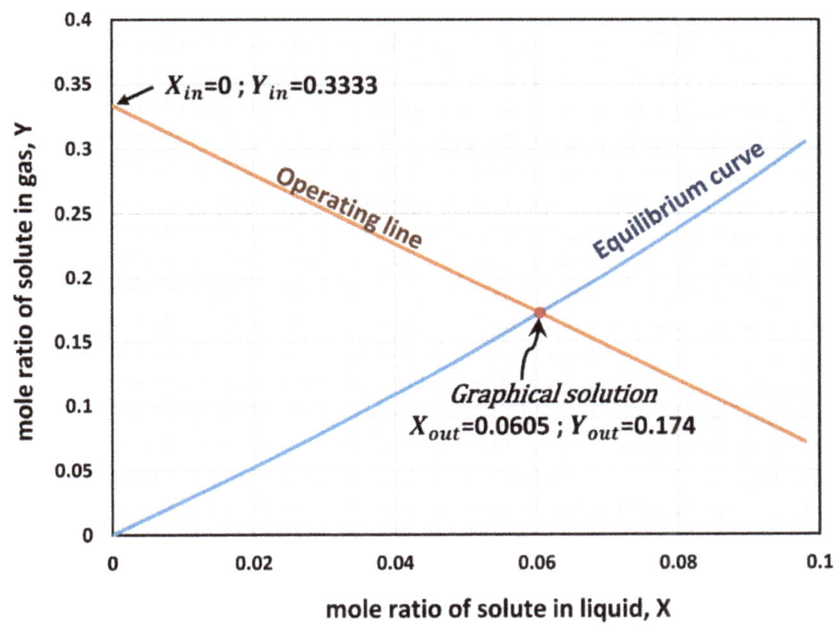

The mole fractions can be calculated from the mole ratios as follows:

$$y_{out} = \frac{Y_{out}}{1 + Y_{out}} = \frac{0.174}{1 + 0.174} = 0.21065$$

$$x_{out} = \frac{X_{out}}{1 + X_{out}} = \frac{0.0605}{1 + 0.0605} = 0.0644$$

(b) the molar flow rates of the exit gas and liquid streams?

$$L_{out} = \frac{L'}{1 - x_{out}} = \frac{500}{1 - 0.0644} = 534.42\; kmol/h$$

Thus L has increased from 500 to 534.42 kmol/h.

$$V_{out} = \frac{V'}{1 - y_{out}} = \frac{187.5}{1 - 0.21065} = 237.54\ kmol/h$$

The gas flow rate has decreased from 250 to 237.54 kmol/h.

(c) the percent recovery of solute?

$$\%\ Recovery\ of\ Solute = 100\left(1 - \frac{Y_{out}}{Y_{in}}\right) = 100\left(1 - \frac{0.174}{0.3333}\right) = 47.8$$

Thus the recovery of solute is 47.8% in this single-stage absorption process.

Problem 6

We wish to remove two solutes, H_2S and CO_2, from a gaseous stream using pure water as a solvent in a single-stage process. The carrier gas is air. The gas stream contains 0.3 mole% H_2S and 0.5 mole% CO_2. The molar flow rate of the gas mixture is 50 kmol/h. The water molar flow rate is 250 kmol/h. The process is carried out at $T = 298$ K and $P = 10$ bars. The Henry law constants for H_2S and CO_2 in water are: 552 bars and 1662 bars, respectively. Determine the compositions of the two discharge streams and the percent recovery of solutes H_2S and CO_2 from the gas phase to the liquid phase

Solution

This problem involves multicomponent gas absorption. The multicomponent gas absorption problems involving two or more solutes can be solved using the single solute procedure assuming that: i) the solutes act independently of each other. Thus, the equilibrium relationship for any given solute does not depend on the mole fractions of other solutes, ii) the system is dilute, and iii) the molar flow rates of gas and liquid streams are constant.

Refer to the diagram given below.

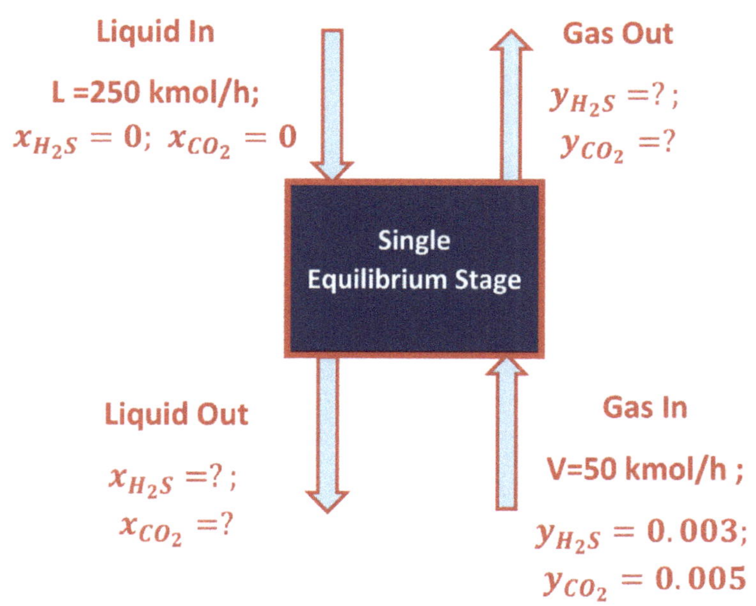

H_2S solute:

The absorption factor A_{H_2S} can be calculated as:

$$A_{H_2S} = \frac{L/V}{K_{H_2S}} = \frac{250/50}{55.2} = 0.09058$$

As shown in the diagram below, $x_{H_2S,in} = 0$. Consequently,

$$\frac{y_{H_2S,out}}{y_{H_2S,in}} = \left(\frac{1}{1+A}\right) = \frac{1}{1+0.09058} = 0.91694$$

$$y_{H_2S,out} = 0.91694 \times 0.003 = 2.7508 \times 10^{-3}$$

$$x_{H_2S,out} = y_{H_2S,out}/K_{H_2S} = 2.7508 \times 10^{-3}/55.2 = 4.983 \times 10^{-5}$$

$$\%\ Recovery\ of\ Solute = 100\left(1 - \frac{y_{H_2S,out}}{y_{H_2S,in}}\right) = 100(1 - 0.91694) = 8.306$$

CO_2 solute:

The absorption factor A_{CO_2} can be calculated as:

$$A_{CO_2} = \frac{L/V}{K_{CO_2}} = \frac{250/50}{166.2} = 0.03008$$

As shown in the diagram below, $x_{CO_2,in} = 0$. Consequently,

$$\frac{y_{CO_2,out}}{y_{CO_2,in}} = \left(\frac{1}{1+A}\right) = \frac{1}{1+0.03008} = 0.9708$$

$$y_{CO_2,out} = 0.9708 \times 0.005 = 4.854 \times 10^{-3}$$

$$x_{CO_2,out} = y_{CO_2,out}/K_{CO_2} = 4.854 \times 10^{-3}/166.2 = 2.9206 \times 10^{-5}$$

$$\%\ Recovery\ of\ Solute = 100\left(1 - \frac{y_{CO_2,out}}{y_{CO_2,in}}\right) = 100(1 - 0.9708) = 2.92$$

Comments:
As the absorption factors are small, the recoveries of solutes are low in this single-stage gas absorption process. We can increase the absorption factors by increasing L/V.

SHORT REVIEW QUESTIONS

1. What are the two types of absorption processes?
2. Give examples of two commercial applications of gas absorption?
3. What is the driving force for mass transfer of solute from gas phase to liquid phase in terms of composition variables?
4. What does an operating line represent?
5. Under what condition is the slope of the operating line constant when the operating line is expressed in terms of the mole fractions of solute?
6. Is it true that the operating line of an equilibrium stage connects a point on the equilibrium curve to a point whose coordinates are the compositions of the streams entering the stage?

7. What is the significance of absorption factor A?
8. What is the advantage of using mole ratios, instead of mole fractions, as composition variables when dealing with concentrated systems with large mass transfer rate of solute?

PRACTICE PROBLEMS

Problem 1

A gas mixture containing 98.5 mole % air and 1.5 mole% acetone is passed through an equilibrium stage so that acetone can be absorbed in water stream. The molar flow rate of gas mixture is 25 kmol/h. The pure water stream enters at a molar flow rate of 100 kmol/h. The Henry law is $y = 2.5x$. Determine the mole fractions of acetone in the exit liquid and gas streams. What is the percent recovery of acetone in the gas phase? You can assume that the molar flow rates of gaseous and liquid streams do not change in passing through the equilibrium stage.

Problem 2

Air containing acetone vapor is being contacted with water to scrub out the acetone. The contactor consists of a single equilibrium stage. The gas stream entering the contactor contains 4 mole percent acetone ($y_{in} = 0.04$). The liquid stream enters the contactor with 0.25 mole percent acetone (x_{in} =0.25/100=0.0025). The flow rate of liquid stream is 1000 kmol/h. The ratio of liquid flow rate to gas flow rate is 15.76. Assume that the flow rates of gas and liquid streams are constant and that the equilibrium relationship for this system at the temperature and pressure of the process is given by $y = 2.5x$, where x and y are mole fractions of acetone in the liquid stream and in the gas stream, respectively. Calculate the following:

i) A, the absorption factor
ii) y_{out}, the mole fraction of acetone in the gas stream leaving the contactor
iii) x_{out}, the mole fraction of acetone in the liquid stream leaving the contactor
iv) the percent recovery of acetone from the gas stream to liquid stream

Problem 3

A gas stream consisting of solute species A and air is to be passed through an equilibrium stage so that the solute A can be absorbed in a water stream. The inlet gas contains 2 mol% A, and has a molar flow rate of 25 kmol/h. The pure water stream has a molar flow rate of 150 kmol/h. The process is to operate isothermally and at constant pressure of 2 bar. The equilibrium relationship (Henry law) for the solute is $P_A = 5.06x_A$ where P_A is in bar. You may assume that the water and air are immiscible and the molar flow rates entering and leaving the equilibrium stage are constant.

a) Determine the concentrations of solute A in the exit streams.
b) Calculate the percent recovery of solute A from the gas phase.
c) Show the graphical solution (operating line, equilibrium line, and solution) on the y-x graph.

Problem 4

A gas mixture consisting of air and sulphur dioxide (SO_2) is contacted in a single-stage contactor with pure water at 2 bar and 295 K. The partial pressure of SO_2 in the original gas is 0.20 bar. The inlet gas molar flow rate is 6 kmol/h. The inlet pure water stream has a molar flow rate 2 kmol/h. The exit gaseous and liquid streams are at equilibrium with each other. The Henry law is $Y = 15X$ where Y is the mole ratio of solute in the gas phase and X is the mole ratio of solute in the liquid phase. The mole ratios are defined as: $Y = y/(1-y)$ and $X = x/(1-x)$. Determine the molar flow rates of exit gaseous and liquid streams. What is the percent recovery of solute in the gas phase? Justify the estimated percent recovery in terms of the absorption factor. The gas and liquid molar rates should not be assumed constant in the present case.

Problem 5

A solute is to be recovered into a liquid solvent from an inert carrier gas using single-stage absorption. The gas enters the absorber at a total molar flow rate of 350 kmol/h with $y = 0.35$ and leaves the absorber with $y = 0.05$. Pure solvent enters the absorber at a molar flow rate of 1200 kmol/h. The carrier gas is insoluble in the solvent and the solvent is nonvolatile.

a) Estimate the mole fraction of solute in the liquid stream leaving the absorber.
b) Calculate the total molar flow rates of the exit streams

Problem 6

A solute is to be recovered into a liquid solvent from an inert carrier gas using single-stage absorption. The gas enters the absorber at a total molar flow rate of 350 kmol/h with $y = 0.35$. Pure solvent enters the absorber at a molar flow rate of 1200 kmol/h. The carrier gas is insoluble in the solvent and the solvent is nonvolatile.

Construct the following plots: (a) y-x plot for the operating line, covering x range of 0 to 0.1; (b) Y-X plot for the operating line, covering x range of 0 to 0.1.

Which coordinates (y-x or Y-X) give a *linear* operating line?

Problem 7

A toxic gaseous species is absorbed from air using pure solvent in a single stage scrubber. The equilibrium relationship for the solute-solvent system is given by : $y = -0.75ln(1 - x)$. The solute concentration is 20 mole% in the entering gas. The molar flow rates of gas and solvent at the entrance are 300 and 625 kmol/h, respectively. Calculate the following:

a) the mole ratios and mole fractions of solute in the exit gas and liquid streams
b) the molar flow rates of the exit gas and liquid streams
c) the percent recovery of solute

Problem 8

We wish to remove two solutes, A and B, from a gaseous stream using pure solvent in a single-stage process. The carrier gas is air. The gas stream contains 1.0 mole% A and 0.5 mole% B. The molar flow rate of the gas mixture is 150 kmol/h. The solvent molar flow rate is 4500 kmol/h. The process is carried out at $T = 298$ K and $P = 1$ bar. The Henry law constants for A and B in solvent are: 15 bars and 10 bars, respectively. Determine the compositions of the two discharge streams and the percent recovery of solutes A and B from the gas phase to the liquid phase

Problem 9

Re-do tutorial problem 6 with $L/V = 100$.

Chapter 3

Single-Stage Liquid Stripping

3.1 LEARNING OUTCOMES

Upon completion of this chapter, students will be able to:

- Describe the process of liquid stripping and its importance
- Explain the direction and driving force for solute transfer in liquid stripping
- Analyze single-stage liquid stripping process for dilute and concentrated systems
- Determine the percent recovery of solute from liquid phase to gas phase

3.2 INTRODUCTION

Stripping is a separation process in which a solute present in the liquid phase is transferred from the liquid phase to the gas phase (see Figure 3.1). In other words, the liquid is stripped of a solute by a gas. Thus the direction of mass transfer of solute in stripping is opposite to that in gas absorption. Stripping is widely used in industrial applications to remove harmful toxic contaminants from waste water. For example, VOCs (volatile organic compounds) are partially soluble in water. They must be removed from water due to environmental consideration and regulations. The VOCs enter water because of industrial, agricultural and commercial activities. Steam and air are often used in stripping operations as stripping gases (Toth and Mizsey, Int. J. Environ. Sci. Technol., Vol. 12, 1321-1330,2015).

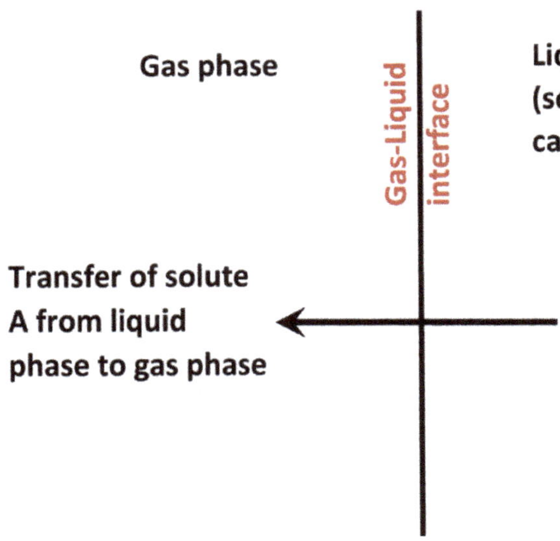

Figure 3.1 Direction of mass transfer of solute in a liquid stripping process.

In a single-stage liquid stripping process shown in Figure 3.2, the liquid phase containing the solute (contaminant or valuable material) enters the stage from the top and the stripping gas phase enters from the bottom. The two streams entering the stage (solute-rich liquid stream and stripping gas stream) are not at phase equilibrium with each other. Hence the transfer of solute from one phase to another is possible thermodynamically. The two input streams undergo intimate mixing on a perforated (sieve) tray due to formation of bubbles of gas. During the mixing process, the solute is transferred from the liquid phase to the gas phase. The treated liquid phase leaves the stage from the bottom and the solute-rich gas comes out of the top. Thus we have counter-current flow of gas and liquid streams as was the case in gas-absorption process discussed in Chapter 2. We will assume that the two streams *leaving the stage* (treated liquid stream and solute-rich gas stream) are in *phase equilibrium* with each other. The conditions necessary to achieve the phase equilibrium are: large interfacial contact area between the two phases, intense mixing, and sufficient residence (contact) time for the two phases.

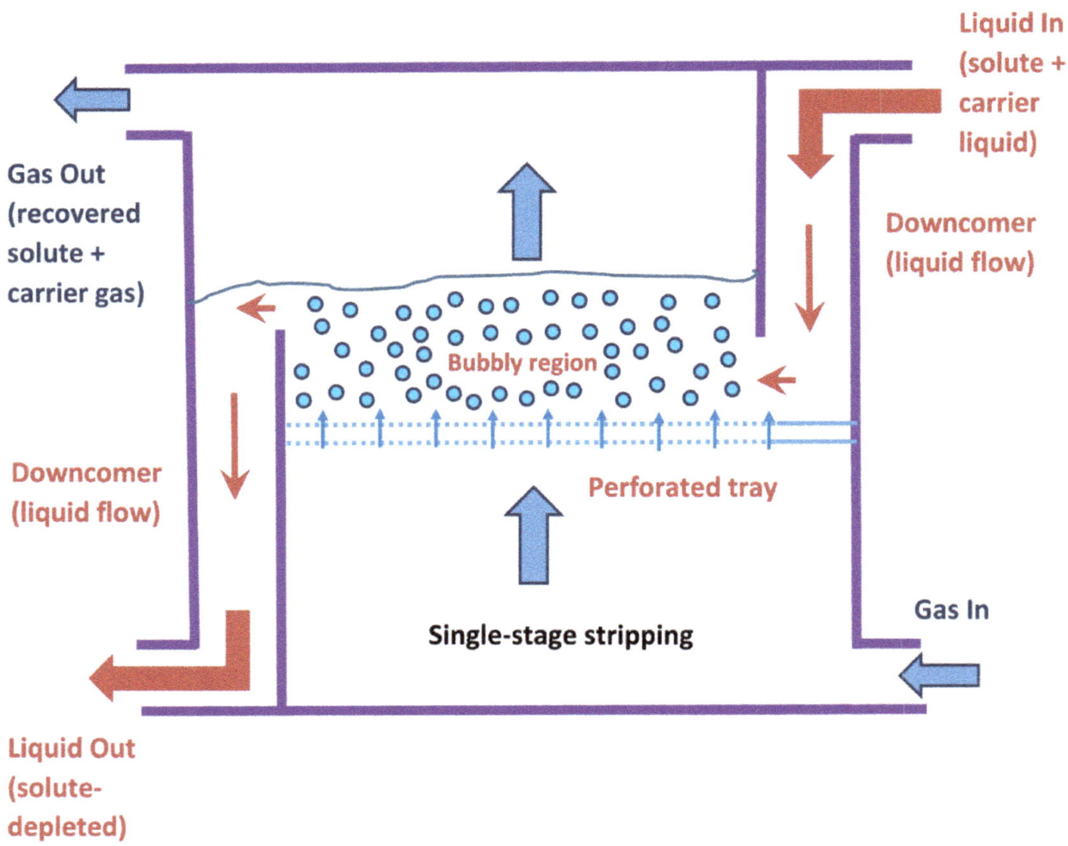

Figure 3.2 Single-stage liquid stripping process.

3.3 DRIVING FORCE FOR MASS TRANSFER IN LIQUID STRIPPING

Let us now consider the driving force for mass transfer of solute from the liquid phase to the gas phase. As noted in Chapters 1 and 2, the driving force for mass transfer is the difference in chemical potentials. The mass flows in the direction of decreasing chemical potential. Thus for solute A to move from the liquid phase to the gas phase, the chemical potential of solute in the liquid phase must be higher than that in the gas phase. Referring to Figure 3.3, the solute (species A) will move from liquid phase to the gas phase provided that:

$$\mu_{A,solu}^{L} > \mu_{A,mix}^{G} \tag{3.1}$$

Thus the driving force for mass transfer (transfer of solute A) is:

$$Driving\ force\ \rightarrow (\mu_{A,solu}^{L} - \mu_{A,mix}^{G}) > 0 \tag{3.2}$$

It can be readily shown (proof given in the tutorial of this chapter) that this driving force for mass transfer of solute A can be further expressed as:

$$Driving\ force \rightarrow (x_A - x_A^*) > 0 \tag{3.3}$$

where x_A^* is the mole fraction of solute in the liquid phase if the liquid phase was in equilibrium with the existing gas phase. If the existing liquid phase is not in equilibrium with the existing gas phase, x_A^* is a fictitious quantity to be determined from an equilibrium relationship. When the existing gas and liquid phases are in equilibrium, $x_A = x_A^*$ and no further transfer of solute is possible from the liquid phase to the gas phase. When the two phases initially come in contact with each other, the driving force is positive $(x_A > x_A^*)$. With continued contact between the two phases, the driving force decreases due to transfer of solute from liquid phase to gas phase. The two phases eventually reach equilibrium if the contact interfacial area between the two phases is large, the mixing is intense, and the residence (contact) time of the two phases in the mixing device is sufficient.

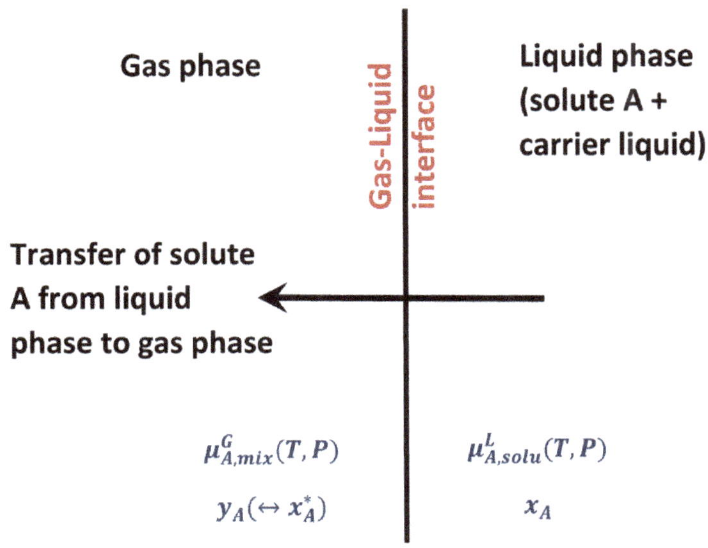

Figure 3.3 Variables governing the driving force for mass transfer of solute from liquid phase to gas phase in a liquid stripping process.

3.4 ANALYSIS OF SINGLE-STAGE LIQUID STRIPPING: DILUTE SYSTEMS

Figure 3.4 shows a schematic diagram of a single equilibrium stage stripping process. The liquid stream molar flow rates are expressed by L and the solute (species A) concentrations in the liquid stream are expressed in terms of mole fraction x. The molar flow rates of gas stream are given in terms of V and the concentrations of solute in the gas stream are given by y. No chemical reaction is involved. As we are dealing with single solute, the subscript "A" is not included in the composition variables.

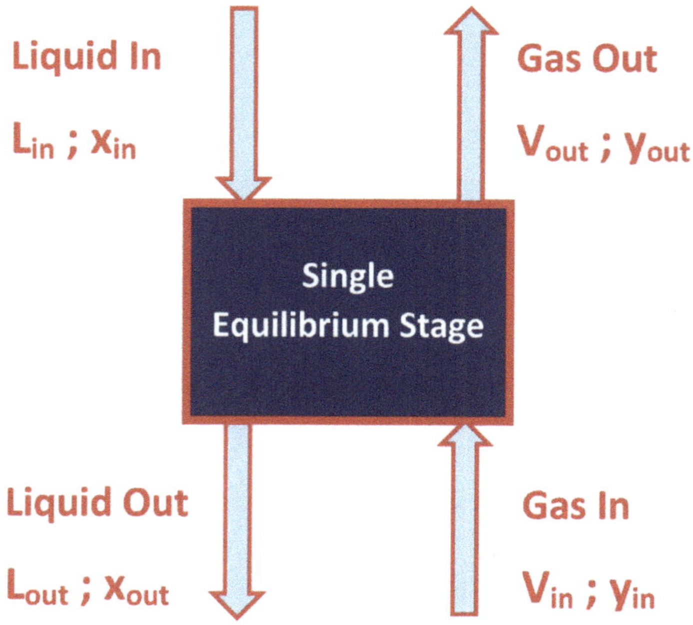

Figure 3.4 Single equilibrium stage liquid stripping.

The overall mass (mole) balance gives:

$$L_{in} + V_{in} = L_{out} + V_{out} \tag{3.4}$$

The solute mole balance gives:

$$x_{in}L_{in} + y_{in}V_{in} = x_{out}L_{out} + y_{out}V_{out} \tag{3.5}$$

Assuming that the system is dilute (concentration of solute in any stream is small) and that the mass transfer rate of solute from one stream to another is low, we can regard:

$$L_{in} = L_{out} = L \tag{3.6}$$

$$V_{in} = V_{out} = V \tag{3.7}$$

Consequently, Equation (3.5) reduces to:

$$y_{out} = y_{in} + (-\tfrac{L}{V})(x_{out} - x_{in}) \tag{3.8}$$

This is the *operating line* of a single-stage liquid stripping process. Note that the equations for the operating lines of single-stage absorption and stripping are exactly the same. However, there are significant differences between the plots of these operating lines. The operating line for stripping process is shown in Figure 3.5. Now the concentration of solute in the liquid phase decreases and increases in the gas phase as stripping proceeds (indicated by arrow on the operating line). This is opposite to what was observed in gas absorption in Chapter 2.

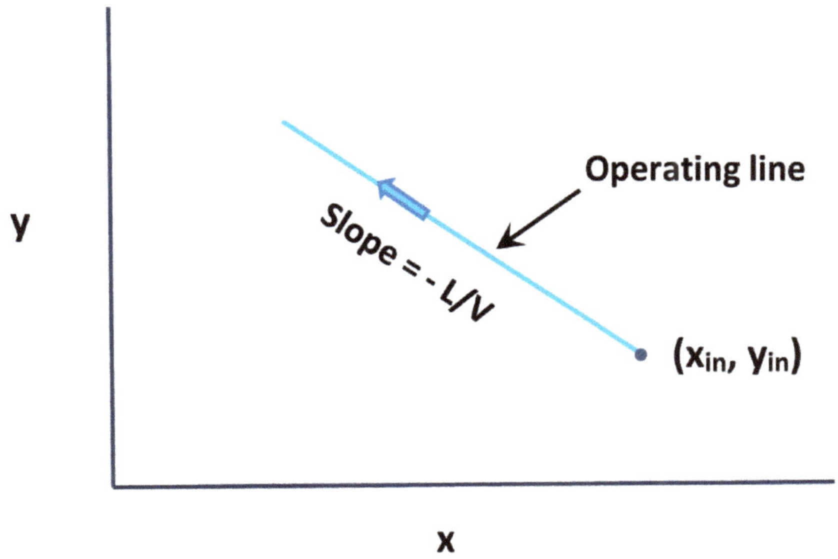

Figure 3.5 Operating line of a single-stage liquid stripping process.

Using the condition of phase equilibrium, that is, the two streams leaving the stage are at equilibrium with each other, we can determine the unknown gas and liquid compositions of the streams leaving the process. Figure 3.6 shows the graphical solution of a single equilibrium stage stripping process. Given the inlet streams compositions and their flow rates, and the equilibrium relationship (Henry law), the intersection of the operating and equilibrium lines gives the unknown compositions (x_{out} and y_{out}).

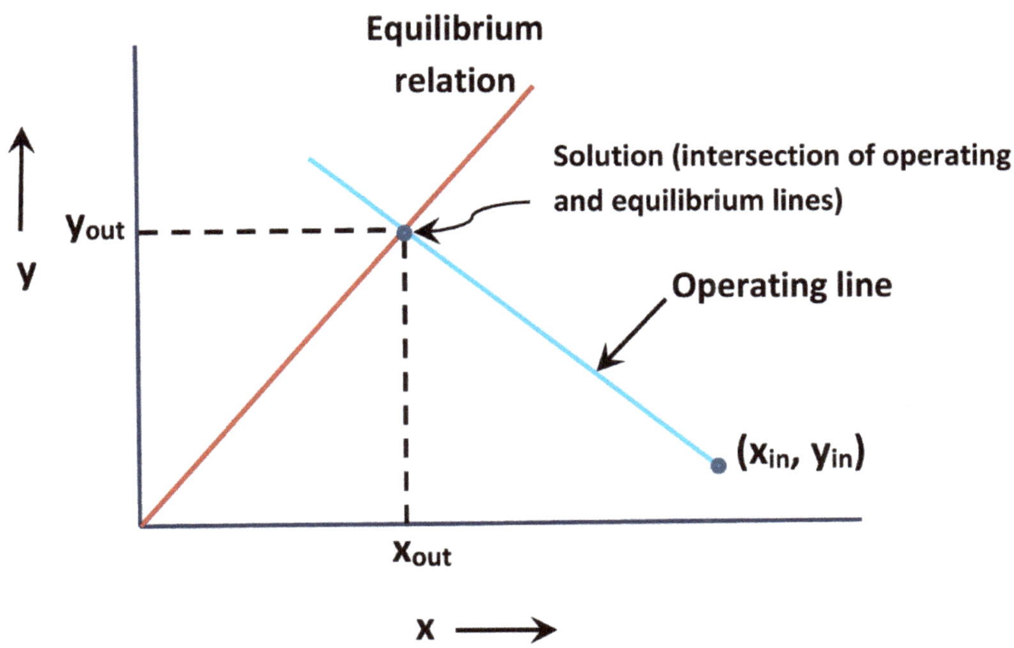

Figure 3.6 Graphical solution of a single equilibrium stage stripping.

The driving force for mass transfer of solute from liquid phase to gas phase is shown in Figure 3.7. When the two phases come in contact with each other, the driving force is initially maximum and as the transfer of solute from liquid phase to gas phase proceeds, the driving force decreases and eventually phase equilibrium is established and consequently, no more transfer of solute from liquid to gas is possible.

The percent recovery of solute from a single equilibrium stage stripping process can be calculated once the compositions of the leaving streams are known:

$$\%\ Recovery\ of\ Solute = 100\left(1 - \frac{x_{out}}{x_{in}}\right) \tag{3.9}$$

An analytical solution of unknown compositions of the two discharge streams can be obtained using the operating line (Equation (3.8)) and the equilibrium relation, that is, Henry law:

$$P_A = K_H x \tag{3.10}$$

$$y = (K_H/P)x = Kx \tag{3.11}$$

where P_A is the partial pressure of the solute in the gas phase, P is the total pressure, K_H is the Henry law constant of solute in units of pressure, and K is the dimensionless Henry law constant or distribution coefficient that depends on temperature and pressure of the system. Since the two leaving streams are at equilibrium with each other, Equation (3.11) can be applied and re-written as:

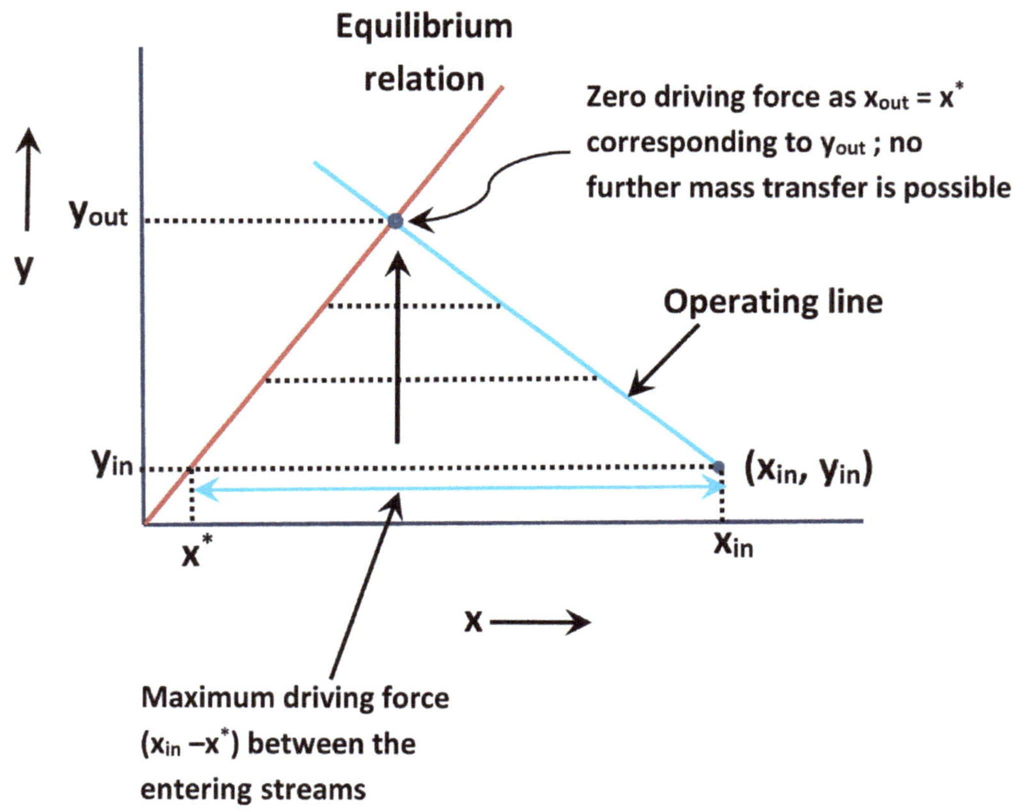

Figure 3.7 The driving force for mass transfer in single stage liquid stripping.

$$y_{out} = Kx_{out} \tag{3.12}$$

From Equations (3.8) and (3.12), the following expression for x_{out} can be derived by eliminating y_{out}

$$\frac{x_{out}}{x_{in}} = \left(\frac{1}{1+S}\right)\left[1 + \frac{V}{L}\left(\frac{y_{in}}{x_{in}}\right)\right] \tag{3.13}$$

where S is called the stripping factor, defined as:

$$S = \frac{\text{slope of equilibrium line}}{|\text{slope of operating line}|} = \frac{K}{L/V} \tag{3.14}$$

The stripping factor "S" is a measure of degree of separation of solute from the liquid phase to the gas phase. To see the significance of "S", consider a single equilibrium stage stripping with solute-free gas entering the device. In this case, Equation (3.13) could be re-written as:

$$\frac{x_{out}}{x_{in}} = \left(\frac{1}{1+S}\right) \tag{3.15}$$

When $S \to 0, x_{out} = x_{in}$ and no separation of solute occurs. When $S \to \infty, x_{out} = 0$, that is, complete separation of solute occurs. Also note that $S = 1/A$.

3.5 ANALYSIS OF SINGLE-STAGE LIQUID STRIPPING: CONCENTRATED SYSTEMS

As noted in Chapter 2, the operating line has a constant slope of *–L/V* only when the system is dilute. In dilute systems, the mass transfer rate of solute from liquid stream to gas stream is small and the flow rates of gas and liquid streams are constant from inlet to outlet. In concentrated systems with large enough mass transfer rates, the flow rates of gas and liquid streams can no longer assumed to be constant from inlet to outlet. When the total flow (carrier + solute) rates of the fluid streams (gas or liquid) are not constant, the operating line is non-linear. The equilibrium relation is also generally non-linear for concentrated systems. As discussed in chapter 2, it is preferred to use mole ratios (instead of mole fractions) as composition variables in concentrated systems. The advantage of using mole ratios is that the operating line for the process turns out to be a linear relation. The mole ratios are defined as follows:

$$X = \frac{\text{moles of solute in liquid stream}}{\text{moles of carrier (solute−free solvent) in liquid stream}} = \frac{x}{1-x} \tag{3.16}$$

$$Y = \frac{\text{moles of solute in gas stream}}{\text{moles of carrier (solute−free gas) in gas stream}} = \frac{y}{1-y} \tag{3.17}$$

Let L' be the carrier (solute-free solvent) flow rate of the liquid stream and V' be the carrier (solute-free gas) flow rate of the gas stream. The carrier fluids of the gas and liquid streams are assumed to be completely immiscible with each other. Only the solute moves from the liquid stream to the gas stream. As the carrier fluids are immiscible, their flow rates do not change. Referring to Figure 3.8, it follows that:

$$L'_{in} = L'_{out} = L' \tag{3.18}$$

$$V'_{in} = V'_{out} = V' \tag{3.19}$$

The solute mole balance gives:

$$X_{in}L'_{in} + Y_{in}V'_{in} = X_{out}L'_{out} + Y_{out}V'_{out} \tag{3.20}$$

From Equations (3.18)-(3.20), it follows that:

$$Y_{out} = Y_{in} + (-\tfrac{L'}{V'})(X_{out} - X_{in}) \tag{3.21}$$

This is the operating line for concentrated systems. It is a linear relation on Y-X scale with a slope of -L'/V'.

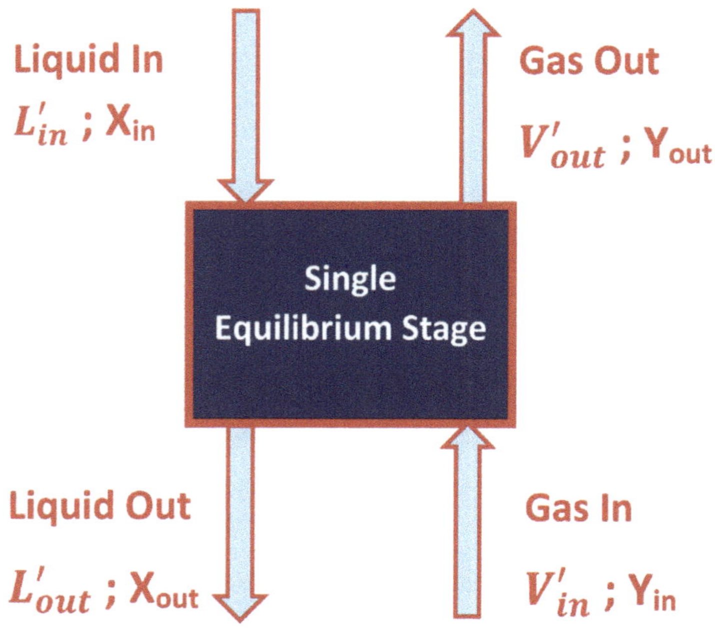

Figure 3.8 Single equilibrium stage stripping.

To solve for Y_{out} and X_{out} , we need the equilibrium relation. In general, the solution is obtained graphically. An analytical expression for Y_{out} can be derived only if the equilibrium relation is assumed to be linear, that is:

$$Y_{out} = K' X_{out} \tag{3.22}$$

From Equations (3.21) and (3.22), it can be readily shown that:

$$\frac{X_{out}}{X_{in}} = \left(\frac{1}{1+S'}\right)\left[1 + \frac{V'}{L'}\left(\frac{Y_{in}}{X_{in}}\right)\right] \tag{3.23}$$

where S' is the stripping factor for concentrated systems defined as:

$$S' = \frac{K'}{L'/V'} \tag{3.24}$$

Note that once X_{out} is known, Y_{out} can be obtained from the equilibrium relation, Equation (3.22). Figure 3.9 shows how to obtain the solution graphically.

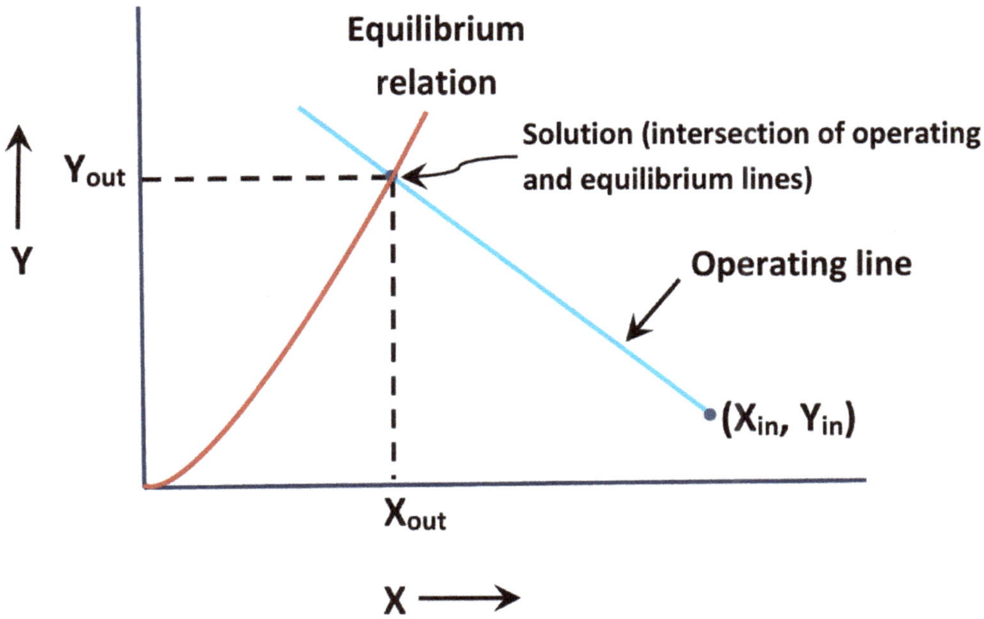

Figure 3.9 Graphical solution of single stage stripping for concentrated systems.

NOTATION

A	absorption factor, equal to $1/S$ where S is the stripping factor defined in Equation (3.14)
K	dimensionless Henry law constant or distribution coefficient of solute, defined in Equation (3.11)
K'	distribution coefficient of solute, defined in Equation (3.22)
K_H	Henry law constant of solute defined in Equation (3.10)
L	molar flow rate of liquid stream (solvent + solute)
L'	molar flow rate of solute-free liquid stream, that is, molar flow rate of solvent
P	total pressure of gas
P_A	partial pressure of solute (species A) in the gas phase
S	stripping factor, defined in Equation (3.14)
S'	stripping factor for concentrated systems, defined in Equation (3.24)
V	molar flow rate of gas stream (carrier gas + solute)
V'	molar flow rate of solute-free gas stream, that is, molar flow rate of carrier gas
$VOCs$	abbreviation for volatile organic compounds
x, x_A	mole fraction of solute A in liquid phase
x^*	mole fraction of solute A in the liquid phase if the liquid phase was in equilibrium with the existing gas phase
X	mole ratio of solute in the liquid phase, defined in Equation (3.16)
y, y_A	mole fraction of solute A in gas phase
Y	mole ratio of solute in the gas phase, defined in Equation (2.18)

Greek Letters

$\mu^{G}_{A,mix}$	chemical potential of solute (species A) in the gas mixture
$\mu^{L}_{A,solu}$	chemical potential of solute (species A) in the liquid solution

TUTORIAL PROBLEMS WITH SOLUTION

Problem 1

Starting from the following expression for the driving force for mass transfer of solute in stripping,

$$Driving\ force \rightarrow (\mu^L_{A,solu} - \mu^G_{A,mix}) > 0$$

show that this driving force for mass transfer can also be expressed as:

$$Driving\ force \rightarrow (x_A - x_A^*) > 0$$

Solution

This problem is similar to the tutorial problem of Chapter 2. Referring to the figure given below and assuming that the gas mixture behaves as an ideal gas mixture and that the liquid solution behaves as an ideal liquid solution, one can derive the following expression (refer to tutorial section of Chapter 2):

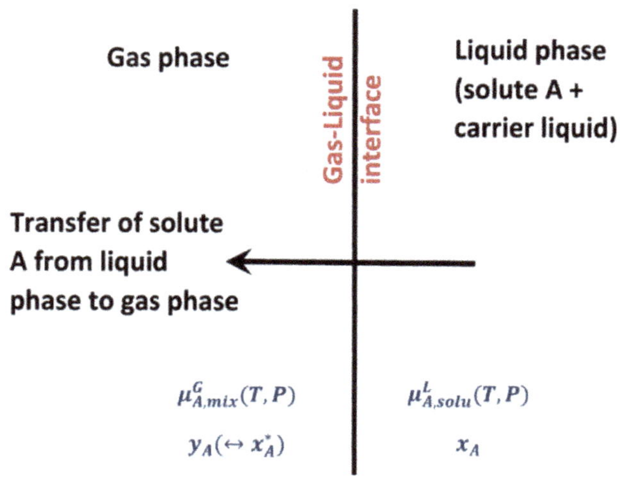

$$\mu^L_{A,solu}(T,P) - \mu^G_{A,mix}(T,P) = RT \ln \frac{x_A P_A^{Sat}}{y_A P}$$

The driving force for mass transfer of species A from liquid phase to gas phase is:

$$\mu^L_{A,solu}(T,P) - \mu^G_{A,mix}(T,P) > 0$$

This is equivalent to the following condition:

$$RT \ln \frac{x_A P_A^{Sat}}{y_A P} > 0$$

Therefore, the driving force for mass transfer of species A from liquid phase to gas phase is:

$$x_A > \frac{y_A P}{P_A^{Sat}}$$

The mole fraction of species A in the liquid phase if the liquid phase is in equilibrium with the gas phase is

$$x_A^* = \frac{y_A P}{P_A^{Sat}}$$

From these two expressions, it follows that the condition for transfer of species A from liquid phase to gas phase is

$$x_A > x_A^*$$

Thus, the driving force for mass transfer of species from liquid phase to gas phase is:

$$Driving\ force\ \rightarrow (x_A - x_A^*) > 0$$

Problem 2

A toxic chemical is stripped from water with air using a single-stage stripping process. The air entering the process is free of toxic chemical. Determine the percentage removal of toxic chemical for the following values of the stripping factor: a) $S = 0.1$, b) $S = 1.0$, c) $S = 2$, and d) $S = 10$. Any comments on your results?

Solution

a) Given $y_{in} = 0, S = 0.1$. From Equation (3.13):

$$\frac{x_{out}}{x_{in}} = \left(\frac{1}{1+S}\right) = \frac{1}{1+0.1} = 0.9091$$

The percentage removal is given by Equation (3.9) as:

$$\%\ Removal\ of\ Solute = 100\left(1 - \frac{x_{out}}{x_{in}}\right) = 100(1 - 0.9091) = 9.09\%$$

b) Repeat the above calculation with $y_{in} = 0$ and S value of 1:

$$\frac{x_{out}}{x_{in}} = \left(\frac{1}{1+S}\right) = \frac{1}{1+1} = 0.5$$

$$\%\ Removal\ of\ Solute = 100\left(1 - \frac{x_{out}}{x_{in}}\right) = 100(1 - 0.5) = 50\%$$

c) Repeat the above calculation with $y_{in} = 0$ and S value of 2:

$$\frac{x_{out}}{x_{in}} = \left(\frac{1}{1+S}\right) = \frac{1}{1+2} = 0.3333$$

$$\%\ Removal\ of\ Solute = 100\left(1 - \frac{x_{out}}{x_{in}}\right) = 100(1 - 0.3333) = 66.7\%$$

d) Repeat the above calculation with $y_{in} = 0$ and S value of 10:

$$\frac{x_{out}}{x_{in}} = \left(\frac{1}{1+S}\right) = \frac{1}{1+10} = 0.09091$$

$$\%\ Removal\ of\ Solute = 100\left(1 - \frac{x_{out}}{x_{in}}\right) = 100(1 - 0.09091) = 90.91\%$$

Obviously, a high value of stripping factor is desirable as it leads to a larger percentage removal of toxic chemical from water.

Problem 3

A toxic chemical is stripped from waste water with air using a single stage stripping process. The air entering the process is free of toxic chemical. The mole fraction of toxic chemical in the waste water stream entering the process is 0.05. The waste water molar flow rate is 10 times the air molar flow rate. The equilibrium relation is $y = 25x$, where y and x are mole fractions of toxic chemical in air and waste water, respectively.

i) What is the value of the stripping factor?
ii) Determine the mole fractions of toxic chemical in the exit streams.
iii) Calculate the percent removal of toxic chemical from the waste water stream

Solution

Refer to the diagram given below.

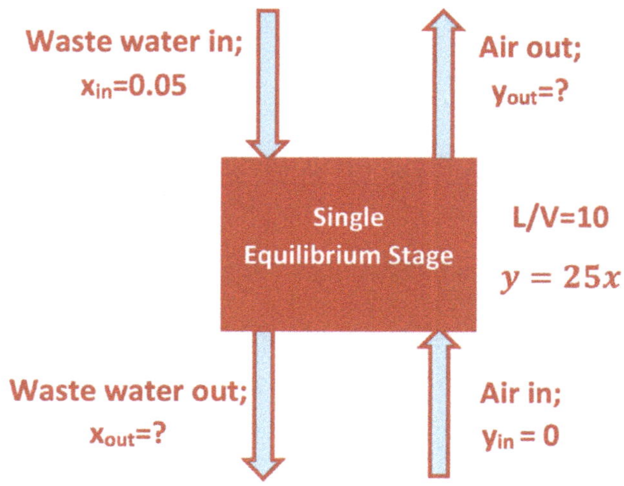

i) The stripping factor is:

$$S = \frac{K}{L/V} = 25/10 = 2.5.$$

ii) As $y_{in} = 0$, we can calculate the mole fraction in the exit liquid stream from:

$$\frac{x_{out}}{x_{in}} = \left(\frac{1}{1+S}\right) = \left(\frac{1}{1+2.5}\right) = 0.2857$$

Thus,

$$x_{out} = 0.2857 \times 0.05 = 0.014285$$

$$y_{out} = 25x_{out} = 25 \times 0.014285 = 0.3571$$

iii) The percent removal of toxic chemical from the waste water stream is as follows:

$$\%\ Removal\ of\ Solute = 100\left(1 - \frac{x_{out}}{x_{in}}\right) = 100(1 - 0.2857) = 71.43\%$$

Note that the stripping factor S is large in the present case ($S = 2.5$) and therefore, the percent removal of solute (toxic chemical) is high.

Problem 4

A hydrocarbon mixture consisting of 2 mole% propane and 98 mole% non-volatile hydrocarbon is being stripped by pure superheated steam using a single stage stripping process. The hydrocarbon molar flow rate is 13 times the steam molar flow rate. The equilibrium relation is $P_{propane} = 60x$, where $P_{propane}$ is in bar. The process is to operate isothermally and at constant pressure of 2 bar. You can assume steam is an inert gas and it does not condense.

a) Determine the mole fractions of propane in the exit streams.
b) What is the value of stripping factor?
c) Calculate percent removal of propane from the hydrocarbon mixture
d) Show the graphical solution (operating line, equilibrium line, and solution) on the y-x graph.

Solution

Refer to the diagram given below.

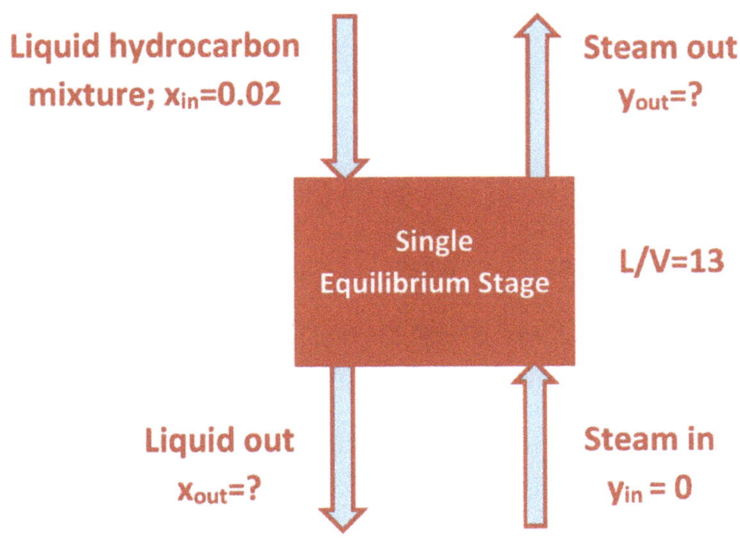

$$K = y/x = 60/P = 60/2 = 30$$

a) As $y_{in} = 0$, we can calculate the mole fraction in the exit liquid stream from:

$$\frac{x_{out}}{x_{in}} = \left(\frac{1}{1+S}\right)$$

The stripping factor is: $S = \frac{K}{L/V} = 30/13 = 2.3077$. Thus,

$$\frac{x_{out}}{x_{in}} = \left(\frac{1}{1+2.3077}\right) = 0.3023\ ;\ x_{out} = 0.3023 \times 0.02 = 0.006046$$

$$y_{out} = 30x_{out} = 30 \times 0.006046 = 0.18138$$

b) The value of the stripping factor is calculated in part (a) as 2.3077

c) The percent removal of propane from the hydrocarbon mixture is as follows:

$$\% \, Removal \, of \, Solute = 100\left(1 - \frac{x_{out}}{x_{in}}\right) = 100(1 - 0.3023) = 69.77\%$$

d) The graphical solution is shown in the figure below

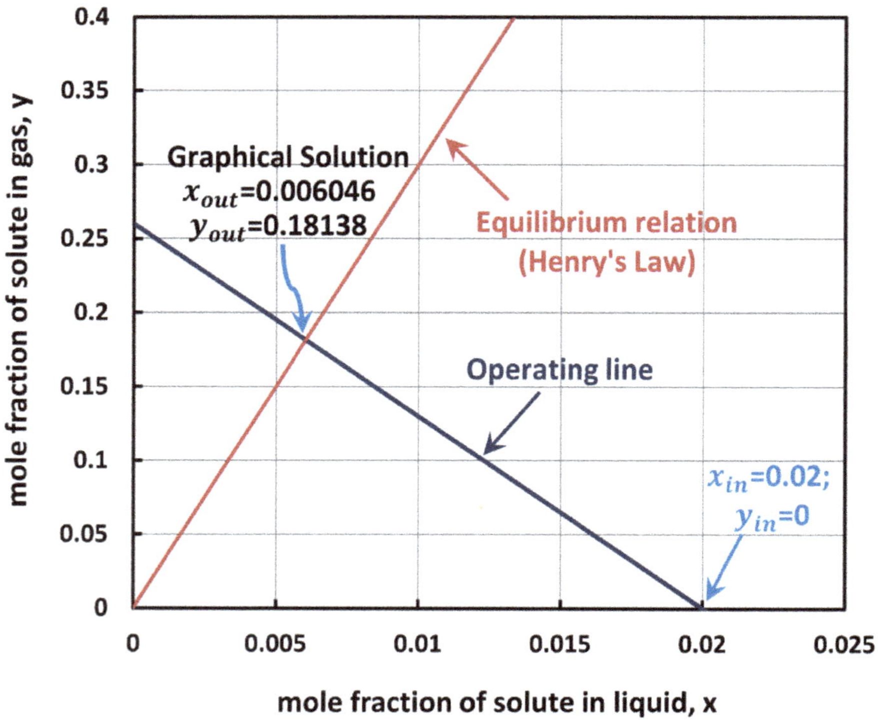

Operating line equation:

$$y_{out} = y_{in} + \left(-\frac{L}{V}\right)(x_{out} - x_{in})$$

$$y_{out} = -13(x_{out} - 0.02)$$

Equilibrium relation:

$$y = 30x$$

Problem 5

A solute is stripped from a concentrated aqueous solution containing solute with a mole ratio of 0.1 by contact with pure air in a single-stage stripping process. The equilibrium relationship in terms of mole ratios is given as $Y = 0.85X$. The molar flow rate of pure air is 1.5 times that of carrier (solute-free) liquid. a) Find the mole fractions of solute in the exit streams, b) Determine the percent removal of solute, c) What percentage removal of solute is obtained if $V'/L' = 3$.

Solution

a) Given $X_{in} = 0.1$, $Y_{in} = 0, V'/L' = 1.5$. From Equation (3.24), the stripping factor S' is:

$$S' = \frac{K'}{L'/V'} = 0.85 \times 1.5 = 1.275$$

From Equation (3.23):

$$\frac{X_{out}}{X_{in}} = \left(\frac{1}{1 + S'}\right) = \frac{1}{1 + 1.275} = 0.4396$$

$$X_{out} = 0.4396 \times 0.1 = 0.04396$$

$$x_{out} = \frac{X_{out}}{1+X_{out}} = \frac{0.04396}{1+0.04396} = 0.04211$$

From Equation (3.22):

$$Y_{out} = K'X_{out} = 0.85 \times 0.04396 = 0.03737$$

$$y_{out} = \frac{Y_{out}}{1+Y_{out}} = \frac{0.03737}{1+0.03737} = 0.03602$$

b) $\% \; Removal \; of \; Solute = 100\left(1 - \frac{X_{out}}{X_{in}}\right) = 100(1 - 0.4396) = 56.04\%$

c) $V'/L' = 3$. The stripping factor S' is:

$$S' = \frac{K'}{L'/V'} = 0.85 \times 3 = 2.55$$

From Equation (3.23):

$$\frac{X_{out}}{X_{in}} = \left(\frac{1}{1+S'}\right) = \frac{1}{1+2.55} = 0.2817$$

$$\% \; Removal \; of \; Solute = 100\left(1 - \frac{X_{out}}{X_{in}}\right) = 100(1 - 0.2817) = 71.83\%$$

As expected, the percent removal of solute from liquid phase to gas phase increases with the increase in stripping factor.

SHORT REVIEW QUESTIONS

1. Give an example of practical application of liquid stripping?
2. What is the driving force for mass transfer of solute from liquid phase to gas phase in terms of composition variables?
3. How does the operating line in single-stage stripping differ from that in single-stage absorption?
4. What is the significance of stripping factor *S*?

PRACTICE PROBLEMS

Problem 1

A hydrocarbon mixture consisting of 5 mole% propane and 95 mole% non-volatile hydrocarbon is being stripped by pure superheated steam using a single stage process. The ratio of steam molar flow rate to hydrocarbon molar flow rate is 1/10, that is, $V/L = 0.1$. The equilibrium relation is $y = 30x$, where y and x are mole fractions of propane in steam and hydrocarbon mixture, respectively.

a) Determine the mole fractions of propane in the exit streams.
b) What is the value of stripping factor?
c) Calculate percent removal of propane from the hydrocarbon mixture

You can assume steam is an inert gas and it does not condense.

Problem 2

A single-stage stripping process is being used to remove ammonia from wastewater using pure air as the stripping gas. The water entering the process contains 0.05 mole fraction ammonia. The ammonia mole fraction in the exit water stream is 0.01. The equilibrium relation is $y = 1.5x$, where y and x are mole fractions of ammonia in air and water streams at equilibrium, respectively. Determine the ratio V/L. You can assume that the molar flow rates of gas and liquid streams are constant.

Problem 3

A single-stage stripping process is being used to remove a chemical solute from water using pure air as the stripping gas. The mole fraction of solute in the water stream entering the process is 0.15. The molar flow rate of pure air is 2 times that of solute-free water. The output streams leaving the stripper are at equilibrium with each other. The equilibrium relationship (Henry's law) in terms of mole ratios is $Y = 1.5\,X$. Determine the following:

a) the mole ratios and mole fractions of solute in the exit gas and liquid streams
b) the percent removal of solute from liquid phase to gas phase

Problem 4

A solute is stripped from a concentrated aqueous solution containing solute with a mole ratio of 0.21 by contact with pure air in a single-stage stripping process. The equilibrium relationship in terms of mole fractions is given as $y = -0.75\ln(1 - x)$. The molar flow rate of pure air is 2.5 times that of carrier (solute-free) liquid.

a) Find the mole ratios and mole fractions of solute in the exit streams
b) Determine the percent removal of solute,

Use graphical approach to solve this problem. The equilibrium relationship should be converted from mole fractions to mole ratios.

Problem 5

We would like to remove CH_4 and H_2S from an aqueous stream using a single stage stripping process. The mole fractions of CH_4 and H_2S in the aqueous stream entering the stripper are 5×10^{-5} and 10^{-4}, respectively. The molar flow rate of aqueous stream is 590 kmol/h and the molar flow rate of stripping gas is 5 kmol/h. The stripping process occurs at 285 K and 1 bar.

a) Estimate the Henry law constants for CH_4 and H_2S at 285 K using the Harvey correlation (see tutorial problem 7, Chapter 1)
b) Determine the compositions of the two discharge streams
c) Determine the percent recovery of CH_4 and H_2S from the liquid phase to gas phase.

Chapter 4

Multi-Stage Gas Absorption

4.1 LEARNING OUTCOMES

Upon completion of this chapter, students will be able to:

- Analyze counter-current multistage absorption process
- Determine the number of equilibrium stages required for a given separation in a multi-stage absorption process using the graphical McCabe-Thiele method
- Develop and apply the Kremser equations for the estimation of number of equilibrium stages analytically
- Explain the importance of minimum liquid flow rate and pinch condition in a multistage gas absorption process

4.2 INTRODUCTION

In Chapter 2, we analyzed a single-stage gas absorption process. The degree of separation of solute in a single stage process is generally quite limited. For large recoveries of solute, it is necessary to utilize a multi-stage contacting of the two fluid streams. Figure 4.1 shows a typical multi-stage arrangement of

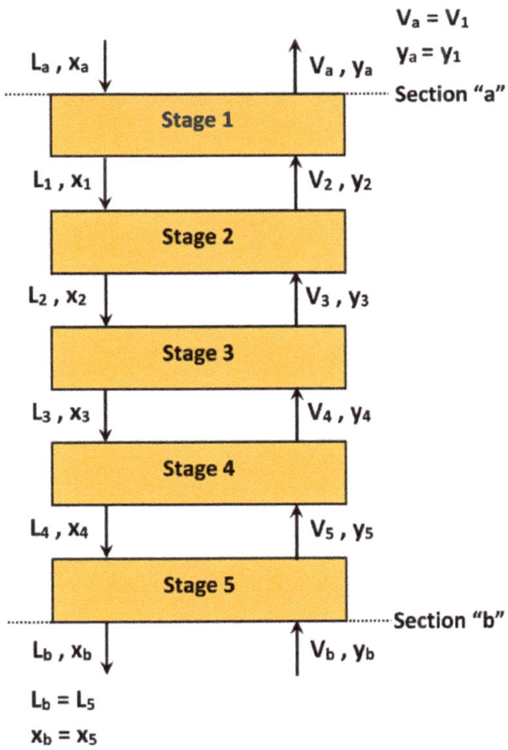

Figure 4.1 A counter-current cascade of five stages.

individual stages. The flow arrangement shown in Figure 4.1 is referred to as "*counter-current*" flow as the two contacting phases flow in the opposite directions. The entering fluid streams flow sequentially through all the stages. The lighter phase (gas) flows in the upward direction and the heavier phase (liquid) flows downward in the direction of gravity. Such an interconnected assembly of individual stages is also called *cascade*. In each stage of a cascade, the entering fluid streams are brought into contact, mixed thoroughly, and then separated. The streams *leaving* any stage of a cascade are assumed to be in equilibrium with each other in an equilibrium stage operation.

Although it is not necessary that the two fluid streams move in a counter-current manner, it is the most effective flow arrangement and is used widely in industrial applications. Figure 4.2 shows a column with multi-stage contacting of two phases.

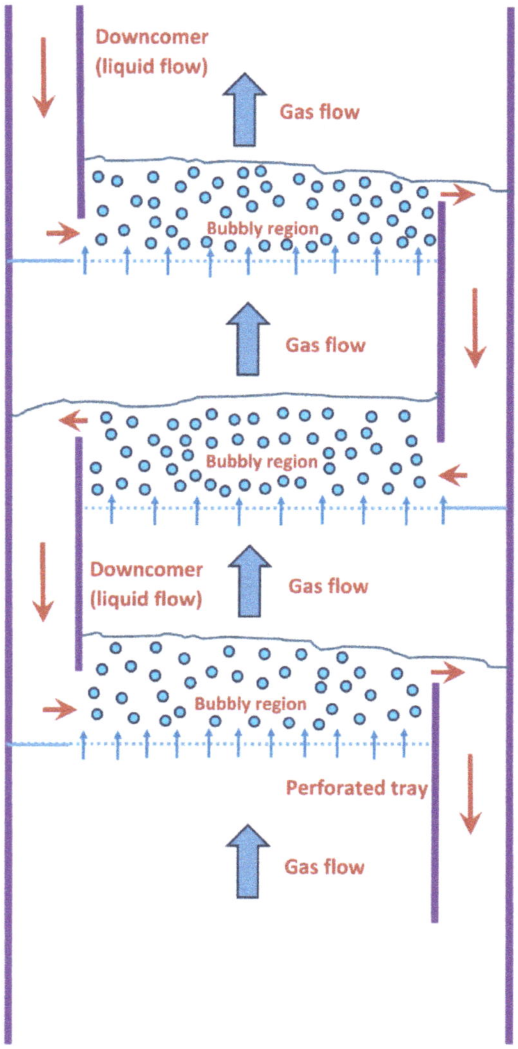

Figure 4.2 Multi-stage counter-current gas absorption column.

4.3 ANALYSIS OF COUNTER-CURRENT MULTI-STAGE ABSORPTION PROCESS

Before we do any mass (mole) balance, it is important to describe the nomenclature that we will be using in analyzing the multi-stage absorption and stripping processes. Following the same notation for liquid and gas flow rates as used in the single –stage processes discussed in Chapters 2 and 3, L will be used for the total (solute + carrier) liquid flow rate and V will be used for the total (solute + carrier) gas flow rate. The subscript of any flow or composition variable (x and y) will indicate the stage from which the stream is emerging. For example, L_2 is the total flow rate of liquid stream leaving stage 2, V_2 is the total flow rate of gas stream leaving stage 2, x_2 is the mole fraction of solute in the liquid stream leaving stage 2, and y_2 is the mole fraction of solute in the gas stream leaving stage 2 (see Figure 4.1).

Consider a counter-current cascade of N stages shown in Figure 4.3. We will do mole balance over the envelope indicated by blue dashed boundary, assuming that no chemical reactions are involved. The overall balance gives:

$$L_a + V_{n+1} = L_n + V_a \tag{4.1}$$

The solute balance gives

$$x_a L_a + y_{n+1} V_{n+1} = x_n L_n + y_a V_a \tag{4.2}$$

Equation (4.2) can be re-written as:

$$y_{n+1} = \left(\frac{L_n}{V_{n+1}}\right) x_n + \frac{y_a V_a - x_a L_a}{V_{n+1}} \tag{4.3}$$

This is the equation of the operating line of counter-current cascade of Figure 4.3. The operating line relates the composition y_{n+1} in the gas stream (*V* stream) with the composition x_n in the liquid stream (*L* stream) passing it at the same level (see Figure 4.3). When Equation (4.3) is plotted on a y-x plot, the plot is generally non-linear since the slope L_n/V_{n+1} varies from one stage to another if the liquid and gas flow rates are not constant. However, for *dilute* systems, the liquid and gas flow rates can be assumed to be constant and therefore, the slope L_n/V_{n+1} is constant leading to a linear operating line.

In this chapter we will assume, for the most part, that the systems are dilute. Concentrated systems where mass transfer rates of solute are large enough to affect the flow rates of gaseous and liquid streams are treated as a special case in the tutorial section of the chapter. Note that in the analysis of concentrated systems we need to use mole ratios of solute as the composition variables and solute-free gas and liquid flow rates in order to obtain a linear operating line. The temperature and pressure are also assumed to be constant, that is, the absorption process is isothermal and isobaric.

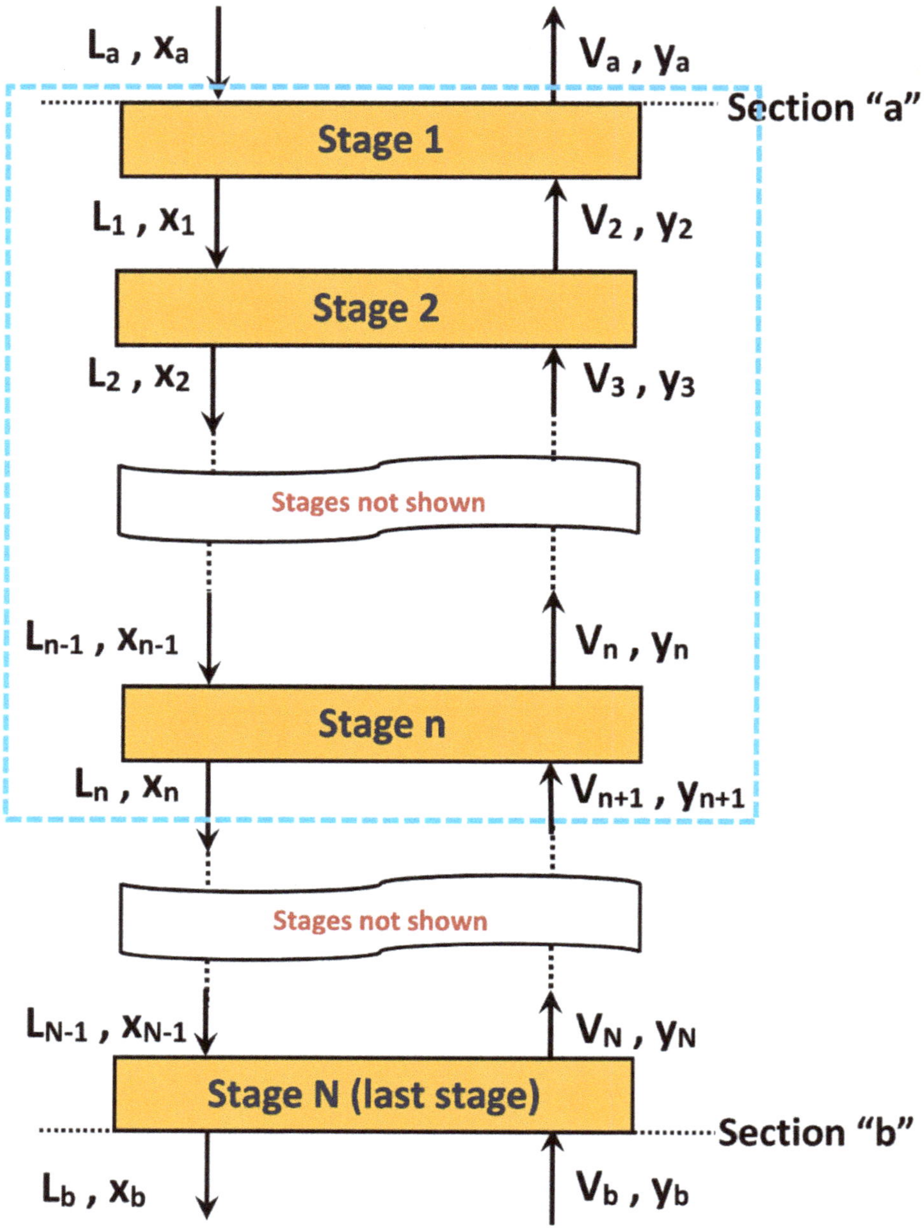

Figure 4.3 A counter-current cascade of N stages.

For dilute systems, one can write:

$$L_a = L_1 = \cdots = L_n = \cdots = L_b = L \tag{4.4}$$

$$V_a = V_1 = \cdots = V_n = \cdots = V_b = V \tag{4.5}$$

Consequently, Equation (4.3) reduces to:

$$y_{n+1} = y_a + (\frac{L}{V})(x_n - x_a) \tag{4.6}$$

This is the *operating line* of a multi-stage absorption cascade assuming dilute systems. It a straight line of constant positive slope (L/V) passing through the terminal compositions (x_a, y_a) and (x_b, y_b).

In a gas absorption process (single-stage or multi-stage), the driving force for mass transfer of solute from the gas phase to the liquid phase is proportional to $y\text{-}y^*$ as discussed in Chapter 2, where y is the mole fraction of solute in the existing gas phase and y^* is the mole fraction of solute in the gas phase if the gas phase was in equilibrium with the existing liquid phase. This is illustrated in Figure 4.4. Note that y^* is a hypothetical quantity (non-existent in the actual system) if the existing gas phase is not in equilibrium with the existing liquid phase.

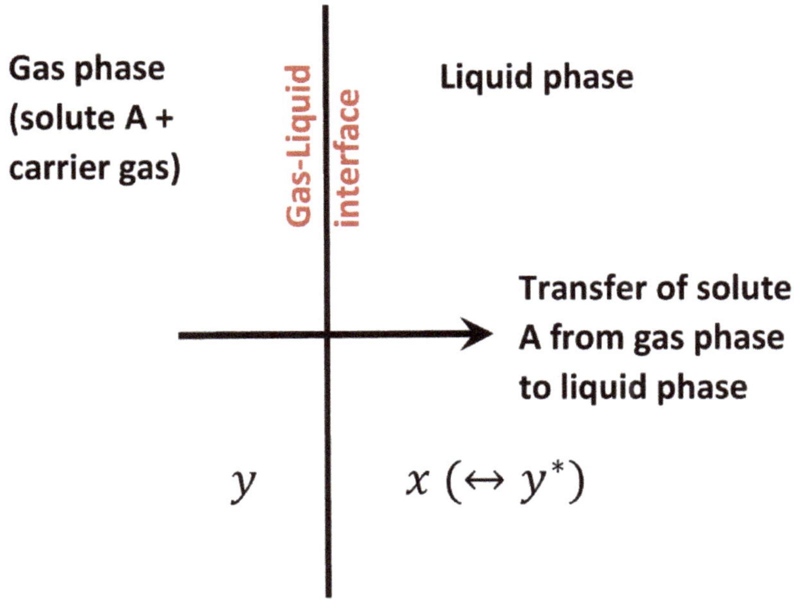

Figure 4.4 Direction and driving force of mass transfer of solute in a gas absorption process.

Figure 4.5 shows the plots of the operating and equilibrium lines for a multi-stage absorption process. The figure also shows the driving force for mass transfer at any location within the absorption column. The equilibrium line represents the Henry law for dilute solutions of solute given as:

$$P_A = K_H x \tag{4.7}$$

$$y = (K_H/P)x = Kx \tag{4.8}$$

where P_A is the partial pressure of the solute in the gas phase, P is the total pressure, K_H is the Henry law constant of solute in units of pressure, and K is the dimensionless Henry law constant or distribution coefficient that depends on temperature and pressure.

The following points should be noted from Figure 4.5: (a) the operating and equilibrium lines are linear; (b) the liquid solution and the gas mixture both become more concentrated in solute if we move in the downward direction from the top of the multi-stage absorption column to the bottom of the column. This is reflected in the operating line shown in Figure 4.5 where the top end of the column is shown near the origin and the bottom end of the column is shown at the upper corner; (c) the operating line is steeper than the equilibrium line provided that the absorption factor (ratio of operating line slope to equilibrium line slope) is greater than unity; (d) the driving force for mass transfer of solute from gas phase to liquid phase, that is, $y\text{-}y^*$, increases from top to bottom of the column. This is to be expected when absorption factor is greater than unity; and (e) the operating line falls above the equilibrium line. This is necessary in gas absorption where solute is transferred from gas phase to liquid phase and the driving force $y\text{-}y^*$ is positive.

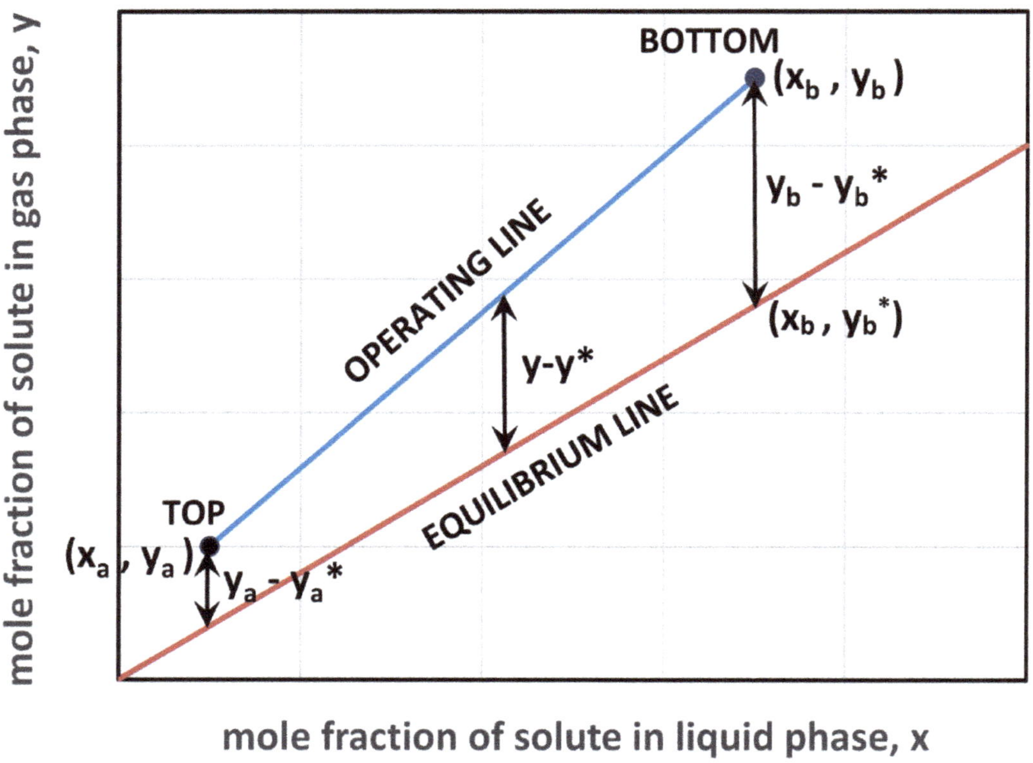

Figure 4.5 Plots of operating and equilibrium lines for multi-stage absorption of dilute systems.

For further explanation of the mass transfer driving force $(y - y^*)$ and its link to an equilibrium stage of a multi-stage gas absorption cascade, we refer to Figure 4.6. For the liquid stream of composition $x_n (= x)$ leaving the stage, the corresponding change (decrease) in the gas composition with respect to the entering

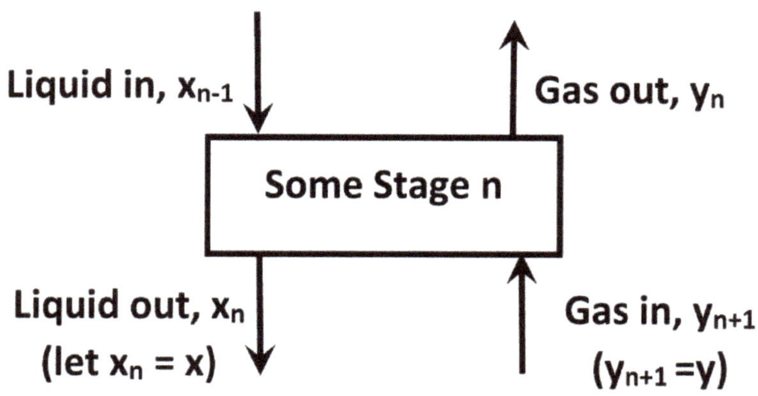

Figure 4.6 Some n[th] stage of a gas absorption cascade.

gas is $y_{n+1} - y_n$. As y_{n+1} is the composition of the gas stream passing the liquid stream (of composition x), we can take $y_{n+1} = y$ and re-write the change in gas composition as $y - y_n$. The driving force for mass transfer of solute from gas phase to liquid phase for a given x is $y - y^*$, where y^* is the composition of the gas stream in phase equilibrium with the liquid stream leaving the stage. This is the maximum change in gas composition possible in the stage under consideration. If an equilibrium is established between the liquid and gas streams leaving the stage, this maximum possible change in gas composition is actually realized as $y_n = y^*$. If equilibrium is not established between the two streams leaving the stage, $y_n > y^*$ as less solute is removed from the gas phase and therefore, $y - y_n < y - y^*$. In this case the full potential of the driving force for mass transfer of solute from gas phase to liquid phase is not actually realized in the given stage.

4.4 DESIGN OF MULTI-STAGE ABSORPTION TOWERS

The key calculation in the design of gas absorption towers is the determination of the number of stages required to carry out the given separation duty. Once the number of stages required are known, one can specify the height of the absorption tower by selecting a suitable spacing between the stages. The number of stages in a multi-stage absorption column can be determined graphically as well as analytically.

4.4.1 Graphical Approach: McCabe-Thiele Method

The graphical approach to determine the number of equilibrium stages is based on the following points: (1) the two fluid streams leaving any stage are at equilibrium with each other. Hence the compositions of the leaving gas and liquid streams from any stage must fall on the equilibrium line; and (2) the compositions of the passing gas and liquid streams at any level are related by the operating line. Hence they should fall on the operating line. These points are illustrated in Figure 4.7 for the nth stage. The nth stage is represented graphically as a triangle with the following two points of the triangle (x_{n-1}, y_n) and (x_n, y_{n+1}) falling on the operating line and the following point of the triangle (x_n, y_n) falling on the equilibrium line.

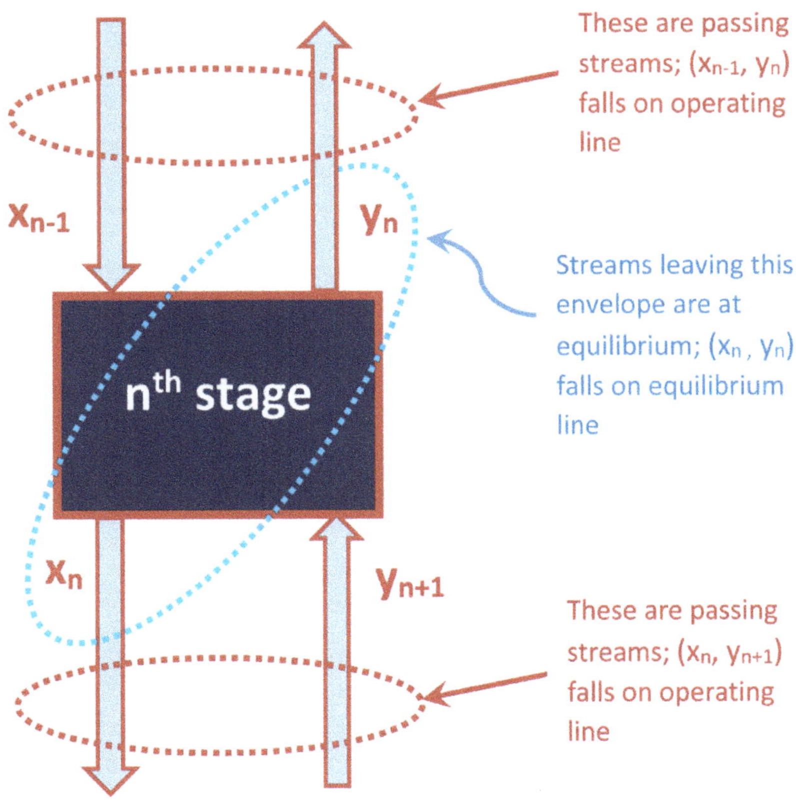

Figure 4.7 nth stage of a gas absorption cascade.

Figure 4.8 shows the graphical method of determining the number of equilibrium stages in a multi-stage absorption process. We begin counting the number of stages from the top of the column as shown in Figure 4.1. To construct the first stage graphically, draw a horizontal line from point (x_a, y_a) on the operating line to the equilibrium line as shown in the figure. The point on the equilibrium line is (x_1, y_1) or (x_1, y_a) as $y_1 = y_a$. From point (x_1, y_1), draw a vertical line to the operating line. This gives the point (x_1, y_2) on the operating line. This completes the construction of stage 1 shown in Figure 4.9 and represented by a triangle in Figure 4.8 with two points, (x_a, y_a) and (x_1, y_2), on the operating line and one point, (x_1, y_1), on the equilibrium line. This procedure of construction of triangular stages by drawing horizontal and vertical lines between the operating and equilibrium lines is continued until the terminal composition at the other end (bottom) of the column is reached.

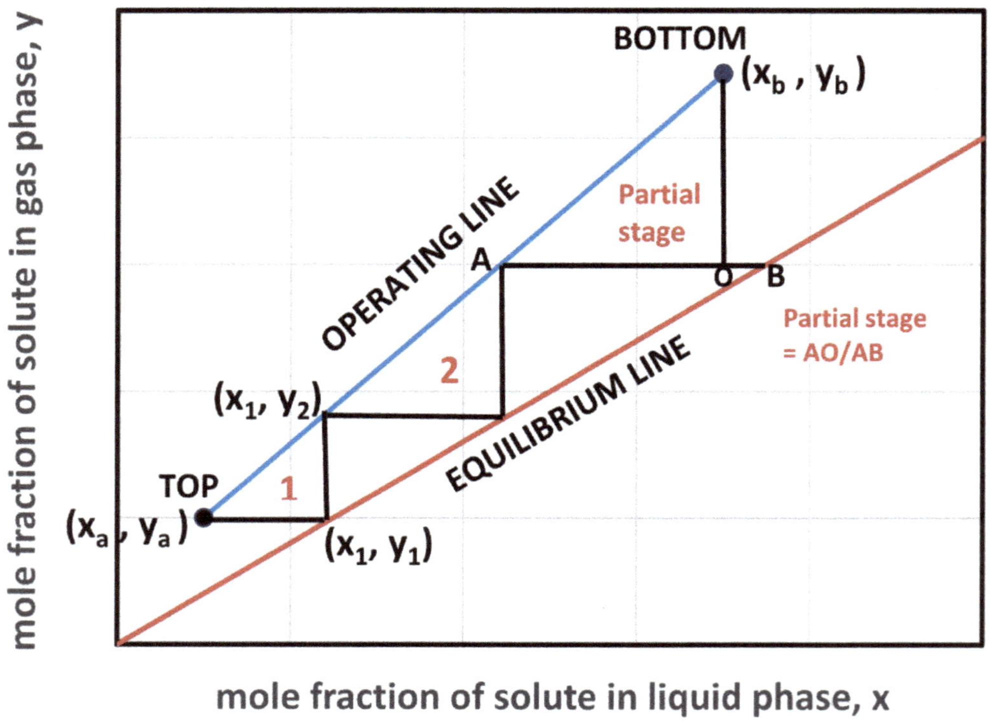

Figure 4.8 McCabe-Thiele graphical method of determining the number of stages in a multi-stage absorption process.

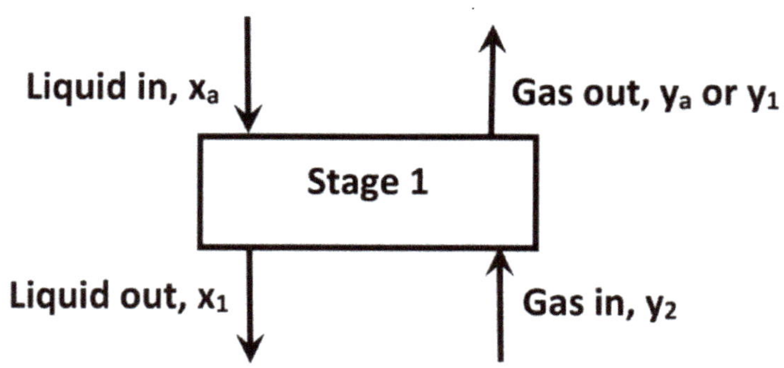

Figure 4.9 Compositions of streams entering and leaving stage 1.

Note that the final stage may be a partial stage as shown in the figure. This graphical construction of stages using the operating and equilibrium lines alternately is referred to as the *McCabe-Thiele method* of determining the number of stages as McCabe and Thiele (W.L. McCabe, E.W. Thiele, Industrial & Engineering Chemistry, Vol. 17, 605, 1925) were the first ones to apply this approach to the design of distillation columns.

4.4.2 Analytical Approach: Kremser Equations

The number of equilibrium stages required in a multi-stage gas absorption process can be calculated from the Kremser equations (A. Kremser, National Petroleum News, Vol. 22, 42, 1930). However, the Kremser equations can be applied only if the operating and equilibrium lines are linear. Before giving the derivations of the Kremser equations, let us first write the Kremser equations and see how they can be applied to predict the number of equilibrium stages. Referring to Figure 4.10, the Kremser equations are given as follows:

$$N = \frac{ln\left(\frac{y_b - y_b^*}{y_a - y_a^*}\right)}{ln(A)} \quad when\ A \neq 1 \tag{4.9}$$

$$N = \frac{y_b - y_a}{y_b - y_b^*} = \frac{y_b - y_a}{y_a - y_a^*} \quad when\ A = 1 \tag{4.10}$$

where A is the absorption factor defined as:

$$A = \frac{L/V}{K} = \frac{y_b - y_a}{y_b^* - y_a^*} \tag{4.11}$$

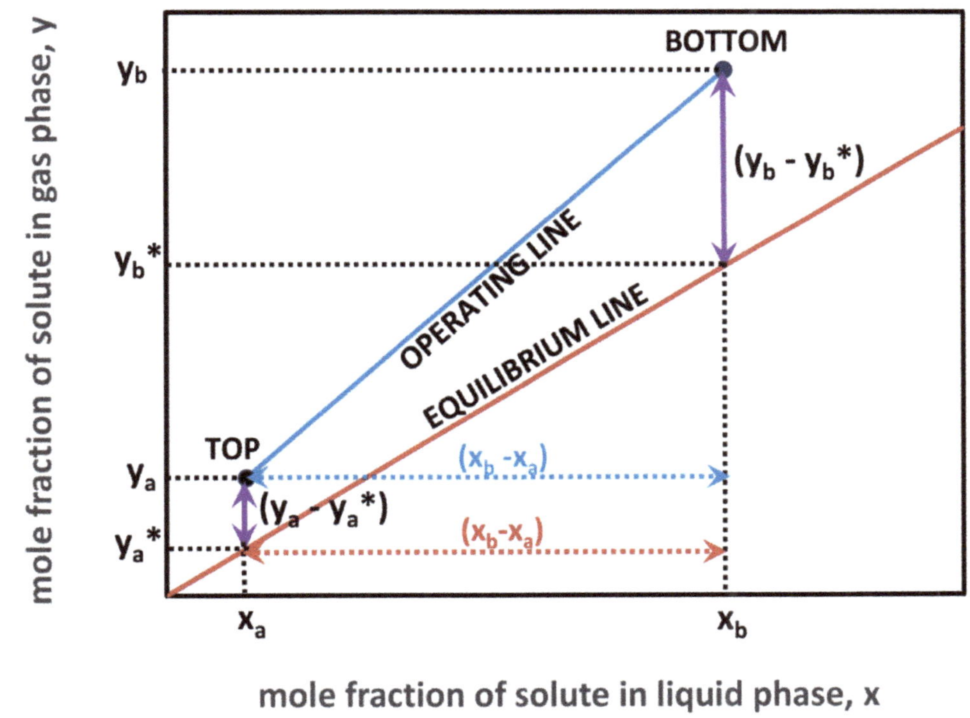

Figure 4.10 Diagram to illustrate the applicability of Kremser equations.

Note that A is the ratio of operating line slope (L/V) to equilibrium line slope (K). Referring to Figure 4.10, the slope of the operating line is $(y_b - y_a)/(x_b - x_a)$ and the slope of the equilibrium line is $(y_b^* - y_a^*)/(x_b - x_a)$. Therefore, $A = (y_b - y_a)/(y_b^* - y_a^*)$.

When $A \neq 1$, that is, the operating line is not parallel to the equilibrium line, Equation (4.9) gives the number of stages. This case is more commonly encountered in practical applications. In the numerator natural log-term on the right side of Equation (4.9), $(y_b - y_b^*)$ is the driving force for mass transfer of solute from gas phase to liquid phase at the bottom end of the column (see Figure 4.10) and $(y_a - y_a^*)$ is the driving force for mass transfer of solute at the top end of the column.

When $A = 1$, that is, the operating line is parallel to the equilibrium line, Equation (4.10) gives the number of stages. In this special case, the number of stages is simply the ratio of overall change in the gas composition $(y_b - y_a)$ in going from top to bottom of the column to the driving force $(y - y^*)$ for mass transfer of solute. Note that $(y - y^*)$ is constant in this case as the operating line and equilibrium line are parallel.

Derivation of Kremser Equations

We begin with the equation for the operating line, Equation (4.6). As the gas and liquid streams leaving any n[th] stage are in phase equilibrium with each other (see Figure 4.11), x_n and y_n are related to each other by the equilibrium relationship (Henry law). Consequently, the operating line equation becomes:

$$y_{n+1} = y_a + (\tfrac{L}{V})(\tfrac{y_n}{K} - x_a) \tag{4.12}$$

Using the quantities shown in Figure 4.11, Equation (4.12) could be re-written as:

$$y_{n+1} = y_1 + (\tfrac{L}{KV})(y_n - y_a^*) \tag{4.13}$$

Note that $y_a = y_1$ and $y_a^* = Kx_a$. Since $A = L/KV$, we can re-write Equation (4.13) as:

$$y_{n+1} = y_1 + Ay_n - Ay_a^* \tag{4.14}$$

Let us now apply this equation to different stages, starting with the first stage from the top of the column $(n = 1)$:

$$y_2 = y_1 + Ay_1 - Ay_a^* \tag{4.15}$$

When $(n = 2)$:

$$y_3 = y_1 + Ay_2 - Ay_a^* \tag{4.16}$$

Upon substitution of y_2 from Equation (4.15) into Equation (4.16), we get:

$$y_3 = y_1(1 + A + A^2) - y_a^*(A + A^2) \tag{4.17}$$

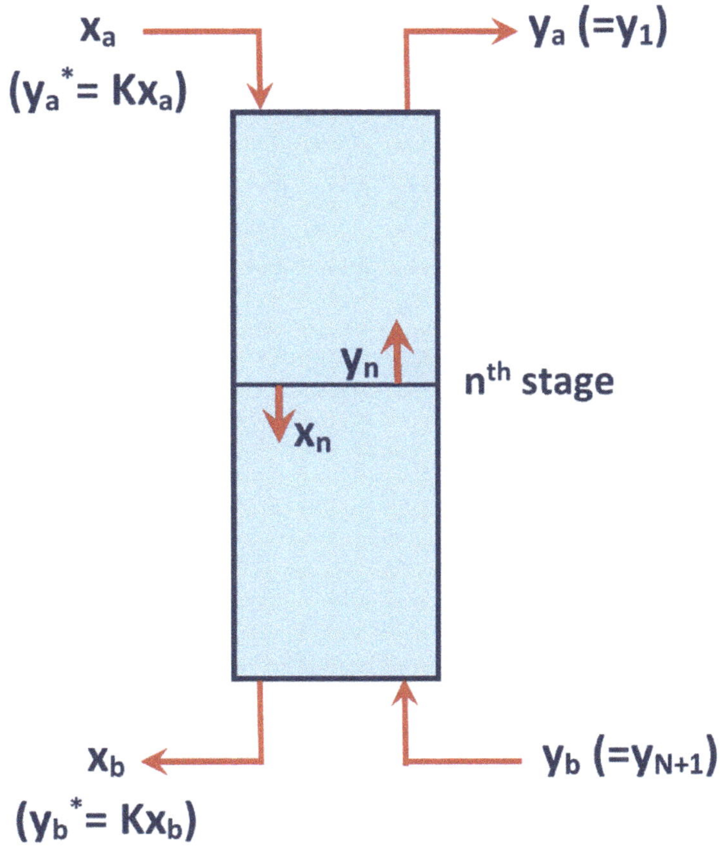

Figure 4.11 Gas absorption column indicating the quantities needed in the derivation of the Kremser equations.

When ($n = 3$), Equation (4.14) gives:

$$y_4 = y_1 + Ay_3 - Ay_a^* \tag{4.18}$$

Upon substitution of y_3 from Equation (4.17) into Equation (4.18), we get:

$$y_4 = y_1(1 + A + A^2 + A^3) - y_a^* (A + A^2 + A^3) \tag{4.19}$$

This result can now be extended to the Nth stage (last stage of the column):

$$y_{N+1} = y_b = y_a(1 + A + A^2 + A^3 + \cdots + A^N) - y_a^* (A + A^2 + A^3 + \cdots + A^N) \tag{4.20}$$

Note that we have replaced y_1 by y_a as they are the same (refer to Figure 4.11). In Equation (4.20), we have two geometric series:

$$(1 + A + A^2 + A^3 + \cdots + A^N) \; and \; (A + A^2 + A^3 + \cdots + A^N) \tag{4.21}$$

The sum of a geometric series is given as follows:

$$(a_1 + a_2 + a_3 + a_4 + \cdots + a_n) = \frac{a_1(1-r^n)}{1-r} \; where \; r = \frac{a_2}{a_1} = \frac{a_3}{a_2} = \cdots \neq 1 \tag{4.22}$$

Consequently, when $A \neq 1$:

$$1 + A + A^2 + A^3 + \cdots + A^N = \frac{1-A^{N+1}}{1-A} \tag{4.23}$$

$$A + A^2 + A^3 + \cdots + A^N = \frac{A(1-A^N)}{1-A} \tag{4.24}$$

When $A = 1$:

$$1 + A + A^2 + A^3 + \cdots + A^N = N + 1 \tag{4.25}$$

$$A + A^2 + A^3 + \cdots + A^N = N \tag{4.26}$$

When $A \neq 1$, Equation (4.20) can now be re-written as follows:

$$y_b = \frac{y_a(1-A^{N+1})}{1-A} - \frac{y_a^* A(1-A^N)}{1-A} \quad where\ A \neq 1 \tag{4.27}$$

Upon re-arrangement, Equation (4.27) gives:

$$A^{N+1}(y_a - y_a^*) = A(y_b - y_a^*) - (y_b - y_a) \quad where\ A \neq 1 \tag{4.28}$$

Substituting $(y_b - y_a) = A(y_b^* - y_a^*)$ from Equation (4.11) into Equation (4.28) and re-arranging, we get the following expression for the number of stages in an absorption column:

$$N = \frac{ln\left(\frac{y_b - y_b^*}{y_a - y_a^*}\right)}{ln(A)} \quad when\ A \neq 1 \tag{4.29}$$

When $A = 1$, Equation (4.20) can be re-written as follows:

$$y_b = y_a(N + 1) - y_a^*(N) \tag{4.30}$$

Upon re-arrangement, Equation (4.30) gives the following expression for the number of stages in an absorption column when $A = 1$:

$$N = \frac{y_b - y_a}{y_a - y_a^*} \quad when\ A = 1 \tag{4.31}$$

Note that $(y_a - y_a^*) = (y_b - y_b^*)$ when $A = 1$.

4.4.3 Minimum L/V Requirement and Pinch Condition

From a practical point of view, it is important to know the minimum solvent flow rate needed to operate an absorption column for a given gas feed stream to be treated. The gas flow rate and its inlet and outlet compositions are specified. The solute concentration in the solvent entering the column is also specified (solvent entering the column is usually solute-free). The minimum liquid flow rate for a given gas flow rate corresponds to minimum *L/V* ratio. Thus the slope *L/V* of the operating line is minimum. The smallest slope of the operating line that one can have is when the operating line just touches the equilibrium line, as shown in Figure 4.12. The point where the two lines meet each other is called *pinch point*. The pinch condition usually occurs at the bottom of the absorption tower. The minimum *L/V* ratio can be calculated as:

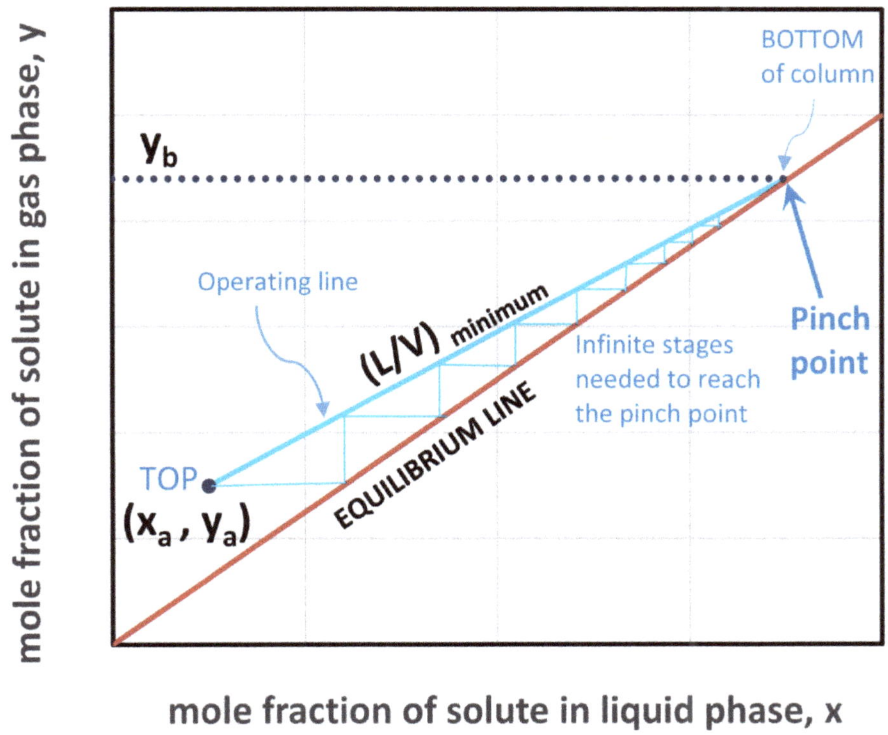

Figure 4.12 Minimum *L/V* and pinch point in an absorption column.

$$(L/V)_{minimum} = \frac{y_p - y_a}{x_p - x_a} \tag{4.32}$$

where (x_p, y_p) are the coordinates of the pinch point (see Figure 4.12). Note that $y_p = y_b$ and $x_p = y_b/K$. An infinite number of stages are required to reach the pinch point where the driving force for mass transfer is zero (see Figure 4.12). This also follows from the Kremser equation, Equation (4.9). When *L/V* is minimum, $y_b - y_b^*$ is zero and $A < 1$. Under this condition, the Kremser equation gives $N \to \infty$. In the design of an absorption column, the designer has to select an appropriate liquid flow rate. The column cannot be operated at a minimum *L/V* ratio as the height of the tower required at minimum *L/V* ratio is infinite. The optimum *L/V* ratio is about 1.2 to 1.5 times the minimum *L/V* ratio. More often than not, actual L/V is taken to be 1.5 times the minimum L/V.

NOTATION

A	absorption factor, defined in Equation (4.11)
K	distribution coefficient of solute, defined in Equation (4.8)
K_H	Henry law constant of solute defined in Equation (4.7)
L	molar flow rate of liquid stream (solvent + solute)
L_a	molar flow rate of liquid stream entering the column at the top, see Figure 4.3
L_b	molar flow rate of liquid stream leaving the column at the bottom, see Figure 4.3
L_n	molar flow rate of liquid stream leaving the nth stage of the column, see Figure 4.3
N	total number of stages in the column
P	total pressure of gas
P_A	partial pressure of solute (species A) in the gas phase
V	molar flow rate of gas stream (carrier gas + solute)

V_a molar flow rate of gas stream leaving the column at the top, see Figure 4.3
V_b molar flow rate of gas stream entering the column at the bottom, see Figure 4.3
V_n molar flow rate of gas stream leaving the nth stage of the column, see Figure 4.3
x mole fraction of solute in the liquid phase
x_a mole fraction of solute in liquid stream entering the column at the top, see Figure 4.3
x_b mole fraction of solute in liquid stream leaving the column at the bottom, see Figure 4.3
x_n mole fraction of solute in liquid stream leaving the nth stage of the column, see Figure 4.3
y mole fraction of solute in the gas phase
y_a mole fraction of solute in gas stream leaving the column at the top, see Figure 4.3
y_b mole fraction of solute in gas stream entering the column at the bottom, see Figure 4.3
y_n mole fraction of solute in gas stream leaving the nth stage of the column, see Figure 4.3
y^* mole fraction of solute in the gas phase if the gas phase was in equilibrium with the existing liquid phase
y_a^* mole fraction of solute in the gas phase in equilibrium with the liquid stream entering the column at the top, see Figures 4.10 and 4.11.
y_b^* mole fraction of solute in the gas phase in equilibrium with the liquid stream leaving the column at the bottom, see Figures 4.10 and 4.11.

TUTORIAL PROBLEMS WITH SOLUTION

Problem 1

A multi-stage countercurrent absorption tower is to be designed to absorb ethyl alcohol vapor from an inert gas stream containing 2.5 mole percent ethyl alcohol. Pure water is to be used as a solvent. The process operates isothermally (303 K) at a total pressure of 101.3 kPa. The molar flow rate of the gas stream is 50 kmol/h. The liquid stream molar flow rate is 45 kmol/h. It is desired to absorb 90% of the alcohol. The equilibrium relationship for this system at the temperature and pressure of the process is given as: $y = 0.7x$, where y and x are mole fractions of ethyl alcohol in gas and liquid phases, respectively at equilibrium.

i) Determine the number of equilibrium stages required using the graphical McCabe-Thiele method.
ii) Determine the number of equilibrium stages required using the Kremser equation.

Solution

Refer to the column schematic shown below.

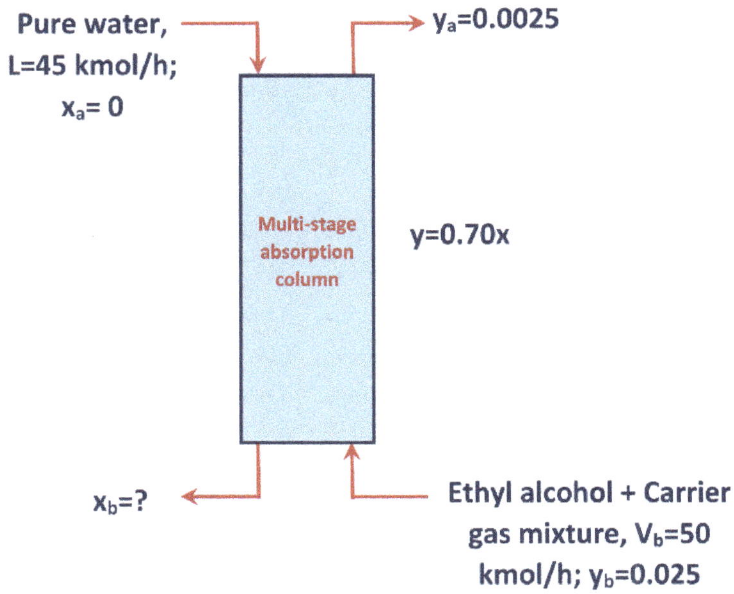

As 90% of entering ethyl alcohol is removed, the mole fraction of ethyl alcohol in the gas leaving the column is as follows:

$$y_a = 0.025 - (0.90 \times 0.025) = 0.0025$$

The mole fraction of ethyl alcohol in the liquid leaving the column at the bottom, that is, x_b , can be calculated from the operating line using the terminal points:

$$y_b = y_a + (\frac{L}{V})(x_b - x_a)$$

$$0.025 = 0.0025 + (45/50)(x_b - 0)$$

$$x_b = 0.025$$

i) McCabe-Thiele Graphical Method to estimate the number of equilibrium stages:

- Plot the equilibrium line, y = 0.70 x, as shown on the graph
- Plot the operating line: straight line that passes through the terminal compositions of the column (0, 0.0025) and (0.025, 0.025)
- Construct stages as shown in the graph below

The McCabe graphical method gives the total number of equilibrium stages N =4.4.

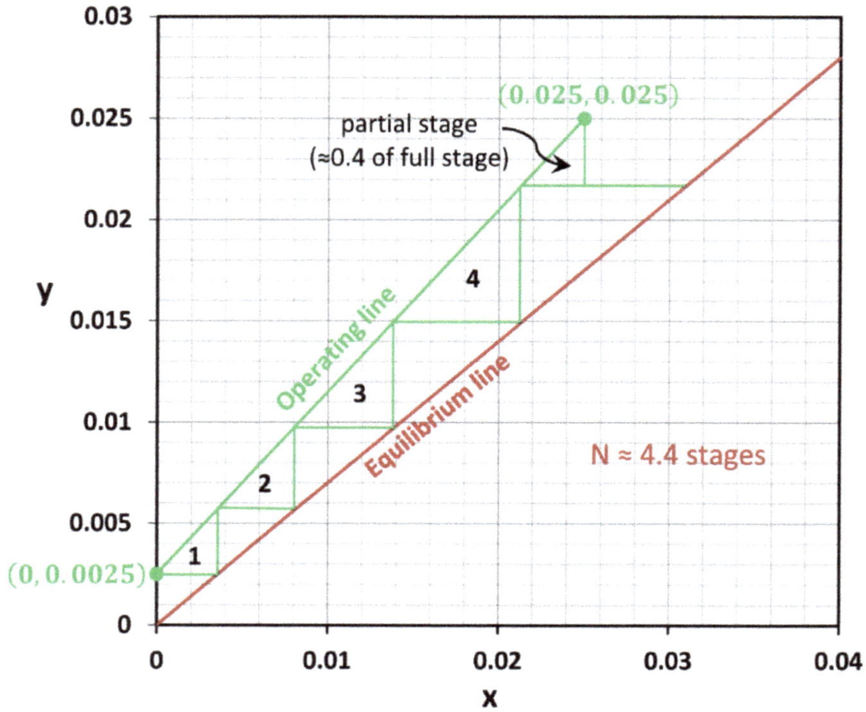

ii) Let us now apply the Kremser equation to estimate the number of equilibrium stages. The absorption factor is

$$A = (L/V)/K = (45/50)/0.70 = 1.2857$$

As $A \neq 1$, the following form of Kremser equation is applicable:

$$N = \frac{ln\left(\frac{y_b - y_b^*}{y_a - y_a^*}\right)}{ln(A)} \ when\ A \neq 1$$

$$y_b = 0.025, y_b^* = Kx_b = 0.7 \times 0.025 = 0.0175$$

$$y_a = 0.0025, y_a^* = Kx_a = 0.7 \times 0 = 0$$

Thus,

$$N = \frac{ln\left(\frac{y_b - y_b^*}{y_a - y_a^*}\right)}{ln(A)} = \frac{ln\left(\frac{0.025 - 0.0175}{0.0025 - 0}\right)}{\ln{(1.2857)}} = 4.3716$$

This value is very close to the value obtained graphically using the McCabe-Thiele method.

Problem 2

Acetone is being absorbed from acetone-air mixture in a multi-stage absorption tower using pure water as a solvent. The gas entering the tower contains 1.5 mole percent acetone and 80 percent of it is to be absorbed. The molar flow rate of water is 2.4 times the gas molar flow rate. The process operates isothermally (300 K) at a total pressure of 1bar. Assume that the equilibrium relationship for this system at the temperature and pressure of the process is given as: $P_{acetone} = K_H x$, where K_H is 2.5 bars.

a) Determine the number of equilibrium stages required using the McCabe-Thiele graphical method.

b) Determine the number of equilibrium stages using the Kremser equation

You can assume that the molar flow rates of gas and liquid streams are constant as the system in dilute in solute.

Solution

(a) Refer to the column schematic shown below.

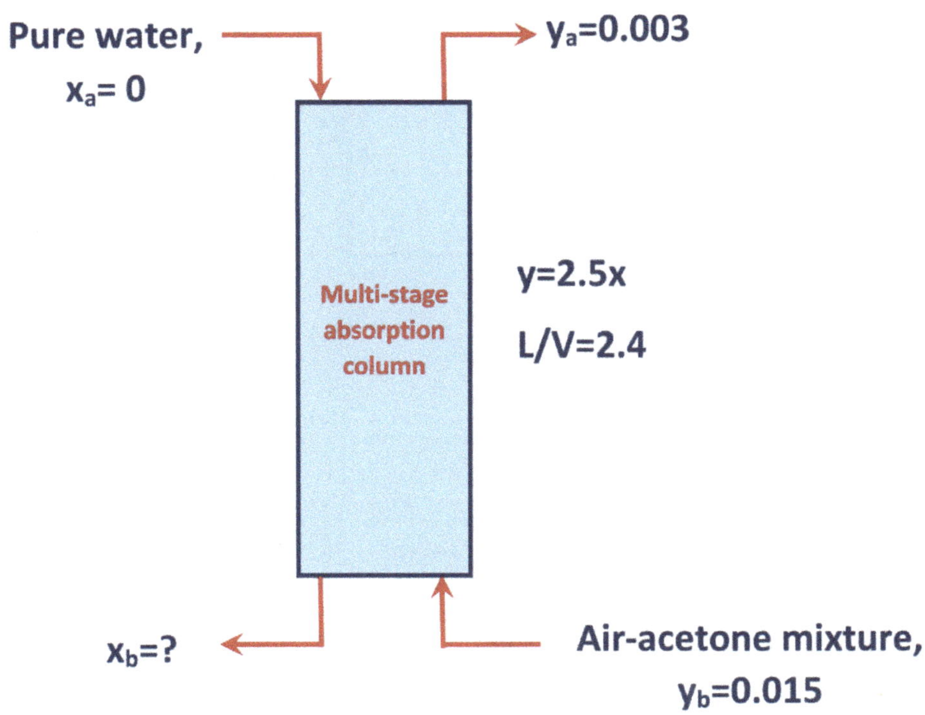

As 80% of entering acetone is removed, the mole fraction of acetone in the gas leaving the column is as follows:

$$y_a = 0.015 - (0.80 \times 0.015) = 0.003$$

The mole fraction of acetone in the liquid leaving the column at the bottom, that is, x_b , can be calculate from the operating line using the terminal points:

$$y_b = y_a + (\tfrac{L}{V})(x_b - x_a)$$

$$0.015 = 0.003 + (2.4)(x_b - 0)$$

$$x_b = 0.005$$

Thus the operating line passes through the points (0, 0.003) and (0.005, 0.015). The following figure shows the plots of the operating and equilibrium lines. Note that the equilibrium line is given as: $y = P_{acetone}/P = K_H x/P = 2.5x$. The stepping off the stages using the McCabe-Thiele method is also shown in the figure. The number of equilibrium stages required is 4.6.

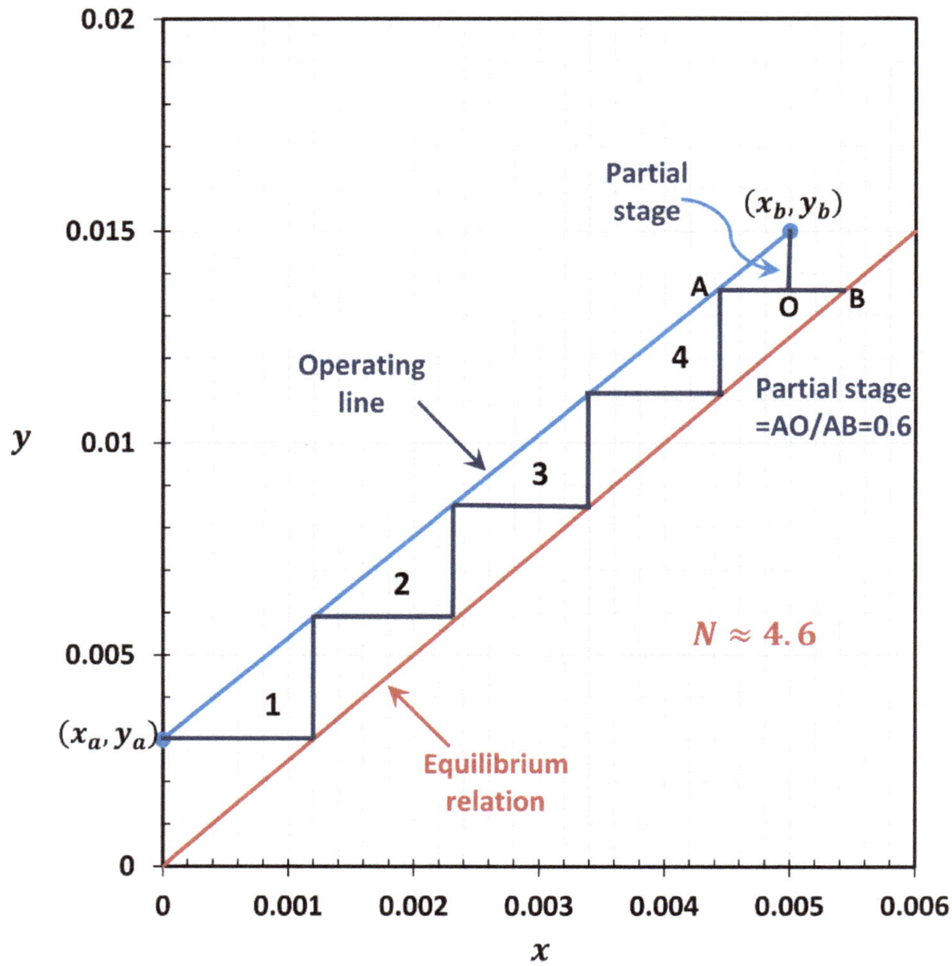

b) Let us now apply the Kremser equation to estimate the number of equilibrium stages. The absorption factor is

$$A = (L/V)/K = (2.4)/2.5 = 0.96$$

As $A \neq 1$, the following form of Kremser equation is applicable:

$$N = \frac{ln\left(\frac{y_b - y_b^*}{y_a - y_a^*}\right)}{ln(A)} \text{ when } A \neq 1$$

$$y_b = 0.015, y_b^* = Kx_b = 2.5 \times 0.005 = 0.0125$$

$$y_a = 0.003, y_a^* = Kx_a = 2.5 \times 0 = 0$$

Thus,

$$N = \frac{ln\left(\frac{y_b - y_b^*}{y_a - y_a^*}\right)}{ln(A)} = \frac{ln\left(\frac{0.015 - 0.0125}{0.003 - 0}\right)}{\ln(0.96)} = 4.466$$

This value is very close to the value obtained graphically using the McCabe-Thiele method

Problem 3

Acetone is being absorbed from acetone-air mixture in a multi-stage absorption column using a non-volatile oil. The gas entering the column contains 25 mole percent acetone and 95 percent of it is to be absorbed. The oil entering the column is acetone free. The liquid leaving at the bottom of the column contains 8 mole percent acetone. Assume that the equilibrium relationship for this system at the temperature and pressure of the process is given as: $y = 2x$, where y and x are mole fractions of acetone in gas and liquid phases, respectively at equilibrium.

Determine the number of equilibrium stages required using the McCabe-Thiele graphical method. Also calculate V_a, L_a, L_b, V' and L'

You can assume that the molar flow rates of acetone-free oil and acetone-free air are constant. Choose $1\ kmol/h$ of entering gas ($V_b = 1kmol/h$) as a basis for your calculations.

Solution

Refer to the column schematic shown below. Let us first consider the *gas entering* at the bottom of the column:

$$y_b = 0.25, V_b = 1\ kmol/h\ , Acetone\ molar\ flow\ rate = 0.25\ \ kmol/h$$

As the solute concentration is high, we will use mole ratios $X\ and\ Y$ and solute-free molar flow rates $L' and\ V'$ in our analysis. These quantities are defined in Chapters 2 and 3.

$$V' = V_b(1 - y_b) = 1\ (1 - 0.25) = 0.75\ kmol/h$$

$$, Y_b = y_b/(1 - y_b) = 0.25/(1 - 0.25) = 0.3333$$

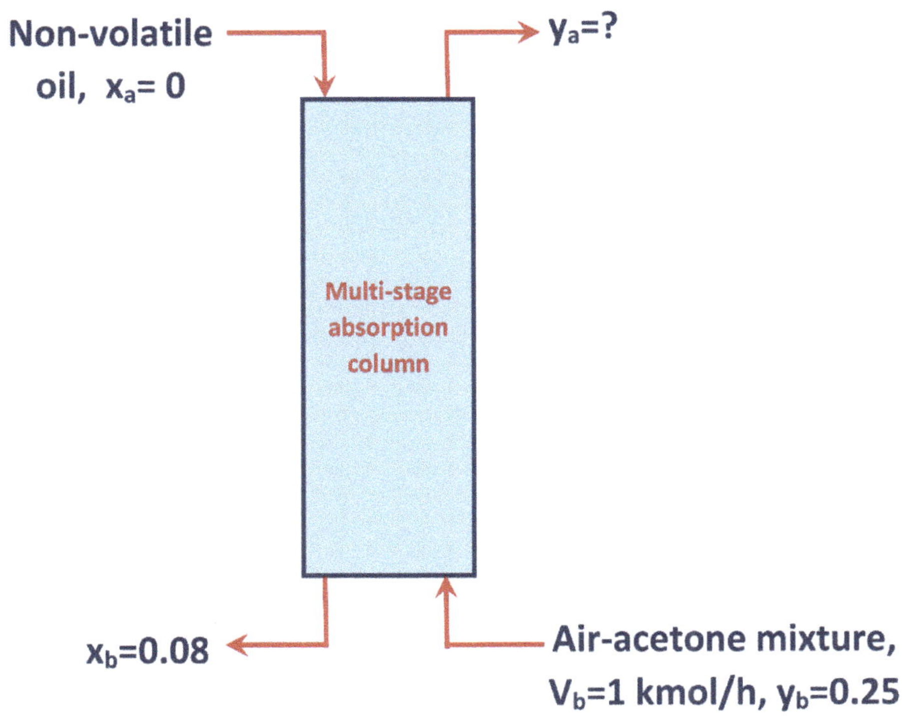

Consider now the *gas leaving* at the top of the column:

$$V' = 0.75\ kmol/h$$

$$Acetone\ molar\ flow\ rate = 0.25 - (0.95 \times 0.25) = 0.0125\ kmol/h$$

$$y_a = 0.0125/(0.75 + 0.0125) = 0.0164$$

$$Y_a = y_a/(1 - y_a) = 0.0164/(1 - 0.0164) = 0.0167$$

$$V_a = 0.75 + 0.0125 = 0.7625\ kmol/h$$

Next we consider *liquid leaving* at the bottom of the column:

$$Acetone\ molar\ flow\ rate = x_b L_b = 0.25 - 0.0125 = 0.2375\ kmol/h$$

$$L_b = 0.2375/x_b = 0.2375/0.08 = 2.969\ kmol/h$$

$$L' = L_b(1 - x_b) = 2.969\,(1 - 0.08) = 2.731\ kmol/h$$

$$L_a = L' = 2.731\ kmol/h$$

In summary, the values of the unknown quantities are:

$V_a = 0.7625\ kmol/h$
$L_a = 2.731\ kmol/h$
$L_b = 2.969\ kmol/h$
$V' = 0.75\ kmol/h$
$L' = 2.731\ kmol/h$

Estimation of Equilibrium Stages

The equilibrium relationship given in terms of mole fractions can be re-written in terms of mole ratios as:

$$\frac{Y}{1+Y} = 2\left(\frac{X}{1+X}\right)$$

The operating line is a straight line on Y-X plot and it passes through the terminal points: (X_a, Y_a) and (X_b, Y_b). The values of the terminal mole ratios are as follows:

$$X_a = 0,\ Y_a = 0.0167$$

$$X_b = x_b/(1 - x_b) = 0.08/(1 - 0.08) = 0.087$$

$$Y_b = 0.3333$$

The plots of the equilibrium and operating lines are shown in the figure below. The operating line is linear whereas the equilibrium relationship on Y-X plot is slightly nonlinear. The stepping off the stages using the McCabe-Thiele method is also shown in the figure. The number of equilibrium stages required is 4.2.

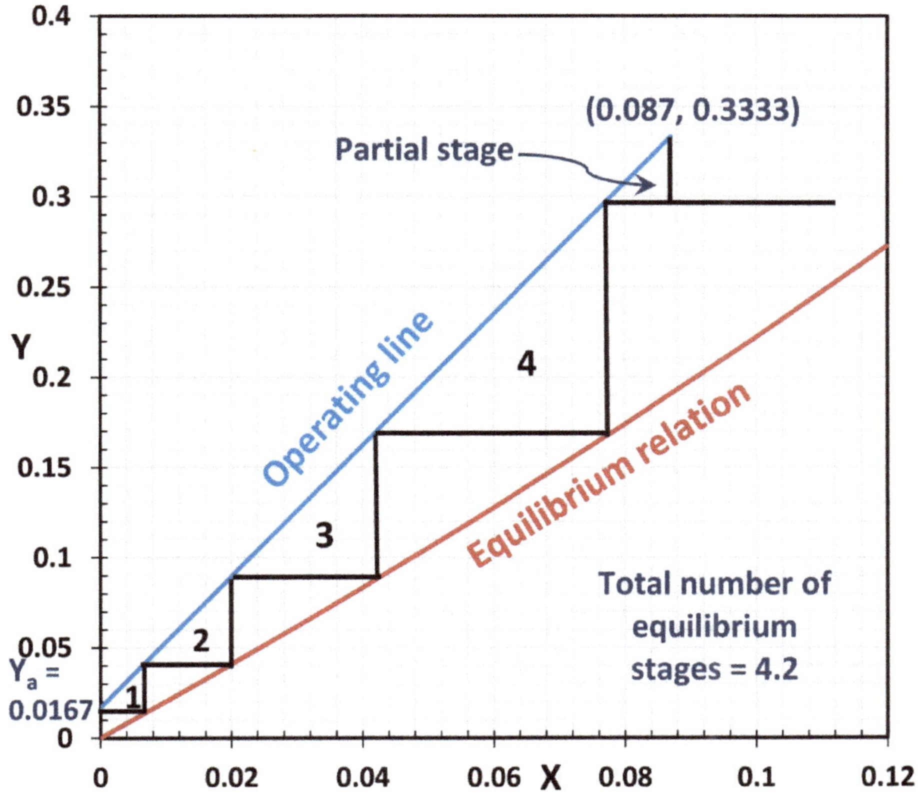

Problem 4

Carbon dioxide is being absorbed from $CO_2 - N_2$ gas mixture in a multi-stage countercurrent absorption tower using pure water as a solvent. The gas entering the tower contains 2 mole percent CO_2 and 92 percent of it is to be absorbed. The actual liquid molar flow rate is 1.5 times the minimum liquid molar flow rates. Assume that the equilibrium relationship for this system at the temperature and pressure of the process is given as: $y = 87.6x$, where y and x are mole fractions of carbon dioxide in gas and liquid phases, respectively at equilibrium.

a) Determine the number of equilibrium stages required using the McCabe-Thiele graphical method.
b) Determine the number of stages using the Kremser equation

You can assume that the molar flow rates of gas and liquid streams are constant as the system in dilute in solute.

Solution

a) Refer to the column schematic shown below. As 92% of entering CO_2 is removed, the mole fraction of CO_2 in the gas leaving the column is as follows:

$$y_a = 0.02 - (0.92 \times 0.02) = 0.0016$$

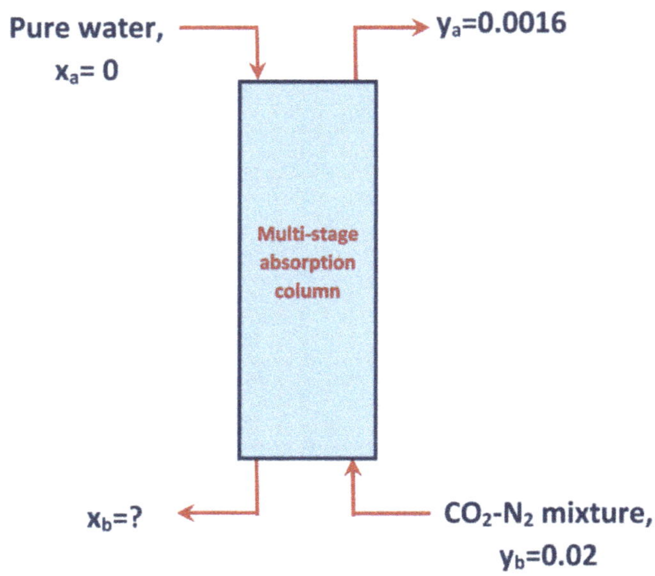

The minimum slope of the operating line is when the operating line just touches the equilibrium line, as shown in the figure below. Thus,

$$Minimum\ slope = (L/V)_{min} = \frac{y_b - y_a}{x_b^* - x_a} = \frac{0.02 - 0.0016}{0.0002283} = 80.6$$

Note that $x_b^* = y_b/K = 0.02/87.6 = 0.0002283$. The actual L/V is $1.5(L/V)_{min}$. Consequently

$$L/V = 1.5 \times 80.6 = 120.9$$

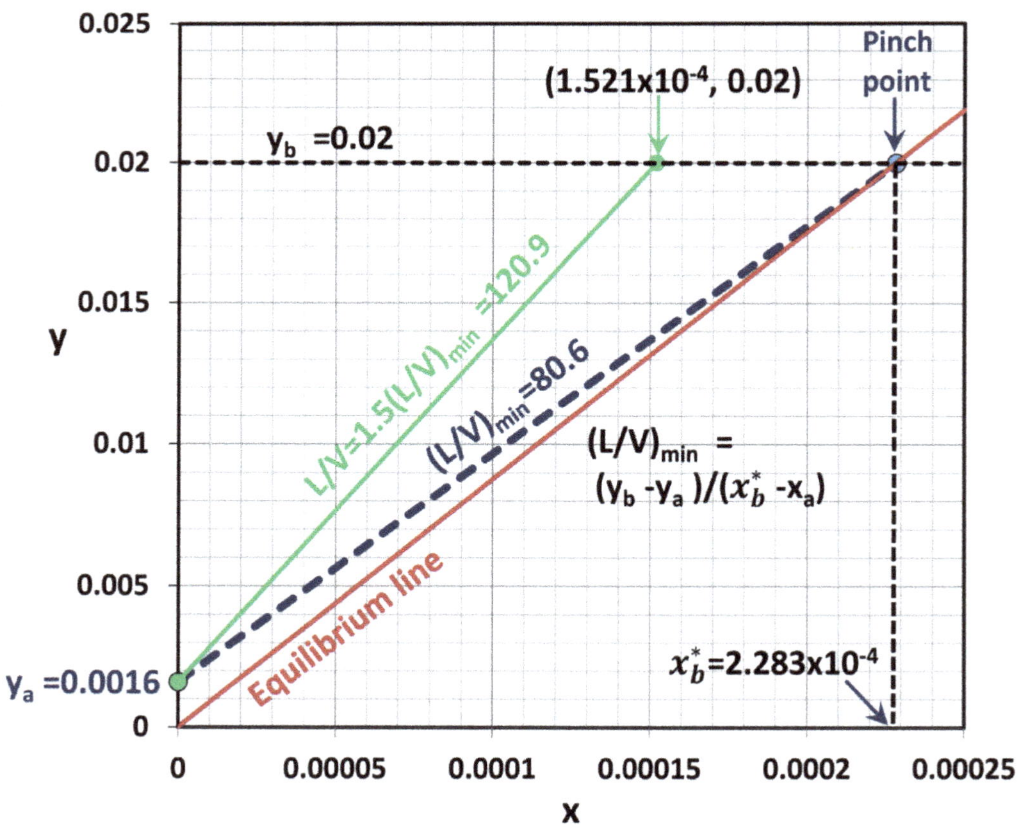

The mole fraction of CO_2 in the liquid leaving the column at the bottom, that is, x_b , can be calculate from the operating line using the terminal points:

$$y_b = y_a + (\tfrac{L}{V})(x_b - x_a)$$

$$0.02 = 0.0016 + (120.9\,)(x_b - 0)$$

$$x_b = 0.0001521$$

Thus the actual operating line passes through the points (0, 0.0016) and (0.0001521, 0.02). The following figure shows the plots of the operating and equilibrium lines. The stepping off the stages using the McCabe-Thiele method is also shown in the figure. The number of equilibrium stages required is 4.6.

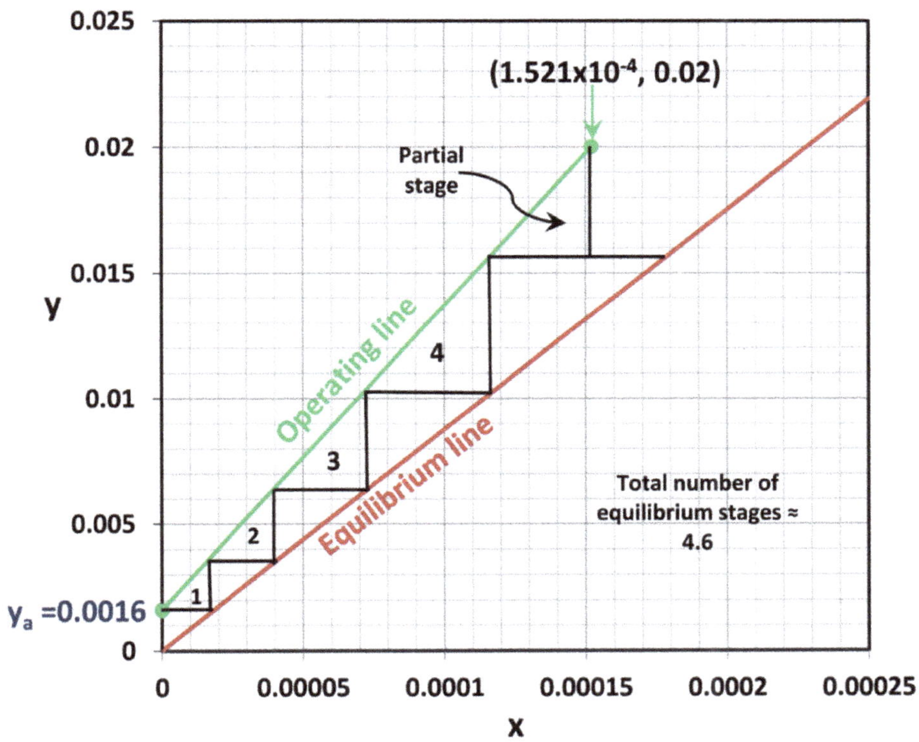

b) Let us now apply the Kremser equation to estimate the number of equilibrium stages. The absorption factor is

$$A = (L/V)/K = 120.9/87.6 = 1.3801$$

As $A \neq 1$, the following form of Kremser equation is applicable:

$$N = \frac{ln\left(\frac{y_b - y_b^*}{y_a - y_a^*}\right)}{ln(A)} \; when\; A \neq 1$$

$$y_b = 0.02, y_b^* = Kx_b = 87.6 \times 0.0001521 = 0.013324$$

$$y_a = 0.0016, y_a^* = Kx_a = 87.6 \times 0 = 0$$

Thus,

$$N = \frac{ln\left(\frac{y_b - y_b^*}{y_a - y_a^*}\right)}{ln(A)} = \frac{ln\left(\frac{0.02 - 0.013324}{0.0016 - 0}\right)}{\ln(1.3801)} = 4.43$$

This value is close to the value obtained graphically using the McCabe-Thiele method.

Problem 5

We wish to remove two solutes, H_2S and CO_2, from a gaseous stream using pure water as a solvent in a multistage absorption process. The carrier gas is air. The gas stream contains 0.3 mole% H_2S and 0.5 mole% CO_2. We desire a recovery of 99% of H_2S from the gas phase to the liquid phase. The ratio of liquid molar flow rate to gas molar flow rate is 100. The process is carried out at $T = 298$ K and $P = 10$ bars. The Henry law constants for H_2S and CO_2 in water are: 552 bars and 1662 bars respectively.

a) Determine the mole fractions of H_2S in the two discharge streams
b) Determine the number of equilibrium stages needed
c) Determine the mole fractions of CO_2 in the two discharge streams
d) What is the percent recovery of CO_2 from the gas phase to the liquid phase?

Solution

This problem involves multicomponent gas absorption. As noted in Chapter 2, the multicomponent gas absorption problems involving two or more solutes can be solved using the single solute procedure assuming that: i) the solutes act independently of each other. Thus, the equilibrium relationship for any given solute does not depend on the mole fractions of other solutes, ii) the system is dilute, and iii) the molar flow rates of gas and liquid streams are constant.

Refer to the diagram given below.

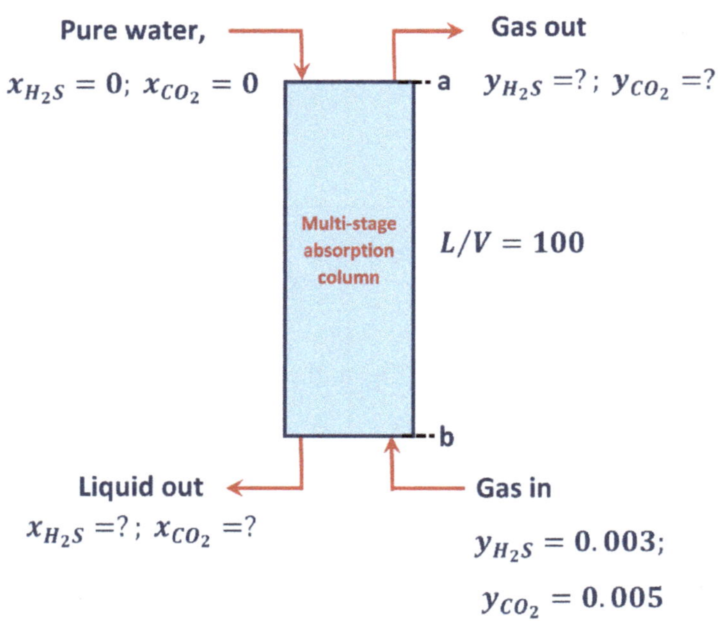

a) H_2S mole fractions:

As the recovery of H_2S is 99%, the mole fraction of H_2S in the gas leaving the absorber is:

$$y_{H_2S,a} = 0.003 - (99/100) \times 0.003 = 3 \times 10^{-5}$$

The mole fraction of H_2S in the liquid leaving the column at the bottom, that is, $x_{H_2S,b}$, can be calculated from the operating line using the terminal points:

$$y_{H_2S,b} = y_{H_2S,a} + (L/V)(x_{H_2S,b} - x_{H_2S,a})$$

$$0.003 = 3 \times 10^{-5} + (100\ )(x_{H_2S,b} - 0)$$

$$x_{H_2S,b} = 2.97 \times 10^{-5}$$

b) Number of equilibrium stages needed

The absorption factor A_{H_2S} can be calculated as:

$$A_{H_2S} = \frac{L/V}{K_{H_2S}} = \frac{100}{55.2} = 1.811594$$

We can now apply the Kremser equation to estimate the number of equilibrium stages required. As $A_{H_2S} \neq 1$, the following form of Kremser equation is applicable:

$$N = \frac{ln\left(\frac{y_{H_2S,b} - y^*_{H_2S,b}}{y_{H_2S,a} - y^*_{H_2S,a}}\right)}{ln(A_{H_2S})} \quad when\ A_{H_2S} \neq 1$$

$$y_{H_2S,b} = 0.003, y^*_{H_2S,b} = Kx_{H_2S,b} = 55.2 \times 2.97 \times 10^{-5} = 1.6394 \times 10^{-3}$$

$$y_{H_2S,a} = 3 \times 10^{-5}, y^*_{H_2S,a} = Kx_{H_2S,a} = 55.2 \times 0 = 0$$

Thus,

$$N = \frac{ln\left(\frac{y_{H_2S,b} - y^*_{H_2S,b}}{y_{H_2S,a} - y^*_{H_2S,a}}\right)}{ln(A_{H_2S})} = \frac{ln\left(\frac{0.003 - 1.6394 \times 10^{-3}}{3 \times 10^{-5} - 0}\right)}{ln(1.811594)} = 6.42$$

c) CO_2 mole fractions:

The absorption factor A_{CO_2} is:

$$A_{CO_2} = \frac{L/V}{K_{CO_2}} = \frac{100}{166.2} = 0.601685$$

Apply Kremser equation to CO_2, given that $N = 6.42$:

$$6.21 = \frac{ln\left(\frac{y_{CO_2,b} - y^*_{CO_2,b}}{y_{CO_2,a} - y^*_{CO_2,a}}\right)}{ln(A_{CO_2})} = \frac{ln\left(\frac{y_{CO_2,b} - y^*_{CO_2,b}}{y_{CO_2,a} - y^*_{CO_2,a}}\right)}{ln(0.601685)}$$

$$y_{CO_2,b} - y^*_{CO_2,b} = (y_{CO_2,a} - y^*_{CO_2,a}) \times 0.038331$$

$$y^*_{CO_2,a} = K_{CO_2} \times x_{CO_2,a} = 0$$

$$0.005 - y^*_{CO_2,b} = (y_{CO_2,a}) \times 0.038331$$

Apply the operating line equation for CO_2 using the terminal points:

$$y_{CO_2,b} = y_{CO_2,a} + (L/V)(x_{CO_2,b} - x_{CO_2,a})$$

$$y_{CO_2,b} = y_{CO_2,a} + (100)(x_{CO_2,b} - 0) = y_{CO_2,a} + (100)(y^*_{CO_2,b}/K_{CO_2})$$

$$0.005 = y_{CO_2,a} + (100)(y^*_{CO_2,b}/166.2)$$

$$100y^*_{CO_2,b} + 166.2y_{CO_2,a} = 0.831$$

There are two unknowns in the equations obtained from the Kremser equation and the operating line: $y_{CO_2,a}$ and $y^*_{CO_2,b}$. Solve for $y_{CO_2,a}$ and $y^*_{CO_2,b}$:

$$y_{CO_2,a} = 2.0386 \times 10^{-3}\ , y^*_{CO_2,b} = 4.92186 \times 10^{-3}$$

$$x_{CO_2,b} = y^*_{CO_2,b}/K_{CO_2} = 4.92186 \times 10^{-3}/166.2 = 2.9614 \times 10^{-5}$$

d) CO_2 recovery:

The percent recovery of CO_2 from gas phase to liquid phase can be calculated as:

$$\%Recovery = \left(1 - \frac{y_{CO_2,a}}{y_{CO_2,b}}\right) \times 100 = \left(1 - \frac{2.0386 \times 10^{-3}}{0.005}\right) \times 100 = 59.228$$

SHORT REVIEW QUESTIONS

1. Define a cascade of stages.
2. What is the difference between counter-current and parallel cascades? Which one is more effective?
3. Is the operating line of a single stage identical to the operating line of a cascade? If not, what are the differences?
4. The operating line of a cascade relates the compositions of two flow streams. Which flow streams are those?
5. Under what condition is the slope of the operating line constant when the operating line is expressed in terms of the mole fractions of solute?
6. What is the main assumption made in the McCabe-Thiele graphical method of estimating the number of equilibrium stages?
7. For McCabe-Thiele graphical method to be applicable, should the equilibrium relation be linear?
8. Under what conditions is the Kremser equation applicable?
9. An absorber consists of 5 equilibrium stages. The mole fraction of solute in the feed gas is 0.05 and the feed liquid is free of solute. What is the mole fraction of solute in the gas leaving the top of the absorber if the absorption factor is 1.3 and the Henry law constant is 10?

10. Define pinch condition. Does it generally occur at the bottom or top of the absorption column?
11. How many equilibrium stages are needed to operate an absorption column at minimum L/V corresponding to pinch condition?

PRACTICE PROBLEMS

Problem 1

Acetone is being absorbed from acetone-air mixture in a multi-stage absorption tower using pure water as a solvent. The gas entering the tower contains 1 mole percent acetone and 80 percent of it is to be absorbed. The molar flow rates of gas and water entering the tower are 25 kmol/h and 85 kmol/h, respectively. The process operates isothermally (300 K) at a total pressure of 101.3 kPa. Assume that the equilibrium relationship for this system at the temperature and pressure of the process is given as: $y = 2.5x$, where y and x are mole fractions of acetone in gas and liquid phases, respectively at equilibrium.

a) Determine the number of equilibrium stages required using the McCabe-Thiele graphical method.
b) Determine the number of stages using the Kremser equation

You can assume that the molar flow rates of gas and liquid streams are constant as the system in dilute in solute.

Problem 2

A multi-stage countercurrent absorption tower is absorbing ethyl alcohol from an inert gas stream containing 2 mole percent ethyl alcohol. Pure water is being used as a solvent. The process operates isothermally (303 K) at a total pressure of 101.3 kPa. The molar flow rate of gas stream is 90 kmol/h. It is desired to remove 85% of the alcohol. Assume that the equilibrium relationship for this system at the temperature and pressure of the process is given as: $y = 0.7x$, where y and x are mole fractions of ethyl alcohol in gas and liquid phases, respectively at equilibrium. The actual liquid molar flow rate is 1.6 times the minimum liquid molar flow rate.

a) Determine the number of equilibrium stages required using the McCabe-Thiele graphical method.
b) Determine the number of stages using the Kremser equation

You can assume that the molar flow rates of gas and liquid streams are constant as the system in dilute in solute.

Problem 3

An absorber consists of 5 equilibrium stages. The mole fraction of the solute in the feed gas is 0.05 and the feed liquid is free of solute. The Henry law for the system is given as: $y = 10x$. What are the mole fractions of solute in the gas and liquid streams leaving the absorber if the absorption factor is 1.3? What is the percent recovery of the solute from gas phase to liquid phase? Re-do the problem if the absorption factor is decreased from 1.3 to 0.65. Does the percent recovery increase or decrease when the absorption faction is reduced to one-half from 1.3 to 0.65?

Problem 4

It is desired to absorb 90% of the solute in a gas containing 25 mol% solute in a countercurrent stage tower. The total inlet gas flow to the tower is 30 kmol/h, and the total inlet pure solvent flow rate is 90 kmol/h. The process is to operate isothermally at 300 K and a total pressure of 101.3 kPa. The equilibrium relation for the solute is $y = 2.53 ln(1 - x)$. Determine the number of equilibrium stages required using the graphical McCabe-Thiele method. Note that the gas and liquid molar flow rates in the absorption tower cannot be assumed to be constant.

Problem 5

We wish to remove two solutes, SO_2 and Cl_2, from a gaseous stream using pure water as a solvent in a multistage absorption process. The carrier gas is air. The gas stream contains 0.3 mole% SO_2 and 0.5 mole% Cl_2. We desire a recovery of 99% of SO_2 from the gas phase to the liquid phase. The ratio of liquid molar flow rate to gas molar flow rate is 100. The process is carried out at $T = 298$ K and $P = 10$ bars. The Henry law constants for SO_2 and Cl_2 in water are: 440 bars and 615 bars respectively.

a) Determine the mole fractions of SO_2 in the two discharge streams
b) Determine the number of equilibrium stages needed
c) Determine the mole fractions of Cl_2 in the two discharge streams
d) What is the percent recovery of Cl_2 from the gas phase to the liquid phase?

Problem 6

Acetone is being absorbed from acetone-air mixture in a multi-stage absorption tower using pure water as a solvent. The gas entering the tower contains 2 mole percent acetone and 90 percent of it is to be absorbed. The molar flow rates of gas and water entering the tower are 35 kmol/h and 85 kmol/h, respectively. The process operates isothermally (300 K) at a total pressure of 1 bar. Assume that the equilibrium relationship for this system is given as: $P_{acetone} = K_H x$, where K_H is 2.5 bars.

a) Determine the number of equilibrium stages required using the McCabe-Thiele graphical method.
b) Determine the number of equilibrium stages using the Kremser equation
c) What would be the number of equilibrium stages required if the pressure of the absorption tower is increased from 1 bar to 2 bars?

You can assume that the molar flow rates of gas and liquid streams are constant as the system in dilute in solute.

Chapter 5

Multi-Stage Liquid Stripping

5.1 LEARNING OUTCOMES

Upon completion of this chapter, students will be able to:

- Analyze counter-current multi-stage liquid stripping process
- Determine the number of equilibrium stages required for a given separation in a multi-stage stripping process using the graphical McCabe-Thiele method
- Develop and apply the Kremser equations for the estimation of number of equilibrium stages analytically
- Explain the importance of minimum gas flow rate and pinch condition in a multi-stage stripping process

5.2 INTRODUCTION

The analysis and design of a multi-stage stripping column is similar to that of a multi-stage gas absorption column discussed in Chapter 4. The counter-current cascade (and nomenclature) shown in Figure 4.1 and the contacting device (perforated/sieve tray column) shown in Figure 4.2 of Chapter 4 are applicable to a multi-stage liquid stripping process. The mass balance carried out in section 4.3 of Chapter 4 over the top n stages (see Figure 4.3 of Chapter 4) is also applicable. Thus the following equation of the operating line derived in Chapter 4 is applicable to multi-stage liquid stripping process:

$$y_{n+1} = y_a + \left(\frac{L}{V}\right)(x_n - x_a) \tag{5.1}$$

The operating line is linear of positive slope (L/V) passing through the terminal compositions (x_a, y_a) and (x_b, y_b). In arriving at this equation of the operating line, it was assumed that the system is dilute enough that the liquid and gas flow rates do not change from one end of the column to the other end of the column. However, there are significant differences between multi-stage absorption and stripping processes.

First of all, let us consider the driving force for mass transfer of solute from liquid phase to gas phase. Referring to Figure 5.1 given below, the driving force for mass transfer in liquid stripping process at any location of the contacting device (perforated tray column) is $(x - x^*)$, where x is the mole fraction of solute in the existing liquid phase and x^* is the mole fraction of solute in the liquid phase if the liquid phase was in phase equilibrium with the existing gas phase. If the two phases are in phase equilibrium with each other $(x - x^*) = 0$. If the existing gas phase is not in equilibrium with the existing liquid phase, x^* is a hypothetical quantity (non-existent in the actual system). For solute to move from liquid phase to gas phase, the driving force $(x - x^*)$ must be positive greater than zero. For $(x - x^*) > 0$, it is necessary that the operating line is located below the equilibrium line as shown in Figure 5.2 given below. Note that in the case of multi-stage gas absorption the operating line was located above the equilibrium line (see Figure 4.5 of Chapter 4) as mass transfer of solute was taking place from gas phase to liquid phase and $(y - y^*) > 0$.

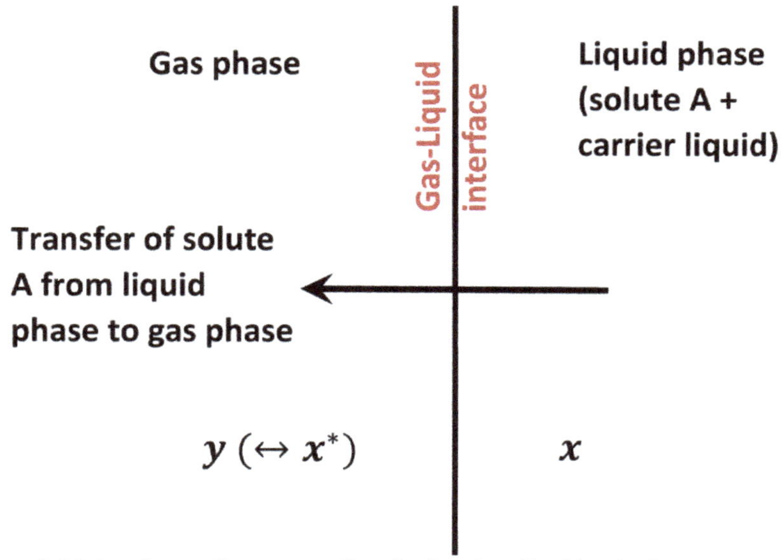

Figure 5.1 Direction and driving force of mass transfer of solute in a liquid stripping process.

Information revealed by Figure 5.2 can be summarized as follows: (a) the operating and equilibrium lines are linear as expected for dilute systems; (b) the liquid solution and the gas mixture both become more concentrated in solute if we move in the upward direction from the bottom of the multi-stage stripping column to the top of the column. This is opposite of what was observed in the case of multi-stage gas absorption; (c) the equilibrium line is steeper than the operating line provided that the stripping factor (ratio of equilibrium line slope to operating line slope) is greater than unity; (d) the driving force for mass transfer of solute from liquid phase to gas phase, that is, x-x^*, increases from bottom to top of the column. This is to be expected when stripping factor is greater than unity; and (e) the operating line falls below the equilibrium line, as noted earlier. This is necessary in liquid stripping where solute is transferred from liquid phase to gas phase and the driving force x-x^* has to be positive.

For further explanation of mass transfer driving force $(x - x^*)$ and its link to an equilibrium stage of a multi-stage liquid stripping cascade, we refer to Figure 5.3 given below. For the gas stream of composition $y_n (= y)$ leaving the stage, the corresponding change (decrease) in the liquid composition with respect to the entering liquid is $x_{n-1} - x_n$. As x_{n-1} is the composition of the liquid stream passing the gas stream (of composition y), we can take $x_{n-1} = x$ and re-write the change in liquid composition as $x - x_n$. The driving force for mass transfer of solute from liquid phase to gas phase for a given y is $x - x^*$, where x^* is the composition of the liquid stream in phase equilibrium with the gas stream leaving the stage. This is the maximum change in liquid composition possible in the stage under consideration. If an equilibrium is established between the liquid and gas streams leaving the stage, this maximum possible change in liquid composition is actually realized as $x_n = x^*$. If equilibrium is not established between the two streams leaving the stage, $x_n > x^*$ as less solute is removed from the liquid phase and therefore, $x - x_n < x - x^*$. In this case the full potential of the driving force for mass transfer of solute from liquid phase to gas phase is not actually realized in the given stage.

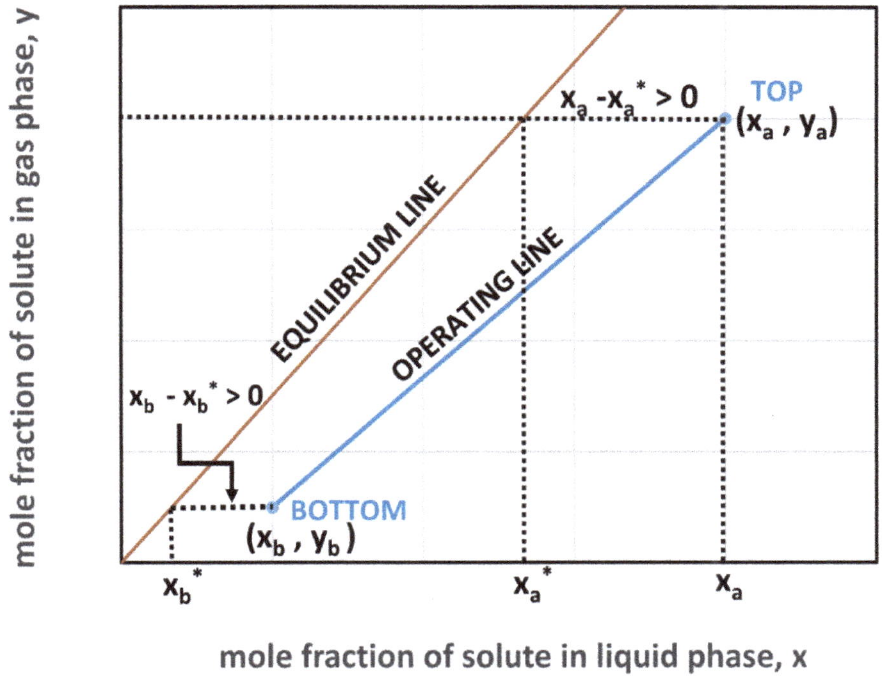

Figure 5.2 Positioning of operating and equilibrium lines in multi-stage liquid stripping process.

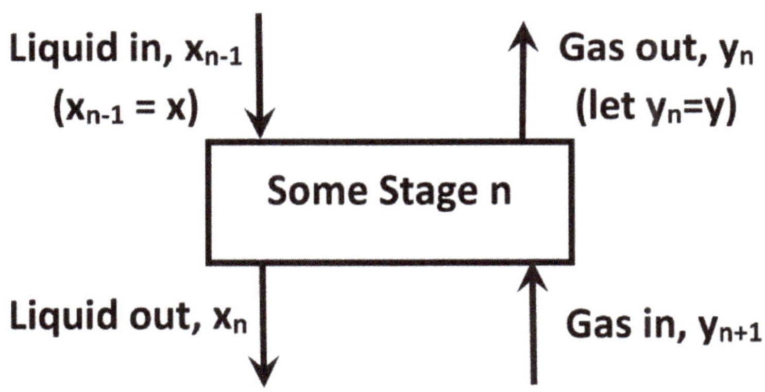

Figure 5.3 Some n[th] stage of a liquid stripping cascade.

5.3 DESIGN OF MULTI-STAGE LIQUID STRIPPING TOWERS

The key calculation in the design of liquid stripping towers is the determination of the number of stages required to carry out the given separation duty. Once the number of stages required are known, one can specify the height of the stripping tower by selecting a suitable spacing between the stages. The number of stages in a multi-stage liquid stripping column can be determined graphically as well as analytically in a manner similar to multi-stage gas absorption. However there are significant differences between absorption and stripping multi-stage processes as far as the graphs and equations are concerned.

5.3.1 Graphical Approach: McCabe-Thiele Method

As discussed in Chapter 4, the graphical approach to determine the number of equilibrium stages is based on the following points: (1) the two fluid streams leaving any stage are at equilibrium with each other. Hence the compositions of the leaving gas and liquid streams from any stage must fall on the equilibrium line; and (2) the compositions of the passing gas and liquid streams at any level are related by the operating line. Figure 5.4 shows the graphical method of determining the number of equilibrium stages in a multi-stage liquid stripping process.

5.3.2 Analytical Approach: Kremser Equations

The number of equilibrium stages required in a multi-stage liquid stripping process can be calculated from the Kremser equations provided that the operating and equilibrium lines are linear. The Kremser equations for multi-stage liquid stripping are given as follows:

$$N = \frac{\ln\left(\frac{x_a - x_a^*}{x_b - x_b^*}\right)}{\ln(S)} \quad when\ S \neq 1 \tag{5.2}$$

$$N = \frac{x_a - x_b}{x_b - x_b^*} = \frac{x_a - x_b}{x_a - x_a^*} \quad when\ S = 1 \tag{5.3}$$

where S is the stripping factor defined as:

$$S = \frac{K}{L/V} = \frac{x_a - x_b}{x_a^* - x_b^*} \tag{5.4}$$

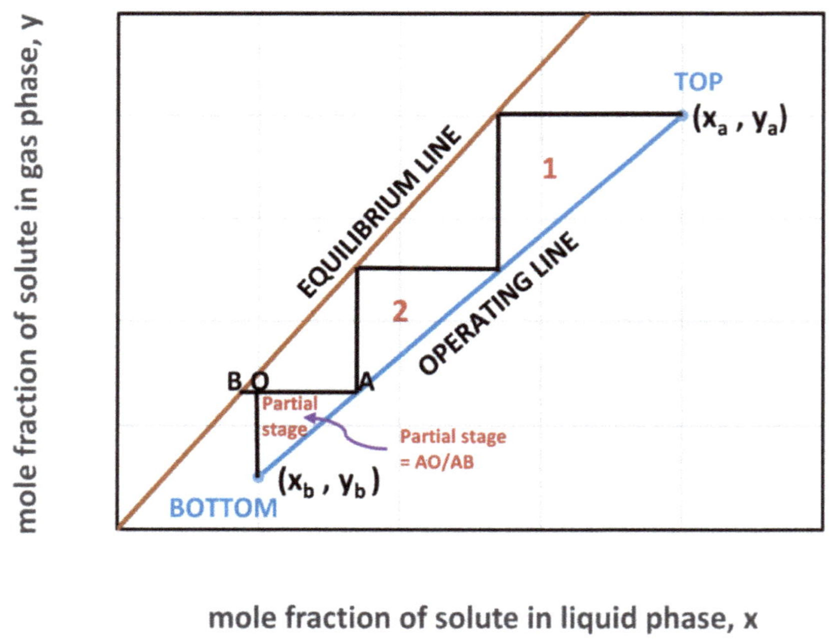

Figure 5.4 McCabe-Thiele graphical method of determining the number of stages in a multi-stage liquid stripping process.

Note that S is the ratio of equilibrium line slope (K) to operating line slope (L/V). Referring to Figure 5.2, the slope of the equilibrium line is $(y_a - y_b)/(x_a^* - x_b^*)$ and the slope of the operating line is $(y_a - y_b)/(x_a - x_b)$ and therefore, $S = (x_a - x_b)/(x_a^* - x_b^*)$.

When $S \neq 1$, that is, the operating line is not parallel to the equilibrium line, Equation (5.2) gives the number of stages. This case is more commonly encountered in practical applications. In the numerator natural log-term on the right side of Equation (5.2), $(x_a - x_a^*)$ is the driving force for mass transfer of solute from liquid phase to gas phase at the top of the column (see Figure 5.2) and $(x_b - x_b^*)$ is the driving force for mass transfer of solute at the bottom end of the column.

When $S = 1$, that is, the operating line is parallel to the equilibrium line, Equation (5.3) gives the number of stages. In this special case, the number of stages is simply the ratio of overall change in the liquid composition $(x_a - x_b)$ in going from bottom to top of the column to the driving force $(x - x^*)$ for mass transfer of solute. Note that $(x - x^*)$ is constant in this case as the operating line and equilibrium line are parallel.

Derivation of Kremser Equations

The Kremser equations derived in Chapter 4 are not restricted to multi-stage gas absorption. They are equally applicable to multi-stage liquid stripping as long as the operating and equilibrium lines are linear. Starting from the Kremser equations of Chapter 4 (Equations (4.9)-(4.10) of Chapter 4), we can derive the Kremser equations for multi-stage liquid stripping (Equations (5.2)-(5.3) of this chapter) by replacing gas phase compositions (y's) by liquid phase compositions (x's). The Kremser equations of Chapter 4 are as follows:

$$N = \frac{ln\left(\frac{y_b - y_b^*}{y_a - y_a^*}\right)}{ln(A)} \quad when\ A \neq 1 \tag{5.5}$$

$$N = \frac{y_b - y_a}{y_b - y_b^*} = \frac{y_b - y_a}{y_a - y_a^*} \quad when\ A = 1 \tag{5.6}$$

where A is the absorption factor defined as:

$$A = \frac{L/V}{K} = \frac{y_b - y_a}{y_b^* - y_a^*} \tag{5.7}$$

Note that by definition, $A = 1/S$.

Referring to Figure 5.5, we can replace the y variables with x variables as follows:

$$y_a = Kx_a^*\ ;\ y_a^* = Kx_a\ ;\ y_b = Kx_b^*\ ;\ y_b^* = Kx_b \tag{5.8}$$

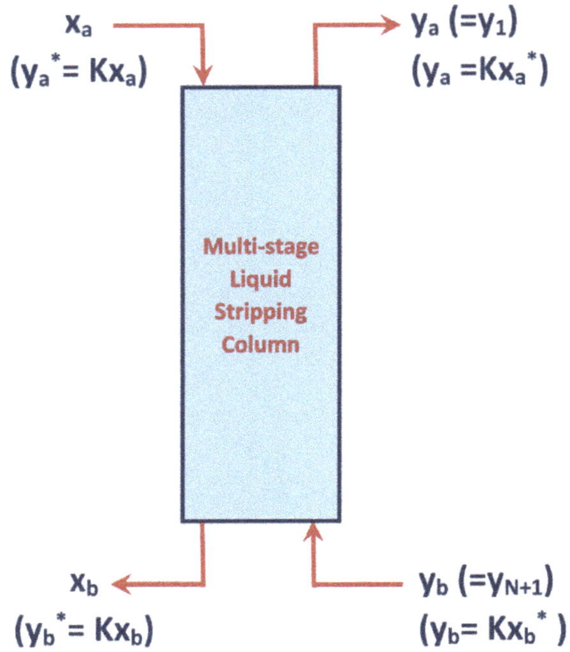

Figure 5.5 Multi-stage liquid stripping column indicating the quantities involved in the Kremser equations.

Thus, Equation (5.5) becomes:

$$N = \frac{ln\left(\frac{x_b - x_b^*}{x_a - x_a^*}\right)}{ln(1/S)} = \frac{\ln\left(\frac{x_a - x_a^*}{x_b - x_b^*}\right)}{\ln(S)} \quad when\ S \neq 1 \tag{5.9}$$

In a similar manner, Equation (5.6) reduces to:

$$N = \frac{x_a^* - x_b^*}{x_b - x_b^*} = \frac{x_a - x_b}{x_b - x_b^*} = \frac{x_a - x_b}{x_a - x_a^*} \quad when\ S = 1 \tag{5.10}$$

Note that when $S = 1, x_a - x_b = x_a^* - x_b^*, x_a - x_a^* = x_b - x_b^*$.

5.3.3 Minimum V/L (Maximum L/V) Requirement and Pinch Condition

It is important from a practical point of view to determine the minimum gas flow rate needed to treat a given liquid stream. In other words, what is the minimum gas flow rate for given L, x_a and x_b and entering gas composition y_b? Note that a minimum in gas flow rate for given liquid flow rate corresponds to a maximum slope L/V of the operating line. The maximum slope of the operating line possible is shown in Figure 5.6. It corresponds to a situation where the operating line just touches the equilibrium line. The point where the operating line touches the equilibrium line is the *pinch point*. In a liquid stripping column, the pinch condition usually occurs at the top of the column. Note that an infinite number of stages is needed to operate the column at a maximum value of L/V. In actual practice, L/V is taken to be about 0.7 times the maximum L/V.

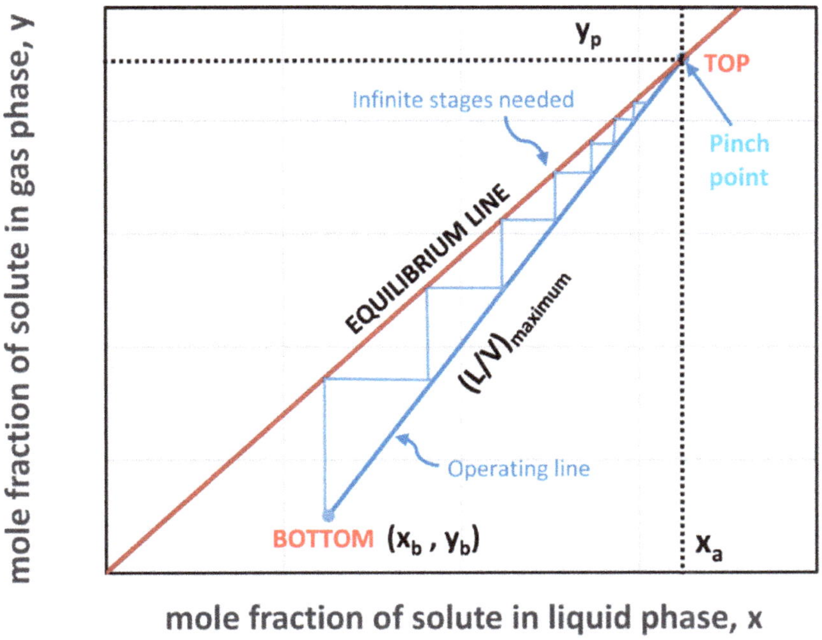

Figure 5.6 Minimum *V/L* (maximum *L/V*) and pinch point in a liquid stripping column.

Note that the maximum *L/V* ratio can be calculated as:

$$(L/V)_{maximum} = \frac{y_p - y_b}{x_a - x_b} \tag{5.11}$$

where (x_a, y_p) are the coordinates of the pinch point (see Figure 5.6) and $y_p = Kx_a$.

NOTATION

A	absorption factor, defined in Equation (5.7)
K	distribution coefficient of solute, defined as $K = y/x$ (also referred to as equilibrium constant)
L	molar flow rate of liquid stream (solvent + solute)
N	total number of stages in the column
S	stripping factor, defined in Equation (5.4)
V	molar flow rate of gas stream (carrier gas + solute)
x	mole fraction of solute in the liquid phase
x_a	mole fraction of solute in liquid stream entering the column at the top, see Figure 5.5
x_b	mole fraction of solute in liquid stream leaving the column at the bottom, see Figure 5.5
x_n	mole fraction of solute in liquid stream leaving the nth stage of the column, see Figure 5.3
x^*	mole fraction of solute *A* in the liquid phase if the liquid phase was in equilibrium with the existing gas phase
x_a^*	mole fraction of solute in the liquid phase in equilibrium with the gas stream leaving the column at the top, see Figure 5.5
x_b^*	mole fraction of solute in the liquid phase in equilibrium with the gas stream entering the column at the botttom, see Figure 5.5
y	mole fraction of solute in the gas phase
y_a	mole fraction of solute in gas stream leaving the column at the top, see Figure 5.5

y_b mole fraction of solute in gas stream entering the column at the bottom, see Figure 5.5

y_n mole fraction of solute in gas stream leaving the nth stage of the column, see Figure 5.3

y_p mole fraction of solute in gas stream at pinch point, see Figure 5.6

y^* mole fraction of solute in the gas phase if the gas phase was in equilibrium with the existing liquid phase

y_a^* mole fraction of solute in the gas phase in equilibrium with the liquid stream entering the column at the top, see Figure 5.5

y_b^* mole fraction of solute in the gas phase in equilibrium with the liquid stream leaving the column at the bottom, see Figure 5.5

TUTORIAL PROBLEMS WITH SOLUTION

Problem 1

A countercurrent stripping tower is to be used to remove a certain chemical solute from the liquid stream. The concentration of solute in the liquid stream is 25 ppm. Pure steam is used as a stripping gas. The ratio of liquid mass flow rate to steam mass flow rate is 2.5. It is desired to reduce the concentration of the solute in the liquid stream to 2 ppm. The equilibrium relation for the solute is $y = 12x$, where y is ppm of solute in the steam and x is ppm of solute in the liquid phase.

Determine the number of equilibrium stages needed using the Kremser equation. You can assume that none of the steam condenses in the stripping.

Solution

$$S = \frac{K}{L/V} = \frac{12}{2.5} = 4.8$$

The ppm of solute in the gas stream leaving the column at the top, that is, y_a , can be calculated from the operating line using the terminal points:

$$y_b = y_a + (\tfrac{L}{V})(x_b - x_a)$$

$$0 = y_a + (2.5)(2 - 25)$$

$$y_a = 57.5 \text{ ppm}$$

Use Kremser equation:

$$N = \frac{ln\left[\frac{x_a - x_a^*}{x_b - x_b^*}\right]}{ln(S)} = \frac{ln\left[\frac{25 - 4.7917}{2 - 0}\right]}{\ln(4.8)} = 1.47$$

Note that: $x_a^* = y_a/12 = 57.5/12 = 4.7917$ ppm and $x_b^* = 0/12 = 0$

In this problem, we have used mass fractions (equivalently ppm) and mass flow rates instead of mole fractions and molar flow rates. The Kremser equations can be applied with any consistent set of units: mass fractions and mass flow rates or mole fractions and molar flow rates.

Problem 2

Ammonia is stripped from a concentrated aqueous solution containing ammonia with a mole ratio of 0.10 by contact with pure air in a multistage countercurrent stripping column consisting of six sieve trays. The equilibrium relationship in terms of mole ratios is given as: Y = 0.85 X. When the molar flow rate of pure air is 1.5 times that of the carrier (ammonia-free) liquid solution, 90 percent of the ammonia is removed. How many equilibrium stages are required, using graphical McCabe-Thiele method? How many equilibrium stages are required, using Kremser equation? What is the stage efficiency?

Solution

As the solute (ammonia) concentration is high, we will use mole ratios X *and* Y and solute-free molar flow rates L' *and* V' in our analysis. These quantities are defined in Chapters 2 and 3.

Refer to the column schematic shown below. Let us first consider the *liquid entering* at the top of the column:

$$X_a = 0.10, \text{ Ammonia molar flow rate} = 0.10\ L'$$

As 90% of the ammonia is being stripped from the liquid phase, the ammonia in the discharge liquid stream can be calculated as follows:

$$0.10(0.10L') = L'X_b$$
$$X_b = 0.01$$

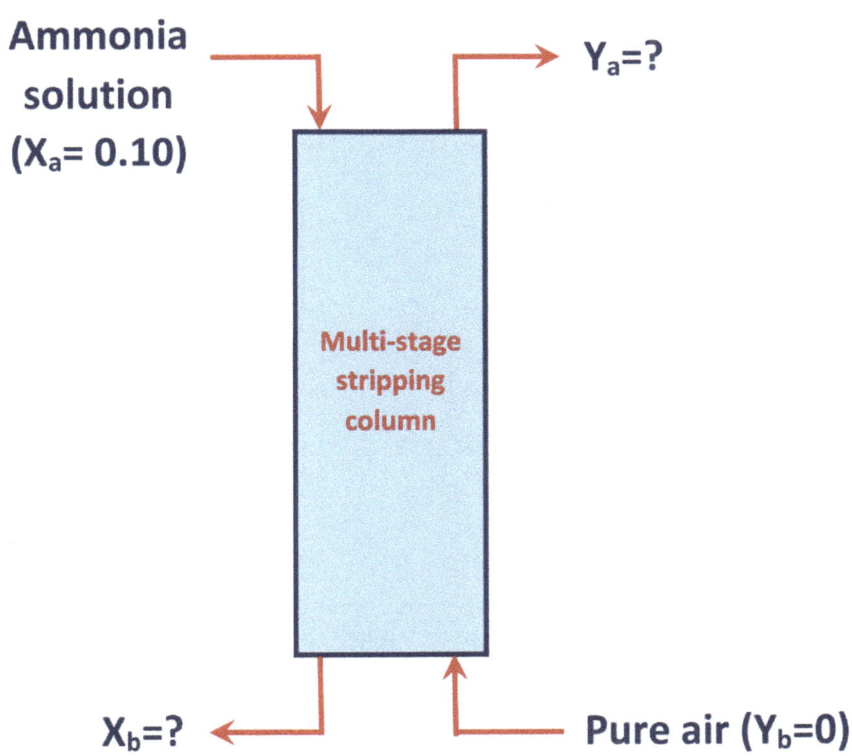

Consider now the *gas leaving* at the top of the column. The amount of ammonia in the discharge gas stream is:

$$Y_aV' = (0.90)X_aL'$$

$$Y_a = 0.90(L'/V')X_a = 0.90 \times \frac{0.10}{1.5} = 0.06$$

Estimation of Equilibrium Stages

The equilibrium relationship given in terms of mole ratios is:

$$Y = 0.85X$$

The operating line is a straight line on Y-X plot and it passes through the terminal points: (X_a, Y_a) and (X_b, Y_b). The values of the terminal mole ratios are as follows:

$$X_a = 0.1,\ Y_a = 0.06$$

$$X_b = 0.01,\ Y_b = 0$$

The plots of the equilibrium and operating lines are shown in the figure below. The stepping off the stages using the McCabe-Thiele method is also shown in the figure. The number of equilibrium stages required is 4.5.

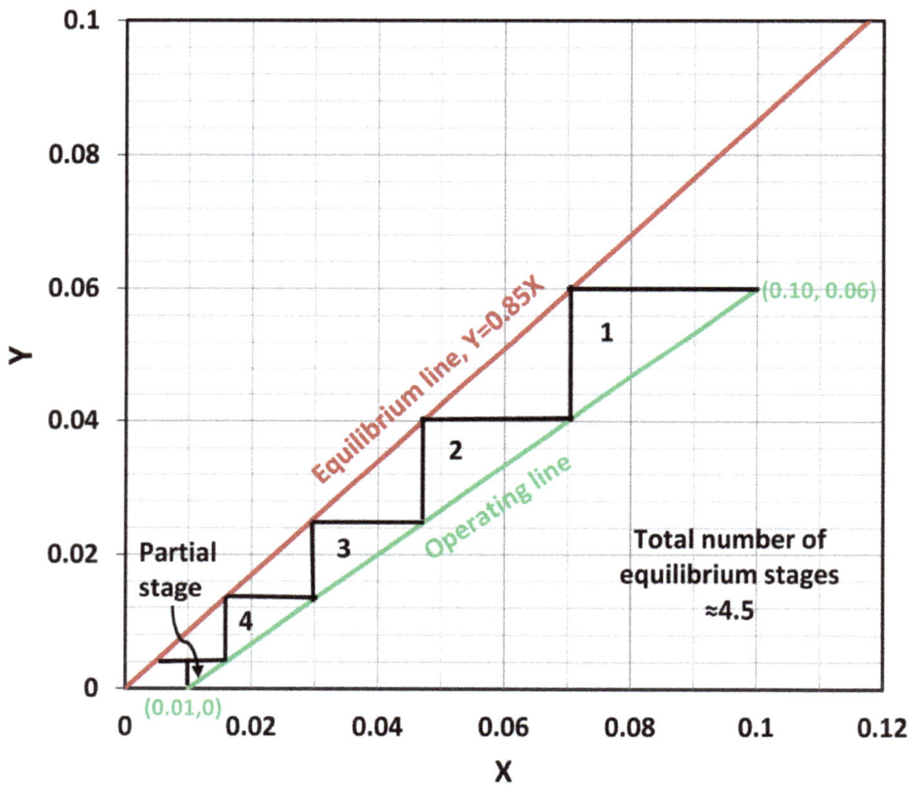

Let us now apply the Kremser equation to estimate the number of equilibrium stages. The stripping factor is

$$S' = K'/(L'/V') = 0.85 \times 1.5 = 1.275$$

As $S' \neq 1$, the following form of the Kremser equation is applicable:

$$N = \frac{\ln\left(\frac{X_a - X_a^*}{X_b - X_b^*}\right)}{ln(S')} \quad when\ S' \neq 1$$

$$X_a = 0.10, X_a^* = Y_a/K' = 0.06/0.85 = 0.0706$$

$$X_b = 0.01, X_b^* = Y_b/K' = 0/0.85 = 0$$

Thus,

$$N = \frac{ln\left(\frac{0.10-0.0706}{0.01-0}\right)}{\ln(1.275)} = 4.44$$

This value is close to the value obtained graphically using the McCabe-Thiele method.

Finally, the stage efficiency can be calculated as:

$$\text{Stage efficiency} = N_{equilibrium}/N_{actual} = 4.5/6 = 0.75$$

Problem 3

A liquid stripping tower with 5 equilibrium stages is utilized for stripping ammonia from waste water with the inlet concentration of 0.2 mol% using air. Determine the concentrations of ammonia in the exit waste water and air streams if the air entering the tower is ammonia free and the molar flow rate of air is 1.5 times that of waste water. The equilibrium equation is y = 1.45 x, where y and x are mole fractions of propane in steam and hydrocarbon mixture, respectively.

Solution

Refer to the column schematic shown below. From the given information, the stripping factor S can be calculated as:

$$S = \frac{K}{L/V} = 1.45 \times 1.5 = 2.175$$

The stripping factor can also be expressed as (see Equation (5.4)):

$$S = \frac{x_a - x_b}{x_a^* - x_b^*} = \frac{0.002 - x_b}{x_a^* - 0} = 2.175$$

Note that $x_a = 0.002$, $y_b = 0$ and $x_b^* = y_b/K = 0$. Upon rearrangement the above equation becomes:

$$0.002 - x_b = 2.175x_a^*$$

The number of stages is $N = 5$. We can now use the Kremser equation (Equation (5.9)) as:

$$N = \frac{\ln\left(\frac{x_a - x_a^*}{x_b - x_b^*}\right)}{\ln(S)} = \frac{\ln\left(\frac{0.002 - x_a^*}{x_b - 0}\right)}{\ln(2.175)} = 5$$

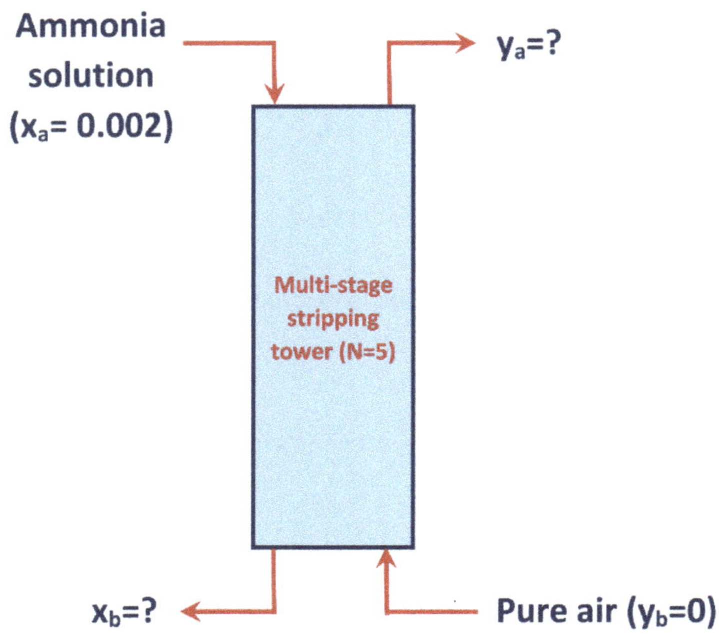

Upon rearrangement, the Kremser equation gives:

$$0.002 - x_a^* = 48.674x_b$$

Thus we have two equations and two unknowns (x_b *and* x_a^*). We can solve for the unknowns and get the following values:

$$x_b = 2.241 \times 10^{-5} \; and \; x_a^* = 9.092 \times 10^{-4}$$

Now we can calculate the concentration of ammonia in the exit gas stream as:

$$y_a = 1.45x_a^* = 1.45 \times 9.092 \times 10^{-4} = 1.318 \times 10^{-3}$$

Problem 4

A toxic chemical is stripped from waste water containing 5 mol% toxic chemical using pure air as the stripping gas. The equilibrium relation is $y = 0.85x$, where y and x are mole fractions of toxic chemical in air and water streams at equilibrium, respectively. The molar flow rate of air is 1.6 times the molar flow rate of wastewater. It is desired to remove 98% of the toxic chemical.

a) How many equilibrium stages are required using the graphical McCabe-Thiele method?
b) How many equilibrium stages are required using the Kremser equation?

You can assume that the molar flow rates of gas and liquid streams are constant.

Solution

a) Refer to the column schematic shown below. Let us first consider the *liquid entering* at the top of the column:

$$x_a = 0.05, \qquad \text{Solute molar flow rate} = 0.05L$$

As 98% of the solute (toxic chemical) is being stripped from the liquid phase, the solute in the discharge liquid stream can be calculated as follows:

$$0.02(0.05L) = Lx_b$$

$$x_b = 0.001$$

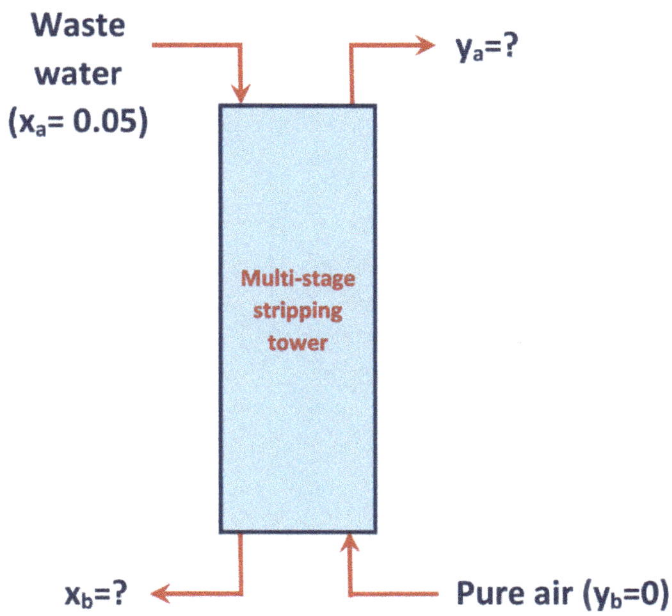

The mole fraction of solute in the gas stream leaving the column at the top, that is, y_a , can be calculated from the operating line using the terminal points:

$$y_b = y_a + (\frac{L}{V})(x_b - x_a)$$

$$0 = y_a + (1/1.6)(0.001 - 0.05)$$

$$y_a = 0.0306$$

Thus the operating line passes through the points (0.001, 0) and (0.05, 0.0306). The following figure shows the plots of the operating and equilibrium lines. The stepping off the stages using the McCabe-Thiele method is also shown in the figure. The number of equilibrium stages required is 8.5.

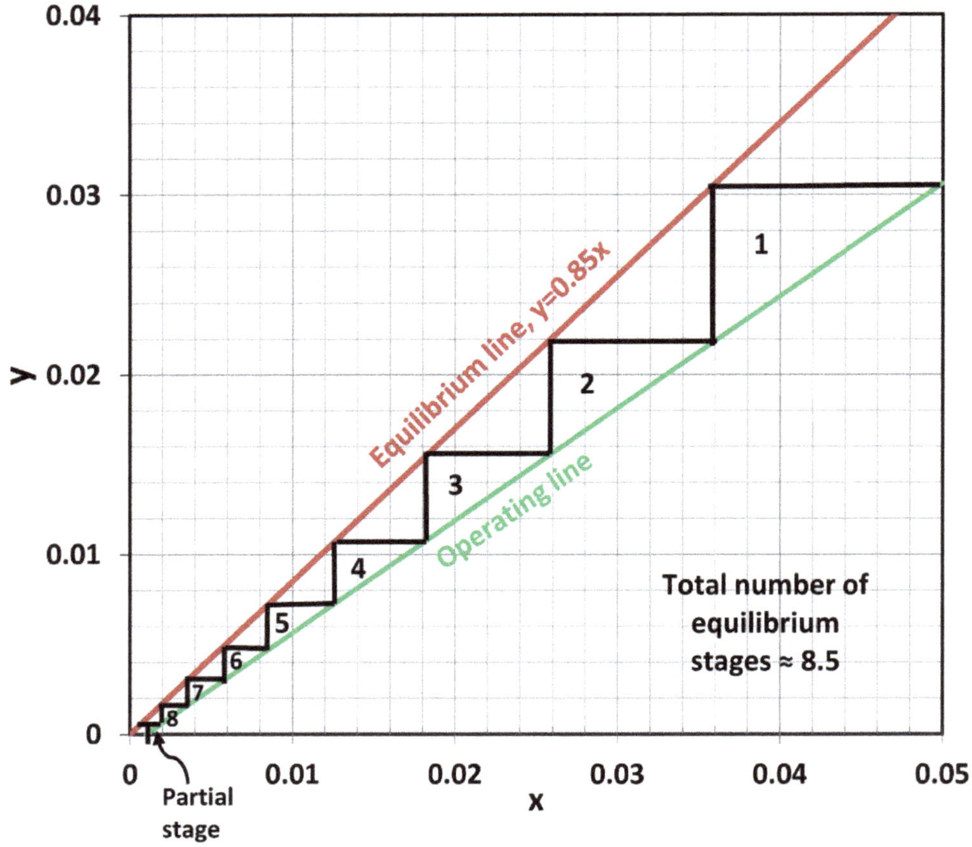

b) Let us now apply the Kremser equation to estimate the number of equilibrium stages. The stripping factor is

$$S = K/(L/V) = 0.85 \times 1.6 = 1.36$$

As $S \neq 1$, the following form of Kremser equation is applicable:

$$N = \frac{\ln\left(\frac{x_a - x_a^*}{x_b - x_b^*}\right)}{\ln(S)} = \frac{\ln\left(\frac{0.05-0.036}{0.001-0}\right)}{\ln(1.36)} = 8.58$$

Note that:

$$x_a^* = y_a/K = 0.0306/0.85 = 0.036$$

$$x_b^* = y_b/K = 0/0.85 = 0$$

The number of equilibrium stages estimated from the Kremser equation is very close to the value obtained graphically from the McCabe-Thiele method.

Problem 5

We would like to remove two solutes A and B from an aqueous stream using a multistage stripping process. The mole fractions of A and B in the aqueous stream entering the stripper are 6×10^{-5} and 2×10^{-4}, respectively. We desire a recovery of 97% of solute A from liquid phase to gas phase. The molar flow rate of aqueous stream is 1000 kmol/h and the molar flow rate of pure stripping gas is 5 kmol/h. The stripping process occurs at 285 K and 1 bar. The Henry law constants of A and B are 250 bars and 100 bars, respectively.

a) Determine the mole fractions of solute *A* in the two discharge streams
b) Determine the number of equilibrium stages needed
c) Determine the mole fractions of solute *B* in the two discharge streams
d) What is the percent recovery of solute *B* from liquid phase to gas phase?
e) What would be the percent recovery of solute *B* from liquid phase to gas phase if the Henry law constant of solute was 150 bars instead of 100 bars?

Solution

This problem involves multicomponent stripping. The multicomponent stripping problems involving two or more solutes can be solved using the single solute procedure assuming that: i) the solutes act independently of each other. Thus, the equilibrium relationship for any given solute does not depend on the mole fractions of other solutes, ii) the system is dilute, and iii) the molar flow rates of gas and liquid streams are constant.

Refer to the diagram given below.

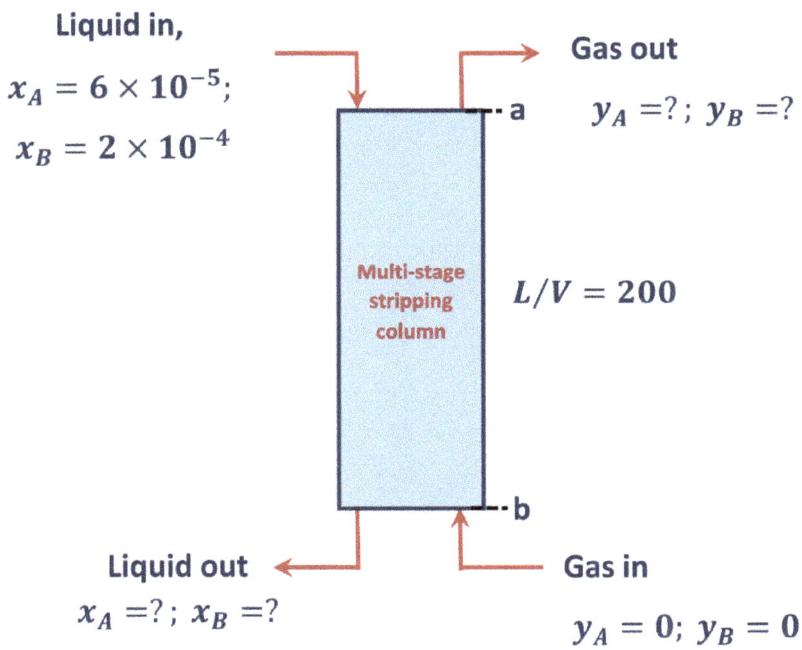

a) *Solute A mole fractions:*

As the recovery of *A* is 97%, the mole fraction of *A* in the liquid leaving the absorber is:

$$x_{A,b} = 6 \times 10^{-5} - (97/100) \times 6 \times 10^{-5} = 1.8 \times 10^{-6}$$

The mole fraction of *A* in the gas leaving the column at the top, that is, $y_{A,a}$, can be calculated from the operating line using the terminal points:

$$y_{A,b} = y_{A,a} + (L/V)(x_{A,b} - x_{A,a})$$

$$0 = y_{A,a} + (200)(1.8 \times 10^{-6} - 6 \times 10^{-5})$$

$$y_{A,a} = 0.01164$$

b) Number of equilibrium stages needed

The stripping factor S_A can be calculated as:

$$S_A = \frac{K_A}{L/V} = \frac{250}{200} = 1.25$$

We can now apply the Kremser equation to estimate the number of equilibrium stages required. As $S_A \neq 1$, the following form of Kremser equation is applicable:

$$N = \frac{ln\left(\frac{x_{A,a}-x^*_{A,a}}{x_{A,b}-x^*_{A,b}}\right)}{ln(S_A)} \quad when\ S_A \neq 1$$

$$x_{A,a} = 6 \times 10^{-5}, x^*_{A,a} = y_{A,a}/K_A = 0.01164/250 = 4.656 \times 10^{-5}$$

$$x_{A,b} = 1.8 \times 10^{-6}, x^*_{A,b} = y_{A,b}/K_A = 0$$

Thus,

$$N = \frac{ln\left(\frac{x_{A,a} - x^*_{A,a}}{x_{A,b} - x^*_{A,b}}\right)}{ln(S_A)} = \frac{ln\left(\frac{6 \times 10^{-5} - 4.656 \times 10^{-5}}{1.8 \times 10^{-6} - 0}\right)}{ln(1.25)} = 9.0$$

c) Solute B mole fractions:

The stripping factor S_B is:

$$S_B = \frac{K_B}{L/V} = \frac{100}{200} = 0.50$$

Apply Kremser equation to solute B, given that $N = 9.0$:

$$x_{B,a} - x^*_{B,a} = (x_{B,b} - x^*_{B,b}) \times exp[9 \times ln(0.50)] = 1.9531 \times 10^{-3} \times (x_{B,b} - x^*_{B,b})$$
$$x^*_{B,b} = y_{B,b}/K_B = 0$$

$$2 \times 10^{-4} - x^*_{B,a} = 1.9531 \times 10^{-3} \times x_{B,b}$$

Apply the operating line equation for solute B using the terminal points:

$$y_{B,b} = y_{B,a} + (L/V)(x_{B,b} - x_{B,a})$$
$$0 = y_{B,a} + (200)\left(x_{B,b} - 2 \times 10^{-4}\right) = K_B x^*_{B,a} + (200)\left(x_{B,b} - 2 \times 10^{-4}\right)$$

$$0.04 = 100x^*_{B,a} + 200x_{B,b}$$

There are two unknowns in the equations obtained from the Kremser equation and the operating line: $x_{B,b}$ and $x^*_{B,a}$. Solve for $x_{B,b}$ and $x^*_{B,a}$:

$$x_{B,b} = 1.001 \times 10^{-4}, x^*_{B,a} = 1.998 \times 10^{-4}$$

$$y_{B,a} = K_B x^*_{B,a} = 100 \times 1.998 \times 10^{-4} = 0.01998$$

d) Solute B recovery:

The percent recovery of solute B from liquid phase to gas phase can be calculated as:

$$\%Recovery = \left(1 - \frac{x_{B,b}}{x_{B,a}}\right) \times 100 = \left(1 - \frac{1.001 \times 10^{-4}}{2.0 \times 10^{-4}}\right) \times 100 = 49.95$$

e) Solute B recovery with K_B=150 bars:

We need to repeat part (c) calculations with $K_B = 150$ bars. Now the stripping factor is:

$$S_B = \frac{K_B}{L/V} = \frac{150}{200} = 0.75$$

As the stripping factor is increased, we expect a higher recovery. The application of the Kremser equation gives:

$$2 \times 10^{-4} - x^*_{B,a} = 0.075085 \times x_{B,b}$$

The application of the operating line gives:

$$0.04 = 150x^*_{B,a} + 200x_{B,b}$$

Solve these equations for $x_{B,b}$ and $x^*_{B,a}$:

$$x_{B,b} = 5.2984 \times 10^{-5}\ ,\ x^*_{B,a} = 1.9602 \times 10^{-4}$$

$$y_{B,a} = K_B x^*_{B,a} = 150 \times 1.9602 \times 10^{-4} = 0.029403$$

The percent recovery of solute B from liquid phase to gas phase can be calculated as:

$$\%Recovery = \left(1 - \frac{x_{B,b}}{x_{B,a}}\right) \times 100 = \left(1 - \frac{5.2984 \times 10^{-5}}{2.0 \times 10^{-4}}\right) \times 100 = 73.508$$

Thus, the recovery increases from 49.95 to 73.508 percent with the increase in Henry law constant of the solute from 100 to 150 bars.

SHORT REVIEW QUESTIONS

1. Is it true that the liquid solution and the gas mixture both become more concentrated in solute if we move in the upward direction from the bottom of the multi-stage stripping column to the top of the column?
2. When stripping factor is greater than unity, the driving force for mass transfer of solute from liquid phase to gas phase, that is, x-x^*, increases from bottom to top of the column. Is this true?
3. Explain why the operating line in multi-stage stripping process falls below the equilibrium line.
4. Define pinch condition. Does it generally occur at the bottom or top of the stripping column?

5. How many equilibrium stages are needed to operate a stripping column at maximum L/V corresponding to the pinch condition? Use the Kremer equation to determine the number of equilibrium stages.

PRACTICE PROBLEMS

Problem 1

A hydrocarbon mixture consisting of 5 mole% propane and 95 mole% non-volatile hydrocarbon is being stripped by pure superheated steam using a multi-stage countercurrent stripping column. The ratio of steam molar flow rate to hydrocarbon molar flow rate is 1/30, that is, $V/L = 1/30$. The propane content of the discharge liquid stream is 0.1 mole%. The equilibrium relation is $y = 30x$, where y and x are mole fractions of propane in steam and hydrocarbon mixture, respectively.

a) Determine the mole fraction of propane in the exit gas stream
b) How many equilibrium stages are required using the graphical McCabe-Thiele method?
c) How many equilibrium stages are required using the Kremser equation?

You can assume steam is an inert gas and it does not condense.

Problem 2

A multi-stage countercurrent stripping tower is being used to remove ammonia from wastewater using pure air as the stripping gas. The water entering the process contains 0.05 mole fraction ammonia. The ammonia mole fraction in the exit water stream is 0.01. The equilibrium relation is $y = 1.5x$, where y and x are mole fractions of ammonia in air and water streams at equilibrium, respectively. The molar flow rate of air is 1.5 times the molar flow rate of wastewater.

a) Determine the mole fraction of ammonia in the exit gas stream
b) How many equilibrium stages are required using the graphical McCabe-Thiele method?
c) How many equilibrium stages are required using the Kremser equation?

You can assume that the molar flow rates of gas and liquid streams are constant.

Problem 3

A toxic chemical is stripped from waste water containing 5 mol% toxic chemical using pure air as the stripping gas. The equilibrium relation is $y = 0.85x$, where y and x are mole fractions of toxic chemical in air and water streams at equilibrium, respectively. The molar flow rate of air is 1.25 times the molar flow rate of wastewater. It is desired to remove 90% of the toxic chemical.

a) Determine the mole fraction of toxic chemical in the exit gas stream
b) How many equilibrium stages are required using the graphical McCabe-Thiele method?
c) How many equilibrium stages are required using the Kremser equation?

You can assume that the molar flow rates of gas and liquid streams are constant.

Problem 4

We would like to remove a solute from an aqueous stream using a multistage stripping process. The mole fraction of the solute in the entering aqueous stream is 10^{-5}. We desire a recovery of 70% of the solute from liquid phase to gas phase. The stripping process occurs at 300 K and 1 bar. The Henry law constant of the solute is 225 bars. The stripping gas entering the column is free of solute. The actual L/V ratio used in the process is 0.65 times the maximum L/V.

a) Determine the number of equilibrium stages required using the McCabe-Thiele graphical method.

b) Determine the number of equilibrium stages using the Kremser equation.

Problem 5

A solute is stripped from a concentrated aqueous solution containing solute with a mole ratio of 0.15 by contact with solute-free pure gas in a multistage countercurrent stripping column. The equilibrium relationship in terms of mole fractions is given as: $y = -0.85\ln(1 - x)$. When the molar flow rate of gas (solute free basis) is 1.4 times that of the carrier (solute-free) liquid solution, 90 percent of the solute is removed. How many equilibrium stages are required, using graphical McCabe-Thiele method?

Problem 6

We would like to remove three solutes A, B and C from an aqueous stream using a multistage stripping process. The mole fractions of A, B and C in the aqueous stream entering the stripper are 8×10^{-5}, 5×10^{-4}, and 10^{-4} respectively. We desire a recovery of 95% of solute A from liquid phase to gas phase. The molar flow rate of aqueous stream is 1200 kmol/h and the molar flow rate of pure stripping gas is 6 kmol/h. The stripping process occurs at 295 K and 1 bar. The Henry law constants of A, B and C are 300 bars, 200 bars and 100 bars, respectively.

a) Determine the mole fractions of solute A in the two discharge streams
b) Determine the number of equilibrium stages needed
c) Determine the mole fractions of solute B in the two discharge streams
d) What is the percent recovery of solute B from liquid phase to gas phase?
e) Determine the mole fractions of solute C in the two discharge streams
f) What is the percent recovery of solute C from liquid phase to gas phase?

Chapter 6

Some Definitions and Basic Concepts related to Distillation

6.1 LEARNING OUTCOMES

Upon completion of this chapter, students will be able to:

- Explain the concept of vapor pressure
- Describe P-V-T behavior and phase diagrams for pure substances
- Estimate vapor pressure of pure substances as a function of temperature using Clausius-Clapeyron and Antoine equations
- Estimate distribution coefficients (K_i-values) of species in a multi-component vapor-liquid system at equilibrium
- Define relative volatility and appreciate its importance in distillation
- Define bubble point and dew point pressures and temperatures
- Estimate bubble point pressure, dew point pressure, bubble point temperature, and dew point temperature for multicomponent systems using the Raoult law
- Estimate bubble and dew points for multicomponent systems using K_i-values
- Explain the importance of bubble point and dew point calculations in distillation
- Construct P-x-y and T-x-y diagrams and y-x equilibrium curve for binary systems using Raoult's law and K_i-values

6.2 INTRODUCTION

As noted in Chapter 1, distillation involves separation of components of a liquid solution by a process of vaporization/boiling of liquid and condensation of vapor. For separation to take place, it is important that the boiling points of the components are different. If the boiling points are the same, that is, we have an azeotropic liquid solution, then the vapor coming out of the liquid solution has the same composition as that of the liquid solution itself and hence, no separation of components takes place by boiling of liquid and condensation of vapor. In some of the distillation processes (batch distillation and flash distillation) the vaporization of the liquid solution and the condensation of the produced vapors are done separately. First the liquid solution is boiled and then the produced vapors are subsequently condensed using some means to cool and condense the vapors. However, in a multi-stage distillation process, vaporization and condensation take place simultaneously at each stage. Simultaneous vaporization and condensation at each stage leads to high enrichment of vapor of more volatile components and high enrichment of liquid of less volatile components. Hence a good degree of separation of a feed liquid mixture into more volatile and less volatile fractions is achieved in a distillation column.

In this chapter, some of the basic concepts and definitions relevant to distillation equilibrium stage operations are discussed. For example, what is vapor pressure of a pure substance and how does it vary with the temperature. The relations relevant to vapor pressure include Clapeyron, Clausius-Clapeyron, and Antoine equations. What is relative volatility and what is its significance in distillation? How to construct P versus composition and T versus composition phase diagrams and y-x equilibrium curve using Raoult's law? The definitions of the bubble and dew points as well as their relevance in distillation and calculation procedures are discussed.

6.3 VAPOR PRESSURE

Vapor pressure of a pure substance, at a given temperature, is defined as the pressure exerted by a vapor in equilibrium with the pure liquid phase of the substance in a closed system, as shown in Figure 6.1. It is often denoted as P^{Sat}. The vapor pressure of a substance is a strong function of temperature; it increases non-linearly with the increase in the temperature. A substance with higher vapor pressure at any given temperature is more volatile as compared with a substance with lower vapor pressure at the same temperature. For example, liquid butane is much more volatile as compared with liquid water at room temperature: $P_{water}^{Sat} = 2.3\ kPa\ and\ P_{butane}^{Sat} = 220\ kPa\ at\ T =\ 20^oC$.

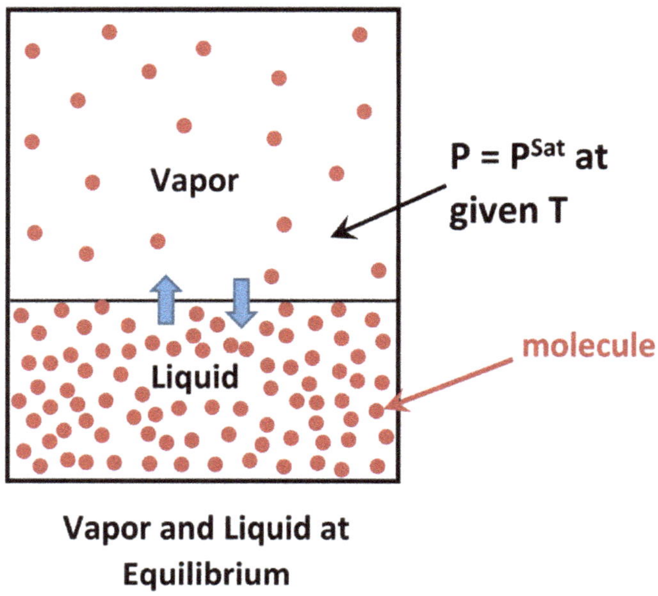

Figure 6.1 Vapor and liquid of a pure substance in equilibrium at a given temperature.

Let us now consider a closed vessel of fixed volume consisting of pure liquid and its vapor in equilibrium with each other at some temperature (say the vessel is placed in a constant temperature bath whose temperature can be varied), as shown in Figure 6.1. Let the initial temperature be T_1 and the corresponding equilibrium vapor pressure be P_1^{Sat}. When the vapor and liquid are at equilibrium with each other, they are also at "saturation condition", that is, the liquid is at its boiling point temperature and vapor is at its dew point temperature. Note that when vapor is at saturation condition (dew point), it tends to condense with a small decrease in temperature. Likewise, a liquid at saturation condition (boiling point) tends to vaporize (boil) with an incremental increase in temperature. For pure substances, the boiling point and dew point temperatures are the same. What happens when the temperature of the vapor-liquid system under consideration is raised to say T_2 ? When the temperature of the system is raised above T_1, boiling of liquid occurs (bubbles are formed inside the bulk liquid phase) leading to an increase in the amount of the vapor phase and a decrease in the amount of the liquid phase. Eventually a new equilibrium is established between the vapor and liquid phases where the vapor pressure is P_2^{Sat} corresponding to temperature T_2. If the temperature of the system is increased further to T_3 such that the entire liquid phase is vaporized but the vapor is still at saturation condition, then the pressure of the saturated vapor is P_3^{Sat} corresponding to vapor pressure at temperature T_3 . If the temperature is raised beyond T_3 to T_4 then the vapor is no longer at saturation condition; it is now a superheated vapor where the pressure P of the vapor is lower than P_4^{Sat}. The process of raising the temperature of vapor-liquid system just described is illustrated schematically in Figure 6.2. The changes are also shown on the P-V-T phase diagram in Figure 6.3. Figure 6.3 clearly shows that the fraction of vapor in the two-phase system increases as the temperature is raised. The P^{Sat} also increases with the increase in temperature. Also in the final state 4, the vapor is superheated with $P_4 < P_4^{Sat}$.

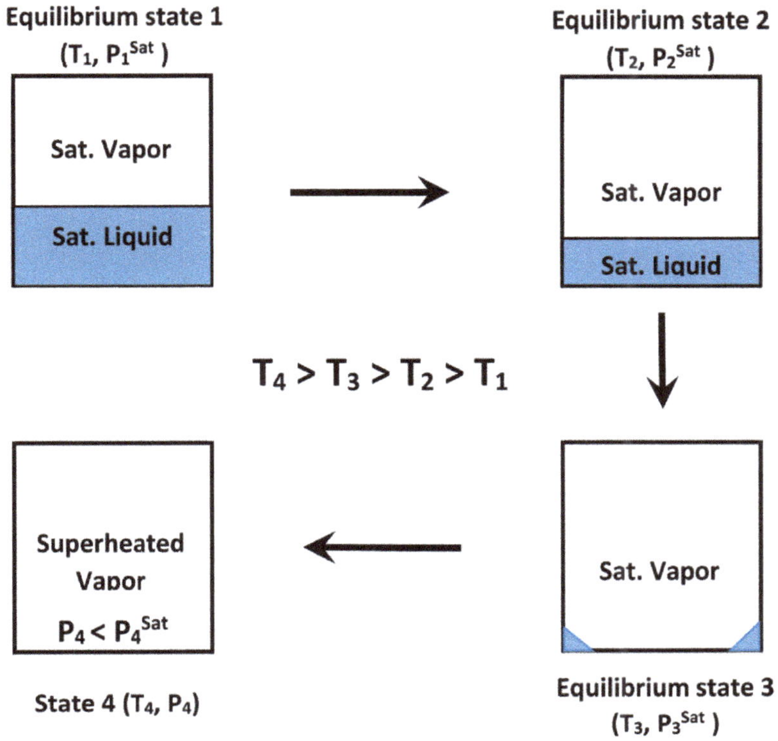

Figure 6.2 Changes in vapor-liquid two-phase closed system (fixed volume) when temperature is raised.

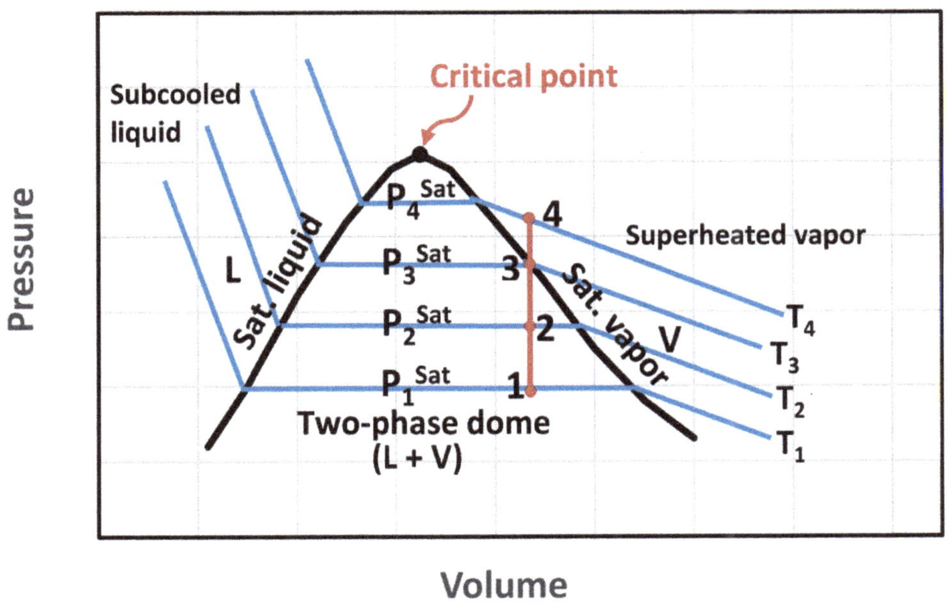

Figure 6.3 P-V-T phase diagram showing the changes in vapor-liquid system when temperature is raised from T_1 to T_4 along the path 1-2-3-4.

We now consider another important experiment where vapor-liquid system is heated at constant pressure. This experiment could be done by heating the material contained in a piston/cylinder assembly with constant external pressure (say 1 atm) imposed on it as shown in Figure 6.4. We assume that the piston can slide up and down freely. Thus the volume of the material enclosed in the piston/cylinder assembly is adjustable. The isobaric heating process, starting with subcooled liquid, is shown in Figure 6.5 with the help of a P-V-T phase diagram. Referring to Figure 6.5, we have subcooled liquid in state 1 ($P_1 > P_1^{Sat}$ $where\ P_1^{Sat}\ is\ vapor\ pressure\ at\ T_1$). Upon heating at constant pressure P $(external\ pressure)$, state 2 corresponding to saturated liquid is reached. In state 2, the temperature is T_2 and the pressure is $P_2^{sat} = P$. Upon further heating at constant pressure, the liquid begins to boil (vapor bubbles are formed in the bulk liquid phase). The heat added to the system goes into phase change from liquid to vapor. The temperature remains constant at T_2 until all of the liquid is converted into vapor phase and state 3 is reached where we have only saturated vapor and no liquid. Upon isobaric heating of the saturated vapor, the temperature rises and state 4 is reached where the steam is in superheated state. In the superheated state 4, $P_4(= P) < P_3^{Sat}$ where P_3^{Sat} is the vapor pressure at temperature T_3 .

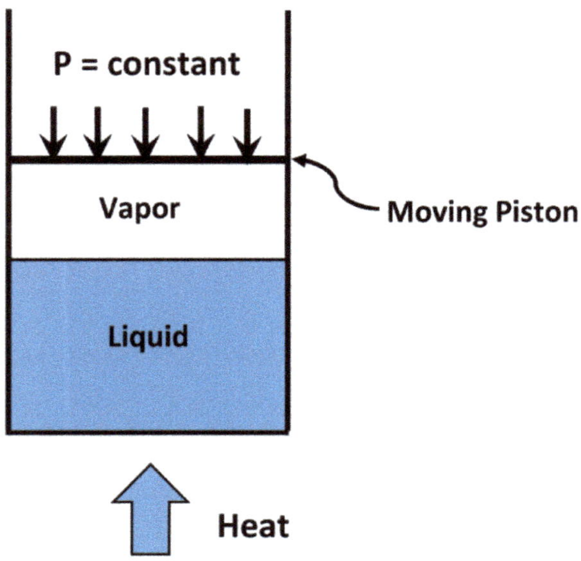

Figure 6.4 Constant pressure heating of fluids.

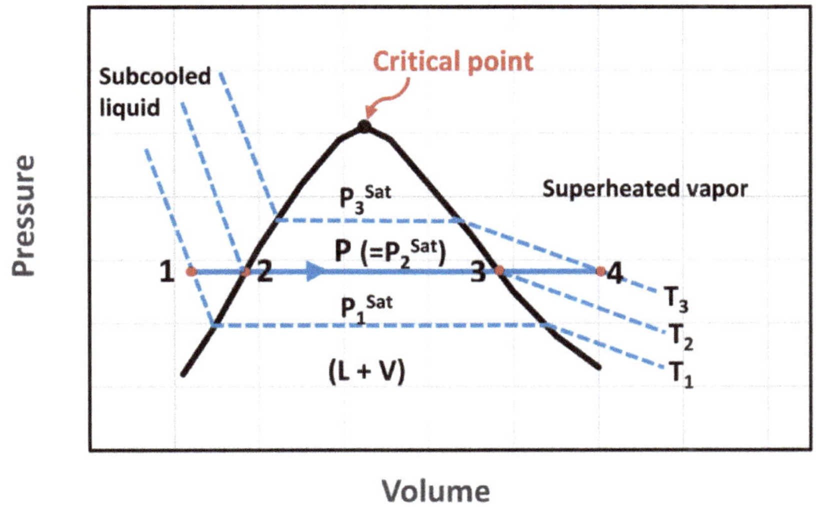

Figure 6.5 P-V-T phase diagram showing the changes in vapor-liquid system when the system is heated at constant pressure along the path 1-2-3-4.

6.3.1 Clapeyron and Clausius-Clapeyron Equations

Starting from the condition of phase equilibrium between vapor and liquid phases of a pure substance, one can easily derive the following *Clapeyron* equation:

$$\frac{dP^{Sat}}{dT} = \frac{\Delta H^{vap}}{T\Delta V^{vap}} \tag{6.1}$$

The Clapeyron equation gives the rate of change of vapor pressure with temperature. ΔH^{vap} is the molar latent heat of vaporization (or enthalpy of vaporization), T is the absolute temperature, and ΔV^{vap} is the molar volume change of vaporization.

If we treat vapor to be an ideal gas and neglect the molar volume of liquid in comparison with the molar volume of vapor, we get to the *Clausius-Clapeyron* equation (Equation (6.4)) as shown below:

$$V^{vap} = \frac{RT}{P^{Sat}} \tag{6.2}$$

$$\Delta V^{vap} = V^{vap} - V^{liq} \approx V^{vap} \tag{6.3}$$

$$\frac{d\ln P^{Sat}}{dT} = \frac{\Delta H^{vap}}{RT^2} \tag{6.4}$$

Upon integration, the Clausius-Clapeyron equation gives:

$$\ln P^{Sat} = -\frac{\Delta H^{vap}}{R}\left(\frac{1}{T}\right) + A \tag{6.5}$$

where "A" is an integration constant for a given species. This equation assumes that ΔH^{vap} is constant independent of temperature. It can be re-written as:

$$lnP^{Sat} = A - \frac{B}{T} \tag{6.6}$$

where $B = \Delta H^{vap}/R$

6.3.2 Antoine Equation

The Antoine equation is a more general form of the Clausius-Clapeyron equation, Equation (6.6). It describes the vapor pressure data of pure substance more satisfactorily than Equation (6.6). It is given as:

$$lnP^{Sat} = A - \frac{B}{T+C} \tag{6.7}$$

where $A, B, and\ C$ are constants for a given substance. These constants are available in the literature for a large number of substances.

6.4 RAOULT'S LAW

Raoult's law states that for an *ideal* liquid solution in phase equilibrium with an *ideal* vapor mixture, the partial vapor pressure of any species "i" in the vapor mixture is equal to the vapor pressure of the pure component "i" at the mixture temperature multiplied by the mole fraction of component in the liquid solution. Thus:

$$P_i = x_i P_i^{Sat} \tag{6.8}$$

where P_i is the partial pressure of species "i" in the vapor mixture, x_i is the mole fraction of species "i" in the liquid solution, and P_i^{Sat} is the vapor pressure of pure species "i" at the system temperature. Note that the partial pressure of species "i" in an ideal gas mixture is defined as the pressure that species "i" would exert if it alone occupied the total volume of the gas mixture at the same temperature. Also the sum of the partial pressures of various gaseous species in an ideal gas mixture is equal to the total pressure of the gas mixture. This is known as Dalton's law.

For an ideal gas mixture, the partial pressure P_i is given as:

$$P_i = y_i P \tag{6.9}$$

where y_i is the mole fraction of species "i" in the vapor mixture and P is the total pressure of the vapor mixture. Combining Equations (6.8) and (6.9), we can write:

$$y_i P = x_i P_i^{Sat} \tag{6.10}$$

This form of Raoult's law is used extensively in practical applications involving multi-component vapor-liquid phase equilibrium. The restrictions are: the liquid solution is ideal and the vapor mixture is ideal. Equation (6.10) could also be re-written in a more general form as:

$$y_i = K_i x_i \tag{6.11}$$

This equation defines the distribution coefficient K_i of a chemical species regardless of the nature (ideal or non-ideal) of the system. In general, the distribution coefficient K_i of any species is a function of temperature, pressure, and liquid and vapor compositions. Fortunately for a number of systems, the composition dependence is small. In the case of an ideal system (ideal vapor mixture in equilibrium with an ideal liquid solution), K_i is completely independent of the composition. For ideal systems, K_i depends only on temperature and pressure and is given as $K_i = P_i^{Sat}/P$.

The K_i-value of any given species is a measure of the tendency of the species to favor the vapor phase over the liquid phase. The concentration or mole fraction of a species in the vapor phase is higher than that in the liquid phase when $K_i > 1$. When $K_i < 1$, the concentration of the species is higher in the liquid phase. The K_i-value of any species increases with temperature at a given pressure.

There are nomographs available in the literature to estimate the K-values of materials such as light hydrocarbons at any given temperature and pressure. For example, DePriester charts are utilized extensively to estimate K-values of light hydrocarbons at different temperatures and pressures. The DePriester charts are based on experimental data. McWilliams (M.L. McWilliams, Chemical Engineering, Vol. 80, 138, 1973) proposed the following empirical fit to describe the DePriester charts for light hydrocarbons:

$$lnK = a_0 + \frac{a_1}{T^2} + \frac{a_2}{T} + b_0 lnP + \frac{b_1}{P^2} + \frac{b_2}{P} \tag{6.12}$$

where P is in psia and T is absolute temperature in °R. The constants in Equation (6.12) are given in Table 6.1 for various chemical species. This equation is valid in the temperature range of 365.7 to 851.7 °R (-70 to 200 °C) and a pressure range of 14.69 to 870.1 psia (101.3 to 6000 kPa).

Table 6.1: Constants in Equation (6.12) for various chemical compounds

Compound	a_0	a_1	a_2	b_0	b_1	b_2
Methane	8.24450	-292860.0	0	-0.89510	59.8465	0
Ethylene	7.90595	-600076.875	0	-0.84677	42.94594	0
Ethane	7.90699	-687248.25	0	-0.88600	49.02654	0
Propylene	7.71725	-923484.6875	0	-0.87871	47.67624	0
Propane	7.15059	-970688.5625	0	-0.76984	0	6.90224
Isobutane	7.72668	-1166846.0	0	-0.92213	0	0
n-Butane	7.994986	-1280557.0	0	-0.96455	0	0
Isopentane	7.58071	-1481583.0	0	-0.93159	0	0
n-Pentane	7.33129	-1524891.0	0	-0.89143	0	0
n-Hexane	6.96783	-1778901.0	0	-0.84634	0	0
n-Heptane	6.52914	-2013803.0	0	-0.79543	0	0
n-Octane	12.48457	0	-7646.81641	-0.73152	0	0
n-Nonane	5.69313	-255104.0	0	-0.67818	0	0
n-Decane	13.80354	0	-9760.45703	-0.71470	0	0

6.5 VOLATILITY AND RELATIVE VOLATILITY

The volatility of any component in a liquid solution (ideal or non-ideal) is defined as the ratio of equilibrium partial pressure of the component in the vapor phase (assumed ideal) to mole fraction of the component in the liquid phase. In a binary liquid solution consisting of species A and B:

$$\vartheta_A = P_A/x_A \tag{6.13}$$

$$\vartheta_B = P_B/x_B \tag{6.14}$$

where ϑ_A and ϑ_B are the volatilities of species A and B, respectively, P_A and P_B are the equilibrium partial pressures of species A and B, respectively, and x_A and x_B are the mole fractions of species A and B, respectively in the liquid phase. Thus for the same mole fraction in the liquid phase, the more volatile component exerts a higher partial pressure. For an ideal liquid solution, the volatility of any species is equal to the vapor pressure of that species in the pure state at the solution temperature. This follows from the Raoult law (Equation (6.8)). Thus a more volatile component has a higher vapor pressure at the system temperature.

The relative volatility is defined as the volatility of one component of a liquid solution relative to another component of the liquid solution and is given as:

$$\alpha_{AB} = \frac{\vartheta_A}{\vartheta_B} = \frac{p_A/x_A}{p_B/x_B} = \frac{y_A/x_A}{y_B/x_B} \tag{6.15}$$

where α_{AB} is the relative volatility of species "A" with respect to species "B".

Let us now consider the physical significance of relative volatility. Consider a binary system of vapor mixture in equilibrium with the liquid solution. Let the volatility of "A" be the same as that of "B", that is, $\alpha_{AB} = 1$. From Equation (6.15), it follows that:

$$\frac{y_A}{x_A} = \frac{y_B}{x_B} \tag{6.16}$$

If the liquid solution is equimolar, $x_A = x_B = 0.5$, then Equation (6.16) gives $y_A = y_B = 0.5$. Thus no separation of liquid solution takes place upon boiling the solution in this case as vapor composition is the same as liquid composition. If the liquid solution is not equimolar, say $x_A = 2x_B$, then $y_A = 2y_B$. This further implies that $y_A = x_A = 0.67\ and\ y_B = x_B = 0.33$. In this case too, the vapor composition is not different from the liquid composition and hence no separation takes place upon distillation. For separation to take place using distillation, it is necessary that the volatilities of the two components are different. When $\alpha_{AB} = 1$, we have an azeotrope meaning the vapor composition is the same as liquid composition, that is, $y_A = x_A$ and $y_B = x_B$, and no separation is possible using distillation.

Let us now take $\alpha_{AB} = 2$, that is, species "A" is two times more volatile than species "B". In this case, $y_A = 2y_B$ when $x_A = x_B$. Now the vapor phase is much richer in a more volatile component as compared with the equimolar liquid solution and hence, separation of components takes place upon distillation. When $x_A = 2x_B$, then $y_A = 4y_B$. Once again, the vapor phase is much richer in more volatile component as compared with the liquid solution.

A more general expression of relative volatility α_{ij} of species "i" with respect to species "j" is as follows:

$$\alpha_{ij} = \frac{y_i/x_i}{y_j/x_j} = \frac{K_i}{K_j} \tag{6.17}$$

Although the distribution coefficients K_i and K_j are strong functions of temperature, the relative volatility α_{ij} is relatively a weak function of temperature as different *K's* increase with temperature in a similar manner. Also note that this definition of relative volatility where α_{ij} is expressed in the form of Equation (6.17) is not restricted to ideal systems. In the special case of ideal vapor-liquid system (ideal solution and ideal vapor) where the Raoult law is applicable, Equation (6.17) reduces to:

$$\alpha_{ij} = \frac{K_i}{K_j} = \frac{P_i^{Sat}}{P_j^{Sat}} \tag{6.18}$$

For a non-ideal liquid solution in equilibrium with an ideal vapor mixture, Equation (6.17) reduces to the following expression for α_{ij}:

$$\alpha_{ij} = \frac{K_i}{K_j} = \frac{\gamma_i P_i^{Sat}}{\gamma_j P_j^{Sat}} \tag{6.19}$$

where γ_i is the activity coefficient of species i in the non-ideal liquid solution.

6.6 BUBBLE POINT AND DEW POINT

A liquid solution at saturation condition is said to be at its bubble point. With small changes in temperature or pressure, the liquid solution at bubble point begins to boil. A vapor mixture at saturation condition is said to be at its dew point. With small changes in temperature or pressure, the vapor mixture at dew point begins to condense.

6.6.1 Definitions of Bubble and Dew points

Pure Substances

Consider a subcooled liquid at a fixed temperature T_o initially in state 1 as shown on a P-V-T diagram in Figure 6.6. If the pressure of this subcooled liquid is decreased from state 1 along an isotherm, state 2 is reached where the pressure is equal to the vapor pressure (P^{Sat}) of the liquid at T_o . In state 2, the liquid is saturated and begins to form bubbles. The pressure where the first bubble appears at a fixed temperature of T_o is P^{Sat} . This pressure is called *bubble point pressure (BUBLP)*. Also note that the pressure does not change until the entire liquid is vaporized or boiled off.

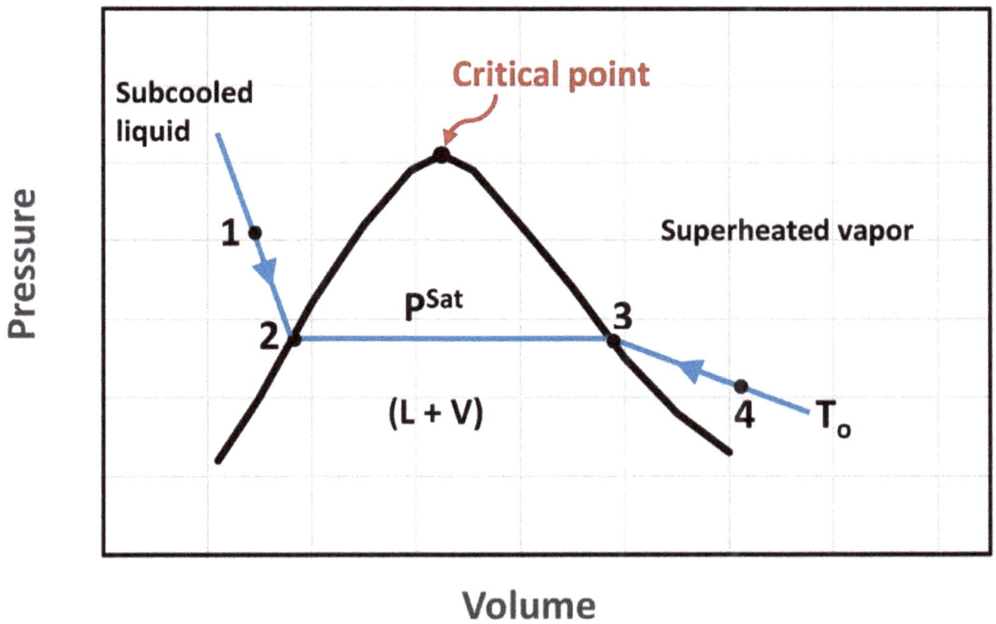

Figure 6.6 Boiling of pure liquid upon decrease of pressure to $P^{Sat}$and condensation of pure vapor upon increase of pressure to P^{Sat}at constant temperature.

Let us now follow the path from state 4 to state 3 along an isotherm (fixed temperature of T_o) as shown in Figure 6.6. We have superheated vapor in state 4. With the increase in pressure of this superheated vapor, we reach state 3 where the vapor becomes saturated vapor. In saturated vapor state 3, the vapor begins to condense. The pressure where the first liquid droplet (also called dew) is formed at a fixed temperature of T_o is P^{Sat}. This pressure is called *dew point pressure (DEWP).* Also note that the pressure does not change until the entire vapor is condensed.

Thus the bubble point pressure and dew point pressure at any given temperature are the same for pure (single-component) substances, that is, $BUBLP = DEWP = P^{Sat}$. This implies that boiling and condensation of pure substances at any given temperature occur at a single pressure corresponding to vapor pressure P^{Sat} at that temperature.

We now consider a different experiment where pressure is fixed and temperature is varied. Referring to Figure 6.7, we heat a subcooled liquid in state 1 at constant pressure P_o until state 2 is reached. In state 2, the liquid is saturated and it begins to boil. The temperature where the first bubble is formed at a fixed pressure of P_o is T^{Sat}. This temperature is called *bubble point temperature (BUBLT)*. The temperature remains constant until the entire liquid is vaporized. What happens when we cool a superheated vapor in state 4 at constant pressure? Upon isobaric cooling, the vapor becomes saturated in state 3 and it begins to condense. The temperature where the first liquid droplet (dew) is formed at a fixed temperature of P_o is T^{Sat}. This temperature is called *dew point temperature (DEWT)*. The temperature remains constant until the entire vapor is condensed.

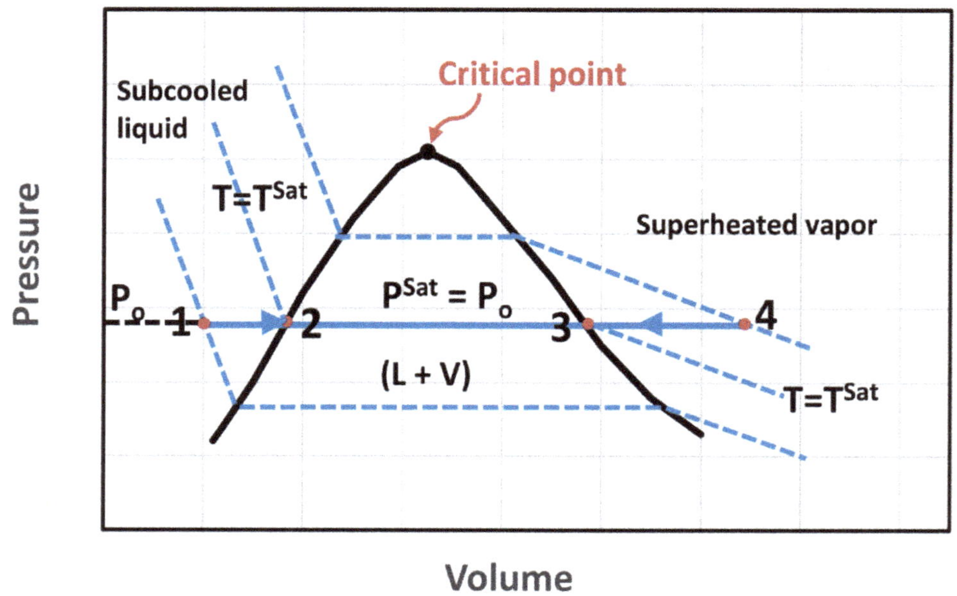

Figure 6.7 Boiling of pure liquid upon increase of temperature to $T^{Sat}$and condensation of pure vapor upon decrease of temperature to T^{Sat}at constant pressure.

Thus we have observed that the bubble point temperature is the same as the dew point temperature at any given pressure for pure (single-component) substances, that is, $BUBLT = DEWT = T^{Sat}$. The boiling and condensation of pure substances at any given pressure occur at a single temperature corresponding to T^{Sat} at that pressure.

Mixtures of Two or More Components

In the case of systems with two or more components, boiling and condensation do not occur at a single pressure at a fixed temperature. That is, *BUBLP* and *DEWP* are not the same. The boiling and condensation take place over a range of pressures at any given temperature. For the same overall composition of the system, *BUBLP* > *DEWP* at a fixed temperature. Likewise, the boiling and condensation do not occur at a single temperature at a fixed pressure. The *BUBLT* and *DEWT* are different and boiling and condensation occur over a range of temperatures. For the same overall composition of the system *DEWT* > *BUBLT* at a fixed pressure. Furthermore, the bubble points (*BUBLP* and *BUBPT*) and dew points (*DEWP* and *DEWT*) vary with the system composition.

As an example, consider a binary system consisting of two chemical species. Figure 6.8 shows the P-x-y phase diagram of a binary system at a fixed temperature. Starting from a subcooled liquid solution of known composition in state 1 (Figure 6.8), we decrease the pressure of the system along the path 1 → 2 → 3. At point 2, the liquid solution begins to boil. The pressure corresponding to state 2 where the first bubble appears is the bubble point pressure *(BUBLP)* of the solution. In order to continue boiling of the liquid solution into the vapor phase, the pressure has to be decreased further from state 2 to state 3. In state 3, the last liquid droplet is vaporized. The pressure corresponding to state 3 is *DEWP*. Note that this boiling behavior is different from what was observed in the case of pure single-component system where boiling occurred at a constant pressure. In the case of multi-component systems, boiling at a fixed temperature occurs over a range of pressures $DEWP \leq P \leq BUBLP$. Also note that *DEWP* where the first liquid droplet (dew) is formed is the same pressure where the last droplet is vaporized. If we begin from superheated state 4 and increase the pressure (keeping temperature fixed) along the path 4 → 3 → 2 , the first liquid droplet is formed in state 3 and to continue condensation further we need to increase the pressure further until the entire vapor is condensed in state 2. Thus condensation at a fixed temperature occurs over a range of pressures $DEWP \leq P \leq BUBLP$.

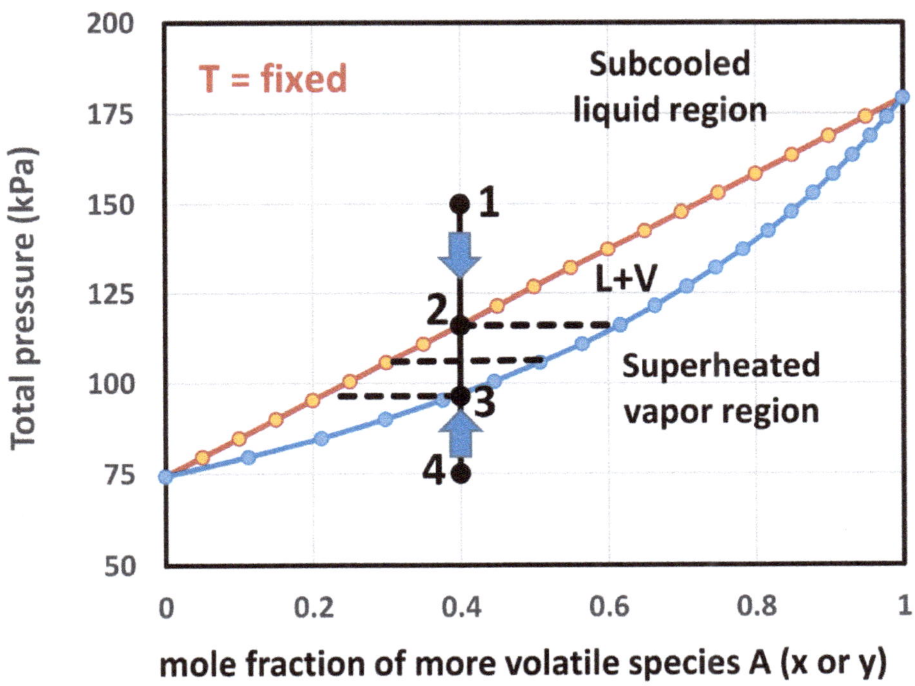

Figure 6.8 P-x-y phase diagram of a binary system at a fixed temperature.

In the preceding discussion, the temperature was held constant and pressure was varied. Now we will keep the pressure fixed and vary the temperature of a binary system. Thus we need T-x-y phase diagram of a binary system at a fixed pressure to define bubble and dew points. Figure 6.9 shows a T-x-y diagram for a binary system consisting of species "*A*" and "*B*" with species "*A*" being more volatile ($T_A^{Sat} \ll T_B^{Sat}$). Starting from a subcooled liquid solution of known composition in state 1 (Figure 6.9) we heat the solution isobarically along the path 1 → 2 → 3. At point 2, the liquid solution begins to boil. The temperature corresponding to state 2 where the first bubble appears is the bubble point temperature *(BUBLT)* of the solution. In order to continue boiling of the liquid solution into the vapor phase, the temperature has to be increased further from state 2 to state 3. In state 3, the last liquid droplet is vaporized. The temperature corresponding to state 3 is *DEWT*. Thus, unlike single-component systems, boiling at a fixed pressure occurs over a range of temperatures $BUBLT \leq T \leq DEWT$ in multi-component systems. If we begin from superheated state 4 and cool the vapor by decreasing the temperature (keeping pressure fixed) along the path 4 → 3 → 2 , the first liquid droplet is formed in state 3 and to continue condensation further we need to decrease the temperature further until the entire vapor is condensed in state 2. Thus condensation, like boiling, at a fixed pressure occurs over a range of temperatures $BUBLT \leq T \leq DEWT$.

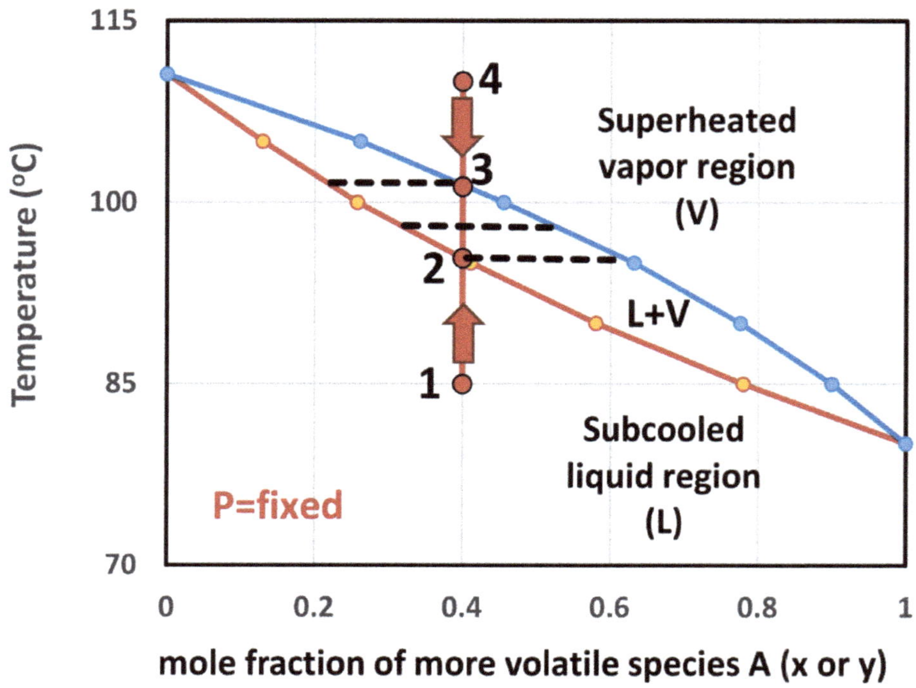

Figure 6.9 T-x-y phase diagram of a binary system at a fixed pressure.

6.6.2 Different Classes of Bubble and Dew Points Calculations

There are four types of bubble and dew points calculations. They are as follows:

- *BUBLP:* Given liquid composition and temperature, calculate the pressure where boiling begins and the composition of the first bubble formed. We can assume that the liquid is initially in subcooled liquid state. If we reduce the pressure (keeping temperature fixed), at what pressure would boiling begin?
- *DEWP:* Given vapor composition and temperature, calculate the pressure where condensation begins and the composition of the first droplet (dew) formed. We can assume that the vapor is initially in superheated vapor state. If we increase the pressure (keeping temperature fixed), at what pressure would condensation begin?
- *BUBLT:* Given liquid composition and pressure, calculate the temperature where boiling begins and the composition of the first bubble formed. We can assume that the liquid is initially in subcooled liquid state. If we increase the temperature (keeping pressure fixed), at what temperature would boiling begin?
- *DEWT:* Given vapor composition and pressure, calculate the temperature where condensation begins and the composition of the first droplet (dew) formed. We can assume that the vapor is initially in superheated vapor state. If we decrease the temperature (keeping pressure fixed), at what temperature would condensation begin?

6.6.3 Bubble and Dew Points Calculations Using Raoult's Law

We will assume that the Raoult law is applicable, that is, the liquid solution and vapor mixture are both ideal in nature.

BUBP Calculation: Given $\{x_i\}$ and T, calculate P and $\{y_i\}$

From Raoult's law:

$$y_i = x_i \frac{P_i^{Sat}}{P} \tag{6.20}$$

As $\sum y_i = 1$, Equation (6.20) gives the bubble point pressure as:

$$P = \sum x_i P_i^{Sat} \tag{6.21}$$

The liquid composition is specified and the vapor pressures of different species can be calculated from the Antoine equation, Equation (6.7), as the temperature is specified. Once the bubble point pressure is known, the composition of the first bubble can be calculated from Equation (6.20).

DEWP Calculation: Given $\{y_i\}$ and T, calculate P and $\{x_i\}$

From Raoult's law:

$$x_i = y_i \frac{P}{P_i^{Sat}} \tag{6.22}$$

As $\sum x_i = 1$, Equation (6.22) gives the dew point pressure as:

$$P = \frac{1}{\sum y_i / P_i^{Sat}} \tag{6.23}$$

The vapor composition is specified and the vapor pressures of different species can be calculated from the Antoine equation, Equation (6.7), as the temperature is specified. Once the dew point pressure is known, the composition of the first liquid droplet (dew) can be calculated from Equation (6.22).

BUBLT Calculation: Given $\{x_i\}$ and P, calculate T and $\{y_i\}$

When temperature is unknown, an iterative procedure is required. The iterative procedure can be developed as follows.

Equation (6.21) could be re-written as:

$$P = P_r^{Sat} \sum x_i P_i^{Sat} / P_r^{Sat} \tag{6.24}$$

where species "*r*" is the "reference species", selected somewhat arbitrarily. Using the definition of relative volatility given in Equation (6.18), Equation (6.24) can be re-arranged as:

$$P_r^{Sat} = P / \sum x_i \alpha_{ir} \tag{6.25}$$

From the Antoine equations for species "*i*" and "*r*", one can write:

$$ln\alpha_{ir} = A_i - A_r + \frac{B_r}{T+C_r} - \frac{B_i}{T+C_i} \tag{6.26}$$

Upon re-arrangement of the Antoine equation for the reference species "*r*", we get:

$$T = \frac{B_r}{A_r - lnP_r^{Sat}} - C_r \tag{6.27}$$

where $A, B, and\ C$ are coefficients in the Antoine equations for species "*i*" and "*r*". The reference species "*r*" is fixed whereas "*i*" refers to any species in a multi-component system.

The iterative procedure for the estimation of *T* is shown in Figure 6.10. The initial guess of temperature can be made as follows:

$$T_{guess} = \sum x_i\, T_i^{Sat} \tag{6.28}$$

where T_i^{Sat} is the boiling point of species i estimated from the Antoine equation at the given total pressure P. Equation (6.28) is not a valid equation and it is to be used only for initial guess purposes. Following the procedure shown in Figure 6.10, an improved value of temperature is calculated from Equation (6.27). If the calculated temperature is within the set tolerance level, the iteration process is stopped otherwise the calculated value of temperature is taken as the new guess value and the iterative process is continued. The percent error between two successive iterations can be calculated and compared with the percent tolerance level ε as follows:

$$\%\ Error = \left|\frac{T_{k+1} - T_k}{T_k}\right| \times 100 \leq \varepsilon \tag{6.29}$$

Once the temperature is estimated, one can use Equation (6.20) to calculate the vapor composition.

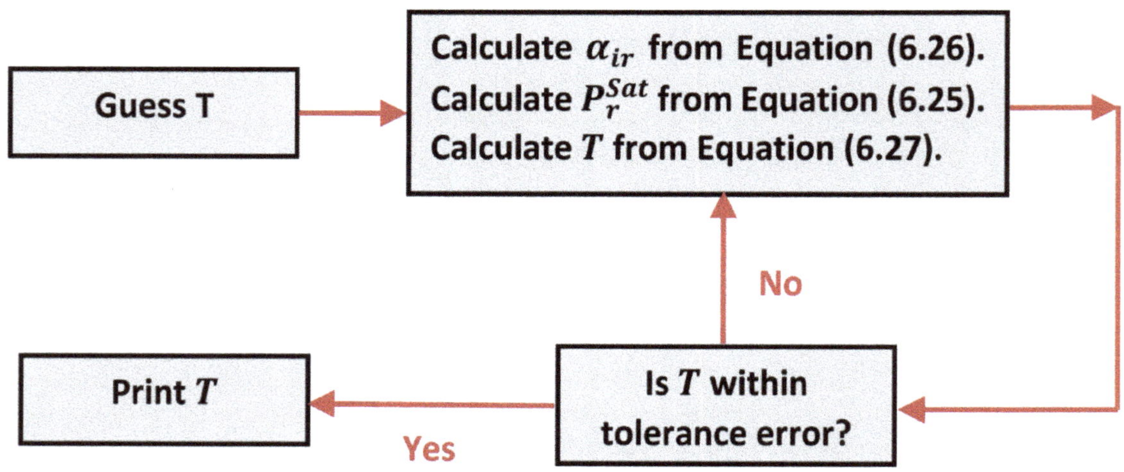

Figure 6.10 Iterative procedure for the calculation of *BUBLT*.

DEWT Calculation: Given $\{y_i\}$ and P, calculate T and $\{x_i\}$

Once again an iterative procedure is required as temperature is unknown. The iterative procedure is similar to that used in *BUBLT* calculation. Equation (6.23) can be re-written as:

$$P = \frac{P_r^{Sat}}{P_r^{Sat} \sum y_i / P_i^{Sat}} \tag{6.30}$$

Using the definition of relative volatility given in Equation (6.18), Equation (6.30) can be re-arranged as:

$$P_r^{Sat} = P \sum y_i/\alpha_{ir} \quad (6.31)$$

Equations (6.26) and (6.27) are applicable to *DEWT* calculation. Thus the iteration scheme for *DEWT* calculation is as shown in Figure 6.11.

The initial guess of temperature can be made as follows:

$$T_{guess} = \sum y_i \, T_i^{Sat} \quad (6.32)$$

where T_i^{Sat} is the boiling point of species i estimated from the Antoine equation at the given total pressure P. Following the procedure shown in Figure 6.11, an improved value of temperature is calculated from Equation (6.27). If the calculated temperature is within the set tolerance level (see Equation (6.29)), the

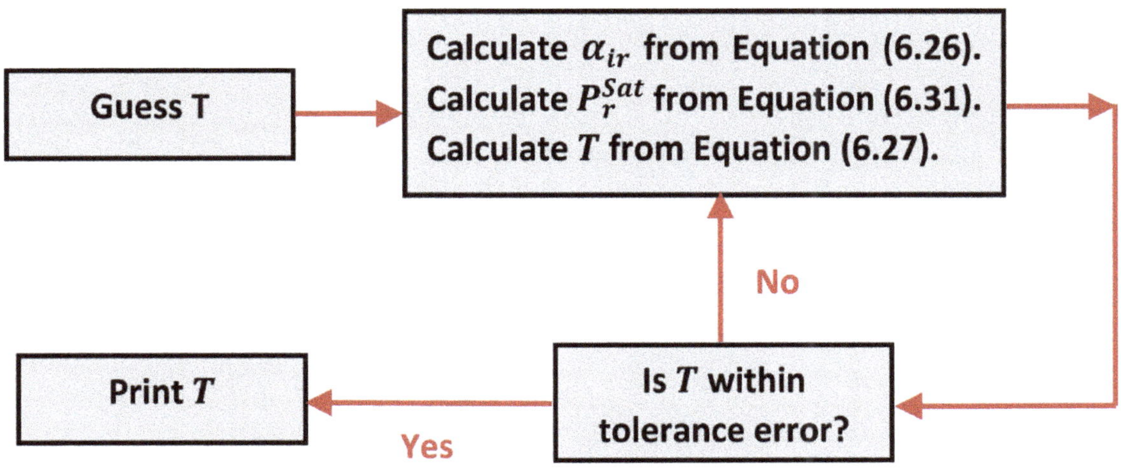

Figure 6.11 Iterative procedure for the calculation of *DEWT*.

iteration process is stopped otherwise the calculated value of temperature is taken as the new guess value and the iterative process is continued. Once the temperature is estimated, one can use Equation (6.22) to calculate the liquid (dew) composition.

6.6.4 Bubble and Dew Points Calculations Using K_i Values

We can use K_i values, available in the form of charts (such as DePriester charts) or equations (such as Equation (6.12)), to carry out the bubble point and dew point calculations. We will assume that K_i is independent of the composition.

BUBLP Calculation: Given $\{x_i\}$ and T, calculate P and $\{y_i\}$

Given T and liquid composition, find P by trial for which the following equation is satisfied:

$$\sum K_i \, x_i = 1 \quad (6.33)$$

The K_i values depend on the choice of P for the given temperature. If P is chosen such that all K_i values are greater than unity, then $\sum K_i \, x_i > 1$ as $\sum x_i = 1$. In this case, P is too low. We need to increase the pressure. Likewise, if P is chosen such that all K_i values are smaller than unity, then $\sum K_i\, x_i < 1$. In this case, P is too high. We need to decrease the pressure. Note that K_i increases with the decrease in pressure. Thus our initial guess of P should be such that some K_i values are greater than unity and some are less than unity.

If at the assumed value of P (which satisfies the criterion that some K_i values are greater than unity and some are less than unity), we still find $\sum K_i\, x_i \neq 1$, we need to either increase the value of P in the next trial if $\sum K_i\, x_i > 1$ or lower the value of P if $\sum K_i\, x_i < 1$. Once P is determined by trial and error, the corresponding K_i values are used to calculate the vapor composition from Equation (6.11).

DEWP Calculation: Given $\{y_i\}$ *and T, calculate P* **and** $\{x_i\}$

For dew point pressure calculations, where T and y_i are known, we have to satisfy the following relation:

$$\sum x_i = \sum \frac{y_i}{K_i} = 1 \tag{6.34}$$

Like in *BUBLP* calculation, we need to make an initial guess of P. As $\sum y_i = 1, \sum y_i/K_i > 1$ if P is chosen such that all K_i values are smaller than unity. If P is chosen such that all K_i values are larger than unity, then $\sum y_i/K_i < 1$. Thus our initial guess value of P should be such that some K_i values are greater than unity and some are less than unity. If at the assumed value of P we still find $\sum y_i/K_i \neq 1$, we need to adjust the value of P (increase P if $\sum y_i/K_i < 1$ or decrease P if $\sum y_i/K_i > 1$) in the next trial. Once P is determined by trial and error, the corresponding K_i values are used to calculate the liquid composition from Equation (6.11).

BUBLT Calculation: Given $\{x_i\}$ *and P, calculate T* **and** $\{y_i\}$

The iterative procedure for *BUBLT* calculation can be developed starting from the following equation:

$$\sum y_i = \sum K_i\, x_i = K_r \sum \alpha_{ir}\, x_i = 1 \tag{6.35}$$

where "*r*" is some reference species, chosen arbitrarily and α_{ir} is the relative volatility of species "*i*" with respect to the reference species "*r*". Equation (6.35) could be re-arranged as:

$$K_r = 1/\sum \alpha_{ir}\, x_i \tag{6.36}$$

The iteration procedure is illustrated in Figure 6.12. The iteration begins with an initial guess of temperature. The initial guessed temperature should be such that some K_i values are greater than unity and some are less than unity. If T is chosen such that all K_i values are greater than unity, then $\sum K_i\, x_i > 1$ as $\sum x_i = 1$. In this case, the guessed temperature is too high. We need to lower the temperature (K_i decreases with the decrease in temperature). Likewise, if T is chosen such that all K_i values are smaller than unity, then $\sum K_i\, x_i < 1$. In this case, we need to adjust the temperature to a higher value in order to increase the K_i values. The initial guess of the temperature could also be made using Equation (6.28) where T_i^{Sat} is the boiling point of pure species i estimated from K_i chart at a given pressure P and $K_i = 1$.
Once T is determined by trial and error, the corresponding K_i values are used to calculate the vapor composition from Equation (6.11). Note that if $\sum K_i\, x_i$ is not exactly equal to unity in the final iteration, the vapor composition should be calculated from the following equation, instead of Equation (6.11), in order to satisfy $\sum y_i = 1$:

$$y_i = K_i x_i/\sum K_i\, x_i = \alpha_{ir} x_i/\sum \alpha_{ir}\, x_i \tag{6.37}$$

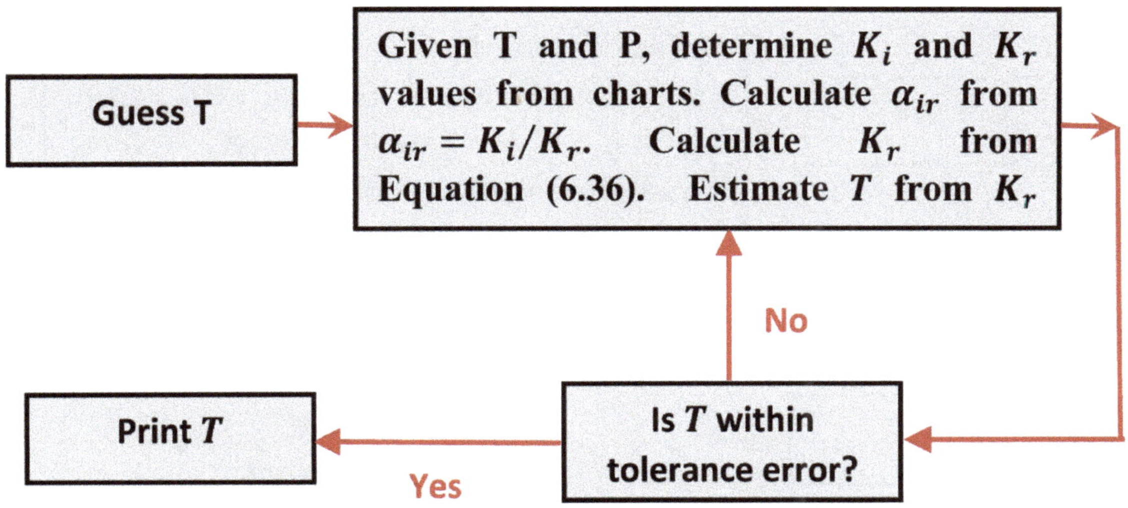

Figure 6.12 Iterative procedure for the calculation of *BUBLT* using K_i values.

DEWT Calculation: Given $\{y_i\}$ *and P, calculate T* and $\{x_i\}$

The iterative procedure for *DEWT* calculation is similar to that used in *BUBLT* calculation. The procedure can be developed starting from the following equation:

$$\sum x_i = \sum y_i/K_i = \sum(y_i/\alpha_{ir})/K_r = 1 \tag{6.38}$$

where "*r*" is some reference species, chosen arbitrarily and α_{ir} is the relative volatility of species "*i*" with respect to the reference species "*r*". Equation (6.38) can be re-arranged as:

$$K_r = \sum y_i/\alpha_{ir} \tag{6.39}$$

The iteration procedure is illustrated in Figure 6.13. The iteration begins with an initial guess of temperature. The initial guessed temperature should be such that some K_i values are greater than unity and some are less than unity. If T is chosen such that all K_i values are greater than unity, then $\sum y_i/K_i < 1$ as $\sum y_i = 1$. In this case, the guessed temperature is too high. We need to lower the temperature (K_i decreases with the decrease in temperature). Likewise, if T is chosen such that all K_i values are smaller than unity, then $\sum y_i/K_i > 1$. In this case, we need to adjust the temperature to a higher value in order to increase the K_i values. The initial guess of the temperature can also be made using Equation (6.32) where T_i^{Sat} is the boiling point of pure species i estimated from K_i chart at a given pressure P and $K_i = 1$.

Once T is determined by trial and error, the corresponding K_i values are used to calculate the liquid composition from $x_i = y_i/K_i$. If $\sum y_i/K_i$ is not exactly equal to unity in the final iteration, the liquid composition can be calculated from the following equation, in order to satisfy $\sum x_i = 1$:

$$x_i = (y_i/K_i)/\sum y_i/K_i = (y_i/\alpha_{ir})/\sum y_i/\alpha_{ir} \tag{6.40}$$

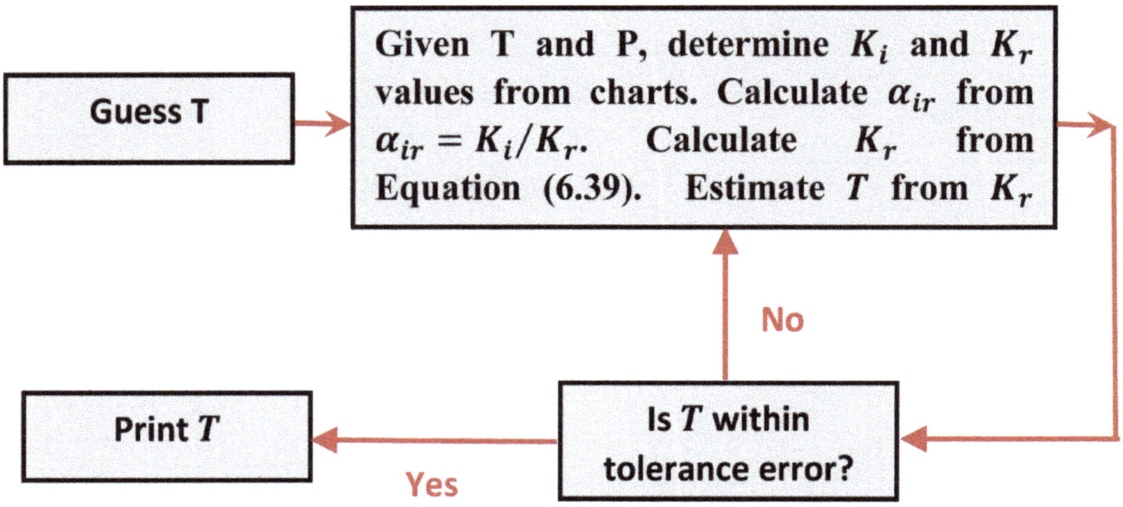

Figure 6.13 Iterative procedure for the calculation of *DEWT* using K_i values.

6.6.5 Importance of Bubble and Dew Points Calculations

Bubble point and dew point calculations are important in the design and analysis of distillation processes. Some practical situations where bubble and dew point calculations are important and necessary are described below:

1) Suppose we are given a feed of known composition at some specified pressure and temperature. However, it is not known in advance if the feed is a subcooled liquid, saturated liquid, saturated vapor, superheated vapor, or a mixture of saturated liquid and saturated vapor. In this situation, bubble point and dew point calculations are made to establish the nature of the feed. At the given T and composition, the bubble point and dew point pressures are calculated and compared with the specified feed pressure P. The feed is subcooled liquid if $P > BUBLP$. The feed is saturated liquid if $P = BUBLP$. If $DEWP < P < BUBLP$, the feed is a two-phase mixture of saturated liquid and saturated vapor. The feed is saturated vapor if $P = DEWP$. If $P < DEWP$, the feed is in superheated vapor state. The K_i values of different species are also helpful in determining the nature of the feed. If all K_i values are greater than unity, the feed is a superheated vapor (above the dew point). If all K_i values are less than unity, the feed is a single-phase subcooled liquid (below the bubble point).

2) Bubble point and dew point calculations are important in flash distillation. For example, we are given the feed composition and pressure of the flash chamber. What should be the flash temperature? In this case, we need to calculate *DEWT* and *BUBLT* of the feed for specified composition and pressure. For flashing (partial vaporization) of feed to occur, it is necessary that $BUBLT < T < DEWT$. Another variation of flash calculations is the specification of flash chamber pressure for the given feed composition and flash temperature. Now we need to calculate *BUBLP* and *DEWP* for the given feed composition and specified temperature. For flashing to occur, $DEWP < P < BUBLP$.

3) In the rigorous analysis and design of distillation columns for multicomponent systems, the calculation procedures require stage-by-stage calculations. Bubble point and dew point calculations are utilized to estimate the temperature of each stage. Bubble point calculations are utilized in the bottom up approach where the pressure at each stage and the composition of the liquid stream leaving the stage are known. The bubble point calculations give the temperature at each stage and the composition of the vapor stream leaving the stage. The liquid and vapor streams leaving the stage are assumed to be at equilibrium with

each other (vapor is at its dew point and liquid is at its bubble point). This bottom up approach requires the specification of the composition of the bottom product of the column (liquid stream leaving the reboiler). Dew point calculations are utilized in the top down approach where the pressure at each stage and the composition of the vapor stream leaving the stage are known. The dew point calculations give the temperature at each stage and the composition of the liquid stream leaving the stage. The liquid and vapor streams leaving the stage are assumed to be at equilibrium with each other. This top down approach requires the specification of the composition of the top product (distillate) of the column.

4) Bubble point and dew point calculations are also helpful in deciding the operating pressure of the distillation column once the top and bottom product compositions are specified. For example, one can determine the dew point temperature of the top vapor product stream over a range of pressures and then select the operating pressure such that the temperature difference between the top vapor stream and cooling water available in the condenser is enough to condense the vapor stream. In a similar manner, one can determine the bubble point temperature of the bottom liquid product stream over a range of pressures and then select the operating pressure such that the temperature difference between the utilities (for example, steam) available for reboiling and bottom liquid stream is large enough to provide the required heating in the reboiler. One can finally select the column operating pressure which meets both the requirements that the overhead vapor must be condensed and the bottom liquid must be reboiled with the available utilities.

6.7 CONSTRUCTION OF P-x-y AND T-x-y PHASE DIAGRAMS AND y-x EQUILIBRIUM CURVE

6.7.1 Construction of Phase Diagrams and Equilibrium Curve Using Raoult's Law

The phase diagrams (P-x-y and T-x-y) and equilibrium y-x curve for binary systems can be constructed using the Raoult law assuming that the liquid solution and vapor mixture are ideal in nature.

Figure 6.8 shows a P-x-y diagram (temperature fixed) of a binary system of species "A" and "B" where species "A" is more volatile. The P-x-y diagram gives the compositions of liquid and vapor phases in equilibrium with each other at different pressures at a fixed temperature. If the liquid solution and vapor mixture are assumed to be ideal, the compositions of the vapor mixture and liquid solution are related by the Raoult law. Thus the Raoult law can be used to construct the P-x-y phase diagram and the corresponding y-x equilibrium curve.

Figure 6.9 shows the T-x-y diagram (pressure fixed) of a binary system of species "A" and "B" where species "A" is more volatile. The T-x-y diagram gives the compositions of liquid and vapor phases in equilibrium with each other at different temperatures at a fixed pressure. Assuming ideal liquid solution and vapor mixture, the Raoult law can be used to construct the T-x-y phase diagram and the corresponding y-x equilibrium curve.

For a binary system of species "A" and "B" (A being more volatile), the governing equations are as follows:

$$P_A = x_A P_A^{Sat} \quad (6.41)$$

$$P_B = x_B P_B^{Sat} = (1 - x_A) P_B^{Sat} \quad (6.42)$$

$$P = P_A + P_B = x_A P_A^{Sat} + (1 - x_A) P_B^{Sat} \quad (6.43)$$

Solving for x_A , Equation (6.43) gives:

$$x_A = \left[\frac{P - P_B^{Sat}}{P_A^{Sat} - P_B^{Sat}}\right] \quad (6.44)$$

The vapor composition y_A can be calculated as:

$$y_A = \frac{x_A P_A^{Sat}}{P} = \left(\frac{P_A^{Sat}}{P}\right)\left[\frac{P - P_B^{Sat}}{P_A^{Sat} - P_B^{Sat}}\right] \tag{6.45}$$

The phase diagrams (P-x-y, T-x-y) and the y-x equilibrium curve can now be constructed using Equations (6.44) and (6.45) along with the Antoine equations for species "A" and "B".

To construct the P-x-y diagram at a given temperature, the vapor pressures P_A^{Sat} and P_B^{Sat} are calculated at the specified temperature using the Antoine equations. Then a number of P values are selected over the pressure range $P_B^{Sat} < P < P_A^{Sat}$ (say 10 values equally spaced between P_B^{Sat} and P_A^{Sat}). The liquid and vapor compositions (x_A and y_A) corresponding to any chosen P are calculated using Equations (6.44) and (6.45). The plots of P vs x_A (bubble point curve) and P vs y_A (dew point curve) give the P-x-y phase diagram. The plot of y_A vs x_A gives the equilibrium curve at constant temperature.

To construct the T-x-y diagram at a given pressure, we need to calculate T_A^{Sat} and T_B^{Sat} corresponding to the specified pressure using the Antoine equations. Then a number of T values are selected over the temperature range $T_A^{Sat} < T < T_B^{Sat}$ (say 10 values equally spaced between T_A^{Sat} and T_B^{Sat}). The vapor pressures of species A and B (P_A^{Sat} and P_B^{Sat}) are calculated at the selected values of T using the Antoine equations. The liquid and vapor compositions (x_A and y_A) corresponding to any chosen T are calculated using Equations (6.44) and (6.45). The plots of T vs x_A (bubble point curve) and T vs y_A (dew point curve) give the T-x-y phase diagram. The plot of y_A vs x_A gives the equilibrium curve at constant pressure. The isobaric y_A vs x_A equilibrium curve is used extensively in the design of the distillation columns.

6.7.2 Construction of Phase Diagrams and Equilibrium Curve Using K_i Values

As noted in the preceding section, the P-x-y diagram gives the compositions of liquid and vapor phases in equilibrium with each other at different pressures at a fixed temperature. The T-x-y diagram gives the compositions of liquid and vapor phases in equilibrium with each other at different temperatures at a fixed pressure.

The governing equations for the construction of P-x-y, T-x-y, and y-x diagrams are as follows:

$$y_A = K_A x_A \tag{6.46}$$

$$y_B = K_B x_B = K_B(1 - x_A) \tag{6.47}$$

$$y_A + y_B = K_A x_A + K_B(1 - x_A) = 1 \tag{6.48}$$

Solving for x_A, we get:

$$x_A = (1 - K_B)/(K_A - K_B) \tag{6.49}$$

To construct the P-x-y diagram at a given temperature, the vapor pressures P_A^{Sat} and P_B^{Sat} are determined from the K_i chart by setting $K_A = K_B = 1$. Then a number of P values are selected over the pressure range $P_B^{Sat} < P < P_A^{Sat}$ (say 10 values equally spaced between P_B^{Sat} and P_A^{Sat}). The values of K_A and K_B are determined at the selected values of P (T is fixed). Using Equation (6.49), the values of x_A are calculated. The corresponding y_A values are determined from Equation (6.46). Thus we can construct the P-x-y diagram and the equilibrium curve (y-x) at a given temperature.

In order to construct the T-x-y diagram at a given pressure, the boiling points T_A^{Sat} and T_B^{Sat} corresponding to the specified pressure are determined from from the K_i chart by setting $K_A = K_B = 1$. Then a number of T values are selected over the temperature range $T_A^{Sat} < T < T_B^{Sat}$ (say 10 values equally spaced between $T_A^{Sat}$and T_B^{Sat}). The values of K_A and K_B are determined at the selected values of T (P is fixed). Using Equation (6.49), the values of x_A are calculated. The corresponding y_A values are determined from Equation (6.46). Thus we can construct the T-x-y diagram and the equilibrium curve (y-x) at a given pressure.

NOTATION

A	constant in Equations (6.5), (6.6) and (6.7)
a_i	constants in Equation (6.12)
B	constant in Equations (6.6) and (6.7)
b_i	constants in Equation (6.12)
C	constant in Equation (6.7)
K_i	distribution coefficient of species *i* defined in Equation (6.11)
P	pressure
P_i^{Sat}	saturation pressure or vapor pressure of pure species *i*
R	universal gas constant
T	temperature
T_i^{Sat}	saturation temperature or boiling point of pure species *i*
V^{liq}	molar volume of liquid
V^{vap}	molar volume of vapor
x_i	mole fraction of species *i* in liquid phase
y_i	mole fraction of species *i* in vapor phase

Greek Letters

α_{ij}	relative volatility of species *i* with respect to species *j*, defined in Equation (6.17)
γ_i	activity coefficient of species *i* in liquid solution
ΔH^{vap}	molar latent heat of vaporization
ΔV^{vap}	molar volume change of vaporization
ϑ_i	volatility of species *i* defined in Equation (6.13)

TUTORIAL PROBLEMS WITH SOLUTION

Problem 1

The vapor pressure of a certain liquid at 10°C is 110 mm Hg. The latent heat of vaporization of this material is 31 kJ/mole. What is the normal boiling point of this liquid?

Solution

$$T = 10^oC, P^{Sat} = 110\ mmHg, \Delta H^{vap} = 31{,}000\ J/mole$$

The Clausius-Clapeyron equation can be written as:

$$ln\left(\frac{P_2^{Sat}}{P_1^{Sat}}\right) = -\frac{\Delta H^{vap}}{R}\left(\frac{1}{T_2} - \frac{1}{T_1}\right)$$

We need to calculate T_2 given that : $T_1 = 10^oC = 283.15\,K, P_1^{Sat} = 110\,mmHg, P_2^{Sat} = 760\,mmHg.$
Thus:

$$ln\left(\frac{760}{110}\right) = -\frac{31000}{8.314}\left(\frac{1}{T_2} - \frac{1}{283.15}\right)$$

The equation gives normal boiling point T_2 equal to 331.86 K (=58.71 °C)

Problem 2

200 g of toluene (MW=92.134 g/mole) is mixed with 15 g of benzene (MW=78.108 g/mole) to form a liquid solution. The normal boiling point of toluene is 383.8 K and that of benzene is 353.2 K. The latent heats of vaporization of toluene and benzene are: 39.2 kJ/mole and 30.8 kJ/mole, respectively.

a) What are the partial pressures of toluene and benzene above the solution at 298 K? What is the total pressure?
b) What is the composition of the vapor mixture above the liquid solution at 298 K?

Assume liquid solution and vapor mixture are ideal in nature.

Solution

The liquid solution consists of 15 g (=15/78.108= 0.192 moles) of benzene and 200 g (=200/92.134=2.171 moles) of toluene. Thus the mole fractions of benzene and toluene in the liquid solution are as follows:

$$x_{benzene} = 0.0813 \text{ and } x_{toluene} = 0.9187$$

The vapor pressures of pure benzene and toluene can be estimated from the Clausius-Clapeyron equation:

$$ln\left(\frac{P_2^{Sat}}{P_1^{Sat}}\right) = -\frac{\Delta H^{vap}}{R}\left(\frac{1}{T_2} - \frac{1}{T_1}\right)$$

For benzene, $T_1 = 353.2\,K, P_1^{Sat} = 1\,atm, T_2 = 298\,K, P_2^{Sat} =?$ Thus:

$$ln\left(\frac{P_2^{Sat}}{1}\right) = -\frac{30800}{8.314}\left(\frac{1}{298} - \frac{1}{353.2}\right)$$

The equation gives $P_2^{Sat} = 0.1433\,atm.$

For toluene, $T_1 = 383.8\,K, P_1^{Sat} = 1\,atm, T_2 = 298\,K, P_2^{Sat} =?$ Thus:

$$ln\left(\frac{P_2^{Sat}}{1}\right) = -\frac{39200}{8.314}\left(\frac{1}{298} - \frac{1}{383.8}\right)$$

The equation gives $P_2^{Sat} = 0.0291\,atm$

a) Partial pressures and total pressure:

$$P_{benzene} = x_{benzene}P_{benzene}^{Sat} = 0.0813 \times 0.1433 = 0.01165 \text{ atm}$$

$$P_{toluene} = x_{toluene}P_{toluene}^{Sat} = 0.9187 \times 0.0291 = 0.02673 \text{ atm}$$

$$total\ pressure = P_{benzene} + P_{toluene} = 0.01165 + 0.02673 = 0.0384 \text{ atm}$$

b) Composition of vapor mixture:

$$y_{benzene}P = P_{benzene} = 0.01165$$

$$y_{benzene} = 0.01165/0.0384 = 0.3034$$

$$y_{toluene}P = P_{toluene} = 0.02673$$

$$y_{benzene} = 0.02673/0.0384 = 0.6961$$

$$\sum y_i = 0.3034 + 0.6961 \approx 1.0$$

Problem 3

A 3-component liquid solution of n-butane, n-pentane, and n-hexane is in equilibrium with its vapor at T=130 °F and P=25 psia. If the mole fraction of n-butane in the liquid solution is 0.15, what are the compositions of the vapor and liquid phases?

Solution

We have a 3-component system: n-butane as species 1, n-pentane as species 2, and n-hexane as species 3. Given that: T=130 °F and P=25 psia, $x_1 = 0.15$. We are required to calculate x_2, x_3, y_1, y_2, y_3.
Number of unknown composition variables = 5. We need the following 5 equations to solve for the unknowns:

$$y_1 = K_1x_1\,; y_2 = K_2x_2;\ y_3 = K_3x_3; \sum x_i = 1.0; \sum y_i = 1.0$$

The distribution coefficients K_i's can be estimated from Equation (6.12) using constants given in Table 6.1. At T=130 +460=590 °R and P=25 psia, Equation (6.12) gives: $K_1 = 3.36, K_2 = 1.08, K_3 = 0.42$.

$$y_1 = K_1x_1 = 3.36 \times 0.15 = 0.504$$

$$K_3 = y_3/x_3 = (1 - y_1 - y_2)/(1 - x_1 - x_2) = (1 - y_1 - K_2x_2)/(1 - x_1 - x_2)$$

From this equation, we can solve for x_2 as:

$$x_2 = (1 - y_1 - K_3 + K_3x_1)/(K_2 - K_3)$$

$$x_2 = (1 - 0.504 - 0.42 + 0.42 \times 0.15)/(1.08 - 0.42) = 0.2106$$

The other unknown composition variables can be calculated as:

$$x_3 = 1 - x_1 - x_2 = 1 - 0.15 - 0.2106 = 0.6394$$

$$y_2 = K_2x_2 = 1.08 \times 0.2106 = 0.2274$$

$$y_3 = K_3x_3 = 0.42 \times 0.6394 = 0.2686$$

Note that: $\sum y_i = y_1 + y_2 + y_3 = 0.504 + 0.2274 + 0.2686 = 1.0$

Problem 4

Estimate the relative volatility of benzene with respect to toluene in a binary benzene-toluene system at different temperatures: a) 80 °C, b) 90 °C, c) 100 °C, and d) 110 °C. Assume that the system is ideal in nature and Raoult's law is applicable. The Antoine equations for benzene and toluene are as follows:

$$lnP_{benzene}^{Sat} = 13.7819 - \frac{2726.81}{t+217.572}$$

$$lnP_{toluene}^{Sat} = 13.932 - \frac{3056.96}{t+217.625}$$

where t is the temperature in °C and the vapor pressures P^{Sat} are in kPa.

Solution

For ideal systems,

$$\alpha_{ij} = \frac{P_i^{Sat}}{P_j^{Sat}}$$

Thus,

$$\alpha_{benzene-toluene} = \frac{P_{benzene}^{Sat}}{P_{toluene}^{Sat}}$$

Using Antoine equations, we get:

T(°C)	$P_{benzene}^{Sat}$(kPa)	$P_{toluene}^{Sat}$(kPa)	$\alpha_{benzene-toluene}$
80	101.33	38.893	2.605
90	136.497	54.31	2.513
100	180.453	74.26	2.43
110	234.532	99.617	2.354

Comments: The relative volatility decreases somewhat with the increase in temperature. However, the decrease is less than 10% over the temperature increase from 80 to 110 °C.

Problem 5

A liquid solution of benzene and toluene is placed in a piston and cylinder assembly. The benzene amount is 6 moles and the toluene amount is 4 moles. The temperature is fixed at 298 K. The vapor pressures of pure benzene and pure toluene at 298 K are 96.4 mmHg and 28.9 mmHg, respectively. The pressure above the liquid solution is initially 760 mmHg. If the pressure above the solution is reduced from 760 mmHg, at what pressure would the first vapor bubble appear? What is the composition of this vapor bubble?

You can assume that the liquid solution and vapor mixture are ideal in nature.

Solution

Binary liquid solution of benzene (species 1) and toluene (species 2). Given that: $x_1 = 0.60$, $x_2 = 0.40$, T=298 K, $P_1^{Sat} = 96.4\ mmHg$, $P_2^{Sat} = 28.9\ mmHg$. The vapor pressure of the liquid solution is:

$$P = x_1P_1^{Sat} + x_2P_2^{Sat} = 0.60 \times 96.4 + 0.40 \times 28.9 = 69.4\ mmHg$$

There is no boiling initially when the pressure above the liquid solution is 760 mmHg.

The first vapor bubble is formed at 69.4 mmHg. The composition of the first vapor bubble is determined from Raoult's law as:

$$y_1 = x_1 P_1^{Sat}/P = 0.60 \times 96.4/69.4 = 0.8334$$

$$y_2 = x_2 P_2^{Sat}/P = 0.40 \times 28.9/69.4 = 0.1666$$

Problem 6

A 4-component vapor mixture containing 25 mole% species 1, 25 mole% species 2, 25 mole% species 3, and 25 mole% species 4 has a total pressure of 1 bar. For this system, the vapor pressures of the pure species are given by the Antoine equation:

$$lnP_i^{Sat} = A_i - \frac{B_i}{T+C_i}$$

where T is in K and the vapor pressures are in bars. The Antoine coefficients are given in the following table:

Species	1	2	3	4
A_i	9.2131	9.1325	9.2164	9.2535
B_i	2477.07	2766.63	2697.55	2911.32
C_i	-39.94	-50.50	-48.78	-56.51

a) Determine the dew point temperature of this vapor.
b) What is the composition of the first drop of liquid?

You can assume that the vapor mixture and liquid solution are ideal in nature. In your calculations, use species 3 as the reference species.

Solution

This problem involves DEWT calculation. We are given vapor composition and P, we are asked to determine T and dew (first liquid drop) composition.

a)Estimation of Dew point temperature:

$$T_{guess} = \sum y_i T_i^{Sat}$$

At P = 1 bar, Antoine equations give: $T_1^{Sat} = 308.9039\ K, T_2^{Sat} = 353.4433\ K, T_3^{Sat} = 341.4702\ K,$

$T_4^{Sat} = 371.1283\ K$. Thus, $T_{guess} = 343.7114\ K$.

At $T = 343.7114\ K, \alpha_{13} = 2.6878, \alpha_{23} = 0.6886, \alpha_{33} = 1.0, \alpha_{43} = 0.3854$. Using Equation (6.31):

$$P_3^{Sat} = P \sum y_i/\alpha_{i3} = 1 \times (y_1/\alpha_{13} + y_2/\alpha_{23} + y_3/\alpha_{33} + y_4/\alpha_{43}) = 1.3547 \text{ bar}$$

Using Equation (6.27), we obtain:

$$T = \frac{B_3}{A_3 - lnP_3^{Sat}} - C_3 = 351.44\ K$$

Since T of 351.44 K is significantly off from the assumed value of 343.7114 K, we need another iteration.

At $T = 351.44\ K, \alpha_{13} = 2.605, \alpha_{23} = 0.6946, \alpha_{33} = 1.0, \alpha_{43} = 0.398$. Using Equation (6.31):

$$P_3^{Sat} = P \sum y_i/\alpha_{i3} = 1 \times (y_1/\alpha_{13} + y_2/\alpha_{23} + y_3/\alpha_{33} + y_4/\alpha_{43}) = 1.334 \text{ bar}$$

Using Equation (6.27), we obtain:

$$T = \frac{B_3}{A_3 - lnP_3^{Sat}} - C_3 = 350.92\ K$$

This value of T is very close to the assumed value of 351.44 K. Thus T=350.92 K

b) Composition of liquid droplet: The composition of the first liquid drop is determined from Raoult's law as:

$$x_1 = y_1 P/P_1^{Sat} = 0.25 \times 1/3.4824 = 0.0718$$

$$x_2 = y_2 P/P_2^{Sat} = 0.25 \times 1/0.9262 = 0.27$$

$$x_3 = y_3 P/P_3^{Sat} = 0.25 \times 1/1.3341 = 0.1874$$

$$x_4 = y_4 P/P_4^{Sat} = 0.25 \times 1/0.5298 = 0.4718$$

Note that at T=350.92 K, the vapor pressures of pure species are determined from the Antoine equations.

Problem 7

Consider a 3-component mixture of methane (10 mole%), ethane (20 mole %), and propane (70 mole%) at 50 °F. The K_i values of different species can be estimated from Equation (6.12) using Table 6.1 data at any given temperature and pressure.

a) Assuming that the mixture is in vapor state, estimate the dew point pressure. Take initial guess of dew pressure as 120 psia.
b) What is the composition of the first liquid drop formed?
c) Assuming that the mixture is in liquid state, estimate the bubble point pressure. Take initial guess of bubble point pressure as 375 psia.
d) What is the composition of the first vapor bubble formed?

Solution

a) Given T=50°F, y_1=0.10, y_2=0.20, y_3=0.70, estimate DEWP. Procedure: Guess P, calculate K_i, check if $\sum y_i/K_i = 1$. If $\sum y_i/K_i < 1$, increase P. If $\sum y_i/K_i > 1$, decrease P.

The following table shows the calculations using the initial guess value of pressure P=120 psia. Equation (6.12) is used to estimate the K_i values.

species	P=120 psia			P=140 psia		P=130 psia	
	y_i	K_i	y_i/K_i	K_i	y_i/K_i	K_i	y_i/K_i
methane (1)	0.1	17.07	0.0059	14.86	0.0067	15.88	0.0063
ethane (2)	0.2	2.79	0.0717	2.43	0.0822	2.60	0.077
propane (3)	0.7	0.81	0.8631	0.71	0.9799	0.76	0.922
	$\sum y_i/K_i = 0.9407$			$\sum y_i/K_i = 1.0688$		$\sum y_i/K_i = 1.0053$	

Thus the dew point pressure is approximately 130 psia.

b) The composition of the first liquid drop is given in the last column of the above table, that is, x_1=0.0063, x_2=0.077, x_3=0.922

c) Given T=50°F, x_1=0.10, x_2=0.20, x_3=0.70, estimate BUBLP. Procedure: Guess P, calculate K_i, check if $\sum K_i x_i = 1$. If $\sum K_i x_i < 1$, decrease P. If $\sum K_i x_i > 1$, increase P.

The following table shows the calculations using the initial guess vale of pressure P=375 psia. Equation (6.12) is used to estimate the K_i values.

species	P=375 psia			P=390 psia		P=393 psia	
	x_i	K_i	K_ix_i	K_i	K_ix_i	K_i	K_ix_i
methane (1)	0.1	6.133	0.6133	5.922	0.592	5.88	0.5881
ethane (2)	0.2	1.014	0.2028	0.979	0.196	0.973	0.1945
propane (3)	0.7	0.324	0.2271	0.315	0.220	0.313	0.2188
	$\sum K_i x_i = 1.0432$			$\sum K_i x_i = 1.008$		$\sum K_i x_i = 1.001$	

Thus the bubble point pressure is approximately 393 psia.

d) The composition of the first vapor bubble is given in the last column of the above table, that is, y_1=0.5881, y_2=0.1945, y_3=0.2188

Problem 8

Consider a binary system of species 1 and 2 at a total pressure of 101.32 kPa. Assuming that the system is ideal in nature and follows the Raoult law, construct the T-x-y and y-x diagrams for the system at 101.32 kPa. For this system, the vapor pressures of the pure species are given by the Antoine equations as:

$$lnP_1^{Sat} = 13.7819 - \frac{2726.81}{t+217.572}$$

$$lnP_2^{Sat} = 13.932 - \frac{3056.96}{t+217.625}$$

where t is in °C and the vapor pressures are in kPa. Species 1 is more volatile.

Solution

The boiling points of pure species 1 and 2 at 101.32 kPa are calculated from the given Antoine equations as: $T_1^{Sat} = 80^oC$ and $T_2^{Sat} = 110.6^oC$. Table given below shows at different temperatures the values of vapor pressures of pure species and mole fractions of more volatile component (species 1) in the vapor and liquid phases at equilibrium with each other. The Antoine equations are used to calculate the vapor pressures. Equations (6.44) and (6.45) are used to calculate the mole fractions in liquid and vapor phases at equilibrium.

The T-x-y and y-x diagrams for the system at 101.32 kPa are developed using the data shown in the table. Note that x and y refer to mole fractions of more volatile species (species 1). The diagrams are shown in the figures given below.

T (°C)	$P_1^{Sat}(kPa)$	$P_2^{Sat}(kPa)$	x	y
80	101.32	38.89	1.00	1.00
83	111.02	43.09	0.86	0.94
86	121.44	47.64	0.73	0.87
89	132.60	52.57	0.61	0.80
92	144.53	57.91	0.50	0.71
95	157.28	63.66	0.40	0.62
98	170.88	69.86	0.31	0.53
101	185.37	76.53	0.23	0.42
104	200.78	83.70	0.15	0.30
107	217.15	91.38	0.08	0.17
110	234.52	99.61	0.01	0.03
110.6	238.11	101.32	0.00	0.00

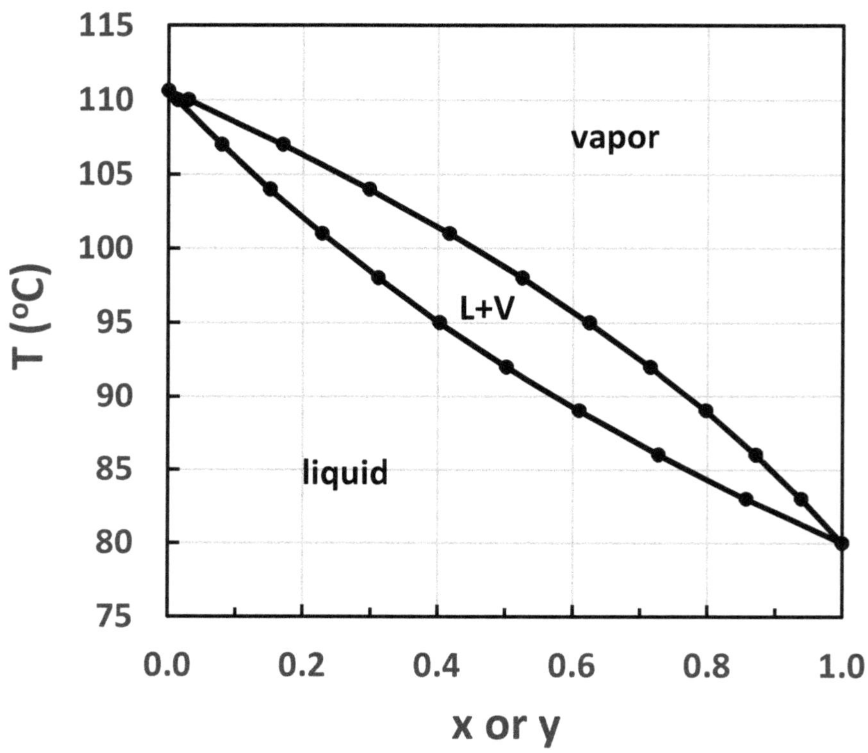

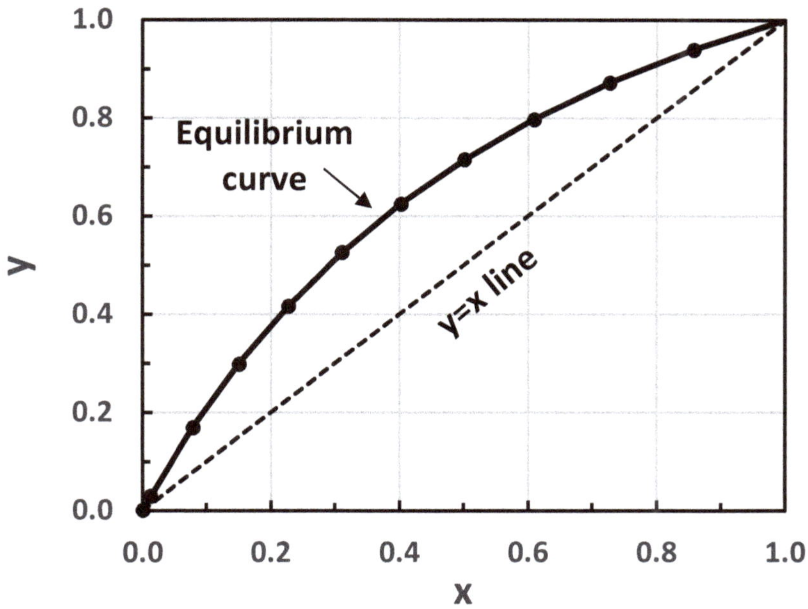

SHORT REVIEW QUESTIONS

1. How does the vapor pressure of a pure substance vary with temperature?
2. What assumptions are made in the derivation of Clausius-Clapeyron equation?
3. What are the limitations of Raoult law?
4. Define volatility and relative volatility
5. Define azeotrope
6. The bubble point and dew point are the same for a pure substance. Is this true?
7. Define bubble point of a multicomponent liquid solution
8. Define dew point of a multicomponent vapor mixture
9. What is the importance of bubble point and dew point calculations in distillation?

PRACTICE PROBLEMS

Problem 1

The normal boiling point of a certain liquid is 35°C. Calculate the vapor pressure at 30°C if the material follows the Trouton rule. Note that according to the Trouton rule, $\Delta H_n^{vap} = 10.5RT_n$ where T_n is the normal boiling point, ΔH_n^{vap} is the latent heat of vaporization at normal boiling point, and R is the universal gas constant. You can assume that the latent heat of vaporization is constant independent of temperature.

Problem 2

100 g of n-hexane (MW=86.18 g/mole) is mixed with 50 g of n-octane (MW=114.23 g/mole) to form a liquid solution. The normal boiling point of n-hexane is 68.7°C and that of n-octane is 125.7°C. The latent heats of vaporization of n-hexane and n-octane are: 335 kJ/kg and 303 kJ/kg, respectively.

a) What are the partial pressures of n-hexane and n-octane above the solution at 25°C? What is the total pressure?
b) What is the composition of the vapor mixture above the liquid solution at 25°C ?

Assume liquid solution and vapor mixture are ideal in nature.

Problem 3

Estimate the relative volatility of n-hexane with respect to n-octane in a binary n-hexane/n-octane system at different temperatures: a) 80 °C, b) 90 °C, c) 100 °C, and d) 120 °C. Assume that the system is ideal in nature and Raoult's law is applicable. The Antoine equations for n-hexane and n-octane are as follows:

$$lnP_{hexane}^{Sat} = 13.8193 - \frac{2696.04}{t+224.317}$$

$$lnP_{octane}^{Sat} = 13.9346 - \frac{3123.13}{t+209.635}$$

where t is the temperature in °C and P^{Sat} is vapor pressure in kPa.

Problem 4

Consider a 4-component liquid solution containing 25 mole% species 1, 25 mole% species 2, 25 mole% species 3, and 25 mole% species 4 is at a pressure of 1 bar. For this system, the vapor pressures of the pure species are given by the Antoine equation:

$$lnP_i^{Sat} = A_i - \frac{B_i}{T+C_i}$$

where T is in K and the vapor pressures are in bars. The Antoine coefficients are given in the following table:

Species	1	2	3	4
A_i	9.2131	9.1325	9.2164	9.2535
B_i	2477.07	2766.63	2697.55	2911.32
C_i	-39.94	-50.50	-48.78	-56.51

a) Determine the bubble point temperature of this liquid solution.
b) What is the composition of the first vapor bubble that is formed?

You can assume that the liquid solution and vapor mixture are ideal in nature. In your calculations, use species 3 as the reference species.

Problem 5

Consider a 3-component system species 1, 2 and 3. For this system, the vapor pressures of the pure species are given by the Antoine equations as:

$$lnP_1^{Sat} = 13.7819 - \frac{2726.81}{t+217.572}$$

$$lnP_2^{Sat} = 13.932 - \frac{3056.96}{t+217.625}$$

$$lnP_3^{Sat} = 13.9726 - \frac{3259.93}{t+212.30}$$

where t is in °C and the vapor pressures are in kPa. Species 1 is more volatile.

Assuming that the system is ideal in nature and follows the Raoult law, do the following calculations:

a) BUBLP calculation. Given that $t = 120^oC, x_1 = 0.25, x_2 = 0.35, x_3 = 0.40$
b) DEWP calculation. Given that $t = 110^oC, y_1 = 0.40, y_2 = 0.35, y_3 = 0.25$
c) BUBLT calculation. Given that $P = 95\ kPa, x_1 = 0.45, x_2 = 0.20, x_3 = 0.35$
d) DEWT calculation. Given that $P = 100\ kPa, y_1 = 0.50, y_2 = 0.30, y_3 = 0.20$

Problem 6

Consider a 4-component mixture of n-butane (35 mole%), n-pentane (30 mole %), n-hexane (15 mole%) and n-heptane (20 mole%) at a pressure of 500 kPa. The K_i values of different species can be estimated from Equation (6.12) using Table 6.1 data at any given temperature and pressure.

Assuming that the mixture is in liquid state, estimate the bubble point temperature of the liquid? What is the composition of the first vapor bubble formed?

Problem 7

Consider a binary system of species 1 and 2 at a total pressure of 101.32 kPa. Assuming that the system is ideal in nature and follows the Raoult law, construct the T-x-y and y-x diagrams for the system at 101.32 kPa. For this system, the vapor pressures of the pure species are given by the Antoine equations as:

$$lnP_1^{Sat} = 13.8193 - \frac{2696.04}{t+224.317}$$

$$lnP_2^{Sat} = 13.9346 - \frac{3123.13}{t+209.635}$$

where t is in °C and the vapor pressures are in kPa. Species 1 is more volatile.

When a liquid solution containing 25 mole% species 1 is heated at a constant pressure of 101.32 kPa, at what temperature would the first vapor bubble form and what is its composition?

Problem 8

Consider a binary system of species 1 and 2 at a total pressure of 1.5 bar. For this system, the vapor pressures of the pure species are given by the Antoine equations as:

$$lnP_1^{Sat} = 13.8193 - \frac{2696.04}{t+224.317}$$

$$lnP_2^{Sat} = 13.9346 - \frac{3123.13}{t+209.635}$$

where t is in °C and the vapor pressures are in kPa. Species 1 is more volatile. The system is ideal in nature and follows the Raoult law.

a) Fill in the blank information of the following table (total pressure is 1.5 bar):

t^oC	P_1^{Sat}(kPa)	P_2^{Sat}(kPa)	x_1	y_1
81.75	150	24.937	1.00	1.00
93.00				
105.00				
117.00				
132.00				
140.336	617.613	150	0.00	0.00

b) Construct the T-x-y diagram for the system at 1.5 bar.

c) When a liquid solution containing 20 mole % species 1 is heated at a constant pressure of 150 kPa, at what temperature would the first bubble form and what is its composition? Use the T-x-y diagram to obtain this information.

d) If the vapor formed in part (c) is condensed to a saturated liquid and a small amount of vapor is formed again from this condensate, what would be the vapor composition? Show this process on the T-x-y diagram.

Chapter 7

Mass and Energy Balances for Open (Flow) Systems

7.1 LEARNING OUTCOMES

Upon completion of this chapter, students will be able to:

- Do mass balance on a control volume with multiple inputs and multiple outputs
- Do energy balance on a control volume with multiple inputs and multiple outputs taking into consideration heat and work flows into or out of the control surface
- Estimate enthalpy of pure substances in vapor or liquid state with respect to reference state
- Estimate enthalpy of multicomponent ideal gas mixtures and ideal liquid solutions

7.2 INTRODUCTION

Mass and energy balances on open (flow) systems are needed in the analysis and design of continuous separation processes such as single-stage flash distillation and multi-stage column distillation. According to the *law of conservation of mass*, mass can neither be created nor destroyed. Like mass, energy is also conserved. The *law of conservation of energy* (also referred to as the first law of thermodynamics) states that energy cannot be created or destroyed but can be transformed from one form to another.

In flow systems, the laws of conservation of mass and energy are applied to a *control volume*. Control volume is a region of space bounded by a control surface. Both mass and energy can pass through the control surface. Typically the control volume is chosen to be the volume of the vessel itself with inlet and outlet, as shown in Figure 7.1. Mass passes through the control surface at inlets and outlets of the control volume. Energy passes through the boundaries of the control volume in three different ways: heat transfer, work transfer, and advection energy transfer due to exchange of mass. Note that mass entering and leaving the control volume carries with it energy.

For a control volume, the general balance equation for any quantity (mass or energy), regardless of the conservation laws, can be written as:

$$(INPUT) - (OUTPUT) + (GENERATION) = (STORAGE\ OR\ ACCUMULATION) \tag{7.1}$$

Applying the conservation laws (no generation of mass or energy), Equation (7.1) simplifies to:

$$(INPUT) - (OUTPUT) = (STORAGE\ OR\ ACCUMULATION) \tag{7.2}$$

In terms of rates, the mass and energy balances for a control volume can be written as:

$$\begin{pmatrix} Rate\ at \\ which\ mass \\ enters\ CV \end{pmatrix} - \begin{pmatrix} Rate\ at \\ which\ mass \\ leaves\ CV \end{pmatrix} = \begin{pmatrix} Rate\ of \\ accumulation \\ of\ mass \end{pmatrix} \tag{7.3}$$

$$\begin{pmatrix} Rate\ at \\ which\ energy \\ enters\ CV \end{pmatrix} - \begin{pmatrix} Rate\ at \\ which\ energy \\ leaves\ CV \end{pmatrix} = \begin{pmatrix} Rate\ of \\ accumulation \\ of\ energy \end{pmatrix} \tag{7.4}$$

where "*CV*" refers to control volume.

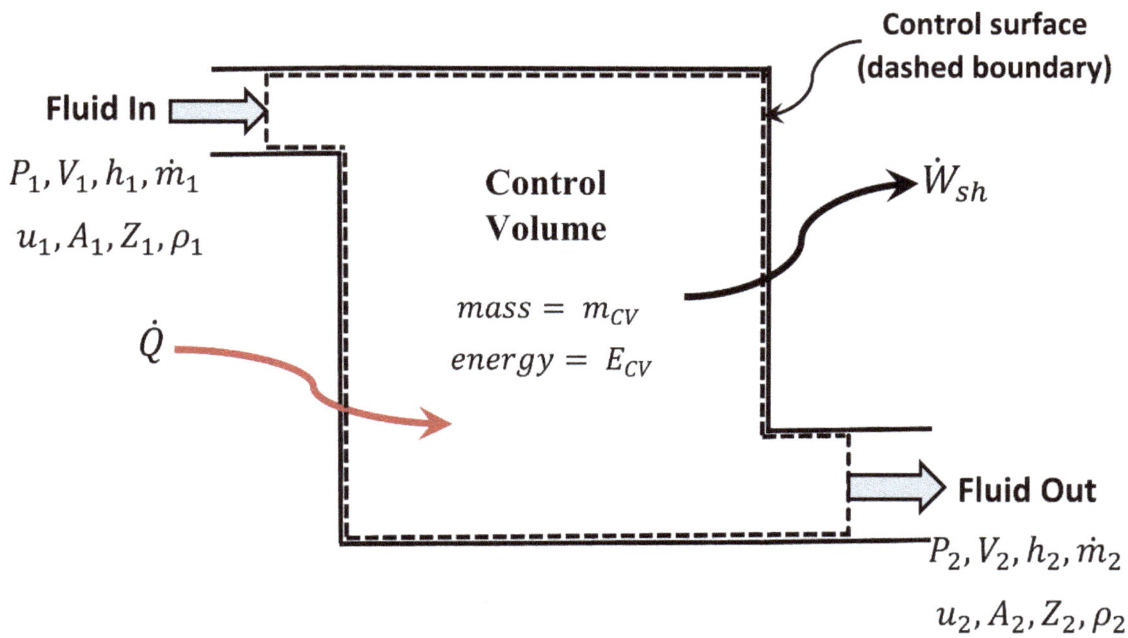

Figure 7.1 Control volume with one inlet and one outlet (P is pressure, V is velocity, h is specific enthalpy, $\dot{m}$ is total mass flow rate, u is specific internal energy, A is cross-sectional area, Z is elevation, ρ is density, $\dot{Q}$ is rate of heat transfer, $\dot{W}_{sh}$ is rate of shaft work, m is mass, and E is total energy).

7.3 MASS BALANCE

Referring to control volume shown in Figure 7.1, the mass balance can be expressed as:

$$\dot{m}_1 - \dot{m}_2 = \frac{d(m)_{CV}}{dt} \tag{7.5}$$

where $\dot{m}$ is the total mass flow rate and m is mass of fluid. At steady state, the storage or accumulation of mass inside the control volume is zero. Thus,

$$\dot{m}_1 = \dot{m}_2 = \dot{m} = constant \tag{7.6}$$

This steady state mass balance can be generalized to control volume with multiple inlets and multiple outlets as follows:

$$\sum_{INLETS} (\dot{m}_i) = \sum_{OUTLETS} (\dot{m}_i) \tag{7.7}$$

In the absence of chemical reactions, mass balance is equivalent to mole balance. Hence,

$$\sum_{INLETS} (F_i) = \sum_{OUTLETS} (F_i) \tag{7.8}$$

where F_i is the total molar flow rate of a fluid stream at *i*th opening (inlet/outlet) of the control volume.

Steady state mass and mole balances on individual species of multi-component mixtures, in the absence of chemical reactions, can be written as:

$$\sum_{INLETS} (\omega_{ji}\dot{m}_i) = \sum_{OUTLETS} (\omega_{ji}\dot{m}_i) \tag{7.9}$$

$$\sum_{INLETS} (x_{ji}F_i) = \sum_{OUTLETS} (x_{ji} F_i) \tag{7.10}$$

where "*j*" refers to a particular chemical species, "*i*" refers to inlet or outlet, ω_j is mass fraction of a species, and x_j is mole fraction of a species. In the case of one inlet and one outlet, Equations (7.9) – (7.10) reduce to:

$$\omega_{j1}\dot{m}_1 = \omega_{j2}\dot{m}_2 \tag{7.11}$$

$$x_{j1}F_1 = x_{j2}F_2 \tag{7.12}$$

Note that $\dot{m}$ and F are the total mass and molar flow rates, respectively at any inlet or outlet of the control volume.

As an example, consider a distillation column shown in Figure 7.2. The distillation column is separating a binary mixture. The feed mass flow rate and composition (mass fraction of more volatile component) are specified. The distillate product withdrawn from the top of the column is also specified in terms of mass flow rate and mass fraction of more volatile component. What are the mass flow rate and composition of the bottoms product?

From overall mass balance, Equation (7.7), it follows that:

$$\dot{m}_1 = \dot{m}_2 + \dot{m}_3 \tag{7.13}$$

From component mass balance on more volatile species (species *A*):

$$\omega_{A1}\dot{m}_1 = \omega_{A2}\dot{m}_2 + \omega_{A3}\dot{m}_3 \tag{7.14}$$

From the data indicated in Figure 7.2, the overall and component mass balances (Equations (7.13)-(7.14) give:

$$\dot{m}_3 = \dot{m}_1 - \dot{m}_2 = 100 - 10 = 90\ kg/min \tag{7.15}$$

$$\omega_{A3} = \frac{\omega_{A1}\dot{m}_1 - \omega_{A2}\dot{m}_2}{\dot{m}_3} = \frac{0.1 \times 100 - 0.60 \times 10}{90} = 0.0445 \tag{7.16}$$

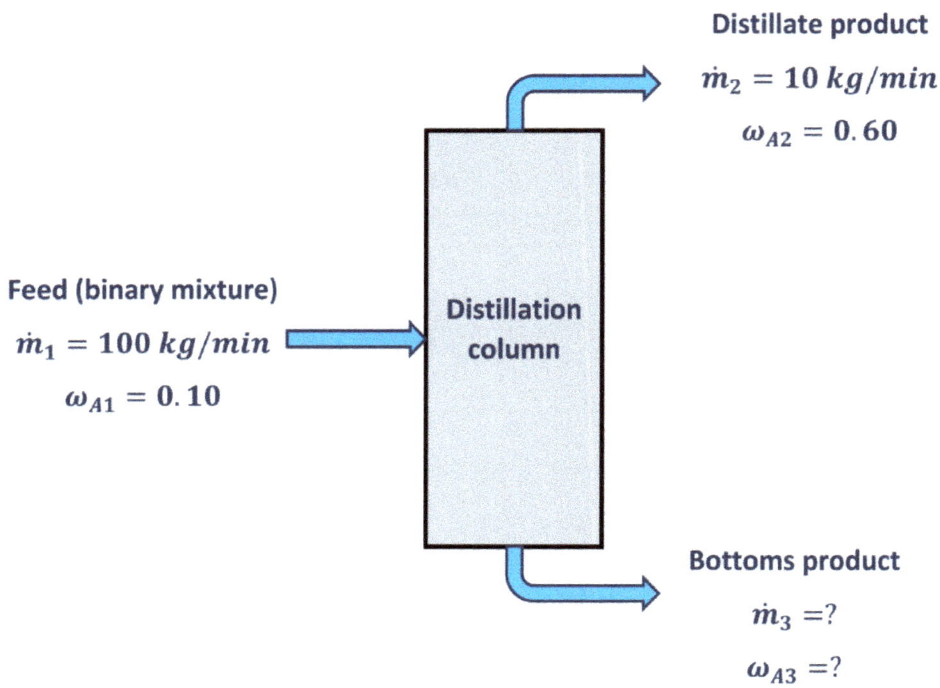

Figure 7.2 External mass balance on a distillation column.

7.4 ENERGY BALANCE

Consider the flow process shown in Figure 7.1. Fluid enters the control volume at some condition and leaves the control volume at a different condition. The variables relevant to energy balance are indicated at the inlet and exit conditions of the control volume shown in Figure 7.1. The rate of heat input to the control volume is $\dot{Q}$. The control volume is producing shaft work at a rate of $\dot{W}_{sh}$.

Referring to energy balance given in Equation (7.4), the various terms in energy balance equation can be written as:

$$\begin{pmatrix} \textit{Rate at} \\ \textit{which energy} \\ \textit{enters CV} \end{pmatrix} = \dot{m}_1\left[u_1 + \frac{1}{2}V_1^2 + gZ_1\right] + \dot{Q} + (P_1A_1V_1) \tag{7.17}$$

$$\begin{pmatrix} \textit{Rate at} \\ \textit{which energy} \\ \textit{leaves CV} \end{pmatrix} = \dot{m}_2\left[u_2 + \frac{1}{2}V_2^2 + gZ_2\right] + \dot{W}_{sh} + (P_2A_2V_2) \tag{7.18}$$

$$\begin{pmatrix} \textit{Rate of} \\ \textit{accumulation of} \\ \textit{energy inside CV} \end{pmatrix} = \frac{dE_{CV}}{dt} = \frac{d(mu)_{CV}}{dt} \tag{7.19}$$

where variables u, V, Z, A, P refer to specific internal energy, fluid velocity, elevation, cross-sectional area of flow, and pressure, respectively. Note that the last term on the right hand side of Equations (7.17) and (7.18), that is, the PAV term, is the flow work (work needed to push the fluid into and out of the control volume).

Thus the energy balance, Equation (7.4), becomes:

$$\left\{\dot{m}_1\left[u_1+\frac{1}{2}V_1^2+gZ_1\right]+\dot{Q}+(P_1A_1V_1)\right\}-\left\{\dot{m}_2\left[u_2+\frac{1}{2}V_2^2+gZ_2\right]+\dot{W}_{sh}+(P_2A_2V_2)\right\}=\frac{d(mu)_{CV}}{dt} \quad (7.20)$$

In this energy balance equation, $\dot{Q}$ is positive if it enters the control volume and negative if it leaves the control volume. $\dot{W}_{sh}$ is positive if it flows out of the control volume and negative if it flows into the control volume.

Assuming steady state, Equation (7.20) can be written as:

$$\dot{m}\left[(u_2-u_1)+\frac{1}{2}(V_2^2-V_1^2)+g(Z_2-Z_1)+\left(\frac{P_2}{\rho_2}-\frac{P_1}{\rho_1}\right)\right]=\dot{Q}-\dot{W}_{sh} \quad (7.21)$$

Making use of the definition of specific enthalpy: $h=u+P/\rho$, Equation (7.21) can be expressed as:

$$\dot{m}\left[(h_2-h_1)+\frac{1}{2}(V_2^2-V_1^2)+g(Z_2-Z_1)\right]=\dot{Q}-\dot{W}_{sh} \quad (7.22)$$

In the absence of shaft work and negligible changes in kinetic and potential energies, Equation (7.22) simplifies to:

$$\dot{m}[(h_2-h_1)]=\dot{Q} \quad (7.23)$$

For multiple inlets and outlets, Equation (7.23) can be generalized as follows:

$$\Sigma_{INLETS}\ (\dot{m}_i\, h_i)+\Sigma\,\dot{Q}=\Sigma_{OUTLETS}\ (\dot{m}_i\, h_i) \quad (7.24)$$

where $\Sigma\,\dot{Q}$ is the total rate of heat input to the control volume (heat may be added at different locations of the control surface). In the absence of heat exchange, energy balance reduces to enthalpy balance:

$$\Sigma_{INLETS}\ (\dot{m}_i\, h_i)=\Sigma_{OUTLETS}\ (\dot{m}_i\, h_i) \quad (7.25)$$

This equation is valid under the restrictions: steady state, negligible kinetic and potential energy changes, no shaft work and no heat exchange.

Equations (7.24)–(7.25) could also be written in terms of molar flow rates and molar enthalpies as:

$$\Sigma_{INLETS}\ (F_i\, h_i)+\Sigma\,\dot{Q}=\Sigma_{OUTLETS}\ (F_i\, h_i) \quad (7.26)$$

$$\Sigma_{INLETS}\ (F_i\, h_i)=\Sigma_{OUTLETS}\ (F_i\, h_i) \quad (7.27)$$

where F_i is the total molar flow rate of the fluid stream at any given inlet or outlet and h_i is the molar enthalpy of the fluid stream. Equation (7.27) assumes that no heat transfer is involved between the control volume and its surroundings.

As an example of energy balance, consider a distillation column shown in Figure 7.3. The values of various quantities are indicated in the figure itself. We are required to calculate the condenser and reboiler heat duties (that is, the heat transfer rates $\dot{Q}_C$ and $\dot{Q}_R$). This problem is an extension of mass balances carried out earlier in section 7.3, see Figure 7.2.

Using energy balance with multiple inlets and outlets, Equation (7.24), one can write:

$$\dot{m}_1 h_1 + \dot{Q}_C + \dot{Q}_R = \dot{m}_2 h_2 + \dot{m}_3 h_3 \tag{7.28}$$

The only unknowns in this equation are the condenser and reboiler heat duties. We need another energy balance in order to calculate the two unknowns. Now we do energy balance around the condenser:

$$(10 + 30) \times 350 + \dot{Q}_C = (10 + 30) \times 50 \tag{7.29}$$

Note that the vapor flow rate into the condenser is 40 kg/min and the liquid reflux specific enthalpy is the same as that of liquid distillate product. From Equation (7.29), the condenser heat duty is $\dot{Q}_C = -12{,}000\ kcal/min$. A negative value of heat transfer rate indicates that heat is rejected by condenser, that is, heat flows out of the control volume (condenser). The only unknown left in Equation (7.28) is the reboiler heat duty. Thus:

$$\dot{Q}_R = \dot{m}_2 h_2 + \dot{m}_3 h_3 - \dot{m}_1 h_1 - \dot{Q}_C$$

$$= (10 \times 50) + (90 \times 100) - (100 \times 25) - (-12000)$$

$$= 19{,}000\ kcal/min \tag{7.30}$$

The sign of reboiler heat transfer rate is positive as this heat flows into the process control volume.

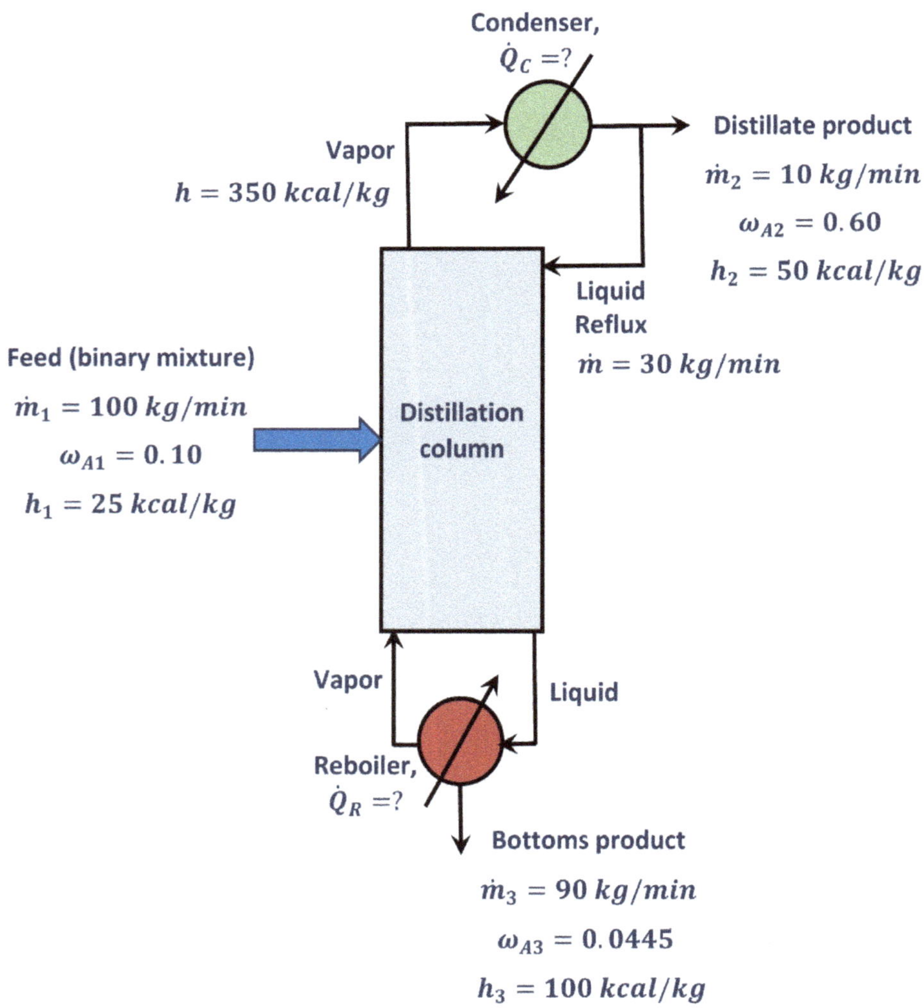

Figure 7.3 External energy balance on a distillation column.

7.5 ESTIMATION OF ENTHALPY OF FLUIDS

As noted in the preceding section, the enthalpies of fluids (single-component and multi-component vapor and liquid phases) are needed in order to carry out the energy balances. First we discuss how to estimate the enthalpies of pure single-component materials and then deal with multi-component gas mixtures and liquid solutions.

7.5.1 Pure Single-Component Substances

It is not possible to calculate the absolute values of enthalpies of materials. The absolute values of enthalpy of materials given in charts and tables in handbooks and textbooks are always relative to some *reference state*. The reference state of a material is any convenient state, selected arbitrarily. The enthalpy of the material is arbitrarily set to zero in the reference state. To define the reference state, we must specify reference temperature (T_R), reference pressure (P_R), and actual state of aggregation of material (gas, liquid or solid) at reference conditions.

Example:

As an example, we consider water. We can take the reference state as " $T_R = 20^oC, P_R = 1\ bar, h_R = 0$, subcooled liquid water". Given this reference state, we can calculate the absolute value of enthalpy of superheated vapor state at a specified set of T and P values (say, $T = 300^oC$ and $P = 1\ bar$). The computational path for the calculation of absolute value of enthalpy of superheated vapor at T and P is shown in Figure 7.4 given below.

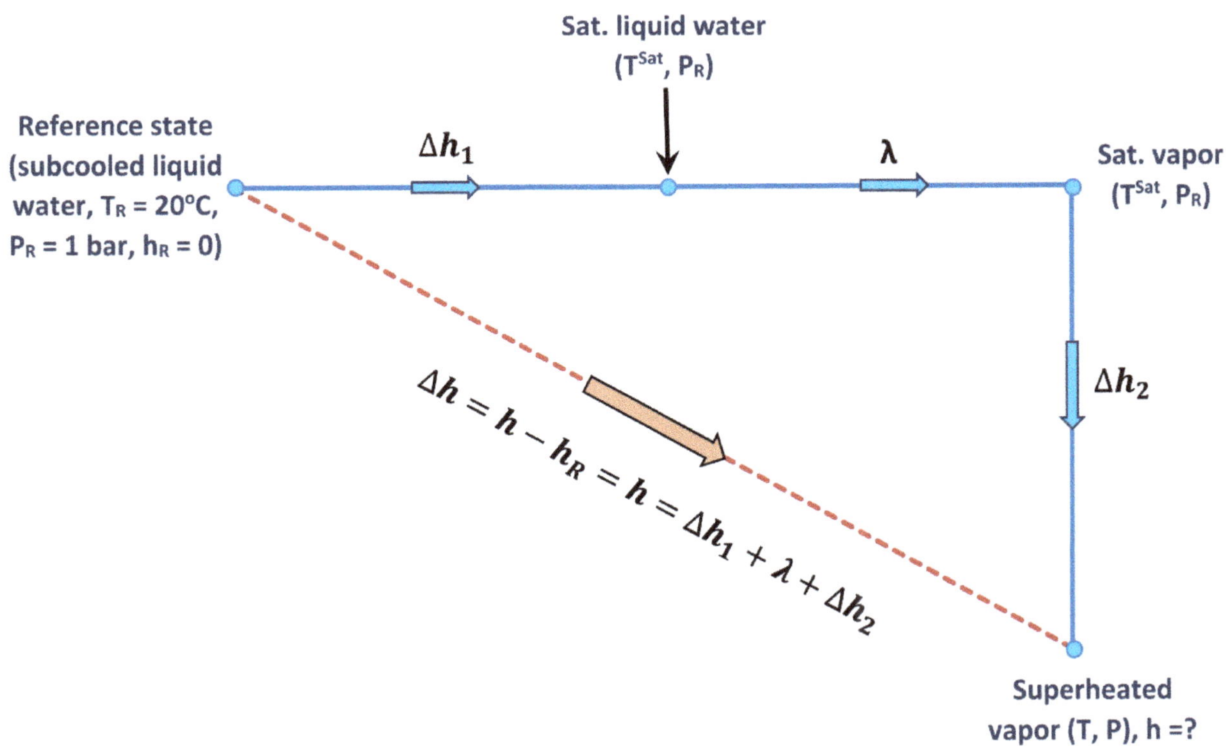

Figure 7.4 Computational path to estimate the absolute value of enthalpy of superheated steam at T and P.

As enthalpy is a state property, the change in enthalpy is the same for different paths connecting the same initial and final states. Therefore,

$$\Delta h = \Delta h_1 + \lambda + \Delta h_2 \tag{7.31}$$

$$\Delta h = h - h_R = h \tag{7.32}$$

$$h = \Delta h_1 + \lambda + \Delta h_2 \tag{7.33}$$

where λ is the latent heat of vaporization. Equation (7.33) gives the absolute value of enthalpy of superheated steam at T and P. Equation (7.33) can be re-written as:

$$h = C_{pL}(T^{Sat} - T_R) + \lambda + C_{pG}(T - T^{Sat}) \tag{7.34}$$

where C_{pL} is constant pressure heat capacity of liquid and C_{pG} is constant pressure heat capacity of vapor (gas). In Equation (7.34), it is assumed that vapor is an ideal gas and consequently the pressure-dependence of enthalpy is negligible. Thus, we need constant pressure heat capacities of liquid and vapor phases and latent heat of vaporization λ to estimate the absolute enthalpy of superheated vapor.

Estimation of Latent Heat of Vaporization (λ):

The *Trouton rule* can be used to make a rough estimation of λ_n at normal boiling point (P^{Sat} = 1 atm). According to the Trouton rule, the entropy change of vaporization at normal boiling point ΔS_n^{Vap} is almost the same for different substances:

$$\frac{\Delta S_n^{Vap}}{R} \approx 10.5 \tag{7.35}$$

where R is the universal gas constant (8.314 J/mol.K). Since,

$$\Delta S_n^{Vap} = \frac{\lambda_n}{T_n} \tag{7.36}$$

it follows that:

$$\frac{\lambda_n}{RT_n} \approx 10.5 \tag{7.37}$$

$$\lambda_n \approx 10.5(RT_n) \tag{7.38}$$

A more accurate prediction of λ_n is given by the empirical Riedel equation:

$$\frac{\lambda_n}{RT_n} = \frac{1.092(lnP_c - 1.013)}{0.93 - T_{rn}} \tag{7.39}$$

where P_c is the critical pressure of the substance in bars and T_{rn} is the reduced normal boiling point, defined as:

$$T_{rn} = \frac{T_n}{T_c} \tag{7.40}$$

where T_c is the absolute critical temperature of the substance and T_n is the absolute normal boiling point. Once the latent heat of vaporization at normal boiling point is determined, the value of λ at any other temperature (reduced temperature T_r) can be estimated from the Watson equation:

$$\frac{\lambda}{\lambda_n} = \left[\frac{1 - T_r}{1 - T_{rn}}\right]^{0.38} \tag{7.41}$$

7.5.2 Multi-Component Gas Mixtures and Liquid Solutions

The enthalpy of multicomponent mixtures (gas mixture or liquid solution) is given as:

$$h(T,P) = \sum x_j \, \bar{h}_j(T,P) \tag{7.42}$$

where $h(T,P)$ is the molar enthalpy of mixture and $\bar{h}_j(T,P)$ is the partial molar enthalpy of any species j in the solution, all at the same T and P, and x_j is the mole fraction of species "*j*". For *ideal solutions* (ideal gas mixture or ideal liquid solution), the partial molar enthalpy of a species in solution is equal to the molar enthalpy of pure species at the same T and P:

$$\bar{h}_j^{id}(T,P) = h_j(T,P) \tag{7.43}$$

Note that there is no bar on top of a pure species molar enthalpy. From Equations (7.42) and (7.43), it follows that for *ideal solutions*:

$$h^{id}(T,P) = \sum x_j \, h_j(T,P) \tag{7.44}$$

When individual components are mixed at constant T and P and the solution/mixture formed is ideal in nature, the heat of mixing is zero, that is, no heat is released or absorbed. The heat of mixing (enthalpy of mixing) of different gases or different liquids is defined as:

$$\Delta h_{mixing} = h(T,P) - \sum x_j \, h_j(T,P) \tag{7.45}$$

This is the general definition of heat of mixing, valid for both ideal and non-ideal solutions. In the case of ideal solutions:

$$\Delta h_{mixing} = h^{id}(T,P) - \sum x_j \, h_j(T,P) \tag{7.46}$$

From Equations (7.44) and (7.46), it follows that:

$$\Delta h_{mixing} = 0 \tag{7.47}$$

As an example, Figure 7.5 shows mixing of three chemical species, all at the same T and P, to form one mole of an ideal solution at the same T and P. In this case, Equation (7.46) gives:

$$\Delta h_{mixing} = h^{id}(T,P) - [x_1 h_1(T,P) + x_2 h_2(T,P) + x_3 h_3(T,P)] \tag{7.48}$$

The enthalpy of ideal solution formed is given by Equation (7.44) as:

$$h^{id}(T,P) = [x_1 h_1(T,P) + x_2 h_2(T,P) + x_3 h_3(T,P)] \tag{7.49}$$

From Equations (7.48) and (7.49), it follows that:

$$\Delta h_{mixing} = 0 \tag{7.50}$$

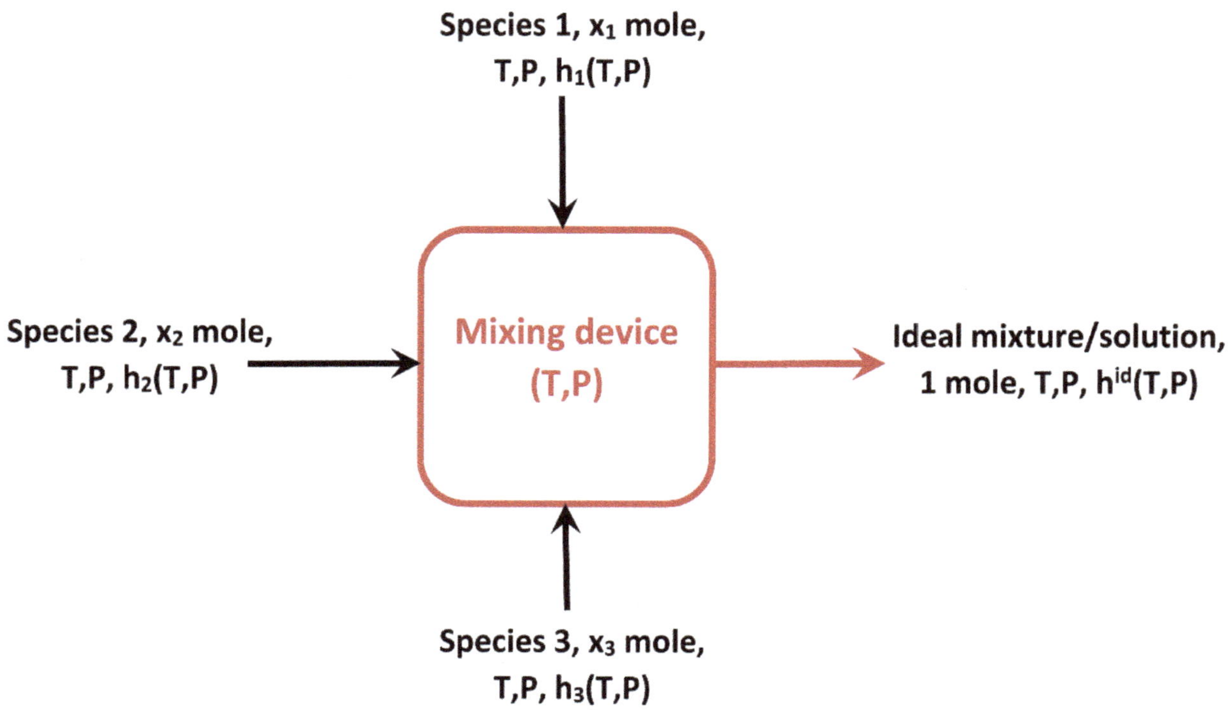

Figure 7.5 Mixing of three different chemical species to form an ideal solution/mixture, all at the same T and P.

NOTATION

A	cross-sectional area of opening (inlet/outlet) of control volume (CV)
C_{PG}	constant pressure molar heat capacity of gas
C_{PL}	constant pressure molar heat capacity of liquid
E_{CV}	total energy inside the control volume
F_i	molar flow rate of fluid stream at *i*th opening (inlet/outlet) of control volume
g	acceleration due to gravity
h	specific enthalpy (per unit mass or unit mole)
h_i	specific enthalpy of fluid stream (per unit mass or unit mole) at *i*th opening (inlet/outlet) of control volume (CV)
h_j	molar enthalpy of pure species *j*
h_R	molar enthalpy of pure species in reference state
h^{id}	molar enthalpy of ideal gas mixture or ideal liquid solution
$\bar{h}_j$	partial molar enthalpy of species *j* in a gas mixture or liquid solution
$\bar{h}_j^{id}$	partial molar enthalpy of species *j* in an ideal gas mixture or ideal liquid solution
m_{CV}	mass inside the control volume
$\dot{m}$	mass flow rate
$\dot{m}_i$	mass flow rate at *i*th opening (inlet/outlet) of control volume (CV)
P	pressure
P_C	critical pressure
P_R	reference state pressure for enthalpy calculations
$\dot{Q}$	heat transfer rate
$\dot{Q}_C$	condenser heat duty (heat transfer rate in condenser)
$\dot{Q}_R$	reboiler heat duty (heat transfer rate in reboiler)
R	universal gas constant

T	temperature
T^{Sat}	saturation temperature
T_C	absolute critical temperature
T_n	normal boiling point temperature
T_r	reduced temperature ($= T/T_C$)
T_R	reference state temperature for enthalpy calculations
T_{rn}	reduced normal boiling point temperature ($= T_n/T_C$)
u	specific internal energy (per unit mass)
V	velocity of fluid stream
$\dot{W}_{sh}$	rate of shaft work
x_j	mole fraction of species *j* in gas mixture or liquid solution
x_{ji}	mole fraction of species *j* in the fluid stream at *i*th opening (inlet/outlet) of control volume (CV)
Z	elevation of the opening (inlet/outlet) of the control volume

Greek Letters

Δh_{mixing}	enthalpy of mixing at same *T* and *P*
ΔS_n^{vap}	molar entropy of vaporization at normal boiling point
λ	molar latent heat of vaporization
λ_n	molar latent heat of vaporization at normal boiling point
ρ	density of fluid
ω_{ji}	mass fraction of species *j* in the fluid stream at *i*th opening (inlet/outlet) of control volume (CV)

TUTORIAL PROBLEMS WITH SOLUTION

Problem 1

Under steady state condition, the fluid velocities in pipe sections A and C are 3 m/s and 7 m/s, respectively. What is the fluid velocity in section B? The diameters of the pipe sections are: $D_A = D_B = 20$ mm and $D_C = 15$ mm.

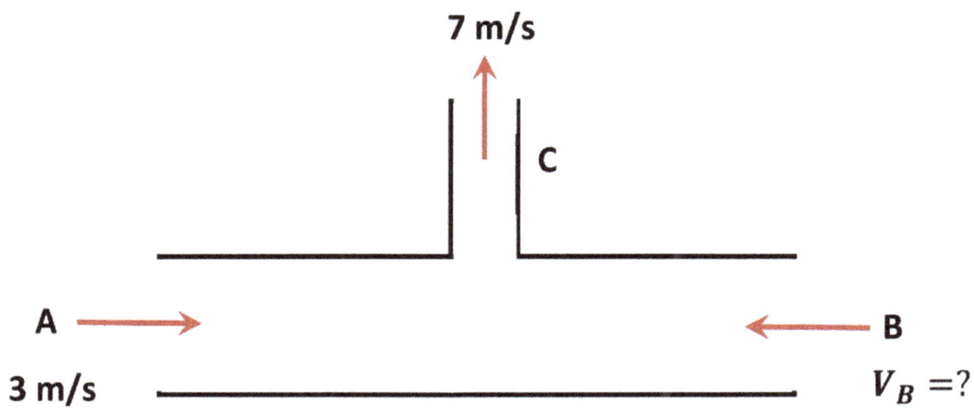

Solution

From mass balance, Equation (7.7), we get:

$$\sum_{outlets} \dot{m}_i = \sum_{inlets} \dot{m}_i$$

$$\dot{m}_C = \dot{m}_A + \dot{m}_B$$

The mass flow rate can also be expressed as:

$$\dot{m} = \rho V A = \rho V \left(\frac{\pi}{4} D^2\right)$$

where ρ is fluid density, V is fluid velocity, and D is pipe diameter. Thus the mass balance gives:

$$V_C D_C^2 = V_A D_A^2 + V_B D_B^2$$

$$V_B = V_C \left(\frac{D_C}{D_B}\right)^2 - V_A \left(\frac{D_A}{D_B}\right)^2 = 7\left(\frac{15}{20}\right)^2 - 3\left(\frac{20}{20}\right)^2 = 0.9375 \text{ m/s}$$

Problem 2

A tank of volume 2 m^3 is being filled with air. The air enters the tank through a pipe of diameter 15 mm at a velocity of 10 m/s. The temperature and pressure of air at the entrance to the tank are: 25°C and 100 kPa.

a) Determine the rate of air mass accumulation inside the tank
b) Determine the rate at which the density of air inside the tank is increasing

Solution

a) Mass balance over the tank volume (control volume CV) can be expressed as:

$$\frac{d(m)_{CV}}{dt} = \dot{m}_{in}$$

$$\dot{m}_{in} = \rho V A = \rho V \left(\frac{\pi}{4} D^2\right)$$

Assuming air to be an ideal gas, the density of air at the entrance to the tank is:

$$\rho = P/RT = 100{,}000/(286.9 \times 298.15) = 1.169 \text{ kg/m}^3$$

Note that $R = 286.9$ J/kg.K for air.

Thus the rate of air mass accumulation inside the tank is:

$$\frac{d(m)_{CV}}{dt} = \rho V \left(\frac{\pi}{4} D^2\right) = 1.169 \times 10 \left(\frac{\pi}{4}\right)\left(\frac{15}{1000}\right)^2 = 2.0658 \times 10^{-3} \text{ kg/s}$$

b) The rate at which the density of air inside the tank is increasing can be determined as follows:

$$\frac{d(m)_{CV}}{dt} = (\text{volume of tank})\frac{d\rho}{dt} = 2.0658 \times 10^{-3} \text{ kg/s}$$

$$\frac{d\rho}{dt} = \frac{2.0658 \times 10^{-3}}{2} = 1.0329 \times 10^{-3} \text{ kg/m}^3\text{.s}$$

Problem 3

Water flows through a horizontal tube heated from the outside by hot gas. The water temperature is increased from 80 °C to 120 °C. Neglecting kinetic energy changes, determine the amount of heat transferred per unit mass of water. Given that $h_1 = 340\ kJ/kg$ and $h_2 = 2700\ kJ/kg$

Solution

The process is shown in the figure below. Using energy balance, Equation (7.23), we can calculate the heat transferred on a unit mass basis as:

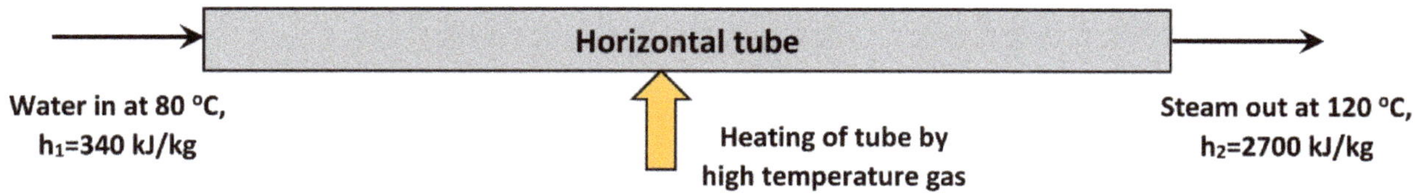

$$Q = h_2 - h_1 = 2700 - 340 = 2360\ kJ/kg$$

Problem 4

Water stream at 20 °C (h_1=84 kJ/kg) is mixed adiabatically with saturated steam stream at 140 °C (h_2=2730 kJ/kg) to form a water stream at 65 °C (h_3=270 kJ/kg) flowing at a mass flow rate of 5 kg/s. What are the mass flow rates of water and saturated steam streams entering the mixing device?

Solution

The process is shown in the figure below.

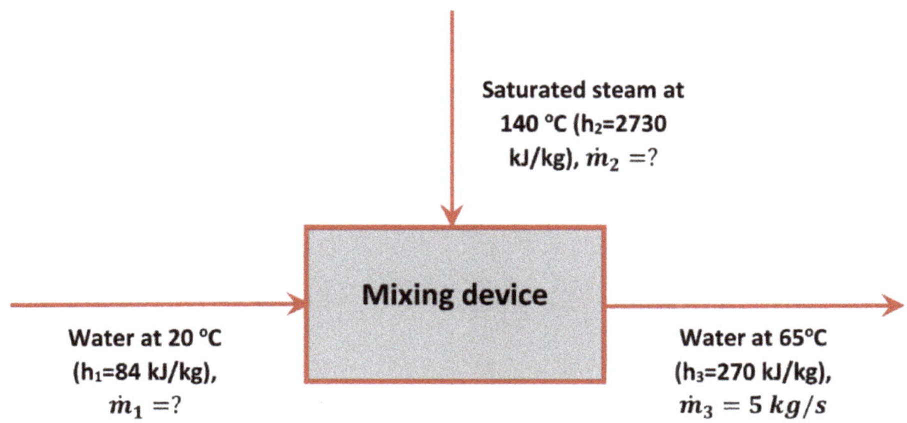

Energy balance, Equation (7.25), gives:

$$\sum_{outlets} \dot{m}_i\, h_i - \sum_{inlets} \dot{m}_i\, h_i = 0$$

$$\dot{m}_3 h_3 - (\dot{m}_1 h_1 + \dot{m}_2 h_2) = 0$$

From mass balance, Equation (7.7), we get:

$$\sum_{outlets} \dot{m}_i = \sum_{inlets} \dot{m}_i$$

$$\dot{m}_3 = \dot{m}_1 + \dot{m}_2 = 5$$

The energy balance can now be written as:

$$5(270) - \dot{m}_1(84) - (5 - \dot{m}_1)(2730) = 0$$

Solution of this equation gives:

$$\dot{m}_1 = 4.65\ kg/s$$

The mass flow rate $\dot{m}_2 = 0.35\ kg/s$.

Problem 5

A liquid flows through a horizontal heating coil which is heated from the outside by superheated steam. The inlet velocity of liquid is 2.6 m/s. As the liquid passes through the coil, it undergoes phase change from liquid to vapor. The fluid leaving the coil is superheated vapor and its velocity is 165 m/s. What is the heat transferred to the fluid per unit mass of fluid? The enthalpies of the fluid at the inlet and outlet are given as: $h_1 = 245$ kJ/kg and $h_2 = 2545$ kJ/kg.

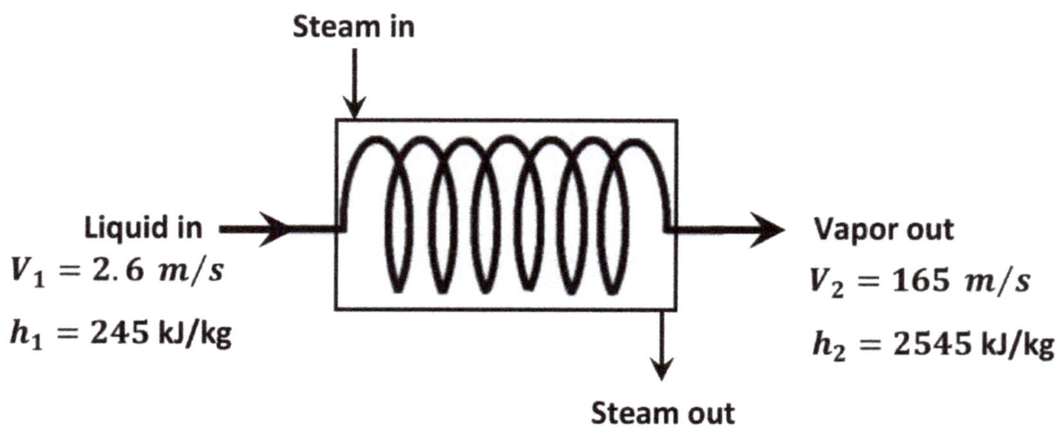

Solution

Energy balance, Equation (7.22), on a unit mass basis gives:

$$\left[\Delta h + \frac{1}{2}\Delta V^2 + g\Delta Z\right] = Q - W_{sh}$$

As $W_{sh} = 0$, and the system is horizontal $\Delta Z = 0$, we get:

$$Q = \Delta h + \frac{1}{2}(V_2^2 - V_1^2) = (2545 - 245) \times 10^3 + \frac{1}{2}(165^2 - 2.6^2) = 2300000 + 13609.12$$
$$= 2313609.12\,\text{J/kg} = 2313.61\,\text{kJ/kg}$$

SHORT REVIEW QUESTIONS

1. Under what conditions is mole balance equivalent to mass balance?
2. From a fundamental point of view, is the mass conserved or the moles conserved?
3. Under what conditions is the enthalpy balance equivalent to energy balance for flow systems?
4. What must be specified in order to define the reference state of a substance?
5. Explain Trouton's rule
6. For ideal solutions, the partial molar enthalpy of a species in solution is equal to the molar enthalpy of pure species at the same T and P. Is this true?

7. The enthalpy of mixing of different gases at the same T and P is zero provided that the mixture is ideal. Is this true?

PRACTICE PROBLEMS

Problem 1

A fluid having a density of 850 kg/m^3 is flowing through a pipe junction shown in the figure below. The pipes are of the same diameter, 5 cm. The mass flow rates of fluid entering the junction are: $\dot{m}_A = 1.5$ kg/s and $\dot{m}_D = 2.0$ kg/s. The velocity in pipe C is 1 m/s. Calculate the mass flow rate in pipe B. You can assume steady state.

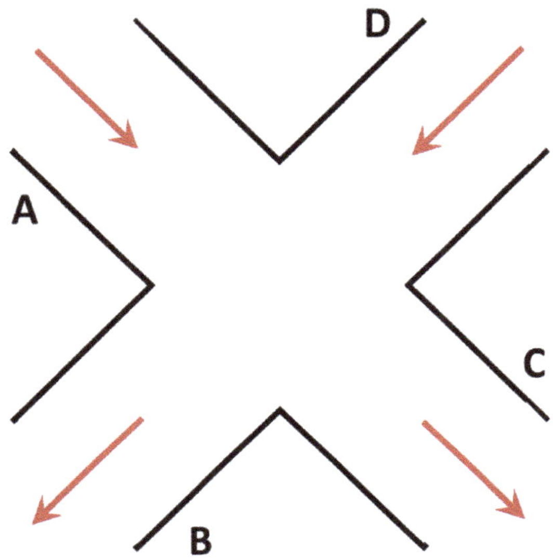

Problem 2

Air is heated in a heater as shown below. Assuming steady state, calculate the discharge velocity of air. Air can be treated as an ideal gas.

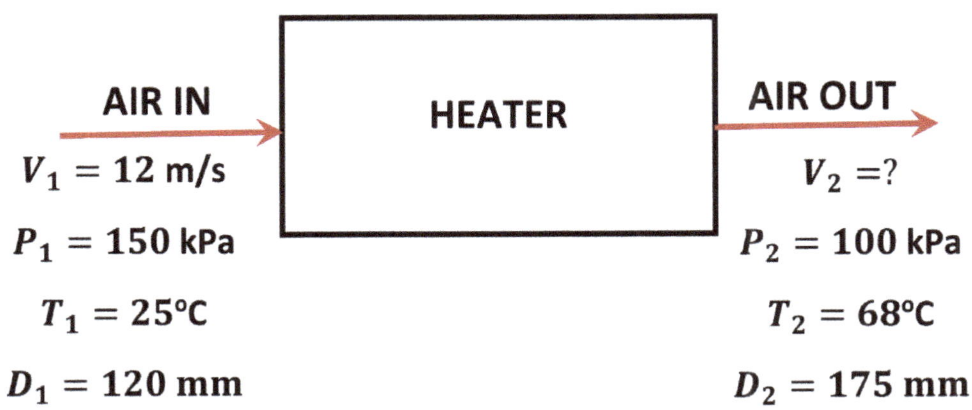

Problem 3

Water is being added to a cylindrical tank of diameter 9 ft at a flow rate of 9 ft^3/min. The tank is initially empty. Water is also leaving the tank from an opening at the bottom of the tank at a flow rate that is dependent on the height of liquid in the tank as:

$$Q_{out} = 2.5\ h\ \text{ft}^3/\text{min}$$

where h is the height of liquid in the tank in ft. Develop the expression for the height of liquid in the tank as a function of time. What is the steady state height of liquid in the tank?

Problem 4

Water stream at 40 °C (h_1=170 kJ/kg) is mixed adiabatically with superheated steam stream at 325 °C (h_2=3063 kJ/kg) to form a saturated vapor stream (h_3=2801 kJ/kg) flowing at a mass flow rate of 9 kg/s. What are the mass flow rates of water and superheated steams entering the mixing device?

Problem 5

Estimate the enthalpy of pure benzene in superheated vapor state at 100 °C and 1 atm pressure with respect to the following reference state: "$T_R = 80.1^oC$, $P_R = 1$ atm, $h_R = 0$, saturated liquid benzene". Given that: $\Delta H^{vap} = 30{,}820$ kJ/kmol at normal boiling point of 80.1 °C, $C_{pG} = 96.3$ kJ/kmol. K.

Chapter 8

Single-Stage Flash Distillation

8.1 LEARNING OUTCOMES

Upon completion of this chapter, students will be able to:

- Explain flash distillation
- Carry out mass and energy balances over a flash distillation unit
- Analyze binary flash distillation and obtain solutions graphically and analytically
- Analyze multicomponent flash distillation
- Solve Rachford-Rice flash equation numerically

8.2 INTRODUCTION

Flash distillation is a single-stage continuous distillation process shown schematically in Figure 8.1. Liquid feed is pumped through a heater to raise its temperature. Liquid at high temperature and pressure is passed through a throttling valve where a sharp reduction in pressure occurs. The throttling valve is located right at the entry into a flash drum. A sharp reduction in pressure of a hot liquid causes flashing (rapid vaporization) in the flash drum. The vapor enriched in more volatile components is taken off overhead while the liquid enriched in less volatile components is withdrawn from the bottom of the drum. Assuming that vapor and liquid leaving the flash drum are at equilibrium with each other, the flash drum can be treated as a single equilibrium stage.

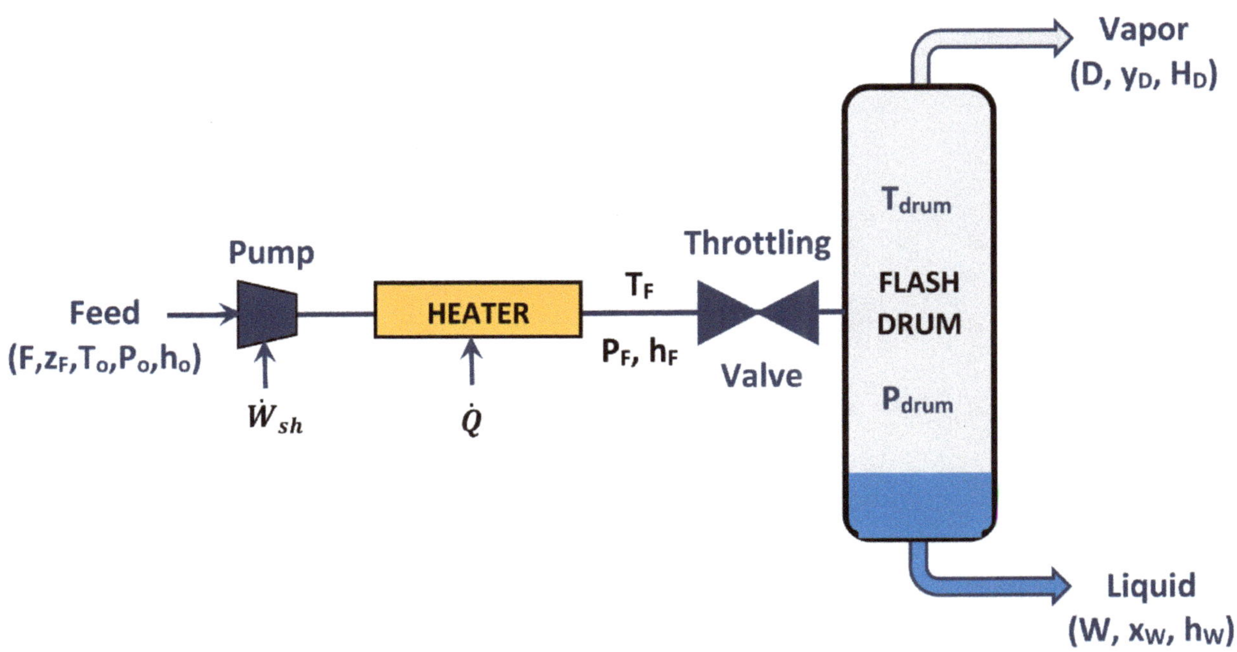

Figure 8.1 Flash distillation process (feed is assumed to be a binary mixture here).

Flash distillation is used widely in petroleum refining operations. The solute recovered in an absorption process is often separated from solvent using flash distillation. Flash distillation is also used extensively in desalination of sea water. In many applications, flash distillation is used as an auxiliary operation to "pre-flash" the feed stream before further processing and separation.

The analysis of flash distillation process generally involves mass balances (component and overall), vapor-liquid equilibrium relation, and energy balances. We will first consider binary systems and then multicomponent systems.

8.3 ANALYSIS OF BINARY FLASH DISTILLATION

Referring to Figure 8.1, the binary flash distillation problems often encountered are as follows:

1) Given feed variables (F, z_F, T_o, P_o), P_{drum}, and fraction of feed that vaporizes ($\psi = D/F$), determine the vapor-distillate and bottoms-liquid flow rates and compositions (D, y_D, W, x_W). Also determine T_{drum}, H_D, h_W, h_F, h_o, T_F, P_F, and $\dot{Q}$. We will refer to this problem as "$P, \psi - Flash$" calculation where P is the flash drum pressure and ψ is the fraction of feed that vaporizes in the flash drum.

2) Given feed variables (F, z_F, T_o, P_o), P_{drum}, and T_{drum}, determine the vapor-distillate and bottoms-liquid flow rates and compositions (D, y_D, W, x_W). Also determine H_D, h_W, h_F, h_o, T_F, P_F, and $\dot{Q}$. This type of problem is often referred to as "$P, T - Flash$" calculation where P and T are pressure and temperature, respectively in the flash drum.

In Figure 8.1, F is the feed flow rate, z_F is the mole fraction of more volatile component in the feed, T_o and P_o are the temperature and pressure of feed before the pump, T_F and P_F are the temperature and pressure of feed after the heater, D is the flow rate of vapor-distillate product, y_D is the mole fraction of more volatile component in the vapor-distillate product, W is the flow rate of bottoms-liquid, x_W is the mole fraction of more volatile component in the bottoms-liquid, H_D is the specific enthalpy of vapor product, h_o, h_F, and h_W are the specific enthalpies of liquid phase at different locations of the process flow sheet (see Figure 8.1), $\dot{Q}$ is the rate of heat transfer in the heater, and $\dot{W}_{sh}$ is the shaft work of the pump.

8.3.1 P, ψ −Flash Distillation

Mass Balances

Referring to Figure 8.1, the overall and component material balances can be written as:

$$F = D + W \tag{8.1}$$

$$z_F F = y_D D + x_W W \tag{8.2}$$

From Equations (8.1) and (8.2), it follows that:

$$y_D = -\left(\frac{1-\psi}{\psi}\right) x_W + \frac{z_F}{\psi} \tag{8.3}$$

where $\psi = D/F$. Equation (8.3) represents the operating line for flash distillation. It relates the composition of the vapor stream to the composition of the bottoms-liquid stream leaving the flash drum. The operating line passes through the feed point (z_F, z_F) where $y = x = z_F$ and has a y-intercept of z_F/ψ. The slope of the operating line is $-W/D\ (= -\ (1-\psi)/\psi)$. Note that the operating line is based on just material balance. We need another relation in order to solve for the compositions of the vapor and liquid streams leaving the flash drum. This other relation is the equilibrium relation. The vapor and liquid streams leaving the flash drum are at equilibrium with each other.

Graphical Solution

Figure 8.2 shows the graphical solution of a P, ψ – flash problem. The y-x equilibrium curve is plotted at the specified drum pressure P_{drum}. The operating line passes through the feed point $y = x = z_F$ ($z_F = 0.60$ in the present example) and has a slope of $-(1-\psi)/\psi$ ($\psi = 0.40$ in the present example). When $y = 1$, the operating line gives $x = 0.333$. Thus, the operating line passes through the point (0.333, 1). The intersection of the operating line with the equilibrium curve gives the compositions of the vapor and liquid streams leaving the flash drum. In the present example, $y_D \approx 0.85$ and $x_W \approx 0.43$. Note that this P, ψ – flash problem is similar to the single-stage stripping problem (Chapter 3) in that the operating lines in both cases relate the compositions of the two discharge streams. Also the slope of the operating lines in both cases is negative, $-W/D$ or $-L/V$ where L refers to liquid flow rate and V refers to gas flow rate, and the operating lines fall below the equilibrium curve.

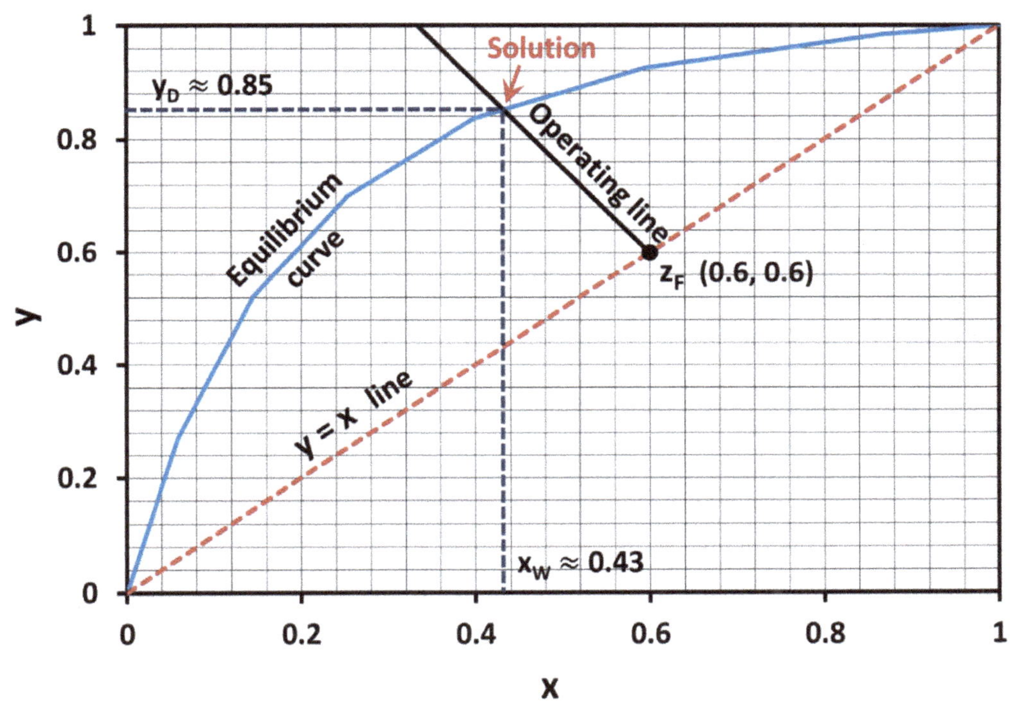

Figure 8.2 Graphical solution of P, ψ – flash problem.

Once the compositions of the vapor and liquid streams leaving the flash drum are known, we can determine the flash drum temperature, T_{drum}. There are 3 ways to determine T_{drum} : a) use T-x-y diagram at the flash drum pressure; b) given P_{drum} and composition of bottoms-liquid stream leaving the flash drum, estimate the bubble point temperature at liquid composition; and c) given P_{drum} and composition of vapor-distillate stream leaving the flash drum, estimate the dew point temperature at vapor composition. Figure 8.3 illustrates the usage of T-x-y diagram to determine T_{drum}. The horizontal line connecting the compositions of the vapor and liquid streams gives T_{drum}.

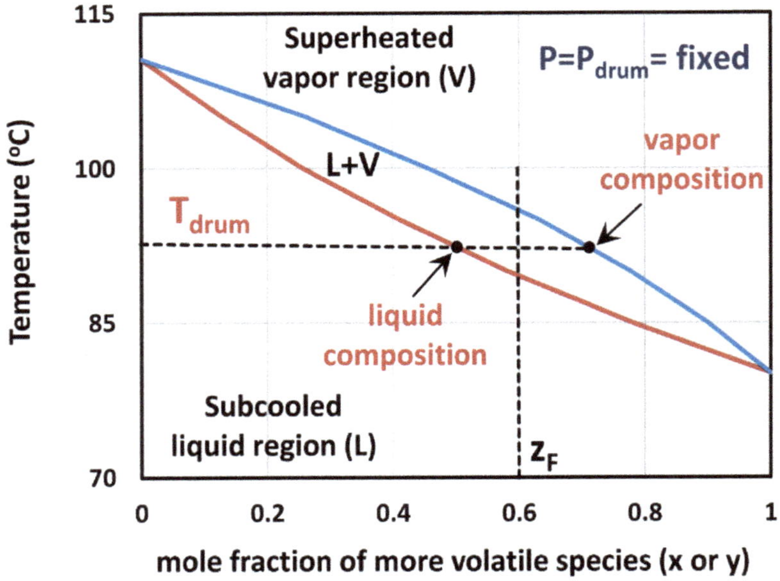

Figure 8.3 Determination of flash temperature from known values of vapor and liquid compositions.

Figure 8.4 shows the limits of operation of P, ψ – flash distillation. When $\psi = 0$, the slope of the operating line, given as $-(1-\psi)/\psi$, is $-\infty$ and therefore, the operating line is simply a vertical line passing through the feed point. In this case, the vapor-distillate flow rate is zero ($D = F\psi = 0$) and all the feed goes to the bottoms-liquid stream ($W = F(1-\psi) = F$). The composition of the trace amount of vapor produced in

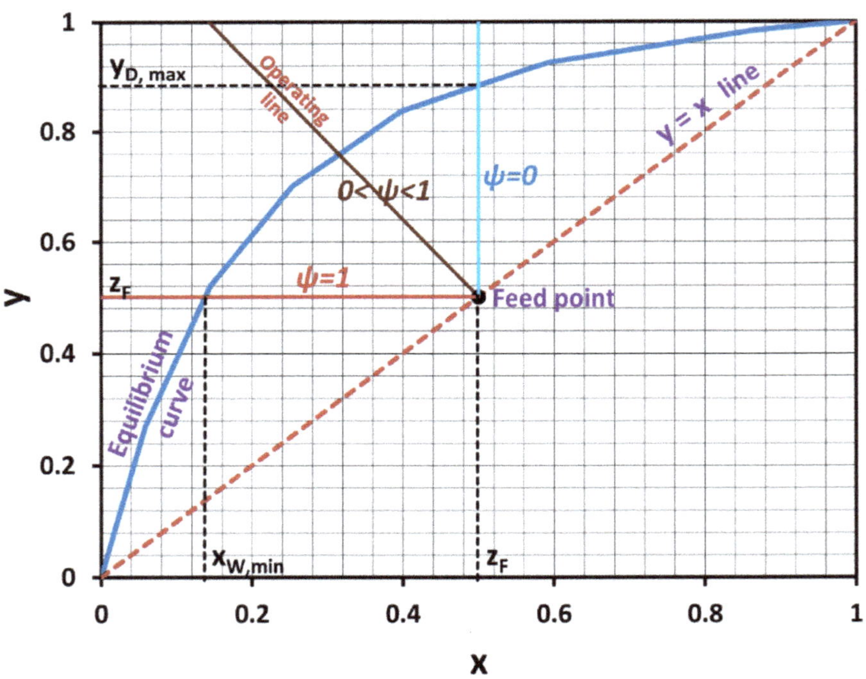

Figure 8.4 Limits of operation of P, ψ – flash (binary) distillation.

this flashing is given by the intersection of the operating line with the equilibrium curve as shown in Figure 8.4. The mole fraction of more volatile species in this trace amount of vapor is the maximum possible ($y_{D,max}$) for the given feed composition. This kind of flashing where $\psi = 0$ occurs when flashing is done at the bubble point temperature of the feed at flash drum pressure, that is, $T_{drum} = BUBLT$. As illustrated in Figure 8.5, the amount of vapor produced (although very rich in more volatile component) is negligible when the flash drum temperature is equal to the bubble point temperature of the feed at the given flash drum pressure (note that one can apply the lever rule to determine the amounts of vapor and liquid in a two-phase mixture at equilibrium).

When $\psi = 1$, the slope of the operating line is zero and therefore, the operating line is a horizontal line passing through the feed point. In this case, all the feed goes to the vapor-distillate stream ($D = F\psi = F$) and bottoms-liquid stream flow rate is zero ($W = F(1-\psi) = 0$). The composition of the trace amount of liquid produced in this flashing is given by the intersection of the operating line with the equilibrium curve as shown in Figure 8.4. The mole fraction of more volatile species in this trace amount of liquid is the minimum possible ($x_{W,min}$) for the given feed composition. This kind of flashing where $\psi = 1$ occurs when flashing is done at the dew point temperature of the feed at flash drum pressure, that is, $T_{drum} = DEWT$. As illustrated in Figure 8.6, the amount of liquid produced (although very lean in more volatile component) is negligible when the flash drum temperature is equal to the dew point temperature of the feed at the given flash drum pressure.

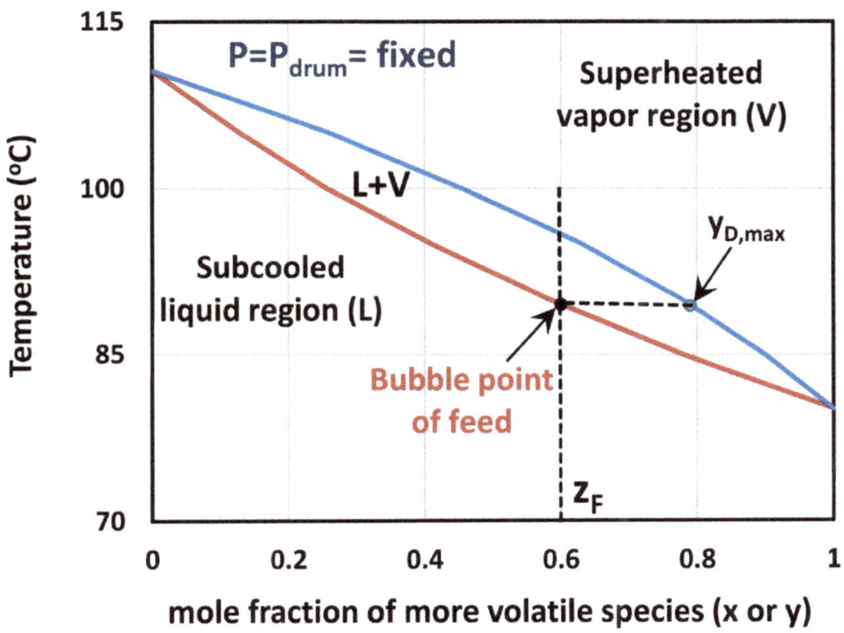

Figure 8.5 Bubble point temperature of feed at flash drum pressure. When $T_{drum} = BUBLT$ of feed, all the feed goes to the bottoms-liquid stream ($\psi = 0$). The composition of trace amount of vapor is $y_{D,max}$.

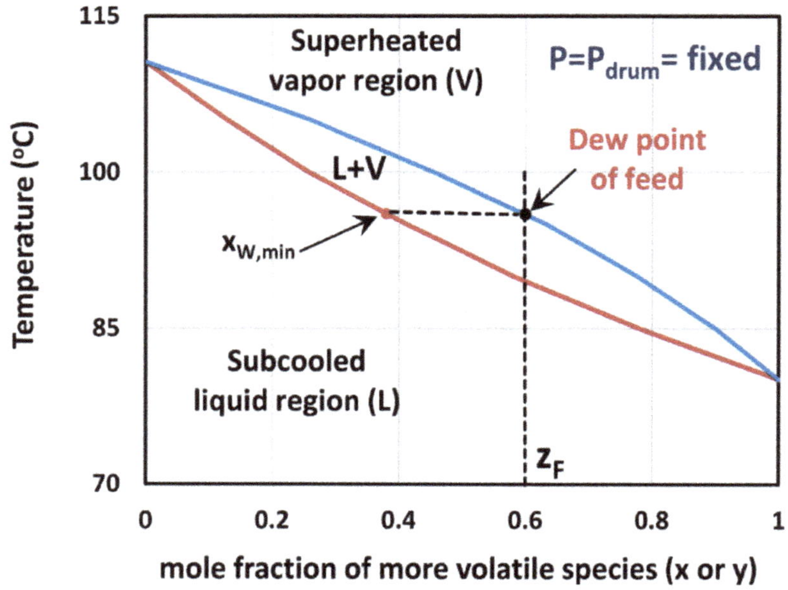

Figure 8.6 Dew point of feed at flash drum pressure. When $T_{drum} = DEWT$ of feed, all the feed goes to the vapor-distillate stream ($\psi = 1$). The composition of trace amount of liquid is $x_{W,min}$.

When $0 < \psi < 1$, the slope of the operating line is negative in the range $-\infty < slope < 0$ and the operating line falls in the second quadrant (see Figure 8.4). The feed is split into vapor-distillate stream ($0 < D(= F\psi) < F$) and bottoms-liquid stream ($0 < W(= F - F\psi) < F$). This kind of flashing occurs when $BUBLT < T_{drum} < DEWT$ where the bubble point and dew point are based on feed composition at flash drum pressure. Figure 8.7 shows this flashing on the T-x-y diagram.

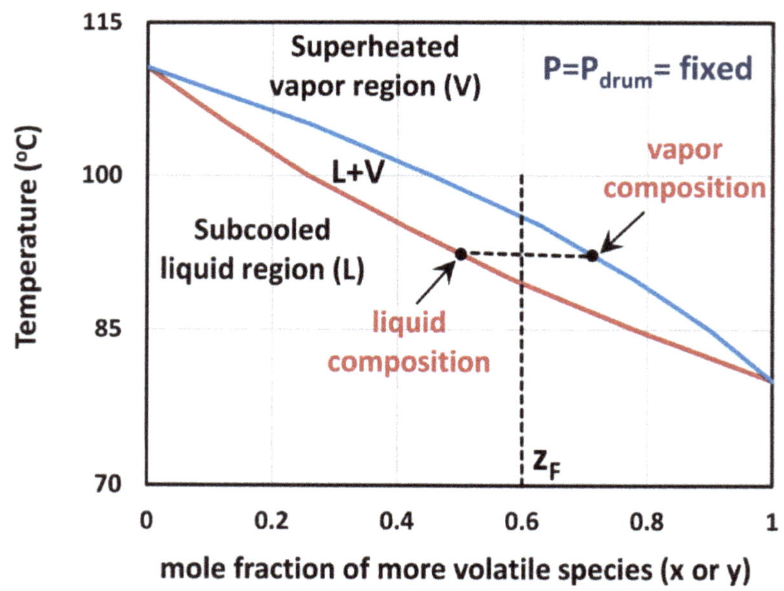

Figure 8.7 Flashing with $BUBLT < T_{drum} < DEWT$ where bubble and dew points refer to feed composition. Part of the feed goes to vapor-distillate stream and part of it goes to bottoms-liquid stream ($0 < \psi < 1$).

Analytical Solution

We can also solve the P, ψ – flash problem analytically, instead of graphically, provided that the equilibrium curve can be expressed in the form of an equation. Assuming that the relative volatility of more volatile species with respect to the less volatile species, denoted by α, is constant (independent of temperature), the equilibrium curve can be represented by the following equation:

$$y = \alpha x/[1 + (\alpha - 1)x] \tag{8.4}$$

This equation simply follows from the definition of relative volatility α:

$$\alpha = (y/x)/[(1 - y)/(1 - x)] \tag{8.5}$$

From Equations (8.3) and (8.4), it follows that:

$$\alpha x_W/[1 + (\alpha - 1)x_W] = -\left(\frac{1-\psi}{\psi}\right)x_W + \frac{z_F}{\psi} \tag{8.6}$$

Upon rearrangement, Equation (8.6) gives:

$$\left(\frac{1-\psi}{\psi}\right)(\alpha - 1)x_W^2 + \left[\alpha + \left(\frac{1-\psi}{\psi}\right) - \left(\frac{\alpha-1}{\psi}\right)z_F\right]x_W - \frac{z_F}{\psi} = 0 \tag{8.7}$$

This is a quadratic equation with the following solution:

$$x_W = \frac{-b+\sqrt{b^2-4ac}}{2a} \tag{8.8}$$

where

$$a = \left(\frac{1-\psi}{\psi}\right)(\alpha - 1) \tag{8.9}$$

$$b = \left[\alpha + \left(\frac{1-\psi}{\psi}\right) - \left(\frac{\alpha-1}{\psi}\right)z_F\right] \tag{8.10}$$

$$c = -\frac{z_F}{\psi} \tag{8.11}$$

Energy Balances

Energy balance is important in order to determine the heat transfer rate in the heater and the shaft work needed to pressurize the feed. For multiple inlets and outlets, energy balance (assuming steady state, negligible kinetic and potential energy changes, see Chapter 7) over a given control volume can be written as:

$$\Sigma_{INLETS}\ (F_i\, h_i) + \Sigma\,\dot{Q} = \Sigma_{OUTLETS}\ (F_i\, h_i) + \Sigma\,\dot{W}_{sh} \tag{8.12}$$

where F_i is the molar flow rate and h_i is the molar enthalpy of fluid stream at *i*th opening (inlet/outlet) of the control volume. We will apply this energy balance over different portions of the flash distillation process shown in Figure 8.1.

First we apply energy balance on the portion of the flash distillation process indicated by dotted envelope in Figure 8.8:

$$Fh_F = DH_D + Wh_W \tag{8.13}$$

Given the flash drum conditions, one can estimate the molar enthalpies of saturated vapor (H_D) and saturated liquid (h_W) and hence calculate the enthalpy of feed h_F prior to throttling using Equation (8.13).

In the next step, we do energy balance on the portion of the process flow sheet including pump and heater as shown in Figure 8.9:

$$Fh_o + \dot{Q} = Fh_F + \dot{W}_{sh} \tag{8.14}$$

The energy balance around the pump (assuming feed liquid to be incompressible and negligible changes in kinetic energy, potential energy, and feed temperature across the pump) gives the pump shaft work as follows:

$$\dot{W}_{sh} = -F\Delta h = -FV_F(P_F - P_o) \tag{8.15}$$

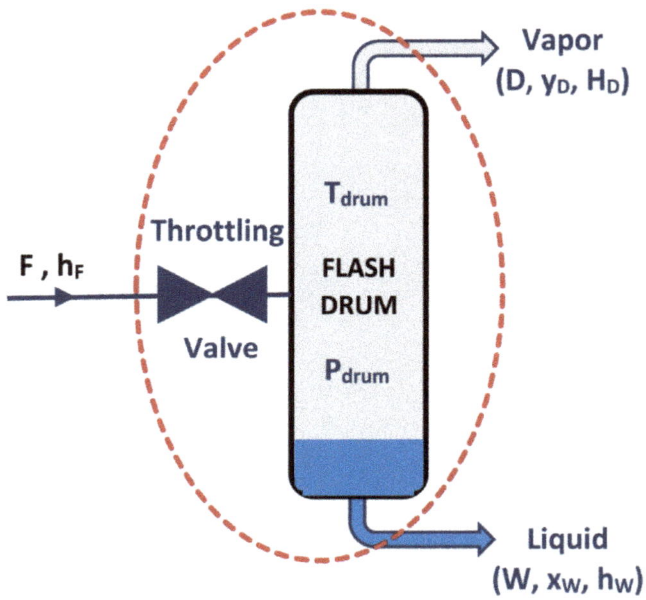

Figure 8.8 Envelope over which energy balance is carried out.

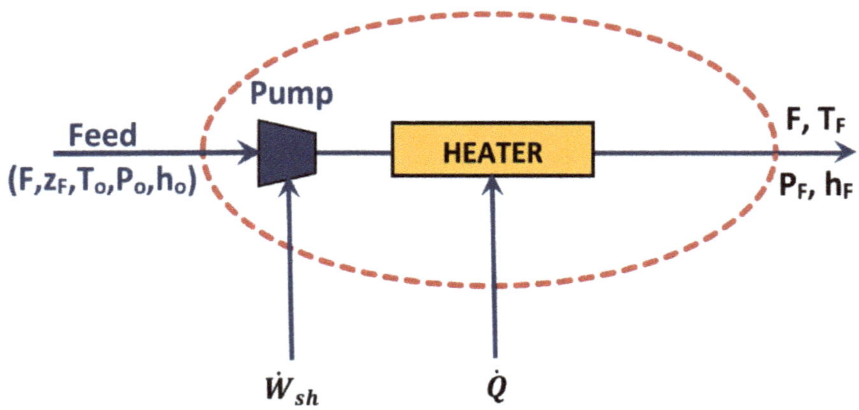

Figure 8.9 Energy balance over the envelope including pump and heater.

where V_F is the molar volume of liquid estimated at T_o and P_o. The pressure at the exit of the pump is assumed to be the same at that prior to the throttling valve (pressure drop across the heat exchanger is neglected).

From Equations (8.14) and (8.15), it follows that:

$$\dot{Q} = F[(h_F - h_o) - V_F(P_F - P_o)] \tag{8.16}$$

Equation (8.16) can be used to calculate the heat transfer rate provided that P_F is specified. This pressure can be selected such that the feed is above the bubble point pressure prior to throttling, that is, $P_F \geq BUBLP$ at $T = T_F$.

Estimation of Enthalpies

In order to perform energy balances, we need the enthalpies of vapor and liquid phases. Referring to Figure 8.1, we need to estimate the enthalpies of vapor mixture (H_D) and liquid solutions (h_W, h_F, and h_o). As noted in Chapter 7, we have to define the reference state of a material in order to calculate the absolute values of enthalpies. As we are dealing with mixtures of two species (A and B), we need to define reference states for both species. For convenience, we can take the reference temperature (T_R) equal to the boiling point of more volatile species (species A) at the flash drum pressure, that is:

$$T_R = T_A^{Sat} \quad \text{and} \quad P_R = P_{drum} \tag{8.17}$$

Thus, the reference states of the two species (A and B) are as follows:

$$\textit{Species A}\text{: Saturated liquid, } T_R = T_A^{Sat}, P_R = P_{drum}, h_R = 0 \tag{8.18}$$

$$\textit{Species B}\text{: Subcooled liquid, } T_R = T_A^{Sat}, P_R = P_{drum}, h_R = 0 \tag{8.19}$$

Note that the reference temperature, pressure, and enthalpy are the same for both the species. At these reference conditions, species A is in saturated liquid state whereas species B is in subcooled liquid state. With the reference states specified in Equations (8.18) and (8.19), we can estimate the molar enthalpies of pure substances A and B and their mixtures in liquid and vapor states. We will assume that the mixtures/solutions are ideal in nature.

The molar enthalpy H_D of vapor-distillate product leaving the flash drum (see Figure 8.1) can be estimated as:

$$H_D = y_D H_A(T_{drum}, P_{drum}) + (1 - y_D) H_B(T_{drum}, P_{drum}) \tag{8.20}$$

Assuming that pure vapors of species A and B behave like ideal gases, we can re-write Equation (8.20) as:

$$H_D = y_D H_A(T_{drum}) + (1 - y_D) H_B(T_{drum}) \tag{8.21}$$

Note that the enthalpy of an ideal gas is independent of pressure and y_D is the mole fraction of more volatile species (species A) in the vapor-distillate product.

Figure 8.10 shows the computational path to estimate H_A. We need to calculate λ_A and ΔH, as shown in the computational path. Thus H_A can be calculated as follows:

$$H_A = \lambda_A + C_{pG,A}(T_{drum} - T_R) \tag{8.22}$$

where the reference temperature is $T_R = T_A^{Sat}$ at P_{drum}, λ_A is the molar latent heat of vaporization of pure species A at T_A^{Sat}, and $C_{pG,A}$ is the constant-pressure molar heat capacity of pure vapor A. The second term on the right hand side of Equation (8.22) is ΔH of pure vapor A.

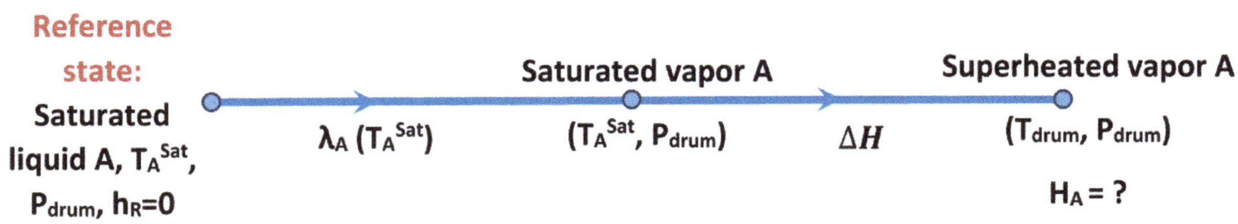

Figure 8.10 Computational path to estimate enthalpy H_A of pure vapor A with respect to the given reference state.

The computational path for H_B calculation is shown in Figure 8.11. Three steps are involved in the computational path. Referring to Figure 8.11, H_B can be calculated as follows:

$$H_B = \Delta h_1 + \lambda_B + \Delta H_1 = C_{pL,B}\left(T_B^{Sat} - T_R\right) + \lambda_B\left(at\ T_B^{Sat}\right) + C_{pG,B}\left(T_{drum} - T_B^{Sat}\right) \tag{8.23}$$

Now we consider the estimation of enthalpies of liquid solutions (h_W, h_F, and h_o). Assuming that the liquid solutions are ideal, the enthalpy h_W of bottoms-liquid product leaving the flash drum (see Figure 8.1) can be estimated from:

$$h_W = x_W h_A(T_{drum}, P_{drum}) + (1 - x_W) h_B(T_{drum}, P_{drum}) \tag{8.24}$$

In order to use this equation, we need the enthalpies of pure liquids A and B. Based on the reference states for species A and B defined in Equations (8.18) and (8.19), we can calculate h_A and h_B as follows:

$$h_A = C_{pL,A}\left(T_{drum} - T_A^{Sat}\right) \tag{8.25}$$

$$h_B = C_{pL,B}\left(T_{drum} - T_A^{Sat}\right) \tag{8.26}$$

Note that the reference temperature is the same for species A and B, that is, $T_R = T_A^{Sat}$ at P_{drum}.

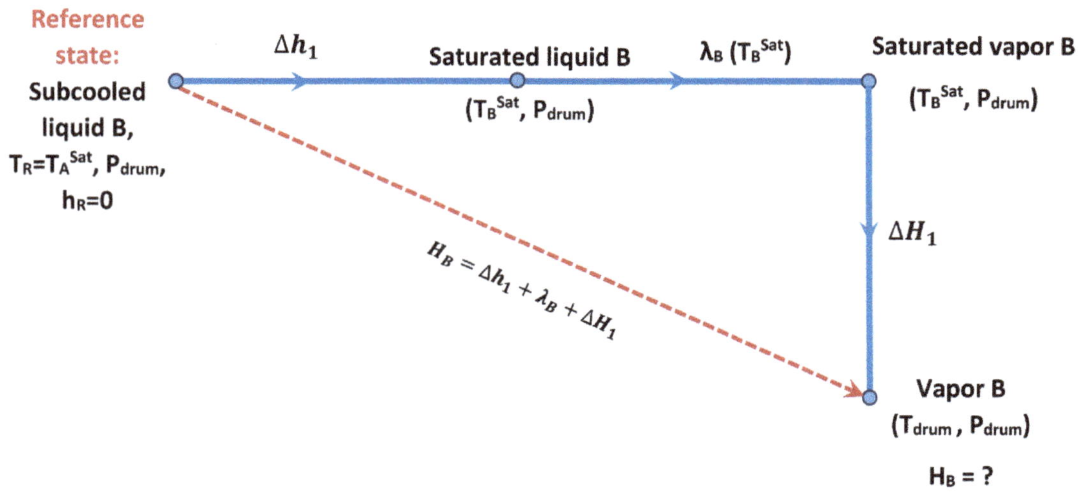

Figure 8.11 Computational path to estimate enthalpy H_B of pure vapor B with respect to the given reference state. Note that the final state of pure vapor B is hypothetical as $T_{drum} < T_B^{Sat}$.

The enthalpy h_F of feed solution prior to throttling (see Figure 8.8) is obtained from energy balance expressed in Equation (8.13). This known value of h_F can be used to estimate T_F, the temperature of liquid feed prior to throttling. Assuming the feed solution to be ideal, its molar enthalpy h_F prior to throttling can be expressed as:

$$h_F = z_F h_A(T_F, P_F) + (1 - z_F) h_B(T_F, P_F) \tag{8.27}$$

Neglecting the effect of pressure on pure liquid enthalpy, we can write:

$$h_A = C_{pL,A}\left(T_F - T_A^{Sat}\right) \tag{8.28}$$

$$h_B = C_{pL,B}\left(T_F - T_A^{Sat}\right) \tag{8.29}$$

Solving for T_F from Equations (8.27)-(8.29), we get:

$$T_F = T_A^{Sat} + \frac{h_F}{z_F C_{pL,A} + (1 - z_F) C_{pL,B}} \tag{8.30}$$

The liquid feed solution enthalpy h_o is needed in the energy balance around the heater (see Figure 8.9 and Equation (8.16)). It can be calculated as:

$$h_o = z_F h_A(T_o, P_o) + (1 - z_F) h_B(T_o, P_o) \tag{8.31}$$

Neglecting the effect of pressure on pure liquid enthalpy, we can calculate h_A and h_B as follows:

$$h_A = C_{pL,A}\left(T_o - T_A^{Sat}\right) \tag{8.32}$$

$$h_B = C_{pL,B}\left(T_o - T_A^{Sat}\right) \tag{8.33}$$

8.3.2 P, T – Flash Calculation

In $P, T - Flash$ calculation, we are given feed variables (F, z_F, T_o, P_o) and flash drum pressure P_{drum} and temperature T_{drum}. We need to determine the vapor-distillate and bottoms-liquid flow rates and compositions (D, y_D, W, x_W), in addition to H_D, h_W, h_F, h_o, T_F, P_F, and $\dot{Q}$.

The component mass balance on more volatile species (species A) gives:

$$z_A F = y_A D + x_A W \tag{8.34}$$

where $z_A = z_F$, $y_A = y_D$, and $x_A = x_W$. Using overall mass balance and the equilibrium relation $y_i = K_i x_i$, we can re-write Equation (8.34) as:

$$y_A = \frac{K_A z_A}{1+\psi(K_A - 1)} \tag{8.35}$$

In a similar manner, we can do component mass balance on species B and obtain:

$$y_B = \frac{K_B z_B}{1+\psi(K_B - 1)} \tag{8.36}$$

where $z_B = 1 - z_A = 1 - z_F$. As $y_A + y_B = 1$, it follows from Equations (8.35) and (8.36) that:

$$1 = \frac{K_A z_A}{1+\psi(K_A - 1)} + \frac{K_B z_B}{1+\psi(K_B - 1)} \tag{8.37}$$

The only unknown in this equation is ψ (=D/F). Given T_{drum} and P_{drum}, we can obtain the values of K_A and K_B from DePriester charts or empirical equations. The feed composition (z_A and z_B) is given. Once ψ is known, one can calculate the vapor-distillate flow rate ($D = F\psi$) and bottoms-liquid flow rate ($W = F(1 - \psi)$). The vapor-distillate composition y_D is then obtained from Equation (8.35) and the bottoms-liquid composition x_W is obtained from the equilibrium relation $y_D = K_A x_W$. The remaining calculations, that is, energy balances and associated enthalpy calculations, are the same as in $P, \psi - Flash$ calculation.

8.4 ANALYSIS OF MULTI-COMPONENT FLASH DISTILLATION

In multi-component flash distillation, the number of components present in the feed are more than two. As the computational procedure for multicomponent flash distillation is not straightforward, we will restrict ourselves to $P, T - Flash$ calculation where we are given the feed variables (F, z_i, T_o, P_o) and flash drum pressure P_{drum} and temperature T_{drum}. We need to determine the vapor-distillate and bottoms-liquid flow rates and compositions (D, y_i, W, x_i), in addition to H_D, h_W, h_F, h_o, T_F, P_F, and $\dot{Q}$.

As a first step, one should check if flashing would occur or not at the given flash drum pressure P_{drum} and temperature T_{drum}. For this purpose, one needs to calculate the bubble point and dew point pressures at the overall feed composition and flash drum temperature, that is: Given $\{x_i\} = \{z_i\}$ and $T = T_{drum}$, calculate *BUBLP*; Given $\{y_i\} = \{z_i\}$ and $T = T_{drum}$, calculate *DEWP*. For flashing to occur, $DEWP < P_{drum} < BUBLP$.

The component mass balance on species "i" gives:

$$z_i F = y_i D + x_i W \tag{8.38}$$

Using overall mass balance and the equilibrium relation $y_i = K_i x_i$, we can re-write Equation (8.38) as:

$$y_i = \frac{K_i z_i}{1+\psi(K_i - 1)} \tag{8.39}$$

Alternatively, we can write:

$$x_i = \frac{z_i}{1+\psi(K_i-1)} \tag{8.40}$$

Because $\sum y_i = 1$ and $\sum x_i = 1$, Equations (8.39) and (8.40) yield:

$$\sum \frac{K_i z_i}{1+\psi(K_i-1)} = 1 \tag{8.41}$$

$$\sum \frac{z_i}{1+\psi(K_i-1)} = 1 \tag{8.42}$$

The distribution constant $K_i(T,P)$ of a chemical species can be determined from DePriester-type of nomographs or appropriate equations at the given flash temperature and pressure. The feed composition is given. The only unknown in Equations (8.41) and (8.42) is ψ. Either of these two equations (Equation (8.41) or (8.42)) can be used to estimate ψ by trial. Unfortunately, Equations (8.41) and (8.42) do not have good convergence properties. However the combination of Equations (8.41) and (8.42), that is, $\sum(y_i - x_i) = 0$, is found to have good convergence properties. Thus:

$$\sum \frac{(K_i-1)z_i}{1+\psi(K_i-1)} = 0 \tag{8.43}$$

Equation (8.43) has excellent convergence properties and can be used to estimate the unknown ψ. Equation (8.43) is often referred to as the Rachford-Rice flash equation.

One can use the Newton-Raphson method (also called Newton method) to find the root of the Rachford-Rice equation. Equation (8.43) can be re-written as:

$$f(\psi) = \sum \frac{(K_i-1)z_i}{1+\psi(K_i-1)} = 0 \tag{8.44}$$

The Newton-Raphson method is illustrated schematically in Figure 8.12. It requires an initial guess of the root. Let x_o be the initial guess of the root of $f(x) = 0$. In order to estimate an improved value of the root, draw a tangent to the curve $y = f(x)$ at $(x_o, f(x_o))$. The intersection of the tangent line with the x-axis gives the improved estimate x_1. The successive values of x generated in this manner converge to the root of $f(x) = 0$.

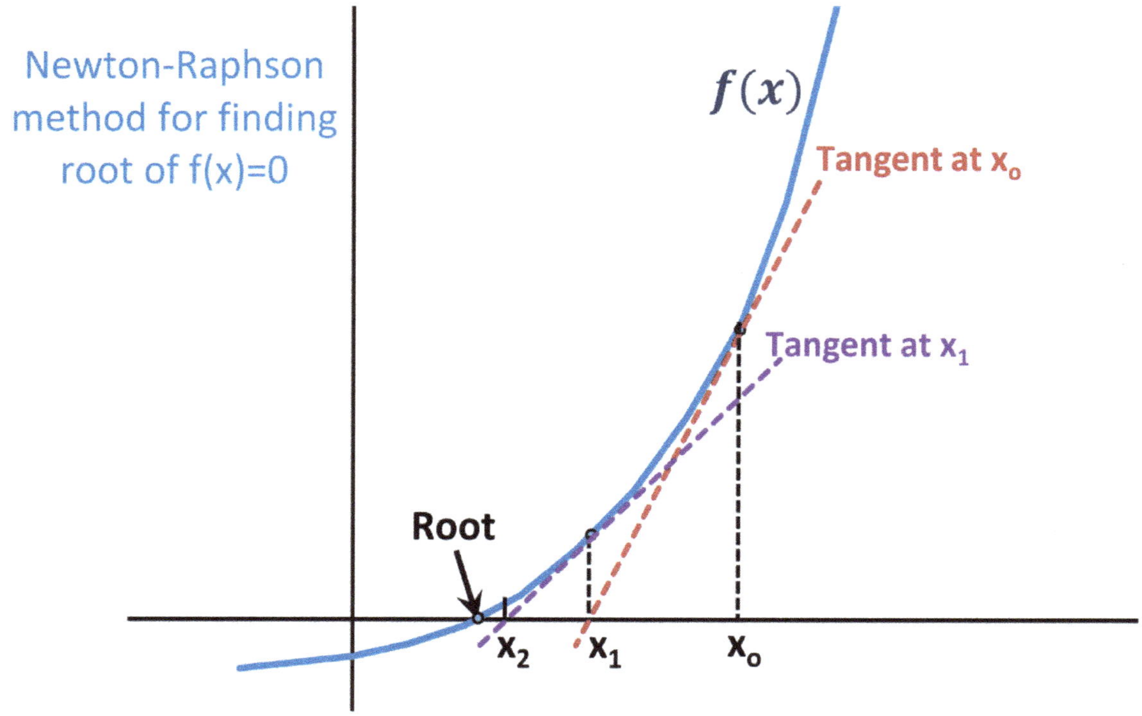

Figure 8.12 Newton-Raphson method for finding root of $f(x) = 0$.

The formula for estimating the improved value of the root (x_{k+1}) over the previous value of the root (x_k) can be derived by expressing the slope of the tangent line as:

$$slope = f'(x_k) = \frac{f(x_k)-0}{(x_k - x_{k+1})} \tag{8.45}$$

Thus,

$$x_{k+1} = x_k - \frac{f(x_k)}{f'(x_k)} \tag{8.46}$$

The derivative of $f(\psi)$ in Rachford-Rice equation, Equation (8.44), is given as:

$$f'(\psi) = -\sum \frac{(K_i-1)^2 z_i}{[1+\psi(K_i-1)]^2} \tag{8.47}$$

Hence we can write:

$$\psi_{k+1} = \psi_k - \frac{f(\psi_k)}{f'(\psi_k)} \tag{8.48}$$

Equation (8.48), along with Equation (8.47), can be used to estimate ψ by successive substitution. The range of ψ is $0 \le \psi \le 1$. The initial guess could be taken as $\psi_o = 0.50$. The iteration process can be terminated when the deviation between two successive values is within the tolerance level, $[|(\psi_{k+1} - \psi_k)|/\psi_k] \times 100 \le \varepsilon$.

Once ψ is known, the vapor composition can be calculate from Equation (8.39) and the liquid composition can be calculated from the equilibrium relation: $x_i = y_i/K_i$. The vapor-distillate flow rate is $D = \psi F$ and the bottoms-liquid flow rate is $W = (1 - \psi)F$. Thus the conditions in the flash drum are known. In order to determine the heat transfer rate in the heater and the shaft work needed to pressurize the feed, one can then carry out energy balance as described in $P, \psi - Flash$ calculation for binary systems, assuming vapor mixtures and liquid solutions to be ideal.

NOTATION

$C_{PG,i}$	constant pressure molar heat capacity of pure gas species i
$C_{PL,i}$	constant pressure molar heat capacity of pure liquid species i
D	vapor-distillate molar flow rate
F	feed molar flow rate
F_i	molar flow rate of fluid stream at ith opening (inlet/outlet) of control volume
h	molar enthalpy
h_A	molar enthalpy of pure species A in liquid state
h_B	molar enthalpy of pure species B in liquid state
h_F	molar enthalpy of feed prior to throttling (see Figure 8.1)
h_i	molar enthalpy of fluid stream at ith opening (inlet/outlet) of the control volume (CV)
h_o	molar enthalpy of feed prior to pumping (see Figure 8.1)
h_R	molar enthalpy of pure species in reference state
h_W	molar enthalpy of bottom liquid stream of flash drum (see Figure 8.1)
H_A	molar enthalpy of pure species A in vapor state
H_B	molar enthalpy of pure species B in vapor state
H_D	molar enthalpy of top vapor distillate stream of flash drum (see Figure 8.1)
K_i	distribution coefficient of species i, defined as $K_i = y_i/x_i$
L	molar flow rate of liquid stream
P	pressure
P_F	pressure of feed stream prior to throttling (see Figure 8.1)
P_o	pressure of feed stream prior to pumping (see Figure 8.1)
P_R	reference state pressure for enthalpy calculations
$\dot{Q}$	heat transfer rate
T	temperature
T_F	temperature of feed stream prior to throttling (see Figure 8.1)
T_o	temperature of feed stream prior to pumping (see Figure 8.1)
T_R	reference state temperature for enthalpy calculations
T_i^{Sat}	saturation temperature of pure species i
V	molar flow rate of vapor stream
V_F	molar volume of liquid feed
W	molar flow rate of bottoms liquid stream of flash drum (see Figure 8.1)
$\dot{W}_{sh}$	rate of shaft work
x	mole fraction of more volatile species in liquid phase (also used as independent variable in Newton-Raphson method)
x_i	mole fraction of species i in liquid phase
x_W	mole fraction of more volatile species in bottoms liquid stream of flash drum (see Figure 8.1)
$x_{W,min}$	mole fraction of more volatile species in bottoms liquid stream of flash drum when flashing is done at dew point temperature of the feed (see Figure 8.6)
y	mole fraction of more volatile species in vapor phase (also used as dependent variable in Newton-Raphson method)
y_D	mole fraction of more volatile species in vapor-distillate stream of flash drum (see Figure 8.1)
y_i	mole fraction of species i in vapor phase

$y_{D,max}$ mole fraction of more volatile species in vapor distillate of flash drum when flashing is done at bubble point temperature of the feed (see Figure 8.5)
z_F mole fraction of more volatile species in the binary feed (see Figure 8.1)
z_i mole fraction of species i in the feed

Greek Letters

α relative volatility
Δh molar enthalpy change of liquid
ΔH molar enthalpy change of vapor
λ_A molar latent heat of vaporization of pure species A
λ_B molar latent heat of vaporization of pure species B
ψ ratio of vapor distillate molar flow rate (D) to feed molar flow rate (F)

TUTORIAL PROBLEMS WITH SOLUTION

Problem 1

An equimolar binary mixture of n-pentane and n-heptane undergoes flash distillation at 1 atm. If $\psi = 0.33$, calculate the compositions of vapor distillate and bottoms liquid streams. The equilibrium data at 1 atm, in terms of mole fractions of more volatile component (n-pentane), is as follows:

x	y
1	1
0.867	0.984
0.594	0.925
0.398	0.836
0.254	0.701
0.145	0.521
0.059	0.271
0	0

Solution

For ψ=0.33 and z_F =0.50, the equation for the operating line, Equation (8.3), is:

$$y_D = -2.03\, x_W + 1.515$$

The intersection of this operating line with the equilibrium curve gives the solution. Figure given below shows the graphical solution. The vapor distillate composition is $y_D \approx 0.795$ and the liquid composition is $x_W \approx 0.355$.

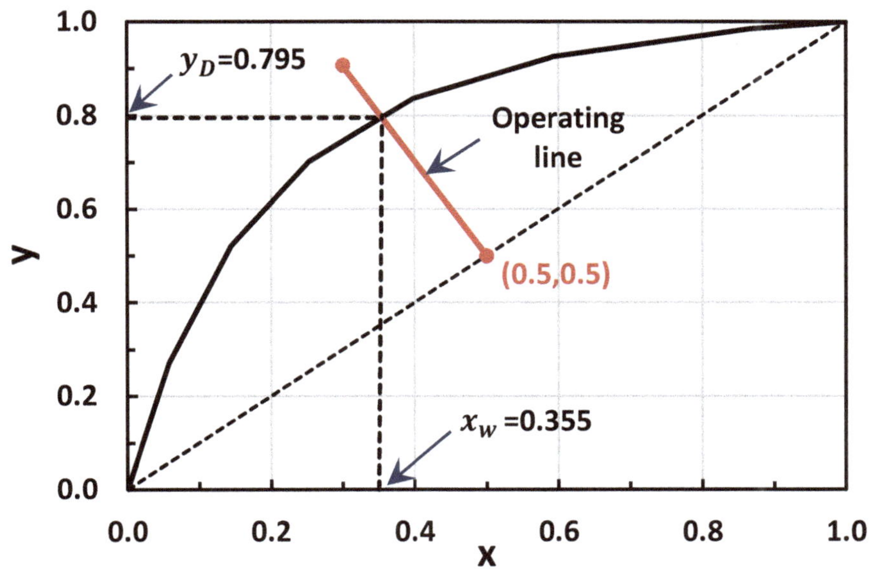

Problem 2

An ethanol-water mixture containing 32 mol% ethanol is subjected to flash distillation at a separator pressure of 1 atm.

Determine graphically the vapor and liquid compositions leaving the separator for each of the following values of ψ : 0, 0.15, 0.30, 0.45, 0.60, 0.75, 0.90, 1.0. How does the liquid composition (that is, x) vary with the increase in ψ? How does the vapor composition (that is, y) vary with the increase in ψ? Does the temperature decrease, increase or remain constant with the increase in ψ?

The equilibrium data for ethanol-water system, in terms of mole fractions of ethanol, at 1 atm are given below:

x	y
0	0
0.019	0.17
0.0721	0.3891
0.0966	0.4375
0.1238	0.4704
0.1661	0.5089
0.2377	0.5445
0.2608	0.558
0.3273	0.5826
0.3965	0.6122
0.5079	0.6564
0.5198	0.6599
0.5732	0.6841
0.6763	0.7385
0.7472	0.7815
0.8943	0.8943
1	1

Solution

Figure given below shows the equilibrium curve and various operating lines corresponding to different values of ψ. The intersection of the operating line with the equilibrium curve gives the compositions of vapor and liquid streams leaving the separator.

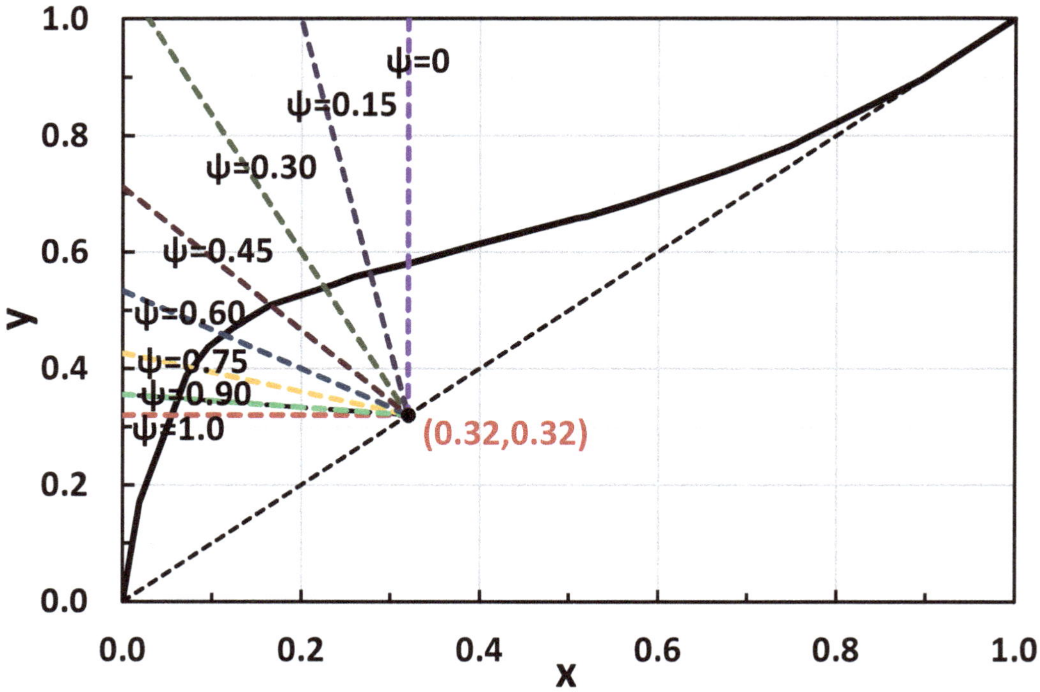

The vapor and liquid compositions corresponding to different values of ψ are summarized in the following table.

ψ	y_D	x_W
0	0.58	0.32
0.15	0.56	0.28
0.30	0.54	0.225
0.45	0.51	0.17
0.60	0.46	0.11
0.75	0.40	0.08
0.90	0.33	0.06
1.0	0.32	0.055

The plots of y_D vs ψ and x_W vs ψ are shown in the figure given below. With the increase in ψ, both y_D and x_W decrease. When ψ=0, the vapor phase has the maximum value of y_D . When ψ=1, the liquid phase has the lowest value of x_W. Note that when ψ=0, all the feed goes into the bottoms liquid product with negligible vapor product. This condition corresponds to the bubble point temperature of feed at the separator pressure. When ψ=1, all the feed goes into the vapor product with negligible bottoms liquid product. This condition corresponds to the dew point temperature of feed at the separator pressure. Thus the temperature rises from bubble point to dew point when ψ is increased from 0 to 1.

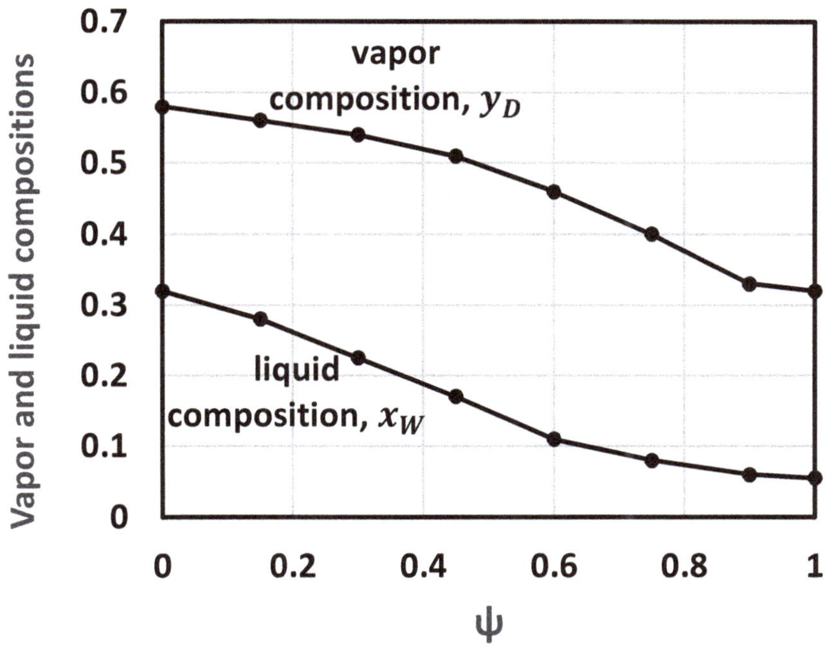

Problem 3

A flash chamber operating at 1 atm is separating a binary mixture of benzene and toluene. The feed mixture is 50 mole % benzene. If ψ=0.33, what are the product compositions. Solve the problem analytically. The relative volatility of benzene with respect to toluene is 2.50.

Solution

The analytical solution for the bottoms liquid product composition is:

$$x_W = \frac{-b+\sqrt{b^2-4ac}}{2a}$$

where

$$a = \left(\frac{1-\psi}{\psi}\right)(\alpha - 1)$$

$$b = \left[\alpha + \left(\frac{1-\psi}{\psi}\right) - \left(\frac{\alpha-1}{\psi}\right) z_F\right]$$

$$c = -\frac{z_F}{\psi}$$

Given that ψ=0.33, $\alpha = 2.50$, $z_F = 0.50$, we can calculate a, b, c and hence x_W. The values are:

a=3.045455, b=2.257576, c=-1.51515, x_W =0.426154. The top vapor product composition is given as:

$$y_D = \alpha x_W/[1 + (\alpha - 1)x_W] = 2.50 \times 0.426154/[1 + (2.50 - 1) \times 0.426154] = 0.64993$$

Thus, the product compositions are: x_W =0.426154 and y_D = 0.64993

Problem 4

A flash chamber operating at 80 °C and 110 kPa is separating 100 kmol/h of a feed that is 3-component mixture of species 1, 2 and 3. The feed composition is z_1=0.42, z_2 = 0.38 and z_3=0.20. Would flashing occur at given temperature (80 °C) and pressure (110 kPa))? If yes, find the product compositions and flow rates.

You can assume that Raoult's law is applicable. The vapor pressures of the pure species at 80 °C are as follows: $P_1^{Sat} = 195.75\ kPa,\ P_2^{Sat} = 97.84\ kPa,\ P_3^{Sat} = 50.32\ kPa.$

Solution

The problem deals with PT-flash calculation of multicomponent systems described in section 8.4. For flashing to occur, $P_{dew} < P < P_{bubble}$. The bubble and dew pressures are as follows:

$$P_{bubble} = 0.42(P_1^{Sat}) + 0.38(P_2^{Sat}) + 0.20(P_3^{Sat}) = 129.46\ kPa$$

$$P_{dew} = 1/[(0.42/P_1^{Sat}) + (0.38/P_2^{Sat}) + (0.20/P_3^{Sat})] = 99.96\ kPa$$

As the pressure in the chamber (110 kPa) falls in the range $P_{dew} < P < P_{bubble}$, flashing will occur at the given chamber conditions.

We now need to solve the Rachford-Rice equation, Equation (8.43), using the Newton-Raphson method, given that: z_1=0.42, z_2 = 0.38, z_3=0.20, $K_1 = P_1^{sat}/P = 1.7795$, $K_2 = P_2^{sat}/P = 0.8895$, $K_3 = P_3^{sat}/P = 0.4575$

The Rachford-Rice equation in the present case can be written as:

$$f(\psi) = \sum \frac{(K_i - 1)z_i}{1 + \psi(K_i - 1)} = 0$$

$$f(\psi) = \frac{0.32729}{1 + \psi(0.7795)} - \frac{0.04199}{1 - \psi(0.1105)} - \frac{0.1085}{1 - \psi(0.5425)} = 0$$

The derivative of $f(\psi)$ is as follows:

$$f'(\psi) = -\sum \frac{(K_i - 1)^2 z_i}{[1 + \psi(K_i - 1)]^2}$$

$$f'(\psi) = -\left[\frac{0.2552}{(1 + \psi(0.7795))^2} + \frac{4.64 \times 10^{-3}}{(1 - \psi(0.1105))^2} + \frac{0.05886}{(1 - \psi(0.5425))^2}\right]$$

Equation (8.48) gives the iteration scheme to solve for ψ as:

$$\psi_{k+1} = \psi_k - \frac{f(\psi_k)}{f'(\psi_k)}$$

The following table gives the iterations starting with the initial guess of ψ_o=0.50

ψ_k	ψ_{k+1}
0.50	0.669938
0.669938	0.666925
0.666925	0.666922
0.666922	0.666922

Thus the solution is ψ=0.666922≈0.667. Given that F=100 kmol/h, D=ψF=66.7 kmol/h, and W=33.3 kmol/h.

The vapor composition follows from Equation (8.39):

$$y_i = \frac{K_i z_i}{1+\psi(K_i-1)}$$

The liquid composition follows from the equilibrium relation:

$$x_i = y_i/K_i$$

The values of y_i and x_i are summarized below.

species	y_i	x_i
1	0.4917	0.2763
2	0.3649	0.4102
3	0.1434	0.3135

Problem 5

A 4-component feed mixture of species 1 to 4 with composition z_1=0.10, z_2 = 0.20, z_3=0.30 and z_4=0.40 undergoes flash distillation at 50 °C and 250 kPa.

a) Verify that flashing occur at the given conditions?
b) What is the fraction of feed that goes to the vapor distillate product?
c) What are the compositions of the exit streams?

You can assume that Raoult's law is applicable. The vapor pressures of the pure species at 50 °C are as follows: $P_1^{Sat} = 1400\ kPa$, $P_2^{Sat} = 480\ kPa$, $P_3^{Sat} = 160\ kPa$, $P_4^{Sat} = 60\ kPa$

Solution:

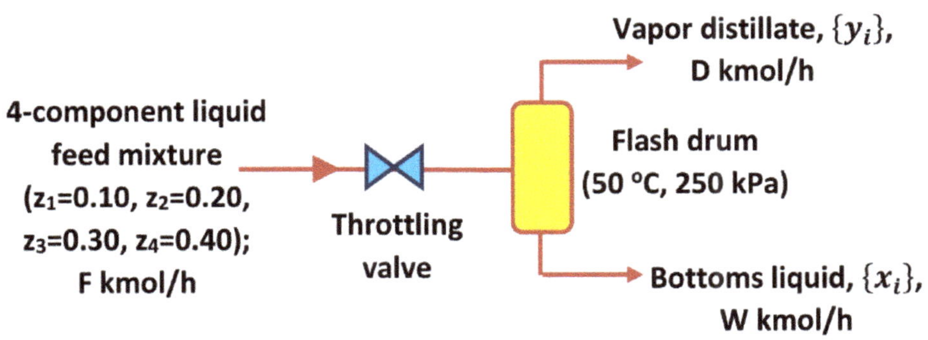

a) For flashing to occur, $P_{dew} < P < P_{bubble}$. The bubble and dew pressures are as follows:

$$P_{bubble} = 0.10(P_1^{Sat}) + 0.20(P_2^{Sat}) + 0.30(P_3^{Sat}) + 0.40(P_4^{Sat}) = 308\ kPa$$

$$P_{dew} = 1/[(0.10/P_1^{Sat}) + (0.20/P_2^{Sat}) + (0.30/P_3^{Sat}) + (0.40/P_4^{Sat})] = 110.745\ kPa$$

As the pressure in the chamber (250 kPa) falls in the range $P_{dew} < P < P_{bubble}$, flashing will occur at the given chamber conditions.

b) We now need to solve the Rachford-Rice equation using the Newton-Raphson method. Given that:

z_1=0.10, z_2 = 0.20, z_3=0.30, z_4=0.40, $K_1 = P_1^{sat}/P = 5.6$, $K_2 = P_2^{sat}/P = 1.92$, $K_3 = P_3^{sat}/P = 0.64$, $K_4 = P_4^{sat}/P = 0.24$

The Rachford-Rice equation in the present case can be written as:

$$f(\psi) = \sum \frac{(K_i - 1)z_i}{1+\psi(K_i - 1)} = 0$$

$$f(\psi) = \frac{0.46}{1+\psi(4.6)} + \frac{0.184}{1+\psi(0.92)} - \frac{0.108}{1-\psi(0.36)} - \frac{0.304}{1-\psi(0.76)} = 0$$

The derivative of $f(\psi)$ is as follows:

$$f'(\psi) = -\sum \frac{(K_i - 1)^2 z_i}{[1+\psi(K_i - 1)]^2}$$

$$f'(\psi) = -\left[\frac{2.116}{(1+\psi(4.6))^2} + \frac{0.16928}{(1+\psi(0.92))^2} + \frac{0.03888}{(1-\psi(0.36))^2} + \frac{0.23104}{(1-\psi(0.76))^2}\right]$$

The iteration scheme to solve for ψ is:

$$\psi_{k+1} = \psi_k - \frac{f(\psi_k)}{f'(\psi_k)}$$

The following table gives the iterations starting with the initial guess of ψ_o=0.50

ψ_k	ψ_{k+1}
0.50	0.117613
0.117613	0.13089
0.13089	0.131221
0.131221	0.131221

Thus the solution is ψ=0.131221≈0.1312. The fraction of the feed that goes to vapor distillate product is 0.1312.

c) Compositions of exit streams:

Vapor distillate composition (top product)

$$y_i = \frac{K_i z_i}{1+\psi(K_i - 1)}$$

Bottoms liquid top product: $x_i = y_i/K_i$

y_1=0.34921	x_1=0.062359
y_2=0.342636	x_2=0.178456
y_3=0.20152	x_3=0.314875
y_4=0.106635	x_4=0.444311
$\sum y_i = 1.0$	$\sum x_i = 1.0$

SHORT REVIEW QUESTIONS

1. Explain flash distillation
2. What is the purpose of the throttling valve?
3. Show on a P-V diagram of a pure substance that a sharp reduction in pressure of saturated liquid causes vaporization of liquid.
4. For flashing to occur at a given temperature, the separator pressure must be within the range: $DEWP < P_{drum} < BUBLP$. Is this true?
5. What happens when the separator pressure is equal to the bubble point pressure of feed at flashing temperature?
6. What happens when the separator pressure is equal to the dew point pressure of feed at flashing temperature?
7. For flashing to occur at a given pressure, the separator temperature must be within the range: $BUBLT < T_{drum} < DEWT$. Is this true?
8. What happens when the flashing temperature is equal to the bubble point temperature of feed at separator pressure?
9. What happens when the flashing temperature is equal to the dew point temperature of feed at separator pressure?
10. Derive the Rachford-Rice flash equation for multicomponent systems. Which numerical technique can be used to solve this equation?

PRACTICE PROBLEMS

Problem 1

A flash chamber operating at 1 atm is separating a binary mixture of benzene and toluene entering at flow rate of 1000 kmol/h. The feed mixture is 50 mole % benzene. If ψ=0.33, what are the product compositions and flow rates. Solve the problem graphically.

The equilibrium data for benzene-toluene system, in terms of mole fractions of benzene, at 1 atm are given below:

x	y
0	0
0.13	0.261
0.258	0.456
0.411	0.632
0.581	0.777
0.78	0.90
1	1

Problem 2

A mixture of 50 mole percent benzene and 50 mole percent toluene is subjected to *flash distillation* at a separator pressure of 1 atm.

Mole fraction benzene (Liquid phase), x	Mole fraction benzene (Vapor phase), y	Temperature, ^{0}C
0	0	110.6
0.13	0.261	105
0.258	0.456	100
0.411	0.632	95
0.581	0.777	90
0.78	0.9	85
1	1	80.1

a) Determine graphically the vapor and liquid compositions leaving the separator for each of the following values of D/F (where D is the distillate vapor flow rate and F is the feed flow rate): 0, 0.2, 0.4, 0.6, 0.8, 1.0. Show the operating lines and solutions for vapor and liquid compositions on the same graph.

(b) Based on your results, Does the liquid composition (that is, x) decrease, increase or remain constant with the increase in D/F ratio?

c) Based on your results, Does the vapor composition (that is, y) decrease, increase or remain constant with the increase in D/F ratio?

d) Based on your results, Does the temperature decrease, increase or remain constant with the increase in D/F ratio?

The equilibrium data for benzene-toluene system at 1 atm is given above.

Problem 3

A flash chamber operating at 1 atm is separating a binary mixture of pentane and hexane. The feed mixture is 50 mole % pentane. If ψ=0.50, what are the product compositions. Solve the problem analytically. The relative volatility of pentane with respect to hexane is 2.95. Repeat calculations for ψ values of 0.25 and 0.75.

Problem 4

A flash chamber operating at 80 °C and 100 kPa is separating 200 kmol/h of a feed that is 3-component mixture of species 1, 2 and 3. The feed composition is z_1=0.30, z_2 = 0.35 and z_3=0.35. Would flashing occur at given temperature (80 °C) and pressure (100 kPa))? If yes, find the product compositions and flow rates.

You can assume that Raoult's law is applicable. The vapor pressures of the pure species at 80 °C are as follows: $P_1^{Sat} = 195.75\ kPa$, $P_2^{Sat} = 97.84\ kPa$, $P_3^{Sat} = 50.32\ kPa$.

Problem 5

A 4-component feed mixture of species 1 to 4 with composition z_1=0.25, z_2 = 0.15, z_3=0.20 and z_4=0.40 undergoes flash distillation at 50 °C and 200 kPa. First verify that flashing will occur at the given conditions. What is the fraction of feed that goes to the vapor distillate product and what are the product compositions?

You can assume that Raoult's law is applicable. The vapor pressures of the pure species at 50 °C are as follows: $P_1^{Sat} = 1400\ kPa$, $P_2^{Sat} = 480\ kPa$, $P_3^{Sat} = 160\ kPa$, $P_4^{Sat} = 60\ kPa$.

Chapter 9

Multi-Stage Column Distillation

9.1 LEARNING OUTCOMES

Upon completion of this chapter, students will be able to:

- Devise different single-stage distillation schemes to separate a mixture
- Devise different multi-stage distillation cascades to separate a mixture
- Describe a multi-stage distillation column with its various components and explain its working
- Explain the meaning of constant molal overflow (*CMO*)
- Do external mass and energy balances on a multi-stage distillation column
- Do internal material balances over different sections of the column
- Develop equations for the operating lines of different sections (enriching/rectifying section, stripping section, feed stage)
- Determine the number of equilibrium stages required for a given separation of binary mixture using the McCabe-Thiele graphical method.
- Determine the optimum feed stage location in a distillation column
- Explain the influence of reflux ratio on the performance of a distillation column
- Explain different pinch conditions: feed pinch and tangent pinch
- Explain the concepts of total reflux, minimum reflux, and optimum reflux, and their implications on the performance of a distillation column
- Explain the influence of different thermal conditions of the feed on the performance of a distillation column
- Explain the difference between partial and total condensers
- Explain the difference between partial and total reboilers
- Describe special applications of distillation column including multiple feeds and side-stream draw-offs. Determine the number of equilibrium stages required in these applications.
- Describe the operation of a distillation column with direct steam injection and estimate the number of stages required
- Explain pressure-swing distillation of azeotropes and determine the number of stages required
- Estimate the number of equilibrium stages required for a given separation of binary mixture using the analytical approach developed by Smoker

9.2 INTRODUCTION

Single-stage flash distillation, although a very simple continuous process, is quite limited in terms of the degree of separation of components. It is mainly used for the separation of components with widely different boiling points. For separation of components with comparable volatility, single-stage flash distillation is not enough as the degree of separation between the components is poor. Therefore, a cascade consisting of a number of interconnected stages is required in order to improve the degree of separation between different components.

9.3 CONTINUOUS SINGLE-STAGE DISTILLATION PROCESSES

We first consider different single-stage schemes for continuous separation of feed mixture stream into different fractions based on the difference in the volatilities of components. Figures 9.1a-d show different single-stage distillation schemes. Figure 9.1a shows a single-stage flash distillation process where partial vaporization of the feed is achieved by adiabatic throttling of the liquid feed mixture. Figure 9.1b shows a single-stage distillation scheme where partial vaporization of liquid feed is achieved by heating the feed in a heat exchanger. The exit stream of the heat exchanger is a two-phase vapor/liquid mixture which is separated into vapor and liquid phases in the separator. In Figure 9.1c is shown a single-stage distillation process where partial condensation of the feed vapor is achieved by cooling the feed vapor mixture in a condenser. In Figure 9.1d, the feed vapor mixture is partially condensed by raising the pressure of the feed

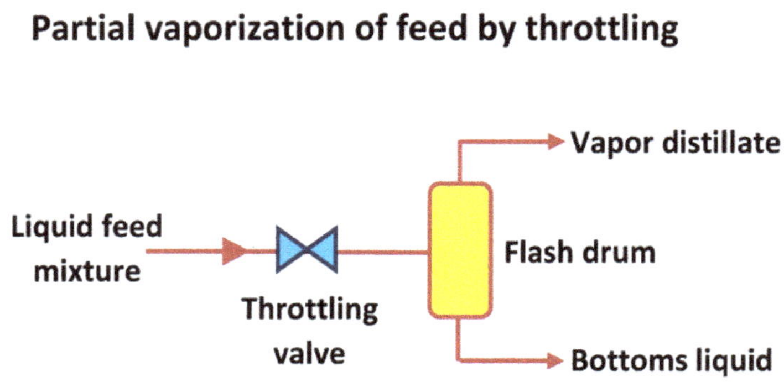

Figure 9.1a Single-stage flash distillation caused by adiabatic throttling of feed liquid mixture.

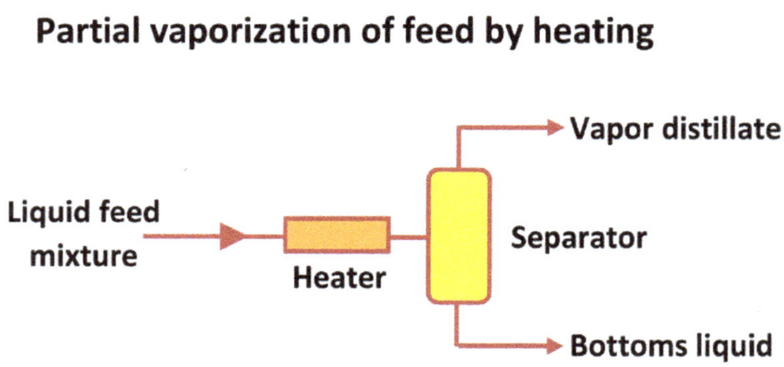

Figure 9.1b Single-stage distillation involving partial vaporization of feed liquid mixture by heating.

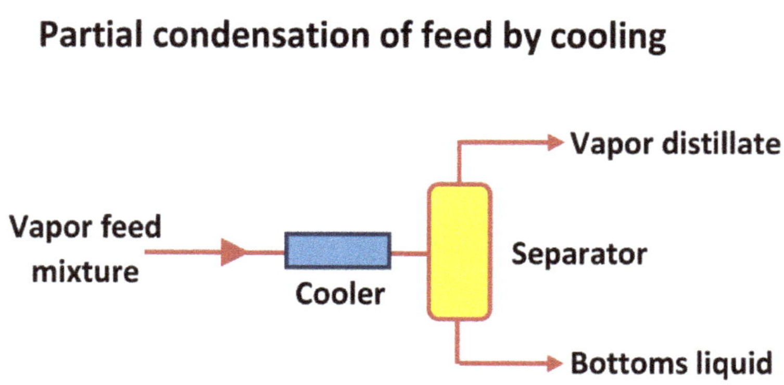

Figure 9.1c Single-stage distillation involving partial condensation of feed vapor mixture by cooling.

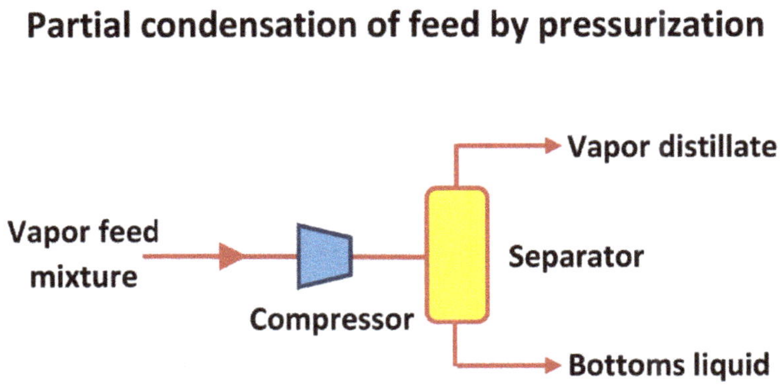

Figure 9.1d Single-stage distillation involving partial condensation of feed vapor mixture by raising the pressure.

using a compressor. In all these schemes, a single-phase feed (liquid or vapor) is transformed into two phases (liquid and vapor phases) which are separated in the separator. The vapor phase emerging from the separator is rich in more volatile components and the bottoms liquid stream is rich is less volatile components. The schemes shown in Figures 9.1b-d can be explained further with the aid of the phase diagrams if the feed mixture is assumed to be binary consisting of only two components and the vapor and liquid phases leaving the separator are at equilibrium with each other. Figure 9.2 shows partial vaporization of subcooled liquid feed upon heating at constant pressure. When the liquid feed is heated such that $BUBLT < T_{Separator} < DEWT$, two phases are formed as shown in the figure: vapor phase, rich in more volatile component and liquid phase, lean in more volatile component.

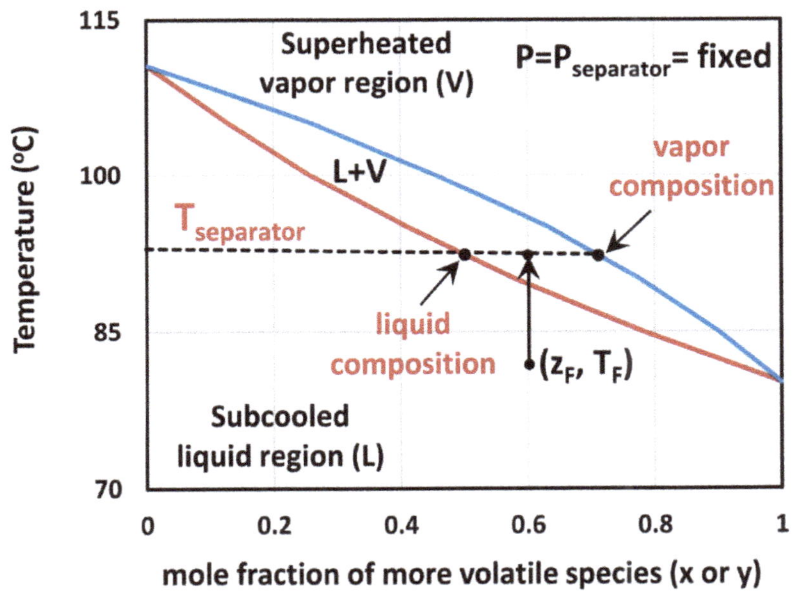

Figure 9.2 Single-stage distillation involving partial vaporization of feed liquid mixture by heating at constant pressure.

Figure 9.3 shows partial condensation of superheated vapor feed upon cooling at constant pressure. When the vapor feed is cooled such that $BUBLT < T_{Separator} < DEWT$, two phases are formed as shown in the figure: vapor phase, rich in more volatile component and liquid phase, lean in more volatile component.
Figure 9.4 shows partial condensation of superheated vapor feed upon pressurization at constant temperature. When the vapor feed is pressurized such that $DEWP < P_{Separator} < BUBLP$, two phases are

formed as shown in the figure: vapor phase, rich in more volatile component and liquid phase, lean in more volatile component.

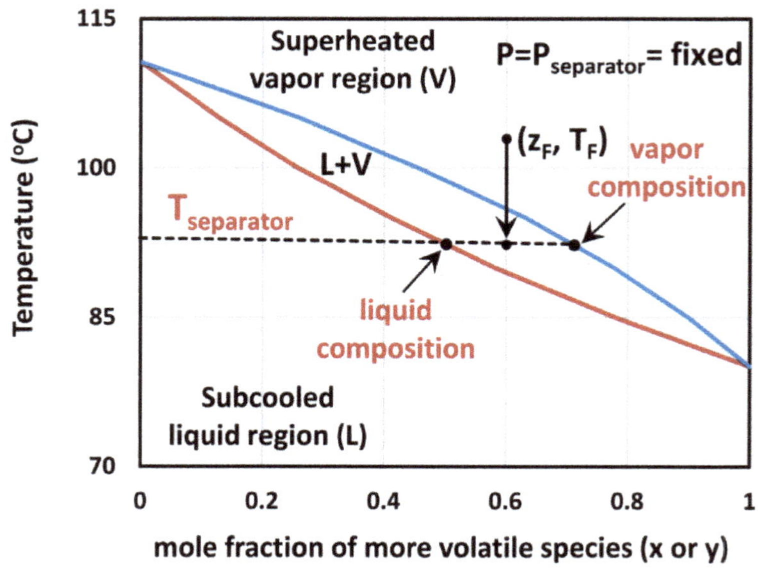

Figure 9.3 Single-stage distillation involving partial condensation of superheated vapor feed mixture by cooling at constant pressure.

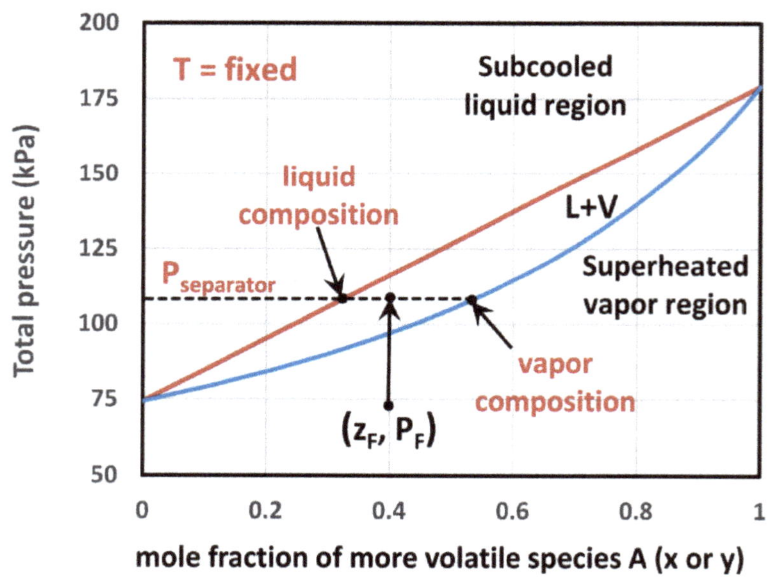

Figure 9.4 Single-stage distillation involving partial condensation of superheated vapor feed mixture by raising the pressure at constant temperature.

9.4 CASCADING OF STAGES

Figure 9.5 shows a distillation cascade consisting of five interconnected stages. The liquid feed mixture enters the cascade through stage 3. The feed is throttled to achieve partial vaporization in chamber 3. The vapor stream leaving chamber 3 is identified as V_3 and the liquid stream leaving chamber 3 is identified as L_3. The symbol V is used to indicate vapor flow rate and the symbol L indicates the liquid flow rate. The subscript on these variables indicates the stage from which the stream is emerging. The vapor and liquid streams emerging from chamber 3 are further processed in subsequent stages.

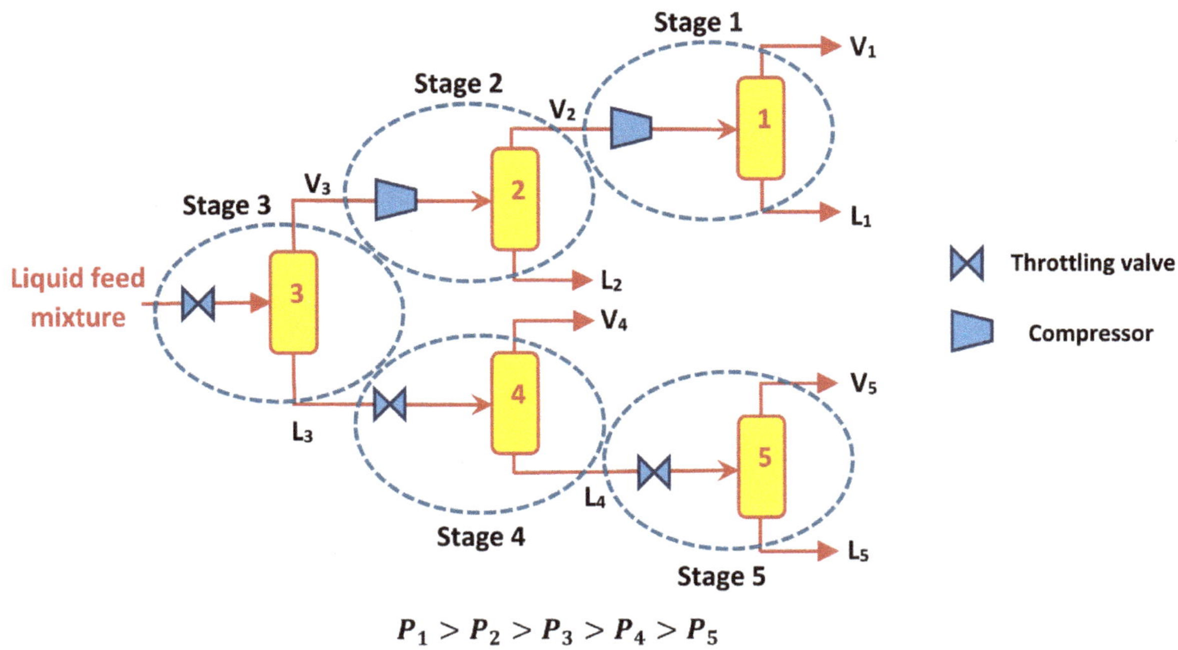

Figure 9.5 Distillation cascade consisting of five interconnected stages.

The vapor stream V_3 is compressed in a compressor resulting in partial condensation in chamber 2. Stage 2 produces a vapor stream V_2 and a liquid stream L_2. Note that when a vapor stream undergoes partial condensation, it is the less volatile component that is knocked out in the liquid phase and hence the vapor stream becomes richer in more volatile component upon partial condensation. The vapor stream V_2 is further compressed in stage 1 resulting in partial condensation and further enrichment of vapor stream with more volatile component. Assuming vapor phase to be a binary mixture with y representing the mole fraction of more volatile component, it follows that $y_1 > y_2 > y_3 > z_F$ where z_F is feed composition (mole fraction of more volatile component in the binary feed).

The liquid stream L_3 is throttled in stage 4 resulting in partial vaporization in chamber 4. Stage 4 produces a vapor stream V_4 and a liquid stream L_4. The liquid stream L_4 becomes richer in less volatile component as more volatile component is vaporized. The liquid stream L_4 is further throttled in stage 5 resulting in partial vaporization and further enrichment of liquid stream with less volatile component. Assuming liquid phase to be a binary mixture with x representing the mole fraction of more volatile component, it follows that $z_F > x_3 > x_4 > x_5$. Thus the cascade shown in Figure 9.5 has an increasing mole fraction of more volatile component in the vapor stream as we go from the feed stage to the top stage and a decreasing mole fraction of more volatile component in the liquid stream as we go from the feed stage to the bottom stage. Also note that pressure increases from bottom stage to the top stage of the cascade, that is, $P_1 > P_2 > P_3 > P_4 > P_5$.

The cascade shown in Figure 9.5 generates a number of intermediate streams which can be sent back to the preceding stage, as shown in Figure 9.6. However, the intermediate liquid stream emerging from a high pressure stage needs to be throttled to reduce the pressure before sending it back to the low-pressure preceding stage. Likewise the intermediate vapor stream emerging from a low pressure stage needs to be pressurized before sending it back to the preceding high-pressure stage. The cascade shown in Figure 9.6 has no intermediate products. Also, the cascade has counter-current flow arrangement, that is, vapor and liquid streams flow in opposite directions. This can be seen more clearly in Figure 9.7 where the same cascade is redrawn.

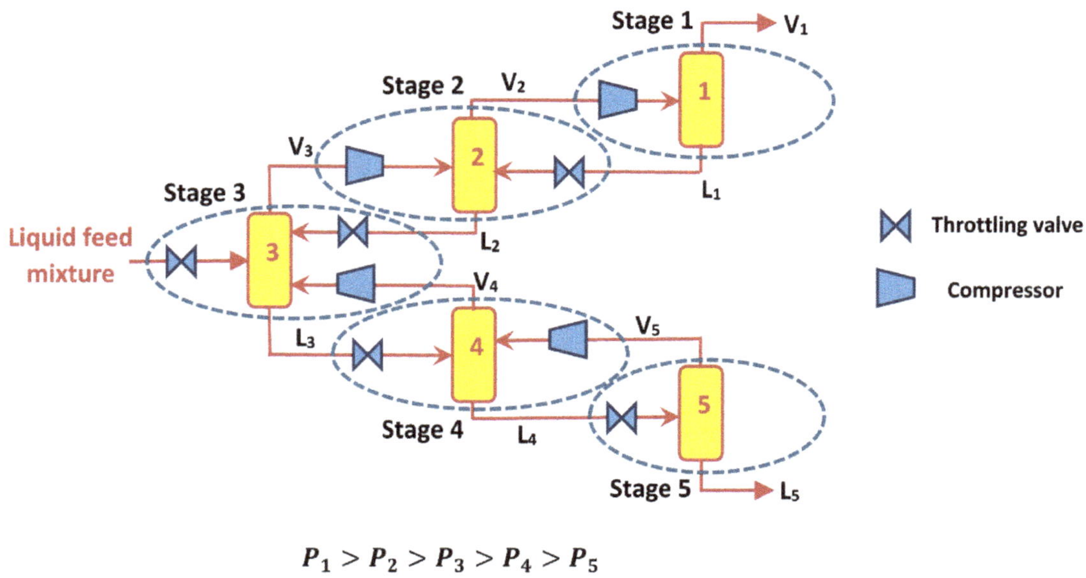

Figure 9.6 Distillation cascade of five stages with no intermediate streams.

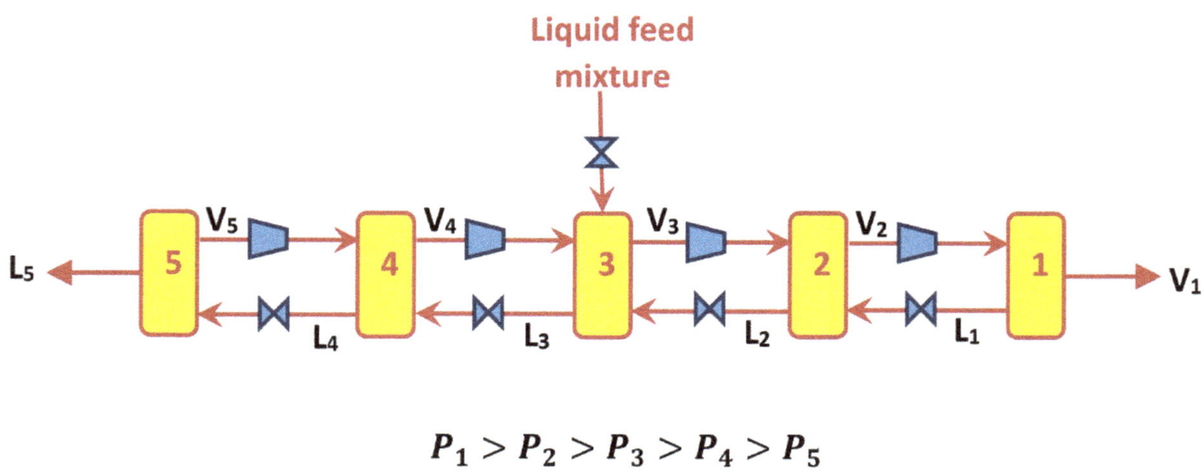

Figure 9.7 Counter-current distillation cascade of five stages.

At steady state, the feed stream splits into vapor-product stream V_1 and liquid-product stream L_5. The vapor-product stream is highly enriched in more volatile component and the liquid-product stream is highly enriched in less volatile component. Thus multi-stage cascading has the potential to achieve a very high degree of separation.

So far we have described *variable-pressure* distillation cascade. The pressure increases from one end of the cascade to the other while the temperature is kept constant. This type of cascade requires a large number of compressors and throttling valves. Thus the capital cost of such a cascade is high. Interestingly, the same degree of separation can be achieved using *variable-temperature* cascade (pressure nearly constant) which is much cheaper.

Figure 9.8 shows a *variable-temperature* distillation cascade consisting of five interconnected stages. The liquid feed mixture enters the cascade through stage 3. The feed is heated to achieve partial vaporization in chamber 3. The vapor stream V_3 emerging from chamber 3 is cooled resulting in partial condensation in chamber 2. Stage 2 produces a vapor stream V_2 and a liquid stream L_2. The vapor stream V_2 is further cooled in stage 1 resulting in partial condensation and further enrichment of vapor stream with more volatile component. Thus, $y_1 > y_2 > y_3 > z_F$ where y is the mole fraction of more volatile component in the vapor stream and z_F is the mole fraction of more volatile component in the binary feed.

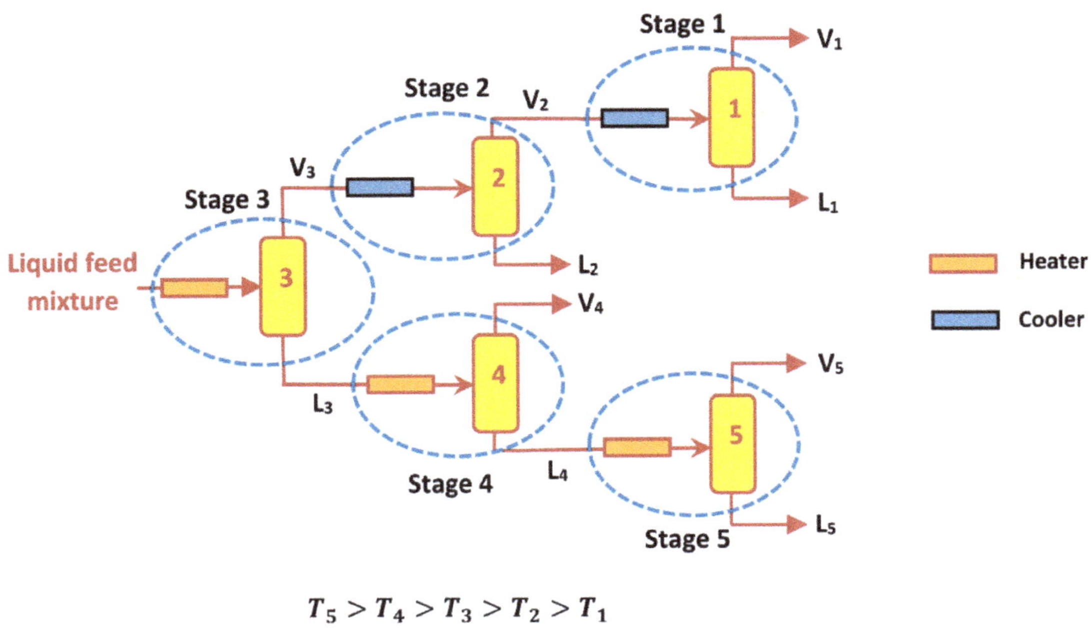

Figure 9.8 Variable-temperature distillation cascade consisting of five interconnected stages.

The liquid stream L_3 is heated in stage 4 resulting in partial vaporization in chamber 4. Stage 4 produces a vapor stream V_4 and a liquid stream L_4. The liquid stream L_4 becomes richer in less volatile component as more volatile component is vaporized. The liquid stream L_4 is further heated in stage 5 resulting in partial vaporization and further enrichment of liquid stream with less volatile component. Thus, $z_F > x_3 > x_4 > x_5$, where x is mole fraction of more volatile component in the liquid stream. Like the variable-pressure cascade, the variable-temperature cascade shown in Figure 9.8 has an increasing mole fraction of more volatile component in the vapor stream as we go from the feed stage to the top stage and a decreasing mole fraction of more volatile component in the liquid stream as we go from the feed stage to the bottom stage. However, the pressure is nearly constant and temperature increases from top stage to the bottom stage of the cascade, that is, $T_5 > T_4 > T_3 > T_2 > T_1$.

Like the variable-pressure cascade, the variable-temperature cascade shown in Figure 9.8 generates a number of intermediate streams which can be sent back to the preceding stage, as shown in Figure 9.9. The cascade shown in Figure 9.9 has no intermediate products. Also, the cascade has counter-current flow arrangement, that is, vapor and liquid streams flow in opposite directions. This can be seen more clearly in Figure 9.10 where the same cascade is redrawn. At steady state, the feed stream splits into vapor-product stream V_1 and liquid-product stream L_5. The vapor-product stream is highly enriched in more volatile component and the liquid-product stream is highly enriched in less volatile component. Thus the variable-temperature multi-stage cascade, like the variable-pressure multi-stage cascade, has the potential to achieve a very high degree of separation.

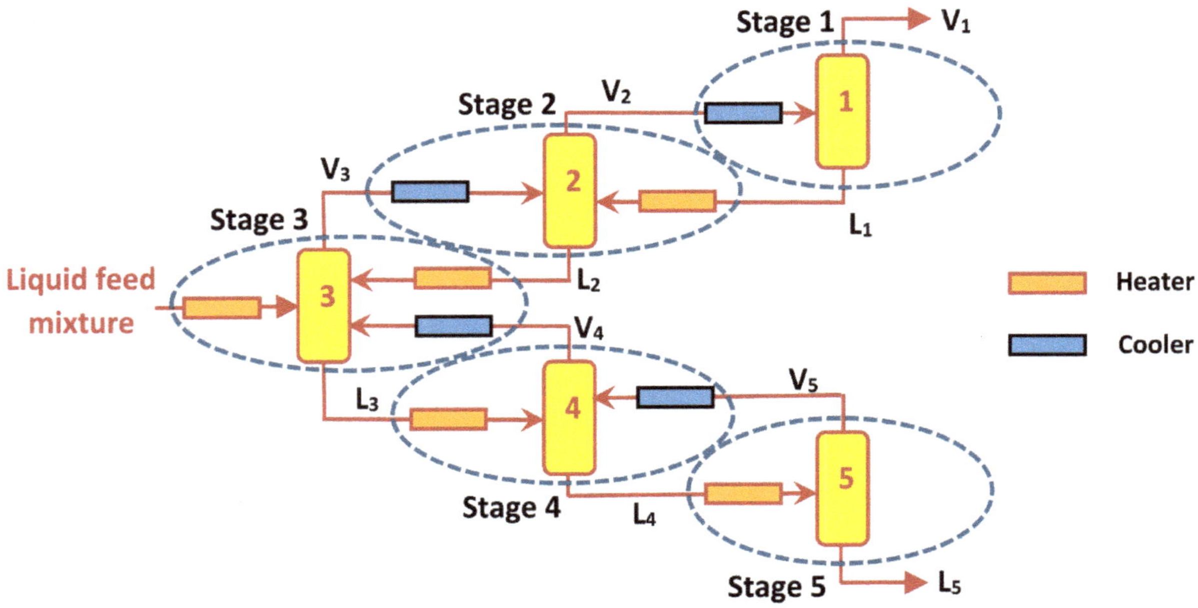

$$T_5 > T_4 > T_3 > T_2 > T_1$$

Figure 9.9 Variable-temperature distillation cascade of five stages with no intermediate streams.

It is important to recognize that in both variable-pressure and variable-temperature cascades the vapor stream entering any stage undergoes partial condensation and hence is enriched with more volatile component. At the same time, the liquid stream entering the stage undergoes partial vaporization and hence is stripped of more volatile component. If one mole of vapor is condensed for every mole of liquid vaporized, the molar flow rates V and L do not change from one stage to another at steady state. This assumption of constant molar flow rates of vapor and liquid streams is referred to as "*constant molal overflow (CMO)*" assumption. This is the key assumption made in the analysis and design of distillation columns using the McCabe-Thiele method.

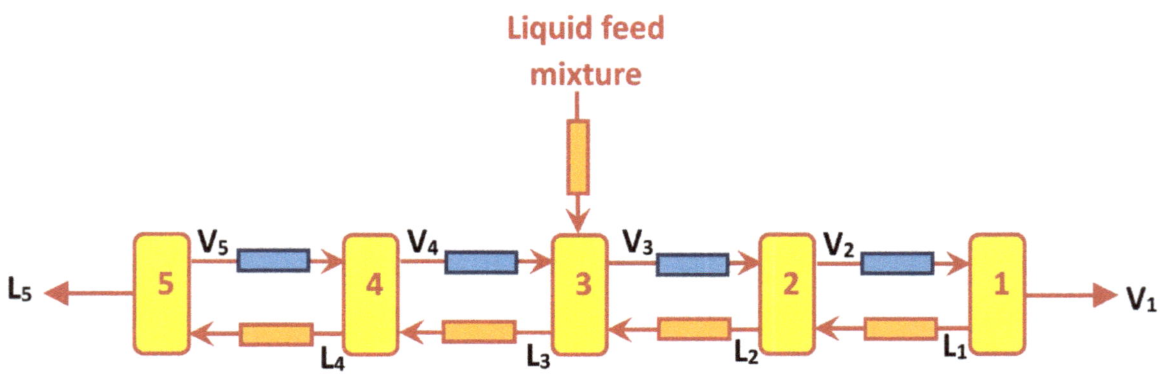

$$T_5 > T_4 > T_3 > T_2 > T_1$$

Figure 9.10 Counter-current variable-temperature distillation cascade of five stages.

9.5 MULTI-STAGE DISTILLATION COLUMN

Actual multi-stage distillation process is based on the modified form of the counter-current variable-temperature cascade shown in Figure 9.10. The modifications carried out are as follows: a) the heating of liquid stream and the cooling of vapor stream is carried out by direct contact and mixing of the streams at each stage. Thus there are no separate intermediate heaters or coolers involved; b) all of the heat required for distillation is supplied through the reboiler installed at the bottom end of the cascade; c) all of the cooling required is done through a condenser installed at the top of the cascade; and d) the entire cascade is built into a single cylindrical column constructed from some metallic material.

Figure 9.11 shows a schematic diagram of the internal view of a multi-stage distillation column. The rising hot vapor comes in contact with relatively cold liquid flowing in the downward direction. The two streams undergo intense mixing on the perforated/sieve tray due to mixing action of vapor bubbles. The bubbles are formed as the vapor stream emerges from tiny holes or perforations on the tray. Upon interaction of hot vapor bubbles with relatively cold liquid, some of the vapor phase condenses and simultaneously some of the liquid phase vaporizes at any given stage due to exchange of heat between the vapor and liquid streams. Simultaneous condensation of vapor and vaporization of liquid at any stage leads to enrichment of the vapor phase with more volatile components and stripping of the liquid phase of more volatile components. Simultaneous condensation and vaporization occur at each stage and due to this repeated condensations and vaporizations in a multi-stage column, the vapor stream becomes highly concentrated in more volatile components as we go up the column and the liquid stream becomes highly concentrated in less volatile components as we go down the column.

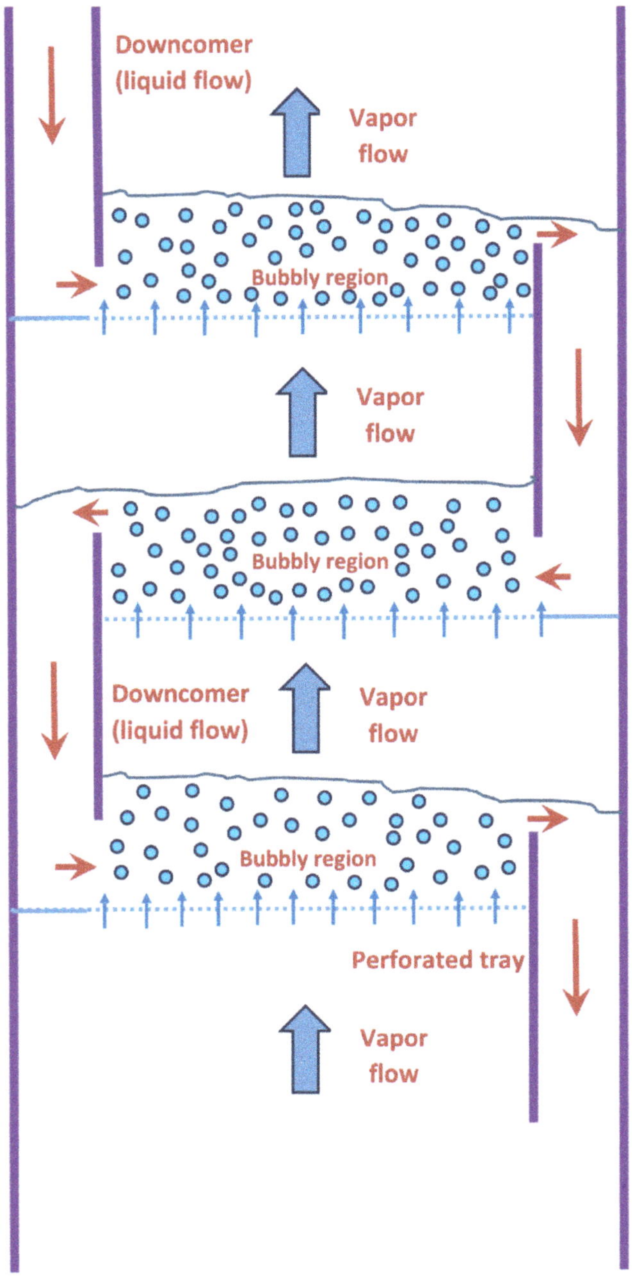

Figure 9.11 Schematic diagram of the internal view of a multi-stage distillation column.

Figures 9.12 – 9.13 further explain what happens at any nth stage of a distillation column. We assume that the stages are numbered from top down. Cold liquid stream with mole fraction of more volatile component x_{n-1} enters the nth stage from the stage immediately above the nth stage, that is, from (n-1)th stage, as shown in Figure 9.12. Hot vapor stream with mole fraction of more volatile component y_{n+1} enters the nth stage from the stage immediately below the nth stage, that is, from (n+1)th stage. The vapor and liquid streams entering the nth stage are not at equilibrium with each other. Upon intense mixing on the stage, the hot vapor stream undergoes partial condensation knocking out some of the less volatile component from the vapor phase and at the same time, the relatively cold liquid stream undergoes partial vaporization depleting the liquid stream of more volatile component. The heat required for vaporization comes from the heat released in condensation. The vapor and liquid streams leaving the nth stage are in equilibrium with each other having same temperature and equilibrium compositions. The changes in compositions of the vapor and liquid streams occurring on nth stage as the streams move from non-equilibrium state to equilibrium state are shown clearly on the T-x-y phase diagram in Figure 9.13. The vapor stream leaving the nth stage is enriched of more volatile component, that is, $y_n > y_{n+1}$ whereas the liquid stream leaving the nth stage is depleted of more volatile component, that is, $x_{n-1} > x_n$. This enrichment of vapor phase and depletion of liquid phase of more volatile component takes place at each stage of the column. As we go up the column, the concentration of more volatile component increases in both phases ($y_n > y_{n+1}$ and $x_{n-1} > x_n$) and the temperature decreases. Also note that the vapor streams within the column are at their dew points and the liquid streams are at their bubble points.

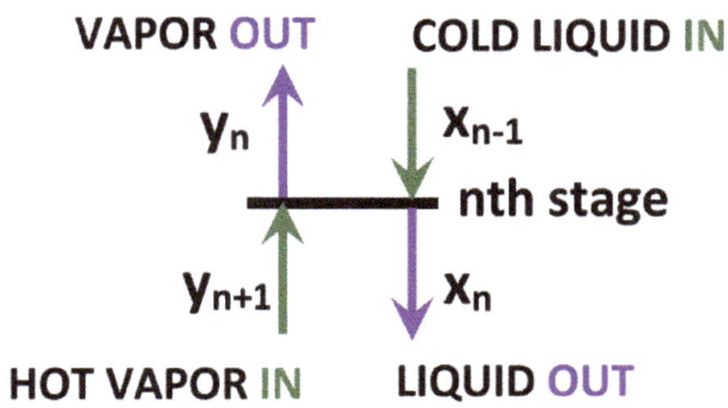

Figure 9.12 Fluid streams entering and leaving any nth stage of a distillation column.

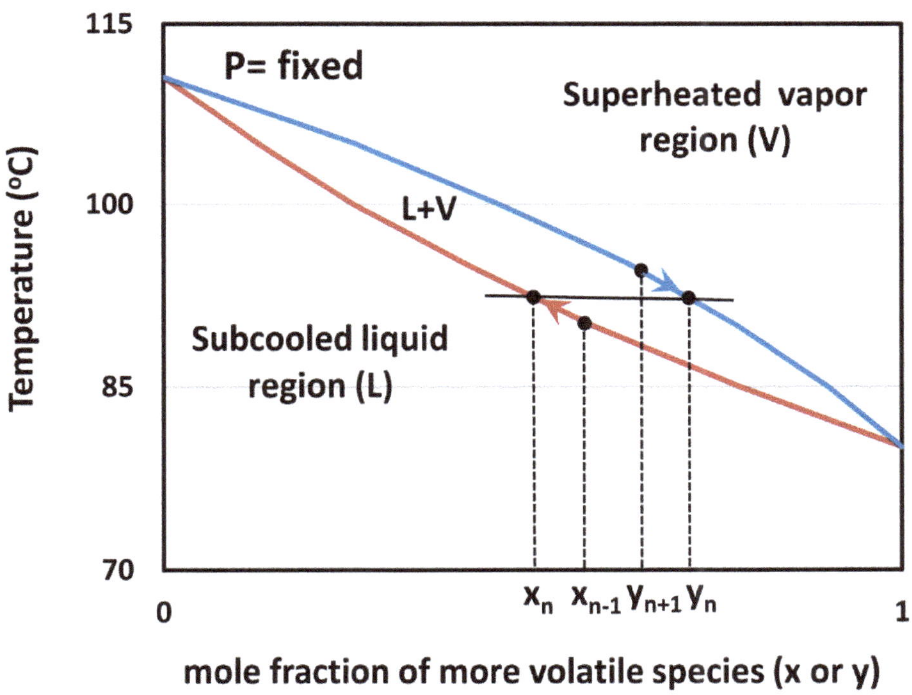

Figure 9.13 Enrichment of vapor stream and depletion of liquid stream of more volatile component occurring on any nth stage of a distillation column.

Figure 9.14 shows a schematic diagram of the full multi-stage distillation column. The section of the column below the feed stage is known as *stripping* section as the liquid stream is being stripped of more volatile components by the rising vapor. The section above the feed stage is called *enriching* or *rectifying* section as the rising vapor is being enriched with more volatile components.

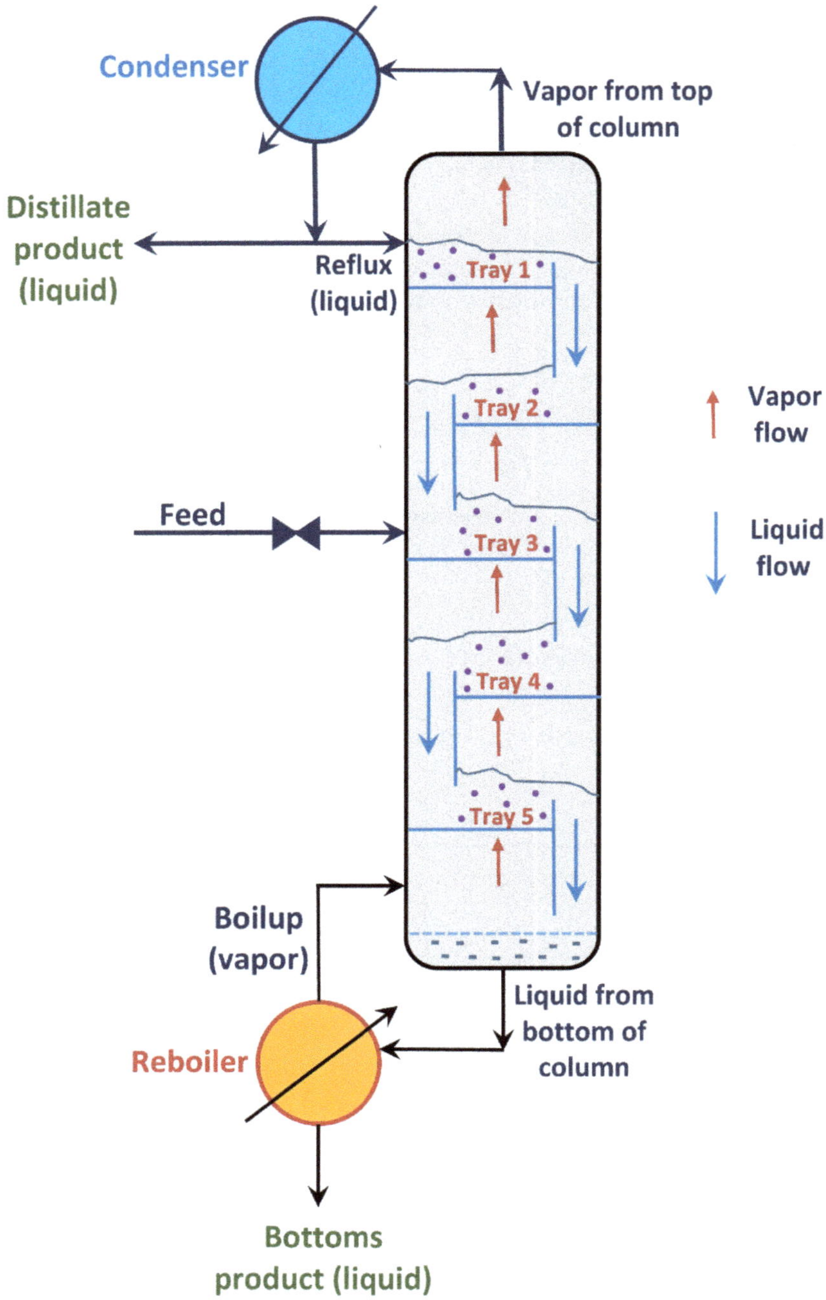

Figure 9.14 Schematic diagram of the full multi-stage distillation column.

The vapor stream emerging from the top of the column is cooled and condensed in a condenser. A portion of the condensed vapor is sent back to the column as liquid reflux. This liquid reflux is necessary in order to provide the downward flowing liquid in the enriching section to interact with the rising vapor and to enrich the vapor of more volatile components. If there is no reflux, no enrichment of vapor would occur in the enriching section and the vapor composition emerging at the top of the column will be the same as that of vapor rising from the feed plate.

The liquid stream emerging from the bottom of the column is heated in a reboiler. The vapor phase generated in the reboiler is sent back to the column as boil-up and some of the liquid is withdrawn from the reboiler as the bottoms-liquid product. The boilup vapor is necessary in order to provide the rising vapor in the stripping section to interact with the downward flowing liquid. If there is no rising hot vapor in the stripping section, the downward flowing liquid in the stripping section would not be stripped of the more volatile components. Consequently the composition of the liquid stream emerging from the bottom of the column will be the same as that of liquid leaving the feed plate.

9.6 ANALYSIS AND DESIGN OF MULTI-STAGE DISTILLATION COLUMN

The analysis and design of a multi-stage distillation column requires component and overall mass balances as well as energy balance. In this chapter, we will restrict ourselves to *binary distillation* involving separation of two-component mixtures (species A and B, with species A being more volatile). Multi-component distillation is covered in a separate chapter. We will assume "*constant molal overflow (CMO)*" in each section of the distillation column, that is: 1) the vapor molar flow rate in enriching section is constant from one stage to another; 2) the liquid molar flow rate in enriching section is constant from one stage to another; 3) the vapor molar flow rate in stripping section is constant from one stage to another; 4) the liquid molar flow rate in stripping section is constant from one stage to another. Note that in any given section (enriching or stripping), the vapor molar flow rate is generally different from the liquid molar flow rate. Also the molar flow rate of any stream (vapor or liquid) changes from one section to another due to feed entering the column. The assumption of *CMO* is valid if the molar latent heat of vaporization of the substance is constant independent of the composition, that is, the molar latent heats of vaporization (or condensation) of all the mixtures (regardless of the composition) are equal over the temperature and pressure operating ranges of the column. Consequently, every mole of condensing vapor vaporizes exactly one mole of liquid. This is possible only if the molar latent heat of vaporizations of pure components are nearly equal and that the heats of mixing of different components are negligible.

Before we do any balances, it is important to describe the notation to be used in the analysis of the distillation column. Figure 9.15 shows a schematic diagram of a distillation column with notation. The description of the notation used in Figure 9.15 is given in Table 9.1.

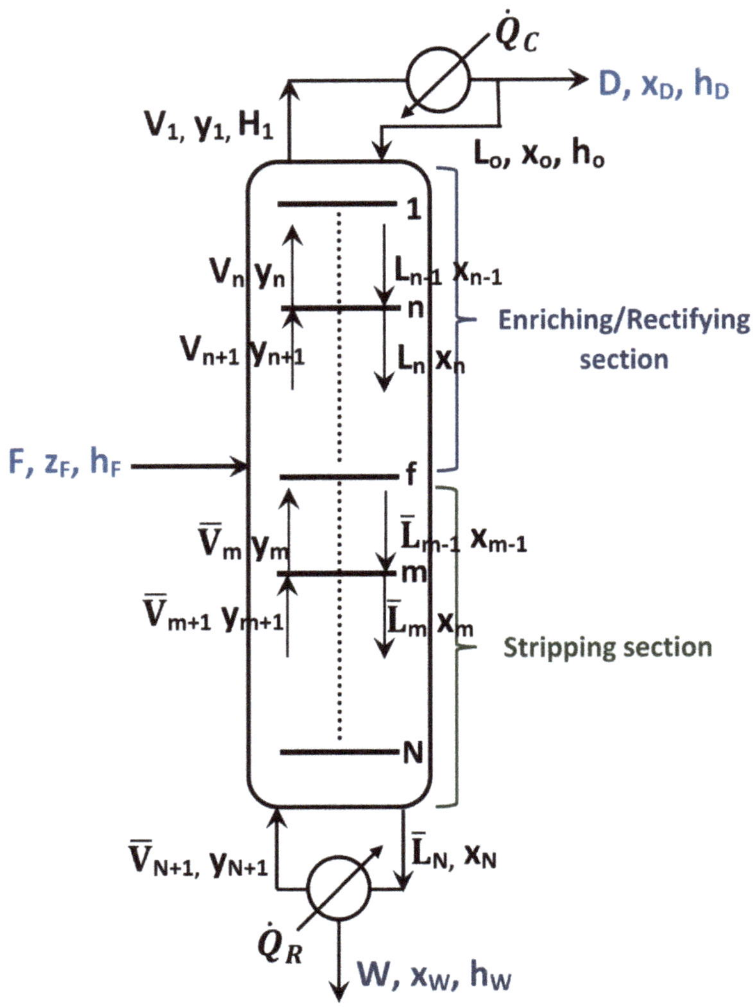

Figure 9.15 Schematic diagram of a distillation column indicating the notation used to describe various quantities in the analysis.

Table 9.1 Description of notation used in the analysis of distillation columns

Symbol	Description
D	Distillate liquid product molar flow rate
F	Feed molar flow rate
W	Bottoms liquid product molar flow rate
L	Molar flow rate of liquid stream in enriching or rectifying section
L_n	Molar flow rate of liquid stream leaving any n^{th} stage in the enriching section (note that subscript of a quantity indicates the number of the stage from which the stream is emerging)
$\bar{L}$	Molar flow rate of liquid stream in stripping section
$\bar{L}_m$	Molar flow rate of liquid stream leaving any m^{th} stage in the stripping section
V	Molar flow rate of vapor stream in enriching or rectifying section
V_n	Molar flow rate of vapor stream leaving any n^{th} stage in the enriching section (note that subscript of a quantity indicates the number of the stage from which the stream is emerging)
$\bar{V}$	Molar flow rate of vapor stream in stripping section

$\bar{V}_m$	Molar flow rate of vapor stream leaving any m[th] stage in the stripping section
h	Molar enthalpy of liquid phase
h_D	Molar enthalpy of distillate liquid product
h_F	Molar enthalpy of feed
h_o	Molar enthalpy of reflux liquid
h_W	Molar enthalpy of bottoms liquid product
H	Molar enthalpy of vapor phase
$\dot{Q}_C$	Condenser heat transfer rate
$\dot{Q}_R$	Reboiler heat transfer rate
x	Mole fraction of more volatile component in the liquid phase
x_n	Mole fraction of more volatile component in the liquid stream leaving n[th] stage
x_D	Mole fraction of more volatile component in the distillate liquid product
x_W	Mole fraction of more volatile component in the bottoms liquid product
y	Mole fraction of more volatile component in the vapor phase
y_n	Mole fraction of more volatile component in the vapor stream leaving n[th] stage
z_F	Mole fraction of more volatile component in the feed
f	Indicates feed tray/stage
N	Total number of stages inside the column

9.6.1 External Balances

Referring to Figure 9.16, the overall material balance at steady state is:

$$F = D + W \tag{9.1}$$

The component (more volatile species) material balance at steady state is:

$$z_F F = x_D D + x_W W \tag{9.2}$$

Eliminating W from Equations (9.1) and (9.2) gives:

$$\frac{D}{F} = \frac{z_F - x_W}{x_D - x_W} \tag{9.3}$$

Eliminating D from Equations (9.1) and (9.2) gives:

$$\frac{W}{F} = \frac{x_D - z_F}{x_D - x_W} \tag{9.4}$$

Given the feed flow rate and compositions of the feed, distillate and bottoms product, Equations (9.3) and (9.4) can be used to calculate the distillate and bottoms product flow rates.

Next we consider energy balance on the column. For multiple inlets and outlets, energy balance (assuming steady state, negligible kinetic and potential energy changes, see Chapter 7) over a given control volume can be written as:

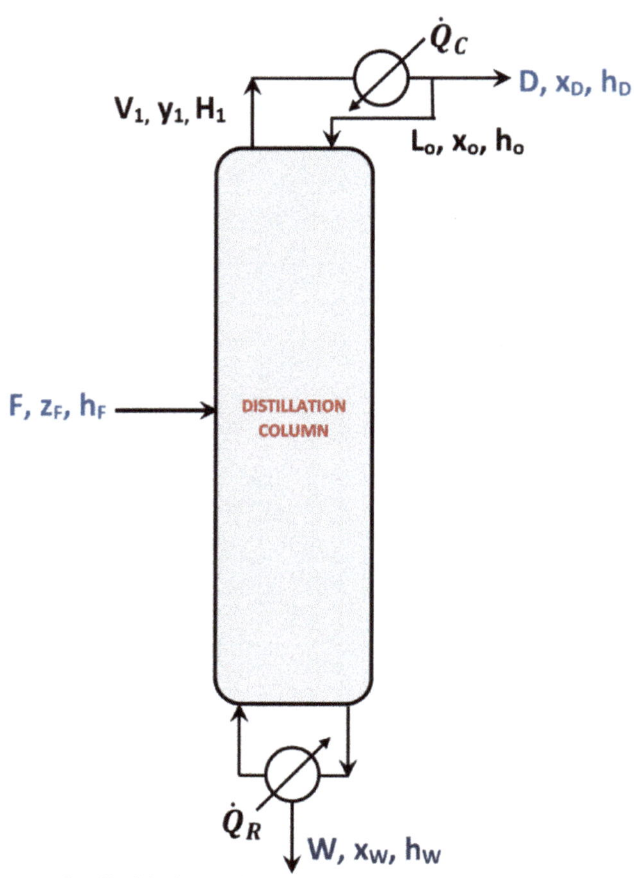

Figure 9.16 Schematic diagram of a distillation column for external material and energy balances.

$$\Sigma_{INLETS}\ (F_i\, h_i) + \Sigma\,\dot{Q} = \Sigma_{OUTLETS}\ (\, F_i\, h_i) + \Sigma\,\dot{W}_{sh} \tag{9.5}$$

where F_i is the molar flow rate of any fluid stream, $\dot{Q}$ is the rate of heat transfer, $\dot{W}_{sh}$ is the rate of shaft work, and h_i is the molar enthalpy of the fluid stream. When applied to the distillation column (Figure 9.16), Equation (9.5) reduces to

$$Fh_F + \dot{Q}_C + \dot{Q}_R = Dh_D + Wh_W \tag{9.6}$$

Note that there is no shaft work involved. Given the specific enthalpies of feed, distillate and bottoms product, Equation (9.6) can be used to estimate the heat duty $\dot{Q}_R$ of the reboiler provided that the heat duty of the condenser $\dot{Q}_C$ is known. We need to do mass and energy balances around the condenser in order to estimate the heat duty $\dot{Q}_C$ of the condenser.

Material balance around the condenser gives:

$$V_1 = D + L_o \tag{9.7}$$

Equation (9.7) can be re-written in terms of the reflux ratio *R*:

$$V_1 = (1 + R)D \tag{9.8}$$

where the reflux ratio *R* is defined as:

$$R = L_o/D \tag{9.9}$$

The application of energy balance (Equation (9.5)) around the condenser gives:

$$V_1 H_1 + \dot{Q}_C = D h_D + L_o h_o \quad (9.10)$$

From Equations (9.8)-(9.10), it can be readily shown that:

$$D(1+R)H_1 + \dot{Q}_C = D(h_D + R h_o) \quad (9.11)$$

As $h_o = h_D$, Equation (9.11) yields

$$\dot{Q}_C = -D(1+R)(H_1 - h_D) \quad (9.12)$$

The condenser heat duty $\dot{Q}_C$ is negative as heat flows out of the control volume. Heat is removed from the vapor emerging from the top of the distillation column using cold water. Also note that:

$$\lambda_D = (H_1 - h_D) \quad (9.13)$$

where λ_D is the molar latent heat of vaporization of distillate product. Thus Equation (9.12) can be re-written as:

$$\dot{Q}_C = -D(1+R)\lambda_D \quad (9.14)$$

Equation (9.12) allows us to calculate the heat duty of the condenser. Once $\dot{Q}_C$ is known, the reboiler heat duty $\dot{Q}_R$ can be calculated from the following version of Equation (9.6):

$$\dot{Q}_R = D h_D + W h_W - F h_F - \dot{Q}_C \quad (9.15)$$

9.6.2 Internal Balances

One important issue in the design of distillation columns is the number of equilibrium stages needed to carry out the specified separation. For example, given the feed information (flow rate, feed composition, feed thermal condition), distillate and bottoms product compositions, and the reflux ratio, how many equilibrium stages are needed to carry out the distillation? Also, which is the optimum stage to feed the column? It is also useful to be able to plot temperature and composition profiles for the column. These profiles show variations in temperature and composition as we go up or down the column. In order to address all these issues, it is essential to carry out internal material balances. However, internal energy balances are not needed if we assume "*constant molal overflow (CMO)*" in each section of the distillation column.

Enriching Section

Consider a material balance envelope covering the top portion of the distillation column including n stages of the enriching section, as shown in Figure 9.17. The overall mass balance over the envelope gives:

$$V_{n+1} = L_n + D \quad (9.16)$$

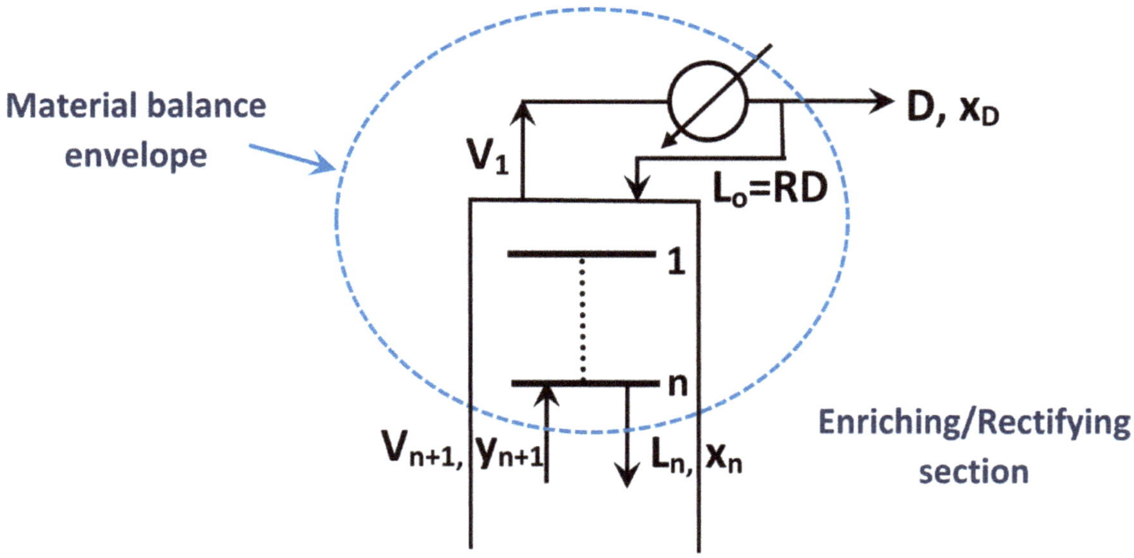

Figure 9.17 Material balance over the top portion of the distillation column including first *n* stages of the enriching section.

The net flow rate of material (including both components) in the upward direction is $D\ (= V_{n+1} - L_n)$. Although the flow rates of vapor and liquid streams may change from one stage to another within the enriching section, their difference is constant at steady state and is equal to D.
The component mass balance on more volatile species gives:

$$y_{n+1}V_{n+1} = x_nL_n + x_DD \tag{9.17}$$

The net flow rate of more volatile component in the upward direction is constant at steady state and is equal to $x_DD\ (= y_{n+1}V_{n+1} - x_nL_n)$.

From Equation (9.17), it follows that:

$$y_{n+1} = \left(\frac{L_n}{V_{n+1}}\right)x_n + \left(\frac{D}{V_{n+1}}\right)x_D \tag{9.18}$$

If we now assume *CMO,* it implies that:

$$V_1 = V_2 = \cdots = V_{n+1} = \cdots = V_f = V = constant \tag{9.19}$$

$$L_o = L_1 = \cdots = L_n = \cdots = L_{f-1} = L = constant \tag{9.20}$$

where subscript "*f*" indicates the feed plate. Thus, Equation (9.18) can be re-written as:

$$y_{n+1} = \left(\frac{L}{V}\right)x_n + \left(\frac{D}{V}\right)x_D \tag{9.21}$$

This is the equation of the *operating line* in the enriching section. It relates the compositions of the passing vapor and liquid streams at the same level in the enriching section of the column. For example, the vapor stream V_{n+1} in Figure 9.17 is passing the liquid stream L_n at the same level in the enriching section of the column. Hence their compositions are related by the operating line Equation (9.21). Using the overall material balance, Equation (9.16), and $L_o = L = RD$, the equation for the operating line, Equation (9.21), can be expressed as:

$$y_{n+1} = \left(\frac{R}{1+R}\right)x_n + \left(\frac{x_D}{1+R}\right) \quad (9.22)$$

Thus the operating line for the enriching section is a straight line of slope of $R/(1+R)$ and an intercept of $x_D/(1+R)$. Note that slope of enriching section operating line is less than unity. Only when $R \to \infty$, the slope is unity. Another important characteristic of the enriching section operating line is that it passes through the point (x_D, x_D), that is when $x_n = x_D$, $y_{n+1} = x_D$.

Stripping Section

Consider a material balance envelope shown in Figure 9.18 covering the bottom portion of the distillation column including stages numbered (m+1) to last stage N of the stripping section where m is any mth stage in the stripping section. The stages in the column are numbered from top down. The overall mass balance over the envelope gives:

$$\bar{L}_m = \bar{V}_{m+1} + W \quad (9.23)$$

The net flow rate of material (including both components) in the downward direction is W $(= \bar{L}_m - \bar{V}_{m+1})$. Although the flow rates of liquid and vapor streams may change from one stage to another within the stripping section, their difference is constant at steady state and is equal to W.

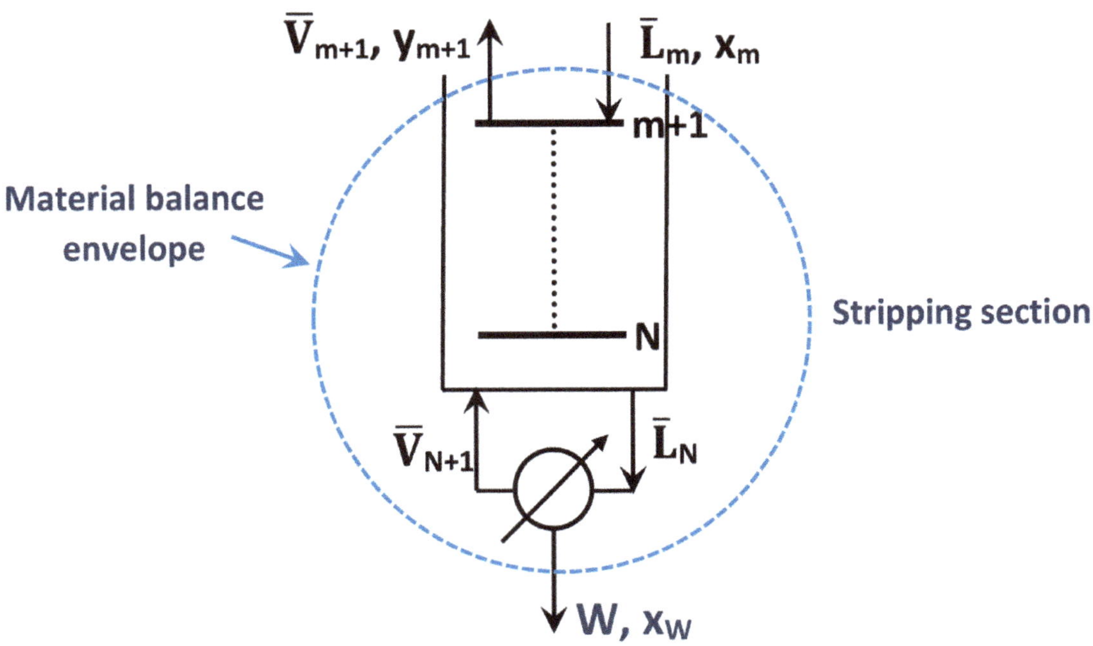

Figure 9.18 Material balance over the bottom portion of the distillation column including stages numbered (m+1) to N of the stripping section (m is any mth stage in the stripping section and N is the last stage).

The component mass balance on more volatile species gives:

$$x_m\bar{L}_m = y_{m+1}\bar{V}_{m+1} + x_W W \quad (9.24)$$

The net flow rate of more volatile component in the downward direction is constant at steady state and is equal to $x_W W \; (= x_m \bar{L}_m - y_{m+1}\bar{V}_{m+1})$.

From Equations (9.23) and (9.24), it follows that:

$$y_{m+1} = \left(\frac{\bar{L}_m}{\bar{V}_{m+1}}\right) x_m - \left(\frac{W}{\bar{V}_{m+1}}\right) x_W \tag{9.25}$$

If we now assume *CMO,* it implies that:

$$\bar{V}_{f+1} = \bar{V}_{f+2} = \cdots = \bar{V}_{N+1} = \bar{V} = constant \tag{9.26}$$

$$\bar{L}_f = \bar{L}_{f+1} = \cdots = \bar{L}_N = \bar{L} = constant \tag{9.27}$$

where subscript "*f*" refers to the feed plate. Thus, Equation (9.25) can be re-written as:

$$y_{m+1} = \left(\frac{\bar{L}}{\bar{V}}\right) x_m - \left(\frac{W}{\bar{V}}\right) x_W \tag{9.28}$$

This is the equation of the *operating line* in the stripping section. It relates the compositions of the passing vapor and liquid streams at the same level in the stripping section of the column. For example, the vapor stream $\bar{V}_{m+1}$ in Figure 9.18 is passing the liquid stream $\bar{L}_m$ at the same level in the stripping section of the column. Hence their compositions are related by the operating line Equation (9.28). Using the overall material balance, Equation (9.23), the equation for the operating line, Equation (9.28), could also be expressed as:

$$y_{m+1} = \left(\frac{\bar{L}}{\bar{L}-W}\right) x_m - \left(\frac{W}{\bar{L}-W}\right) x_W \tag{9.29}$$

Thus the operating line for the stripping section is a straight line of slope $\bar{L}/\bar{V} \; (= \bar{L}/(\bar{L}-W))$. Note that slope of stripping section operating line is greater than unity. Another important characteristic of the stripping section operating line is that it passes through the point (x_W, x_W), that is when $x_m = x_W, y_{m+1} = x_W$.

Feed Stage

As already noted, the section of the distillation column above the feed stage/tray is the enriching or rectifying section. The section below the feed stage is the stripping section. The enriching section operating line is valid in the enriching section whereas the stripping section operating line is valid in the stripping section. Thus, it is obvious that at the feed stage we have to switch from enriching section operating line to stripping section operating line or vice versa. In fact, the feed stage is identified as the stage where we switch from one operating line to another.

It is important to realize that the feed stage is not necessarily the one where the two operating lines (enriching and stripping) intersect. However, the stage where the enriching and stripping operating lines intersect is the *optimum* feed stage. This point will become clear in the next section dealing with McCabe-Thiele graphical approach for the analysis and design of distillation columns. For now, let us define the *feed line*. Feed line is defined to be one on which all possible intersections of the enriching and stripping operating lines fall. Thus feed line is the locus of the intersection of enriching and stripping operating lines.

The equation for the feed line can be derived as follows. The enriching and stripping operating lines can be re-written as:

$$y_{n+1} V = x_n L + x_D D \tag{9.30}$$

$$y_{m+1}\bar{V} = x_m\bar{L} - x_W W \tag{9.31}$$

At the intersection of the operating lines, *y*'s are equal and *x*'s are equal:

$$x_n = x_m = x \tag{9.32}$$

$$y_{n+1} = y_{m+1} = y \tag{9.33}$$

Consequently,

$$yV = xL + x_D D \tag{9.34}$$

$$y\bar{V} = x\bar{L} - x_W W \tag{9.35}$$

In order to obtain the locus of intersections of these equations, we subtract Equation (9.35) from Equation (9.34):

$$y(V - \bar{V}) = x(L - \bar{L}) + x_D D + x_W W \tag{9.36}$$

This is one form of the equation for feed line. Upon re-arrangement, Equation (9.36) can be re-written as:

$$y = \left(\frac{L-\bar{L}}{V-\bar{V}}\right)x + \left(\frac{x_D D + x_W W}{V-\bar{V}}\right) \tag{9.37}$$

Using Equation (9.2), Equation (9.37) becomes

$$y = \left(\frac{L-\bar{L}}{V-\bar{V}}\right)x + \left(\frac{z_F F}{V-\bar{V}}\right) \tag{9.38}$$

This equation can be further simplified if we consider material balance around the feed stage. Referring to Figure 9.19, the material balance around the feed stage gives:

$$F + L + \bar{V} = V + \bar{L} \tag{9.39}$$

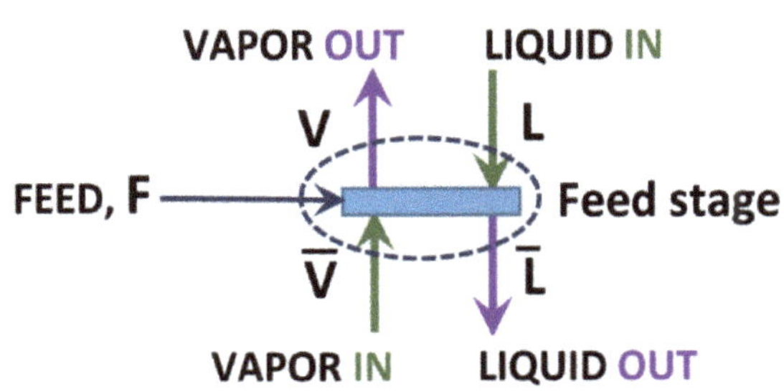

Figure 9.19 Material balance on the feed stage.

Let us now define a new term q as:

$$q = \left(\frac{\bar{L}-L}{F}\right) \tag{9.40}$$

Obviously q is the change in liquid flow rate, per unit feed flow rate, in going from enriching section to stripping section of the column. From Equations (9.39) and (9.40), it follows that:

$$V - \bar{V} = (1 - q)F \tag{9.41}$$

The equation of feed line, Equation (9.38) can now be written as:

$$y = -\left(\frac{q}{1-q}\right)x + \left(\frac{z_F}{1-q}\right) \tag{9.42}$$

This is the final form of the feed line, also referred to as $q - line$. The q-line is a straight line of slope of -q/(1-q) and it passes through the point (z_F, z_F), that is, when $x = z_F$, $y = z_F$.

9.6.3 McCabe-Thiele Method of Analysis and Design of Distillation Columns

McCabe-Thiele method is a graphical procedure for determining the number of equilibrium stages required for a given separation of binary mixtures. Given the feed information (feed composition and feed thermal condition), distillate and bottoms product compositions, and the reflux ratio, the number of equilibrium stages needed to carry out the distillation can be estimated graphically. Also, the optimum stage where to feed the column can be determined, in addition to temperature and composition profiles for the column. The McCabe-Thiele method requires y-x equilibrium curve and assumes *CMO* in each section of the column. The steps involved in the McCabe-Thiele method are as follows:

- Draw the equilibrium y-x curve at the operating pressure of the distillation column, where y and x are mole fractions of the more volatile species of the binary mixture.
- Draw y = x line on the same graph
- Locate the feed point (z_F, z_F), distillate-product composition point (x_D, x_D), and bottoms-product composition point (x_W, x_W). All these points fall on the y = x line.
- Draw the enriching/rectifying operating line. The enriching operating line passes through the point (x_D, x_D) and intersects the y-axis at $x_D/(1 + R)$.
- From the given feed information, determine the q-value and draw the q-line (feed line). The q-line passes through the feed point (z_F, z_F) and has a slope of -q/(1-q) and y-intercept of z_F/(1-q).
- Draw the stripping section operating line. Note that the intersection of the feed line (q-line) and enriching section operating line fixes one point through which the stripping section operating line must pass. The other point through which the stripping section operating line passes is the bottoms-product composition point (x_W, x_W).
- Step off the stages using staircase type construction, as shown in Figure 9.20. The stepping off the stages can begin either at the top of the column or at the bottom of the column. In this book, we will generally begin stepping off stages at the top of the column. We will assume that the condenser at the top of the column is a *total condenser* (*partial condenser*, if used, itself acts as an equilibrium stage). Beginning at the distillate-composition point (x_D, x_D) on the enriching-section operating line, move horizontally to the equilibrium curve (see Figure 9.20). Then move vertically down from the equilibrium curve to the enriching-section operating line. This completes the construction of the first equilibrium stage of the column. The equilibrium stage is represented by a triangle on the graph. The points on the operating line give the compositions of the liquid and vapor streams passing each other at the top and the bottom of the stage. The point on the equilibrium curve gives

the compositions of the liquid and vapor streams leaving the equilibrium stage. Figure 9.21 shows the passing and leaving streams corresponding to stage 1.

- Continue the graphical construction process of triangle-shaped equilibrium stages as shown in Figure 9.20 by moving horizontally from the operating line to the equilibrium curve and vertically from the equilibrium curve to the operating line as long as we are in the enriching section.

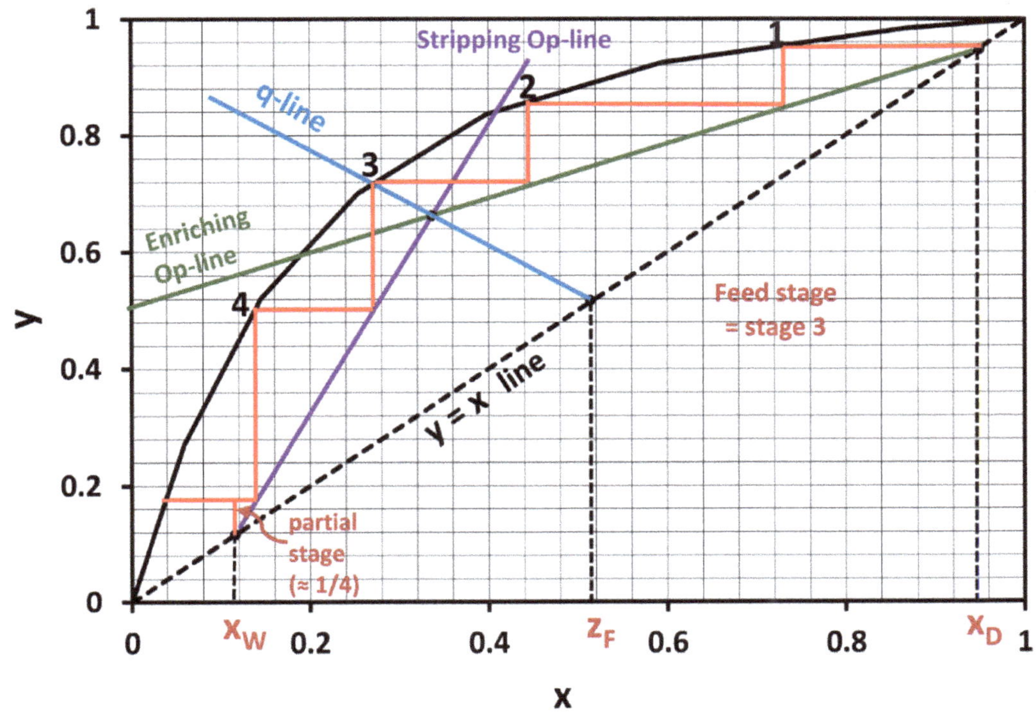

Figure 9.20 McCabe-Thiele diagram of distillation column with optimum feed-stage location.

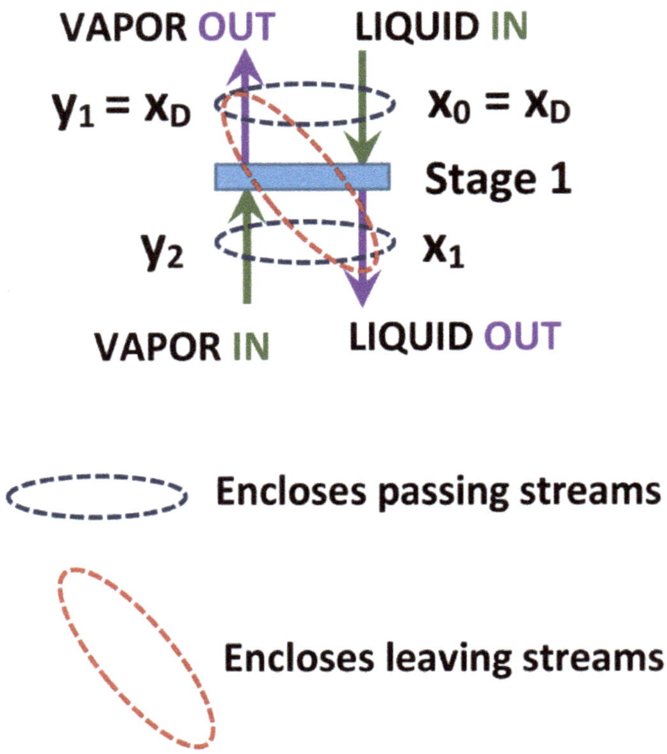

Figure 9.21 Passing and leaving streams on stage 1.

- At the feed stage, we need to switch from the enriching-section operating line to the stripping-section operating line. The feed stage is represented by a triangle that has one point on the enriching-section operating line, one point on the stripping-section operating line, and the third point on the equilibrium curve. For the example shown in Figure 9.20, the feed stage is stage number 3.

- The process of graphical construction of triangular-shaped equilibrium stages is continued after the feed stage by moving horizontally from the stripping-section operating line to the equilibrium curve and vertically from the equilibrium curve to the stripping-section operating line. The construction procedure is stopped once the composition of the bottoms-liquid stream is reached. Note that the last stage is often a partial stage. We can determine the fraction of a full stage that corresponds to the partial stage as follows:

-

$$Fraction = \frac{horizontal\ distance\ from\ operating\ line\ to\ x_W}{total\ horizontal\ distance\ from\ operating\ line\ to\ equilibrium\ curve}$$

  The fractional stage is rounded off to determine the total number of stages required.

- It should be noted that re-boiler acts as an equilibrium stage by itself and is part of the total number of equilibrium stages constructed using the McCabe-Thiele method as the construction process ends at the composition of the bottoms-liquid stream emerging from the re-boiler.

Although the McCabe-Thiele method is popular and is used extensively for the design of distillation columns for binary mixtures, its drawback is that it is graphical in nature. When the relative volatility of more volatile component is only slightly or moderately larger than one, that is, the binary mixture is a close boiling mixture, the number of stages required for a given separation is very large making it impractical to construct a McCabe-Thiele diagram. In such cases an analytical expression developed by Smoker (E.H. Smoker, Trans. AIChE, Vol. 34, 165, 1938) can be used to estimate the number of equilibrium stages. The Smoker equation is discussed in the appendix of this chapter. The key assumptions in the derivation of the Smoker equation are: the relative volatility is constant (McCabe-Thiele method does not assume constant relative volatility); the mixture is binary; the column has one feed stream and no side product streams; and *CMO* is valid in each section (rectifying and stripping) of the column.

Optimum Feed-stage Location

The optimum feed stage is one which includes the point of intersection of the enriching and stripping operating lines. For example, in Figure 9.20 the feed stage (stage number 3) includes the point of intersection of the two operating lines. Hence stage 3 is the optimum feed stage for the separation considered in Figure 9.20.

If the feed stage is located anywhere other than the optimum location, an unnecessarily large number of stages are required to effect the same separation. As an example, Figure 9.22 shows the McCabe-Thiele diagram where the feed stage (stage number 5 in the diagram) is chosen below the optimum location where the intersection of the two operating lines takes place. Now we need 5.67 equilibrium stages for the same separation as in Figure 9.20 where only 4.25 equilibrium stages were required. Note that Figures 9.20 and 9.22 have the same operating lines and the same values of x_D, x_W, z_F and deal with the same binary system.

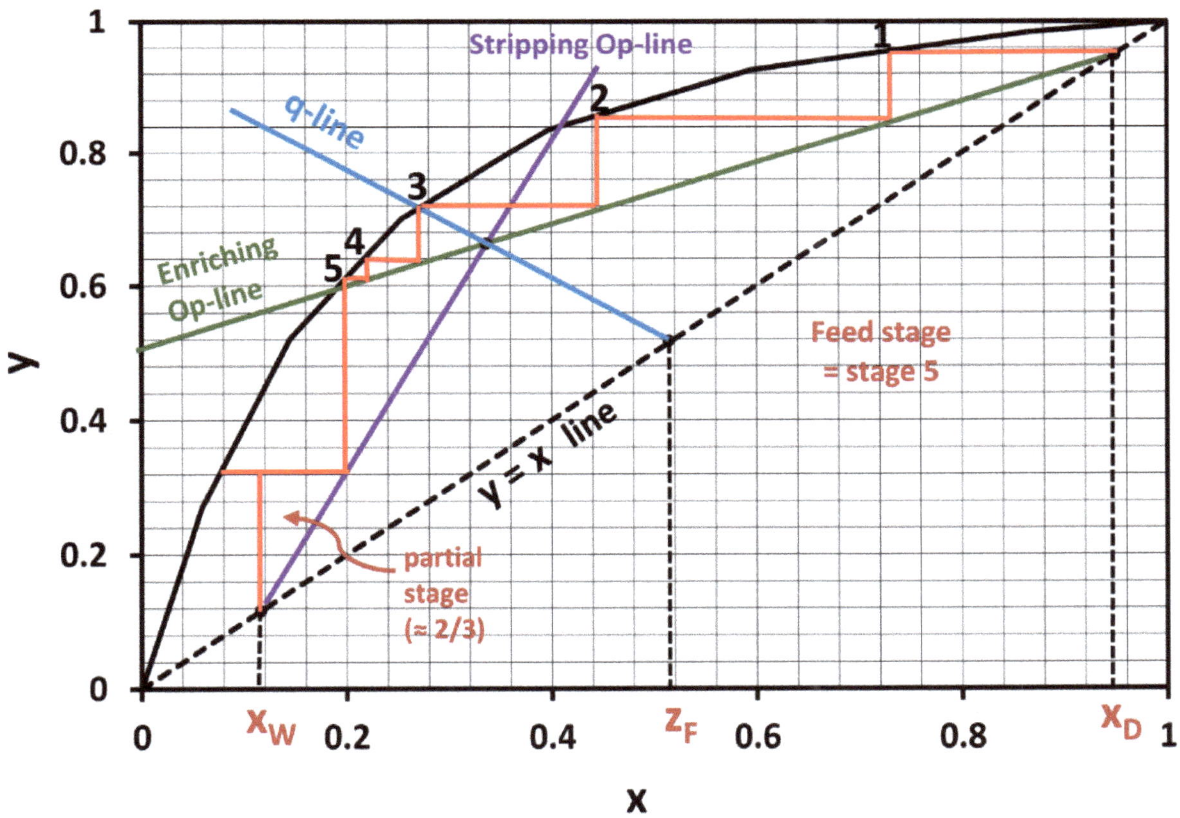

Figure 9.22 McCabe-Thiele diagram of distillation column with non-optimum feed-stage location; feed stage (stage 5) is located below the point of intersection of the operating lines.

Figure 9.23 shows another example of a non-optimum feed stage location. The stages marked in red lines represent the non-optimum case. The feed stage in this case (stage number 2) is located above the optimum point where the intersection of the two operating lines takes place. The stages marked in black dotted lines represent the optimum case. The feed stage in this case (stage number 3) includes the intersection point of the two operating lines. The total number of equilibrium stages required in the non-optimum case is 4.5 whereas the optimum feed-stage case requires less than 4 stages.

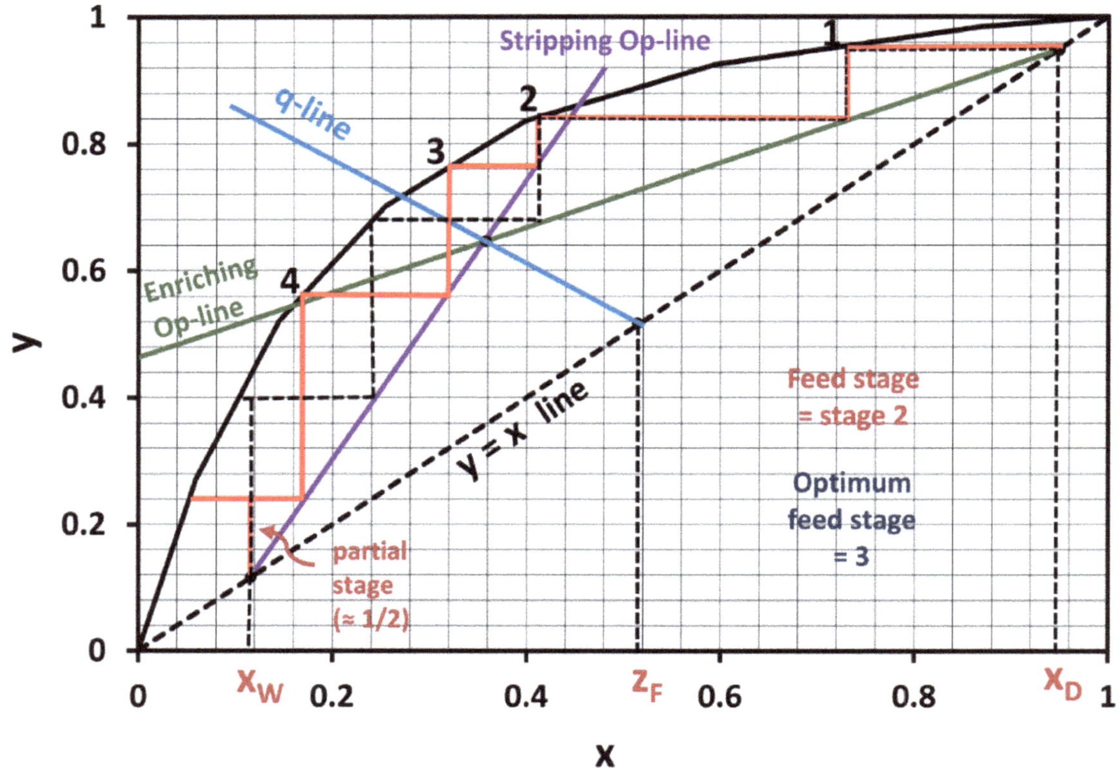

Figure 9.23 McCabe-Thiele diagram of the distillation column with non-optimum and optimum feed-stage locations. The staircase construction of stages with non-optimum feed stage is done in red lines. The feed stage (stage 2) is located above the point of intersection of the operating lines. The staircase construction of stages with optimum feed stage location is done in black dotted lines; in this case the feed stage location is stage 3 which includes the intersection point of the two operating lines.

Reflux Ratio Operating Range and Optimum Value of Reflux Ratio

Reflux ratio (R) has a strong influence on the design and performance of a distillation column. Given the feed flow rate (F), feed composition (z_F), feed thermal condition ($q\ value$), and the desired separation (x_W and x_D values), there is a trade-off involved between the energy requirements (heat duties of condenser and reboiler) and the number of stages needed for the desired separation.

When F, z_F, x_W, x_D and q are all fixed and the reflux ratio R is varied, $D\ and\ W$ do not change (see Equations (9.3) and (9.4)). However, the liquid flow rate in the enriching section changes ($L = L_o = RD$) upon varying R. The vapor flow rate in the enriching section also changes according to $V = V_1 = (1 + R)D$. The reflux ratio also affects the *boilup* ratio (s) in the stripping section defined as $s = \bar{V}/W$. The relationship between s and R is given as:

$$s = q - 1 + (q + R)\left[\frac{z_F - x_W}{x_D - z_F}\right] \tag{9.43}$$

This relation follows from the definition of q (Equation (9.40)) and Equations (9.3) and (9.4). For fixed F, z_F, q, x_W and x_D, varying R changes s and hence boilup vapor flow rate $\bar{V}$. The heat duties of condenser and reboiler are given as:

$$\dot{Q}_C = V_1\lambda_D = -D(1+R)\lambda_D \tag{9.44}$$

$$\dot{Q}_R = -\bar{V}\lambda_W \tag{9.45}$$

where λ_W is the molar latent heat of vaporization of bottoms product.

If R is decreased, the heat duties of the condenser and reboiler decrease as the load on the distillation column decreases. When R is decreased for fixed F, z_F, q, x_W and x_D , the vapor flow rates ($V$and $\bar{V}$) and the liquid flow rates (L and $\bar{L}$) in both enriching section and stripping section decrease. Therefore, less amount of vapor needs to be condensed in the condenser and less amount of vapor (boil-up) needs to be generated in the reboiler. However, we now need more number of stages to achieve the same degree of separation. If R is increased, an opposite behavior is expected in that the heat duties of the condenser and reboiler are expected to go up whereas the number of equilibrium stages required for the same separation are expected to come down.

Figure 9.24 illustrates the effect of variation of R on the number of equilibrium stages required for a fixed degree of separation (z_F, q, x_W, x_D fixed). As R increases, we need a smaller number of equilibrium stages. In the example considered in Figure 9.24, the number of stages required is more than 4 when $R = R_1$. Upon increasing the reflux ratio to $R = R_2$, we need only about 3 stages for the same degree of separation. Also note that the optimum feed stage location changes with the change in R.

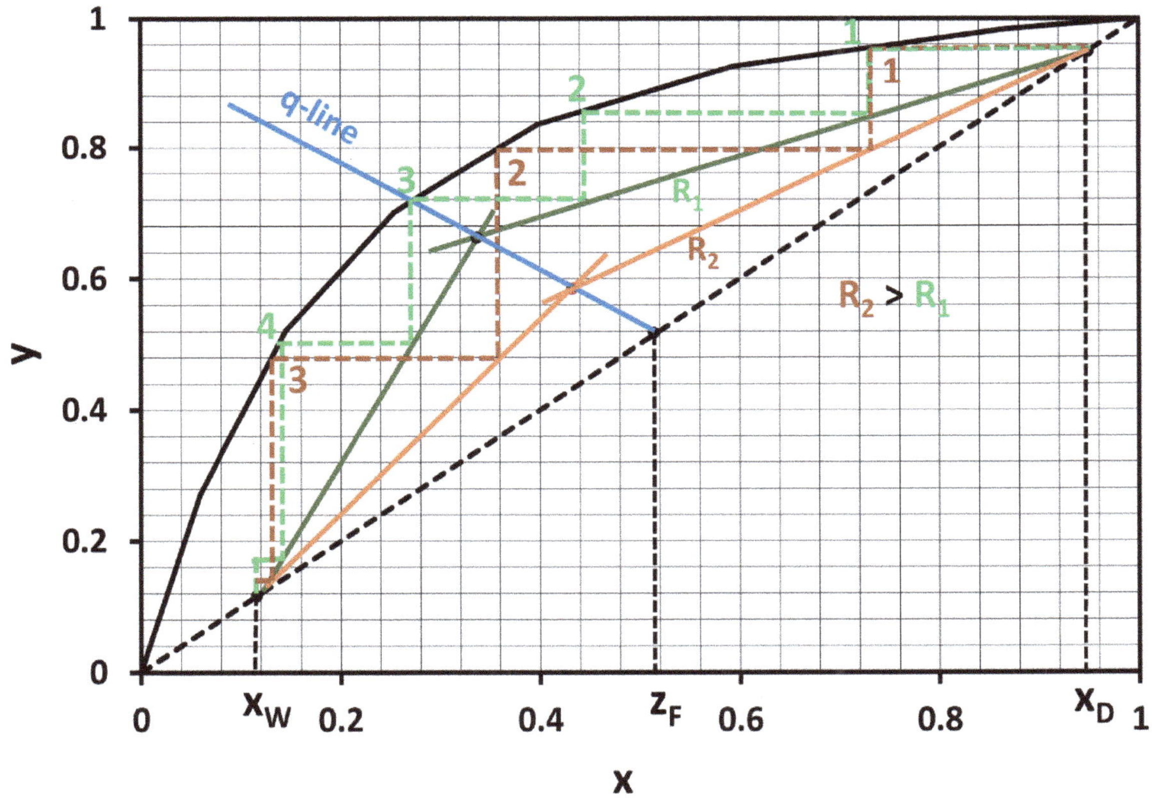

Figure 9.24 Effect of reflux ratio on the number of equilibrium stages required for the same separation.

The reflux ratio could be varied in the range of R_{min} to infinity, that is, $R_{min} \leq R \leq \infty$. The slope of the enriching section operating line is $R/(1+R)$. As R is decreased the enriching section operating line becomes less steeper, that is, the enriching section operating line rotates clockwise around the point (x_D, x_D). If we continue to decrease R, at some R value two things can happen depending on the shape of the equilibrium curve. If the equilibrium curve is concave downwards throughout the concentration range (typical of *ideal* mixtures), the operating line passes through the point of intersection of q-line and equilibrium curve. This intersection point of q-line and enriching section operating line on the equilibrium curve is called *feed pinch point*. The reflux ratio corresponding to pinch condition is R_{min}. One cannot decrease R any further as no point on the operating lines can lie above the equilibrium curve. When $R = R_{min}$, an infinite number of stages are needed in both enriching and stripping sections to achieve the given separation. Figure 9.25 illustrates the pinch condition occurring at the feed plate.

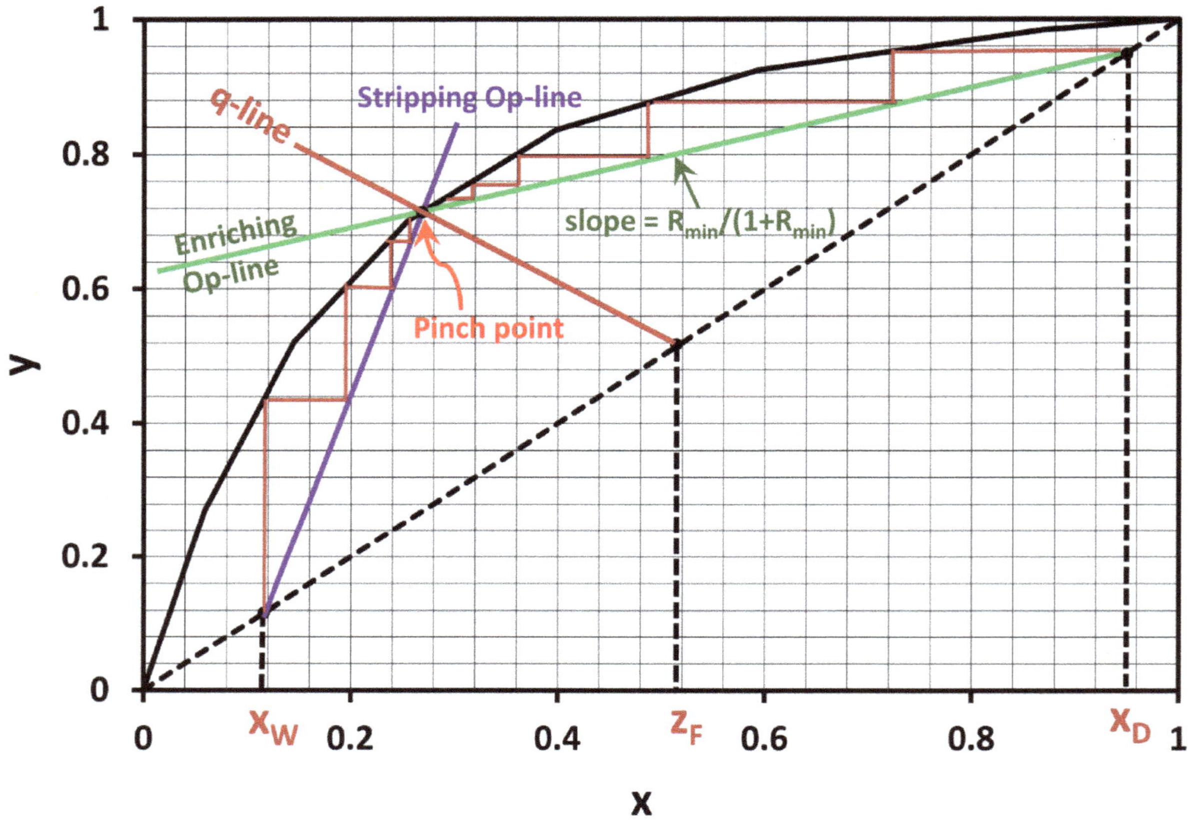

Figure 9.25 Feed pinch point where $R = R_{min}$ and all three lines (q-line, enriching and stripping operating lines) intersect on the equilibrium curve.

If the equilibrium curve exhibits inflection in that a portion of the equilibrium curve is concave downwards and a portion of it is concave upwards (typical of *non-ideal* mixtures), the minimum reflux ratio is determined from the slope of the enriching section operating line that is tangent to the equilibrium curve, as illustrated in Figure 9.26. Note that the enriching section operating line cannot cross the equilibrium curve and therefore, the smallest slope one can have is when the operating line becomes tangent to the equilibrium curve. The point where the enriching section operating line becomes tangent to the equilibrium curve is referred to as *tangent pinch* point. This pinch condition occurs in the enriching section. Although the stripping section requires a finite number of stages, the enriching section requires an infinite number of stages to achieve the given separation.

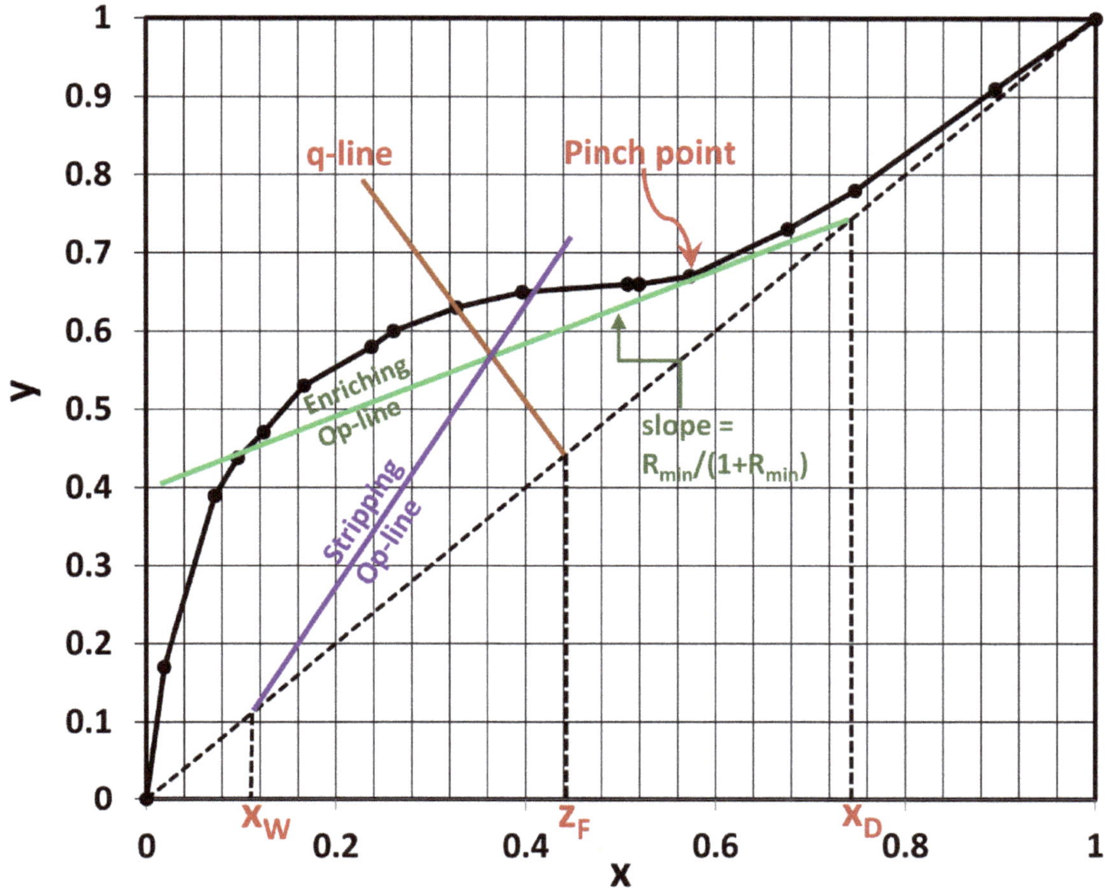

Figure 9.26 Tangent pinch point where $R = R_{min}$ and the enriching section operating line becomes tangent to the equilibrium curve at some point in the enriching section.

What happens when $R < R_{min}$ for given z_F, x_D, x_W and q ? Figure 9.27 illustrates what happens when $R < R_{min}$. The point of intersection of the two operating lines and the feed line (q-line) lies above the equilibrium curve. It is *impossible* to achieve the given separation (x_D and x_W) when $R < R_{min}$ as this would require operation beyond the equilibrium curve. We will never reach the feed stage even if we employ an infinitely large number of stages in stripping and enriching sections. It is impossible to achieve any compositional change once equilibrium is established. Thus all points on the operating lines must lie below the equilibrium curve.

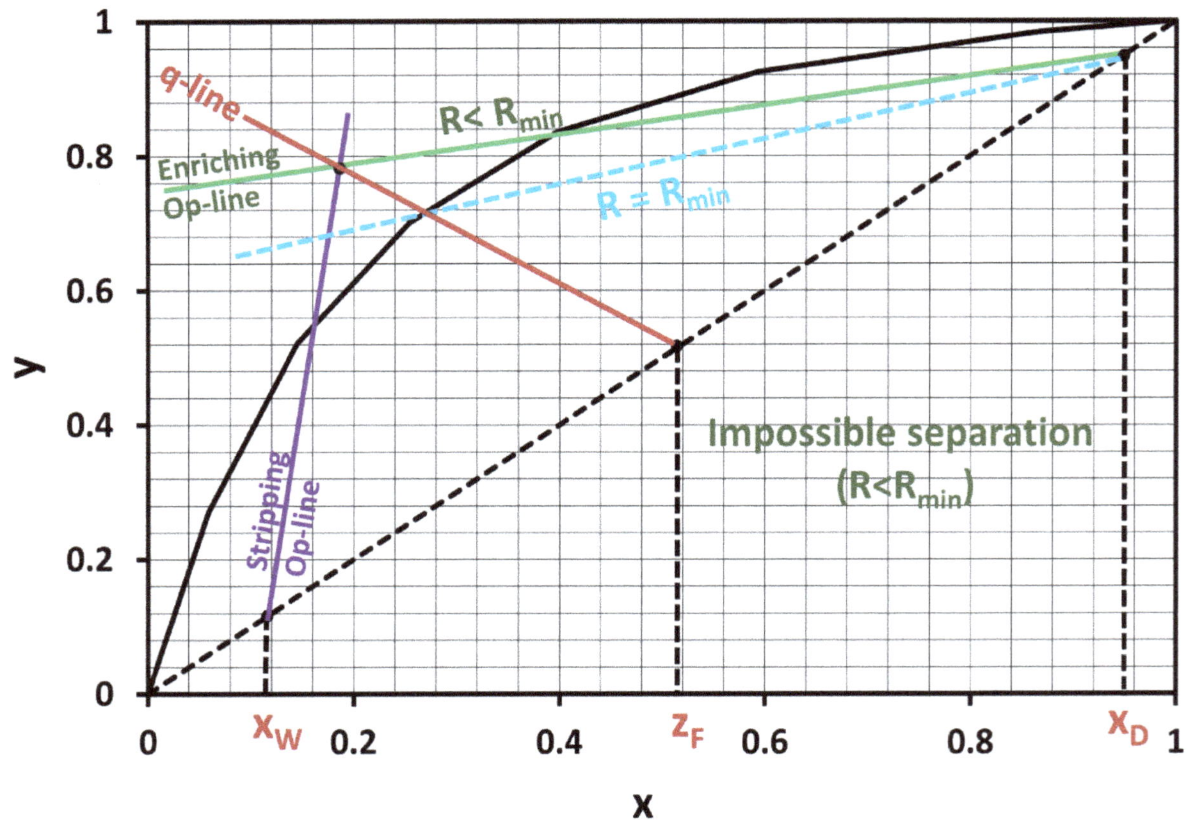

Figure 9.27 When $R < R_{min}$, it is impossible to achieve the given separation.

As long as x_D and x_W are fixed for given feed composition and feed thermal condition (q-value), we cannot operate the column with $R < R_{min}$. However we can operate the column with $R < R_{min}$ if we reduce the x_D value. As an example, consider Figure 9.28. The minimum reflux ratio R_{min} depends on the x_D value. Let $R = R_{min}$ corresponding to x_D of x_{D2}. We need an infinite number of stages to achieve x_D of x_{D2} due to pinch condition on the feed tray. If we keep R fixed but increase x_D from x_{D2} to x_{D1}, it is impossible to achieve this separation even with an infinite number of stages as the intersection of the operating line and q-line lies above the equilibrium curve. Also note that when $x_D = x_{D1}$, $R < R_{min}$ where R_{min} corresponds to x_{D1}. For the same R, if we decrease x_D from x_{D2} to x_{D3}, the separation is readily achieved using a finite number of stages as now $R > R_{min}$ where R_{min} corresponds to x_{D3}. Clearly the value of R_{min} increases with the increase in x_D value.

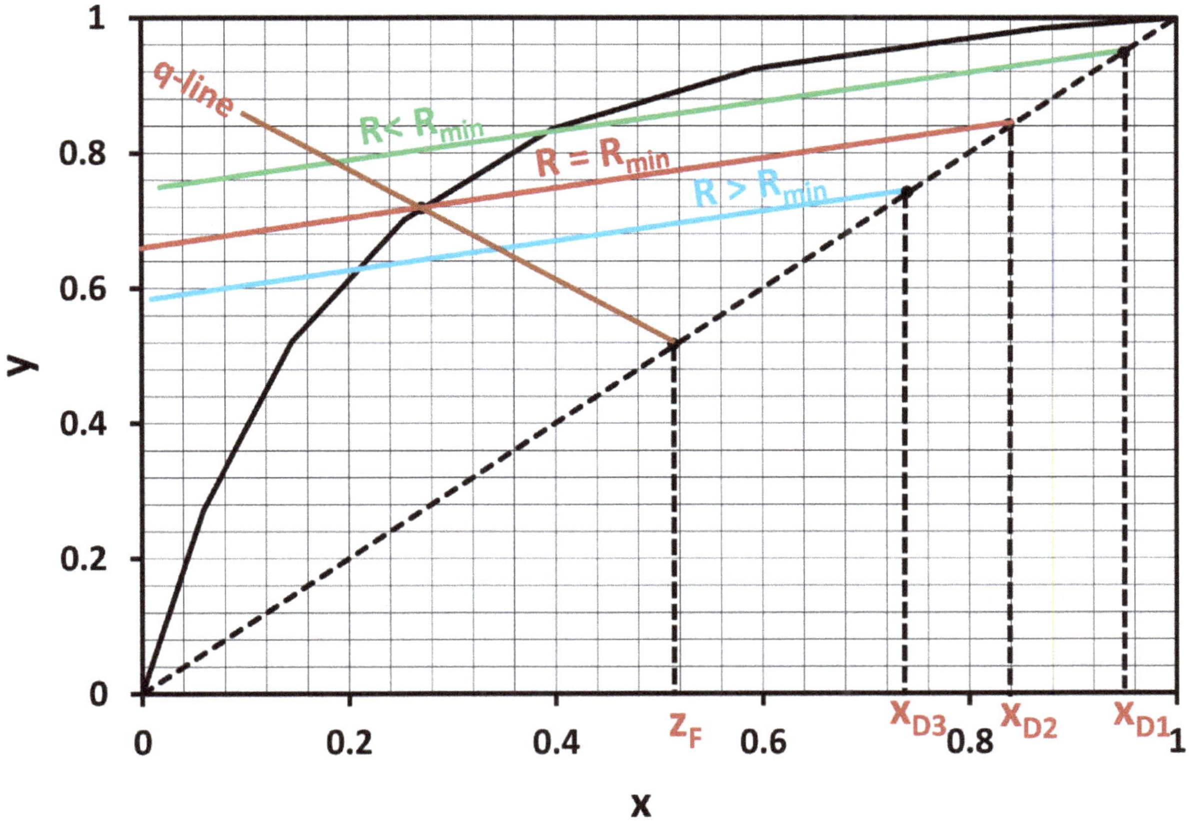

Figure 9.28 Effect of x_D variation on R_{min} keeping z_F and q fixed.

As noted earlier, the reflux ratio could be varied in the range of R_{min} to infinity, that is, $R_{min} \leq R \leq \infty$. We just considered one extreme case, that of *minimum reflux* where $R = R_{min}$. Let us now consider the other extreme case, that of *total reflux* where $R \to \infty$. The reflux ratio of infinity, also referred to as total reflux, occurs when all the vapor emerging from the top of the column is condensed and sent back to the column. Thus no distillate product is withdrawn, $D = 0$. When $D = 0$, $R = L/D \to \infty$ and the slope of the enriching section operating line, that is, $R/(1 + R)$, becomes unity. At total reflux, no bottoms product is withdrawn ($W = 0$) and the boilup ratio ($s = \bar{V}/W$) becomes infinity. All the liquid emerging from the bottom of the column is vaporized in the reboiler and sent back to the column. When $W = 0$, $s = \bar{V}/W \to \infty$ and the slope of the stripping section operating line, that is, $(s + 1)/s$, becomes unity. Thus the operating lines at total reflux coincide with the 45° diagonal line ($y = x$ line). Also note that the feed flow rate is zero ($F = 0$) at total reflux. Figure 9.29 shows the conditions that prevail when distillation column is operating at total reflux.

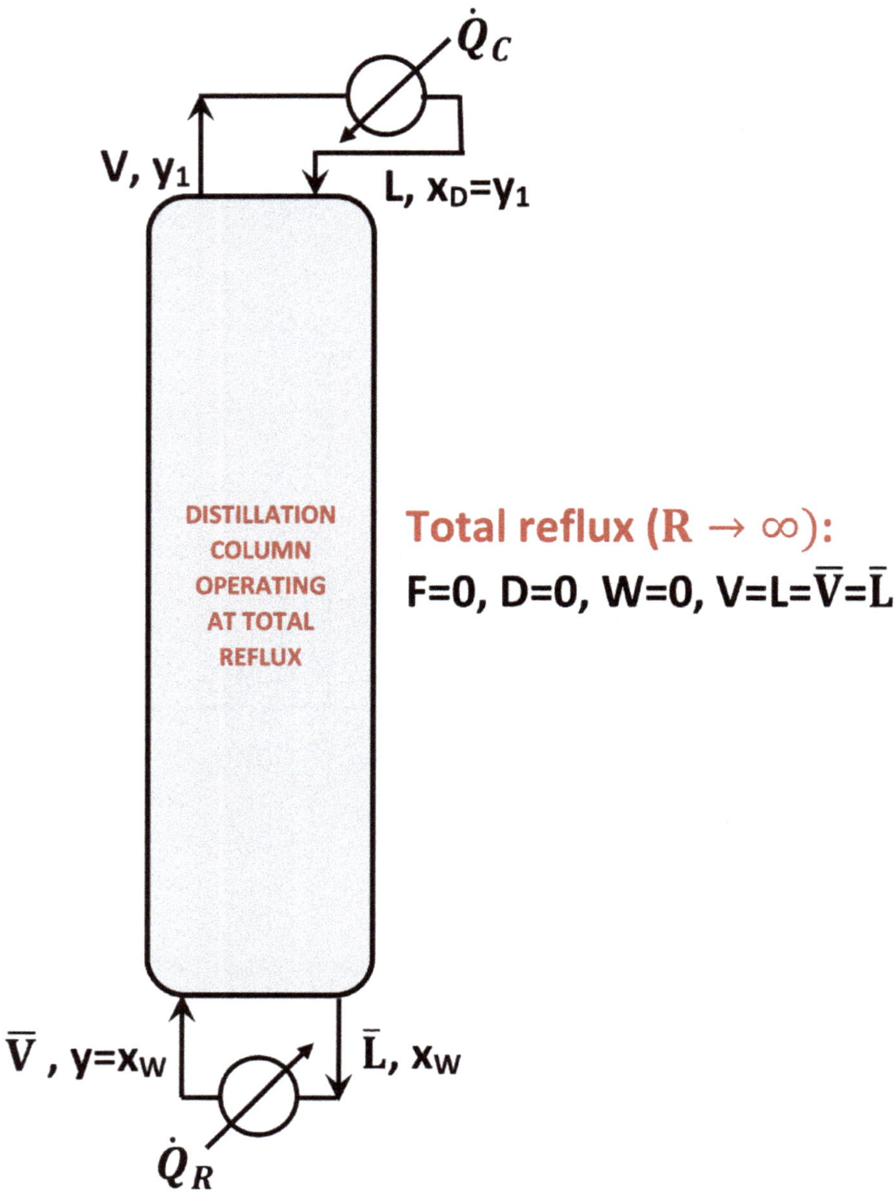

Figure 9.29 Distillation column operating at total reflux.

The industrial distillation columns are usually operated in total reflux mode during start-up. The column is fed with the mixture to be separated. The feed accumulates at the base of the column. The reboiler and condenser are activated with feed, distillate and bottom product lines closed. After some time, steady state is reached and the column operates under total reflux condition. Assuming no heat loss, the heat duties of reboiler and condenser are the same, that is, $\dot{Q}_R + \dot{Q}_C = 0$. The vapor and liquid flow rates within the column are varied by adjusting the heat duties of reboiler and condenser.

One important aspect of operating a column at total reflux is that the number of stages/trays required is *minimum* for a given separation (x_D and x_W). This minimum number of trays can be found by constructing stages using the McCabe-Thiele graphical approach, as shown in Figure 9.30. Note that $y = x$ line is used as the operating line for the entire column as enriching and stripping operating lines coincide with $y = x$ line.

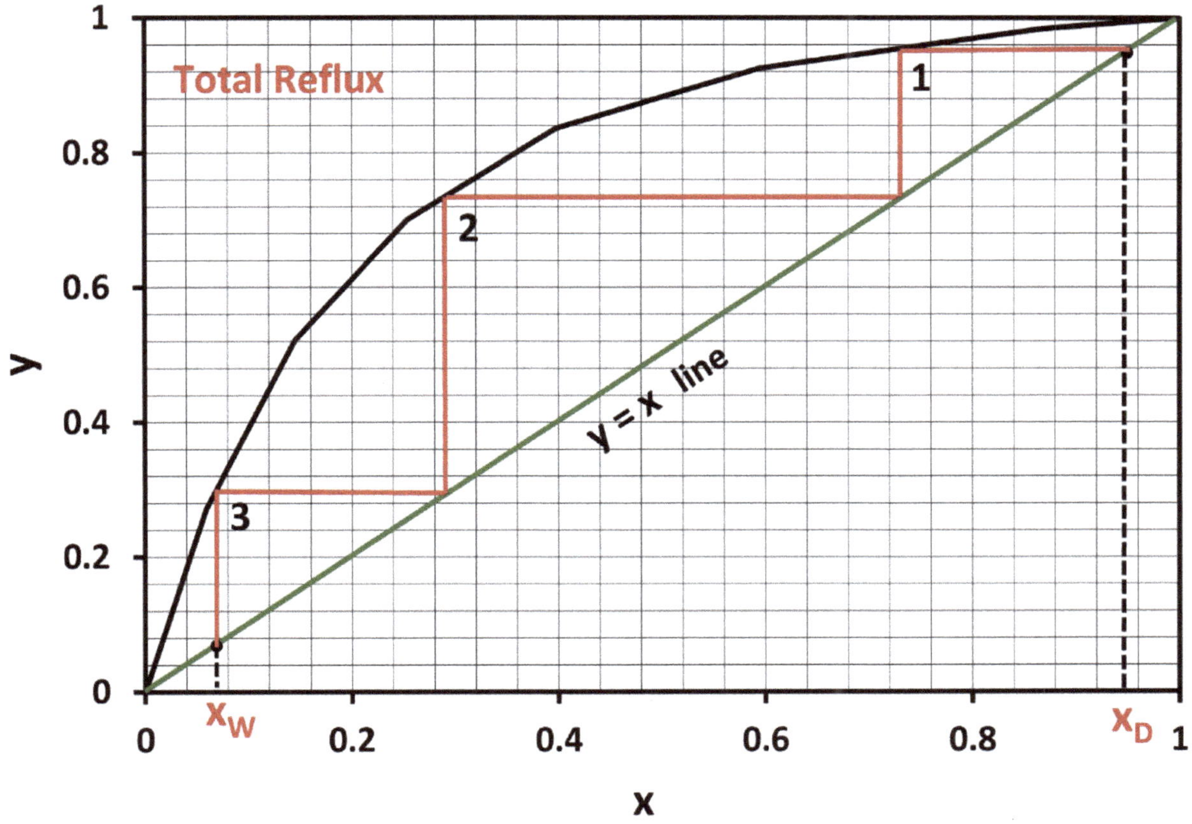

Figure 9.30 Graphical construction of stages under total reflux.

Figure 9.31 shows how the variation of reflux ratio affects the capital, operating, and total costs of a distillation column. When R/R_{min} is close to unity, the capital cost is high as a large number of trays (hence very tall column) are needed to carry out the given separation. The operating cost is low as the heating (reboiler) and cooling (condenser) costs are low. With the increase in R/R_{min}, the capital cost falls rapidly as the number of trays needed are reduced. The operating cost increases gradually with the increase in R/R_{min}. At high values of R/R_{min}, the capital cost begins to rise as large sized reboiler and condenser are needed to provide the heating and cooling effects. The optimum total cost is usually in the range 1.1 $< R/R_{min} <$ 1.5 (often taken as 1.25).

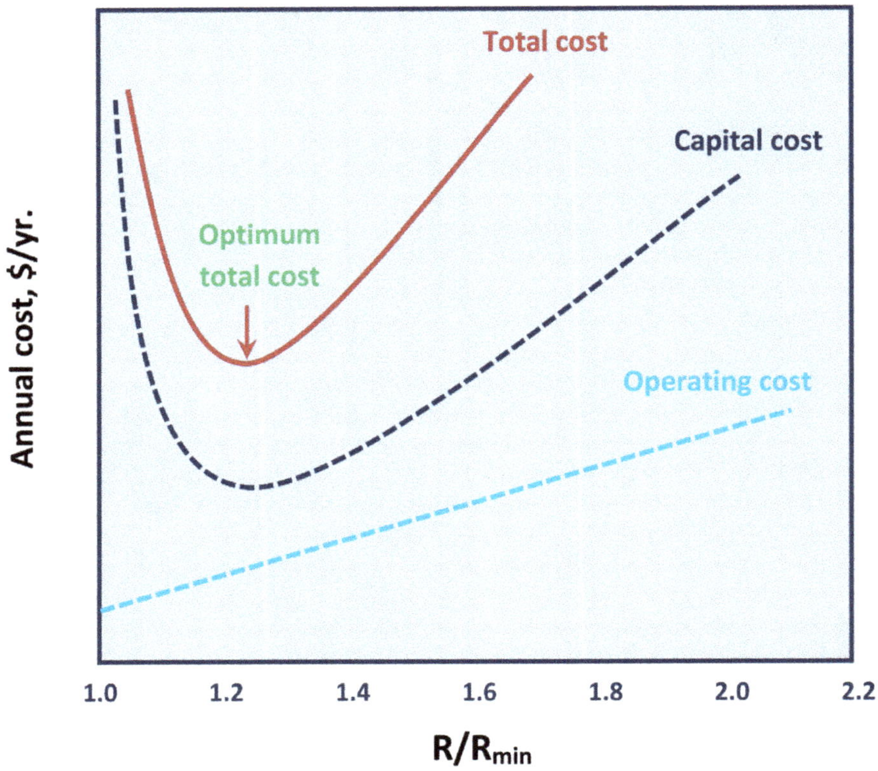

Figure 9.31 Effect of R/R_{min} on annual cost of operating a distillation column.

Different Thermal Conditions of Feed

Figure 9.32 shows the plots of q-line (feed line) for various feed conditions. As noted earlier, the slope of the q-line is $-q/(1-q)$. For feed as subcooled liquid, $q > 1$ and the slope of q-line is positive greater than unity. For feed as saturated liquid, $q = 1$ and the slope of q-line is infinity (q-line is vertical). For feed as two-phase vapor-liquid mixture, $0 < q < 1$ and the slope of q-line is negative. When feed is saturated vapor, $q = 0$ and the slope of q-line is zero (q-line is horizontal). For feed as superheated vapor, $q < 0$ and the slope of q-line is positive in the range $0 < slope < 1$.

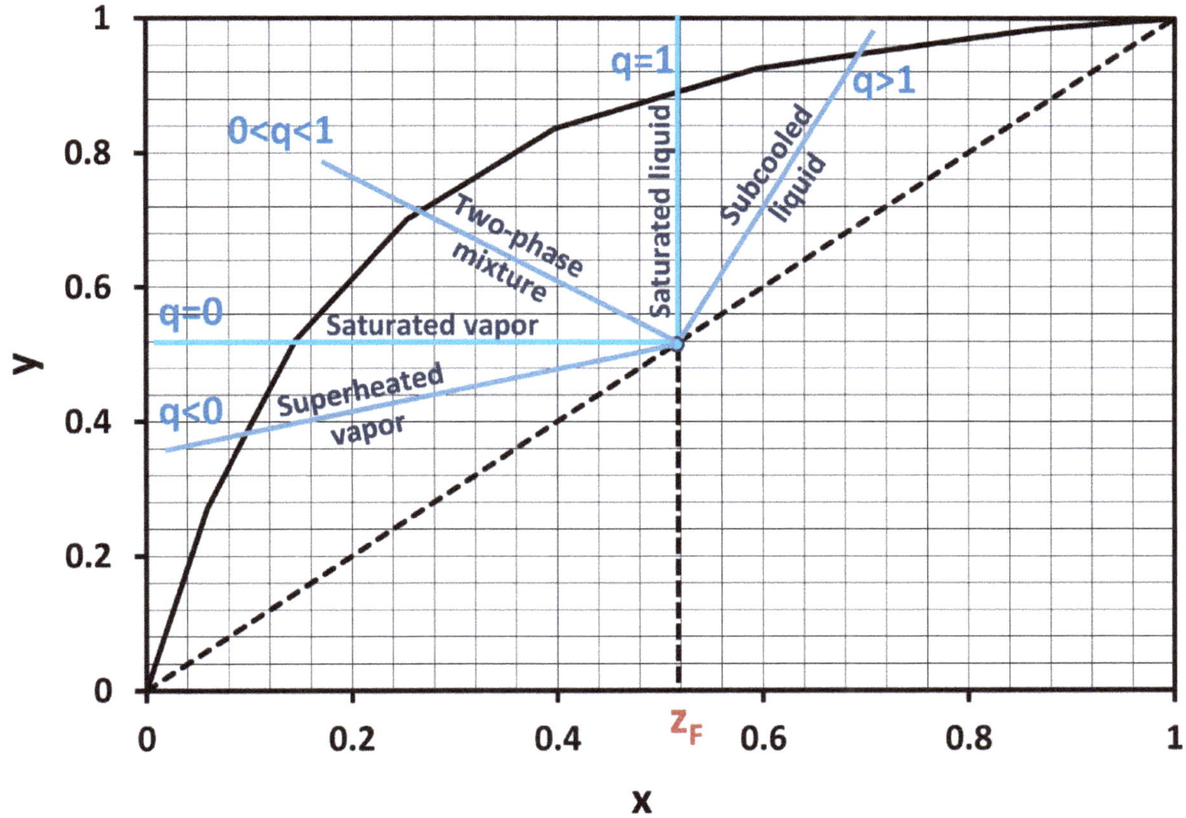

Figure 9.32 Feed or q-lines for different feed conditions.

Figures 9.33-9.37 show schematically the vapor and liquid streams entering and leaving the feed tray for different feed conditions. When the feed is subcooled liquid (Figure 9.33), part of the vapor stream condenses to heat the subcooled liquid to the bubble point. The condensed vapor and the entire feed stream join the liquid stream flowing down the column as illustrated in Figure 9.33. Consequently, $\overline{V} > V$ and $\overline{L} - L > F$ and $q > 1$. When the feed is saturated liquid (Figure 9.34), no condensation of vapor stream takes place. The entire feed stream joins the liquid stream flowing down the column. Thus, $\overline{V} = V$ and $\overline{L} - L = F$ and $q = 1$. When feed is two-phase vapor-liquid mixture (Figure 9.35), the vapor portion of the feed joins the vapor stream and the liquid portion of the feed joins the liquid stream. Consequently, $V > \overline{V}$ and $0 < \overline{L} - L < F$ and $0 < q < 1$. When feed is saturated vapor (Figure 9.36), the liquid stream flowing down the column is unaffected. The entire feed stream joins the vapor stream. Thus, $V - \overline{V} = F$ and $\overline{L} = L$ and $q = 0$. Finally, when feed is superheated vapor (Figure 9.37), part of the liquid stream flowing down the column in the enriching section is vaporized as shown in Figure 9.37 to cool the feed vapor to saturated vapor. The entire feed vapor stream and the liquid stream vaporized inside the column join the vapor stream rising through the stripping section. Consequently, $V - \overline{V} > F$ and $\overline{L} < L$ and $q < 0$.

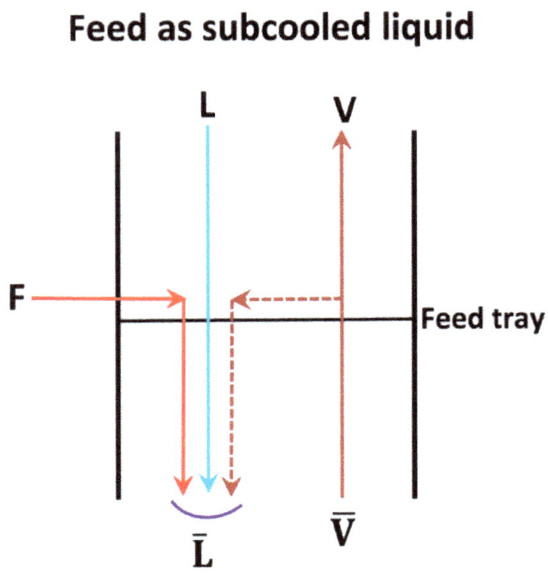

$$\bar{L} > L \,;\, \bar{L} - L > F \,;\, \bar{V} > V \,;$$

$$q = \frac{\bar{L}-L}{F} > 1$$

Figure 9.33 Flow through feed tray with feed as subcooled liquid.

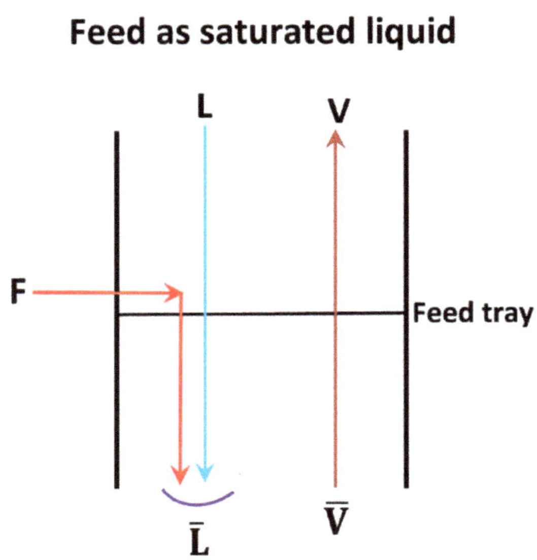

$$\bar{L} > L \,;\, \bar{L} - L = F \,;\, \bar{V} = V \,;$$

$$q = \frac{\bar{L}-L}{F} = 1$$

Figure 9.34 Flow through feed tray with feed as saturated liquid.

Feed as vapor-liquid mixture

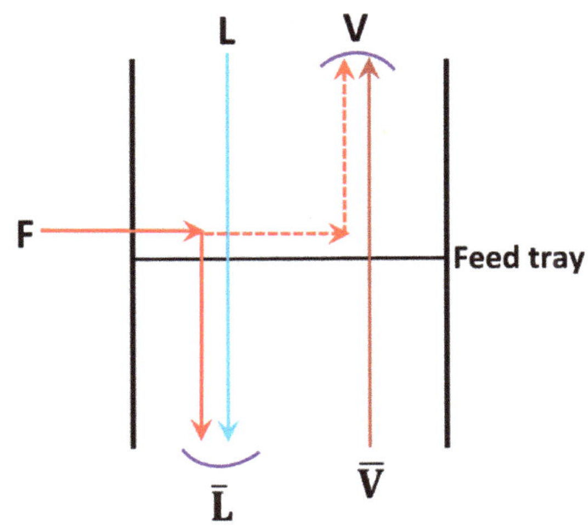

$$0 < \bar{L} - L < F\,;\, V > \bar{V};$$
$$0 < q\left(= \frac{\bar{L}-L}{F}\right) < 1$$

Figure 9.35 Flow through feed tray with feed as two-phase vapor-liquid mixture.

Feed as saturated vapor

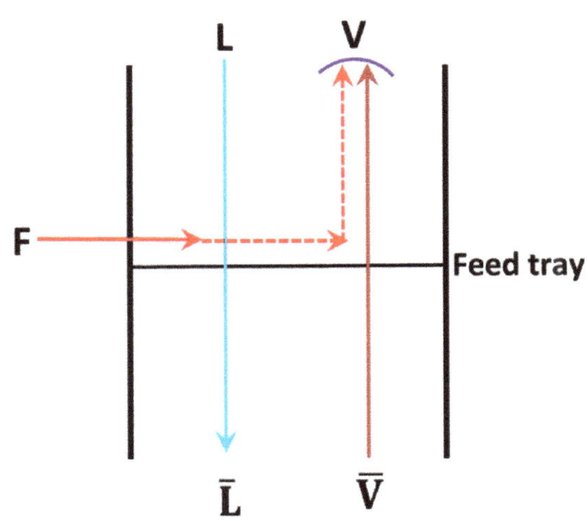

$$\bar{L} = L\,;\, V > \bar{V};$$
$$q\left(= \frac{\bar{L}-L}{F}\right) = 0$$

Figure 9.36 Flow through feed tray with feed as saturated vapor.

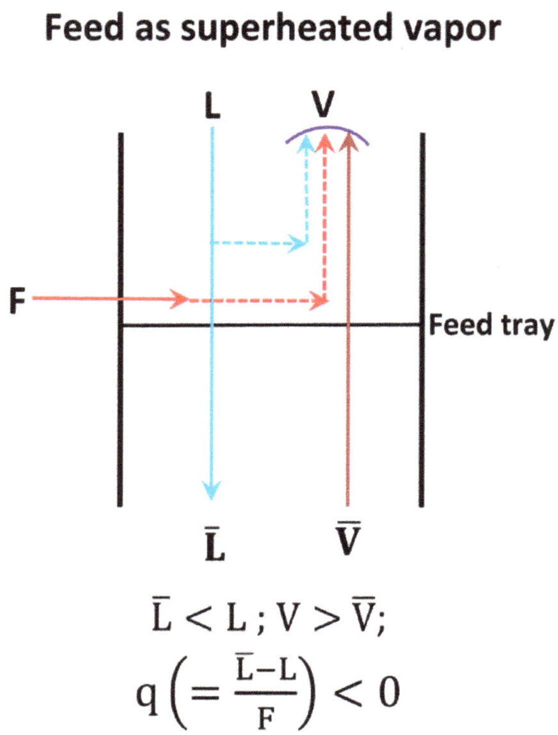

Figure 9.37 Flow through feed tray with feed as superheated vapor.

Energy Balance on the Feed Tray

Earlier we carried out material balance on the feed tray (see Equation (9.39)). The energy balance on the feed tray was not required. Now we will do energy balance on the feed tray to relate q to enthalpies of fluids on the feed tray. Referring to Figure 9.38, the energy balance (Equation (9.5)) on the feed tray, assuming no heat losses, gives:

$$h_F F + H_{f+1}\bar{V} + h_{f-1}L = h_f\bar{L} + H_f V \tag{9.46}$$

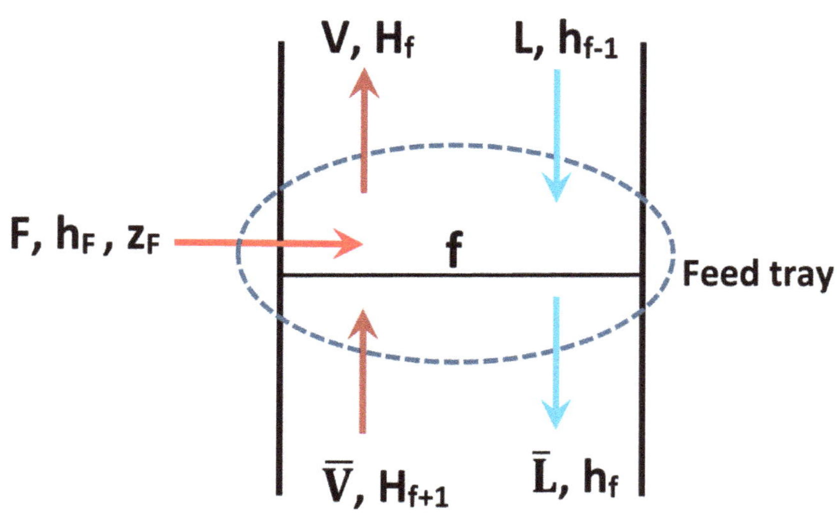

Figure 9.38 Energy balance on the feed tray.

As an approximation, we take:

$$H_{f+1} = H_f \text{ and } h_{f-1} = h_f \tag{9.47}$$

This approximation is a decent approximation. The vapor enthalpies and liquid enthalpies do not change much from one stage to next especially when CMO (constant molal overflow) is valid. Upon making the above approximation (Equation (9.47)), energy balance, Equation (9.46), simplifies to:

$$h_F F + H_f(\bar{V} - V) = h_f(\bar{L} - L) \tag{9.48}$$

From material balance, Equation (9.39), and energy balance, Equation (9.48), it follows that:

$$q = \frac{\bar{L}-L}{F} = \frac{H_f-h_F}{H_f-h_f} \tag{9.49}$$

Thus the "quality q" of the feed can be determined from enthalpies of feed, vapor and liquid streams on the feed plate. Using a slightly different notation for enthalpies, we can re-write Equation (9.49) as:

$$q = \frac{H_V-h_F}{H_V-h_L} \tag{9.50}$$

where H_V is the enthalpy of feed at dew point at the column pressure, h_F is the feed of enthalpy at its entrance conditions, and h_L is the enthalpy of feed at bubble point at the column pressure. Note that $H_V - h_L$ is molar latent heat of vaporization of feed and $H_V - h_F$ is heat needed to vaporize 1 mole of feed at entrance conditions.

Obviously, the feed quality factor q can be estimated from enthalpy data using Equation (9.50). When feed enters the column as *subcooled liquid* ($T_F < T_b$):

$$h_L - h_F = C_{pL}(T_b - T_F) \tag{9.51}$$

where C_{pL} is the heat capacity of liquid feed and T_b is the bubble point of feed at column pressure. Thus,

$$q = \frac{H_V-h_F}{H_V-h_L} = 1 + \frac{h_L-h_F}{H_V-h_L} = 1 + \frac{C_{pL}(T_b-T_F)}{\lambda} \tag{9.52}$$

When feed enters the column as *superheated vapor* ($T_F > T_d$):

$$h_F - H_V = C_{pV}(T_F - T_d) \tag{9.53}$$

where C_{pV} is the heat capacity of vapor feed and T_d is the dew point of feed at column pressure. Thus,

$$q = \frac{H_V-h_F}{H_V-h_L} = -\frac{C_{pV}(T_F-T_d)}{\lambda} \tag{9.54}$$

Note that when feed is *saturated liquid*, $h_F = h_L$ and $q = 1$ from Equation (9.50). When feed is *saturated vapor*, Equation (9.50) gives $q = 0$ as $h_F = H_V$.

Effect of q on Performance of Distillation Column

What happens when q is increased from $q < 0$ (superheated vapor) to $q > 1$ (subcooled liquid) keeping reflux ratio (R), feed composition (z_F), distillate composition (x_D), and bottoms composition (x_W), the same? When q is increased, the slope of the stripping section decreases as shown in Figure 9.39. Also the point of intersection of operating lines and feed line moves closer to the $y = x$ line. The slope of stripping line is given as: $slope = (1 + s)/s = 1 + (1/s)$. A decrease in the slope implies an increase in the boilup ratio s. This is consistent with Equation (9.43). An increase in s indicates an increase in the reboiler duty for a given feed flow rate. This makes sense as it is expected that if the feed is subcooled liquid, vapor boilup flow rate ($\bar{V}$) and hence reboiler duty, should go up.

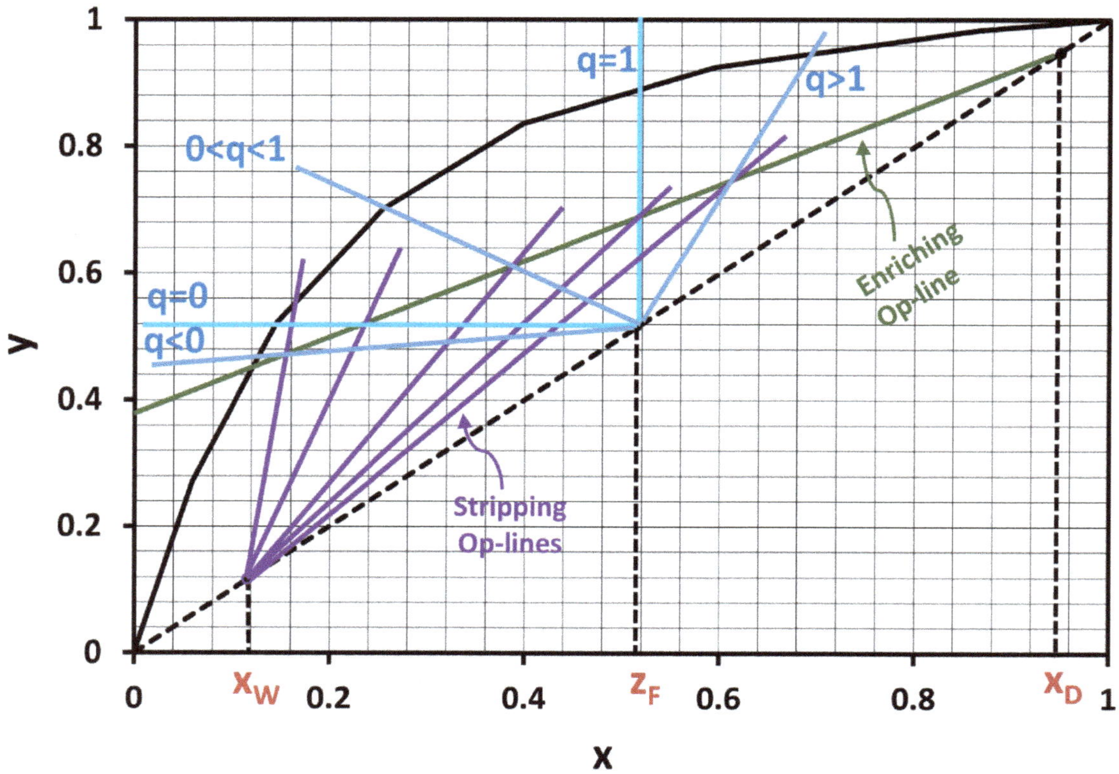

Figure 9.39 Effect of q on distillation performance keeping R, z_F, x_D, and x_W the same.

What happens to the number of plates needed to carry out the same separation? As the point of intersection of operating lines and feed line moves away from the equilibrium curve towards the $y = x$ line, we expect a decrease in number of plates with the increase in q. This is also consistent with our earlier observation that there is trade-off involved between reboiler heat duty and number of plates. Figure 9.40 confirms this observation in the present case. With the increase in q from 0 to 1, the number of plates needed for the same separation reduces from 4 to 3.

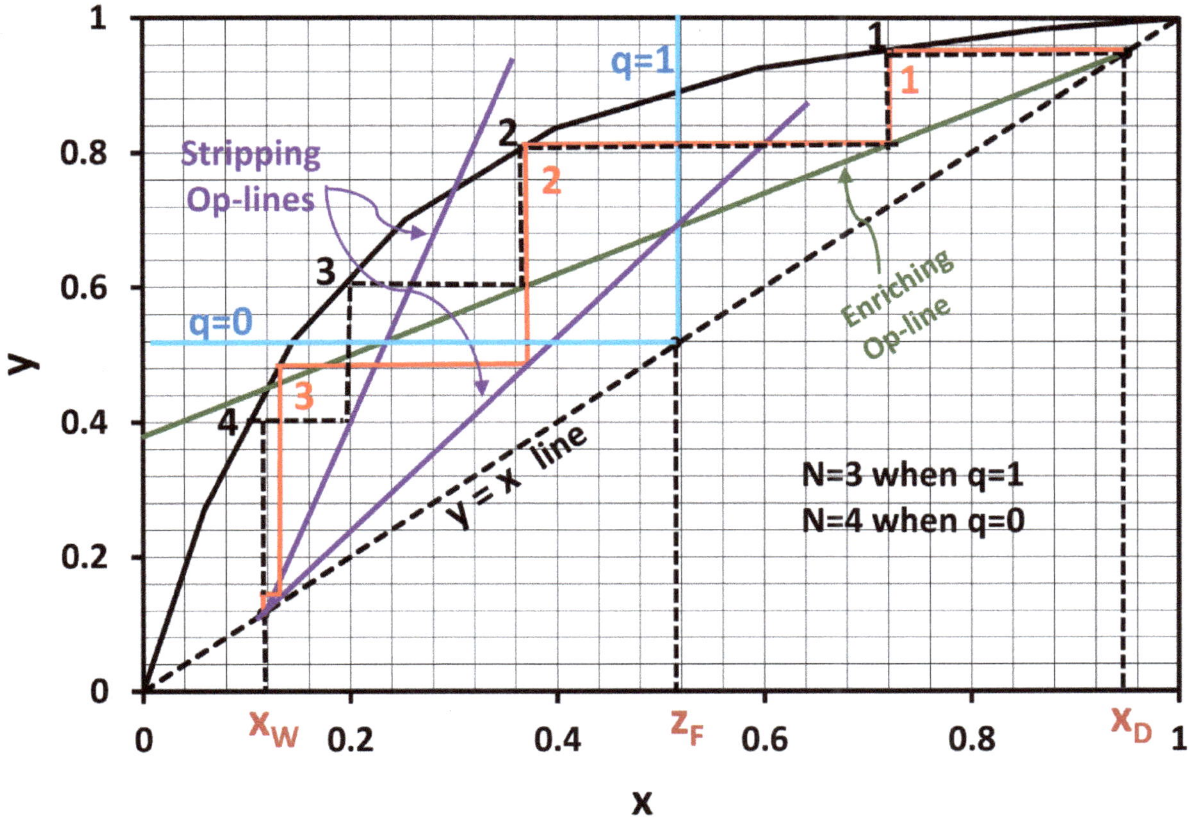

Figure 9.40 Effect of q on the number of plates needed for the same separation.

Partial vs Total Condensers

Thus far we have generally assumed that a *total* condenser is being used in the distillation column. Figure 9.41 shows a total condenser where all the vapor emerging from the top of the column is condensed. As indicated in Figure 9.41, $y_1 = x_D = x_o$, that is, the compositions of the liquid streams leaving the condenser (liquid distillate and reflux liquid) are the same as that of the vapor stream entering the condenser. As total condenser is *not* an equilibrium stage, it does not show up in the McCabe-Thiele diagram (see Figure 9.42) as a triangular stage. The first stage in the McCabe-Thiele diagram represents the top tray inside the column.

Figure 9.43 shows a *partial* condenser. In this case, only a part of the vapor stream emerging from the top of the column and entering the condenser is condensed and returned as liquid reflux to the column. The distillate product is a vapor stream. Now the condenser acts as an *equilibrium stage*.

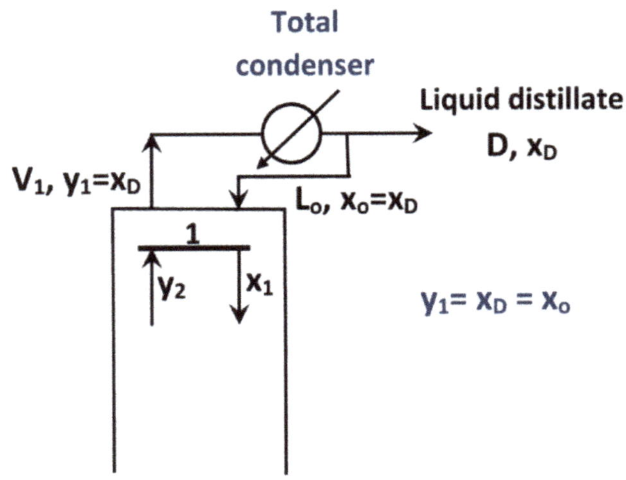

Figure 9.41 Distillation column with total condenser.

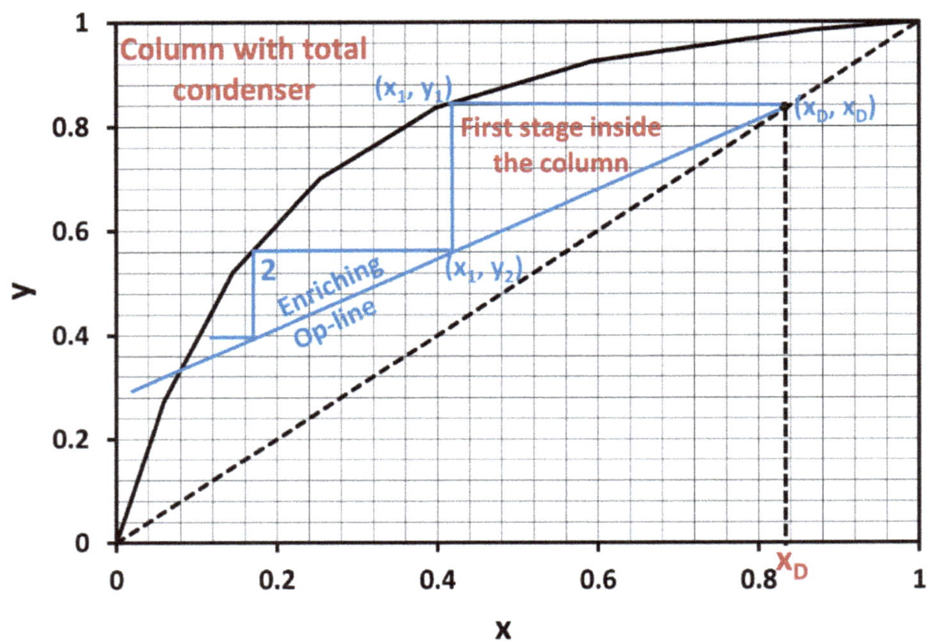

Figure 9.42 McCabe-Thiele diagram of distillation column using total condenser.

The vapor product leaving the condenser is in equilibrium with the liquid reflux returning to the column. As partial condenser is an equilibrium stage, it shows up in the McCabe-Thiele diagram (see Figure 9.44). The first stage of the McCabe-Thiele diagram represents a partial condenser. Partial condensers are generally used when the distillate product desired is in vapor state.

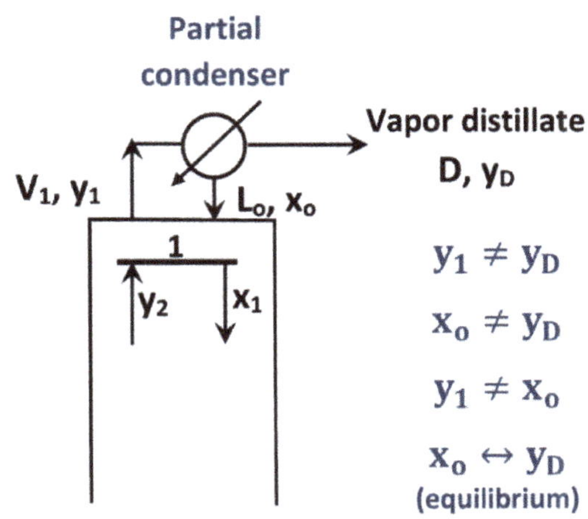

Figure 9.43 Distillation column with partial condenser.

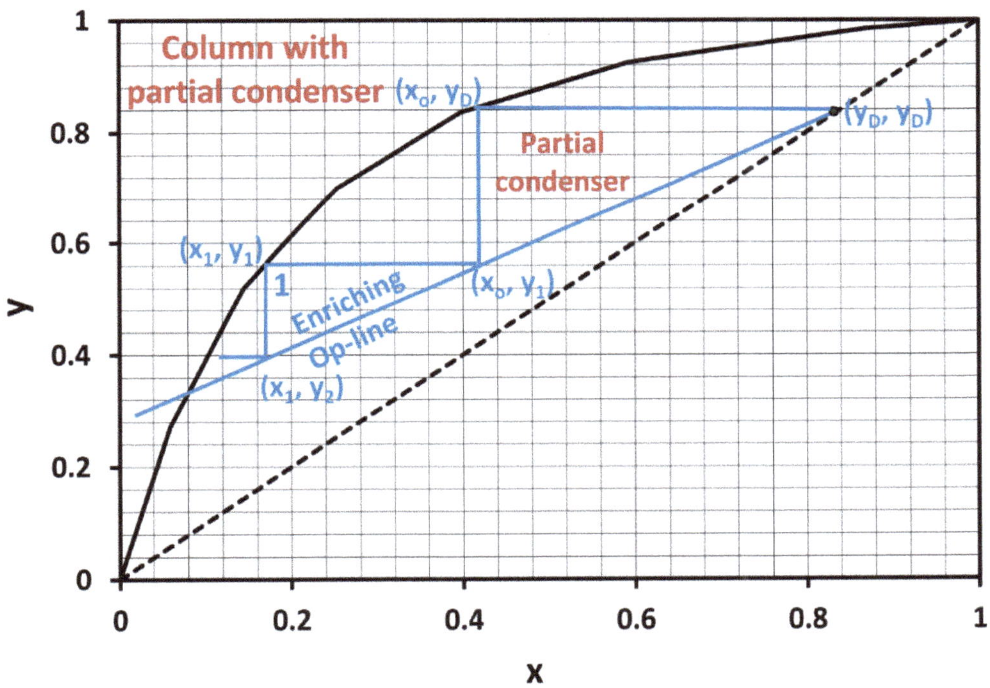

Figure 9.44 McCabe-Thiele diagram of distillation column using partial condenser.

Partial vs Total Reboilers

In this chapter and previous chapters we have assumed that the reboiler installed in a distillation column is a *partial reboiler*. Figure 9.45 shows a partial reboiler. In a partial reboiler, the liquid stream emerging from the bottom of the column and entering the reboiler is only partially vaporized in the reboiler. The boilup vapor stream leaving the partial reboiler (and returning to the column) is in equilibrium with the bottoms liquid product stream. Thus partial reboiler acts as an equilibrium stage. As partial reboiler is an equilibrium stage, it shows up in the McCabe-Thiele diagram (see Figure 9.46). If we construct the McCabe-Thiele diagram starting from distillate product composition to bottoms product composition (x_W), the partial reboiler is the last stage on the McCabe-Thiele diagram as shown in Figure 9.46. Although partial reboilers are more commonly used, some distillation columns may be equipped with total reboilers. Figure 9.47 shows a total reboiler. In a total reboiler, the entire liquid stream entering the reboiler is converted into vapor stream. The vapor composition is the same as that of the liquid bottoms product composition; referring to Figure 9.47, $y_{N+1} = x_N = x_W$. The total reboiler is *not* an equilibrium stage and therefore, it does not show up in the McCabe-Thiele diagram (Figure 9.48) as a triangular stage.

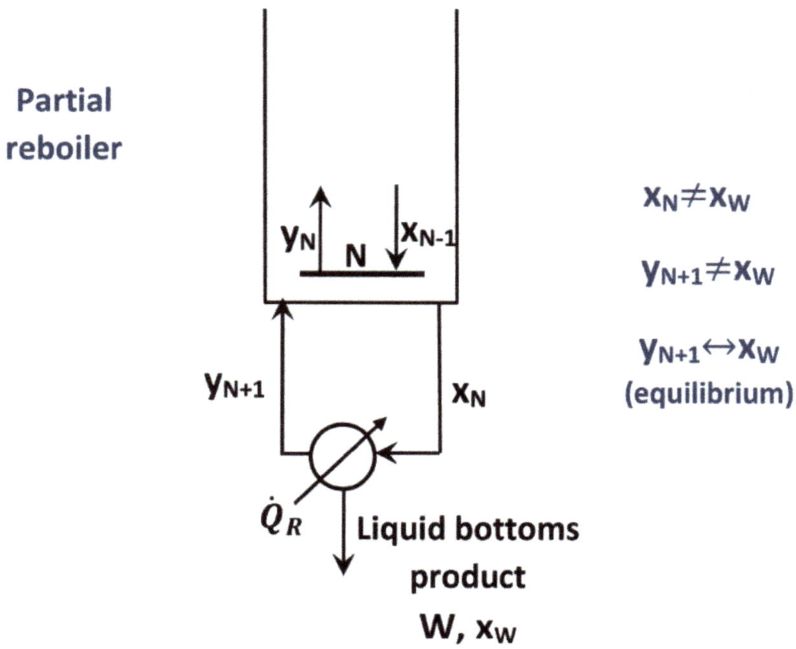

Figure 9.45 Distillation column with partial reboiler.

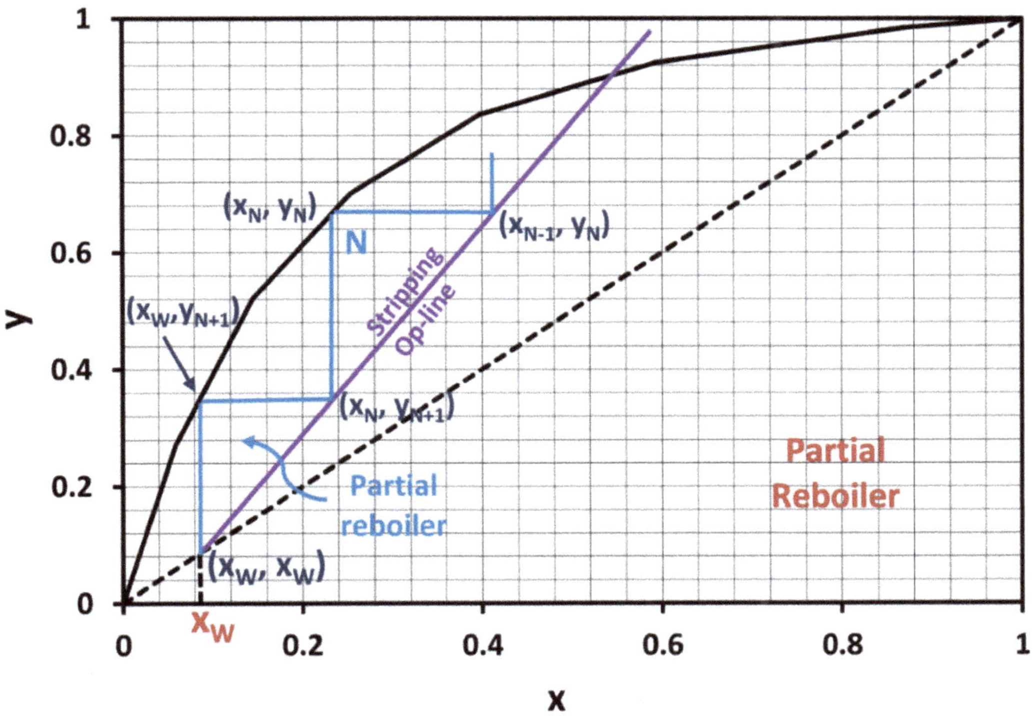

Figure 9.46 McCabe-Thiele diagram of distillation column using partial reboiler.

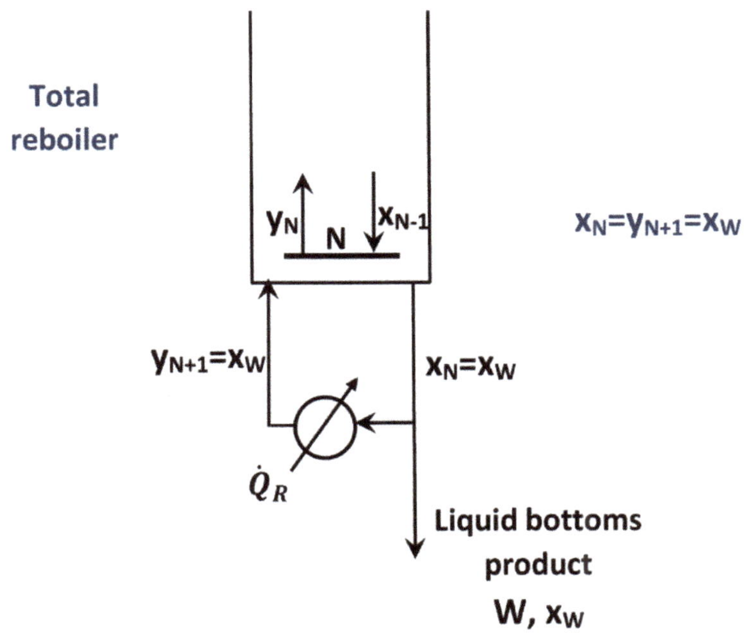

Figure 9.47 Distillation column with total reboiler.

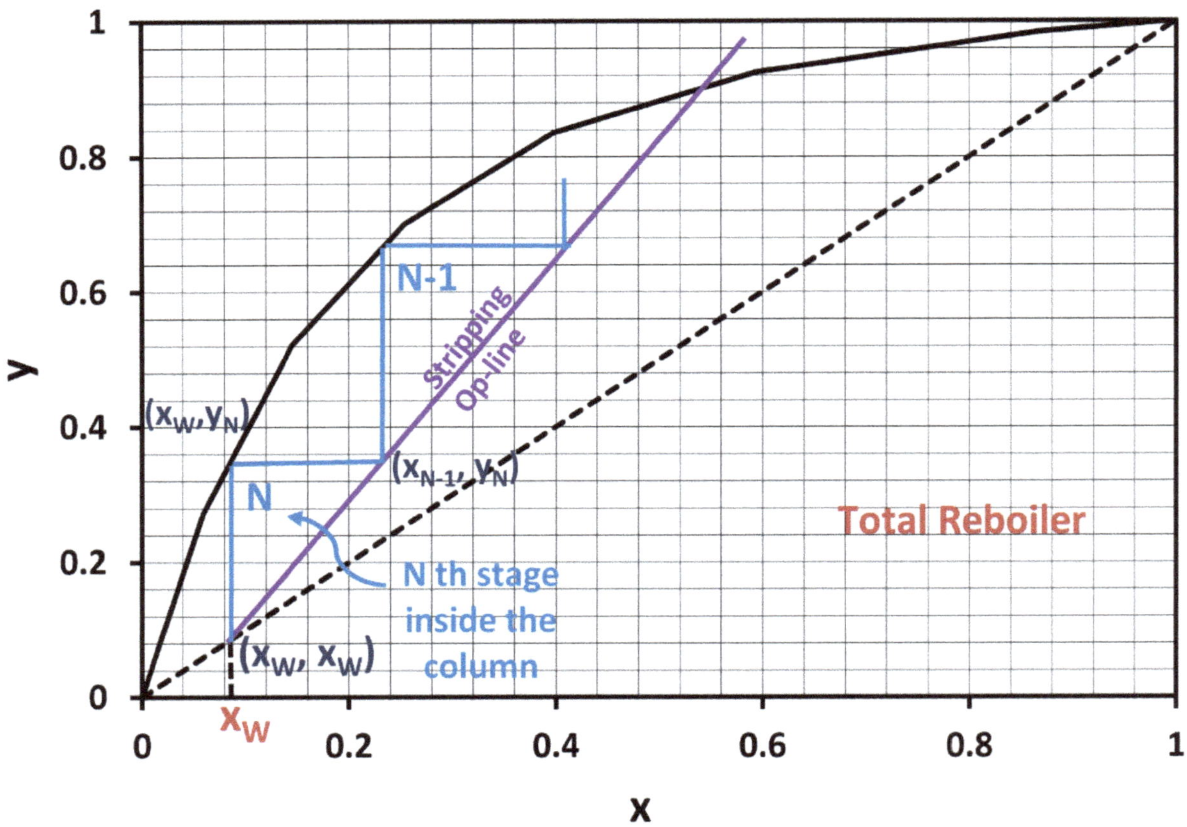

Figure 9.48 McCabe-Thiele diagram of distillation column using total reboiler.

9.6.4 Special Applications of Column Distillation

Enriching Column Distillation

Thus far we have considered complete distillation columns with enriching section above the feed tray and stripping section below the feed tray. In some applications, only an enriching section is involved. In enriching-alone distillation columns, the feed is a superheated or saturated vapor that enters the column from the bottom (see Figure 9.49). The overhead distillate product in enriching columns is generally very rich in the more volatile component. However, the bottoms liquid product is not very different from the feed in composition, being only slightly leaner in more volatile component as compared with the feed.

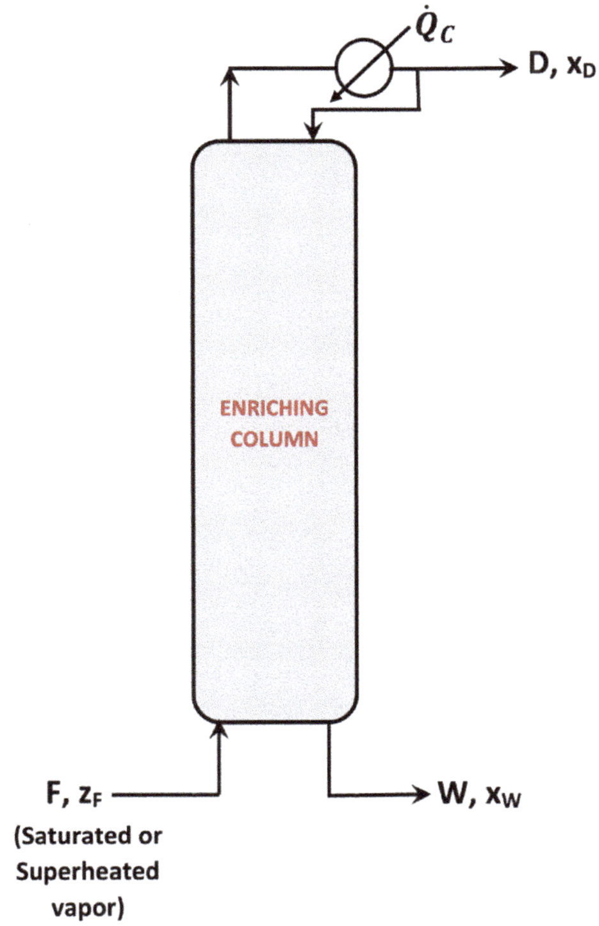

Figure 9.49 Enriching column distillation without stripping section.

For given feed composition (z_F), feed thermal condition (q), reflux ratio (R), distillate product composition (x_D) and bottoms liquid composition (x_W), the number of equilibrium stages can be determined using the McCabe-Thiele graphical approach. Note that there is no reboiler involved in enriching distillation column. Consequently, the vapor boilup rate $\bar{V}$ is zero and the slope of the stripping section operating line is infinite. Although the stripping section is not present, the stripping section operating line is needed in the construction of McCabe-Thiele diagram and stepping off the stages. The stripping section operating line is a vertical line passing through the point (x_W, x_W). Figure 9.50 shows the McCabe-Thiele diagram for enriching column distillation. The feed is assumed to be a saturated vapor ($q = 0$, feed line is horizontal).

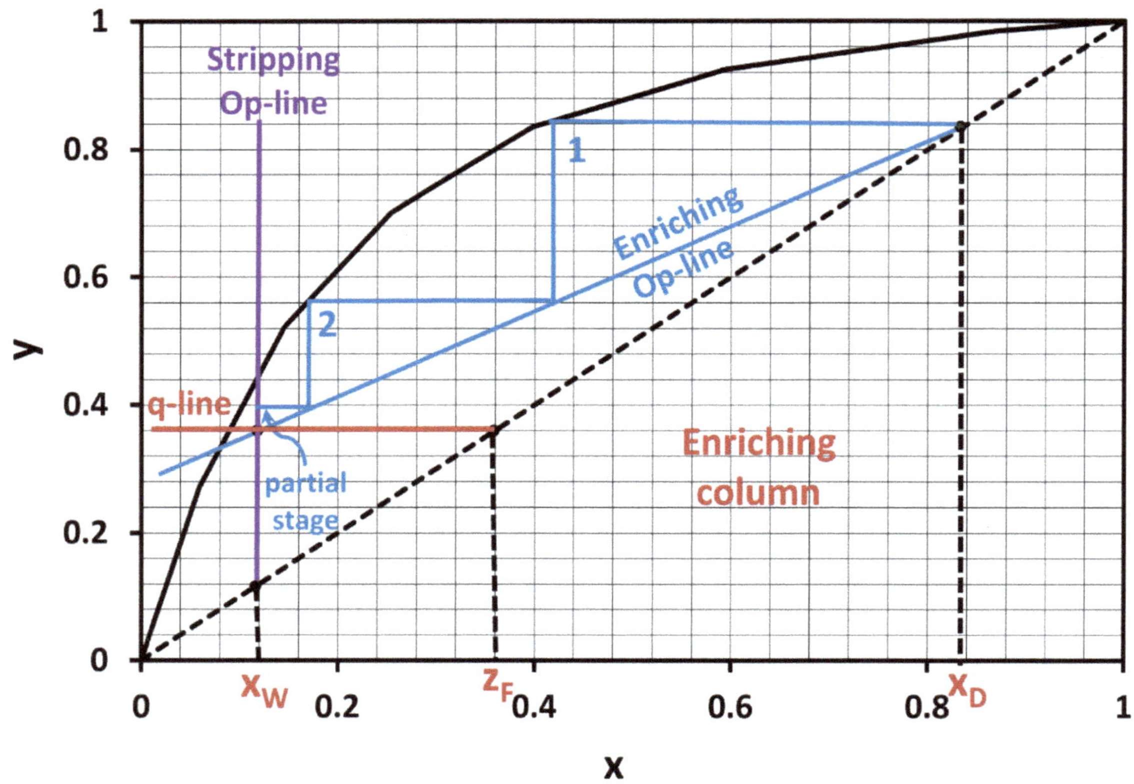

Figure 9.50 McCabe-Thiele diagram of enriching column without any stripping section.

Stripping Column Distillation

In some applications, only a stripping section is needed. The feed in stripping distillation towers is a subcooled or saturated liquid. The feed enters the column from the top (Figure 9.51). No reflux is involved. The bottoms liquid product in stripping towers is nearly a pure liquid very rich in less volatile component. The top vapor distillate product is only slightly richer in more volatile component in comparison with feed composition.

As no reflux is involved in stripping distillation columns, the enriching section operating line is horizontal and is given as $y = y_D$. Figure 9.52 shows the McCabe-Thiele diagram for stripping distillation column. The feed is assumed to be subcooled liquid ($q > 1$).

Figure 9.51 Stripping column distillation without enriching section.

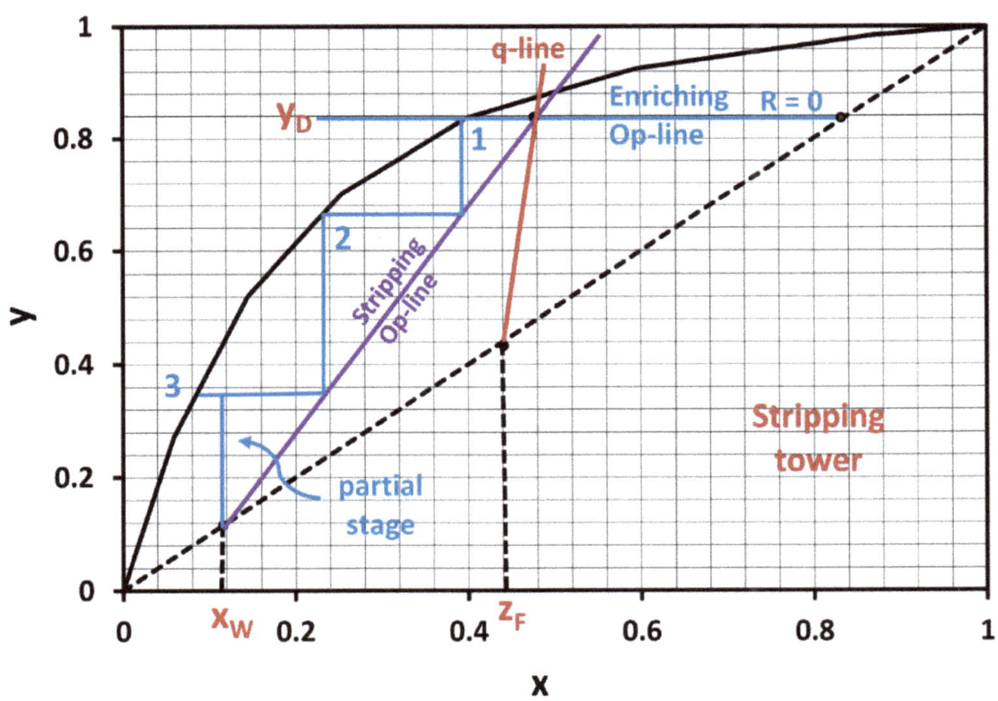

Figure 9.52 McCabe-Thiele diagram of stripping column without any enriching section.

Distillation Column with Multiple Feeds

In some applications, the distillation column is fed with two or more feed streams entering the column on different trays. As an example, consider a distillation column shown in Figure 9.53 with two feed streams. In order to design this two-feed column, we can divide the column into three sections: enriching section above the upper feed stream F_1, middle section between the upper feed stream F_1 and lower feed stream F_2, and stripping section below the lower feed stream F_2. The operating lines are then developed for each section using material balances.

The overall material balance at steady state is:

$$F_1 + F_2 = D + W \tag{9.55}$$

The component (more volatile species) material balance at steady state is:

$$z_1 F_1 + z_2 F_2 = x_D D + x_W W \tag{9.56}$$

Eliminating W from Equations (9.55) and (9.56) gives:

$$D = \frac{z_1 F_1 + z_2 F_2 - (F_1 + F_2) x_W}{x_D - x_W} \tag{9.57}$$

Eliminating D from Equations (9.55) and (9.56) gives:

$$W = \frac{(F_1 + F_2) x_D - z_1 F_1 - z_2 F_2}{x_D - x_W} \tag{9.58}$$

Given the feed flow rates and compositions of the feeds, distillate and bottoms product, Equations (9.57) and (9.58) can be used to calculate the distillate and bottoms product flow rates.

Now consider the enriching section above the upper feed stream F_1. The component (more volatile species) mass balance over the control volume enclosed by a red-dashed oval/circle at the top of the column (Figure 9.53) gives the following equation for the operating line of *enriching section*:

$$y = \left(\frac{L}{V}\right) x + \left(\frac{D}{V}\right) x_D = \left(\frac{R}{1+R}\right) x + \frac{x_D}{1+R} \tag{9.59}$$

Given R and x_D, one can plot the operating line of enriching section. It passes through the point (x_D, x_D) and has an intercept of $x_D/(1+R)$.

The component mass balance over the control volume enclosed by a blue-dashed oval/circle at the top of the column (Figure 9.53) gives the following equation for the operating line of the *middle section*:

$$y = \left(\frac{L'}{V'}\right) x + \left(\frac{x_D D - z_1 F_1}{V'}\right) \tag{9.60}$$

The slope of the middle section operating line (L'/V') can be determined as described below:

$$q_1 = \frac{L' - L}{F_1} \tag{9.61}$$

$$L' = L + q_1 F_1 = RD + q_1 F_1 \tag{9.62}$$

$$V' = L' + D - F_1 = (1 + R)D + (q_1 - 1)F_1 \tag{9.63}$$

As R, q_1, D, F_1 are all known (specified or calculated), L' and V' (hence slope L'/V') can be calculated from Equations (9.62) and (9.63). The middle section operating line can now be plotted keeping in mind that it passes through the point of intersection of q_1-line and enriching section operating line (Equation (9.59)).

The enriching section operating line (Equation (9.59)) and the middle section operating line (Equation (9.60)) are plotted in Figure 9.54. The feeds F_1 and F_2 are saturated liquids with different compositions. We still need to develop the stripping section operating line. The component mass balance over the control volume enclosed by a red-dashed oval/circle at the bottom of the column (Figure 9.53) gives the following equation for the operating line of the *stripping section*:

$$y = \left(\frac{\bar{L}}{\bar{V}}\right)x - \left(\frac{W}{\bar{V}}\right)x_W \tag{9.64}$$

This operating line passes through the point (x_W, x_W) and through the point of intersection of q_2-line and middle section operating line (Equation (9.60)).

Once the three operating lines and two feed lines are plotted, one can step off stages as shown in Figure 9.54. Note that the feed stage locations are optimum as the operating lines are switched upon encountering the intersection of operating lines and feed line. If the feeds are saturated liquids that is, $q_1 = q_2 = 1$, the slopes of the three operating lines follow the relation: $\bar{L}/\bar{V} > L'/V' > L/V$. The vapor flow rate is unaffected due to saturated liquid feed. However, the liquid flow rate below the feed tray increases due to feed addition.

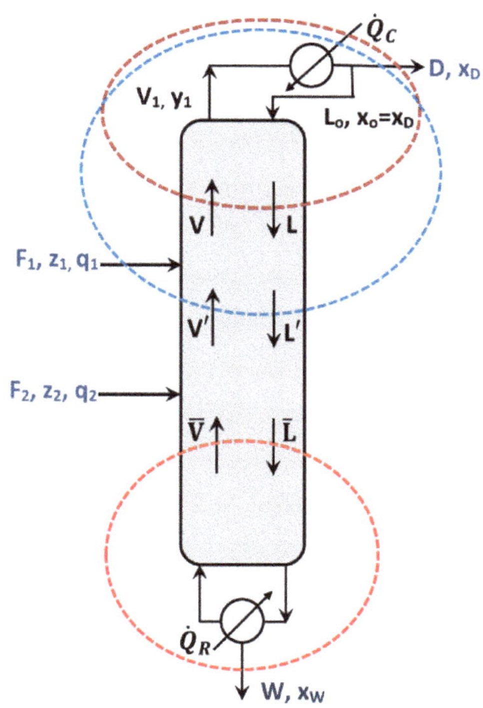

Figure 9.53 Distillation column with two feeds.

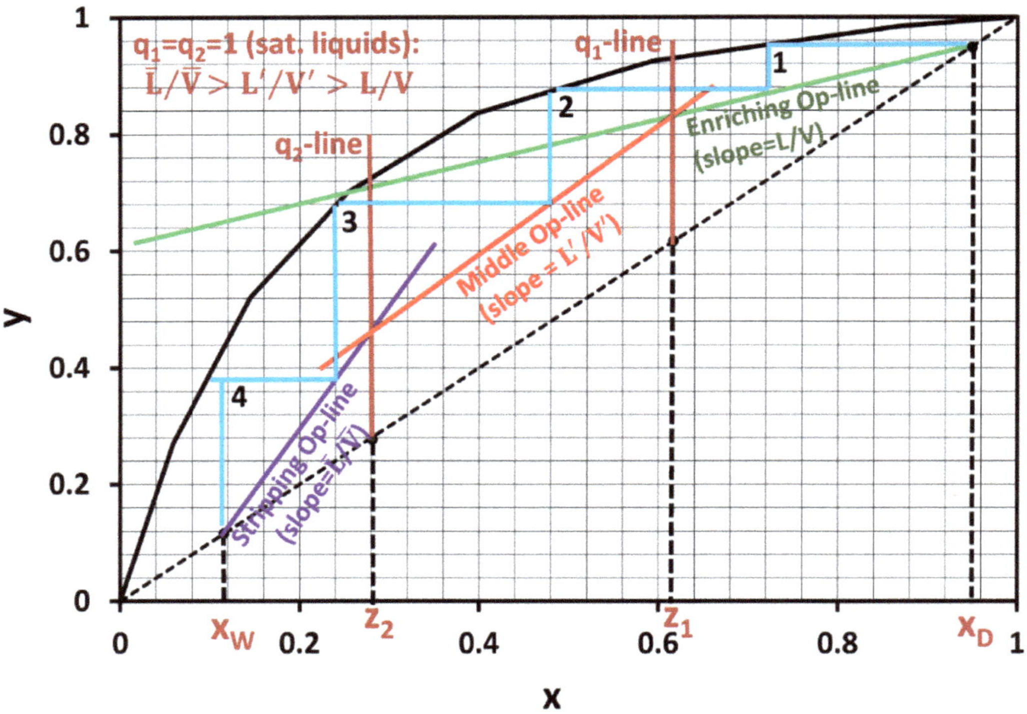

Figure 9.54 McCabe-Thiele diagram of a distillation column with two feeds of different compositions (feeds assumed to be saturated liquids).

Distillation Column with Side-stream Draw-offs

In some applications, a distillation product of intermediate composition is desired. In such situations, a vapor or liquid *side stream* is withdrawn from the column somewhere between the top distillate product and the bottoms product. The location of side draw depends on the composition desired. The flow rate, composition or location of side draw, and state (vapor or liquid) of the side stream are usually specified.

Figure 9.55 shows a distillation column with a side stream located somewhere above the feed tray. The side stream is assumed to be saturated liquid with a flow rate S and composition x_S. The plate from which the side stream is withdrawn is given. We need to develop equations for operating lines of different sections of the column: the top enriching section above the side stream, the middle section between the feed stream and side stream, and the bottom stripping section below the feed tray.

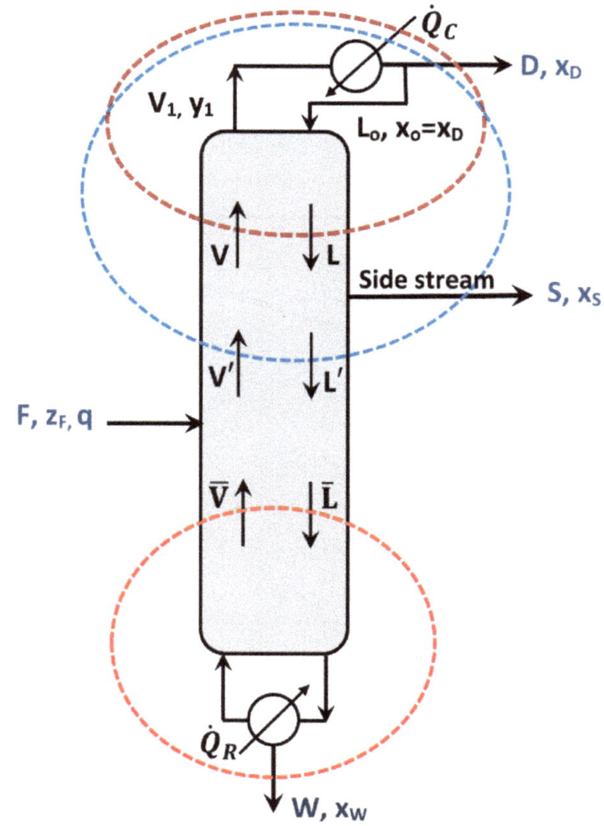

Figure 9.55 Distillation column with a liquid side stream above the feed plate.

The equation for the top enriching section is the same as Equation (9.59). This equation follows from component mass balance over the control volume enclosed by a red-dashed oval/circle at the top of the column (Figure 9.55). Given R and x_D, one can readily plot the enriching section operating line. As noted earlier, this operating line passes through the point (x_D, x_D) and has an intercept of $x_D/(1 + R)$.

After plotting the enriching section operating, we need to plot the *side stream q-line*. The side stream is equivalent to a negative feed of saturated liquid with composition x_S. Thus the side stream q-line is simply a vertical line at $x = x_S$. Note that the usual (positive feed) q-line is vertical when the feed is saturated liquid. Given the tray from which the side stream is withdrawn as a saturated liquid, we can determine x_S from stepping off stages as shown in Figure 9.56. In Figure 9.56, it is assumed that the side stream is withdrawn as a saturated liquid from tray 2. Once x_S is determined, we can plot the side stream q-line as $x = x_S$. The enriching section and middle section operating lines must intersect at this side stream q-line. Before discussing the middle section operating line, let us first see how can we calculate the distillate and bottoms products flow rates D and W. The calculation of these flow rates require external material balances as described below:

$$F = S + D + W \tag{9.65}$$

$$z_F F = x_S S + x_D D + x_W W \tag{9.66}$$

Eliminating W from Equations (9.65) and (9.66) gives:

$$D = \frac{(z_F - x_W)F - (x_S - x_W)S}{x_D - x_W} \tag{9.67}$$

Eliminating D from Equations (9.65) and (9.66) gives:

$$W = \frac{(x_D - z_F)F - (x_D - x_S)S}{x_D - x_W} \tag{9.68}$$

Given the feed and side stream flow rates and compositions of feed, side stream, distillate and bottoms product, Equations (9.67) and (9.68) can be used to calculate the distillate and bottoms product flow rates.

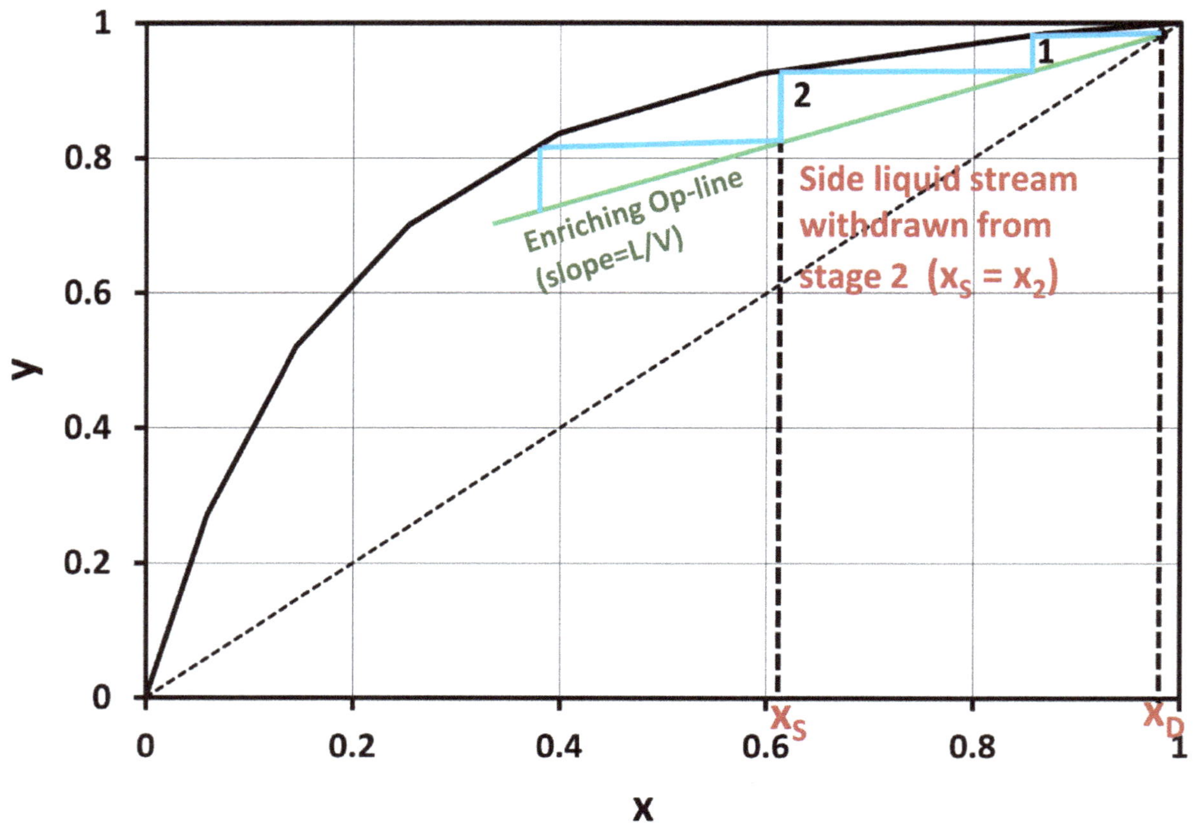

Figure 9.56 Graphical determination of side stream composition. The side stream is withdrawn from tray 2 as saturated liquid.

Let us now develop the operating line for the middle section. The component mass balance over the control volume enclosed by a blue-dashed oval/circle at the top of the column (Figure 9.55) gives the following equation for the operating line of the *middle section*:

$$y = \left(\frac{L'}{V'}\right)x + \left(\frac{x_D D + x_S S}{V'}\right) \tag{9.69}$$

We need L' and V' in order to calculate the slope L'/V' of the middle section operating line. For saturated liquid side stream:

$$L' = L - S = RD - S \tag{9.70}$$

$$V' = V = (1 + R)D \tag{9.71}$$

The middle section operating line can now be plotted readily. It passes through the intersection of the enriching section operating line and the side stream q-line and has a slope of L'/V'.

Figure 9.57 shows the plots of the enriching section and middle section operating lines along with the side stream q-line. Note that the slope of the middle section operating line L'/V' is less than the slope of the enriching section operating line L/V as liquid is being withdrawn through the side stream. In the figure, it is assumed that the feed to the column is saturated liquid. Thus the feed q-line is also vertical $x = z_F$ line. We still need to develop and plot the bottom stripping section operating line. The component mass balance over the control volume enclosed by a red-dashed oval/circle at the bottom of the column (Figure 9.55) gives Equation (9.64) as the operating line of *stripping section*. This operating line passes through the point (x_W, x_W) and through the point of intersection of feed q-line and middle section operating line (Equation (9.69)).

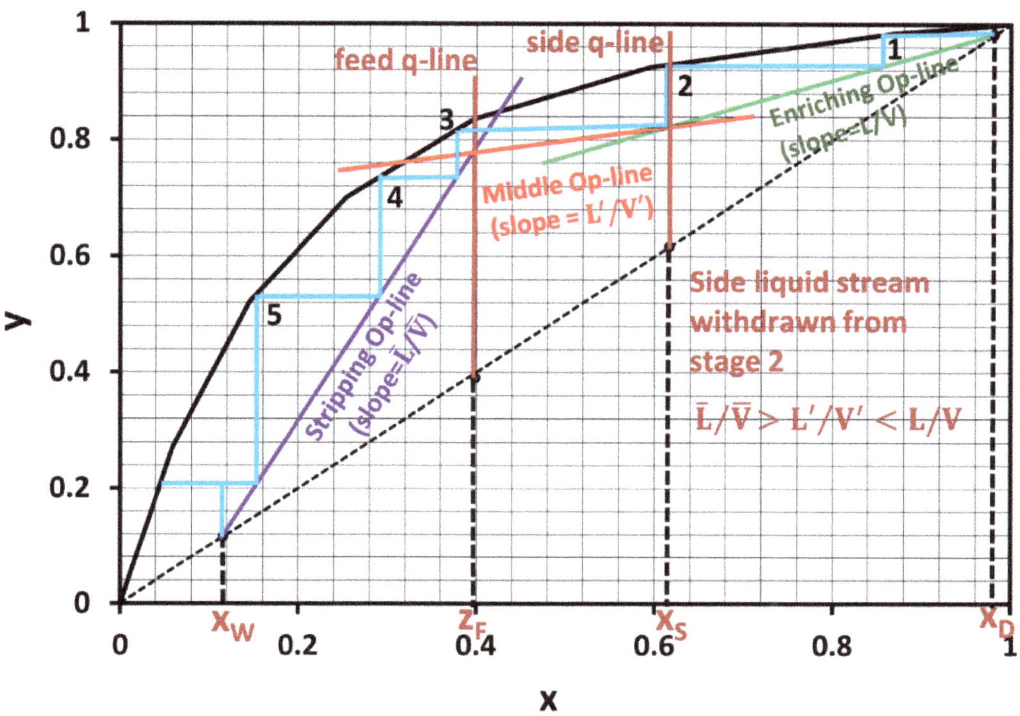

Figure 9.57 McCabe-Thiele diagram of distillation column with side stream. The side stream is withdrawn from tray 2 as saturated liquid. The feed to the column is saturated liquid.

Once the three operating lines, feed q-line and side stream q-line are plotted, one can step off stages as shown in Figure 9.57. If the feed and side stream are saturated liquids, the slopes of the three operating lines follow the relation: $\bar{L}/\bar{V} > L'/V' < L/V$.

In the preceding discussion we have assumed that the side stream is saturated liquid. What happens when the side stream is saturated vapor? Let the side stream be withdrawn as saturated vapor from top of tray 3. Thus the composition of the vapor side stream withdrawn is $y_S = y_3$. This composition can be determined by stepping off two stages (see Figure 9.56) and reading the final y-value on the enriching operating line. The side stream q-line is then simply a horizontal line $y = y_S$. The component mass balance over the control volume enclosed by a blue-dashed oval/circle at the top of the column (Figure 9.55) gives the following equation for the operating line of the *middle section*:

$$y = \left(\frac{L'}{V'}\right)x + \left(\frac{x_D D + y_S S}{V'}\right) \tag{9.72}$$

We need L' and V' in order to calculate the slope L'/V' of the middle section operating line. For saturated vapor side stream:

$$L' = L = RD \tag{9.73}$$

$$V' = V + S = (1 + R)D + S \tag{9.74}$$

The middle section operating line can now be plotted readily. It passes through the intersection of the enriching section operating line and the side stream q-line and has a slope of L'/V'. Note that Equations (9.67) and (9.68) are still valid except that x_S is to be replaced by y_S.

The operating line for the bottom stripping section also remains the same, that is, Equation (9.69). This equation passes through the point (x_W, x_W) and through the point the intersection of middle section operating line and feed line. The feed to the column is a two-phase vapor-liquid mixture.

Once the three operating lines, feed q-line and side stream q-line are plotted, one can step off stages as shown in Figure 9.58. Note that the slopes of the three operating lines in the present case follow the relation: $\bar{L}/\bar{V} > L'/V' < L/V$.

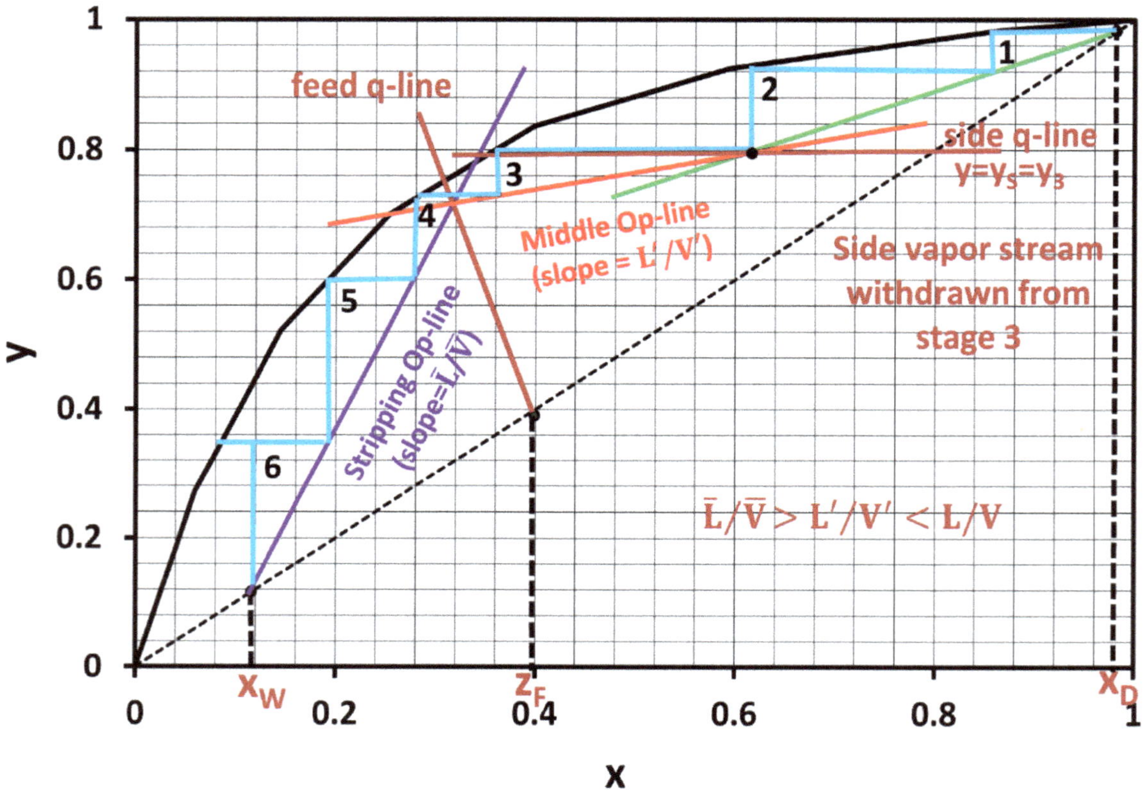

Figure 9.58 McCabe-Thiele diagram of distillation column with side stream. The side stream is withdrawn from tray 3 as saturated vapor. The feed to the column is a two-phase vapor-liquid mixture.

Distillation Column with Direct Steam Injection

In most applications of distillation, the heat required is supplied through the reboiler using steam as the heating fluid in the reboiler. The steam passes through the reboiler tubes without any mixing or direct contact with the column fluids (boiling solution). However, in certain applications of distillation where an aqueous solution of more volatile species A and less volatile water (species B) is being separated, it is advantageous to supply the heat required for distillation directly to the column through steam injection into the base of the column. The steam now comes in direct contact with the boiling solution (steam is bubbled through the solution) and no reboiler is employed.

The only difference between the usual distillation column with reboiler and the distillation column with direct steam injection is the bottom section of the column. The rest of the column with direct steam injection operates in the same way as the usual distillation column with reboiler. Thus the enriching section operating line and the feed q-line remain the same. Only the stripping section operating line is modified. Figure 9.59 shows the bottom section of a distillation column with direct steam injection. The overall and component (more volatile species) material balances give:

$$\bar{L} - \bar{V} = W - S' \tag{9.75}$$

$$x\bar{L} - y\bar{V} = x_W W - y_S' S' \tag{9.76}$$

where S' is the steam molar flow rate and y_S' is the mole fraction of more volatile species in the steam being injected. Assuming steam being injected is pure ($y_S' = 0$) and saturated:

$$\bar{V} = S'; \ \bar{L} = W \tag{9.77}$$

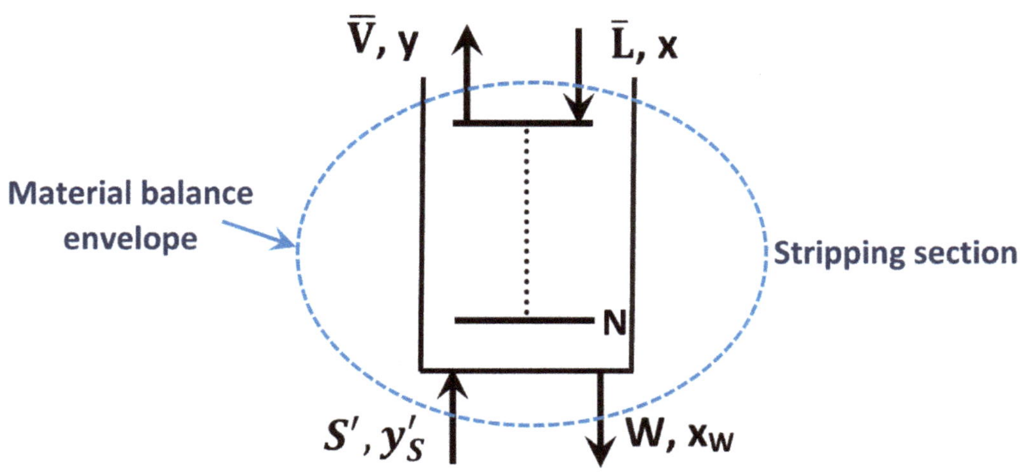

Figure 9.59 Direct steam injection at the base of the column to replace the reboiler.

Thus, the stripping section operating line, Equation (9.76), becomes:

$$y = \left(\frac{W}{S'}\right)(x - x_W) \tag{9.78}$$

This line passes through the point $(x_W, 0)$ and has a slope of $\bar{L}/\bar{V} = W/S'$.

Figure 9.60 shows the McCabe-Thiele diagram for distillation column with direct steam injection. The feed is assumed to be saturated liquid and the stages are stepped off in the usual way. As there is no reboiler involved, all the stages are located inside the column.

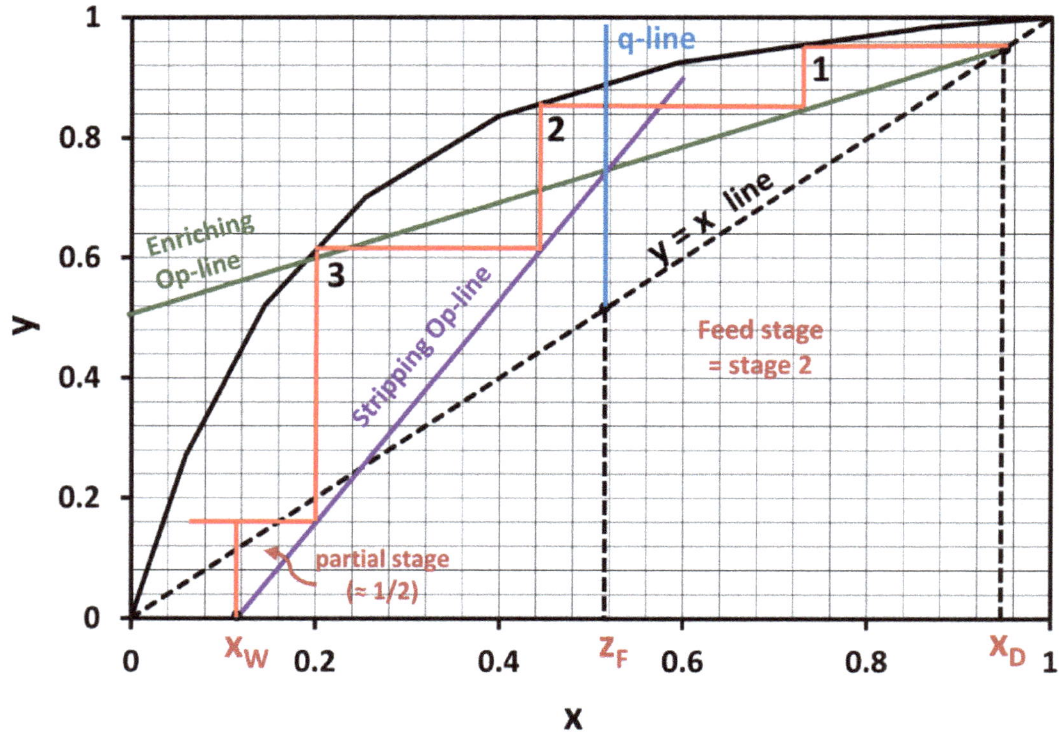

Figure 9.60 McCabe-Thiele diagram of distillation column with direct steam injection without reboiler.

Pressure-swing Distillation of Azeotropes

An azeotrope is a mixture of two or more chemical species (liquids) whose composition cannot be altered by boiling. Thus azeotrope mixtures cannot be separated into components using simple distillation. When an azeotrope liquid mixture/solution is boiled, the vapor produced has the same composition as that of the liquid solution. Also, the boiling point of azeotrope liquid solution does not change as boiling proceeds, that is, the whole liquid solution would boil off at constant temperature because the composition of the solution does not change with boiling. Because the composition and boiling point of the azeotropes do not change upon boiling, they are also referred to as "*constant boiling point*" mixtures.

The boiling point of an azeotrope could be higher (*negative azeotrope*) or lower (*positive azeotrope*) than that of its components. For example, ethanol-water system forms a positive azeotrope at 95.63 % ethanol and 4.37 % water (by mass). The boiling point of ethanol is 78.4 °C and that of water is 100 °C (1 atm pressure), whereas the azeotrope boils at 78.2 °C, which is lower than the boiling points of components. Positive azeotropes are also referred to as *minimum-boiling azeotropes.* An example of negative azeotrope (*maximum-boiling azeotrope*) is hydrochloric acid-water system at 20.2 % acid (by mass) and 79.8 % water. The boiling point of hydrochloric acid is -85.05°C and that of water is 100 °C. The azeotrope boils at 108.6 °C, which is higher than its components. Only non-ideal mixtures of liquids form azeotropes. If the chemical components are dissimilar and the repulsive forces are strong, minimum boiling azeotropes can form. Maximum boiling azeotropes are formed when chemical components of the mixture attract each other.

Azeotropic liquid solutions cannot be separated into components using ordinary distillation. Special techniques are required to separate azeotropic mixtures. Although a number of different techniques are available to separate azeotropic mixtures into components, here we will restrict ourselves to so-called "*pressure-swing distillation (PSD)*" technique to separate a binary azeotropic mixture. In usual distillation of non-azeotropic mixture of two species having different volatilities only one distillation column is required. However, the PSD technique utilizes two distillation columns operating at different pressures in order to separate an azeotropic mixture. The PSD technique is based on the fact that the azeotropic point (composition) of a binary system is sensitive to pressure. By simply changing the pressure of the system, the composition where an azeotrope is formed can be changed. As an example, Figure 9.61 shows the T-x-y phase diagram of a minimum boiling azeotrope. By increasing the pressure from P_1 to P_2 the azeotropic point is shifted to the right towards higher concentration of species A. The shift in azeotrope composition with the increase in pressure can be seen more clearly in Figure 9.62 where equilibrium y-x curves are plotted at low and high pressures. The equilibrium curve crosses the $y = x$ line at azeotropic point. The azeotropic point shifts to higher concentration with the increase in pressure.

Figure 9.63 shows the PSD scheme for separation of a minimum boiling azeotropic binary mixture of species A and B. The T-x-y phase diagrams and equilibrium y-x curves for this azeotropic system have been described in Figures 9.61 and 9.62. Two distillation columns operating at two different pressures P_1 (low pressure) and P_2 (high pressure) are needed to separate the mixture. Nearly pure A is produced as the bottom product stream from the low pressure column and nearly pure B is produced as the bottom product stream from the high pressure column. The distillate product of a low pressure column is used as feed for the second high pressure column. The distillate product stream from the high pressure column is recycled back and mixed with the fresh feed to the low pressure column. Thus the feed that actually goes into the low pressure column is a mixture of two streams (fresh feed stream plus recycled distillate stream from the high pressure column).

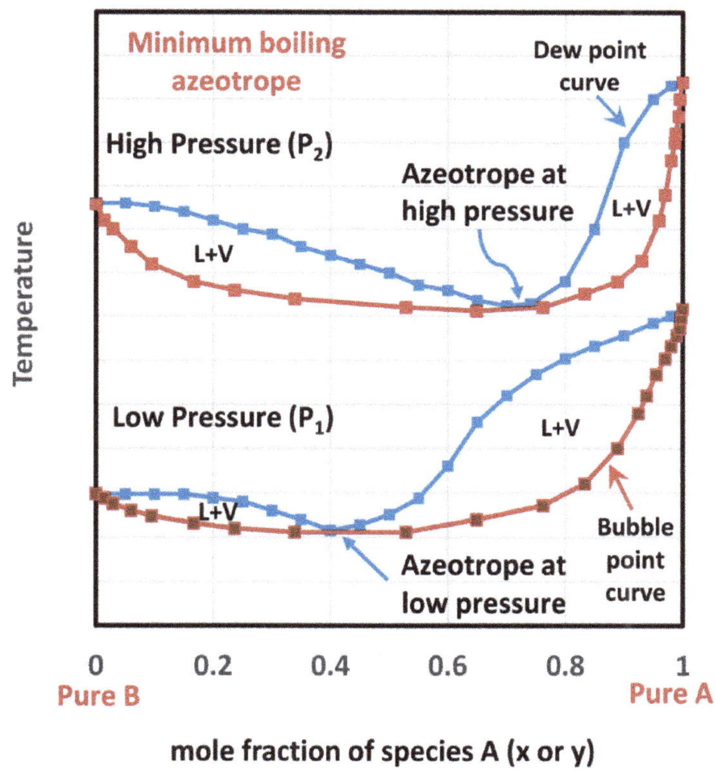

Figure 9.61 T-x-y phase diagrams for minimum boiling azeotrope at different pressures.

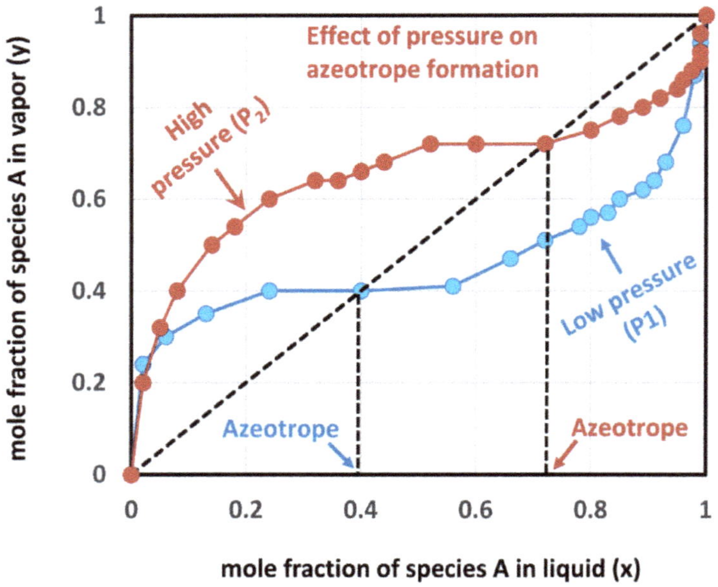

Figure 9.62 Equilibrium y-x curves for minimum boiling azeotrope at different pressures.

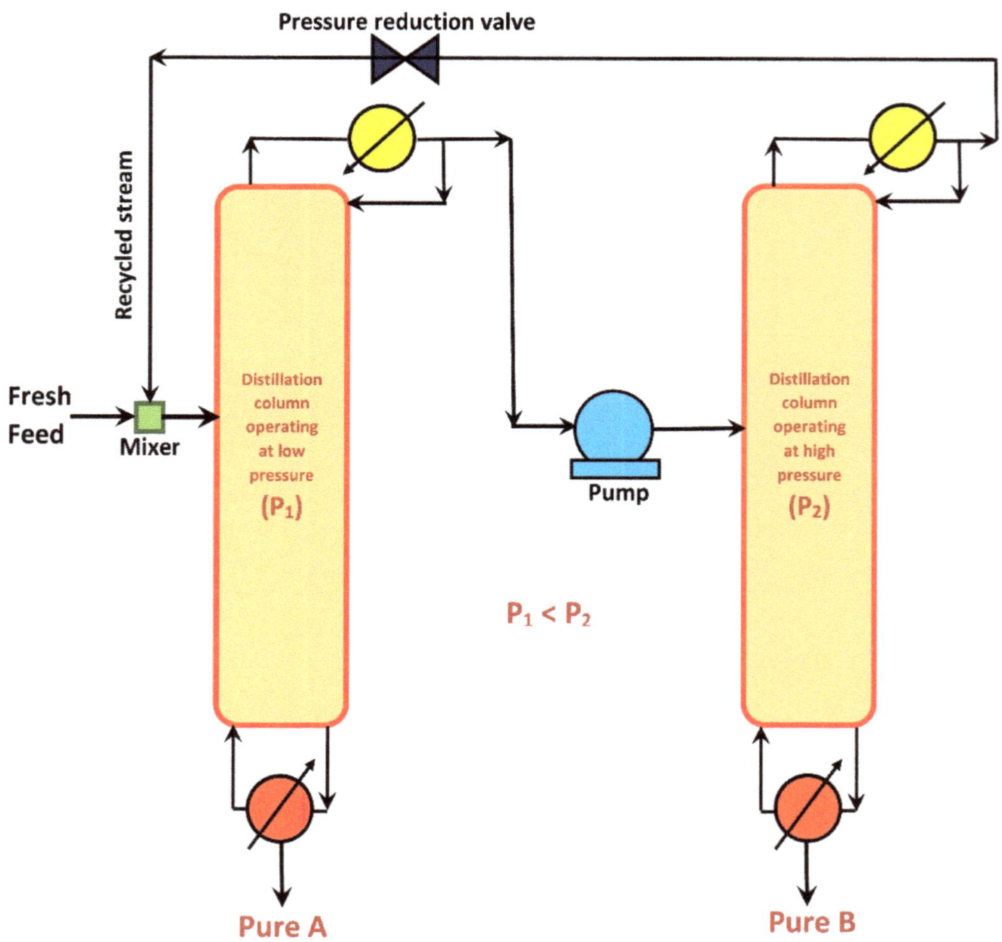

Figure 9.63 PSD scheme for the separation of minimum boiling azeotrope of species A and B.

Figure 9.64 further explains how the PSD scheme works. The feed (mixture of fresh feed plus recycled distillate stream as shown in Figure 9.63) to column 1 at low pressure splits into bottoms product as nearly pure A and distillate product with composition close to azeotrope composition at P_1. Although the distillate product composition of low pressure column is only slightly different (richer in A) compared with the azeotropic composition at low pressure P_1, it is significantly different (leaner in A) compared with the azeotropic composition at high pressure P_2. This distillate product stream of low pressure column 1 is therefore used as feed to the high pressure distillation column operating at P_2. The feed to the high pressure column separates into bottoms product as nearly pure B and distillate product which is significantly richer in A as compared with the azeotropic composition at low pressure P_1. The distillate product from high pressure column 2 is recycled back to low pressure column 1 and mixed with the fresh (azeotropic) feed entering column 1. The mixed feed (fresh feed + recycled stream) actually entering the low pressure column is significantly richer in A as compared with the azeotropic composition at low pressure P_1.

The net effect of the PSD scheme is illustrated in Figure 9.65. The fresh feed enters the PSD box as an azeotropic mixture at P_1 and nearly pure A and pure B streams emerge from the PSD box.

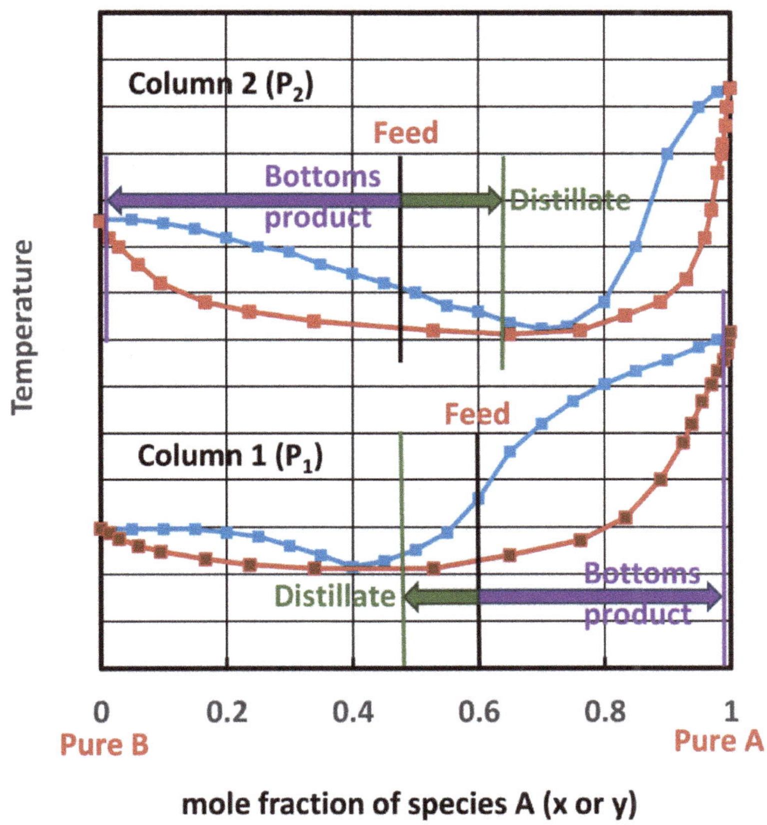

Figure 9.64 Explanation of PSD scheme for separation of minimum boiling azeotrope of species A and B.

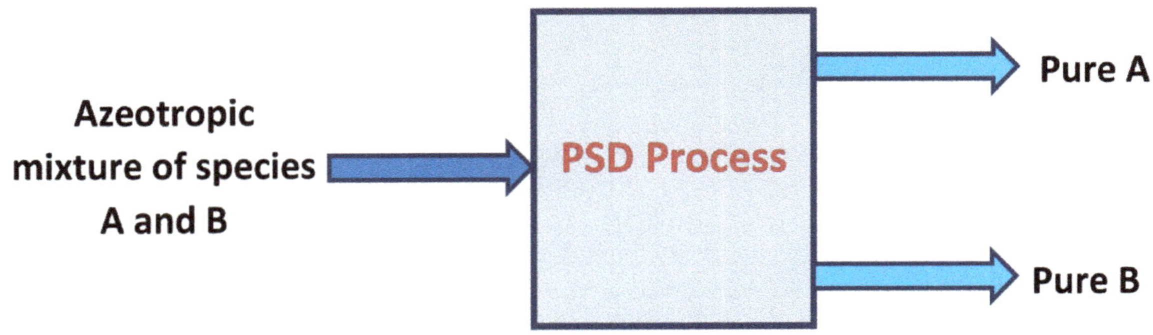

Figure 9.65 Separation of azeotropic mixture of species A and B using the PSD process.

Let us now consider how to design the two columns involved in the PSD scheme. For the low pressure column 1 operating at P_1, it is preferred to plot the equilibrium curve as y_B versus x_B (species A is less volatile compared with the azeotrope formed at P_1). The y_B versus $x_B$equilibrium curve is shown in Figure 9.66. We assume that the mole fraction of species B in the feed after mixing of fresh feed with recycled stream (see Figure 9.63) is known. The distillate and bottoms product compositions of column 1 in terms of mole fractions of species B are specified. The thermal condition of the feed (q value) and the reflux ratio of column 1 are also known. Figure 9.66 shows the McCabe-Thiele diagram of column 1. It is assumed that the feed is a two-phase mixture of vapor and liquid entering column 1. The stages are stepped off as shown in Figure 9.66.

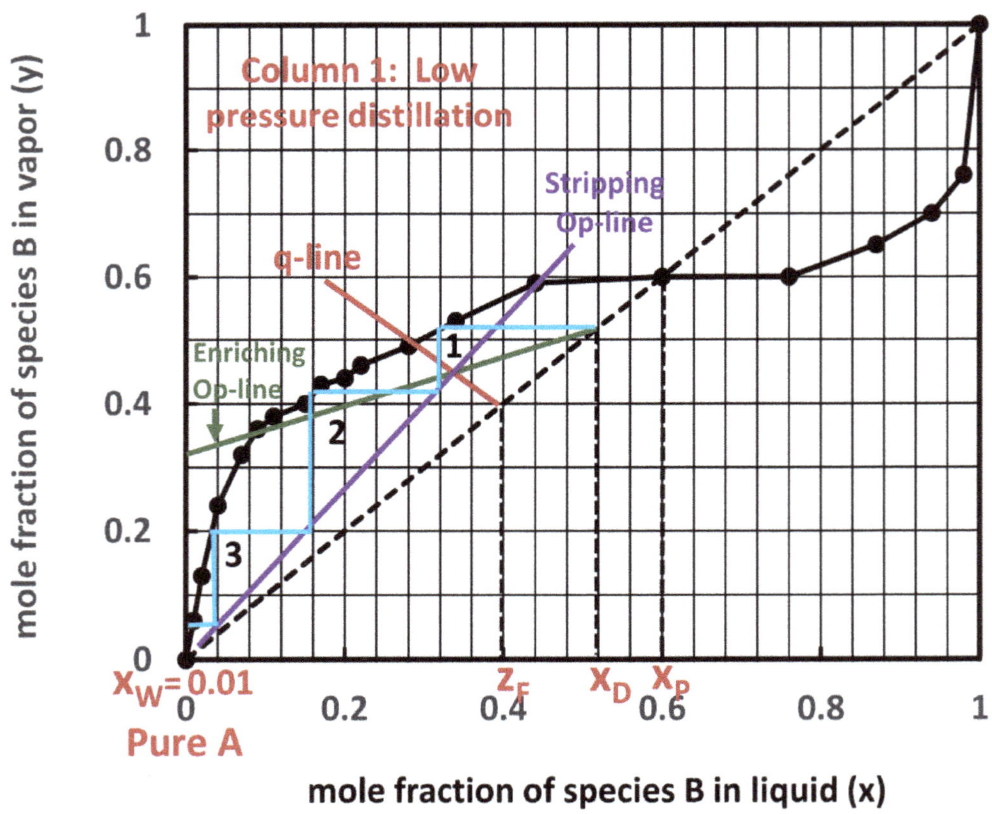

Figure 9.66 McCabe-Thiele diagram of low pressure column 1 of the PSD process.

For the high pressure column 2 operating at P_2, it is preferred to plot the equilibrium curve as y_A versus x_A (species B is less volatile compared with the azeotrope formed at P_2). The y_A versus x_A equilibrium curve is shown in Figure 9.67. We assume that the mole fractions of species A in the feed, distillate and bottoms product of column 2 are all known. The thermal condition of the feed (q value) and the reflux ratio of column 2 are also known. Figure 9.67 shows the McCabe-Thiele diagram of column 2. It is assumed that the feed is subcooled liquid entering column 2. The stages are stepped off as shown in Figure 9.67.

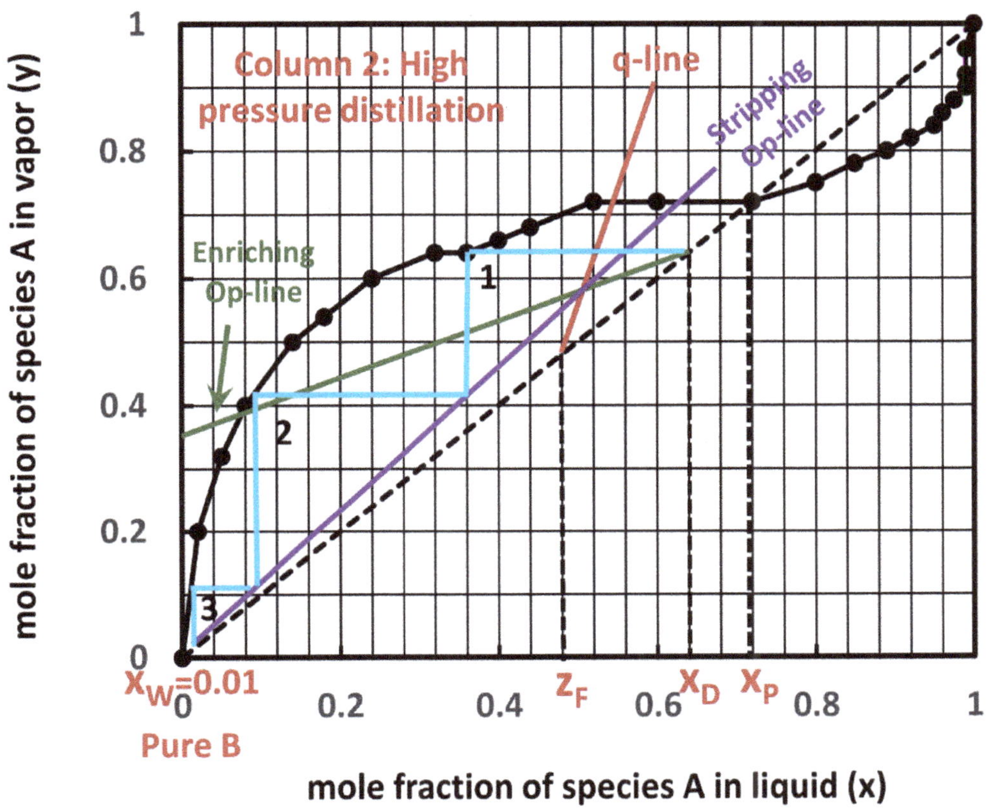

Figure 9.67 McCabe-Thiele diagram of high pressure column 2 of the PSD process.

NOTATION

C_{PL}	constant pressure molar heat capacity of liquid feed
C_{PV}	constant pressure molar heat capacity of vapor feed
CMO	constant molal overflow
D	molar flow rate of distillate liquid product
F	feed molar flow rate
F_i	molar flow rate of fluid stream at ith opening (inlet/outlet) of control volume
h	molar enthalpy of liquid
h_D	molar enthalpy of distillate liquid product
h_F	molar enthalpy of feed
h_f	molar enthalpy of liquid on the feed tray
h_i	molar enthalpy of fluid stream at ith opening (inlet/outlet) of the control volume (CV)
h_L	molar enthalpy of feed at bubble point

h_o	molar enthalpy of reflux liquid
h_W	molar enthalpy of bottoms liquid product
H	molar enthalpy of vapor
H_f	molar enthalpy of vapor on the feed tray
H_V	molar enthalpy of feed at dew point
L	molar flow rate of liquid stream in enriching section
L_n	molar flow rate of liquid stream leaving any *n*th stage in the enriching section
$\bar{L}$	molar flow rate of liquid stream in stripping section
$\bar{L}_m$	molar flow rate of liquid stream leaving any *m*th stage in the stripping section
L'	molar flow rate of liquid stream in the middle section when column has two feed streams or a side stream
N	number of stages in the column; in some situations N includes the partial reboiler and in some situations N is the number of stages inside the column excluding the partial reboiler
P	pressure
P_F	pressure of feed stream
PSD	pressure-swing distillation
q	feed quality, defined in Equation (9.40)
$\dot{Q}$	heat transfer rate
$\dot{Q}_C$	heat transfer rate in condenser
$\dot{Q}_R$	heat transfer rate in reboiler
R	reflux ratio
R_{min}	minimum reflux ratio
s	boilup ratio, defined as the ratio of the amount of liquid that is boiled back as vapor into the column to the amount of liquid leaving the column as the bottoms product ($s = \bar{V}/W$)
S	molar flow rate of side stream withdrawn from the column
S'	molar flow rate of direct steam injection into the column
T	temperature
T_F	temperature of feed stream
T_b	bubble point of feed at column pressure
T_d	dew point of feed at column pressure
V	molar flow rate of vapor stream in enriching section
V_n	molar flow rate of vapor stream leaving any *n*th stage in the enriching section
$\bar{V}$	molar flow rate of vapor stream in stripping section
$\bar{V}_m$	molar flow rate of vapor stream leaving any *m*th stage in the stripping section
V'	molar flow rate of vapor stream in the middle section when column has two feed streams or a side stream
W	molar flow rate of bottoms liquid stream of column
$\dot{W}_{sh}$	rate of shaft work
x	mole fraction of more volatile species in liquid phase
x_D	mole fraction of more volatile species in the distillate liquid product
x_m	mole fraction of more volatile species in the liquid stream leaving any *m*th stage in the stripping section
x_n	mole fraction of more volatile species in the liquid stream leaving any *n*th stage
x_s	mole fraction of more volatile species in the liquid stream withdrawn from the side of the column, that is, the side stream
x_W	mole fraction of more volatile species in bottoms liquid product of distillation column
y	mole fraction of more volatile species in vapor phase
y_m	mole fraction of more volatile species in the vapor stream leaving any *m*th stage in the stripping section
y_n	mole fraction of more volatile species in the vapor stream leaving any *n*th stage

y_S mole fraction of more volatile species in the vapor stream withdrawn from the side of the column, that is, the side stream
y_S' mole fraction of more volatile species in the steam being injected into the column operating under direct steam injection
z_F mole fraction of more volatile species in the feed stream
z_i mole fraction of more volatile species in *i*th feed stream when the column is fed with more than one feed stream

Greek Letters

λ_D molar latent heat of vaporization of distillate product of the column
λ_B molar latent heat of vaporization of bottoms product of the column

TUTORIAL PROBLEMS WITH SOLUTION

Problem 1

A multistage distillation column is operated at 1 atm pressure to separate ethanol-water mixture. The feed composition is 25 wt% ethanol and 75 wt% water and has a mass flow rate of 9,100 kg/h. The desired distillate and bottom compositions are: 79 wt% ethanol and 4 wt% ethanol, respectively. The reflux ratio (R) is 2.5.

Given that:

$h_D = 251\ kJ/kg, h_W = 377\ kJ/kg, h_F = 126\ kJ/kg,$ and $H_1 = 1382\ kJ/kg$

Calculate the distillate and bottom product flow rates. Also estimate the heat duties of the condenser and reboiler.

Solution

A schematic diagram of the column is shown below.

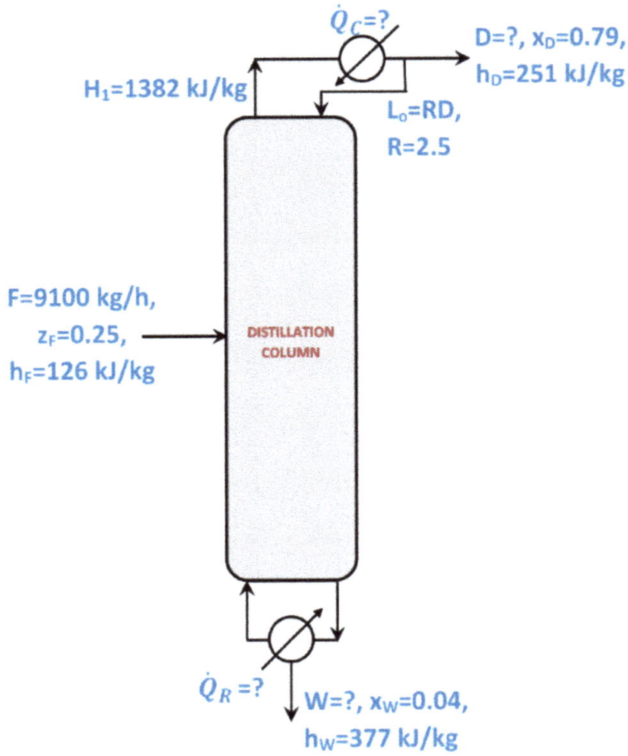

$$\frac{D}{F} = \frac{z_F - x_W}{x_D - x_W} = \frac{0.25-0.04}{0.79-0.04} = 0.28$$

$$D = 0.28F = 0.28 \times 9100 = 2548\ kg/h$$

$$\frac{W}{F} = \frac{x_D - z_F}{x_D - x_W} = \frac{0.79-0.25}{0.79-0.04} = 0.72$$

$$W = 0.72F = 0.72 \times 9100 = 6552\ kg/h$$

Energy balance around condenser gives

$$\dot{Q}_C = -D(1+R)(H_1 - h_D) = -2548(1+2.5)(1382-251) = -10086258\ kJ/h$$

Overall energy balance on the column gives:

$$\dot{Q}_R = Dh_D + Wh_W - Fh_F - \dot{Q}_C$$

$$= 2548 \times 251 + 6552 \times 377 - 9100 \times 126 - (-10086258) = 12049310\ kJ/h$$

Problem 2

Calculate the slope of the feed line (q-line) under the following conditions:

a) Feed entering the distillation column is a two-phase vapor-liquid mixture consisting of 75 mol% vapor and 25 mol% liquid.

b) Feed entering the column is a superheated vapor such that for each mole of feed entering the column, 0.1 mole of liquid will vaporize on the feed stage.

c) Feed entering the column is a subcooled liquid, 41 °F below the bubble point temperature at column pressure. Given that the average heat capacity of liquid is 28 Btu/lb-mol°F and the latent heat of vaporization is 14,225 Btu/lb-mol.

d) Feed entering the column is subcooled liquid. Given that: enthalpy of subcooled liquid feed is 18 kcal/kmol, enthalpy of feed as saturated vapor is 400 kcal/kmol, and enthalpy of feed as saturated liquid is 59 kcal/kmol.

Solution

a)

$$V - \bar{V} = 0.75F$$

$$\bar{L} - L = 0.25F$$

$$q = \frac{\bar{L} - L}{F} = 0.25$$

$$slope = -\frac{q}{1-q} = -\frac{0.25}{1-0.25} = -\frac{1}{3}$$

b) 0.1 mole of liquid is vaporized per mole of feed. For F moles of feed, the liquid vaporized is 0.1F. Thus,

$$V - \bar{V} = F + 0.1F = 1.1F$$

$$\bar{L} - L = -0.1F$$

$$q = \frac{\bar{L} - L}{F} = -0.1$$

$$slope = -\frac{q}{1-q} = -\frac{(-0.1)}{1-(-0.1)} = 0.091$$

c)

$$q = \frac{H_V - h_F}{H_V - h_L} = 1 + \frac{h_L - h_F}{H_V - h_L} = 1 + \frac{C_{pL}(T_b - T_F)}{\lambda} = 1 + \frac{28 \times 41}{14225} = 1.0807$$

$$slope = -\frac{q}{1-q} = -\frac{(1.0807)}{1-(1.0807)} = 13.39$$

d)

$$q = \frac{H_V - h_F}{H_V - h_L} = \frac{400-18}{400-59} = 1.1202$$

$$slope = -\frac{q}{1-q} = -\frac{(1.1202)}{1-(1.1202)} = 9.32$$

Problem 3

Feed entering the distillation column is saturated liquid at column conditions. It consists of a binary mixture of heptane and ethyl benzene. The mole fraction of heptane in the feed is 0.39. The feed flow rate is 250 kmol/h. The distillate product contains 95 mol% heptane and the bottoms product contains 1.25 mol% heptane. The reflux ratio is 2.15.

Do the following:

a) Calculate the distillate product flow rate, D
b) Calculate the bottoms product flow rate, W
c) Determine the optimum number of equilibrium stages using the McCabe-Thiele method
d) Determine the optimum feed tray location
e) Plot Txy diagram
f) Plot composition profile of the column
g) Plot temperature profile of the column

The equilibrium data for heptane-ethyl benzene, in terms of mole fraction of more volatile component heptane, at column pressure (101.32 kPa) is as follows:

x	y	T (°C)
0	0	136.1
0.08	0.23	129.4
0.25	0.514	119.4
0.485	0.73	110.6
0.79	0.904	102.8
1.0	1.0	98.3

Solution

Refer to the diagram given below.

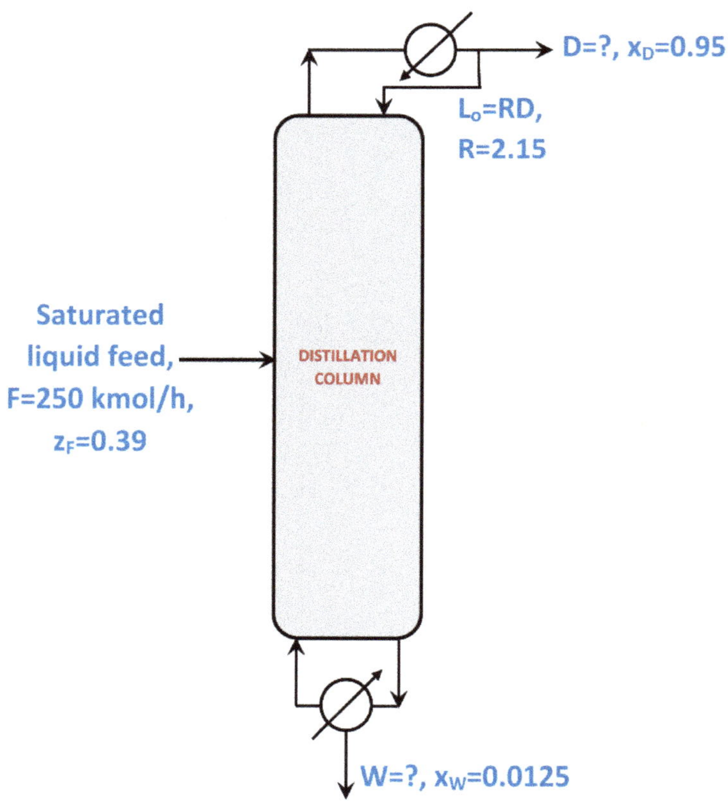

a)

$$\frac{D}{F} = \frac{z_F - x_W}{x_D - x_W} = \frac{0.39 - 0.0125}{0.95 - 0.0125} = 0.4027$$

$$D = 0.4027F = 0.4027 \times 250 = 100.68\ kmol/h$$

b)

$$\frac{W}{F} = \frac{x_D - z_F}{x_D - x_W} = \frac{0.95 - 0.39}{0.95 - 0.0125} = 0.5973$$

$$W = 0.5973F = 0.5973 \times 250 = 149.32\ kmol/h$$

c) As the feed is saturated liquid, $q = 1$ and the slope of q-line is ∞. Thus q-line is a vertical line passing through the point (0.39, 0.39). The enriching section operating line passes through the point (0.95, 0.95) and has a y-intercept of $x_D/(1 + R) = 0.95/(1 + 2.15) = 0.3016$. The figure given below shows the plots of the operating lines and stepping off the stages using the McCabe-Thiele method. The optimum number of equilibrium stages required is 11.

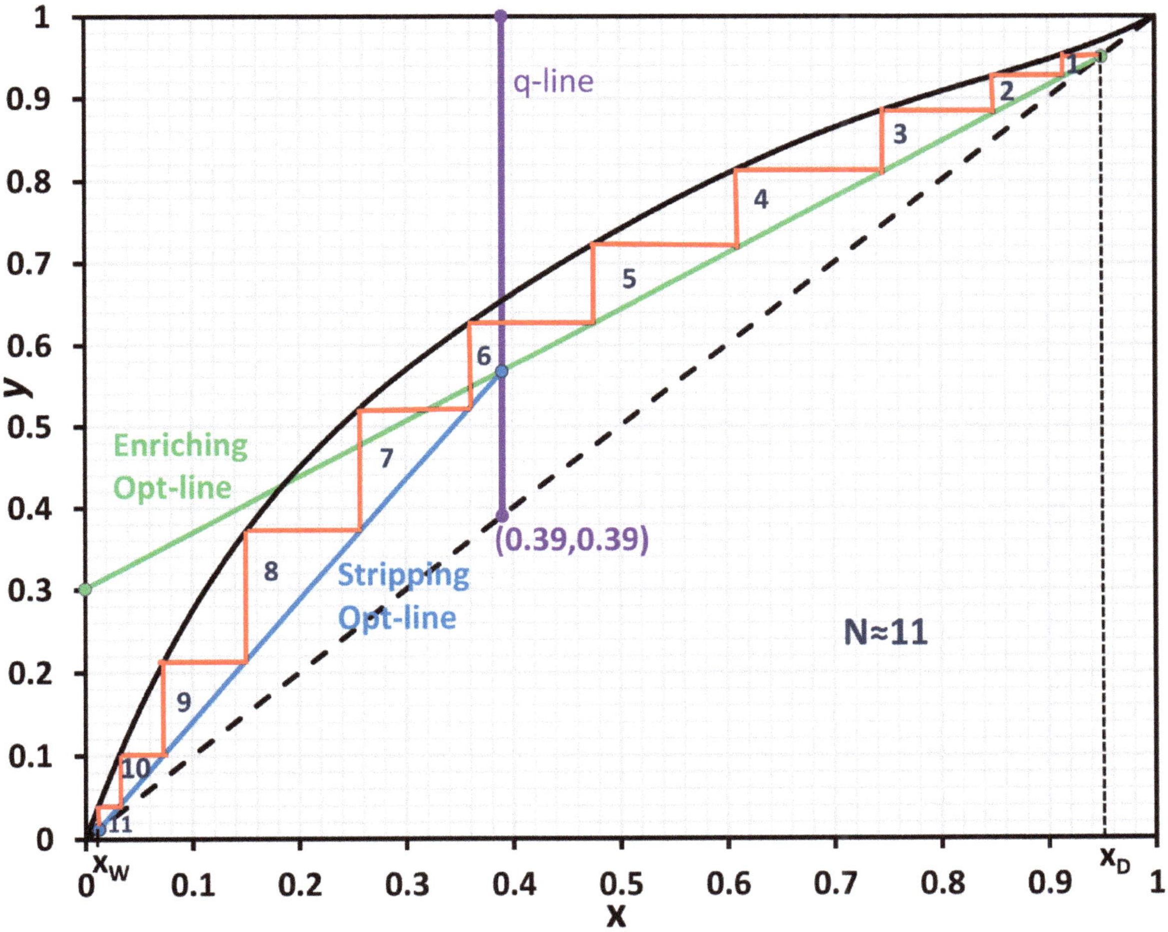

d) The optimum feed location is tray number 6, counting from the top of the column.

e) The T-x-y diagram of the system is shown below.

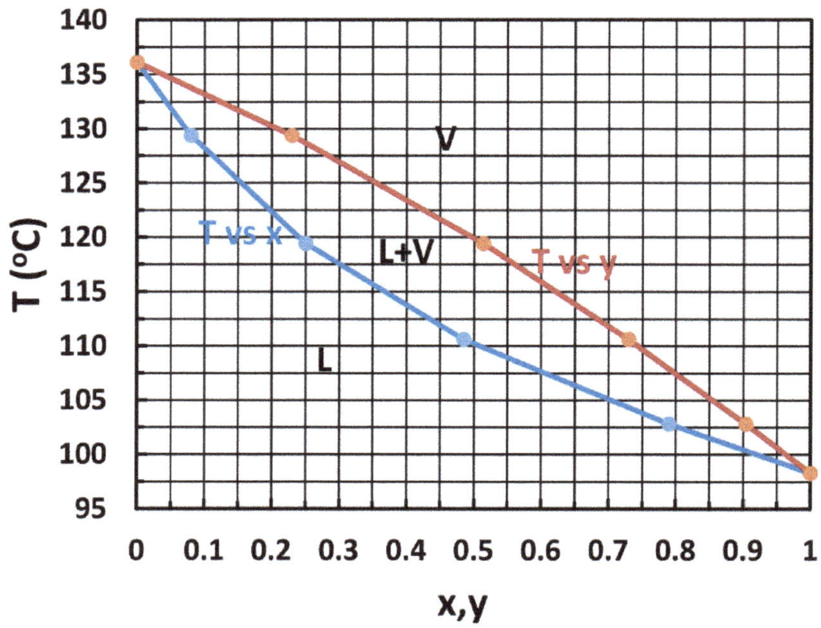

f) The composition profile of the column is shown below. It shows equilibrium composition at each stage.

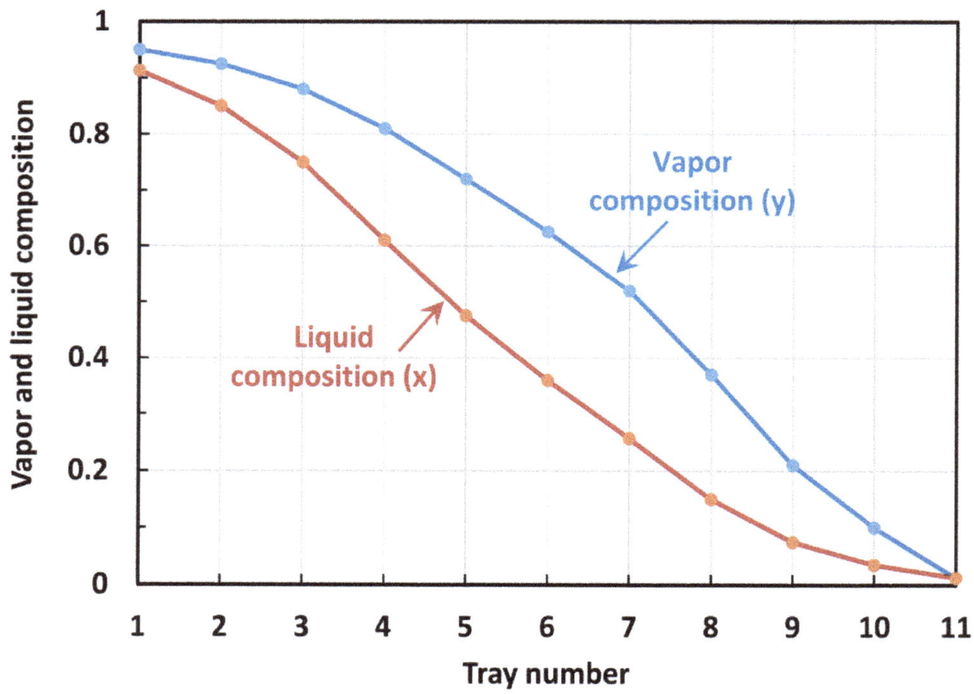

g) The temperature profile of the column is shown below. It shows the equilibrium temperature of vapor and liquid streams leaving each stage. Counting of trays is done from top to bottom. The last stage (stage 11) is reboiler.

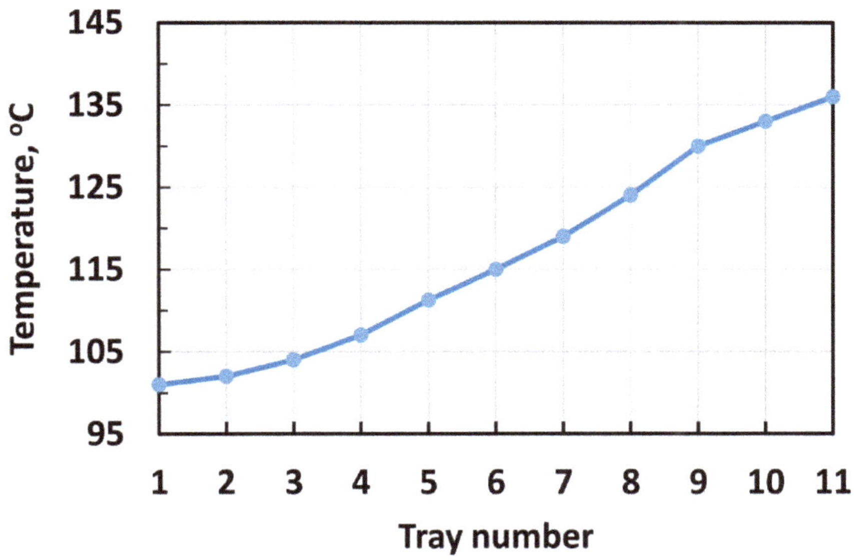

Problem 4

Determine graphically the minimum number of equilibrium stages required to separate a binary feed mixture of heptane and nonane containing 58 mol% heptane into a distillate product of composition 95 mol% heptane and a bottoms product of composition 7 mol% heptane.

The equilibrium data for heptane-nonane, in terms of mole fraction of more volatile component heptane, at column pressure is as follows:

x	y
0	0
0.007	0.025
0.114	0.332
0.257	0.589
0.437	0.773
0.668	0.906
0.968	0.993
1.0	1.0

Solution

The minimum number of equilibrium required is 4 as shown below.

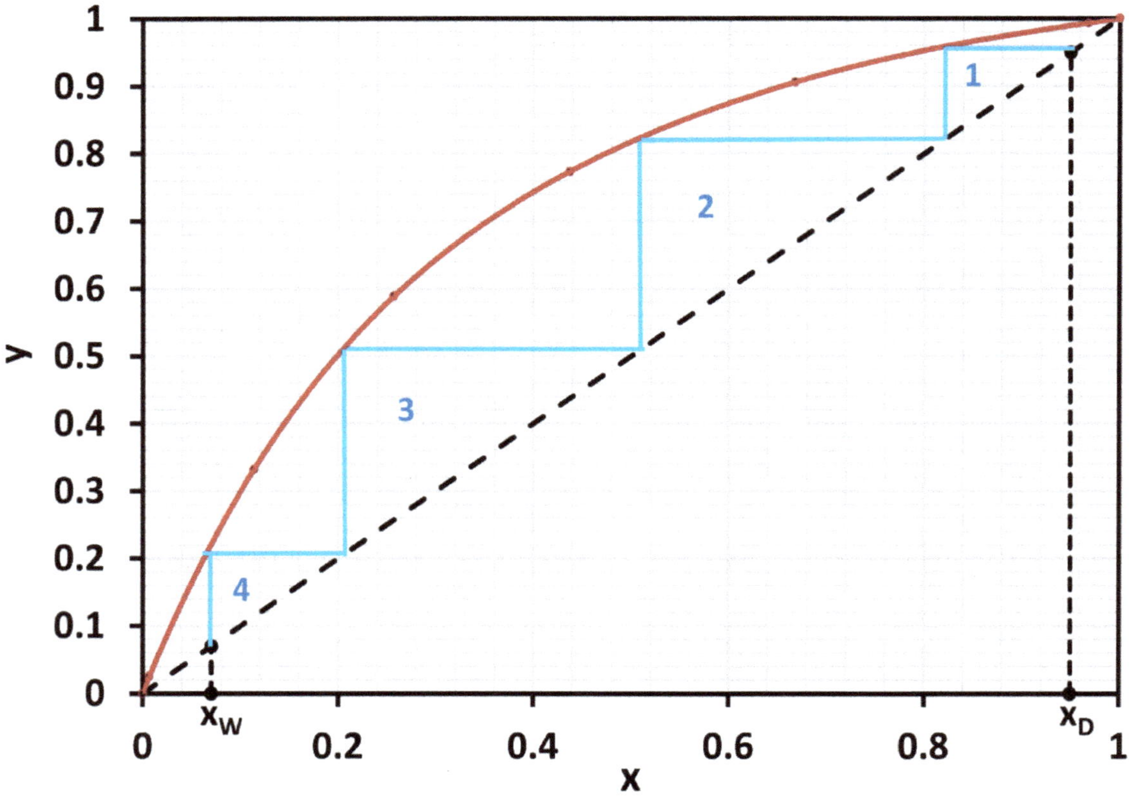

Problem 5

Feed entering the distillation column is a two-phase vapor-liquid mixture such that:

$$\frac{V - \bar{V}}{F} = 0.27$$

The feed consists of two components, namely methanol and water. The overall composition of the feed is 57 mol% methanol and the total feed flow rate is 338 kmol/h. The distillate product contains 96 mol% methanol and the bottoms product contains 7 mol% methanol. The reflux ratio is 2.95. The column is operating at 1 atm pressure with direct steam injection into the bottom of the column (instead of using a reboiler). The steam injected into the column is saturated steam at 1 atm pressure.

Determine the number of equilibrium stages required for the given separation. What is the optimum feed tray location?

The equilibrium data for methanol-water, in terms of mole fraction of more volatile component methanol, at column pressure (1 atm) is as follows:

x	y
0	0
0.02	0.134
0.04	0.23
0.06	0.304
0.08	0.365
0.10	0.418
0.15	0.517

0.20	0.579
0.30	0.665
0.40	0.729
0.50	0.779
0.60	0.825
0.70	0.87
0.80	0.915
0.90	0.958
0.95	0.979
1.0	1.0

Solution

Refer to the schematic diagram of the column shown below.

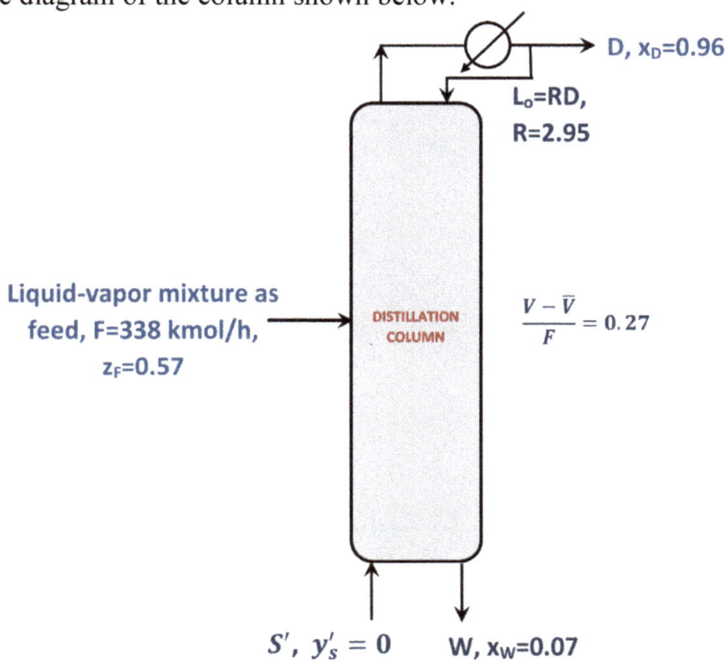

The enriching section operating line is as follows:

$$y_{n+1} = \left(\frac{R}{1+R}\right)x_n + \left(\frac{x_D}{1+R}\right) = \left(\frac{2.95}{1+2.95}\right)x_n + \left(\frac{0.96}{1+2.95}\right) = 0.7468x_n + 0.243$$

This line passes through the point (0.96, 0.96) and has a y-intercept of 0.243.

The feed quality q factor is:

$$q = \frac{\bar{L} - L}{F} = 1 + \frac{\bar{V} - V}{F} = 1 + (-0.27) = 0.73$$

The equation of feed line, Equation (9.42) can now be written as:

$$y = -\left(\frac{q}{1-q}\right)x + \left(\frac{z_F}{1-q}\right) = -\left(\frac{0.73}{1-0.73}\right)x + \left(\frac{0.57}{1-0.73}\right) = -2.7037x + 2.111$$

Thus the q-line (feed line) passes through the points (0.57, 0.57) and (0.411, 1.0).

The stripping section operating line, Equation (9.78), is:

$$y = \left(\frac{W}{S'}\right)(x - x_W) = \left(\frac{\bar{L}}{\bar{V}}\right)(x - x_W)$$

This line passes through the point $(x_w, 0)$ and the point of intersection of enriching section operating line and q-line. Note that $x_w = 0.07$.

The figure given below shows the plots of operating lines and q-line. The stepping off stages is also shown. The total number of stages required is approximately 6 (the sixth stage is partial stage). The optimum feed location is stage number 4.

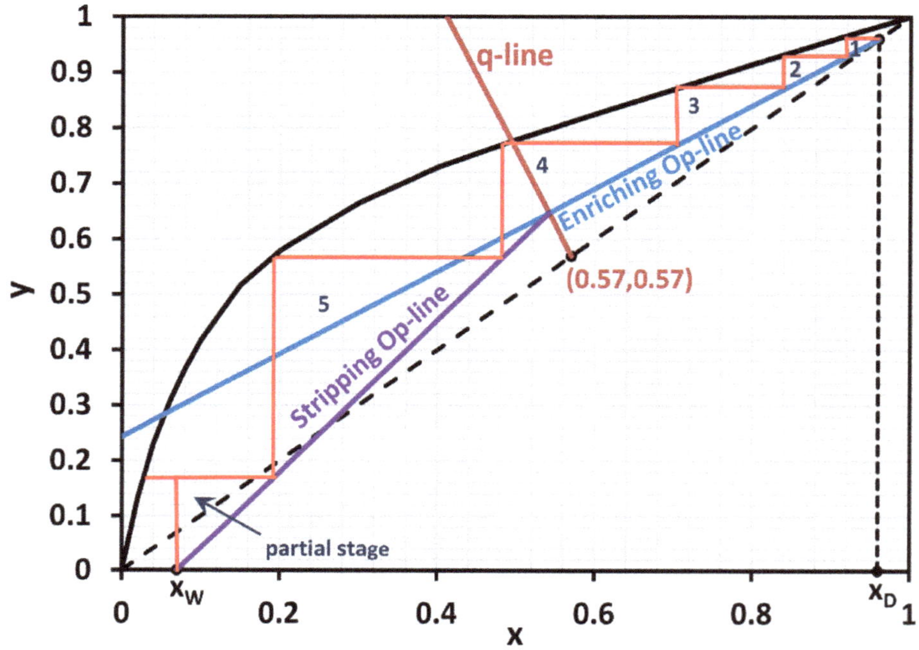

Problem 6

A multistage distillation column is operated at 1 atm pressure to separate acetone-ethanol mixture. The column has two feeds. Feed 1 is a saturated liquid consisting of 58 mol% acetone and has a flow rate of 78 kmol/h. Feed 2 is a two-phase vapor-liquid mixture consisting of 58 % vapor. The acetone content of feed 2 is 38 mol%. The flow rate of feed 2 is 98 kmol/h. The desired distillate and bottom compositions are: 89 mol% acetone and 9 mol% acetone, respectively. The boilup ratio $\bar{V}/W$ is 1.98.

Do the following:

a) Calculate the distillate product flow rate, D
b) Calculate the bottoms product flow rate, W
c) Determine the optimum number of equilibrium stages using the McCabe-Thiele method
d) Determine the optimum feed tray locations

The equilibrium data for acetone-ethanol, in terms of mole fraction of more volatile component acetone, at column pressure (1 atm) is as follows:

x	y	T (K)
0.00	0.00	351.45
0.033	0.111	349.55

0.078	0.216	347.15
0.149	0.345	343.95
0.195	0.410	342.25
0.316	0.534	338.75
0.414	0.614	336.55
0.532	0.697	334.45
0.691	0.796	332.15
0.852	0.896	330.45
1.00	1.00	329.25

Solution

Refer to the schematic diagram of the column shown below.

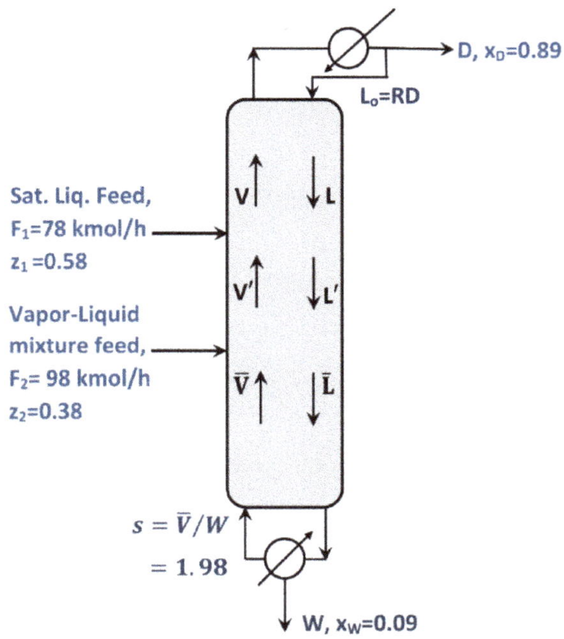

a) The distillate flow rate D is given by Equation (9.57) as:

$$D = \frac{z_1F_1 + z_2F_2 - (F_1 + F_2)x_W}{x_D - x_W} = \frac{(0.58\times0.78) + (0.38\times98) - (78+98)(0.09)}{0.89 - 0.09} = 83.3\ kmol/h$$

b) The bottoms product flow rate W is given by Equation (9.58) as:

$$W = \frac{(F_1 + F_2)x_D - z_1F_1 - z_2F_2}{x_D - x_W} = \frac{(78+98)(0.89) - (0.58\times78) - (0.38\times98)}{0.89 - 0.09} = 92.7\ kmol/h$$

c) The values of q_1 and q_2 corresponding to feed streams 1 and 2 are given as follows:

$$q_1 = \frac{L' - L}{F_1} = 1.0$$

$$q_2 = \frac{\bar{L} - L'}{F_2} = \frac{(L' + 0.42F_2) - L'}{F_2} = 0.42$$

Note that feed stream 2 consists of 42% liquid and 58% vapor. The q_1-line corresponding to feed 1 is a vertical line passing through the point (0.58, 0.58). The q_2-line corresponding to feed 2 passes through the point (0.38, 0.38) and has a y-intercept of $z_2/(1-q_2) = 0.38/(1-0.42) = 0.6552$. The stripping section operating line is as follows:

$$y_{m+1} = \left(\frac{1+s}{s}\right)x_m - \frac{x_W}{s} = \left(\frac{1+1.98}{1.98}\right)x_m - \frac{0.09}{1.98} = 1.505x_m - 0.0455$$

This equation passes through the points (0.09, 0.09) and (0.38, 0.5265). Thus we are in a position to plot the two q lines and the stripping section operating line.

We plot the enriching section operating line as the next step. The enriching section operating line is given as:

$$y = \left(\frac{R}{1+R}\right)x + \frac{x_D}{1+R}$$

We need to determine the reflux ratio R before we can plot this line. The vapor flow rate in the middle section of the column V' is:

$$V' = \bar{V} + 0.58 \times F_2 = sW + 0.58 \times F_2 = 1.98 \times 92.7 + 0.58 \times 98 = 240.386\ kmol/h$$

Using Equation (9.63), we can calculate the reflux ratio R as:

$$V' = L' + D - F_1 = (1+R)D + (q_1 - 1)F_1$$

$$240.386 = (1+R)(83.3) + (1-1)(78)$$

$$R = 1.886$$

Thus the enriching section operating line becomes:

$$y = \left(\frac{R}{1+R}\right)x + \frac{x_D}{1+R} = \left(\frac{1.886}{1+1.886}\right)x + \frac{0.89}{1+1.886} = 0.6535x + 0.3084$$

This equation passes through the points (0.89, 0.89) and (0.58, 0.6874).

Once the q-lines, stripping and enriching sections operating lines are plotted, the middle section operating line is plotted through the intersection points of enriching/stripping operating lines and q-lines as shown in the figure given below.

The optimum number of equilibrium stages needed is approximately 12.

d) The optimum feed locations are: tray 7 for feed 1 stream and tray 10 for feed stream 2, counting stages from the top of the column.

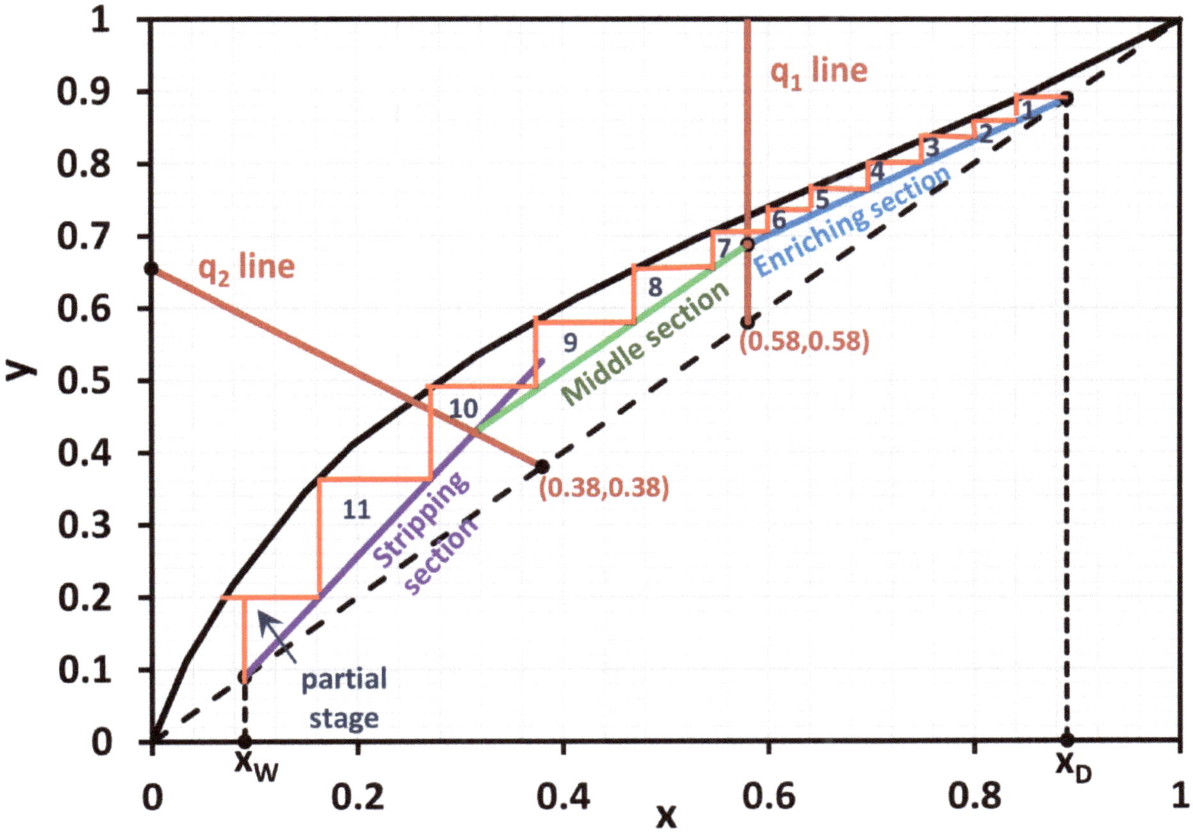

SHORT REVIEW QUESTIONS

1. With the help of process flow sheets, describe different single-stage distillation schemes to separate a feed mixture.
2. Explain partial vaporization of feed liquid mixture by heating at constant pressure using the T-x-y diagram.
3. Explain partial condensation of superheated vapor feed mixture by cooling at constant pressure using the T-x-y diagram.
4. With the help of process flow sheets, explain variable-pressure and variable-temperature multi-stage distillation cascades without any intermediate streams to separate a feed mixture. Which type of multi-stage distillation cascade, variable-pressure or variable temperature, is more relevant in distillation columns?
5. With the help of a schematic diagram, describe the working of a multistage distillation column using sieve trays. What happens at each tray of the column?
6. What is the purpose of using liquid reflux in the operation of a distillation column?
7. What is the purpose of using vapor boil-up in the operation of a distillation column?
8. Define CMO. Under what conditions is CMO valid?
9. Show that the equation for the stripping section operating line, Equation (9.28) can be re-written in terms of the boilup ratio (s) as: $y_{m+1} = \left(\frac{1+s}{s}\right) x_m - \frac{x_W}{s}$. This equation passes through the point (x_W, x_W) and has a slope of $\left(\frac{1+s}{s}\right)$. Thus given x_W and s we can plot the stripping section operating line independently.
10. Define the quality "q" of the feed. What is the q-value when feed is a subcooled liquid, saturated liquid, vapor-liquid mixture, saturated vapor, or superheated vapor?

11. Assuming that the compositions of the vapor and liquid streams do not change on the feed tray, show that the feed line equation, Equation (9.38), follows from component balance around the feed stage (see Figure 9.19).
12. How is the performance of a distillation column affected by varying the reflux ratio R?
13. Define feed pinch and tangent pinch. When do you normally observe these pinch conditions?
14. How many equilibrium stages are needed if the distillation column is operated at a minimum reflux ratio?
15. The number of equilibrium stages needed is minimum if we operate the column at total reflux. Is this true?
16. The optimum reflux ratio falls in between the limiting values of reflux ratio, that is, in between R_{min} and R_{max}. What are the factors that need to be considered to determine the optimum reflux ratio?
17. Show schematically the difference between a partial condenser and total condenser.
18. Show schematically the difference between a partial reboiler and total reboiler.
19. Under certain conditions the distillation columns can be operated either as an enriching column without stripping section or as a stripping column without enriching section. Is this true?
20. Explain pressure swing distillation (PSD) to separate a binary azeotropic mixture.

PRACTICE PROBLEMS

Problem 1

Feed entering the distillation column is saturated liquid at column conditions. It consists of a binary mixture of heptane and ethyl benzene. The mole fraction of heptane in the feed is 0.50. The feed flow rate is 225 kmol/h. The distillate product contains 90 mol% heptane and the bottoms product contains 5.0 mol% heptane. The reflux ratio is 3.0.

Do the following:

a) Calculate the distillate product flow rate, D
b) Calculate the bottoms product flow rate, W
c) Determine the optimum number of equilibrium stages using the McCabe-Thiele graphical met
d) Determine the optimum feed tray location

The equilibrium data for heptane-ethyl benzene, in terms of mole fraction of more volatile component heptane, at column pressure (101.32 kPa) is as follows:

x	y	T (°C)
0	0	136.1
0.08	0.23	129.4
0.25	0.514	119.4
0.485	0.73	110.6
0.79	0.904	102.8
1.0	1.0	98.3

Problem 2

Feed entering the distillation column is saturated liquid at column conditions. It consists of a binary mixture of benzene and toluene. The mole fraction of benzene in the feed is 0.57. The feed flow rate is 29 kmol/h. The distillate product contains 95 mol% benzene and the bottoms product contains 3.0 mol% benzene. The reflux ratio is 3.25.

Do the following:

a) Calculate the distillate product flow rate, D
b) Calculate the bottoms product flow rate, W
c) Determine the optimum number of equilibrium stages using the McCabe-Thiele method
d) Determine the optimum feed tray location
e) Plot Txy diagram
f) Plot composition profile of the column
g) Plot temperature profile of the column

Assume Raoult's law is applicable and the column pressure is 101.32 kPa. The vapor pressures of pure benzene and pure toluene are given by the following Antoine equations:

$$lnP_{benzene}^{Sat} = 13.7819 - \frac{2726.81}{t+217.572}$$

$$lnP_{toluene}^{Sat} = 13.932 - \frac{3056.96}{t+217.625}$$

where t is in °C and the vapor pressures are in kPa.

Problem 3

Consider the separation of methanol from water in a distillation column. The feed is 60 mole% methanol and 40 mole% water and is input as a two-phase mixture so that

$$\frac{V-\overline{V}}{F} = 0.3$$

where V is the vapor flow rate above the feed plate (that is, the enriching section), $\overline{V}$ is the vapor flow rate below the feed plate (that is, stripping section), and F is the feed flow rate. The column is well insulated and has a total condenser. The reflux ratio is 3 (R = 3). We desire a distillate concentration of 92 mole% methanol and a bottoms concentration of 8 mole % methanol. Column pressure is 1 atm.

Determine the optimum number of equilibrium stages and the optimum feed location.
The equilibrium data for methanol-water system, in terms of mole fraction of more volatile component methanol, at 1 atm are given in tutorial problem 5.

Problem 4

Determine graphically the minimum number of equilibrium stages required to separate a binary feed mixture of benzene and toluene containing 55 mol% benzene into a distillate product of composition 98 mol% benzene and a bottoms product of composition 5 mol% benzene.

Assume Raoult's law is applicable and the column pressure is 101.32 kPa. The vapor pressures of pure benzene and pure toluene are given by the following Antoine equations:

$$lnP_{benzene}^{Sat} = 13.7819 - \frac{2726.81}{t+217.572}$$

$$lnP_{toluene}^{Sat} = 13.932 - \frac{3056.96}{t+217.625}$$

where t is in °C and the vapor pressures are in kPa.

Problem 5

Feed entering the distillation column is a two-phase vapor-liquid mixture such that:

$$\frac{\bar{L} - L}{F} = 0.81$$

The feed consists of two components, namely ethanol and water. The overall composition of the feed is 50 mol% ethanol and the total feed flow rate is 300 kmol/h. The distillate product contains 78 mol% ethanol and the bottoms product contains 6 mol% ethanol. The reflux ratio is 2.90. The column is operating at 1 atm pressure with direct steam injection into the bottom of the column (instead of using a reboiler). The steam injected into the column is saturated steam at 1 atm pressure.

Determine the number of equilibrium stages required for the given separation. What is the optimum feed tray location?

The equilibrium data for ethanol-water, in terms of mole fraction of more volatile component ethanol, at column pressure (1 atm) is as follows:

x	y
0	0
0.019	0.17
0.0721	0.3891
0.0966	0.4375
0.1238	0.4704
0.1661	0.5089
0.2377	0.5445
0.2608	0.558
0.3273	0.5826
0.3965	0.6122
0.5079	0.6564
0.5198	0.6599
0.5732	0.6841
0.6763	0.7385
0.7472	0.7815
0.8943	0.8943
1	1

Problem 6

A multistage distillation column is operated at 1 atm pressure to separate acetone-ethanol mixture. The feed to the column is a saturated liquid consisting of 58 mol% acetone and has a flow rate of 98 kmol/h. The liquid side product is withdrawn from second stage (counting from the top of the column) at a flow rate of $S = 13$ kmol/h. The reflux ratio (R) is 3.0. The desired distillate and bottoms product compositions are: 89 mol% acetone and 9 mol% acetone, respectively.

Do the following:

a) Determine the mole fraction of acetone in the side stream x_S
b) Determine the optimum number of equilibrium stages using the McCabe-Thiele method
c) Determine the optimum feed tray location

The equilibrium data for acetone-ethanol, in terms of mole fraction of more volatile component acetone, at column pressure (1 atm) is given in tutorial problem 6.

Problem 7

A saturated liquid mixture of benzene and toluene containing 68 mol% benzene is fed to a stripping column at a flow rate of 388 kmol/h. The column is operating at 101.32 kPa. The bottoms product contains 9 mol% benzene and has a flow rate of 58 kmol/h.

Calculate the composition and flow rate of overhead vapor product. How many equilibrium stages are required?

Assume Raoult's law is applicable. The vapor pressures of pure benzene and pure toluene are given by the following Antoine equations:

$$lnP_{benzene}^{Sat} = 13.7819 - \frac{2726.81}{t+217.572}$$

$$lnP_{toluene}^{Sat} = 13.932 - \frac{3056.96}{t+217.625}$$

where t is in °C and the vapor pressures are in kPa.

APPENDIX: THE SMOKER EQUATIONS

Smoker (E.H. Smoker, Trans. AIChE, Vol. 34, 165, 1938) has developed analytical expressions to estimate the number of equilibrium stages required for a given separation of binary mixtures in a distillation column. It is assumed that the relative volatility is constant and that the *CMO* is valid in each section (rectifying and stripping) of the column. The column has one feed stream and no side product streams.

The operating lines for the enriching and stripping sections are straight lines under the assumption of CMO. They can be represented by an equation of straight line as follows:

$$y = mx + b \tag{9.A1}$$

Assuming constant relative volatility (α), the equilibrium curve can be represented as:

$$y = \frac{\alpha x}{1+(\alpha-1)x} \tag{9.A2}$$

The intersection point of operating line and equilibrium curve can be determined from the following equation:

$$m(\alpha - 1)x^2 + [b(\alpha - 1) + m - \alpha]x + b = 0 \tag{9.A3}$$

This equation is obtained by eliminating y from Equations (9.A1) and (9.A2). In distillation applications, this equation has only one real root, $x = k$, within the range [0 1]. Let the real root for the enriching section be $x = k_1$ and for the stripping section be $x = k_2$. Figures 9.A1 and 9.A2 show the intersection roots for the enriching ($x = k_1$) and stripping ($x = k_2$) sections.

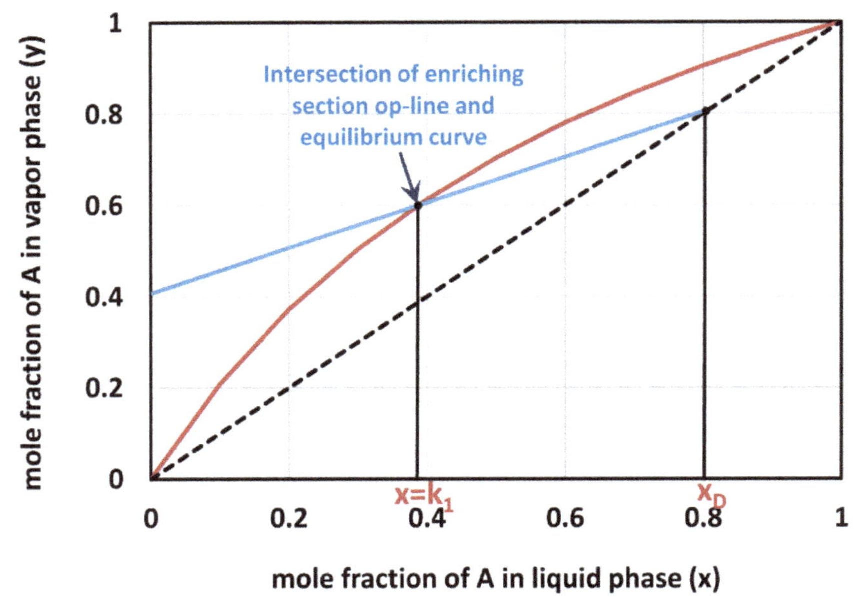

Figure 9.A1 Intersection of enriching section operating line and equilibrium curve.

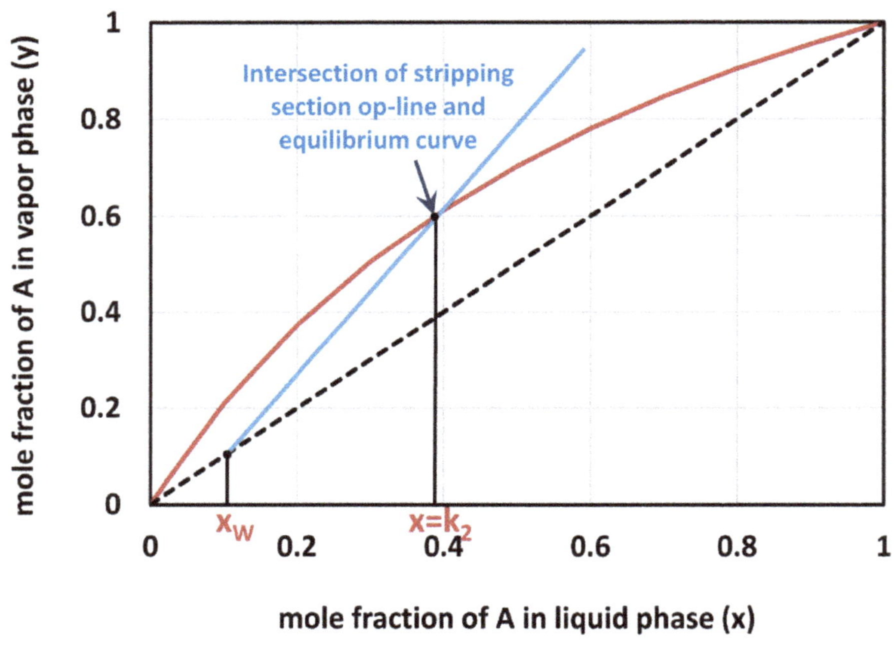

Figure 9.A2 Intersection of stripping section operating line and equilibrium curve.

Slope (m) and Intercept (b) of the Operating Lines

The slope and y-intercept of the enriching section operating line are:

$$m_1 = \frac{R}{1+R};\ b_1 = x_D(1 - m_1) \tag{9.A4}$$

The slope and intercept of the enriching section operating line can be used to determine the point of intersection (x_{int}, y_{int}) of the operating line and feed q line (see Figure 9.A3). The equations for enriching section operating line and q line are as follows:

$$y = m_1 x + x_D(1 - m_1) \tag{9.A5}$$

$$y = -\left(\frac{q}{1-q}\right)x + \left(\frac{z_F}{1-q}\right) \tag{9.A6}$$

Eliminating y from Equations (9.A5) and (9.A6), we get the value of x corresponding to intersection point of operating lines and q line as:

$$x_{int} = \frac{z_F - b_1(1-q)}{1+(m_1-1)(1-q)} \tag{9.A7}$$

The corresponding value of y at the intersection point is:

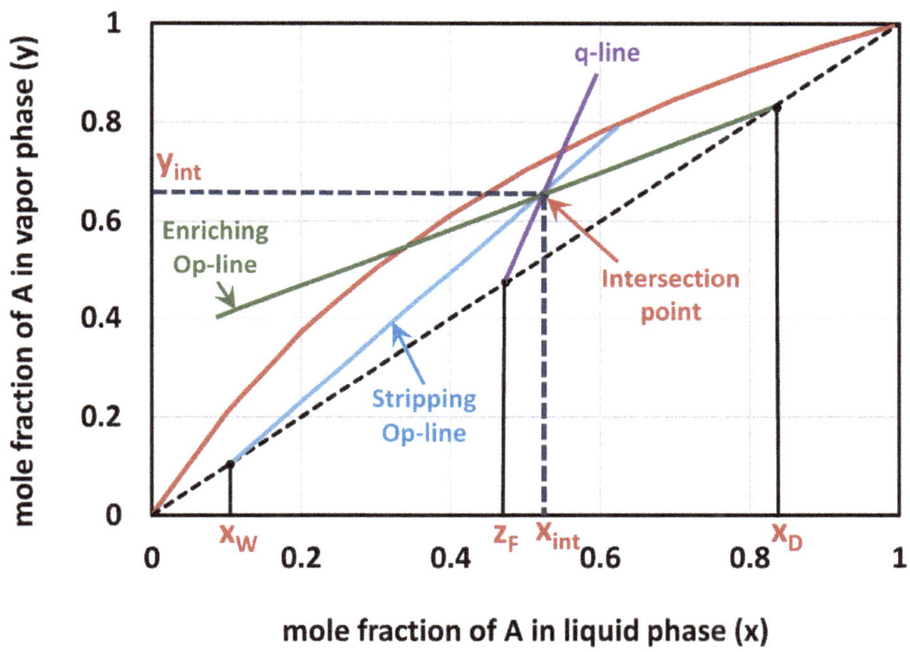

Figure 9.A3 Intersection of operating lines and feed q line.

$$y_{int} = x_D - \frac{R}{1+R}(x_D - x_{int}) \tag{9.A8}$$

We will now express the slope and intercept of the stripping section operating in terms of the intersection point (x_{int}, y_{int}). Referring to Figure 9.A3, the slope of the stripping section operating line can be written as:

$$m_2 = \frac{y_{int} - x_W}{x_{int} - x_W} \tag{9.A9}$$

The y-intercept of the stripping section operating line can be written as:

$$b_2 = x_W(1 - m_2) \tag{9.A10}$$

Substitution of y_{int} from Equations (9.A8) into Equation (9.A9), the slope and intercept of the stripping section operating line can be expressed as:

$$m_2 = \frac{Rx_{int}+x_D-(1+R)x_W}{(1+R)(x_{int}-x_W)} \tag{9.A11}$$

$$b_2 = \frac{(x_{int}-x_D)x_W}{(1+R)(x_{int}-x_W)} \tag{9.A12}$$

Smoker Equations

The Smoker equations can be written separately for the enriching and stripping sections. For the *enriching section*, the number of equilibrium stages is given by the following form of the Smoker equation:

$$N_R = ln\left[\frac{(x_D-k_1)(1-\beta_1(x_{int}-k_1))}{(x_{int}-k_1)(1-\beta_1(x_D-k_1))}\right] \Big/ ln\left[\frac{\alpha}{m_1c_1^2}\right] \tag{9.A13}$$

where

$$c_1 = 1 + (\alpha - 1)k_1 \tag{9.A14}$$

$$\beta_1 = \frac{m_1c_1(\alpha-1)}{\alpha-m_1c_1^2} \tag{9.A15}$$

For the *stripping section*, the number of equilibrium stages (including reboiler) is given by the following Smoker equation:

$$N_S = ln\left[\frac{(x_{int}-k_2)(1-\beta_2(x_W-k_2))}{(x_W-k_2)(1-\beta_2(x_{int}-k_2))}\right] \Big/ ln\left[\frac{\alpha}{m_2c_2^2}\right] \tag{9.A16}$$

where

$$c_2 = 1 + (\alpha - 1)k_2 \tag{9.A17}$$

$$\beta_2 = \frac{m_2c_2(\alpha-1)}{\alpha-m_2c_2^2} \tag{9.A18}$$

Calculation Procedure

Given α, z_F, x_W, x_D, R, and q, we can determine the number of equilibrium stages in the enriching and stripping sections as follows:

- Calculate m_1 and b_1 from Equation (9.A4)
- Estimate k_1 from Equation (9.A3)
- Calculate x_{int} from Equation (9.A7)
- Calculate c_1 form Equation (9.A14) and β_1 from Equation (9.A15)
- Calculate N_R from Equation (9.A13)
- Calculate m_2 and b_2 from Equations (9.A11) and (9.A12), respectively
- Estimate k_2 from Equation (9.A3)
- Calculate c_2 form Equation (9.A17) and β_2 from Equation (9.A18)
- Calculate N_S from Equation (9.A16)

Chapter 10

Approximate Analytical Method of Designing Distillation Columns

10.1 LEARNING OUTCOMES

Upon completion of this chapter, students will be able to:

- Derive Fenske equation for minimum number of theoretical trays (N_{min}) at total reflux
- Derive Underwood equation for minimum reflux ratio (R_{min})
- Explain and use the Fenske-Underwood-Gilliland (FUG) method of designing distillation columns for binary mixtures
- Find optimum location of feed tray using Kirkbride equation
- Use empirical overall column efficiency to estimate the actual number of stages required in the column when equilibrium is not achieved on the trays
- Use Murphree tray efficiency to estimate the actual number of stages required when equilibrium is not achieved on the trays
- Specify spacing between the trays and estimate the height of the column
- Describe different types of trays used in distillation columns (sieve trays, bubble cap trays, and valve trays)
- Estimate the column diameter based on flooding velocity of vapor
- Carry out preliminary mechanical design of the distillation column taking into consideration different loads including wind load. Select the material of construction and estimate the column shell thickness.

10.2 INTRODUCTION

The preliminary design of a distillation column for a given separation involves specification of the following: a) the number of trays/stages required; b) the spacing between the trays; c) the height of the column; and d) the diameter of the column. In this chapter, we will discuss an approximate analytical method (also referred to as shortcut method) of designing distillation columns. Although we will restrict ourselves to distillation of binary mixtures in this chapter, the method presented here could easily be generalized to the design of distillation columns for multicomponent mixtures.

10.3 ESTIMATION OF THE NUMBER OF STAGES

In order to estimate the number of stages required for a given separation, we need to estimate N_{min} and R_{min}. N_{min} is the minimum number of stages required for the given separation and R_{min} is the minimum reflux ratio allowed. Note that N_{min} corresponds to total reflux ($R \to \infty$) and at $R = R_{min}$, an infinite number of stages are required to carry out the given separation.

10.3.1 Fenske Equation for N_{min} at Total Reflux

Consider a binary mixture of species "A" and "B". Let species "A" be more volatile as compared with species "B". The relative volatility α_{AB} is defined as:

$$\alpha_{AB} = \frac{y_A/x_A}{y_B/x_B} \tag{10.1}$$

Upon rearrangement, Equation (10.1) gives:

$$\frac{y_A}{x_A} = \alpha_{AB}\left(\frac{y_B}{x_B}\right) \tag{10.2}$$

Since $y_B = 1 - y_A$ and $x_B = 1 - x_A$, Equation (10.1) further simplifies to:

$$\frac{y_A}{1-y_A} = \alpha_{AB}\left(\frac{x_A}{1-x_A}\right) \tag{10.3}$$

Keeping in mind that y and x are mole fractions of the more volatile species (species A) of the binary mixture and α is the relative volatility of the more volatile species relative to the less volatile species (species B), we can drop the subscripts from variables in Equation (10.3):

$$\frac{y}{1-y} = \alpha\left(\frac{x}{1-x}\right) \tag{10.4}$$

This is an *equilibrium relationship* that relates the compositions of liquid and vapor phases at equilibrium with each other. We will assume that the relative volatility α is constant from the bottom of the column to the top. If the change in the value of α from the bottom to the top of the column is moderate, one can use a geometric mean value defined as $\sqrt{\alpha_{top}\alpha_{bot}}$. Note that the relative volatility of ideal mixtures does not change much over the temperature range encountered in a distillation column. However, the relativity volatility of non-ideal mixtures could change by a large amount over the temperature range encountered in a distillation column.

Figure 10.1 shows the influence on relative volatility on the equilibrium curve. With the increase in relative volatility, the equilibrium curve moves away from the $y = x$ line. The shift of the equilibrium curve away from the $y = x$ line with the increase in α implies that for the same separation, the number of stages required are less. This can be seen more clearly in Figure 10.2 where the number of stages needed are shown for two values of α at total reflux. For the same separation (x_D and x_W values), the number of stages required are 9 at a low α value of 2 and less than 3 at a high α value of 10. Thus separation of mixtures with high value of α require smaller columns.

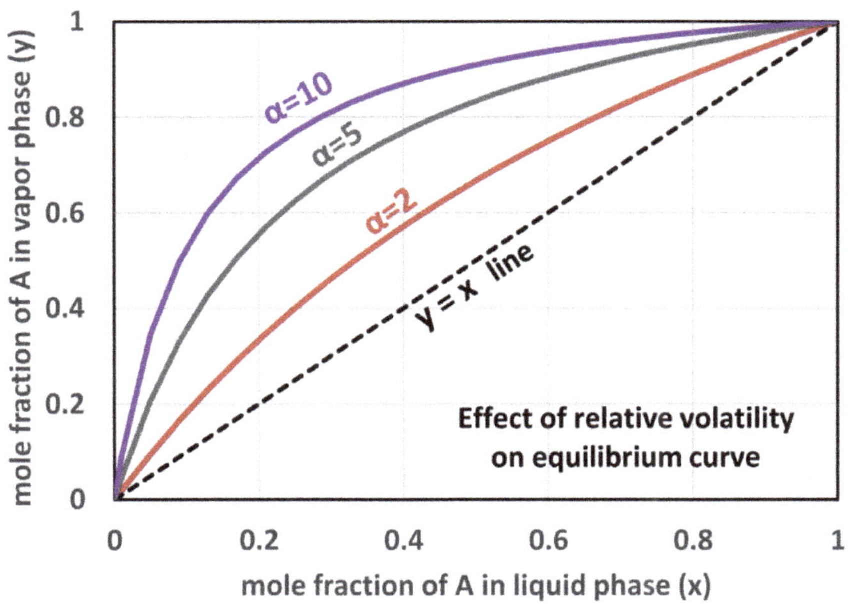

Figure 10.1 Effect of relative volatility on equilibrium curve.

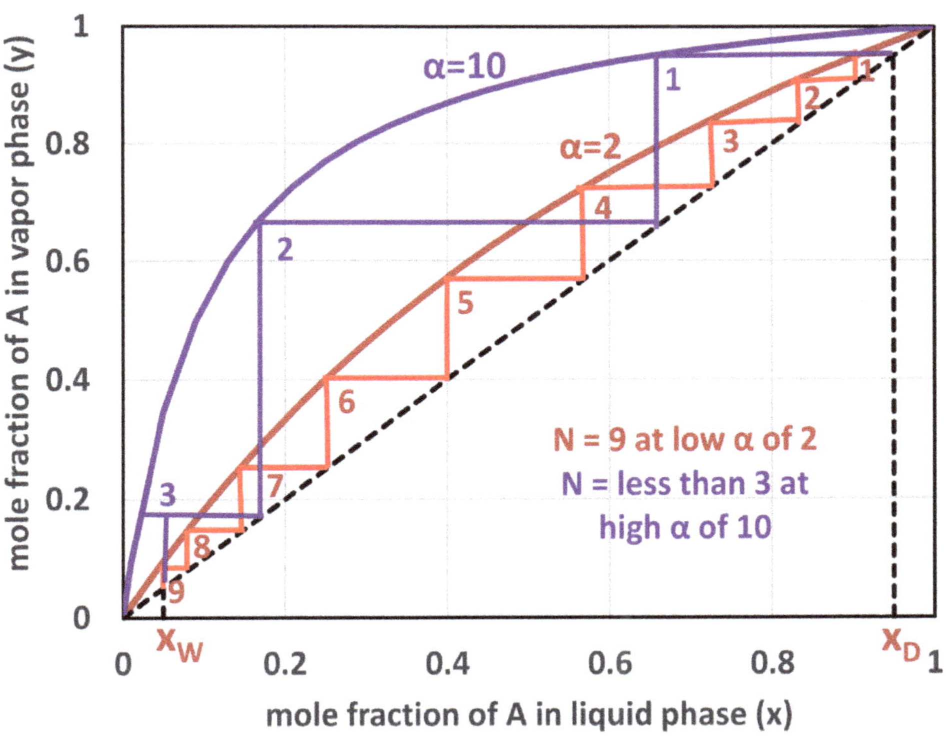

Figure 10.2 Effect of relative volatility on the number of stages required for the separation.

Let us now apply Equation (10.4) to nth stage of a column. Assuming that liquid and vapor streams leaving the nth stage are at *equilibrium* with each other, Equation (10.4) gives:

$$\frac{y_n}{1-y_n} = \alpha\left(\frac{x_n}{1-x_n}\right) \tag{10.5}$$

At total reflux, the compositions of the passing liquid and vapor streams are the same, that is, $y = x$ for the passing streams. Substituting $y_n = x_{n-1}$ (compositions of passing vapor and liquid streams) in Equation (10.5), we obtain:

$$\frac{x_{n-1}}{1-x_{n-1}} = \alpha\left(\frac{x_n}{1-x_n}\right) \tag{10.6}$$

We will now apply this equation to different stages of the distillation column starting with from stage 1 at the top:

$$\frac{x_0}{1-x_0} = \alpha\left(\frac{x_1}{1-x_1}\right) \quad (n = 1) \tag{10.7}$$

$$\frac{x_1}{1-x_1} = \alpha\left(\frac{x_2}{1-x_2}\right) \quad (n = 2) \tag{10.8}$$

$$\vdots$$

$$\frac{x_N}{1-x_N} = \alpha\left(\frac{x_W}{1-x_W}\right) \quad (n = N + 1;\ Reboiler) \tag{10.9}$$

Upon multiplication of Equations (10.7) to (10.9), we obtain:

$$\frac{x_0}{1-x_0} = \frac{x_D}{1-D} = \alpha^{N+1}\left(\frac{x_W}{1-x_W}\right) \tag{10.10}$$

Note that N is the number of stages within the column excluding reboiler. If N includes the reboiler, Equation (10.10) should be modified as:

$$\frac{x_D}{1-x_D} = \alpha^{N}\left(\frac{x_W}{1-x_W}\right) \tag{10.11}$$

As $N = N_{min}$ at total reflux, Equation (10.11) yields:

$$N_{min} = \frac{ln\left[\left(\frac{x_D}{1-x_D}\right)\left(\frac{1-x_W}{x_W}\right)\right]}{ln(\alpha)} \tag{10.12}$$

Another version of this equation is as follows:

$$N_{min} = \left(\frac{1}{ln\alpha}\right) ln\left[\frac{\left(\frac{x_A}{x_B}\right)_{distillate}}{\left(\frac{x_A}{x_B}\right)_{bottoms}}\right] \tag{10.13}$$

where "A" is more volatile species and "B" is less volatile species. This is the well-known Fenske equation for N_{min} at total reflux (M.R. Fenske, Industrial and Engineering Chemistry, Vol. 24, 482-485, 1932). Note that this N_{min} includes the reboiler.

Figure 10.3 shows the variation of N_{min} with α. The plot is generated from Fenske equation (Equation (10.12)) using the following values of distillate and bottoms product compositions: $x_D = 0.95$ and $x_W = 0.05$.

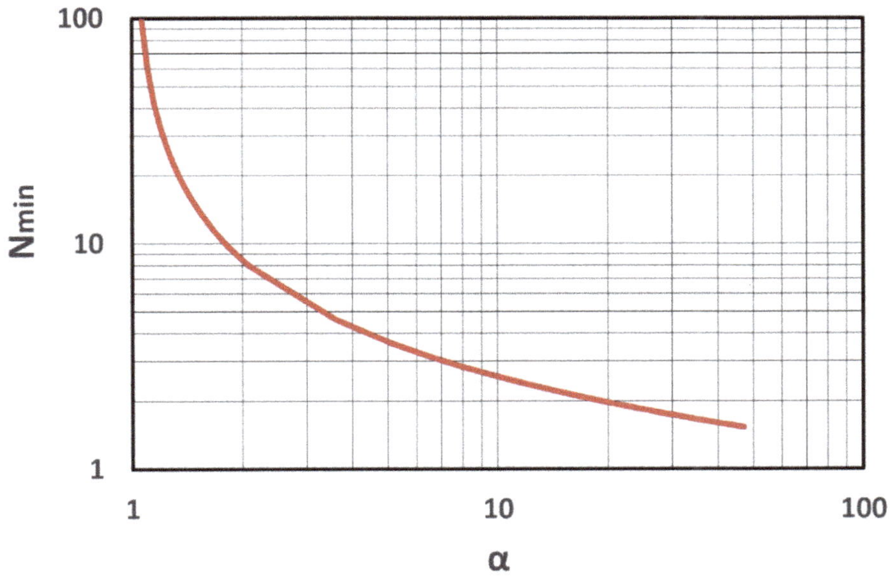

Figure 10.3 Variation of number of theoretical trays with relative volatility at total reflux.

10.3.2 Underwood Equation for R_{min}

The minimum reflux ratio (R_{min}) is determined from the slope of the enriching section operating line passing through the *pinch point*. The equation of the enriching section operating line passing through the pinch point is given as:

$$y = \left(\frac{R_{min}}{1+R_{min}}\right)x + \left(\frac{x_D}{1+R_{min}}\right) \tag{10.14}$$

For *ideal* mixtures, the equilibrium y-x curve is concave downwards throughout the concentration range and the pinch point occurs at the feed plate. In this case, the pinch point, also referred to *feed pinch point*, is that point on the equilibrium curve where the feed line (q-line) and the two operating lines all intersect. However, *non-ideal* mixtures often exhibit inflections in their equilibrium y-x curve, that is, part of the equilibrium curve is concave downwards and part of it concave upwards. Consequently, they exhibit so called *tangent pinch point*. In such cases, the pinch point is determined by the enriching section operating line that is tangent to the equilibrium curve. The location of the tangent pinch point is away from the feed stage where the q-line and operating lines intersect.

Here we will assume that the binary mixture is ideal. It exhibits feed pinch point and the relative volatility is nearly constant over the temperature range encountered in a distillation column.

The equation for the equilibrium curve is as follows:

$$y = \alpha x/[1 + (\alpha - 1)x] \tag{10.15}$$

This equation is a re-arrangement of Equation (10.4). The feed line (q-line) equation is given as:

$$y = -\left(\frac{q}{1-q}\right)x + \frac{z_F}{1-q} \tag{10.16}$$

Elimination of x and y from Equations (10.14) to (10.16) gives the following equation:

$$\frac{R_{min}z_F+qx_D}{R_{min}(1-z_F)+q(1-x_D)} = \frac{\alpha[x_D(q-1)+z_F(R_{min}-1)]}{(R_{min}+1)(1-z_F)+(q-1)(1-x_D)} \quad (10.17)$$

This equation is referred to as *Underwood equation* for a binary system. If the feed is saturated liquid ($q = 1$), Equation (10.17) reduces to:

$$R_{min} = \left(\frac{1}{\alpha-1}\right)\left[\frac{x_D}{z_F} - \frac{\alpha(1-x_D)}{(1-z_F)}\right] \quad (10.18)$$

If the feed is saturated vapor ($q = 0$), Equation (10.17) reduces to:

$$R_{min} = \left(\frac{1}{\alpha-1}\right)\left[\frac{\alpha x_D}{z_F} - \frac{(1-x_D)}{(1-z_F)}\right] - 1 \quad (10.19)$$

10.3.3 Gilliland Correlation for the Estimation of Number of Theoretical Stages

Gilliland proposed a graphical method for estimating the number of theoretical (equilibrium) stages as a function of the reflux ratio for a given separation. He defined the following two new variables (E.R. Gilliland, Industrial and Engineering Chemistry, Vol. 32, 1220-1223, 1940):

$$X_G = \frac{R-R_{min}}{R+1} \quad (10.20)$$

$$Y_G = \frac{N-N_{min}}{N+1} \quad (10.21)$$

Based on a wide variety of distillation data (covering different systems, different number of components, different feed conditions, different operating pressures, different relative volatilities, and different values of R_{min} and N_{min}), he found that a unique curve is obtained when Y_G is plotted as a function of X_G. Note that N and N_{min} include the reboiler.

Several researchers have proposed equations to fit the Gilliland plot. Some of the popular equations describing the Gilliland plot are as follows:

$$Y_G = 1 - exp\left[\left(\frac{X_G-1}{\sqrt{X_G}}\right)\left(\frac{1+54.4X_G}{11+117.2X_G}\right)\right] \quad (10.22)$$

$$Y_G = 0.75\left[1 - X_G^{0.5668}\right] \quad (10.23)$$

Equation (10.22) can be called Gilliland-Molokanov equation (Y.K. Molokanov et al., Industrial and Engineering Chemistry, Vol. 12, 209-212, 1972) and Equation (10.23) can be referred to as Gilliland-Eduljee equation (H.E. Eduljee, Hydrocarbon Processing, Vol. 54, 120-122, 1975).

10.3.4 Fenske-Underwood-Gilliland (FUG) Method of Designing Distillation Columns

Given the feed condition (q) and its composition (z_F), the distillate and bottoms product compositions (x_D and x_W), the reflux ratio (R) and the relative volatility (α), one can estimate the number of theoretical (equilibrium) stages required using the Fenske-Underwood-Gilliland (FUG) method. The FUG method of designing distillation columns involves the following procedure:

- Estimate N_{min} from Fenske equation, Equation (10.12)
- Estimate R_{min} using Underwood equation, Equation (10.17)
- Calculate $X_G = (R - R_{min})/(R + 1)$

- Use Gilliland-Molokanov equation (Equation (10.22)) to calculate Y_G
- Determine the number of theoretical stages from the known value of Y_G using Equation (10.21)

Once the total number of equilibrium stages is estimated, the optimum location of the feed tray can be determined using the following empirical equation proposed by Kirkbride (C.G. Kirkbride, Petroleum Refiner, Vol. 23, 87-102, 1944):

$$\frac{n_F-1}{N-n_F} = \left[\left(\frac{1-z_F}{z_F}\right)\left(\frac{x_W}{1-x_D}\right)^2 \frac{W}{D}\right]^{0.206} \tag{10.24}$$

where n_F is the number (location) of the feed stage counting from top of the column and N is the total number of theoretical stages (including partial reboiler) required.

Note that the FUG method involves two key assumptions: constant molal overflow (CMO) and constant relative volatility. The method is best suited for ideal mixtures.

10.4 ESTIMATION OF ACTUAL (NON-EQUILIBRIUM) NUMBER OF STAGES

Up until now we have assumed that equilibrium is achieved at each stage of the column, that is, the phases leaving any stage are in thermodynamic equilibrium with each other. Such stages are referred to as equilibrium/theoretical/ideal stages. More often than not, the phases leaving a stage are not at equilibrium with each other. The *Murphree tray efficiency* (η_M) is defined as:

$$\eta_M = \frac{y_n-y_{n+1}}{y_n^*-y_{n+1}} \tag{10.25}$$

where y_n is the actual composition of vapor leaving stage n, y_{n+1} is the actual composition of vapor entering stage n, y_n^* is the composition of vapor in equilibrium with the liquid leaving stage n (see Figure 10.4). Thus, Murphree efficiency of a tray is the actual change in vapor composition ($y_n - y_{n+1}$) from inlet to outlet of a stage divided by the change in vapor composition ($y_n^* - y_{n+1}$) if the vapor leaving the stage is in equilibrium with the liquid leaving the stage. Note that if the vapor stream leaving the stage is not at equilibrium with the liquid stream leaving the stage, $y_n < y_n^*$ and therefore, $\eta_M < 1$.

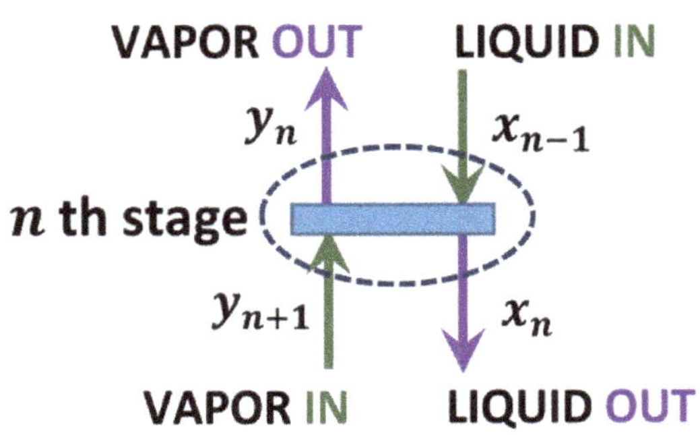

Figure 10.4 nth stage of a distillation column.

The Murphree efficiency of a tray can be readily incorporated into the graphical McCabe-Thiele method of determining the number of actual (not theoretical) stages. The analysis involves the construction of a *pseudo-equilibrium (partial-equilibrium)* curve which gives the actual (not equilibrium) composition of the vapor stream leaving the stage, that is, y_n, for any given composition of the liquid stream leaving the same stage, that is, x_n. The procedure involved in the construction of pseudo-equilibrium curve is as follows:

- Select a number of points (x_n, y_{n+1}) on the operating line for the construction of pseudo-equilibrium curve.
- For any given x_n, determine the equilibrium vapor composition y_n^* from the true equilibrium curve.
- Calculate the actual composition y_n of vapor stream leaving the stage as follows:

$$y_n = y_{n+1} + \eta_M (y_n^* - y_{n+1}) \tag{10.26}$$

The point (x_n, y_n) lies on the pseudo-equilibrium curve (see Figure 10.5).
- Repeat the process for all the selected points on the two operating lines.
- The plot of a series of (x_n, y_n) points gives the pseudo-equilibrium curve.

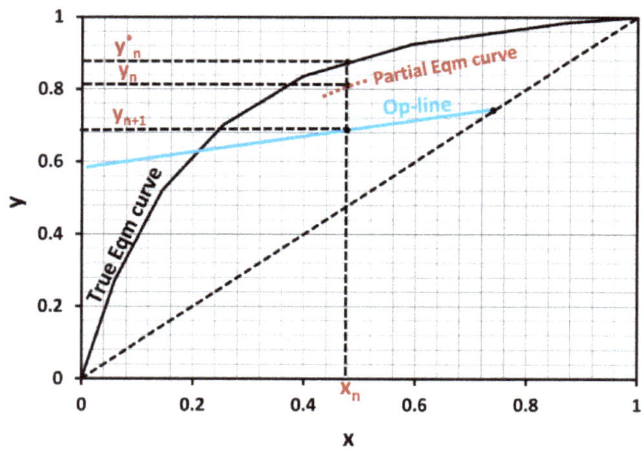

Figure 10.5 Construction of partial (pseudo) equilibrium curve.

Once the pseudo-equilibrium curve is known, the estimation of the actual number of stages can be done by stepping off triangular stages in the usual way alternating between the pseudo-equilibrium curve and

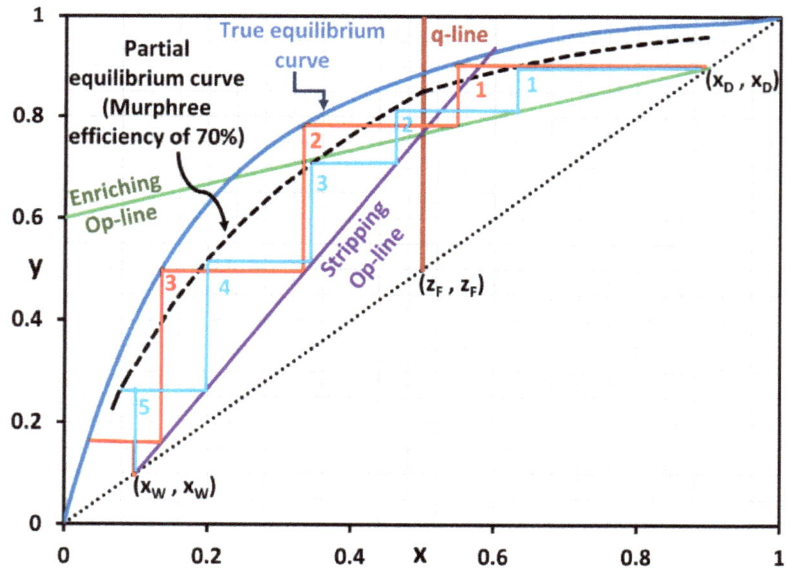

Figure 10.6 Construction of partial (pseudo) equilibrium curve assuming constant Murphree efficiency of 70%. The number of ideal/equilibrium/theoretical stages is approximately 3.4 whereas the actual number of stages with η_M of 0.70 is approximately 4.8.

the operating line. As an example, Figure 10.6 shows the graphical procedure of obtaining the actual number of stages using the pseudo or partial equilibrium curve. The Murphree efficiency of the trays is assumed to be constant, that is, 70%. As can be seen for the same separation, the number of actual stages required is significanltly larger than the theoretical or equilibrium number of stages.

When the Murphree efficiency is not known, one can use the *overall efficiency* η_0 of the column to estimate the actual number of stages needed. The overall efficiency of the column is defined as:

$$\eta_0 = \frac{N_{ideal}}{N_{actual}} \tag{10.27}$$

where N_{ideal} is the total number of ideal or theoretical (equilibrium) stages needed and N_{actual} is the actual number of stages needed for the same separation. Note that partial reboiler is not included in N_{ideal} or N_{actual} in the definition of η_0.

The overall efficiency has the advantage that it is easy to apply. Also it is easily determined experimentally. A distillation column of known number of trays is operated under total reflux condition. The samples of the distillate and bottoms liquid are collected and analyzed for their compositions. For the same separation, the theoretical number of trays are determined either graphically using the McCabe-Thiele method or through the Fenske equation. The overall efficiency of the column is simply the ratio of theoretical number of trays to the actual number of trays.

A number of empirical correlations are available in the literature to predict the overall efficiency (η_0) of distillation columns. Two factors which influence the overall efficiency in a significant way are: relative volatility (α) and liquid viscosity (μ_L). With the increase in relative volatility, the efficiency drops. With the increase in α, less number of equilibrium stages are required for the same separation. In order to achieve the same overall separation with a smaller number of equilibrium trays, the increase in vapor composition at any nth stage $y_n - y_{n+1}$ is large and therefore, a higher amount of mass transfer is required at each stage to reach equilibrium. More often than not, the contact time between the phases and the level of mixing are not sufficient to achieve equilibrium. With the increase in liquid viscosity, the efficiency drops due to decrease in the rate of mass transfer.

The most widely used empirical correlation for overall efficiency of distillation columns is the O'Connell correlation (H.E. O'Connell, Trans. Amer. Inst. Chem. Eng., Vol. 42, 741-775, 1946) given as follows:

$$\eta_0 = 0.52782 - 0.27511\ log_{10}(\alpha\mu_L) + 0.044923\ [log_{10}(\alpha\mu_L)]^2 \tag{10.28}$$

where μ_L is the liquid viscosity at feed composition in units of cP and α is the relative volatility of more volatile species. A simpler version of η_0 correlation is as follows:

$$\eta_0 = 0.492\ (\alpha\mu_L)^{-0.245} \tag{10.29}$$

Both liquid viscosity and relative volatility are evaluated at average tower conditions (average of top and bottom temperatures and average of top and bottom pressures).

Figure 10.7 shows the plots of overall efficiency generated from Equations (10.28) and (10.29). The overall efficiency decreases substantially with the increase in relative volatility-viscosity product ($\alpha\mu_L$) and there is good agreement between the predicted values from Equations (10.28) and (10.29).

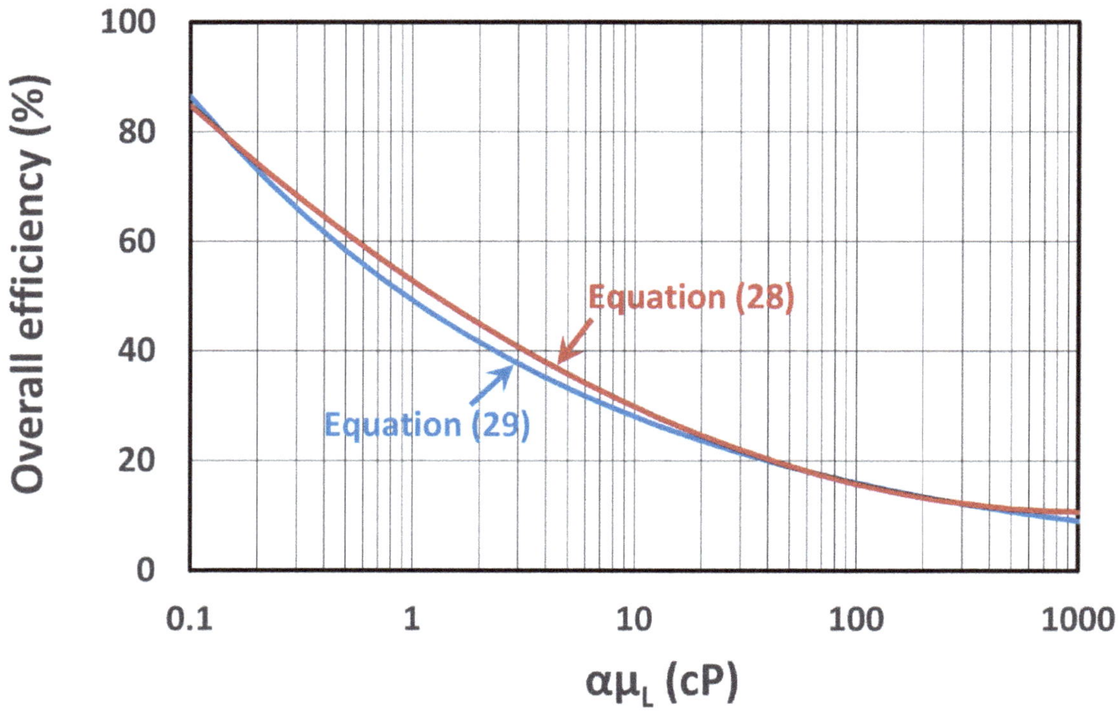

Figure 10.7 Overall column efficiency predicted from Equations (10.28) and (10.29).

Finally it should be noted that the overall efficiency η_0 is related to Murphree stage efficiency η_M. The relationship depends on the ratio of equilibrium line slope (K) and to operating line slope (L/V), that is, KV/L. The equilibrium curve in different sections (enriching and stripping sections) of the column is assumed to be approximately linear. The operating lines are also assumed to be linear. The relationship is given as:

$$\eta_0 = \frac{ln[1+\eta_M((KV/L)-1)]}{ln(KV/L)} \tag{10.30}$$

When $KV/L = 1$, that is, the equilibrium and operating lines are parallel, the two efficiencies are equal: $\eta_0 = \eta_M$. If equilibrium line is steeper than operating line, that is, $KV/L > 1$ which happens in stripping section, $\eta_0 > \eta_M$. If operating line is steeper than equilibrium line, that is, $KV/L < 1$ which happens in enriching section, $\eta_0 < \eta_M$. As distillation columns consist of both stripping and enriching sections, $\eta_0 \approx \eta_M$. Higher value of η_0 in the stripping section offsets the lower value in enriching section.

10.5 ESTIMATION OF TRAY/PLATE SPACING AND HEIGHT OF COLUMN

In order to estimate the height of the column, we need to know the spacing between the trays/plates in addition to the total number of trays required. The spacing between the trays should be considered for maintenance requirements. The workers should be able to access the tray in order to do the maintenance work. For large diameter columns (diameter > 4 ft), the tray spacing of 2 ft is enough for workers to be able to crawl through the column and do the maintenance work on the trays. For small diameter columns (diameter < 4 ft), tray spacing of 1.5 ft is enough as workers can reach the tray for maintenance through the side entry.

It should be noted that the selection of column diameter itself depends on the tray spacing. As the tray spacing increases, the height of the column increases. However, the column diameter is reduced. Thus there is trade off involved between "tray spacing" and "column diameter". With the increase in tray spacing, the vapor velocity where *flooding* of the column occurs increases. Consequently, one can operate the column at higher vapor velocity using a smaller diameter.

The total height of the column can be expressed as:

$$h = (N - 1)h_p + h_{top} + h_{bottom} \tag{10.31}$$

where N is the total number of trays/stages excluding the partial reboiler, h_p is the spacing between the trays/plates, h_{top} is the disengaging height at the top (typically about 5 ft) and h_{bottom} is the liquid reservoir height at the bottom of the column (typically 5 to 10 ft).

Note that tall and skinny columns are not desirable. Probably the tallest distillation column in the world is the one located at Shell Chemical Company complex in Deer Park, Texas. This column is 18 ft in diameter and 338 ft in height. The height to diameter ratio is about 19. The ratio of height to column diameter is important. This ratio should be no more than about 25 to 30. If the height to diameter ratio exceeds these values, the column needs to be redesigned. One can decrease the spacing between the trays or go for two or more columns arranged in series.

10.6 DIFFERENT TYPES OF TRAYS/PLATES

Three types of trays or plates are the most commonly used in distillation columns for contacting vapor and liquid phases. They are: a) Sieve or perforated tray, b) Bubble-cap tray, and c) Valve tray.

Sieve trays: They are basically perforated metal plates with holes drilled in them. Figure 10.8 shows a sieve tray. The number, size, and arrangement (pattern) of the holes are design parameters. The hole size (diameter) is usually selected from within the range of 5 to 12 mm (3/16 to ½ inch), with 5 mm hole diameter being the most commonly used size. Holes are usually arranged in an equilateral triangular pattern and the total area of the holes is about 5 to 15% of the tray area. The vapor flows in the upward direction through the holes of the tray and the liquid flows in the crossflow mode on each tray. The velocity of vapor flowing through the holes into the liquid is high enough to create a bubbly/frothy two-phase mixture of vapor and liquid on the tray with large interfacial area for mass transfer. The depth of the bubbly region on the tray is controlled by the weir height. The vapor pressure drop across the tray is high enough to keep the liquid from dripping down the holes (phenomenon referred to as weeping). Sieve trays are relatively inexpensive and easy to manufacture. However, they have the disadvantage that at low vapor velocities, the liquid begins to drip down from the holes. This phenomenon of weeping of trays deteriorates the efficiency of the distillation column.

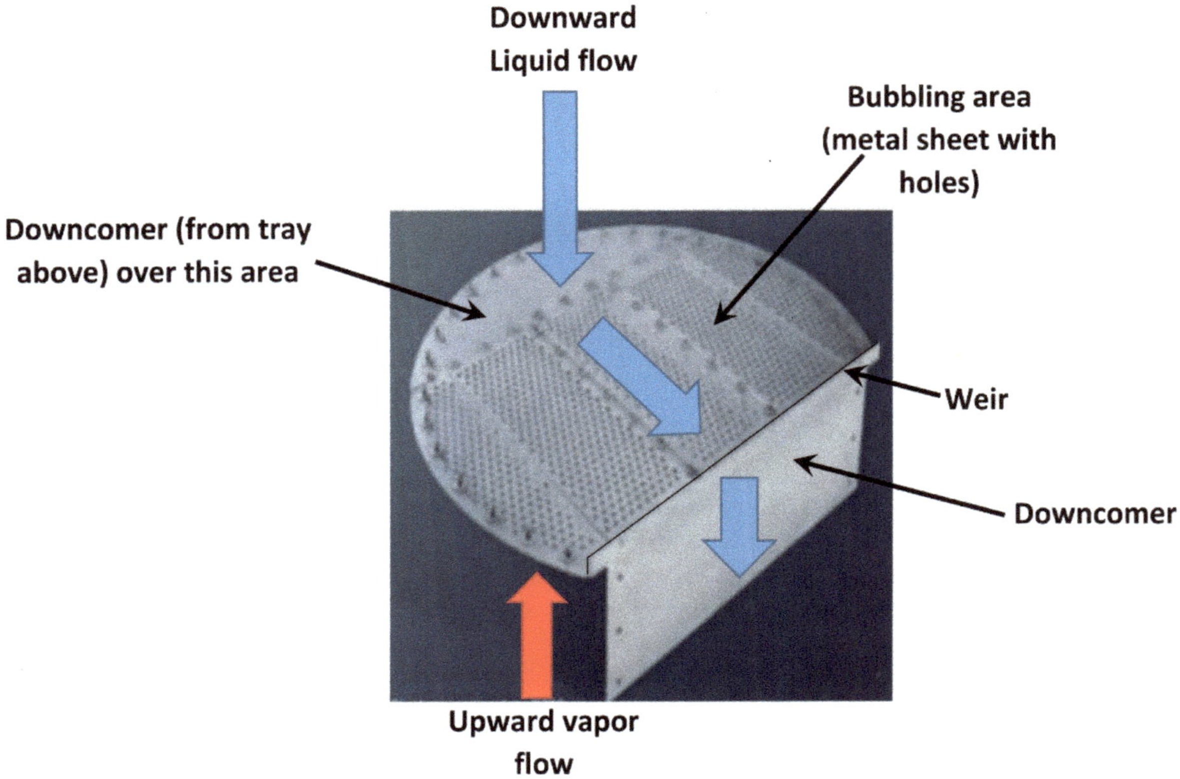

Figure 10.8 Sieve tray.

Bubble-cap trays: In a bubble-cap tray, each hole on the tray is fitted with a riser (see Figure 10.9) through which the vapor rises. The riser is covered with a cap fixed at the top. Some space is left between the riser and the cap to allow the flow of vapor. The upward moving vapor in the riser is directed downward by the cap. The cap has several slots on the side through which the vapor is discharged and bubbled through the liquid present on the tray. The advantage of bubble-cap trays is that weeping of trays is eliminated. As a result the column could be operated over a wide range of vapor flow rates without deterioration of efficiency. However, the disadvantages are: high cost of bubble cap trays and difficult maintenance. Due to these disadvantages, bubble-cap trays are rarely used in the construction of new distillation columns today.

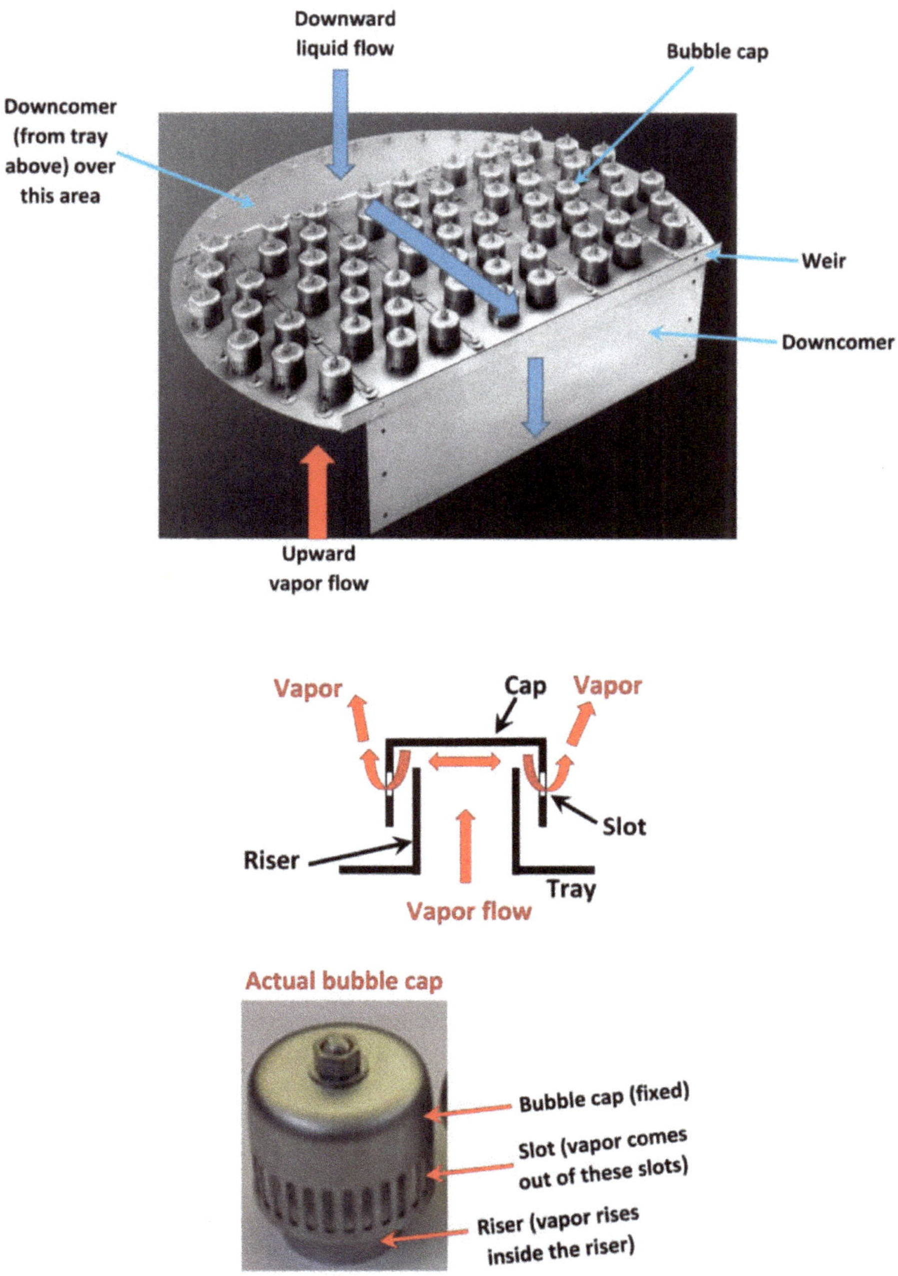

Figure 10.9 Bubble-cap tray.

Valve trays: Valve trays are becoming the most commonly used devices today in distillation columns for contacting the vapor and liquid phases. In valve trays, the holes on the tray are covered by liftable caps as shown in Figure 10.10. As the vapor rises, it lifts the caps and bubbles through the liquid on the tray. Mixing in these devices is better than that in sieve trays as vapor flow is directed horizontally into the liquid. The most important advantage of valve caps is that the flow area of vapor is dependent on the vapor flow rate. At low vapor flow rates, the gap between the valve cap and tray is small. At high vapor flow rates, the gap is large. The velocity of gas through the gap remains nearly the same over a broad range of vapor flow rates. Consequently, the weeping phenomenon encountered in sieve trays at low vapor flow rates is eliminated. While valve trays are more expensive than sieve trays, they are much cheaper than the bubble-cap trays.

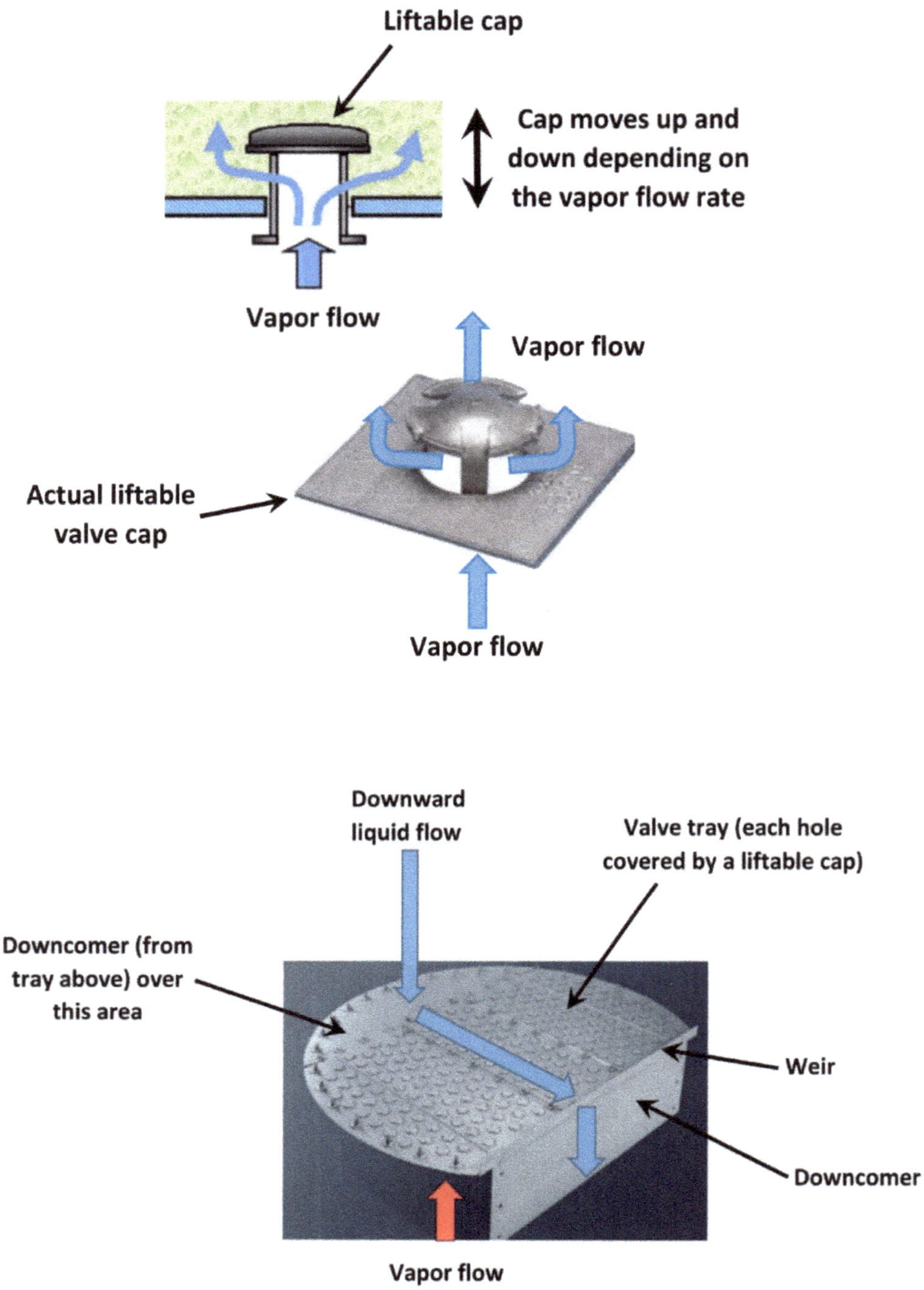

Figure 10.10 Valve tray.

10.7 ESTIMATION OF COLUMN DIAMETER

In a distillation column, vapor flows vertically upward whereas the liquid flows in the downward direction in a cross-flow mode as shown in Figure 10.11. The vapor flows upward through the holes in the tray. The vapor superficial velocity is defined as:

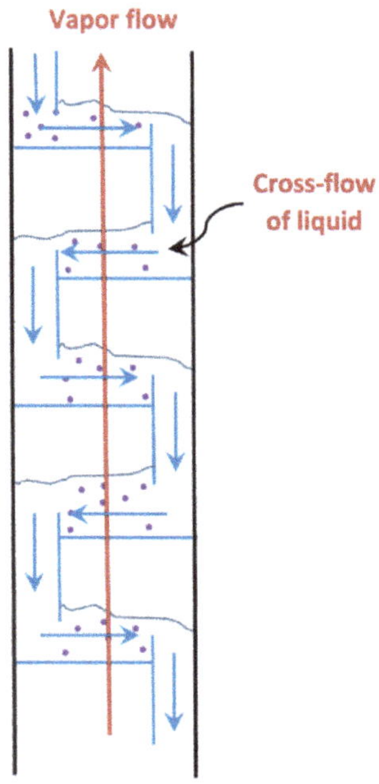

Figure 10.11 Upward flow of vapor and cross-flow of liquid in a distillation column.

$$U = \frac{Q}{(A - A_d)} \tag{10.32}$$

where Q is the volumetric flow rate of vapor, A is the total cross-sectional area of the column, and A_d is the cross-sectional area of one downcomer. This velocity is the actual average velocity of vapor in the space between two adjacent trays. The performance of a distillation column depends on this vapor velocity. At low vapor velocities, the pressure drop of vapor across any tray is not large enough to keep the liquid above the tray when the tray is sieve-type tray. Thus at low vapor velocities "*weeping phenomenon*" occurs on sieve-type trays and the liquid drips through the tray as depicted in Figure 10.12. The weeping phenomenon lowers the separation efficiency of the column significantly. At high vapor velocities, extensive entrainment of liquid droplets occurs in the vapor phase and these liquid droplets are carried upward by vapor to the next plate, located above (see Figure 10.12). This phenomenon whereby a significant amount of liquid is entrained in the form of droplets and is carried upward with vapor is called "*flooding*". At flooding condition, there also occurs backup of liquid in the downcomers as liquid is not able to easily flow over the tray due to high vapor velocity. Flooding of columns should be avoided as it leads to mixing of fluids from one plate to another and hence deterioration of separation efficiency. Typically one should operate the column at a vapor velocity of about 75 to 85% of the flooding velocity.

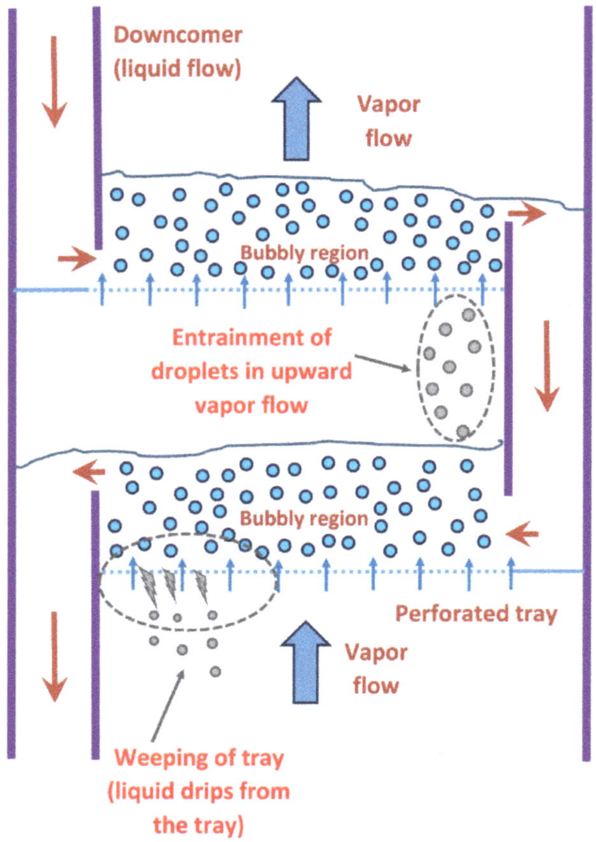

Figure 10.12 "Weeping" of sieve tray at low vapor velocities and "flooding" of column (entrainment and carry over of liquid from one plate to another above) at high vapor velocities.

Keeping the liquid and vapor flow rates constant, varying the column diameter varies the vapor velocity. For example, vapor velocity increases with the decrease in column diameter keeping the liquid and vapor flow rates constant. At some column diameter, the vapor velocity reaches the flooding velocity. This flooding velocity varies with the ratio of liquid to vapor flow rates. It also depends on the physical properties of the fluids involved. As already noted the flooding condition should be avoided in the actual column. Thus one should select a column diameter large enough so that the vapor velocity is no more than 75 to 85% of the flooding velocity.

10.7.1 Estimation of Flooding Velocity of Vapor

At incipient flooding, the droplets of liquid become just suspended in the vapor phase. Thus force balance (see Figure 10.13) gives:

$$F_G = F_B + F_D \tag{10.33}$$

where F_G is the gravitational force (weight) acting on the droplet in the downward direction, F_B is the buoyancy force acting on the droplet in the upward direction, and F_D is the drag force acting on the droplet in the upward direction due to drag exerted by vapor flow around the droplet. These forces can be expressed as follows:

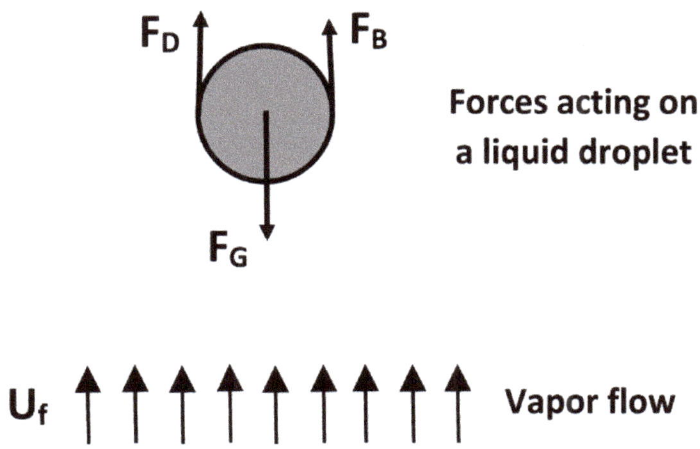

Figure 10.13 Forces acting on a droplet.

$$F_G = \rho_L \left(\frac{\pi d_p^3}{6}\right) g \tag{10.34}$$

$$F_B = \rho_V \left(\frac{\pi d_p^3}{6}\right) g \tag{10.35}$$

$$F_D = C_D \left(\frac{\pi d_p^2}{4}\right)\left(\frac{\rho_V U_f^2}{2}\right) \tag{10.36}$$

where ρ_L and ρ_V are liquid and vapor densities, d_p is droplet diameter, g is acceleration due to gravity, C_D is drag coefficient, and U_f is the vapor velocity at flooding condition. Upon substitution of force terms given in Equations (10.34)-(10.36) into force balance Equation (10.33), we get:

$$\rho_L \left(\frac{\pi d_p^3}{6}\right) g = \rho_V \left(\frac{\pi d_p^3}{6}\right) g + C_D \left(\frac{\pi d_p^2}{4}\right)\left(\frac{\rho_V U_f^2}{2}\right) \tag{10.37}$$

Upon rearrangement, Equation (10.37) can be expressed as:

$$U_f = \sqrt{\left(\frac{4 d_p g}{3 C_D}\right)\left(\frac{\rho_L - \rho_V}{\rho_V}\right)} \tag{10.38}$$

Thus the flooding velocity of vapor is:

$$U_f = C' \sqrt{\left(\frac{\rho_L - \rho_V}{\rho_V}\right)} \tag{10.39}$$

where C' is given as:

$$C' = \sqrt{\left(\frac{4 d_p g}{3 C_D}\right)} \tag{10.40}$$

Parameter C' is an empirical quantity that is determined from experimental data. It is a complicated function of liquid and vapor flow rates, liquid surface tension, and spacing between trays (h_p). Based on experimental observations, C' can be expressed as:

$$C' = C(F_{LV}, h_p)\left(\frac{\gamma}{20}\right)^{0.2} \tag{10.41}$$

where γ is liquid surface tension in units of dynes/cm and F_{LV} is a flow parameter defined as:

$$F_{LV} = \left(\frac{\dot{m}_L}{\dot{m}_V}\right)\sqrt{\frac{\rho_V}{\rho_L}} \tag{10.42}$$

where $\dot{m}_L$ and $\dot{m}_V$ are the mass flow rates of liquid and vapor. In terms of molar flow rates of liquid (L) and vapor (V), the flow parameter is defined as:

$$F_{LV} = \left(\frac{L}{V}\right)\left(\frac{M_L}{M_V}\right)\sqrt{\frac{\rho_V}{\rho_L}} \tag{10.43}$$

where M_L and M_V are the molecular weights liquid and vapor. Note that $M_L = M_V$ if the liquid and vapor compositions are the same.

Combining Equations (10.39) and (10.41), we get:

$$U_f = C(F_{LV}, h_p)\left(\frac{\gamma}{20}\right)^{0.2}\sqrt{\left(\frac{\rho_L-\rho_V}{\rho_V}\right)} \tag{10.44}$$

The dependence of C (also referred to as capacity factor) on F_{LV} and h_p is determined experimentally. Fair and Matthews (J.R. Fair, R.L. Matthews, Petroleum Refiner, Vol. 37, 153, 1958) have presented an empirical correlation for estimating the capacity factor C in ft/s for sieve trays. The correlation was presented in a graphical form as C versus F_{LV} with tray spacing h_p as a parameter. Kessler and Wankat (D.P. Kessler, P.C. Wankat, Chemical Engineering, Vol. 95, 72-74, 1988) fitted the graphical correlation of Fair and Matthews for C (ft/s) with the following expression:

$$log_{10}C = a + b(log_{10}F_{LV}) + c(log_{10}F_{LV})^2 \tag{10.45}$$

where the coefficients a, b and c are dependent on the spacing between the trays. They are given in Table 10.1.

It should be noted that the capacity factor correlation originally proposed for sieve-tray columns is also applicable to bubble-cap and valve trays.

Table 10.1: Values for coefficients a, b and c in Equation (10.45)

$h_p(inches)$	a	b	c
6	-1.1977	-0.53143	-0.18760
9	-1.1622	-0.56014	-0.18168
12	-1.0674	-0.55780	-0.17919
18	-1.0262	-0.63513	-0.20097
24	-0.94506	-0.70234	-0.22618
36	-0.85984	-0.73980	-0.23735

Figure 10.14 shows the plots of capacity factor C (ft/s) versus flow parameter F_{LV} for different values of tray spacing h_p. The plots are generated from Equation (10.45). From the plots it is clear that the capacity factor, and hence flooding velocity, increases with the increase in plate/tray spacing for the same ratio of liquid to vapor flow rates (that is, same F_{LV}). Thus one can operate a distillation column at higher vapor velocities using a smaller diameter column, without flooding when the tray spacing is increased (keeping the ratio of liquid to vapor flow rates constant). However the disadvantage is that we need a taller column when tray spacing is increased.

10.7.2 Estimation of Column Diameter

Once the flooding velocity is estimated, we can determine the diameter of the column by taking the operating vapor velocity U to be some fraction of the flooding velocity. Typically the fraction is 0.80 or so. Thus:

$$U = 0.80(U_f) \tag{10.46}$$

The operating molar flow rate of vapor V can be expressed as:

$$V = (A - A_d)U(\frac{\rho_V}{M_V}) \tag{10.47}$$

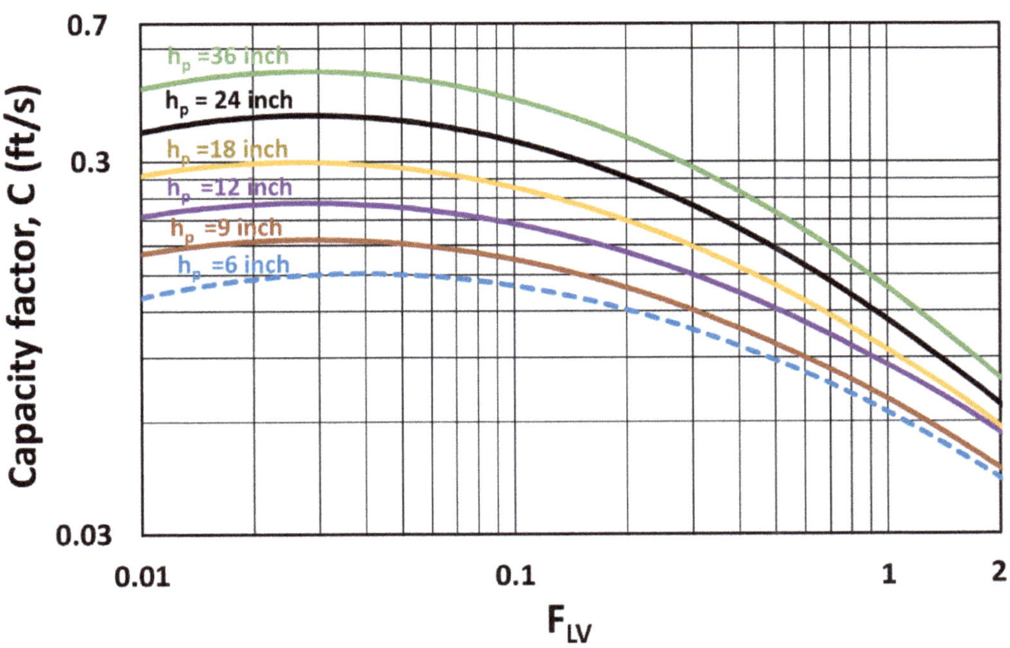

Figure 10.14 Capacity factor C(ft/s) as a function of flow parameter F_{LV} for different values of tray spacing h_p . The plots are generated from Equation (10.45).

where A_d is the cross-sectional area of one downcomer and A is the total cross-sectional area of the column related to column diameter D as:

$$A = \pi D^2/4 \tag{10.48}$$

From Equations (10.46) to (10.48), the column diameter can be expressed as:

$$D = 2\sqrt{\frac{VM_V}{0.80\pi U_f\left(1-\frac{A_d}{A}\right)\rho_V}} \tag{10.49}$$

The area ratio A_d/A is typically 0.1 when $F_{LV} \leq 0.1$. When $F_{LV} \geq 1$, $A_d/A = 0.2$; when $0.1 \leq F_{LV} \leq 1$, $A_d/A = 0.1 + (F_{LV} - 0.1)/9$.

10.8 PRELIMINARY MECHANICAL DESIGN OF A DISTILLATION COLUMN

Thus far we have considered so-called *chemical design* of a distillation column that involved specification of the actual number of trays required, location of optimum feed tray, type of tray and spacing of trays, height of a column, and diameter of a column. The next thing to consider is *mechanical design* of a distillation column. The keys aspects of mechanical design of a column are: specification of material of construction of column and the thickness of the column shell walls.

10.8.1 Selection of Material of Column Construction

The selection of material of construction of column requires careful consideration of process conditions such as temperature, pressure, chemicals involved and their corrosive nature. For example, low temperature can cause some materials to become brittle leading to fracture. High temperature generally decreases the tensile strength of metallic materials. If the pressure inside the column is high, materials with high tensile strength need to be considered. If the process chemicals are corrosive in nature, the engineer should consider corrosion resistant materials; alternatively, one can consider coating or lining of internal walls of the column with appropriate corrosion-resistant material.

Although a wide variety of materials are used in the construction of distillation columns depending upon the specific application, carbon steel and stainless steels are the most commonly used materials of construction. Carbon steel, also known as mild steel, is an alloy of iron and carbon. It is relatively cheap and is readily available. Although carbon steel has good tensile strength, it is not that resistant to corrosion. If corrosion is an issue, stainless steel is preferred. Stainless steel is an alloy of chromium-nickel-iron with low carbon content. Stainless steel with high levels of chromium is strongly resistant to corrosion.

10.8.2 Shell Thickness of Distillation Column

Once the material of construction is selected, shell thickness should be specified. The shell thickness should be such that the column can withstand all kinds of loads/stresses: internal pressure, dead weight, wind load, and seismic load due to earthquakes if the column is located in a region of high seismic activity. Allowance should also be made for corrosion. In the discussion to follow, we will assume that the column is located in a region where seismic activity is negligible and therefore, seismic load is not considered.

Tensile Stresses in a Shell due to Internal Pressure

Consider a cylindrical vessel (column) subjected to an internal pressure of P_i as shown in Figure 10.15. Due to internal pressure, tensile stresses are induced in the walls of the vessel. There are two types of tensile stresses induced: *axial or longitudinal stress* which tends to rupture the vessel longitudinally, and *circumferential or hoop stress* (stress in the direction along the circumference) which tends to rupture the vessel circumferentially.

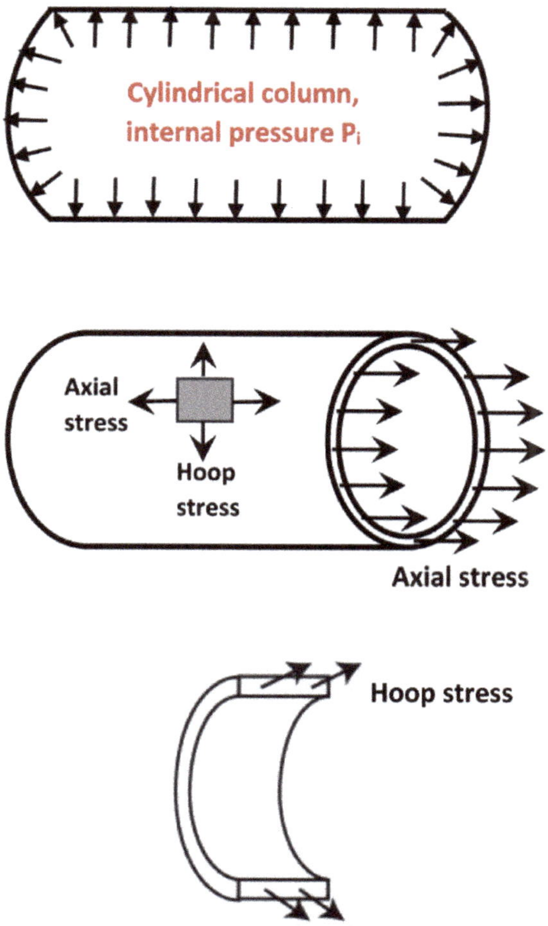

Figure 10.15 Stresses in a cylindrical vessel due to internal pressure.

What should be the minimum shell thickness to withstand the axial and hoop stresses (without any rupturing of vessel) given that the maximum allowable stress or design stress for the material under consideration is σ_{design}?

In order to answer this question, we need to do force balance in axial and circumferential directions. Let the shell thickness be t_s. Let the inside diameter of the cylinder be D_i. We will assume that the shell thickness is small as compared with the shell diameter, that is, $t_s/D_i < 1/20$.

The force tending to rupture the vessel longitudinally (F_{axial}) is given as:

$$F_{axial} = P_i\left(\pi D_i^2/4\right) \quad (10.50)$$

This force is balanced by pressure induced axial stress (σ_p) as follows:

$$F_{axial} = P_i\left(\pi D_i^2/4\right) = \sigma_p(\pi D_i t_s) \quad (10.51)$$

As the maximum allowable stress is σ_{design}, the minimum shell thickness to withstand the axial stress induced by internal pressure is:

$$t_s = P_i D_i/4\sigma_{design} \quad (10.52)$$

This equation assumes the cylindrical vessel is seamless (without any welded joints). Usually the cylindrical vessel is seam welded. Therefore allowance must be made for weakness in the welded joint by introducing a joint efficiency factor J as follows:

$$t_s = P_i D_i / 4J\sigma_{design} \tag{10.53}$$

Some allowance for corrosion should also be made. Thus the minimum shell thickness based on longitudinal stresses is:

$$t_s = \frac{P_i D_i}{4J\sigma_{design}} + c \tag{10.54}$$

where c is corrosion allowance.

Let us now consider force balance in circumferential direction (see Figure 10.16). The pressure induced force on half-cylinder tending to rupture the cylindrical vessel circumferentially is:

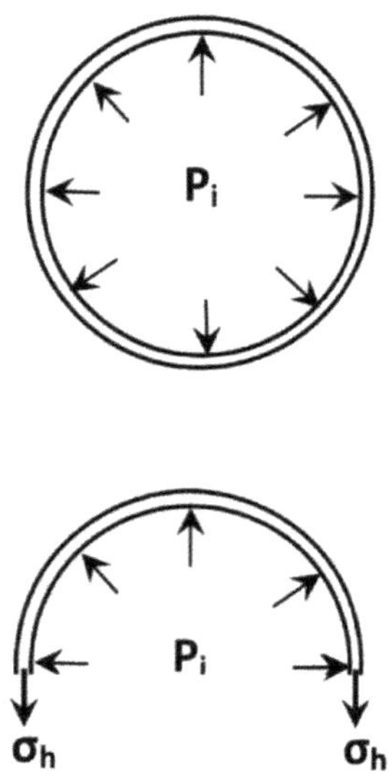

Figure 10.16. Hoop stress in a cylindrical shell of internal diameter D_i, thickness t_s, and length l.

$$F_{circumferential} = P_i l D_i \tag{10.55}$$

where l is the length of cylinder. The resisting force due to hoop stress is $2t_s l\sigma_h$. Thus force balance gives:

$$P_i l D_i = 2t_s l\sigma_h \tag{10.56}$$

As the maximum allowable stress is σ_{design}, the minimum shell thickness to withstand the hoop stress induced by internal pressure is:

$$t_s = P_i D_i / 2\sigma_{design} \tag{10.57}$$

Making allowance for welded joint efficiency and corrosion, Equation (10.57) could be re-written as:

$$t_s = \frac{P_i D_i}{2J\sigma_{design}} + c \tag{10.58}$$

Note that the minimum shell thickness based on hoop stress is larger than the minimum thickness based on axial stress. Thus hoop stress is controlling and the minimum shell thickness is given by Equation (10.58).

However, it is important to recognize that the minimum shell thickness of the column cannot be decided on the basis of hoop stress alone. We need to consider all other loads present on the column before selecting the shell thickness.

Compressive Stresses in a Shell due to Dead Loads

Dead load is defined as the weight of the structure itself including the weight of all fixtures permanently attached to the structure. There is always a possibility that the structure may fail or collapse under its own weight if compressive stresses induced by dead load are not properly taken into account in designing the structure. As dead load stresses increase towards the base of the column, the column needs to be much thicker at the base as compared with the top portion.

The dead loads to be considered are: weight of semi-elliptical head of the column, weight of column shell, weight of trays including the weight of liquid present on the trays, and weight of column insulation. There may be additional dead loads present on the column such as fittings and valves, piping, platforms, ladders, instruments, condenser, etc.

The weight of semi-elliptical head with major to minor axis ratio of 2 (see Figure 10.17) is given as follows:

$$W_{eh} = \frac{\pi}{6}\left[D_0^3 - D_i^3\right]\rho_s g \tag{10.59}$$

where D_i is the inside diameter of the column shell, D_0 is the outside diameter of the column shell, ρ_s is shell material density, and g is acceleration due to gravity (9.81 m/s^2).

The weight of a cylindrical column shell as a function of distance from the top (X) of the column can be expressed as:

$$W_{sh} = \pi D_i X t_s \rho_s g \tag{10.60}$$

where t_s is the shell thickness.

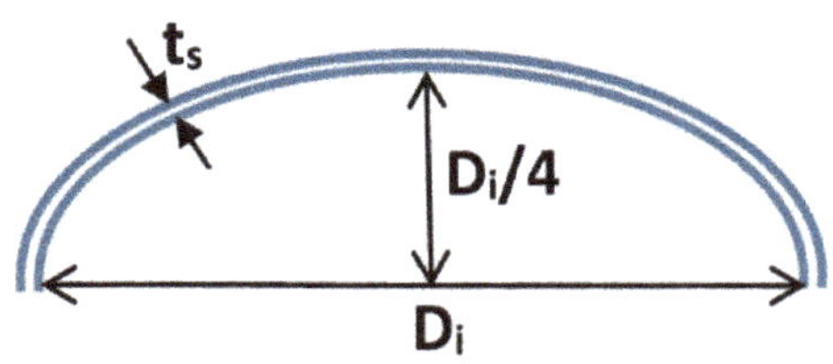

Figure 10.17 Semi-elliptical head of a distillation column.

The number of trays present in the length portion X of the column, where X is measured from the top of the column, is given as:

$$n_X = \frac{X - h_{top}}{h_p} + 1 \tag{10.61}$$

Let the liquid depth on trays be h_L, the liquid density be ρ_L, the area of tray be A_p, and mass of tray/plate be m_p. The weight of a single tray including its liquid content (ω_p) can be expressed as:

$$\omega_p = (A_p h_L \rho_L + m_p)g \tag{10.62}$$

The total weight of trays present in the length portion X of the column, where X is measured from the top of the column, is as follows:

$$W_p = n_X \omega_p = n_X (A_p h_L \rho_L + m_p)g = \left(\frac{X - h_{top}}{h_p} + 1\right)(A_p h_L \rho_L + m_p)g \tag{10.63}$$

The weight of insulation of thickness t_{ins} and density ρ_{ins} as a function of distance from the top (X) of the column is given as:

$$W_{ins} = \pi D_i X t_{ins} \rho_{ins} g \tag{10.64}$$

Note that we are assuming that the shell thickness is small as compared with the shell diameter so that as an approximation $D_i \approx D_0$.

The total dead weight ($W_{eh} + W_{sh} + W_p + W_{ins}$) is supported by the column shell itself. Thus the compressive stress in the column shell due to total dead weight as a function of distance from the top (X) of the column is given as:

$$\sigma_W = -(W_{eh} + W_{sh} + W_p + W_{ins})/\pi D_i t_s \tag{10.65}$$

Note that negative sign is included as compressive stresses are negative.

Compressive and Tensile Stresses in a Shell due to Wind Load

Wind load on the column could be substantial if the column is located in the open air. Its importance increases with the increase in column height. Figure 10.18 shows a schematic diagram of the column subjected to wind load. For the sake of simplicity we will assume that the wind load, defined as force per unit length of the column exerted by wind, is uniform over the entire column length. It should be noted that wind speed is generally not constant. It varies with height. The wind velocity is less near the ground and high at the top of the column.

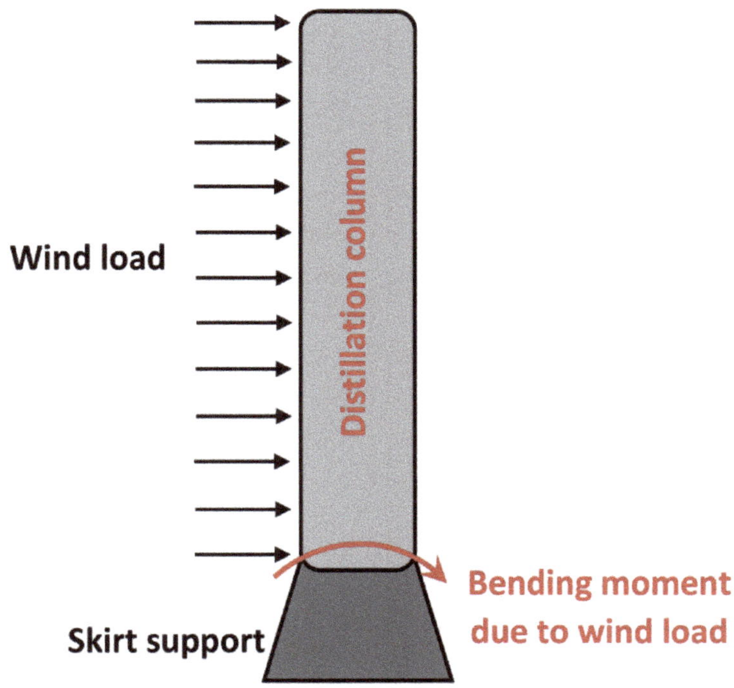

Figure 10.18 Distillation column subjected to wind load.

The column under wind load acts as a cantilever beam (rigid structural element anchored at one end to a support). The wind load generates bending moment on the column as shown in Figure 10.18. Actually bending moment is present on the column at all levels below the top of the column as shown more clearly in Figure 10.19. At any location X measured from the top of the column, the bending moment can be expressed as:

$$M_X = Force \times moment\ arm = (P_W X) \times (X/2) = P_W X^2/2 \qquad (10.66)$$

where P_W is the wind load, that is, force exerted by the wind per unit length of the column. Figure 10.20 shows the bending moment distribution on the column due to wind load. As expected the bending moment increases towards the base of the column.

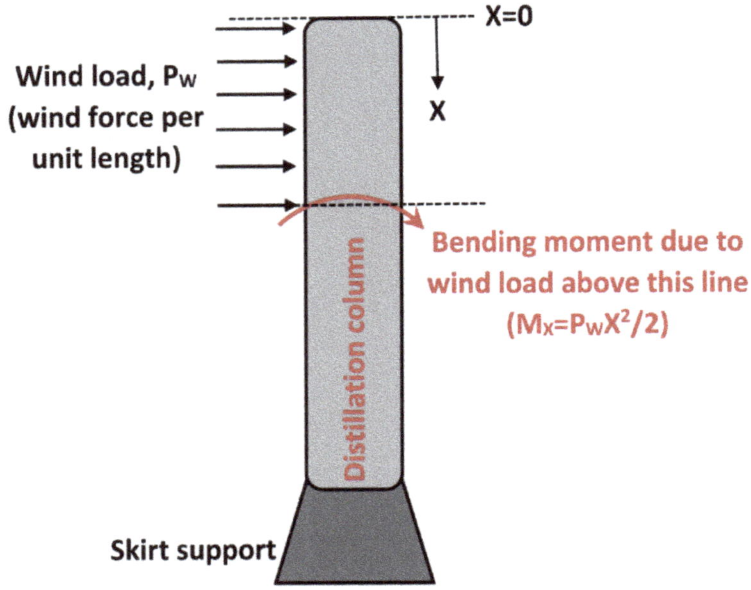

Figure 10.19 Wind induced bending moment at any distance X from the top of the column.

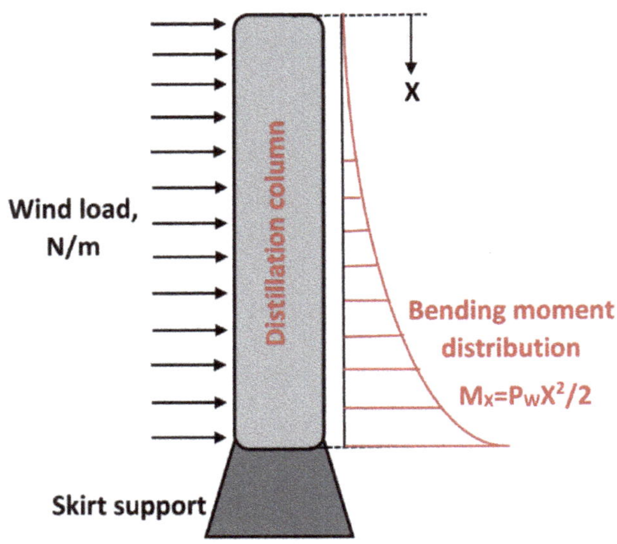

Figure 10.20 Bending moment distribution on the column due to wind load.

The bending moment induces both compressive and tensile stresses in the column depending on the location. The bending stress is tensile on the so-called "upwind side" (column side facing the wind) and compressive on the "downwind side" of the column. The bending stress is given as:

$$\sigma_b = \pm \frac{M}{I}\left(\frac{D_i}{2} + t_s\right) \tag{10.67}$$

where I is second moment of area of vessel about the plane of bending, given as:

$$I = \frac{\pi}{64}\left(D_0^4 - D_i^4\right) \tag{10.68}$$

Note that σ_b = positive when tensile (upwind side) and negative when compressive (downside).

Total Longitudinal Stress in a Shell

The total longitudinal stress in the shell of a distillation column is the sum of all the axial stresses present in the column, that is:

$$\sigma_z = \sigma_p + \sigma_W + \sigma_b \tag{10.69}$$

where $\sigma_z, \sigma_p, \sigma_W, \sigma_b$ are total axial stress, pressure induced axial stress, dead loads induced axial stress, and bending moment induced axial stress, respectively.

Selection of Minimum Shell Thickness

We need to consider the three principal stresses. Neglecting any torsional stress on the column, the three principal stresses are:

$$\sigma_1 = \sigma_h \tag{10.70}$$

$$\sigma_2 = \sigma_z \tag{10.71}$$

$$\sigma_3 \approx -P_i/2 \tag{10.72}$$

Let σ_{design} be the specified design stress for the chosen material of construction at process temperature. The maximum stress at any point in the shell is the numerically (magnitudinal) greatest value of the following:

$$\sigma_1 - \sigma_2 \tag{10.73}$$

$$\sigma_1 - \sigma_3 \tag{10.74}$$

$$\sigma_2 - \sigma_3 \tag{10.75}$$

Note that if we neglect σ_3 (which is negligible as compared with other stresses), the maximum stress in the shell is the numerically greatest value of the following: $\sigma_h - \sigma_z$, σ_h, or σ_z.

The column shell thickness should be selected such that the maximum stress in the shell does not exceed the specified design stress of the material, that is:

$$\sigma_{max} < \sigma_{design} \tag{10.76}$$

Thus the column shell thickness cannot be selected on the basis of hoop stress alone. It is possible that the shell thickness selected on the basis of hoop stress alone does not meet the criterion given in Equation (10.76). One way to select the shell thickness is to first estimate t_s from hoop stress analysis, that is, Equation (10.54). Using this value of t_s, estimate the maximum stress in the structure and see if the criterion given in Equation (10.76) is satisfied. If the criterion is not satisfied, select a higher value of t_s and repeat the calculations.

Finally it should be recognized that the mechanical design of distillation columns presented here is preliminary in nature as several details have been left out for the sake of simplicity. For example, seismic load is not considered. The shell is assumed to be thin (t_s/D_i < 1/20) of constant thickness. The wind load is assumed to be uniform. There may be additional dead loads present on the actual column not considered here.

NOTATION

A cross-sectional area of the column
A_d cross-sectional area of downcomer
A_p area of tray of a column, Equation (10.62)
c corrosion allowance, Equation (10.54)
C capacity factor (ft/s) defined in Equation (10.41), given by Equation (10.45), plotted in Figure 10.14
C' parameter defined in Equation (10.40)
C_D drag coefficient defined in Equation (10.36)
D column inside diameter
D_i inside diameter of the cylindrical column shell, same as D
D_o outside diameter of the cylindrical column shell
d_p liquid droplet diameter
F_B buoyancy force on a droplet
F_D drag force on a droplet
F_G gravity force on a droplet
F_{axial} axial force acting on the shell, force tending to rupture the shell axially or longitudinally
$F_{circumferential}$ circumferential force acting on the shell, force tending to rupture the shell circumferentially
F_{LV} flow parameter defined in Equation (10.42)
FUG Fenske-Underwood-Gilliland
g acceleration due to gravity
h total height of column, see Equation (10.31)
h_{bottom} liquid reservoir height at the bottom of the column
h_L liquid depth on trays
h_p spacing between the trays/plates
h_{top} disengaging height at the top of the column
I second moment of area of vessel about the plane of bending, see Equation (10.68)
J joint efficiency factor
K slope of y versus x equilibrium curve
l length of cylindrical shell
L molar flow rate of liquid in the column
m_p mass of a single tray/plate without any liquid
$\dot{m}_L$ mass flow rate of liquid in the column
$\dot{m}_V$ mass flow rate of vapor in the column
M_L molecular weight of liquid
M_V molecular weight of vapor
M_X bending moment at location X measured from the top of the column, see Equation (10.66).
n nth stage of a column
n_F number (location) of feed stage counting from top of the column
n_X number of trays present in the length portion X of the column, where X is measured from the top of the column, see Equation (10.61)
N number of stages in the column; in some situations N includes the partial reboiler and in some situations N is the number of stages inside the column excluding the partial reboiler
N_{min} minimum number of stages at total reflux (including reboiler)
P_i internal pressure of a vessel /column
P_W wind load, defined as force exerted by the wind per unit length of the column, see Figure 10.19
Q volumetric flow rate of vapor
q feed quality, defined in Chapter 9, see Equation (9.40)

R reflux ratio
R_{min} minimum reflux ratio
t_{ins} thickness of insulation
t_s shell thickness
U vapor superficial velocity, defined in Equation (10.32)
U_f vapor superficial velocity at flooding condition
V molar flow rate of vapor stream in enriching section
W_{eh} weight of semi-elliptical head of the column, Equation (10.59)
W_{ins} weight of insulation, see Equation (10.64)
W_p total weight of trays in the length portion X from the top of the column, see Equation (10.63)
W_{sh} weight of cylindrical column shell as a function of distance from the top, see Equation (10.60)
x mole fraction of more volatile species in liquid phase
x_A mole fraction of species A in liquid phase
x_B mole fraction of species B in liquid phase
x_D mole fraction of more volatile species in the distillate liquid product
x_n mole fraction of more volatile species in the liquid stream leaving any *n*th stage
x_W mole fraction of more volatile species in bottoms liquid product of distillation column
X distance from top of the column
X_G Gilliland parameter, defined in Equation (10.20)
y mole fraction of more volatile species in vapor phase
y_A mole fraction of species A in vapor phase
y_B mole fraction of species B in vapor phase
y_n mole fraction of more volatile species in the vapor stream leaving any *n*th stage
y_n^* mole fraction of more volatile species in the vapor phase in equilibrium with the liquid stream leaving the *n*th stage
Y_G Gilliland parameter, defined in Equation (10.21)
z_F mole fraction of more volatile species in the feed stream

Greek Letters

α relative volatility of more volatile species in a binary solution
α_{AB} relative volatility of species A with respect to species B
α_{bot} relative volatility of more volatile species at the bottom of the column
α_{top} relative volatility of more volatile species at the top of the column
γ surface tension
η_o overall efficiency of the column, Equation (10.27)
η_M Murphree tray efficiency, Equation (10.25)
μ_L viscosity of liquid at feed composition in units of cP
ρ density of material
ρ_{ins} density of insulation material
ρ_L density of liquid
ρ_s density of shell material
ρ_V density of vapor
σ stress
σ_1 principal stress acting in principal direction 1
σ_2 principal stress acting in principal direction 2
σ_3 principal stress acting in principal direction 3
σ_b bending stress
σ_h hoop stress
σ_p pressure induced axial stress
σ_W compressive stress due to dead weight

σ_z total axial stress
σ_{design} design stress (maximum allowable stress)
σ_{max} maximum stress in the shell
ω_p mass of a single tray/plate with its liquid content

TUTORIAL PROBLEMS WITH SOLUTION

Problem 1

Feed entering the distillation column is saturated liquid at column conditions. It consists of a binary mixture of benzene and toluene. The mole fraction of benzene in the feed is 0.57. The distillate product contains 95 mol% benzene and the bottoms product contains 3.0 mol% benzene. The reflux ratio is 3.25. Do the following:

a) Determine the optimum number of equilibrium stages using the McCabe-Thiele method
b) Determine the optimum feed tray location graphically
c) Estimate the number of equilibrium stages using the FUG method. The relative volatility of benzene with respect to toluene is approximately 2.50.
d) Determine the optimum location of the feed tray using the Kirkbride equation.

Assume Raoult's law is applicable and the column pressure is 101.32 kPa. The vapor pressures of pure benzene and pure toluene are given by the following Antoine equations:

$$lnP_{benzene}^{Sat} = 13.7819 - \frac{2726.81}{t+217.572}$$

$$lnP_{toluene}^{Sat} = 13.932 - \frac{3056.96}{t+217.625}$$

where t is in °C and the vapor pressures are in kPa.

Solution: Refer to the column diagram given below.

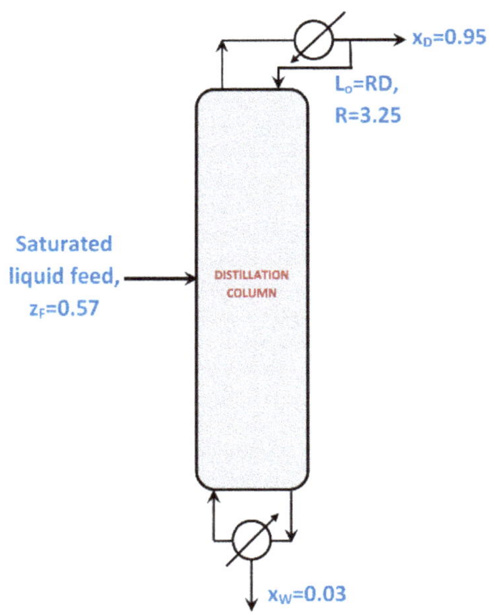

a) As the feed is saturated liquid, $q = 1$ and the slope of q-line is ∞. Thus q-line is a vertical line passing through the point (0.57, 0.57). The enriching section operating line passes through the point (0.95, 0.95) and has a y-intercept of $x_D/(1+R) = 0.95/(1+3.25) = 0.2235$. Before we can estimate the optimum number of equilibrium stages required using the McCabe-Thiele graphical method, we need to generate the equilibrium data.

T (°C)	$P_1^{Sat}(kPa)$	$P_2^{Sat}(kPa)$	x	y
80.0	101.32	38.89	1.00	1.00
82.0	107.72	41.65	0.90	0.96
84.0	114.42	44.57	0.81	0.92
86.0	121.45	47.65	0.73	0.87
88.0	128.80	50.89	0.65	0.82
90.0	136.50	54.31	0.57	0.77
92.0	144.54	57.91	0.50	0.71
94.0	152.95	61.70	0.43	0.66
96.0	161.73	65.68	0.37	0.59
98.0	170.90	69.87	0.31	0.53
100.0	180.45	74.26	0.25	0.45
102.0	190.42	78.87	0.20	0.38
104.0	200.79	83.70	0.15	0.30
106.0	211.60	88.77	0.10	0.21
108.0	222.84	94.07	0.06	0.12
110.6	238.13	101.32	0.00	0.00

The boiling points of pure species 1 (benzene) and 2 (toluene) at 101.32 kPa are calculated from the given Antoine equations as: $T_1^{Sat} = 80^oC$ and $T_2^{Sat} = 110.60^oC$. Table given above shows at different temperatures the values of vapor pressures of pure species and mole fractions of more volatile component (species 1-benzene) in the vapor and liquid phases at equilibrium with each other. The Antoine equations are used to calculate the vapor pressures. Equations (6.44) and (6.45) of Chapter 6 are used to calculate the mole fractions of more volatile component in liquid and vapor phases at equilibrium. The y-x diagram for the system at 101.32 kPa is developed using the data shown in the table. Note that x and y refer to mole fractions of more volatile species (species 1).

The following figure shows the plots of the operating lines and stepping off the stages using the McCabe-Thiele method. The optimum number of equilibrium stages required (including partial reboiler) is about 10.

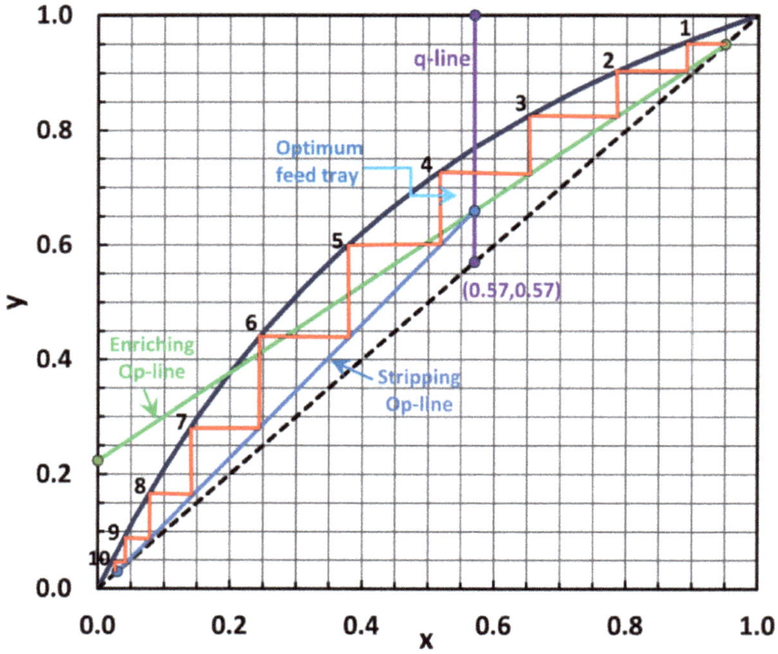

b) The optimum feed tray location is 4 (as shown in the figure above).

c) Estimation of the number of equilibrium stages using the FUG method

The FUG method involves the following procedure:

- Estimate N_{min} from Fenske equation, Equation (10.12)
- Estimate R_{min} using Underwood equation, Equation (10.17)
- Calculate $X_G = (R - R_{min})/(R + 1)$
- Use Gilliland-Molokanov equation (Equation (10.22)) to calculate Y_G
- Determine the number of equilibrium stages from the known value of Y_G using Equation (10.21)

Given: Saturated-liquid feed $z_F = 0.57, x_D = 0.95, x_W = 0.03, \alpha = 2.50, R = 3.25$

$$N_{min} = \frac{ln\left[\left(\frac{x_D}{1-x_D}\right)\left(\frac{1-x_W}{x_W}\right)\right]}{ln(\alpha)} = \frac{ln\left[\left(\frac{0.95}{1-0.95}\right)\left(\frac{1-0.03}{0.03}\right)\right]}{ln(2.50)} = 7.0071$$

$$R_{min} = \left(\frac{1}{\alpha - 1}\right)\left[\frac{x_D}{z_F} - \frac{\alpha(1 - x_D)}{(1 - z_F)}\right] = \left(\frac{1}{2.5 - 1}\right)\left[\frac{0.95}{0.57} - \frac{2.50(1 - 0.95)}{(1 - 0.57)}\right] = 0.91731$$

$$X_G = \frac{(R - R_{min})}{(R + 1)} = \frac{(3.25 - 0.91731)}{(3.25 + 1)} = 0.54887$$

$$Y_G = 1 - exp\left[\left(\frac{X_G - 1}{\sqrt{X_G}}\right)\left(\frac{1 + 54.4X_G}{11 + 117.2X_G}\right)\right] = 1 - exp\left[\left(\frac{0.54887 - 1}{\sqrt{0.54887}}\right)\left(\frac{1 + 54.4 \times 0.54887}{11 + 117.2 \times 0.54887}\right)\right]$$
$$= 0.220773$$

$$Y_G = \frac{N - N_{min}}{N + 1} = 0.220773$$

$$N = 9.276$$

This value of N is in good agreement with the value obtained graphically, that is, $N \approx 10$. Thus the FUG method of estimating the number of equilibrium stages is quite good.

d) The optimum location of the feed tray using the Kirkbride equation

$$\frac{n_F - 1}{N - n_F} = \left[\left(\frac{1 - z_F}{z_F}\right)\left(\frac{x_W}{1 - x_D}\right)^2 \frac{W}{D}\right]^{0.206}$$

$$\frac{W}{D} = \left(\frac{x_D - z_F}{z_F - x_W}\right) = \frac{0.95 - 0.57}{0.57 - 0.03} = 0.7037$$

$$\frac{n_F - 1}{9.276 - n_F} = \left[\left(\frac{1 - 0.57}{0.57}\right)\left(\frac{0.03}{1 - 0.95}\right)^2 0.7037\right]^{0.206} = 0.71112$$

$$n_F = 4.44$$

Thus the optimum feed tray is the 4th tray from the top. The graphical approach also gives the 4th stage as the optimum feed tray.

Problem 2

Estimate the total height of a distillation column, given that: the total number of trays including the partial reboiler is 11, the spacing between trays is 2 ft, the disengaging height at the top of the column is 5 ft, and the liquid reservoir height at the bottom of the column is 6 ft.

Solution

The total height of the column can be calculated using Equation (10.31) as:

$$h = (N - 1)h_p + h_{top} + h_{bottom} = (10 - 1) \times 2 + 5 + 6 = 29 \text{ ft}$$

Note that N is the number of trays excluding the partial reboiler.

Problem 3

A conventional distillation column is being used to separate an equimolar mixture of n-hexane and n-heptane. The feed is saturated liquid of viscosity 0.21 cP at column conditions. The relative volatility is approximately 2.30. Estimate the overall efficiency of the distillation column using the O'Connell correlation. What is the value of overall efficiency if Equation (10.29) is used instead? If the theoretical number of stages is 13 (excluding partial reboiler), how many stages are required in the actual column?

Solution

Using the O'Connell correlation, Equation (10.28), gives:

$$\eta_0 = 0.52782 - 0.27511\, log_{10}(2.30 \times 0.21) + 0.044923\, [log_{10}(2.30 \times 0.21)]^2 = 0.619$$

Using Equation (10.29), we get:

$$\eta_0 = 0.492\, (\alpha\mu_L)^{-0.245} = 0.492(2.30 \times 0.21)^{-0.245} = 0.588$$

The overall column efficiency calculated from a simple expression given by Equation (10.29) is close to the value obtained from the O'Connell correlation.

The actual number of stages required is:

$$N_{actual} = \frac{N_{ideal}}{\eta_o} = \frac{13}{0.619} = 21$$

Problem 4

A sieve-plate distillation column is separating a hydrocarbon mixture feed. The mass flow rate of vapor at the bottom of the column is 19,000 kg/h. The liquid mass flow rate is 17,500 kg/h. The densities of vapor and liquid are 3.5 kg/m^3 and 700 kg/m^3, respectively. Assuming that the column operates at 80% of flooding velocity, estimate the column diameter. The liquid surface tension is 20 dynes/cm. The tray spacing is 24 inches.

Solution

Given: $\dot{m}_V = 19{,}000\ kg/h$, $\dot{m}_L = 17{,}500\ kg/h$, $\rho_V = 3.5\ kg/m^3$, $\rho_L = 700\ kg/m^3$, $\gamma = 20\ dynes/cm$, $h_p = 24\ inches$, $U = 0.80(U_f)$

The flow parameter F_{LV} is:

$$F_{LV} = \left(\frac{\dot{m}_L}{\dot{m}_V}\right)\sqrt{\frac{\rho_V}{\rho_L}} = \frac{17500}{19000}\sqrt{\frac{3.5}{700}} = 0.065$$

Using Equation (10.45), we can calculate C

$$log_{10}C = -0.94506 - 0.70234[log_{10}(0.065)] - 0.22618[log_{10}(0.065)]^2$$

This equation gives a C value of 0.371 ft/s. The flooding velocity is calculated from Equation (10.44):

$$U_f = 0.371\left(\frac{20}{20}\right)^{0.2}\sqrt{\left(\frac{700-3.5}{3.5}\right)} = 5.23\ ft/s$$

Thus, the vapor velocity in the column is 0.80(5.23) = 4.18 ft/s=1.28 m/s.

The column diameter can be estimated from Equation (10.49):

$$D = 2\sqrt{\frac{\dot{m}_V}{0.80\pi U_f\left(1-\frac{A_d}{A}\right)\rho_V}} = 2\sqrt{\frac{19000/3600}{\pi(1.28)(1-0.1)(3.5)}} = 1.291\ m$$

Note that the area ratio A_d/A is 0.1 as $F_{LV} \leq 0.1$.

Problem 5

A thin cylindrical seamless vessel of inside diameter 1 m is subjected to an internal pressure of 1.75 N/mm^2. If the thickness of the vessel is 25 mm, what are the hoop and longitudinal stresses?

Solution

Using Equation (10.56), the hoop stress can be calculated as:

$$\sigma_h = P_iD_i/2t_s = 1.75 \times 1000/(2 \times 25) = 35\ N/mm^2$$

Using Equation (10.51), the longitudinal stress can be calculated as:

$$\sigma_p = P_i D_i / 4t_s = 1.75 \times 1000/(4 \times 25) = 17.5\ N/mm^2$$

Problem 6

A cylindrical vessel of inside diameter 1.5 m is subjected to an internal pressure of 2.0 N/mm^2. If the maximum stress is not to exceed 100 MPa, what should be the thickness of the vessel? Assume that the cylinder is seam welded with joint efficiency of 75%.

Solution

Given: $D_i = 1500\ mm, P_i = 2\ N/mm^2, \sigma_{design} = 100\ MPa, J = 0.75$

When the hoop stress does not exceed 100 MPa, Equation (10.58) gives

$$t_s = \frac{P_i D_i}{2J\sigma_{design}} + c = \frac{2\times1500}{2\times0.75\times100} + 0 = 20\ mm$$

When the longitudinal stress does not exceed 100 MPa, Equation (10.55) gives

$$t_s = \frac{P_i D_i}{4J\sigma_{design}} + c = \frac{2 \times 1500}{4 \times 0.75 \times 100} + 0 = 10\ mm$$

Thus, the minimum wall thickness should be 20 mm.

SHORT REVIEW QUESTIONS

1. For the same separation, the number of theoretical trays required decreases with the increase in relative volatility. Is this true or false?
2. Fenske equation assumes that the relative volatility is constant in the column. Is this true or false?
3. Describe the difference between feed pinch point and tangent pinch point.
4. Describe the FUG method of designing the distillation columns
5. What are the assumptions made in FUG method of designing distillation columns?
6. Define Murphree tray efficiency and overall column efficiency.
7. Why does the overall column efficiency decrease with the increase in liquid viscosity?
8. Why does the overall column efficiency decrease with the increase in relative volatility?
9. Describe three types of trays most commonly used in distillation columns for contacting vapor and liquid phases.
10. Describe weeping phenomenon in distillation columns. How is the efficiency of column affected by weeping?
11. Describe flooding in distillation columns.
12. At what vapor velocities (expressed in terms of flooding velocity) should one operate a distillation column?
13. What are the factors on which the flooding velocity depends?
14. The flooding velocity increases with the increase in tray spacing for the same ratio of liquid to vapor mass flow rates. Is this true or false?
15. The column diameter is selected such that the vapor velocity is well below the flooding velocity. Is this true or false?
16. Give examples of materials used in the construction of distillation columns.
17. What are the factors to be considered in specifying the shell thickness of a distillation column?
18. What are the types of tensile stresses generated when a cylindrical vessel is subjected to an internal pressure?
19. The wind load acting on a distillation column generates a bending moment on the column. Is this true or false?

20. Which type of loads generate compressive stresses in distillation columns?

PRACTICE PROBLEMS

Problem 1

Feed entering the distillation column is saturated liquid at column conditions. It consists of a binary mixture of pentane and hexane. The mole fraction of pentane in the feed is 0.57. The distillate product contains 95 mol% pentane and the bottoms product contains 5.0 mol% pentane. The reflux ratio is 3.0. The column operates at 1 atm. Do the following:

a) Determine the optimum number of equilibrium stages using the McCabe-Thiele method
b) Determine the optimum feed tray location graphically
c) Estimate the number of equilibrium stages using the FUG method. The average relative volatility of pentane with respect to hexane is approximately 2.95.
d) Determine the optimum location of the feed tray using the Kirkbride equation.

The equilibrium data of pentane-hexane system at 1 atm is given in the table below.

T (K)	x	y
342.06	0	0
339.40	0.05	0.12705
336.91	0.10	0.23699
334.58	0.15	0.33263
332.39	0.20	0.41626
330.32	0.25	0.48975
328.38	0.30	0.55462
326.53	0.35	0.61214
324.79	0.40	0.66335
323.14	0.45	0.70911
321.56	0.50	0.75016
320.07	0.55	0.78711
318.64	0.60	0.82048
317.28	0.65	0.85048
315.97	0.70	0.87816
314.72	0.75	0.90317
313.53	0.80	0.92601
312.38	0.85	0.94692
311.28	0.90	0.96610
310.22	0.95	0.98374
309.20	1.00	1.00000

Problem 2

An equimolar binary mixture of hydrocarbons is being separated in a conventional distillation column. The feed is saturated liquid of viscosity 0.30 cP at column conditions. The relative volatility is approximately 2.75. Estimate the overall efficiency of the column using the O'Connell correlation. If the theoretical number of stages (excluding partial reboiler) is 10, how many stages are required in the actual column?

Problem 3

A thin cylindrical vessel of inside diameter 1.15 m is subjected to an internal pressure of 2.1 N/mm^2. Estimate the minimum wall thickness if the longitudinal stress does not exceed 30 MPa. What is the minimum wall thickness if the hoop stress does not exceed 45 MPa?

Problem 4

A seam welded cylindrical vessel of inside diameter 1.75 m is subjected to an internal pressure of 2.5 N/mm^2. What should be the thickness of the vessel if the maximum stress not to exceed is 110 MPa? Assume that the welded joint efficiency is 80%.

Problem 5

A distillation column consisting of sieve plates is separating a hydrocarbon mixture feed. The mass flow rate of vapor at the bottom of the column is 45 lb_m/s. The ratio of liquid mass flow rate to vapor mass flow rate is 0.85. The densities of vapor and liquid are 0.2 lb_m/ft^3 and 40 lb_m/ft^3, respectively. Assuming that the column operates at 80% of the flooding velocity, estimate the column diameter. The liquid surface tension is 15 dynes/cm and the tray spacing is 24 inches.

Chapter 11

Ponchon-Savarit Graphical Method of Designing Distillation Columns

11.1 LEARNING OUTCOMES

Upon completion of this chapter, students will be able to:

- Apply the Ponchon-Savarit graphical method in the analysis and design of distillation columns for binary mixtures
- Estimate the enthalpies of binary liquid solutions at bubble point temperature
- Estimate the enthalpies of binary vapor mixtures at dew point temperature
- Construct enthalpy-composition phase diagram for binary mixtures
- Do material and energy balances over different sections of the column and develop the equations for the operating lines of different sections of the column
- Draw the operating lines on the enthalpy-composition phase diagram
- Construct stages on the enthalpy-composition diagram and estimate the total number of stages required for a given separation
- Estimate the minimum number of stages at total reflux using the enthalpy-composition diagram
- Estimate the minimum reflux ratio using the enthalpy-composition diagram

11.2 INTRODUCTION

In the preceding chapters on multistage column distillation, we have generally assumed CMO (constant molal overflow) meaning that in a given section (enriching or stripping) of the distillation column the vapor and liquid molar flow rates do not vary from one stage to another. CMO is valid only if the molar latent heats of vaporization of the species are equal so that for every mole of liquid that vaporizes exactly 1 mole of vapor condenses. For a number of binary mixtures, the assumption of CMO is appropriate as the molar latent heats of vaporization are close to each other. For example, consider a binary system of benzene and toluene. The latent heat of vaporization of benzene is 7.36 kcal/mol and that of toluene is 7.96 kcal/mol at normal boiling points. For ethanol-water system, the latent heat of vaporization of ethanol is 9.22 kcal/mol and that of water is 9.717 kcal/mol at normal boiling points. However, in some cases the latent heats of vaporization of the two species are significantly different. For example, consider a binary heptane-ethylbenzene system. The latent heat of vaporization of heptane is 7.57 kcal/mol whereas the latent heat of vaporization of ethylbenzene is 8.6 kcal/mol at normal boiling points.

In this chapter, we will relax the assumption of CMO, that is, the vapor and liquid flow rates are allowed to vary from one stage to another in a given section of the column. However, the graphical approach of McCabe-Thiele to determine the number of stages in a distillation column is no longer valid as it is based on the assumption of CMO.

When CMO is not a valid assumption, one can use the so-called *Ponchon-Savarit* graphical method (M. Ponchon, Tech. Moderne, Vol. 13, 20-55, 1921; R. Savarit, Arts et Metiers, pp. 65, 142, 178, 241, 266, 307, 1922) to design a distillation column. The Ponchon-Savarit method involves both material and energy (enthalpy) balances. Like the McCabe-Thiele graphical method, the Ponchon-Savarit method is restricted to binary distillation. We need the enthalpy-composition phase diagram for the binary mixture under consideration in order to apply the Ponchon-Savarit method. Thus we will first discuss enthalpy-composition phase diagram for binary mixtures and how to construct such a diagram before going to the discussion of Ponchon-Savarit method of designing distillation columns.

11.3 ENTHALPY-COMPOSITION DIAGRAM FOR BINARY MIXTURES

Figure 11.1 shows a typical enthalpy-composition phase diagram for binary mixtures at a fixed pressure (column pressure). It consists of plots of h vs x and H vs y, where h is the enthalpy of saturated liquid and H is the enthalpy of saturated vapor. The saturated liquid enthalpy curve (h vs x plot) is referred to as bubble point curve and the saturated vapor enthalpy curve (H vs y plot) is referred to as dew point curve. The region bounded by h vs x and H vs y curves is the two-phase liquid-vapor region. Above the saturated vapor enthalpy curve (H vs y), we have a superheated vapor region and below the saturated liquid enthalpy curve (h vs x), we have a subcooled liquid region.

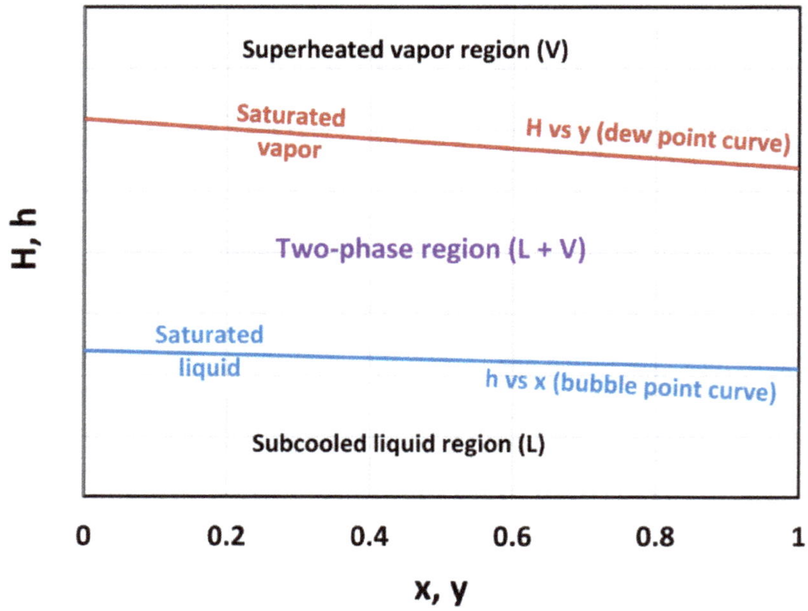

Figure 11.1 Enthalpy-composition phase diagram for a binary mixture.

Figure 11.2 shows the additional features of enthalpy-composition diagram. The vertical line between the dew point curve (H vs y) and the bubble point curve (h vs x) at any given composition represents the latent heat of vaporization of a binary mixture at that composition:

$$\lambda_{mix} = H - h \tag{11.1}$$

When $x = y = 0$, the vertical distance between dew point and bubble point curves gives the latent heat of vaporization of pure species B, that is, λ_B. When $x = y = 1$, the vertical distance between dew point and bubble point curves gives the latent heat of vaporization of species A, that is, λ_A. Figure 11.2 also shows a number of *tie-lines*. The tie-line connects vapor and liquid phases in equilibrium with each other. The vapor phase with mole fraction y of a more volatile component is in equilibrium with liquid phase with mole fraction x of a more volatile component. Note that the two phases at equilibrium have the same temperature meaning the tie-lines are also constant-temperature lines (isotherms). At $x = y = 0$, we have pure species B with a vertical tie-line corresponding to $T = T_B^{Sat}$. At $x = y = 1$, we have pure species A with a vertical tie-line corresponding to $T = T_A^{Sat}$. As $T_B^{Sat} > T_A^{Sat}$ (species A being more volatile with lower boiling point), the temperature of tie-lines decreases as we go from left ($x = y = 0$) to right ($x = y = 1$) of the figure.

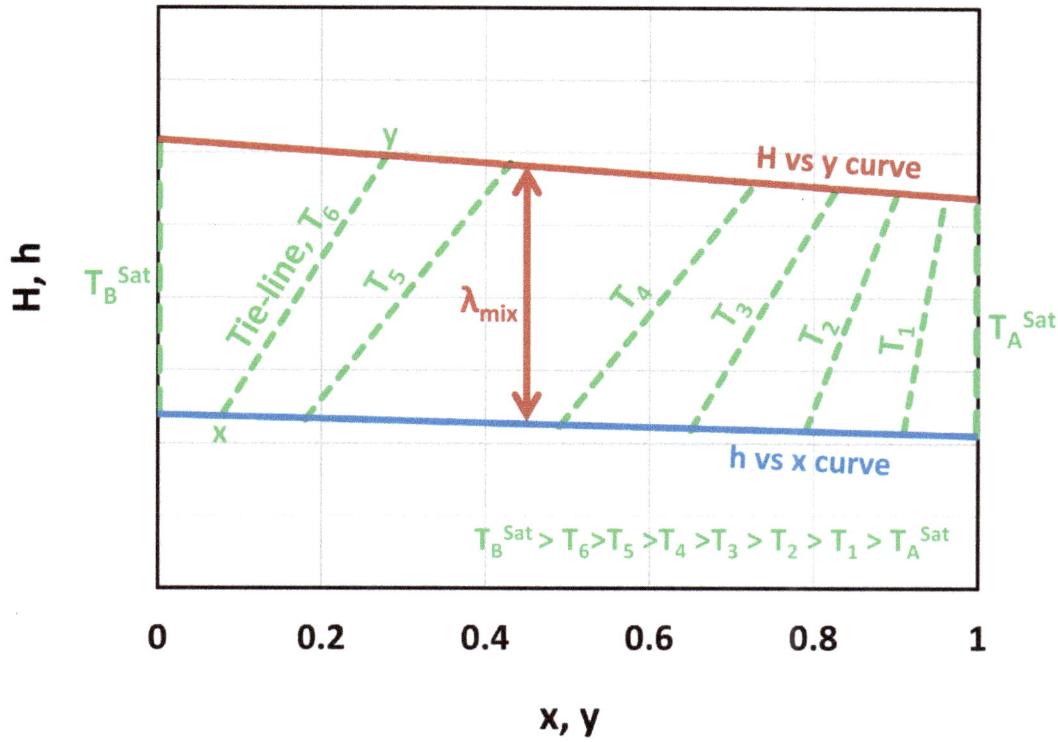

Figure 11.2 Tie-lines on enthalpy-composition phase diagram.

The tie-lines can be drawn readily from the $y - x$ equilibrium curve available for the system at the same pressure. Figure 11.3 demonstrates the connection between the $y - x$ equilibrium curve and the enthalpy-composition diagram ($H - x - y$ diagram). A point on the $y - x$ equilibrium curve is equivalent to a tie-line on the $H - x - y$ diagram.

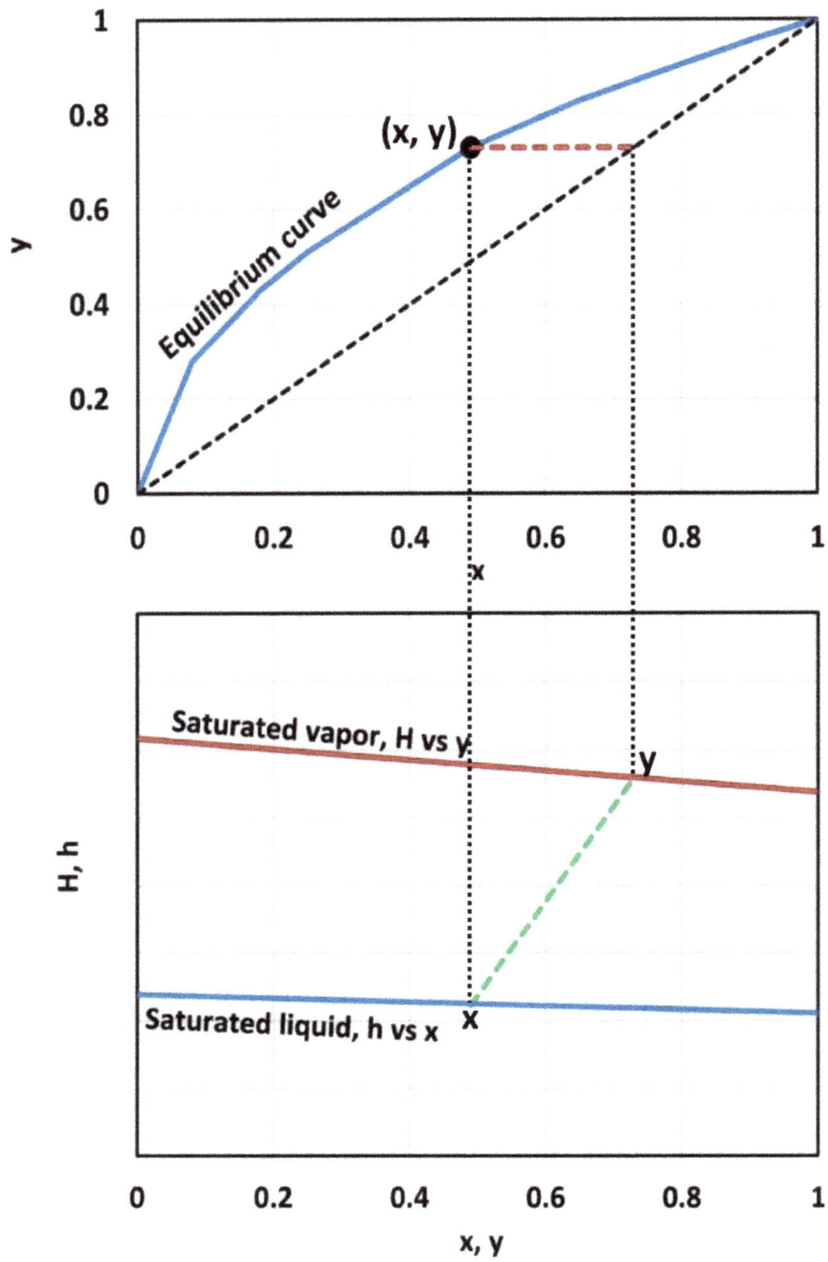

Figure 11.3 Connection between $y - x$ equilibrium curve and enthalpy-composition diagram.

11.4 CONSTRUCTION OF ENTHAPLY-COMPOSITION DIAGRAM FOR BINARY MIXTURES

In order to construct the enthalpy-composition diagram, we need the following data: latent heats of vaporization of species, λ_A and λ_B at boiling points (column pressure); heat capacities of species in liquid state, $C_{pL,A}$ and $C_{pL,B}$; and heat capacities of species in vapor state, $C_{pV,A}$ and $C_{pV,B}$. We will assume that the binary vapor mixtures and liquid solutions are ideal in nature. The vapor-liquid equilibrium data (y vs x data) are either given or could be estimated. We also assume that the dew point and bubble point temperatures of mixtures at column pressure can be estimated.

We have to first define the reference states for the species involved before we can calculate enthalpies. As discussed in Chapter 8, the reference temperature T_R is often taken as equal to the boiling point of the more volatile species (lower boiling point component), that is:

$$T_R = T_A^{Sat} \qquad \text{at } P = P_{column} \tag{11.2}$$

Thus the reference states for the two species (more volatile A and less volatile B) are:

Species A: Saturated liquid; $T_R = T_A^{Sat}$; $P = P_{column}$; $h_R = 0$ (11.3)

Species B: Subcooled liquid; $T_R = T_A^{Sat}$; $P = P_{column}$; $h_R = 0$ (11.4)

Once the reference states are defined, the enthalpies of saturated liquid solutions at bubble point temperature (h vs x) and saturated vapor mixtures at dew point temperature (H vs y) can be estimated.

11.4.1 Enthalpy of Liquid Solutions at Bubble Point Temperature

The enthalpy of pure liquid A at bubble point temperature T_{bubble} can be calculated as:

$$h_A = C_{pL,A}\left(T_{bubble} - T_A^{Sat}\right) \tag{11.5}$$

The enthalpy of pure liquid B at bubble point temperature T_{bubble} can be calculated as:

$$h_B = C_{pL,B}\left(T_{bubble} - T_A^{Sat}\right) \tag{11.6}$$

Assuming liquid solution to be an ideal solution, the enthalpy of solution h can be calculated as:

$$h = xh_A + (1 - x)h_B \tag{11.7}$$

where x is the mole fraction of more volatile species A in liquid solution. One can generate the bubble point curve of enthalpy-composition diagram by calculating h for different values of x.

11.4.2 Enthalpy of Vapor Mixtures at Dew Point Temperature

The enthalpy of pure vapor A at dew point temperature T_{dew} can be calculated as:

$$H_A = \lambda_A + C_{pV,A}\left(T_{dew} - T_A^{Sat}\right) \tag{11.8}$$

The enthalpy of pure vapor B at dew point temperature T_{dew} can be calculated as:

$$H_B = C_{pL,B}\left(T_B^{Sat} - T_A^{Sat}\right) + \lambda_B + C_{pV,B}\left(T_{dew} - T_B^{Sat}\right) \tag{11.9}$$

Assuming vapor mixture to be an ideal gas mixture, the enthalpy of vapor mixture H can be calculated as:

$$H = yH_A + (1 - y)H_B \tag{11.10}$$

where y is the mole fraction of more volatile species A in vapor mixture. One can generate the dew point curve of enthalpy-composition diagram by calculating H for different values of y.

Note that the calculations become quite complex if the liquid solution and vapor mixture cannot be treated as ideal.

11.4.3 Example: Construction of Enthalpy-Composition Diagram for Benzene-Toluene System

Consider a binary system of benzene and toluene. We would like to develop enthalpy-composition phase diagram for this binary system at atmospheric pressure (101.325 kPa). Liquid solutions and vapor mixtures of benzene and toluene can be treated as ideal. The relevant physical property data for this system are given in Table 11.1. The Antoine equation can be used to estimate the vapor pressure of a pure species at any temperature:

$$log_{10}P^{Sat} = A - \frac{B}{t+C} \quad (11.11)$$

where constants A, B, C are given in Table 11.2, t is in °C, and P^{Sat} is vapor pressure in bars.

Table 11.1: Physical property data for benzene-toluene system

Species	Normal boiling point (°C)	Latent heat of vaporization at normal boiling point (kJ/kmol)	Heat capacity of liquid phase (kJ/kmol.K)	Heat capacity of vapor phase (kJ/kmol.K)
Benzene (species A)	80.1	30770	138.2	96.3
Toluene (species B)	110.6	33180	167.5	138.2

Table 11.2: Antoine constants for benzene and toluene

Species	A	B	C
Benzene (species A)	4.02232	1206.53	220.291
Toluene (species B)	4.0854	1348.77	219.976

As a first step, we select a number of temperatures within the range $T_A^{Sat} < T < T_B^{Sat}$. The vapor pressures of pure benzene and toluene are calculated at the selected temperatures using the Antoine equations. The equilibrium compositions of liquid solutions and vapor mixtures are then calculated using the following relations based on Raoult's law:

$$x_A = \frac{P - P_B^{Sat}}{P_A^{Sat} - P_B^{Sat}} \quad (11.12)$$

$$y_A = \frac{x_A P_A^{Sat}}{P} \quad (11.13)$$

where P is total pressure (column pressure), P_A^{Sat} is vapor pressure of species A (benzene) and P_B^{Sat} is vapor pressure of species B (toluene).

Table 11.3 shows the calculated data. For the sake of convenience, the subscripts on the composition variables have been dropped keeping in mind that x, y refer to mole fractions of more volatile species (species A, benzene). Figure 11.4 shows the T-x-y diagram generated using the data of Table 11.3. The equilibrium y vs x curve is shown in Figure 11.5.

Table 11.3 Calculated equilibrium data for benzene (species A)/toluene (species B) system

T (°C)	P_A^{Sat}(bar)	P_B^{Sat}(bar)	x	y
80.1	1.013417	0.389618	0.999732	0.999897
83.1	1.110464	0.431648	0.856788	0.938991
86.1	1.214628	0.477254	0.726899	0.871366
89.1	1.326254	0.52665	0.608551	0.79654
92.1	1.445696	0.580059	0.500431	0.71401
95.1	1.573311	0.63771	0.401389	0.623252
98.1	1.709464	0.699839	0.310423	0.523718
101.1	1.854523	0.766688	0.226654	0.414838
104.1	2.008864	0.838505	0.149309	0.296018
107.1	2.172863	0.915545	0.077709	0.166644
110.1	2.346904	0.998066	0.011257	0.026074
110.6	2.376914	1.012371	0.000644	0.001511

In next step, the enthalpies of liquid solutions of benzene and toluene are calculated to generate h vs x curve. For any given x value in Table 11.3, the corresponding temperature given in the table on the same row is the bubble point temperature. Using x vs T_{bubble} data, the enthalpies of the liquid solutions are calculated from Equations (11.5) to (11.7). The calculated data are given in Table 11.4. The enthalpies of vapor mixtures of benzene and toluene are calculated next. For any given y value in Table 11.3, the corresponding temperature given in the table on the same row is the dew point temperature. Using y vs T_{dew} data, the enthalpies of the vapor mixtures are calculated from Equations (11.8) to (11.10). The calculated data are given in Table 11.5.

Table 11.4 Enthalpies of liquid solutions of benzene and toluene at bubble point temperature

x	T_{bubble} (°C)	h_A (kJ/kmol)	h_B (kJ/kmol)	h (kJ/kmol)
0.999732	80.1	0	0	0
0.856788	83.1	414.6	502.5	427.1883
0.726899	86.1	829.2	1005	877.2112
0.608551	89.1	1243.8	1507.5	1347.025
0.500431	92.1	1658.4	2010	1834.049
0.401389	95.1	2073	2512.5	2336.089
0.310423	98.1	2487.6	3015	2851.283
0.226654	101.1	2902.2	3517.5	3378.04
0.149309	104.1	3316.8	4020	3915.006
0.077709	107.1	3731.4	4522.5	4461.024
0.011257	110.1	4146	5025	5015.105
0.000644	110.6	4215.1	5108.75	5108.174

Table 11.5 Enthalpies of vapor mixtures of benzene and toluene at dew point temperature

y	T_{dew} (°C)	H_A (kJ/kmol)	H_B (kJ/kmol)	H (kJ/kmol)
0.999897	80.1	30770	34073.65	30770.34
0.938991	83.1	31058.9	34488.25	31268.12
0.871366	86.1	31347.8	34902.85	31805.1
0.79654	89.1	31636.7	35317.45	32385.59
0.71401	92.1	31925.6	35732.05	33014.21
0.623252	95.1	32214.5	36146.65	33695.93
0.523718	98.1	32503.4	36561.25	34436.08

0.414838	101.1	32792.3	36975.85	35240.36
0.296018	104.1	33081.2	37390.45	36114.83
0.166644	107.1	33370.1	37805.05	37065.99
0.026074	110.1	33659	38219.65	38100.74
0.001511	110.6	33707.15	38288.75	38281.83

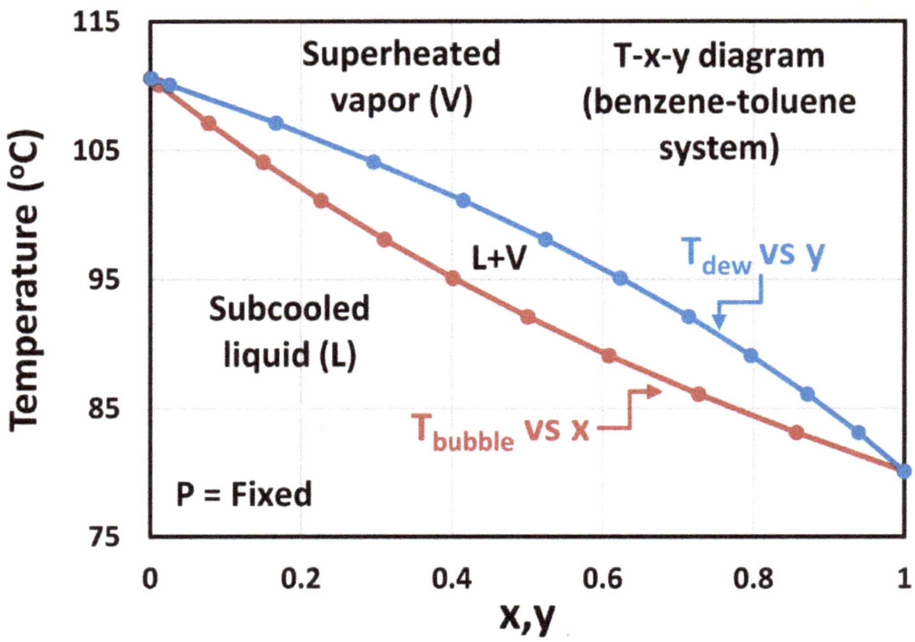

Figure 11.4 T-x-y diagram of benzene-toluene system at atmospheric pressure (101.325 kPa).

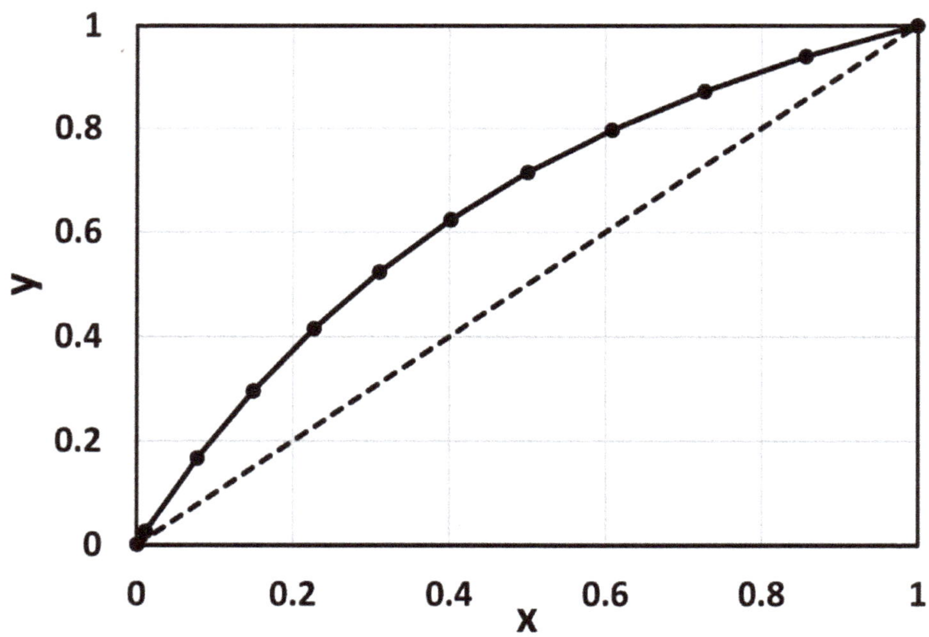

Figure 11.5 The equilibrium y *vs* x curve of benzene-toluene system at atmospheric pressure.

Figure 11.6 shows the enthalpy-composition diagram for benzene-toluene system along with the tie lines. The data given in Tables 11.4 and 11.5 are used to generate this enthalpy-composition diagram. The tie-lines are drawn using the equilibrium y vs x data given in Table 11.3.

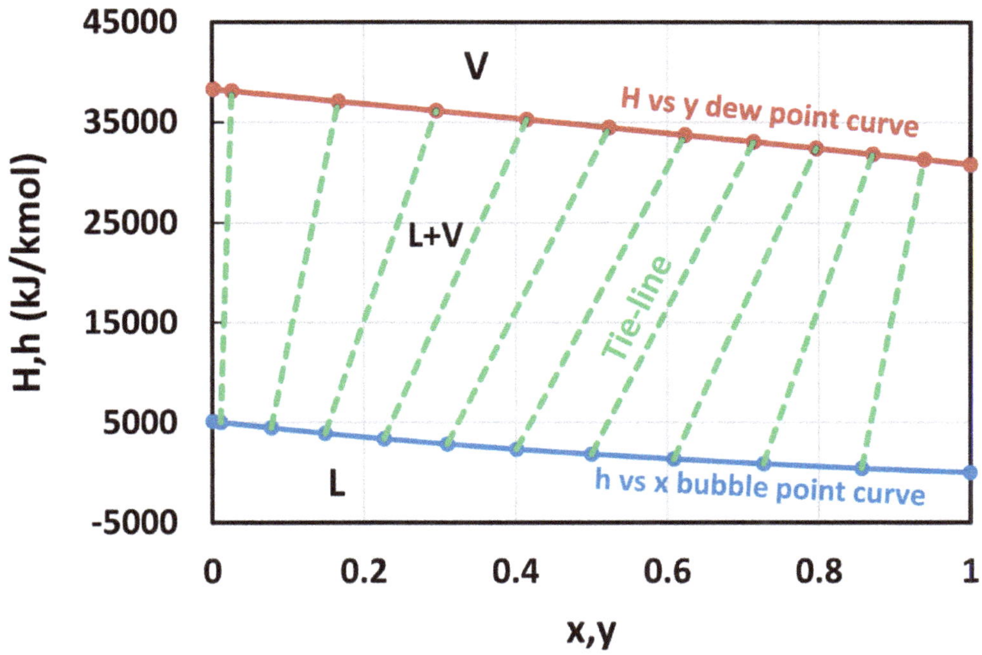

Figure 11.6 Enthalpy-composition diagram of benzene-toluene system at atmospheric pressure.

11.5 PONCHON-SAVARIT GRAPHICAL METHOD OF ANALYSIS AND DESIGN OF DISTILLATION COLUMNS

When CMO was assumed as in the McCabe-Thiele graphical method discussed in the preceding chapters, only material balances were needed to develop the equations for operating lines of different sections of the column. In order to relax the assumption of CMO as done in the Ponchon-Savarit method, both material balances and energy balances are needed to develop equations for the operating lines of different sections of the column.

11.5.1 Enriching Section

Consider a material and energy balance envelope covering the top portion of the distillation column including n stages of the enriching section, as shown in Figure 11.7. The overall mass balance over the envelope gives:

$$V_{n+1} = L_n + D \tag{11.14}$$

The net flow rate of material (including both components) in the upward direction is D $(= V_{n+1} - L_n)$. Although the flow rates of vapor and liquid streams may change from one stage to another within the enriching section, their difference is constant at steady state and is equal to D.

The component mass balance on more volatile species gives:

$$y_{n+1}V_{n+1} = x_nL_n + x_DD \tag{11.15}$$

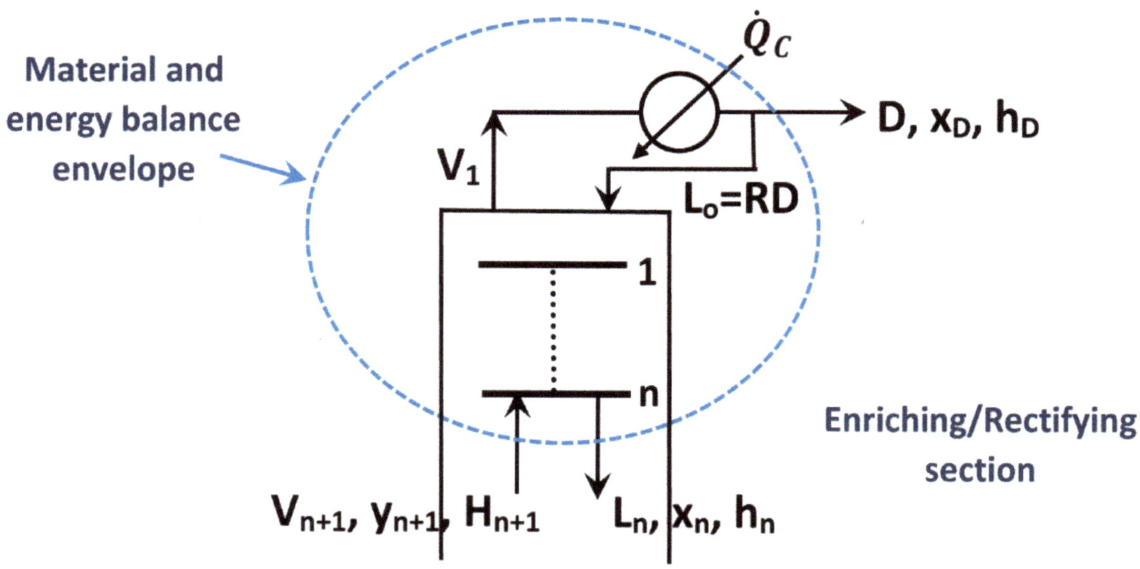

Figure 11.7 Material and energy balance over the top portion of the distillation column including first *n* stages of the enriching section.

The net flow rate of more volatile component in the upward direction is constant at steady state and is equal to $x_D D$ $(= y_{n+1}V_{n+1} - x_n L_n)$.

From Equations (11.14) and (11.15), it follows that:

$$\frac{L_n}{V_{n+1}} = \frac{x_D - y_{n+1}}{x_D - x_n} \tag{11.16}$$

Energy balance over the envelope shown in Figure 11.7 gives:

$$V_{n+1}H_{n+1} + \dot{Q}_C = L_n h_n + D h_D \tag{11.17}$$

Upon re-arrangement, Equation (11.17) could be written as:

$$\frac{V_{n+1}H_{n+1} - L_n h_n}{D} = h_D - \frac{\dot{Q}_C}{D} \tag{11.18}$$

From overall mass balance, Equation (11.14), and Equation (11.18):

$$\frac{V_{n+1}H_{n+1} - L_n h_n}{V_{n+1} - L_n} = h_D - \frac{\dot{Q}_C}{D} \tag{11.19}$$

We now define the so-called *upper difference point* (h_{diff}) as follows:

$$h_{diff} = \frac{difference\ in\ total\ enthalpies\ of\ passing\ streams}{difference\ in\ total\ flow\ rates\ of\ passing\ streams} = \frac{V_{n+1}H_{n+1} - L_n h_n}{V_{n+1} - L_n} \tag{11.20}$$

From Equations (11.19) and (11.20), it follows that:

$$h_{diff} = h_D - \frac{\dot{Q}_C}{D} \tag{11.21}$$

The energy balance around the condenser (assumed to be total condenser) gives:

$$-\frac{\dot{Q}_C}{D} = (1 + R)\lambda_D \tag{11.22}$$

where λ_D is the molar latent heat of vaporization of distillate product.

The upper difference point along with its relation to other enthalpies and condenser heat duty is shown in Figure 11.8 (see envelope marked in green dashed oval).

Equation (11.20) could be re-arranged as:

$$\frac{L_n}{V_{n+1}} = \frac{h_{diff} - H_{n+1}}{h_{diff} - h_n} \tag{11.23}$$

Combining Equations (11.16) and (11.23) gives:

$$\frac{L_n}{V_{n+1}} = \frac{x_D - y_{n+1}}{x_D - x_n} = \frac{h_{diff} - H_{n+1}}{h_{diff} - h_n} \tag{11.24}$$

This equation represents the *operating line* for the enriching section on the enthalpy-composition diagram. According to Equation (11.24), the enriching section operating line on the enthalpy-composition diagram passes through three points: $(x_D, h_{diff}), (y_{n+1}, H_{n+1})$ and (x_n, h_n). The operating line is shown schematically in Figure 11.9. An important point to note is that we now have a family of enriching section operating lines, all radiating from the upper difference point P' as shown in Figure 11.10. Each operating line connects the compositions of *passing* vapor and liquid streams. Also for each stage, there are two operating lines: one connecting the compositions of the passing vapor and liquid streams at the top of the stage and the other one connecting the compositions of the passing vapor and liquid streams at the bottom of the stage. Note that in the McCabe-Thiele graphical method, we had just one operating line for the entire enriching section.

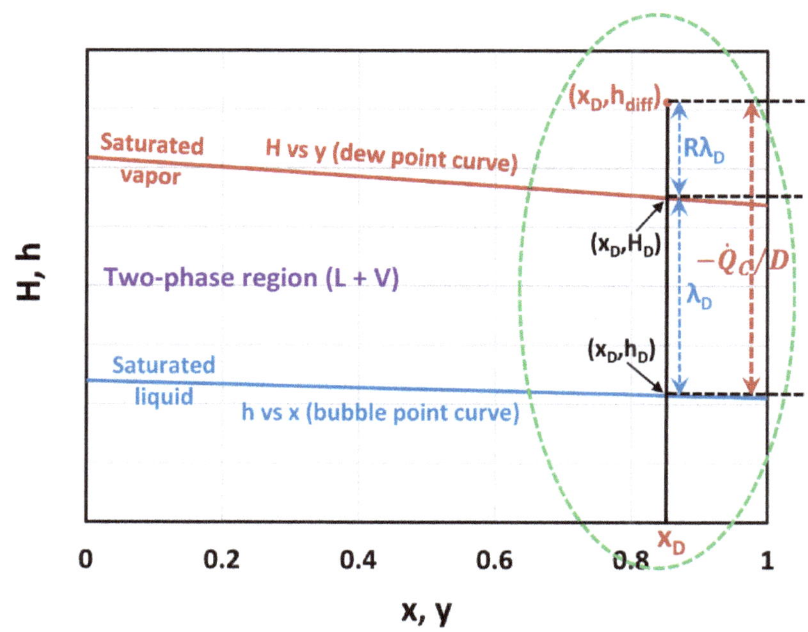

Figure 11.8 Upper difference point and its relation to other enthalpies and condenser heat duty indicated in green dashed oval on the right hand side of the figure.

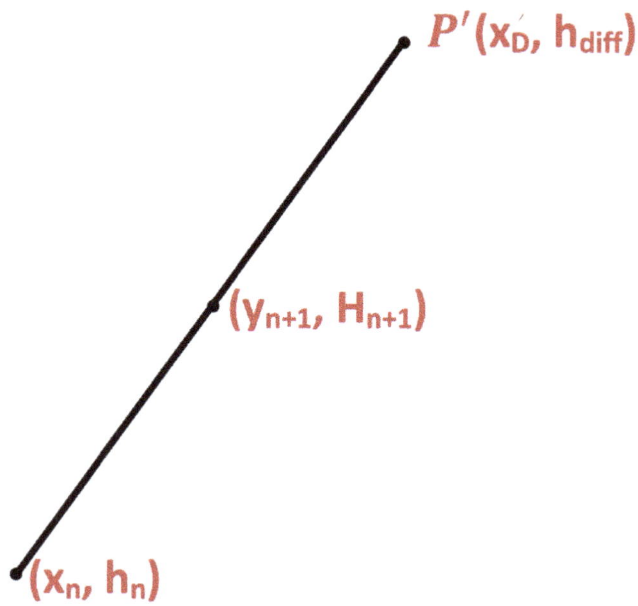

Figure 11.9 Enriching section operating line on enthalpy-composition diagram.

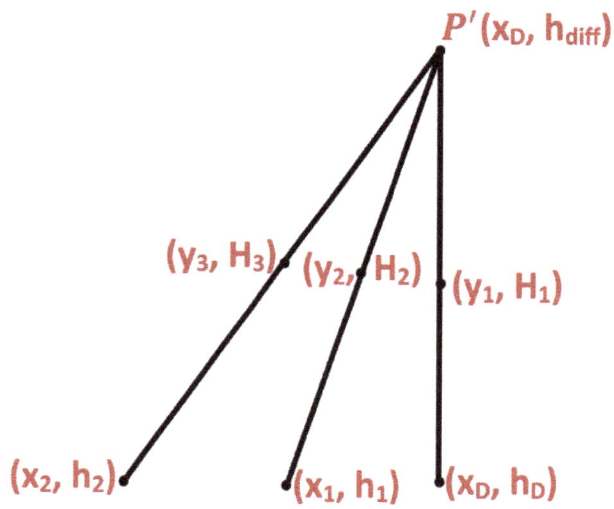

Figure 11.10 Family of enriching section operating lines on enthalpy-composition diagram.

Figure 11.11 illustrates the stepping off stages in the enriching section. Each stage is represented by two operating lines and a tie-line (green line in Figure 11.11). While the operating line connects the compositions of passing vapor and liquid streams at the same level, the tie-line connects the compositions of vapor and liquid streams leaving the stage at equilibrium with each other. The procedure for stepping off stages is as follows: Draw the operating line starting from the upper dfference point and ending on the saturated liquid curve, then draw the tie-line from (y,H) intersection point (intersection of operating line and saturated vapor curve) to the saturated liquid curve, and finally the second operating line connecting (x,h) intersection point (intersection of tie-line and saturated liquid curve) to the upper difference point. This completes the construction of a single stage. The process of construction of stages is continued by drawing the next tie-line from (y,H) intersection point on the saturated vapor curve to the saturated liquid curve.

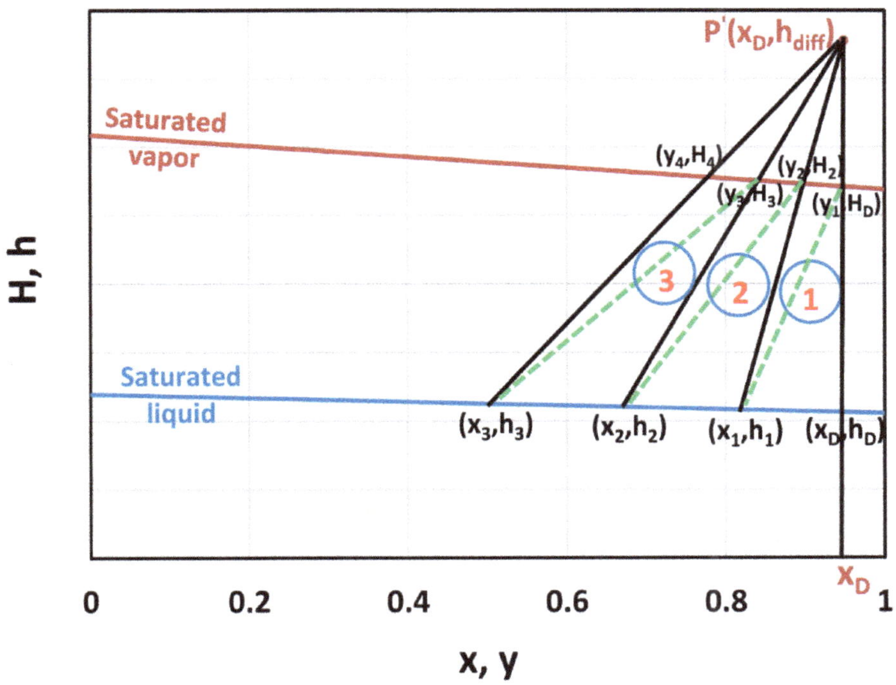

Figure 11.11 Stepping off stages in the enriching section using the enthalpy-composition diagram.

11.5.2 Stripping Section

Consider a material and energy balance envelope shown in Figure 11.12 covering the bottom portion of the distillation column including stages numbered (m+1) to last stage N of the stripping section where m is any m[th] stage in the stripping section. The stages in the column are numbered from top down. The overall mass balance over the envelope gives:

$$\bar{L}_m = \bar{V}_{m+1} + W \tag{11.25}$$

The net flow rate of material (including both components) in the downward direction is W ($= \bar{L}_m - \bar{V}_{m+1}$). Although the flow rates of liquid and vapor streams may change from one stage to another within the stripping section, their difference is constant at steady state and is equal to W.

The component mass balance on more volatile species gives:

$$x_m\bar{L}_m = y_{m+1}\bar{V}_{m+1} + x_W W \tag{11.26}$$

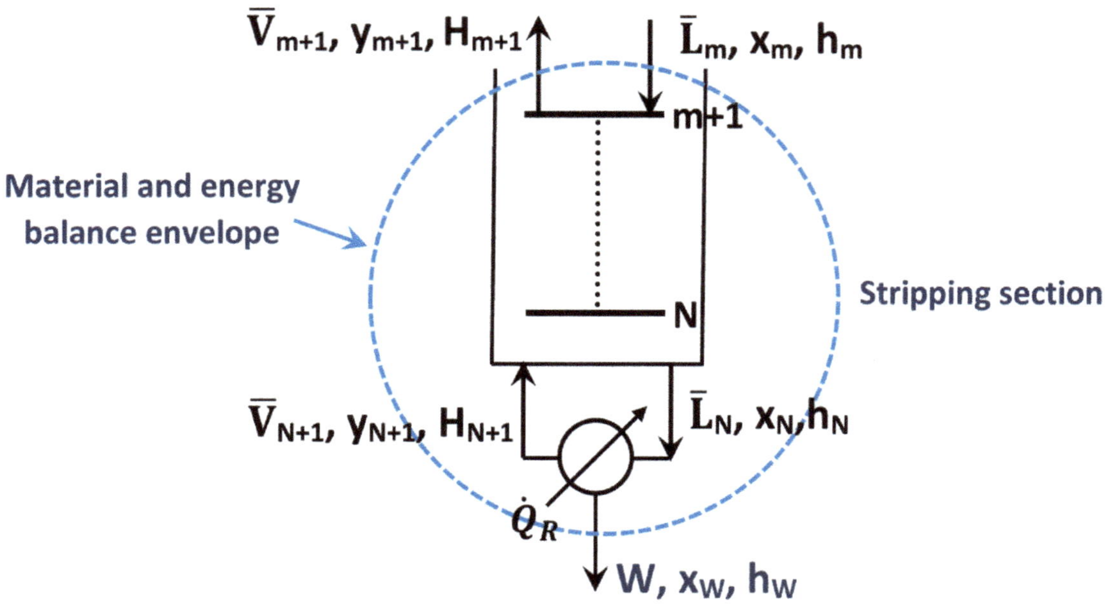

Figure 11.12 Material and energy balance over the bottom portion of the distillation column including stages numbered (m+1) to N of the stripping section.

The net flow rate of more volatile component in the downward direction is constant at steady state and is equal to $x_W W$ $(= x_m \bar{L}_m - y_{m+1}\bar{V}_{m+1})$.

From Equations (11.25) and (11.26), it follows that:

$$\frac{\bar{L}_m}{\bar{V}_{m+1}} = \frac{y_{m+1}-x_W}{x_m-x_W} \tag{11.27}$$

Energy balance over the envelope shown in Figure 11.12 gives:

$$\bar{L}_m h_m + \dot{Q}_R = \bar{V}_{m+1}H_{m+1} + W h_W \tag{11.28}$$

Upon re-arrangement, Equation (11.28) could be written as:

$$\frac{\bar{L}_m h_m - \bar{V}_{m+1}H_{m+1}}{W} = h_W - \frac{\dot{Q}_R}{W} \tag{11.29}$$

From overall mass balance, Equation (11.25), and Equation (11.29):

$$\frac{\bar{L}_m h_m - \bar{V}_{m+1}H_{m+1}}{\bar{L}_m - \bar{V}_{m+1}} = h_W - \frac{\dot{Q}_R}{W} \tag{11.30}$$

We now define the so-called *lower difference point* ($\bar{h}_{diff}$) as follows:

$$\bar{h}_{diff} = \frac{difference\ in\ total\ enthalpies\ of\ passing\ streams}{difference\ in\ total\ flow\ rates\ of\ passing\ streams} = \frac{\bar{L}_m h_m - \bar{V}_{m+1}H_{m+1}}{\bar{L}_m - \bar{V}_{m+1}} \tag{11.31}$$

Note that net flow is in the downward direction in the stripping section.

From Equations (11.30) and (11.31), it follows that:

$$\bar{h}_{diff} = h_W - \frac{\dot{Q}_R}{W} \quad (11.32)$$

This equation allows the calculation of the lower difference point enthalpy from the knowledge of enthalpy of the bottoms liquid product and the heat duty of the reboiler.

Equation (11.31) could be re-arranged as:

$$\frac{\bar{L}_m}{\bar{V}_{m+1}} = \frac{H_{m+1}-\bar{h}_{diff}}{h_m-\bar{h}_{diff}} \quad (11.33)$$

Combining Equations (11.27) and (11.33) gives:

$$\frac{\bar{L}_m}{\bar{V}_{m+1}} = \frac{y_{m+1}-x_W}{x_m-x_W} = \frac{H_{m+1}-\bar{h}_{diff}}{h_m-\bar{h}_{diff}} \quad (11.34)$$

This equation represents the *operating line* for the stripping section on the enthalpy-composition diagram. According to Equation (11.34), the stripping section operating line on the enthalpy-composition diagram passes through three points: $(x_W, \bar{h}_{diff})$, (x_m , h_m) and (y_{m+1}, H_{m+1}). The operating line is shown schematically in Figure 11.13.

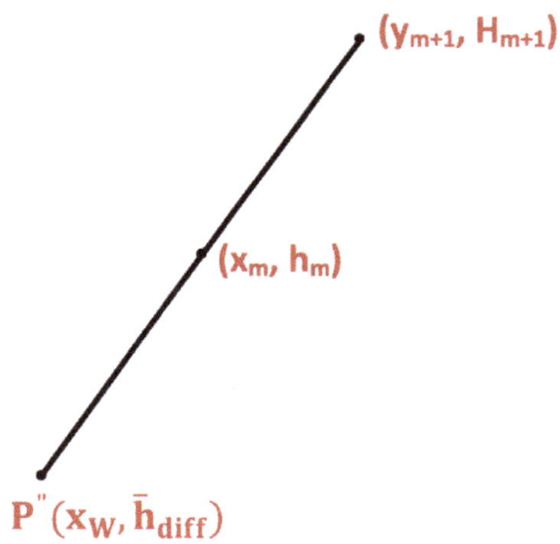

Figure 11.13 Stripping section operating line on enthalpy-composition diagram.

Like enriching section, we have a family of stripping section operating lines, all radiating from the lower difference point P'' as shown in Figure 11.14. Each operating line connects the compositions of *passing* vapor and liquid streams. For each stage, there are two operating lines: one connecting the compositions of the passing vapor and liquid streams at the top of the stage and the other one connecting the compositions of the passing vapor and liquid streams at the bottom of the stage.

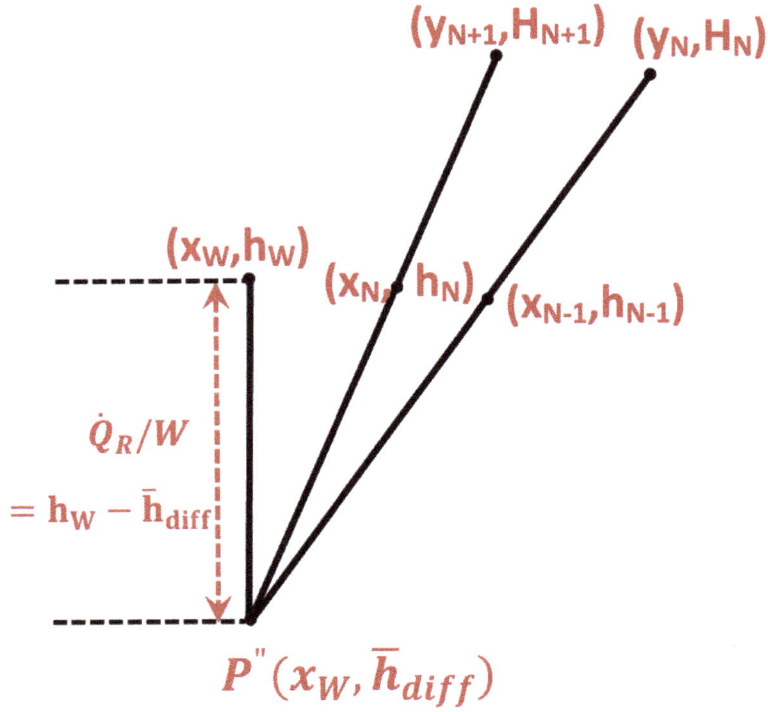

Figure 11.14 Family of stripping section operating lines on enthalpy-composition diagram.

Figure 11.15 illustrates the stepping off stages in the stripping section. The process is similar to enriching section. Like enriching section, each stage in stripping section (except partial reboiler stage marked as R) is represented by two operating lines and a tie-line (green line in Figure 11.15). We are assuming that the counting of stages is done from top to down of the distillation column with the last stage inside the column as Nth stage. After the Nth stage comes the partial reboiler stage which consists of one operating line and a tie-line. The operating line connects the compositions of the vapor boilup stream emerging from the reboiler and the liquid stream entering the reboiler. The tie-line connects the equilibrium compositions of boilup vapor stream and bottoms liquid product stream.

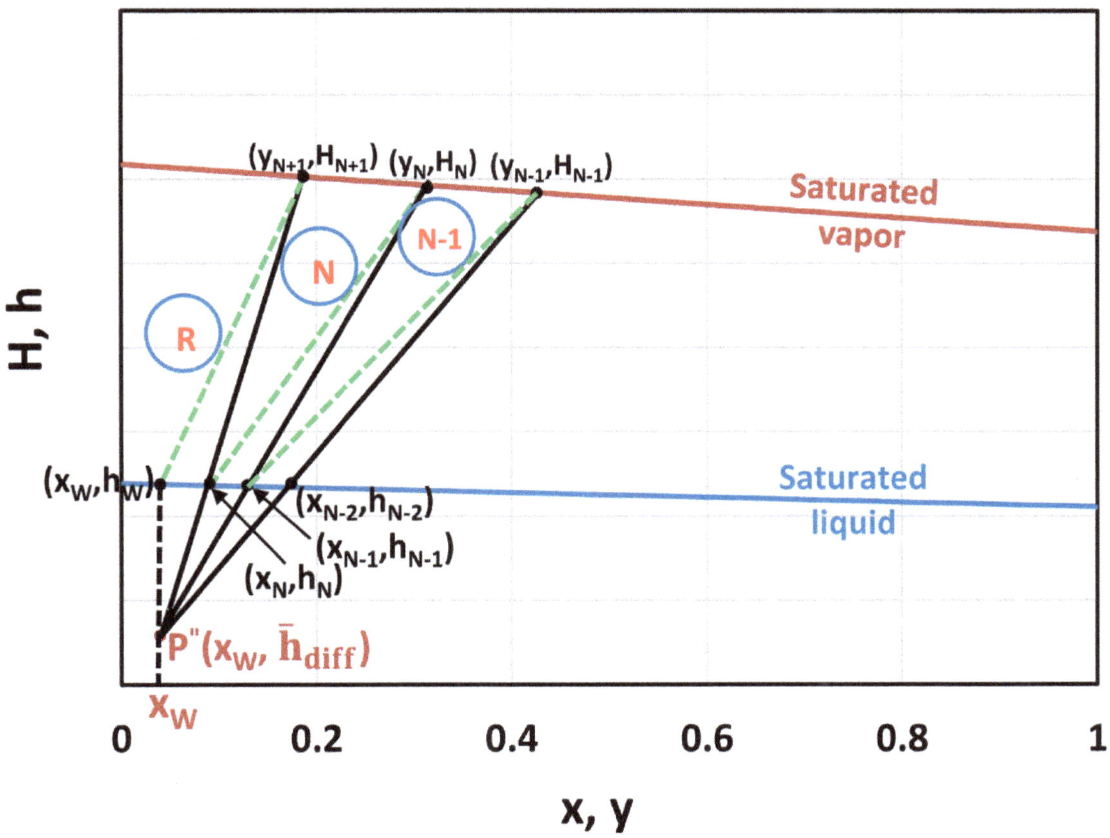

Figure 11.15 Stepping off stages in the stripping section using the enthalpy-composition diagram.

11.5.3 Feed Line

The feed line equation can be developed using overall and component mass balances and overall energy balance along with the definitions of upper and lower difference points.

The overall and component (more volatile species) material balances on the column give:

$$F = D + W \tag{11.35}$$

$$z_F F = x_D D + x_W W \tag{11.36}$$

From Equations (11.35) and (11.36), it can be readily shown that:

$$\frac{W}{D} = \frac{x_D - z_F}{z_F - x_W} \tag{11.37}$$

Overall energy balance on the column gives:

$$F h_F + \dot{Q}_R + \dot{Q}_C = W h_W + D h_D \tag{11.38}$$

Replacing $\dot{Q}_R$ and $\dot{Q}_C$ in Equation (11.38) with upper and lower different point enthalpies given by Equations (11.21) and (11.32), it can be readily shown that:

$$Fh_F = W\bar{h}_{diff} + Dh_{diff} \tag{11.39}$$

Using overall material balance, Equation (11.35), we can re-write Equation (11.39) as:

$$\frac{W}{D} = \frac{h_{diff}-h_F}{h_F-\bar{h}_{diff}} \tag{11.40}$$

Combining Equations (11.37) and (11.40) gives:

$$\frac{W}{D} = \frac{x_D-z_F}{z_F-x_W} = \frac{h_{diff}-h_F}{h_F-\bar{h}_{diff}} \tag{11.41}$$

This equation represents the *feed line* on the enthalpy-composition diagram. According to Equation (11.41), the feed line passes through the following three points on the enthalpy-composition diagram: lower difference point $P^{"}(x_W, \bar{h}_{diff})$, feed point $P_F(z_F, h_F)$ and upper difference point $P' (x_D, h_{diff})$. The feed line is shown schematically in Figure 11.16.

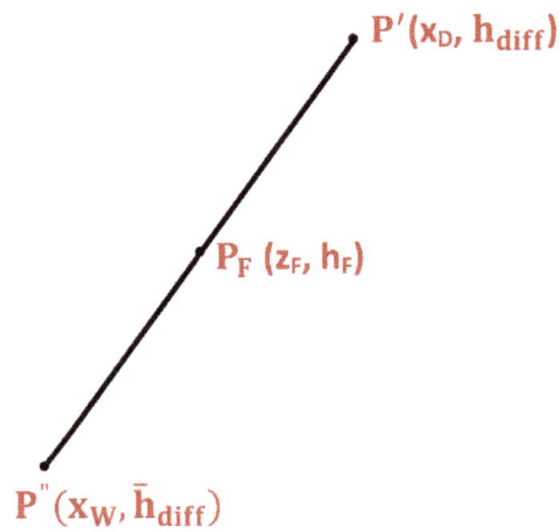

Figure 11.16 Feed line on enthalpy-composition diagram.

11.5.4 Estimation of Total Number of Stages using Ponchon-Savarit Diagram

Figure 11.17 shows the complete Ponchon-Savarit diagram for the estimation of total number of equilibrium stages in a distillation column. The feed is saturated liquid. The key steps in the construction of the Ponchon-Savarit diagram are as follows:

- Locate the upper and lower difference points ($P'$and $P^{"}$) on the enthalpy-composition diagram. Connect these points by a straight line and locate the feed point P_F on this line. The line connecting P' and $P^{"}$ is the feedline, indicated in sky blue color on the Ponchon-Savarit diagram.

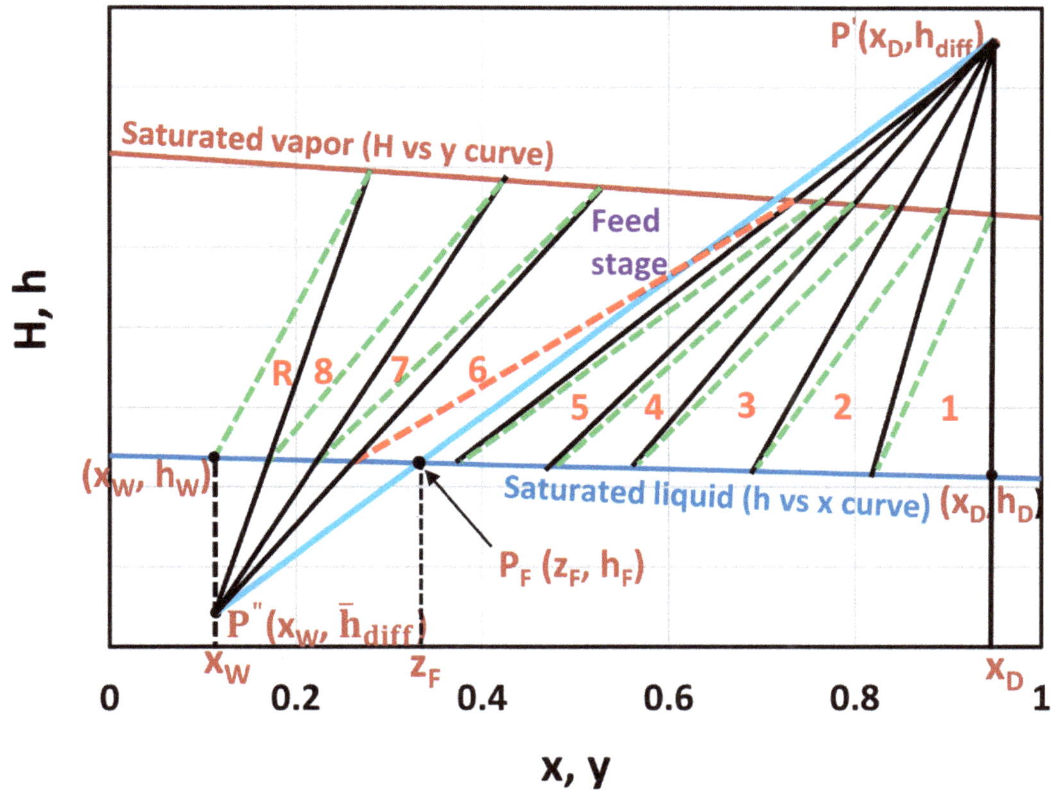

Figure 11.17 Ponchon-Savarit diagram for the estimation of total number of equilibrium stages in a distillation column.

- Starting from the upper difference point P', that is, enriching section, draw the operating lines as indicated in the diagram (Figure 11.17). The operating lines are indicated as black solid lines on the Ponchon-Savarit diagram. The first operating line in the enriching section is a vertical line originating from P'.

- The tie-lines, shown as dashed green lines in Figure 11.17, are used to go from one operating line to the next. When the tie-line crosses the feedline, the operating line is switched from enriching section to stripping section. In the stripping section, the operating lines originate from the lower difference point P''. The tie-line which crosses the feedline is indicated as dashed red line on the Ponchon-Savarit diagram.

- The process of construction of operating and tie lines is continued until the composition of the bottoms liquid product of the column, that is, x_W is reached.

- There are a total of eight equilibrium stages plus a partial reboiler for the separation shown in Figure 11.17. The optimum feed stage is number 6 counting from top of the column. The last equilibrium stage represents the partial reboiler which consists of only one operating line and a tie-line. All other stages consist of two operating lines and a tie-line.

11.5.5 Limiting Reflux Ratios –Total Reflux and Minimum Reflux

As noted in Chapter 9, the reflux ratio can be varied in the range $R_{min} \leq R \leq \infty$. At total reflux corresponding to $R \to \infty$, there is no feed and no distillate and bottoms product are withdrawn from the column, that is, $F = 0, D = 0, W = 0$. In the McCabe-Thiele method, total reflux corresponds to having the operating lines coincide with $y = x$ line. In the Ponchon-Savarit method, total reflux corresponds to difference points going to + and – infinity ($h_{diff} \to \infty$ and $\bar{h}_{diff} \to -\infty$). The operating lines now become vertical lines and the number of stages required for a given separation is *minimum*. Figure 11.18 shows the Ponchon-Savarit diagram at total reflux. The vertical operating lines are connected by tie-lines.

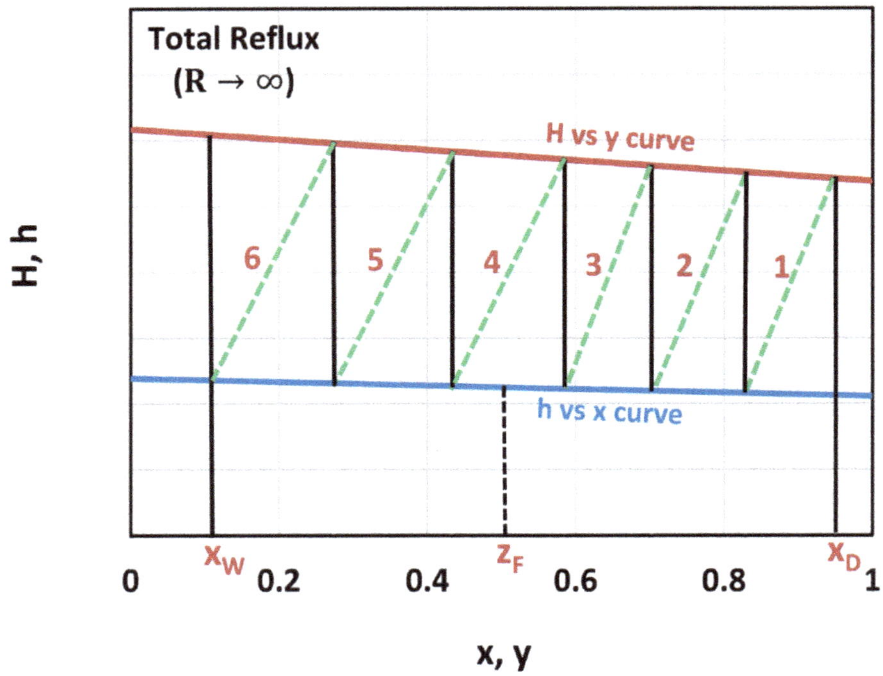

Figure 11.18 Ponchon-Savarit diagram for the estimation of minimum number of equilibrium stages at total reflux.

At minimum reflux ratio, $R = R_{min}$, an infinite number of stages are needed to theoretically achieve a given separation. Assuming that the pinch condition occurs on the feed stage (typical of ideal mixtures), minimum reflux ratio is determined by extending the tie-line passing through the feed point $P_F(z_F, h_F)$ to the intersections with the two vertical lines corresponding to the product compositions ($x = x_D$ vertical line and $x = x_W$ vertical line), as shown in Figure 11.19. From the upper difference point enthalpy estimated graphically, the minimum reflux is calculated using the following relation:

$$R_{min} = \left(h_{diff} - H_D\right)/\lambda_D \tag{11.42}$$

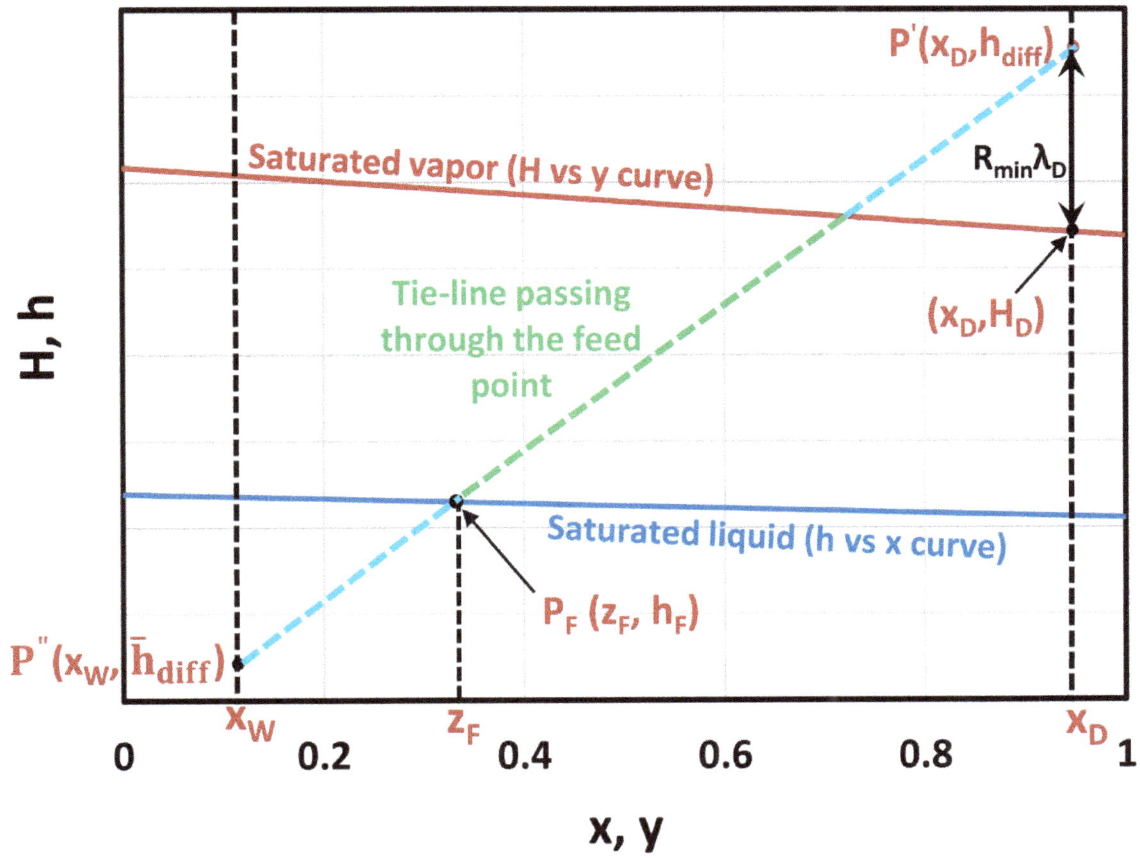

Figure 11.19 Estimation of minimum reflux ratio using Ponchon-Savarit method (pinch condition assumed to occur on the feed stage).

NOTATION

$C_{PL,A}$ constant pressure molar heat capacity of pure species *A* in liquid state
$C_{PL,B}$ constant pressure molar heat capacity of pure species *B* in liquid state
$C_{PV,A}$ constant pressure molar heat capacity of pure species *A* in vapor state
$C_{PV,B}$ constant pressure molar heat capacity of pure species *B* in vapor state
CMO constant molal overflow
A constant in Antoine equation, Equation (11.11)
B constant in Antoine equation, Equation (11.11)
C constant in Antoine equation, Equation (11.11)
D distillate molar flow rate
F feed molar flow rate
h molar enthalpy of liquid solution at bubble point temperature
h_A molar enthalpy of pure liquid species *A* at bubble point temperature of liquid solution
h_B molar enthalpy of pure liquid species *B* at bubble point temperature of liquid solution
h_D molar enthalpy of distillate saturated liquid product
h_F molar enthalpy of feed
h_n molar enthalpy of liquid stream leaving any nth stage of enriching section
h_m molar enthalpy of liquid stream leaving any mth stage of stripping section
h_W molar enthalpy of bottoms saturated liquid product
h_{diff} upper difference point enthalpy, defined in Equation (11.20)

H molar enthalpy of vapor mixture at dew point temperature
H_A molar enthalpy of pure vapor species *A* at dew point temperature of vapor mixture
H_B molar enthalpy of pure vapor species *B* at dew point temperature of vapor mixture
H_D molar enthalpy of saturated vapor distillate
H_{n+1} molar enthalpy of vapor stream leaving (n+1)th stage of enriching section
H_{m+1} molar enthalpy of vapor stream leaving (m+1)th stage of stripping section
$\bar{h}_{diff}$ lower difference point enthalpy, defined in Equation (11.31)
L_n molar flow rate of liquid stream leaving any *n*th stage in the enriching section
$\bar{L}_m$ molar flow rate of liquid stream leaving any *m*th stage in the stripping section
N number of stages in the column; in some situations *N* includes the partial reboiler and in some situations *N* is the number of stages inside the column excluding the partial reboiler
P pressure
P^{Sat} vapor pressure of pure species, see Antoine equation (Equation (11.11))
P_A^{Sat} vapor pressure of pure species A
P_B^{Sat} vapor pressure of pure species B
$\dot{Q}_C$ heat transfer rate in condenser
$\dot{Q}_R$ heat transfer rate in reboiler
R reflux ratio
R_{min} minimum reflux ratio
t temperature in Antoine equation, Equation (11.11)
T temperature
T_{bubble} bubble point temperature of a solution
T_{dew} dew point temperature of vapor mixture
T_R reference temperature for enthalpy calculations
T_A^{Sat} saturation temperature of pure species A
T_B^{Sat} saturation temperature of pure species B
V_{n+1} molar flow rate of vapor stream leaving any (*n*+1)th stage in the enriching section
$\bar{V}_{m+1}$ molar flow rate of vapor stream leaving any (*m*+1)th stage in the stripping section
W molar flow rate of bottoms liquid stream of column
x mole fraction of more volatile species in liquid phase
x_A mole fraction of species *A* in liquid phase
x_D mole fraction of more volatile species in the distillate liquid product
x_m mole fraction of more volatile species in the liquid stream leaving any *m*th stage in the stripping section
x_n mole fraction of more volatile species in the liquid stream leaving any *n*th stage in the enriching section
x_W mole fraction of more volatile species in bottoms liquid product of distillation column
y mole fraction of more volatile species in vapor phase
y_A mole fraction of species A in vapor phase
y_{m+1} mole fraction of more volatile species in the vapor stream leaving any (*m*+1)th stage in the stripping section
y_{n+1} mole fraction of more volatile species in the vapor stream leaving any (*n*+1)th stage in the enriching section
z_F mole fraction of more volatile species in the feed stream

Greek Letters

λ_A molar latent heat of vaporization of pure species A
λ_B molar latent heat of vaporization of pure species B
λ_D molar latent heat of vaporization of distillate product of the column

λ_{mix} molar latent heat of vaporization of a binary mixture

TUTORIAL PROBLEMS WITH SOLUTION

Problem 1

A distillation column is to be operated at 1 atm pressure to separate ethanol-water mixture. The feed composition is 24 wt% ethanol and 76 wt% water. The desired distillate and bottom compositions are: 81 wt% ethanol and 5 wt% ethanol, respectively. The reflux ratio (R) is 2.9.

Given that:

$h_D = 251\ kJ/kg, h_W = 377\ kJ/kg, h_F = 126\ kJ/kg,$ and $H_D = 1382\ kJ/kg$

Estimate the enthalpies corresponding to upper and lower difference points

Solution

The upper difference point is given by Equation (11.21) as:

$$h_{diff} = h_D - \frac{\dot{Q}_C}{D}$$

Using Equation (11.22), we get

$$-\frac{\dot{Q}_C}{D} = (1+R)\lambda_D = (1+2.9)(1382-251) = 4410.9\ kJ/kg$$

Thus:

$$h_{diff} = 377 - (-4410.9) = 4787.9\ kJ/kg$$

For the lower difference point, we use Equation (11.41):

$$\frac{h_{diff}-h_F}{h_F-\bar{h}_{diff}} = \frac{x_D-z_F}{z_F-x_W} = \frac{0.81-0.24}{0.24-0.05} = 3$$

$$\frac{4787.9-126}{126-\bar{h}_{diff}} = 3$$

Solving for $\bar{h}_{diff}$, we get $\bar{h}_{diff} = -1427.97\ kJ/kg$

Problem 2

A multistage distillation column is to be operated at 1 atm pressure to separate a binary mixture of species 1 (more volatile) and species 2 (less volatile). The feed is saturated liquid of composition $z_1 = 0.35$ and $z_2 = 0.65$. The flow rate of feed is 12 kmol/h. The desired distillate composition is $x_1 = 0.95$ and $x_2 = 0.05$. The desired residue (bottoms product) composition is $x_1 = 0.03$ and $x_2 = 0.97$. Assume that column uses a total condenser and a partial reboiler.

The equilibrium data for mixtures of species 1 and 2 at 1 atm pressure are given below.

Mole fraction of more volatile species 1		**Enthalpy of mixture, MJ/kmol**	
Liquid (x)	**Vapor (y)**	**Liquid, h_L**	**Vapor, H_V**
0	0	24.3	61.2
0.08	0.28	24.1	59.6
0.18	0.43	23.2	58.5
0.25	0.51	22.8	58.1
0.49	0.73	22.05	56.5
0.65	0.83	21.75	55.2
0.79	0.90	21.70	54.4
0.91	0.96	21.6	53.8
1.0	1.0	21.4	53.3

Do the following:

a) Plot the equilibrium curve $y\ vs.\ x$ and enthalpy-composition diagram with tie-lines.
b) Estimate the minimum reflux ratio R_{min} using the Ponchon-Savarit method.
c) Estimate the number of stages required at total reflux using the Ponchon-Savarit method.
d) Calculate condenser and reboiler heat duties at a reflux ratio of $1.25R_{min}$.
e) Using the Ponchon-Savarit method, estimate the number of equilibrium stages at $R = 1.25R_{min}$ with optimum feed location.

Solution

a) The plots of equilibrium curve $y\ vs.\ x$ and enthalpy-composition diagram with tie-lines are shown below.

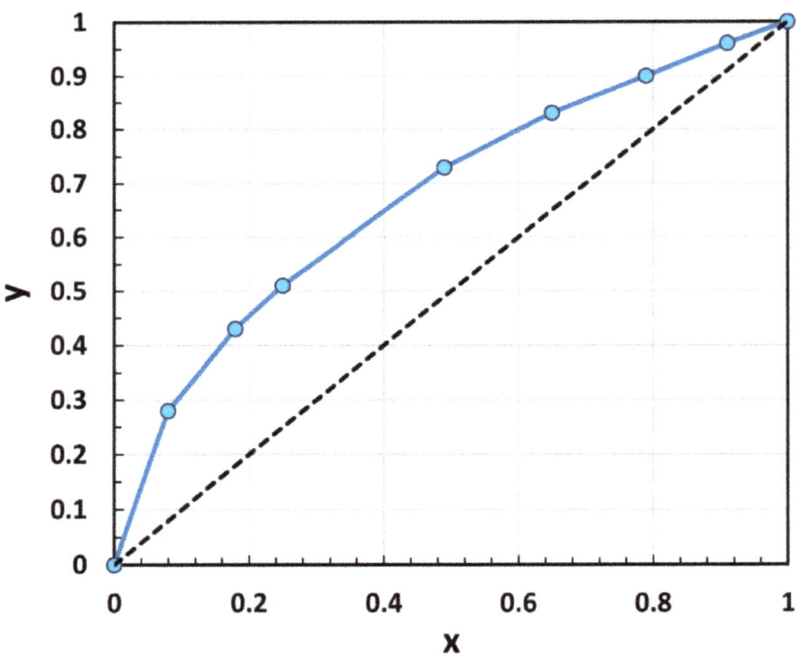

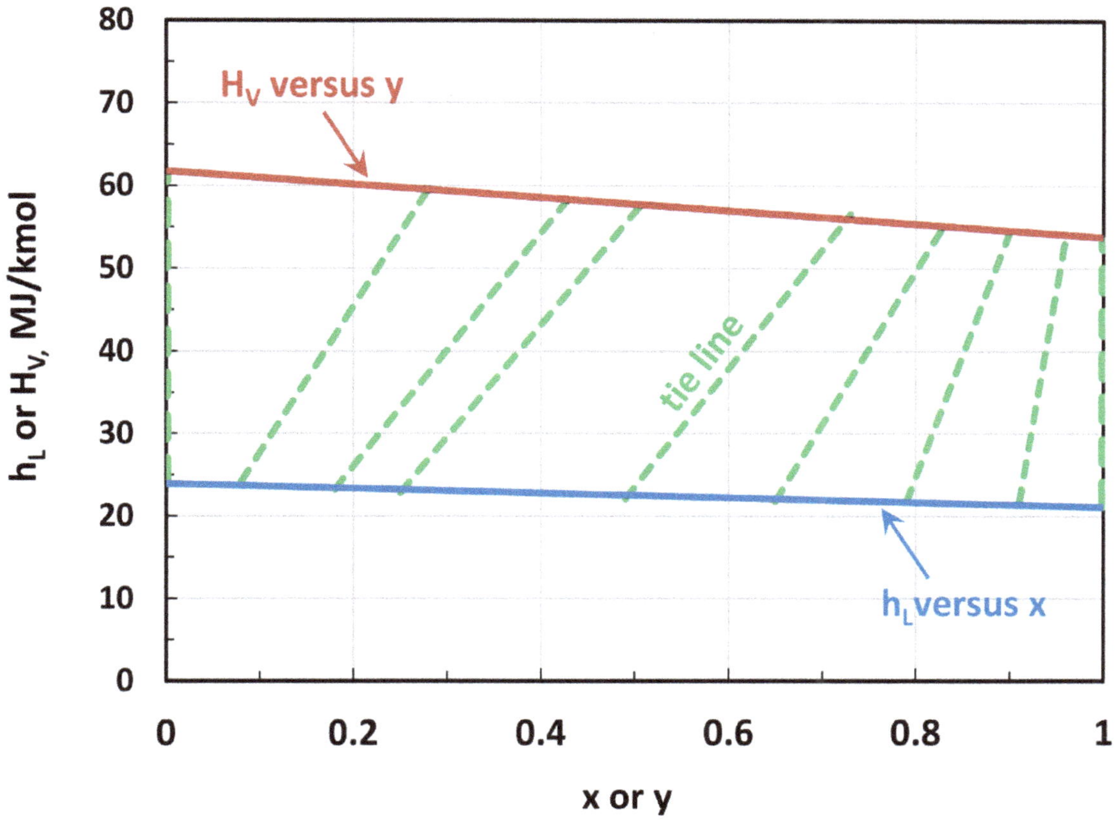

b) Estimation of R_{min}

As discussed in section 11.5.6, minimum reflux ratio is determined by extending the tie-line passing through the feed point $P_F(z_F, h_F)$ to the intersections with the two vertical lines corresponding to the product compositions ($x = x_D$ vertical line and $x = x_W$ vertical line). The extension of tie line passing through the feed point is shown in the figure given below. From the upper difference point enthalpy estimated graphically, the minimum reflux is calculated using Equation (11.42):

$$R_{min} = \left(h_{diff} - H_D\right)/\lambda_D$$

where $h_{diff} = 105\ MJ/kmol$, $H_D = 54\ MJ/kmol$, and $\lambda_D = H_D - h_D = 54 - 22 = 32\ MJ/kmol$. Thus,

$$R_{min} = (105 - 54)/32 = 1.59$$

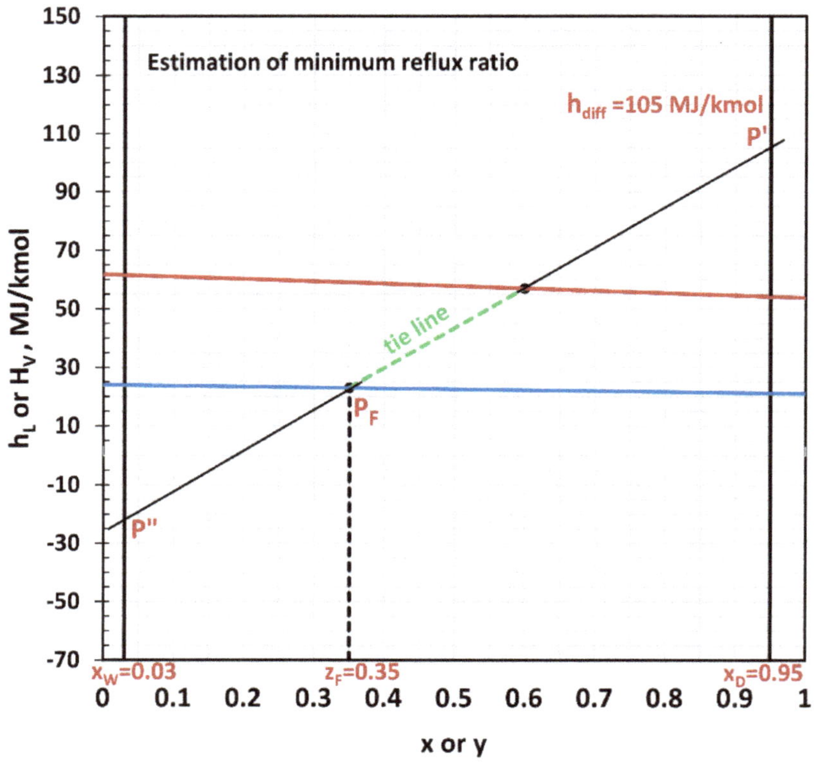

c) Number of stages required at total reflux are estimated graphically as shown below. $N \approx 6$ including partial reboiler.

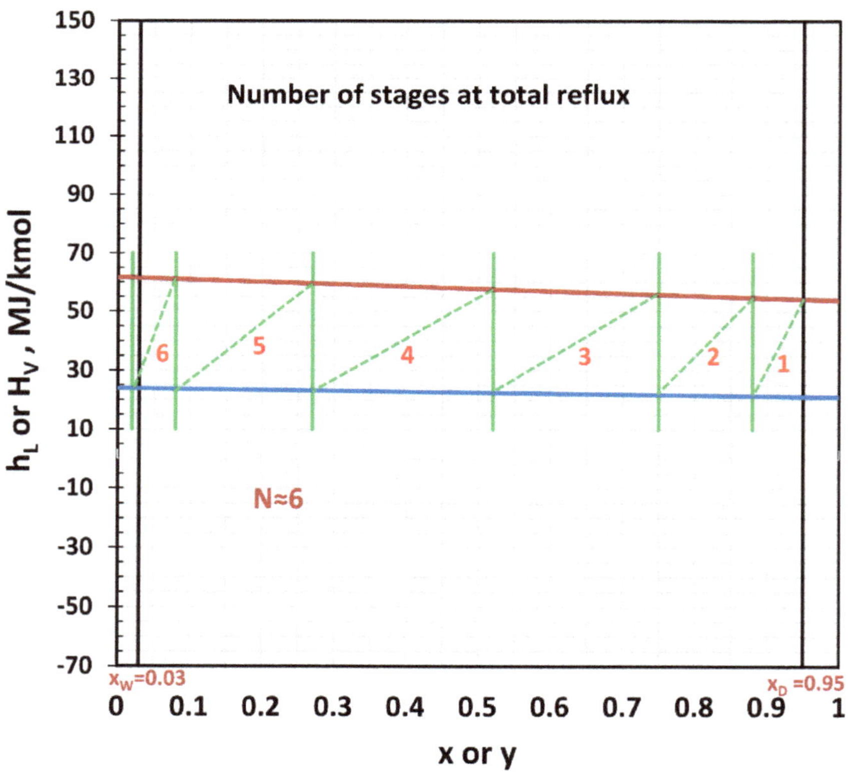

d) Condenser and reboiler heat duties at a reflux ratio of $1.25R_{min}$.

$$R = 1.25 \times 1.59 = 1.9875$$

$$-\frac{\dot{Q}_C}{D} = (1+R)\lambda_D = (1+1.9875)(32) = 95.6\ MJ/kmol$$

We need to calculate the distillate flow rate in order to determine the condenser heat duty. From Equation (11.41):

$$\frac{W}{D} = \frac{x_D - z_F}{z_F - x_W} = \frac{0.95-0.35}{0.35-0.03} = 1.875$$

From overall material balance:

$$F = D + W = D + 1.875D = 2.875D = 12$$

Thus:

$$D = 4.174\ kmol/h \text{ and } W = 7.826\ kmol/h$$

$$\dot{Q}_C = -95.6 \times 4.174 = -399.03\ \ MJ/h = -110.84\ kW$$

The reboiler heat duty is calculated using overall energy balance, Equation (11.38):

$$Fh_F + \dot{Q}_R + \dot{Q}_C = Wh_W + Dh_D$$

$$\dot{Q}_R = Wh_W + Dh_D - Fh_F - \dot{Q}_C = (7.826 \times 24) + (4.174 \times 22) - (12 \times 23) - (-399.03) =$$

$$402.682\ MJ/h = 111.543\ kW$$

e) The number of equilibrium stages at $R = 1.25R_{min}$ with optimum feed location.

To estimate the number of stages graphically, we need to locate the upper and lower difference points and the feed point on the Ponchon-Savarit diagram and construct stages as discussed in section 11.5.4.

$$h_{diff} = h_D - \frac{\dot{Q}_C}{D} = 22 + 95.6 = 117.6\ \ MJ/kmol$$

$$\bar{h}_{diff} = h_W - \frac{\dot{Q}_R}{W} = 24 - (402.682/7.826) = -27.454\ \ MJ/kmol$$

The difference points and the feed point are located on the enthalpy composition diagram and stages are constructed as shown in the diagram below.

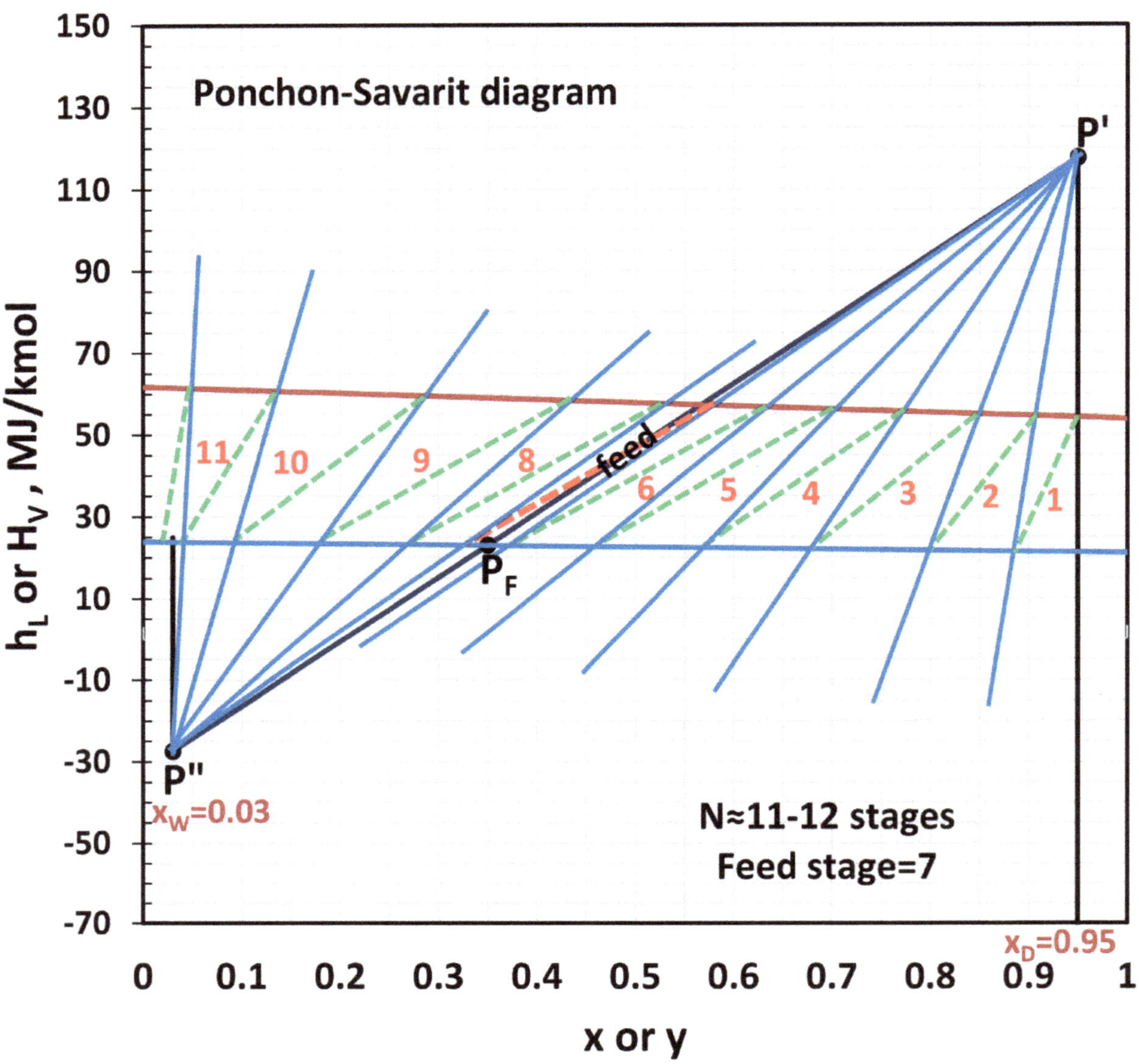

The number of stages required (including partial reboiler) is approximately 11 to 12. The optimum feed tray is 7.

Problem 3

A distillation column is to operate at 1 atm pressure to separate a binary mixture of n-hexane and n-octane. The feed is saturated liquid consisting of 37 mol% n-hexane. The desired distillate composition is 97 mol% n-hexane. The desired residue (bottoms product) composition is 8.5 mol% n-hexane. The column uses a total condenser and the reflux ratio is 1.6.

Using the Ponchon-Savarit method, estimate the number of equilibrium stages required and the location of the optimum feed tray. Assume that column uses a partial reboiler.

The equilibrium data and enthalpy-concentration data for mixtures of n-hexane and n-octane at 1 atm pressure are given below.

Composition, mole fraction of n-hexane		**Enthalpy-concentration data, kJ/mole**		
Liquid (x)	**Vapor (y)**	**x or y**	**Liquid, h_L**	**Vapor, H_V**
0	0	0	29.29	65.69
0.1	0.36	0.1	26.36	64.43
0.3	0.70	0.3	20.92	61.50
0.5	0.85	0.5	17.15	58.16
0.55	0.90	0.7	14.23	53.97
0.70	0.95	0.9	12.97	48.53
1.0	1.0	1.0	12.55	41.84

Solution

Given: Feed as saturated liquid, $R = 1.6$, $z_F = 0.37$, $x_D = 0.97$, $x_W = 0.085$. Estimate the number of equilibrium stages required and the location of the optimum feed tray. Let us first estimate the difference point enthalpies:

$$h_{diff} = (1 + R)\lambda_D + h_D = (1 + 1.6)(46 - 11.5) + 11.5 = 101.2\ MJ/kmol$$

Note that λ_D and h_D are obtained from the enthalpy-composition diagram. We can calculate $\bar{h}_{diff}$ from the following relation:

$$\frac{h_{diff} - h_F}{h_F - \bar{h}_{diff}} = \frac{x_D - z_F}{z_F - x_W} = \frac{0.97 - 0.37}{0.37 - 0.085} = 2.1053$$

$$\frac{101.2 - 21.5}{21.5 - \bar{h}_{diff}} = 2.1053$$

Solving for $\bar{h}_{diff}$, we get $\bar{h}_{diff} = -16.35\ MJ/kmol$

The difference points and the feed point are located on the enthalpy-composition diagram and stages are constructed as shown in the diagram below. The number of stages required (including partial reboiler) is approximately 5. The optimum feed tray is 3.

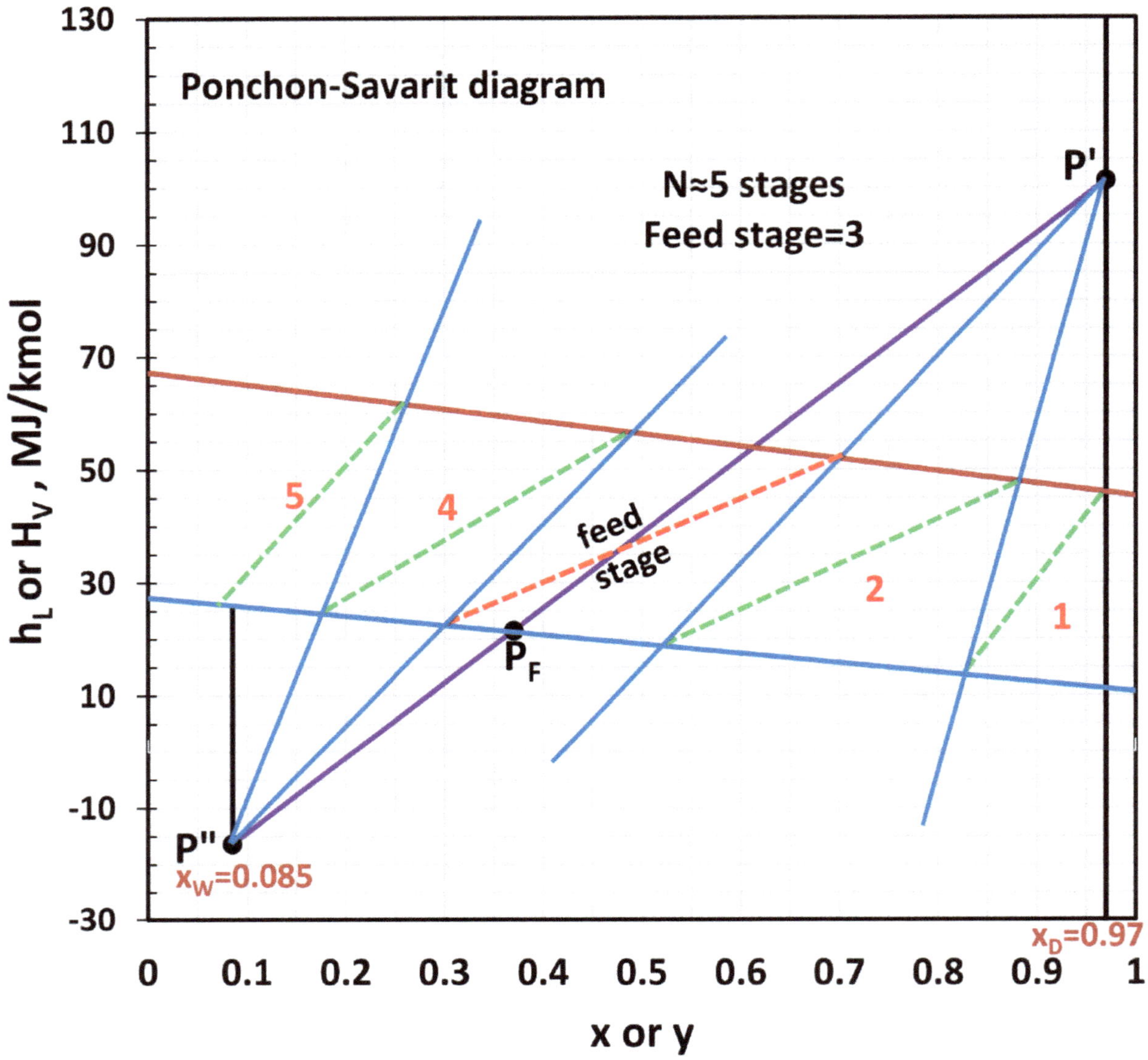

Problem 4

Re-do tutorial problem 3 using the McCabe-Thiele method instead of the Ponchon-Savarit method. Compare the values obtained by the two methods.

Solution

As the feed is saturated liquid, $q = 1$. Other information given: $z_F = 0.37$, $x_D = 0.97$, $x_W = 0.085$, and $R = 1.6$. Based on the given information, the McCabe-Thiele diagram is constructed as shown in the figure given below. Thus, $N = 5$ and the optimum feed location is stage 3. These values are in total agreement with the values obtained through the Ponchon-Savarit method. This is not unexpected as the latent heats of vaporization of n-hexane and n-octane are similar ($\lambda_{hexane} = 29.29$ kJ/mol and $\lambda_{octane} = 36.4$ kJ/mol at normal boiling points) and hence, the assumption of CMO is valid.

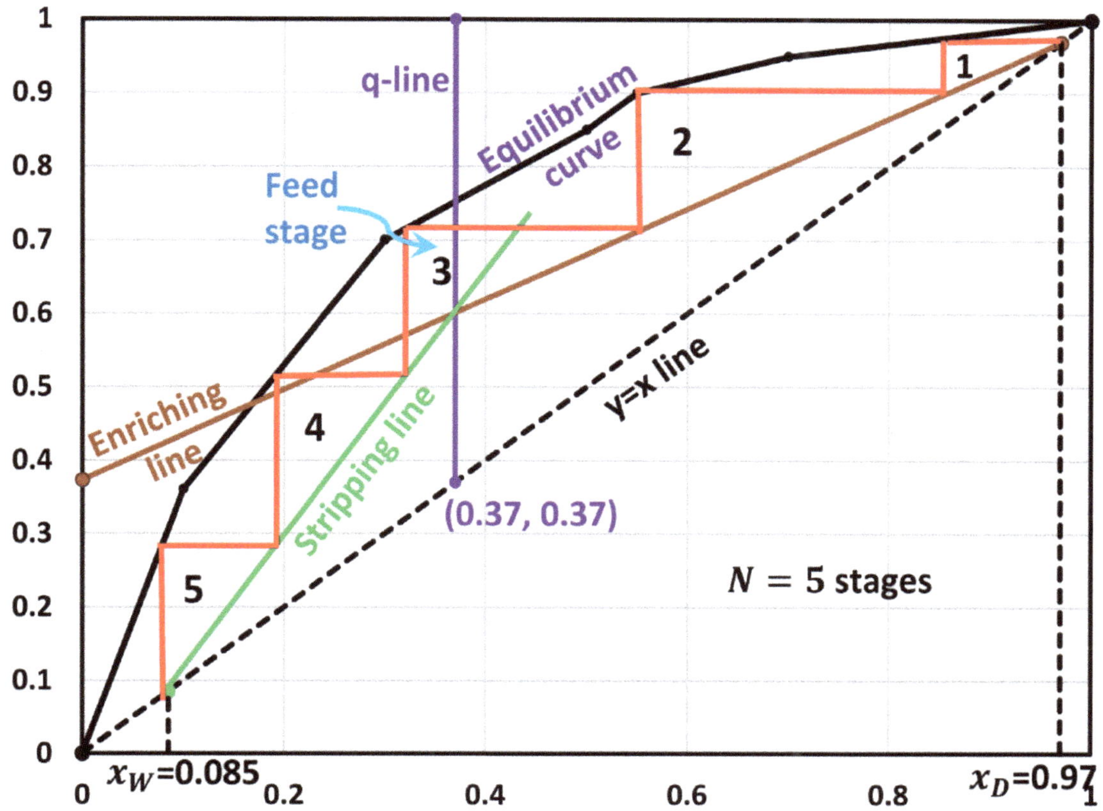

SHORT REVIEW QUESTIONS

1. The assumption of CMO (constant molal overflow) is not required in the Ponchon-Savarit method. Is this true or false?
2. The Ponchon-Savarit method requires both material and energy balances. Is this true or false?
3. What is the enthalpy of mixing of two or more miscible pure liquid species to form an ideal solution, keeping T and P constant?
4. What is the enthalpy of mixing of two or more pure ideal gases to form an ideal gas mixture, keeping T and P constant?
5. A point on a y vs x equilibrium curve becomes a tie-line on enthalpy-composition diagram. Is this true or false?
6. Define the upper and lower difference points involved in the Ponchon-Savarit method.
7. Using energy balance around the condenser, show that $h_{diff} = h_D - (\dot{Q}_C/D)$.
8. Using overall energy balance on the distillation column, show that $\bar{h}_{diff} = h_W - (\dot{Q}_R/W)$.
9. All the enriching section operating lines on enthalpy-composition diagram originate from the upper difference point. Is this true or false?
10. All the stripping section operating lines on enthalpy-composition diagram originate from the lower difference point. Is this true or false?
11. The feed line on enthalpy-composition diagram passes through three points: upper difference point, feed point, and lower difference point. Is this true or false?
12. The Ponchon-Savarit method is restricted to binary systems. Is this true or false?
13. What are the values of the difference point enthalpies (h_{diff} and $\bar{h}_{diff}$) at total reflux where $R \to \infty$.

14. The enriching and stripping operating lines are vertical lines on enthalpy-composition diagram at total reflux. Is this true or false?
15. Assuming that the pinch condition occurs on the feed stage, the minimum reflux ratio is determined by extending the tie-line passing through the feed point to the intersections with the vertical lines $x = x_D$ and $x = x_W$. Is this true or false?

PRACTICE PROBLEMS

Problem 1

A distillation column is to be operated at 1 atm pressure to separate methanol-water mixture. The feed composition is 50 mol% methanol and 50 mol% water. The desired distillate and bottom compositions are: 95 mol% methanol and 5 mol% methanol, respectively. The reflux ratio (R) is 2.0. The feed flow rate is 25 lb-mol/h.

Given that:

$$h_D = 2250 \frac{Btu}{lb-mol}, h_W = 3070 \frac{Btu}{lb-mol}, h_F = 2500 \frac{Btu}{lb-mol}, \text{and } H_D = 17{,}500 \frac{Btu}{lb-mol}$$

a) Estimate the enthalpies corresponding to upper and lower difference points
b) Estimate the condenser and reboiler heat duties

Problem 2

A multistage distillation column is to be operated at 1 atm pressure to separate a binary mixture of species 1 (more volatile) and species 2 (less volatile). The feed is saturated vapor of composition $z_1 = z_2 = 0.50$. The flow rate of feed is 15 kmol/h. The desired distillate composition is $x_1 = 0.93$ and $x_2 = 0.07$. The desired residue (bottoms product) composition is $x_1 = 0.05$ and $x_2 = 0.95$. Assume that column uses a total condenser and a partial reboiler.

The equilibrium data for mixtures of species 1 and 2 at 1 atm pressure are given in tutorial problem 2. Do the following:

a) Estimate the minimum reflux ratio R_{min} using the Ponchon-Savarit method.
b) Estimate the number of stages required at total reflux using the Ponchon-Savarit method.
c) Calculate condenser and reboiler heat duties at a reflux ratio of $1.35R_{min}$.
d) Using the Ponchon-Savarit method, estimate the number of equilibrium stages at $R = 1.35R_{min}$ with optimum feed location.
e) Re-do parts (a), (b), and (d) using the McCabe-Thiele method instead of the Ponchon-Savarit method. Compare the values obtained by the two methods.

Problem 3

A multistage distillation column is to be operated at 1 atm pressure to separate ethanol-water mixture. The feed is saturated liquid containing 25 wt% ethanol. The flow rate of feed is 7,000 kg/h. The desired distillate composition is 82 wt.% ethanol and the bottoms composition is 3 wt.% ethanol. The column uses a total condenser and a partial reboiler.

Using the Ponchon-Savarit method, do the following:

a) Estimate the minimum reflux ratio R_{min}
b) Estimate the number of stages required at total reflux
c) Calculate condenser and reboiler heat duties at a reflux ratio of $1.3R_{min}$.

d) Estimate the number of equilibrium stages at $R = 1.3R_{min}$ with optimum feed location.

The equilibrium data for mixtures of ethanol and water at 1 atm pressure are given below.

Composition, mass fraction of ethanol		**Enthalpy-data, kJ/kg**	
Liquid (x)	**Vapor (y)**	**Liquid, h_L**	**Vapor, H_V**
0	0	418.9	2675
0.02	0.192	-	-
0.04	0.325	-	-
0.05	0.377	393.8	2593.5
0.10	0.527	371.7	2493.5
0.20	0.656	335.6	2355.1
0.30	0.713	314	2193.4
0.50	0.771	285.9	1870.1
0.60	0.794	273.3	1707.3
0.70	0.822	258.4	1544.5
0.80	0.858	241.4	1386.3
0.82	0.868	-	-
0.86	0.888	-	-
0.90	0.912	225	1223.5
1.0	0.978	207	1064.1

Problem 4

A mixture of methanol and water is to be separated using a multistage distillation column at 1 atm. The feed is saturated liquid consisting of 50 mol% methanol. The desired distillate composition is 95 mol% methanol. The residue (bottoms product) composition is 5 mol% methanol. The reflux ratio is 2.0.

Using the Ponchon-Savarit method, estimate the number of equilibrium stages required and the location of the optimum feed tray. Assume that column uses a total condenser and a partial reboiler.

The equilibrium data and enthalpy-concentration data for mixtures of methanol and water at 1 atm pressure are given below.

Composition, mole fraction of methanol		**Enthalpy-concentration data, BTU/lb-mole**		
Liquid (x)	**Vapor (y)**	**x or y**	**Liquid, h_L**	**Vapor, H_V**
0	0	0	3240	20,720
0.02	0.134	0.05	3070	20,520
0.04	0.23	0.10	2950	20,340
0.06	0.304	0.15	2850	20,160
0.08	0.365	0.20	2760	20,000
0.10	0.418	0.30	2620	19,640
0.15	0.517	0.40	2540	19,310
0.20	0.579	0.50	2470	18,970
0.30	0.665	0.60	2410	18,650
0.40	0.729	0.70	2370	18,310
0.50	0.779	0.80	2330	17,980
0.60	0.825	0.90	2290	17,680
0.70	0.870	1.0	2250	17,390
0.80	0.915			

0.90	0.958			
0.95	0.979			
1.0	1.0			

Problem 5

A multistage distillation column is to be used to separate a mixture of benzene and toluene at 1 atm pressure. The feed consists of 40 mol% benzene and is fed to the optimum tray as saturated vapor. The distillate consists of 97 mol% benzene and the bottoms product contains 4 mol% benzene. The reflux ratio is 1.95.

Using the Ponchon-Savarit method, estimate the number of equilibrium stages required and the location of the optimum feed tray. Assume that column uses a total condenser and a partial reboiler.

The equilibrium data and enthalpy-concentration data for mixtures of benzene and toluene at 1 atm pressure are given below.

Composition, mole fraction of benzene		**Enthalpy-concentration data, kJ/kmol**	
Liquid (x)	**Vapor (y)**	**Liquid, h_L**	**Vapor, H_V**
0.999732	0.999897	0	30770.34
0.856788	0.938991	427.1883	31268.12
0.726899	0.871366	877.2112	31805.1
0.608551	0.79654	1347.025	32385.59
0.500431	0.71401	1834.049	33014.21
0.401389	0.623252	2336.089	33695.93
0.310423	0.523718	2851.283	34436.08
0.226654	0.414838	3378.04	35240.36
0.149309	0.296018	3915.006	36114.83
0.077709	0.166644	4461.024	37065.99
0.011257	0.026074	5015.105	38100.74
0.000644	0.001511	5108.74	38281.83

Problem 6

Re-do practice problem 5 using the McCabe-Thiele method instead of the Ponchon-Savarit method. Compare the values obtained by the two methods.

Chapter 12

Batch Distillation-Without and With Reflux

12.1 LEARNING OUTCOMES

Upon completion of this chapter, students will be able to:

- Analyze batch distillation of binary mixtures
- Derive different forms of Rayleigh equation
- Analyze batch distillation with reflux
- Explain the influence of reflux on batch distillation
- Explain and analyze batch steam distillation

12.2 INTRODUCTION

In batch distillation (also referred to as differential distillation), liquid feed is charged to a still pot/kettle. The liquid charge is boiled slowly by supplying heat to the pot. The heat may be supplied by passing steam through a heating coil immersed in the liquid feed. The vapor product is passed through a condenser where it is condensed. The condensed liquid product is collected in a product receiver. Figure 12.1 shows a schematic diagram of the batch distillation process.

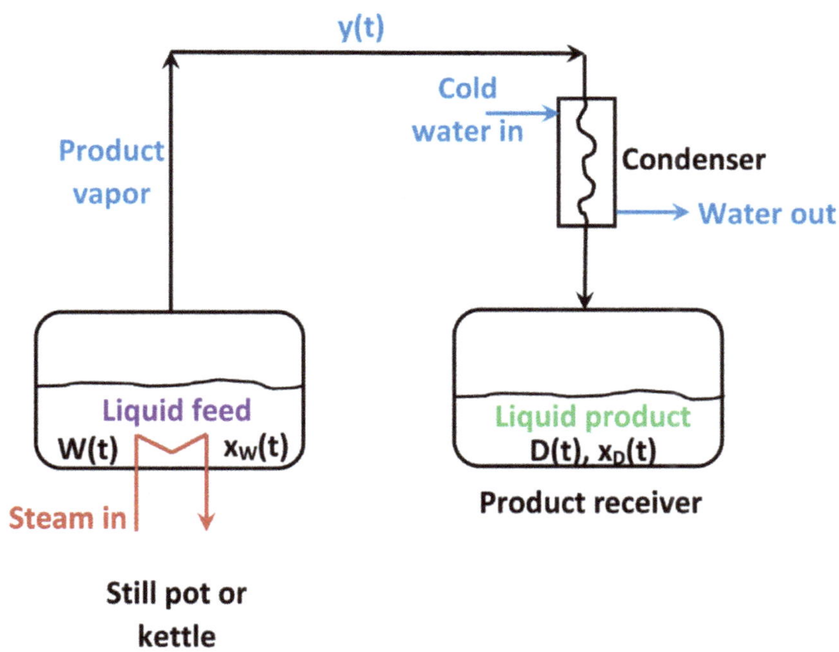

Figure 12.1 Batch distillation apparatus.

The batch distillation process can be explained with the help of a T-x-y diagram given in Figure 12.2. The vapor produced initially is rich in more volatile component. However, as vaporization proceeds, the concentration of more volatile component in the vapor product decreases. The concentration of more volatile component in the liquid left inside the pot also decreases. The boiling temperature in the pot rises as distillation proceeds due to a decrease in the concentration of more volatile component.

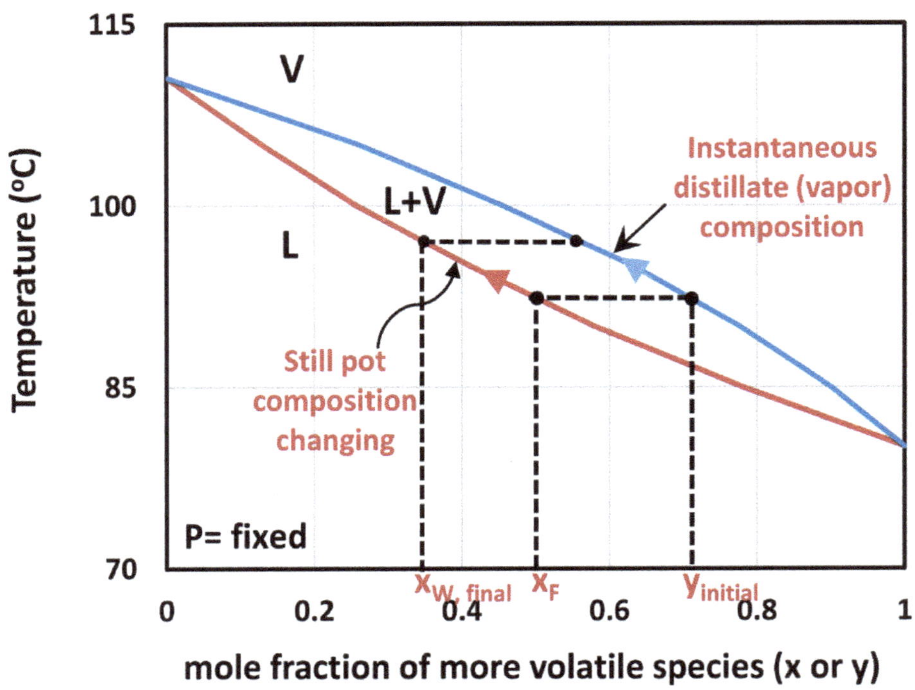

Figure 12.2 Batch distillation process explained on a T-x-y diagram.

Batch distillation is extensively employed in the production of biochemical and pharmaceutical products where the production quantities are usually small but the desired purity of the product is very high. In general, batch distillation is preferred over continuous column distillation when: a) the amounts to be separated are relatively small, b) feed composition varies a lot over a period of time, and c) the separation of materials is required on an irregular basis.

12.3 ANALYSIS OF BATCH DISTILLATION

Referring to Figure 12.1, let the initial amount of feed (assumed binary mixture) charged to a still pot be F moles. Let $W(t)$ be the moles of liquid left inside the still pot at any time t and $D(t)$ be the moles of liquid distillate collected in the product receiver. Let x_F be the mole fraction of more volatile component in the initial feed, $x_W(t)$ be the mole fraction of more volatile component in the liquid left inside the pot at any time t, $y(t)$ be the instantaneous mole fraction of more volatile component in the vapor leaving the pot, and $x_D(t)$ be the mole fraction of more volatile component in the distillate product collected in the receiver.

The overall material balance gives:

$$F = W + D \tag{12.1}$$

The component material balance on more volatile species gives:

$$x_F F = x_W W + x_D D \tag{12.2}$$

The initial feed amount F and the feed composition x_F are usually known. The composition x_W of the residue liquid left in the still pot at the end of the distillation process may also be specified. Given F, x_F and x_W , we are left with three unknown variables (W, D, x_D) but we have two equations so far (Equations (12.1) and (12.2)). Thus we need a third equation which utilizes the fact that the vapor emerging from the still pot is in equilibrium the liquid present inside the pot. This third equation is known as the *Rayleigh equation*.

In order to derive the Raleigh equation, consider a still pot containing a quantity of liquid W at any time t. Let a differential amount of this liquid dW be vaporized. The amount of distillate produced is dD. From overall material balance:

$$-dW = dD \tag{12.3}$$

The amount of more volatile component produced is ydD. Thus component material balance gives:

$$-d(x_W W) = ydD \tag{12.4}$$

From Equations (12.3) and (12.4) it follows that:

$$\frac{dW}{W} = \frac{dx_W}{y - x_W} \tag{12.5}$$

Upon integration from start of distillation ($t = 0, W = F, x_W = x_F$) to end of distillation ($W, x_{W,final}$), we obtain:

$$\int_F^W \frac{dW}{W} = \int_{x_F}^{x_{W,final}} \frac{dx_W}{y - x_W} \tag{12.6}$$

Thus:

$$ln\left(\frac{W}{F}\right) = -\int_{x_{W,final}}^{x_F} \frac{dx}{y - x} \tag{12.7}$$

where the subscript W has been dropped from x_W in the integrand on the right-hand-side of Equation (12.7). This is the Rayleigh equation. We can estimate the amount of liquid left in the still pot at the end of the distillation process using the Raleigh equation provided that the final composition of the left over liquid in the still pot is specified. Once the left over amount W is known, one can readily calculate the amount of distillate collected (D) and its composition (x_D) using overall and component material balances:

$$D = F - W \tag{12.8}$$

$$x_D = (x_F F - x_{W,final} W)/D \tag{12.9}$$

In order to estimate W from the Rayleigh equation (Equation (12.7)), we need to evaluate the integral on the right-hand-side of Equation (12.7). As the vapor coming out of the still pot at any instant is in equilibrium the liquid present in the still pot, y in the integral (Equation (12.7)) is related to x through the equilibrium $y - x$ curve. Using the equilibrium curve, one can determine the y values over the range of x values ($x_{W,final} \leq x \leq x_F$) present in the integral. By plotting the values of $1/(y - x)$ versus x , one can evaluate the integral in the Rayleigh equation graphically as shown in Figure 12.3. Thus:

$$W/F = exp(-area\ under\ the\ curve) \tag{12.10}$$

The area under the curve can be approximated by the area of a trapezoid as shown in Figure 12.4. Referring to Figure 12.4, the area of a single trapezoid is given by the following trapezoid rule:

$$Area\ of\ trapezoid = \frac{1}{2}[f(a) + f(b)](b - a) \tag{12.11}$$

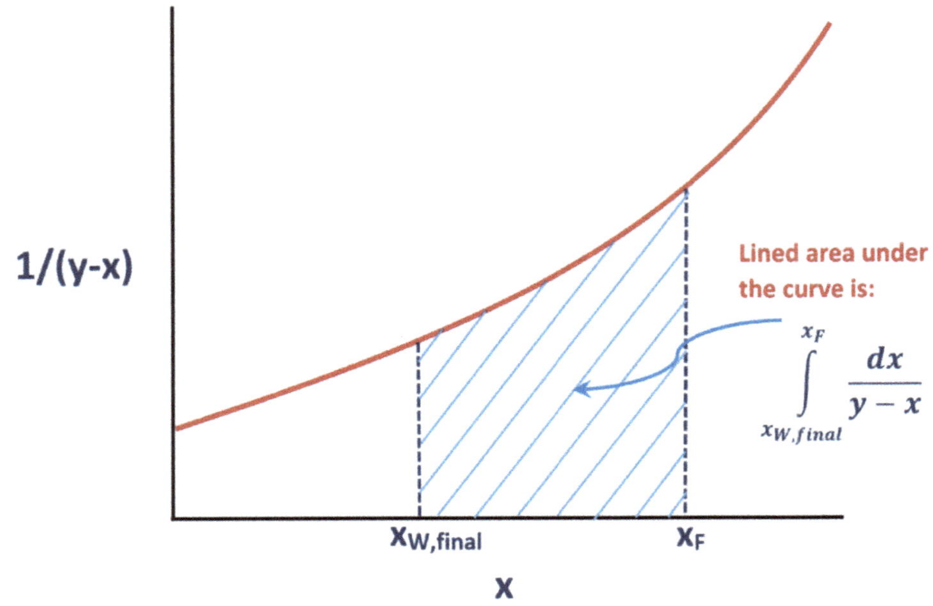

Figure 12.3 Graphical integration of function $1/(y - x)$.

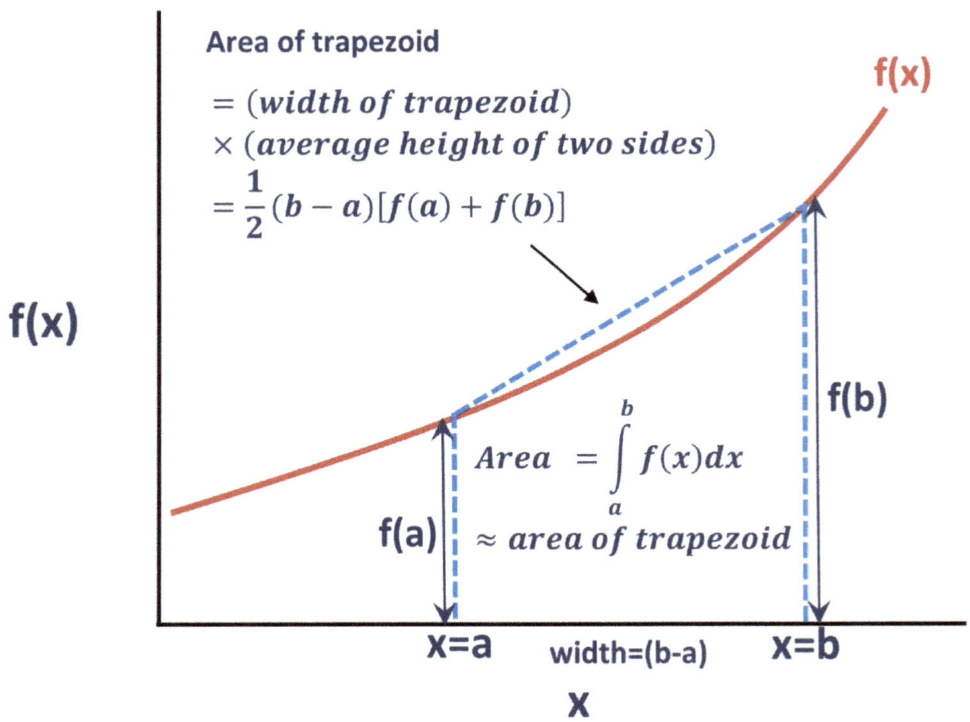

Figure 12.4 Area under the curve approximated by a single trapezoid.

If the area under the curve is approximated by a single trapezoid as shown in Figure 12.4, the error could be large. A better approach is to subdivide the integration limit interval [a b] into a sequence of n sub-intervals of equal width $h = (b - a)/n$. Then use the trapezoid rule on each sub-interval. For example, let $n = 3$ as shown in Figure 12.5. Now the area is given as:

$$Area = \frac{1}{2}h[f(a) + 2\{f(a + h) + f(a + 2h)\} + f(b)] \tag{12.12}$$

where $h = (b - a)/3$. This area is almost the exact area under the curve shown in Figure 12.4 over the interval [a b].

In general, the area under the curve can be expressed as:

$$Area = \frac{1}{2}h[f_0 + 2\{f_1 + f_2 + f_3 + \cdots f_{n-1}\} + f_n] \tag{12.13}$$

where $h = (b - a)/n$ and $f_0 = f(a), f_1 = f(a + h), f_2 = f(a + 2h), \ldots, f_n = f(a + nh) = f(b)$. This is the *composite trapezoid rule* based on equally spaced points.

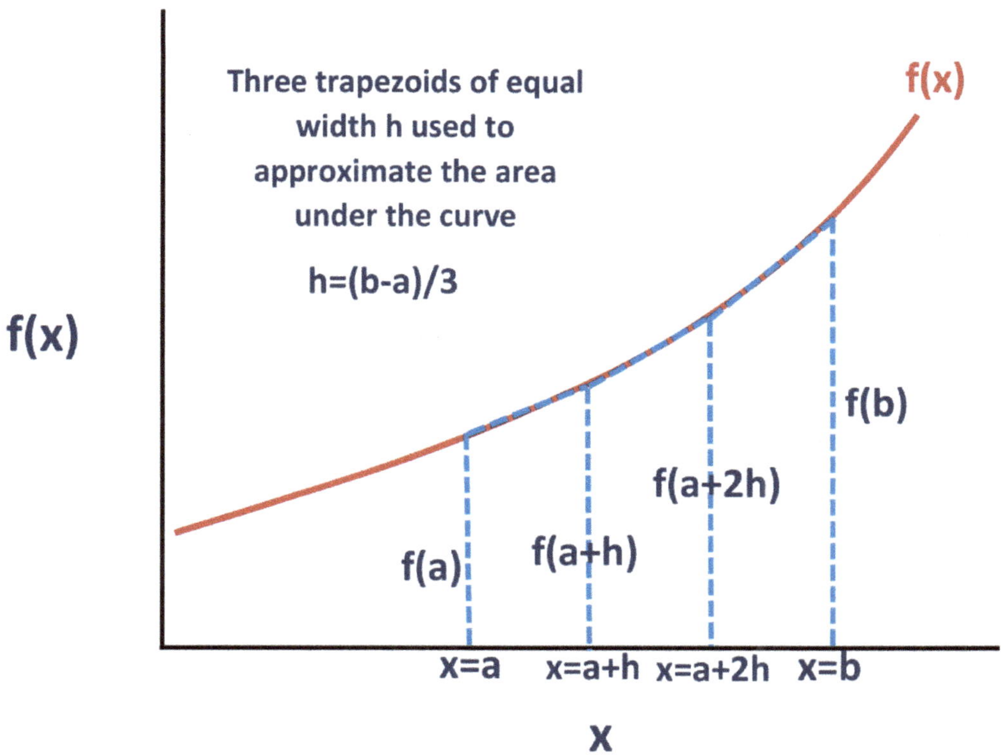

$$Area = \int_a^b f(x)dx \approx \textbf{areas of 3 trapezoids}$$

$$= \frac{1}{2}h[f(a) + 2\{f(a + h) + f(a + 2h)\} + f(b)]$$

Figure 12.5 Area under the curve approximated by three trapezoids.

Analytical Form of the Rayleigh Equation

An analytical form of the Raleigh equation can be derived provided that the relative volatility of more volatile species with respect to the less volatile species, denoted by α, is constant (independent of temperature). As noted in Chapter 8, the equilibrium curve can be represented by the following equation:

$$y = \alpha x/[1 + (\alpha - 1)x] \tag{12.14}$$

Substituting Equation (12.14) into Equation (12.7):

$$ln\left(\frac{W}{F}\right) = -\int_{x_{W,final}}^{x_F} \frac{dx}{\frac{\alpha x}{[1+(\alpha-1)x]} - x} \tag{12.15}$$

The integral on the right-hand-side of Equation (12.15) can be evaluated using partial fractions and taking the limits. Upon integration, we obtain:

$$ln\left(\frac{W}{F}\right) = \frac{1}{(\alpha-1)} ln\left[\frac{x_W(1-x_F)}{x_F(1-x_W)}\right] + ln\left[\frac{1-x_F}{1-x_W}\right] \tag{12.16}$$

The subscript on $x_{W,final}$ has been simplified to x_W. In Equation (12.16), x_W is the mole fraction of more volatile component in the residue at any instant and W is the amount of liquid residue left in the pot at any instant.

Equation (12.16) is the analytical form of the Rayleigh equation. It could readily be recast in another form as:

$$\frac{n_A}{n_{A0}} = \left[\frac{n_B}{n_{B0}}\right]^{\alpha} \tag{12.17}$$

where n_{A0} and n_{B0} are the *initial* number of moles of species A (more volatile) and B in the still pot, and n_A and n_B are the number of moles of species A and B in the still pot at any instant.

12.4 BATCH DISTILLATION WITH REFLUX

In the preceding section, we considered simple batch distillation without any reflux. The separation achieved in a simple batch distillation process is often limited. One way to improve the separation of components is to repeat the batch distillation process using the product of first batch as feed for the next batch. Another important way to sharpen the separation of components is to install a rectifier/enriching column over the still pot and reflux a portion of the distillate through the column and back to the still pot, as shown in Figure 12.6. The downward flow of reflux liquid enriches the rising vapor of more volatile component in a manner similar to that observed in the case of a continuous column distillation. Thus batch distillation with reflux combines the desirable features of simple batch distillation and continuous multi-stage distillation. More often than not, a constant value of reflux ratio (R) is used in operating a batch distillation apparatus with reflux.

For the batch distillation with reflux, we need to modify the original Rayleigh equation (Equation (12.6)). When a differential amount dW of pot liquid is vaporized, the amount of more volatile component produced is now $x_D dD$ (instead of is ydD). Thus the differential mass balance on more volatile component becomes:

$$-d(x_W W) = x_D dD \tag{12.18}$$

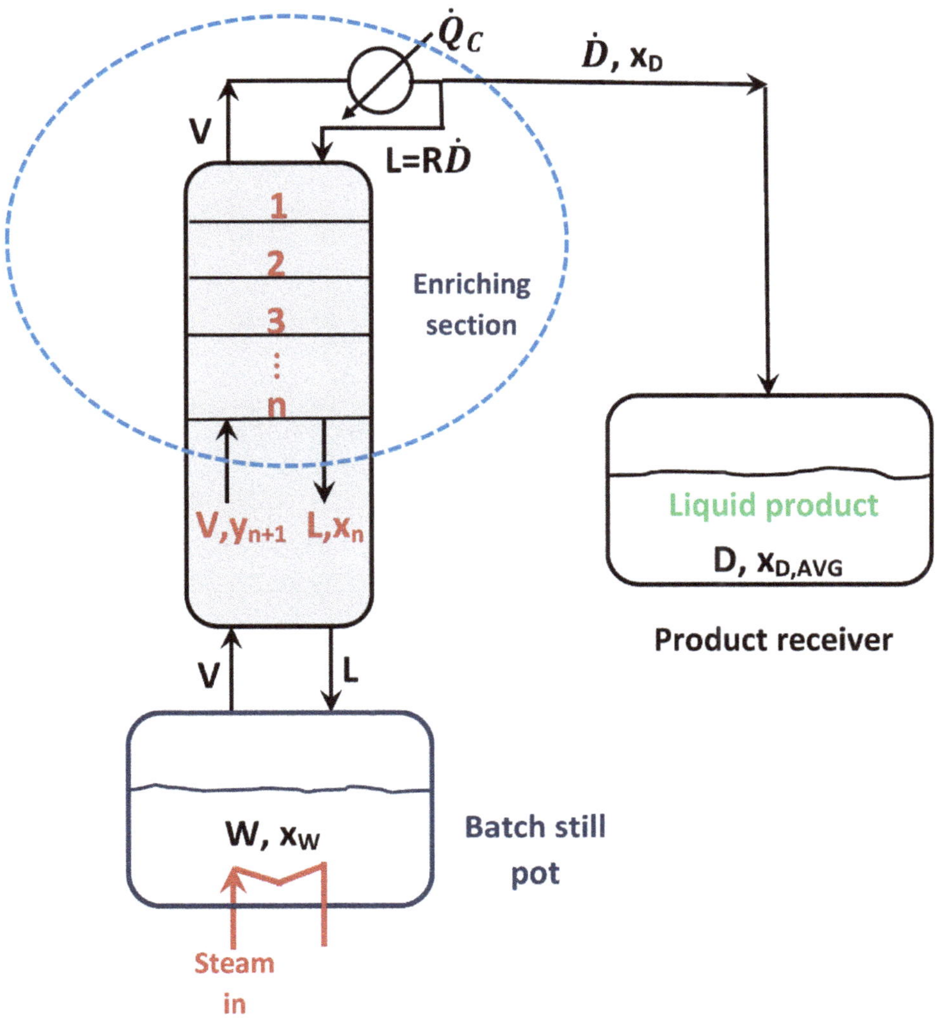

Figure 12.6 Batch distillation with reflux.

Consequently, Equation (12.5) becomes:

$$\frac{dW}{W} = \frac{dx_W}{x_D - x_W} \tag{12.19}$$

Upon integration of Equation (12.19), we get the following modified Rayleigh equation for batch distillation with reflux as:

$$ln\left(\frac{W}{F}\right) = \int_{x_F}^{x_{W,final}} \frac{dx_W}{x_D - x_W} \tag{12.20}$$

Now that the vapor composition variable y in the original Rayleigh equation, Equation (12.6), is replaced by distillate composition variable x_D. To integrate the modified Rayleigh equation, we need the relationship between instantaneous x_D and x_W. In simple batch distillation without reflux, the distillate vapor leaving the still pot at any instant was in equilibrium with the liquid present in the still pot, that is, y and x_W were related through the equilibrium curve. However, in batch distillation with reflux, the distillate product at any instant is not in equilibrium with the pot liquid and hence the instantaneous distillate composition x_D is not related to x_W through any equilibrium relation. So how is instantaneous x_D related to x_W?

The relation between distillate composition x_D and pot liquid composition x_W at any instant is found through the enriching section operating line. Thus we need an equation for the operating line of rectifying section in order to solve the problems related to batch distillation with reflux. To this end, we carry out overall and component material balances over the envelope shown in Figure 12.6:

$$V = L + \dot{D} = (R+1)\dot{D} \tag{12.21}$$

$$y_{n+1}V = x_n L + x_D \dot{D} \tag{12.22}$$

$$y_{n+1} = \left(\frac{L}{V}\right) x_n + x_D\left(\frac{\dot{D}}{V}\right) \tag{12.23}$$

$$y_{n+1} = \frac{R}{R+1} x_n + \frac{x_D}{R+1} \tag{12.24}$$

where $V, L, \dot{D}$ are instantaneous molar flow rates of vapor, liquid, and distillate streams (see Figure 12.6), respectively and R is the reflux ratio. Equation (12.24) is the operating line of the rectification column of the batch distillation process shown in Figure 12.6. It is based on the assumption of *CMO* (constant molal overflow).

The relation between distillate composition x_D and pot liquid composition x_W at any instant is found through the enriching section operating line, Equation (12.24). Assuming that the number of trays in the rectifier column is known and that the reflux ratio R is held constant during the operation, we can obtain x_D as function of x_W and estimate the value of the integral in Equation (12.20) using the following procedure:

- Plot the operating lines (Equation (12.24)) on the McCabe-Thiele diagram for a series of x_D values (R fixed). Note that the operating lines are parallel as shown in Figure 12.7.
- Step off the specified number of stages (trays inside the column plus still pot as a stage) on each operating line starting at x_D to determine the x_W value corresponding to that x_D. This way we obtain x_W values corresponding to a series of x_D values.
- Calculate $1/(x_D - x_W)$ and plot it as a function of x_W as shown in Figure 12.8. The integral in Equation (12.20) is negative of the area under the curve over the x_W range from $x_{W,final}$ to x_F.
- Use the trapezoidal rule discussed earlier to estimate the area under the curve.

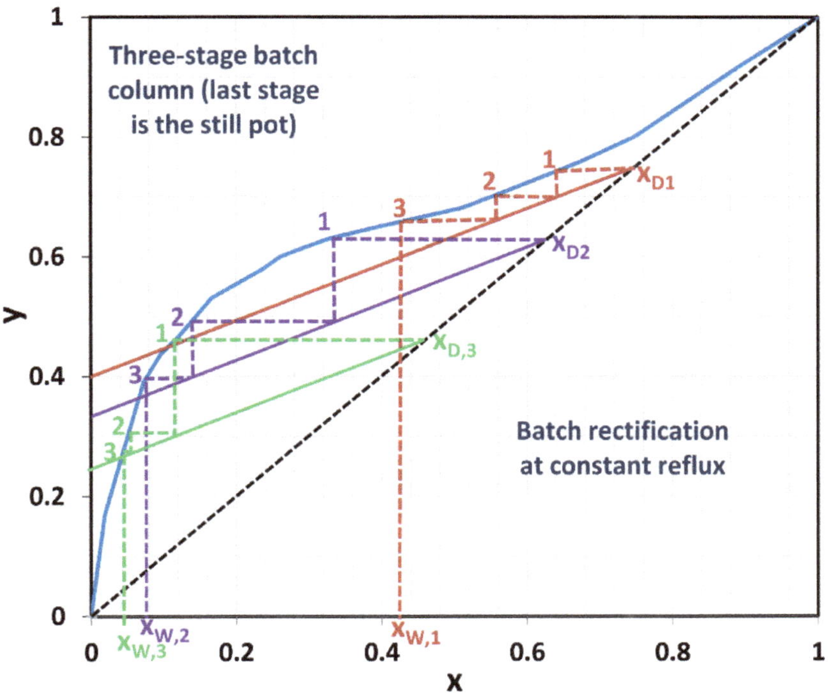

Figure 12.7 Estimation of x_W for given x_D for batch distillation with constant reflux.

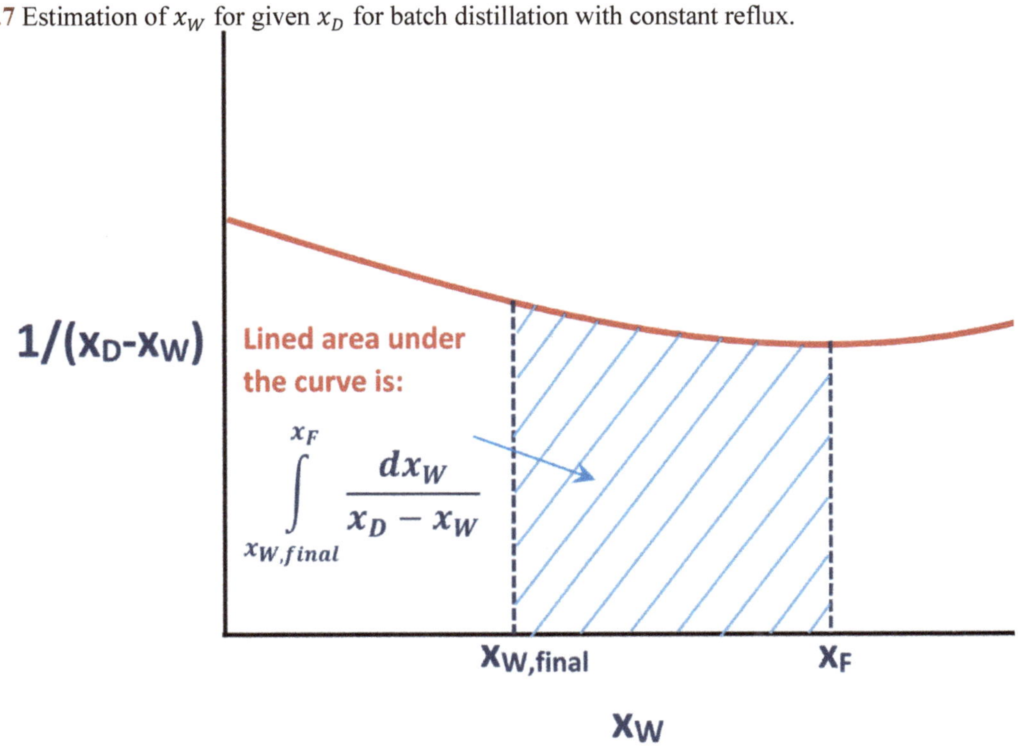

Figure 12.8 Graphical integration of function $1/(x_D - x_W)$.

One type of calculation often involved in batch distillation with reflux is as follows: Given F, x_F, $x_{W,final}$, R and the number of trays in the rectifier column, what is the final charge W left in the still pot and what is the average distillate composition $x_{D,AVG}$ and the amount of distillate D collected? To solve this problem, we can use the modified Rayleigh equation (Equation (12.20)) to first estimate the final charge W left in the still pot. This requires the evaluation of the integral using the procedure just described.

Use Equation (12.8) to calculate D and the following equation to estimate the average composition of the distillate collected:

$$x_{D,AVG} = (x_F F - x_{W,final} W)/D \tag{12.25}$$

12.5 BATCH STEAM DISTILLATION

Batch steam distillation is often used to separate a high-boiling point volatile organic compound (species A) from a mixture of volatile (species A) and non-volatile organic compounds. At atmospheric pressure, the high-boiling point volatile compound is generally not separated from a mixture by ordinary distillation as it may decompose at high temperatures. The high boiling point organic compounds tend to be large molecules which are prone to break-up when heated to high temperatures. In such cases steam distillation is often used. In steam distillation, the feed consists of two immiscible liquids namely water and organic mixture (species A + non-volatile organic compounds). The two phases are agitated and boiled together, as shown in Figure 12.9. The feed may also contain suspended solids. We assume that the feed is exposed to an external pressure of one atmosphere. Thus boiling occurs when the total vapor pressure becomes

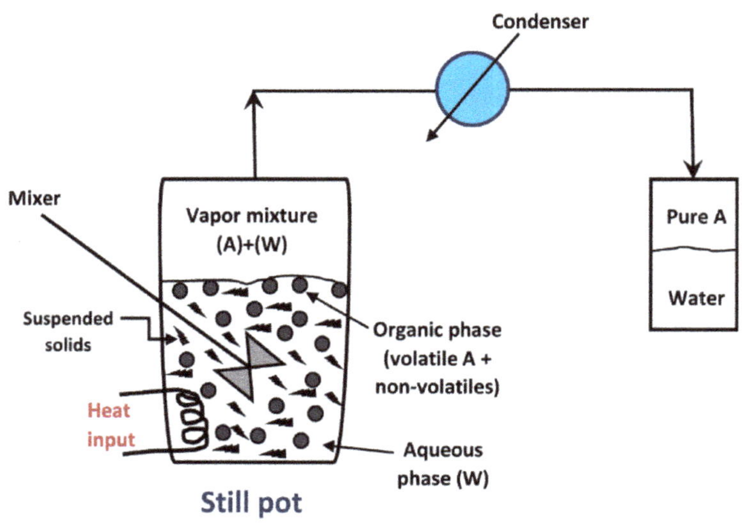

Figure 12.9 Batch steam distillation of an organic mixture of volatile and non-volatile compounds.

equal to atmospheric pressure. As the two liquid phases are immiscible, each liquid phase exerts its own partial pressure at the prevailing temperature, independent of each other. Consequently, the feed mixture boils at a temperature lower than the boiling points of pure water and pure species A, especially when $x_A =$ 1. The combined partial pressures of organic species A and water vapor (steam) reach the external pressure (atmospheric pressure) even though the vapor pressures of the individual components (species A and water) are lower than the external pressure.

The total pressure of the vapor mixture is the sum of partial pressures of species A and water vapor:

$$P = P_A + P_W \tag{12.26}$$

As the water phase in the still pot is pure, the partial pressure P_W is simply the vapor pressure of water P_W^{Sat} at the prevailing temperature. The partial pressure of species A can be estimated from the Raoult law. Thus:

$$P = x_A P_A^{Sat} + P_W^{Sat} \tag{12.27}$$

where x_A is the mole fraction species A in the organic mixture. When P in Equation (12.27) becomes equal to the external pressure (assumed one atmosphere) to which the feed is exposed, boiling of the pot liquid occurs. Thus Equation (12.27) is the boiling (bubble) point condition when P is the known external pressure to which the feed is exposed. The bubble point temperature can be determined graphically as shown in Figure 12.10. It is given by the point of intersection of two plots: $P - x_A P_A^{Sat}$ versus T plot and P_W^{Sat} versus T plot. We need vapor pressure versus temperature data for the two species (volatile organic species A and water) in order to obtain the bubble point.

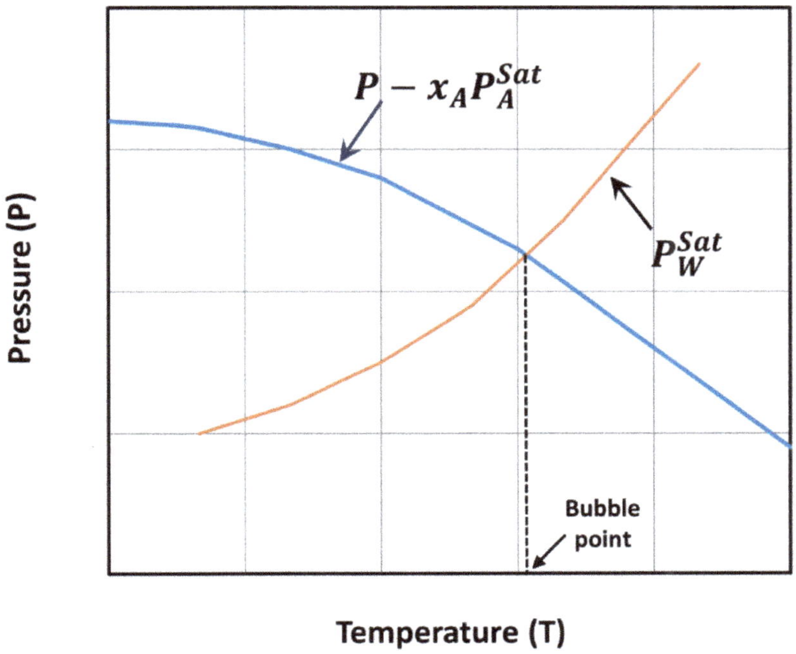

Figure 12.10 Graphical determination of bubble point temperature in steam distillation of an organic mixture of volatile and non-volatile compounds.

As an example, consider steam distillation of phenylamine. For the sake of simplicity we will assume phenylamine to be pure. The normal boiling points of phenylamine (species A) and water are:

$$T_A^{Sat} = 184\ °\text{C}\ ;\ T_W^{Sat} = 100\ °\text{C} \tag{12.28}$$

Phenylamine is immiscible in water. Thus the still pot contains two-phase phenylamine-water mixture at atmospheric pressure. This two-phase system boils at 98 °C as $P_A^{Sat} = 7.07\ kPa$ and $P_W^{Sat} = 94.30\ kPa$ and $P \approx 101.325\ kPa$ (1 atm). Note that phenylamine boils at a much lower temperature than its normal boiling point when it is boiled with water. As long as the two phases (water and phenylamine) are present, the mixture boils at the same temperature giving off vapor of constant composition:

$$y_A = \frac{p_A}{P} = \frac{P_A^{Sat}}{P}\ ;\ y_W = \frac{p_W}{P} = \frac{P_W^{Sat}}{P} \tag{12.29}$$

Let us now estimate the number of moles of steam produced to distill one mole of species A ($n_A = 1$). In the vapor phase leaving the still pot (assuming vapor to be ideal gas):

$$\frac{moles\ of\ A}{moles\ of\ steam} = \frac{n_A}{n_W} = \frac{p_A v/RT}{p_W v/RT} = \frac{x_A P_A^{Sat}}{P_W^{Sat}} \tag{12.30}$$

Upon re-arrangement, Equation (12.30) gives:

$$n_W = \frac{P_W^{Sat}}{x_A P_A^{Sat}} n_A \tag{12.31}$$

Thus, we can find the number of moles of steam produced to distill one mole of species A using Equation (12.31). If the amount of non-volatile compounds in the organic mixture is small, we can take $x_A \approx 1$ at the beginning of processing the feed mixture. The mole fraction of volatile organic in the organic mixture decreases and the still temperature increases as distillation proceeds. At the beginning of the distillation process, the temperature in the still pot is lower than the normal boiling point of water (100°C). However, at the end of the distillation process when all of the volatile organic compound is recovered, the temperature rises to the normal boiling point of water (assuming distillation is carried out at atmospheric pressure).

NOTATION

CMO constant molal overflow
D moles of distillate collected in the product receiver
$\dot{D}$ distillate molar flow rate
F initial moles of feed in the still pot
L molar flow rate of liquid stream
n_A moles of species A present in the still pot at any instant
n_{Ao} moles of species A present initially in the still pot
n_B moles of species B present in the still pot at any instant
n_{Bo} moles of species B initially in the still pot
n_W moles of steam needed to distill n_A moles of species A in batch steam distillation, see Equation (12.31)
P total pressure
P_A partial pressure of species A
P_W partial pressure of water vapor
P_A^{Sat} saturation pressure (vapor pressure) of species A
P_W^{Sat} saturation pressure (vapor pressure) of water
$\dot{Q}_C$ heat transfer rate in condenser
R reflux ratio or Universal gas constant
t time
T temperature
T_A^{Sat} saturation temperature (boiling point) of pure species A
T_W^{Sat} saturation temperature (boiling point) of water
V molar flow rate of vapor stream
W moles of liquid left inside the still pot at any time t
x mole fraction of more volatile species in liquid phase
x_A mole fraction of species A in liquid phase
x_D in batch distillation without reflux, x_D is the average mole fraction of more volatile species present in the distillate liquid product collected in the receiver. In batch distillation with reflux, x_D is the instantaneous mole fraction of more volatile species in the distillate stream, see Figure 12.6
x_F mole fraction of more volatile species in the initial feed
x_n mole fraction of more volatile species in the liquid stream leaving nth stage
x_W mole fraction of more volatile species in the liquid left inside the still pot
$x_{D,AVG}$ average mole fraction of more volatile species in the distillate liquid product collected in batch distillation with reflux
$x_{W,final}$ mole fraction of more volatile species in the liquid left inside the still pot at the end of distillation
y mole fraction of more volatile species in vapor phase
y_A mole fraction of species A in vapor phase

y_{n+1} mole fraction of more volatile species in the vapor stream leaving (n+1)th stage
y_W mole fraction of steam in vapor phase

Greek Letters

α relative volatility
ν molar volume of vapor

TUTORIAL PROBLEMS WITH SOLUTION

Problem 1

100 moles of a binary mixture of methanol and water containing 70 mol% methanol is fed to a batch still and distilled until the still residue contains 40 mol% methanol. The distillation is carried out at 1 atm pressure.

Do the following:

a) Plot the equilibrium data as $1/(y-x)$ vs. x
b) What fraction of the charge remains in the still?
c) What is the amount of distillate collected?
d) What is the composition of distillate collected?

The equilibrium data for methanol-water system at 1 atm pressure are as follows:

Composition, mole fraction of methanol	
Liquid (x)	**Vapor (y)**
0	0
0.02	0.134
0.04	0.23
0.06	0.304
0.08	0.365
0.10	0.418
0.15	0.517
0.20	0.579
0.30	0.665
0.40	0.729
0.50	0.779
0.60	0.825
0.65	0.845
0.70	0.870
0.75	0.895
0.80	0.915
0.90	0.958
0.95	0.979
1.0	1.0

Solution

a) The $1/(y-x)$ versus x data and the corresponding plot are given below.

x	1/(y-x)
0.02	8.77193
0.04	5.26316
0.06	4.09836
0.08	3.50877
0.10	3.14465
0.15	2.7248
0.20	2.63852
0.30	2.73973
0.40	3.03951
0.50	3.58423
0.60	4.44444
0.65	5.12821
0.70	5.88235
0.75	6.89655
0.80	8.69565
0.90	17.2414
0.95	34.4828

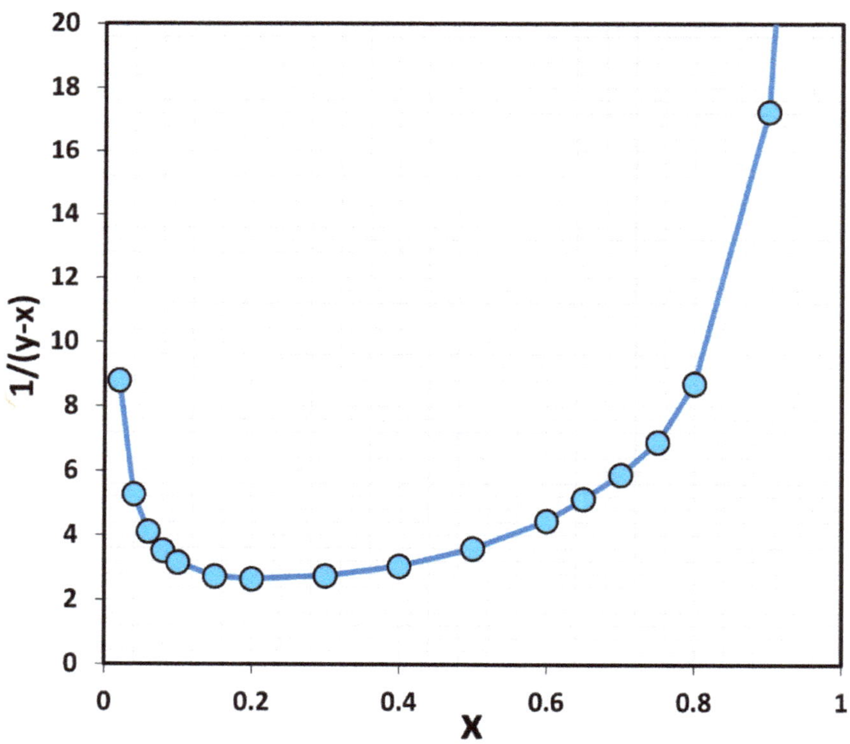

b) Given: $F = 100$ moles, $x_F = 0.70$, $x_W = 0.40$. Using the Rayleigh equation (Equation (12.7))

$$ln\left(\frac{W}{F}\right) = -\int_{0.40}^{0.70} \frac{dx}{y-x} = -Area$$

The area under the curve $1/(y-x)$ versus x curve is determined (see the Figure below) using the trapezoidal rule. The total area under the curve is approximated by 4 trapezoids as shown in the figure below. The total area of 4 trapezoids is as follows:

$$Area = 0.1 \times \left(\frac{3.0395 + 3.5842}{2}\right) + 0.1 \times \left(\frac{3.5842 + 4.4444}{2}\right) + 0.05 \times \left(\frac{4.4444 + 5.1282}{2}\right) + 0.05 \times \left(\frac{5.1282 + 5.8823}{2}\right) = 1.2472$$

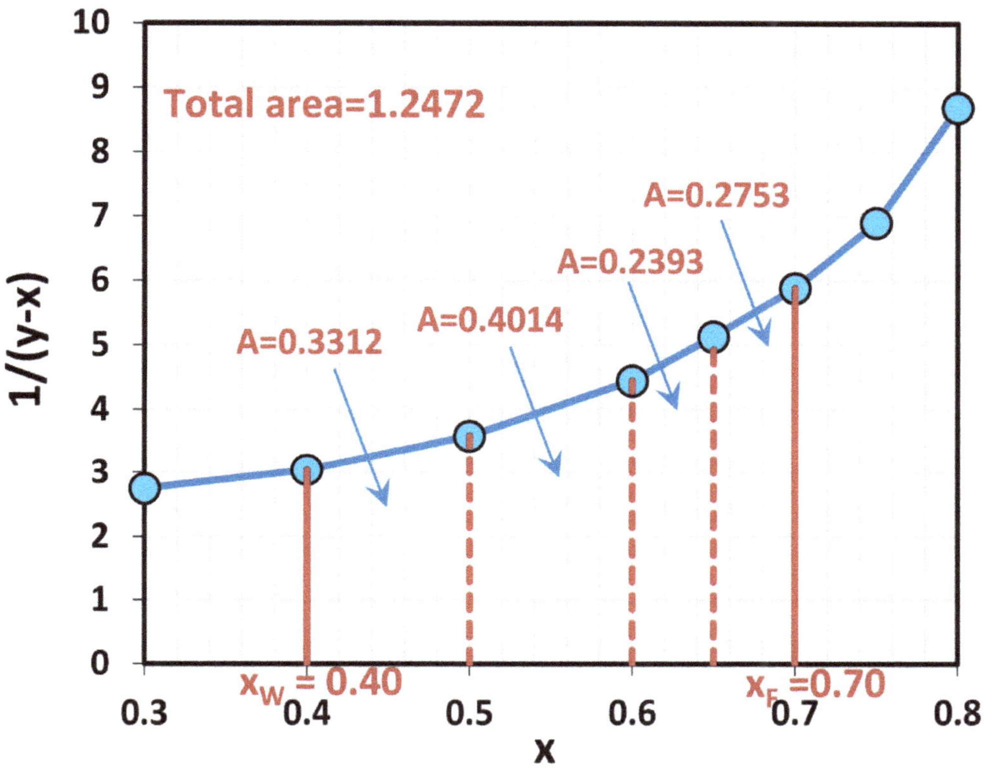

Thus:

$$ln\left(\frac{W}{F}\right) = -1.2472$$

$$\frac{W}{F} = 0.2873$$

$$W = 0.2873 \times 100 = 28.73 \text{ moles}$$

c) The amount of distillate collected is:

$$D = F - W = 100 - 28.73 = 71.27 \text{ moles}$$

d) Using Equation (12.9), we can calculate the distillate composition as:

$$x_D = \frac{x_F F - x_W W}{D} = \frac{0.70 \times 100 - 0.40 \times 28.73}{71.27} = 0.8209$$

Problem 2

A simple batch distillation is used to remove n-butane from crude pentane (mixture of n-pentane and n-butane) containing 13 mol% n-butane and 87 mol% n-pentane. The distillation is carried out at 1 atm pressure until 93 mol% of the n-butane is removed. The average relative volatility of n-butane with respect to n-pentane is 3.5. How much n-pentane is removed? What is the composition of the residue?

Solution

We will take 1 mol feed as the basis for calculations. Let the more volatile species (n-butane) be denoted by "A" and the less volatile species (n-pentane) be denoted by "B". This problem involves the application of Rayleigh equation, Equation (12.17):

$$\frac{n_A}{n_{A0}} = \left[\frac{n_B}{n_{B0}}\right]^{\alpha}$$

We are given: $\alpha = 3.5$, $n_{Ao} = 0.13, n_{Bo} = 0.87$, $n_A = 0.0091$. We need to calculate n_B:

$$\frac{n_B}{n_{B0}} = \left[\frac{n_A}{n_{A0}}\right]^{1/\alpha} = \left[\frac{0.0091}{0.13}\right]^{1/3.5} = 0.4678$$

$$n_B = 0.4678 \times 0.87 = 0.407 \text{ moles}$$

Thus n-pentane removed is $n_{Bo} - n_B = 0.87 - 0.407 = 0.463$ moles. The composition of the residue is:

$$x_W = \frac{n_A}{n_A + n_B} = \frac{0.0091}{0.0091 + 0.407} = 0.0219$$

Problem 3

A batch still is loaded with 100 kmol of a liquid containing an equimolar mixture of benzene and toluene. The liquid boil-up rate (-dW/dt) in the still pot is 10 kmol/hr. The relative volatility of benzene with respect to toluene is 2.41. Determine how long would it take for still pot composition (x_W) to decrease from 0.50 to 0.25. What would be the average distillate composition at that time?

Solution

Given: $F = 100$ kmoles, $x_F = 0.50$, $x_w = 0.25$, $-dW/dt = 10$ kmol/h, $\alpha = 2.41$.

Rayleigh equation:

$$ln\left(\frac{W}{F}\right) = \frac{1}{(\alpha-1)} ln\left[\frac{x_W(1-x_F)}{x_F(1-x_W)}\right] + ln\left[\frac{1-x_F}{1-x_W}\right]$$

$$\left(\frac{W}{F}\right) = \left[\left(\frac{x_W}{x_F}\right)\left(\frac{1-x_F}{1-x_W}\right)^{\alpha}\right]^{\frac{1}{(\alpha-1)}}$$

$$\left(\frac{W}{100}\right) = \left[\left(\frac{0.25}{0.50}\right)\left(\frac{1-0.50}{1-0.25}\right)^{2.41}\right]^{\frac{1}{(2.41-1)}} = 0.30586$$

$$W = 30.586 \text{ kmol}$$

Upon integration of $-dW/dt = 10$ using the initial condition that $W = F = 100$ kmoles, the following result is obtained:

$$W = 100 - 10t$$

We can calculate t when $W = 30.586$ kmol. Thus,

$$t = (100 - W)/10 = (100 - 30.586)/10 = 6.94\ h$$

The average distillate composition is given by:

$$x_{D,avg} = \frac{x_F F - x_W W}{D} = \frac{0.5 \times 100 - 0.25 \times 30.586}{100 - 30.586} = 0.61$$

We can re-do this problem using the other form of the Rayleigh equation, that is,

$$\frac{n_A}{n_{A0}} = \left[\frac{n_B}{n_{B0}}\right]^{\alpha}$$

Substitute $n_{A0} = 50$ kmoles, $n_{B0} = 50$ kmoles, $n_A = x_W W$, $n_B = (1 - x_W)W$ and solve for W given that $x_W = 0.25$:

$$\frac{0.25W}{50} = \left[\frac{0.75W}{50}\right]^{2.41}$$

$$W = 30.586\ \text{kmol}$$

$$W = 100 - 10t = 30.586$$

$$t = 6.94\ h$$

Problem 4

Batch distillation with reflux is used to separate a mixture of methanol and water. The distillation system consists of a still pot, a column with two equilibrium stages, and a total condenser. The feed to the still pot is an equimolar mixture of methanol and water. The amount of initial feed is 100 kmoles. The reflux ratio is 1.75. The system is operated at 1 atm pressure. The final residue composition in the still pot is 10 mol% methanol.

a) Calculate the amount of residue left in the still pot.
b) Calculate the amount of distillate collected.
c) Calculate the average composition of distillate

The equilibrium data for methanol-water system at 1 atm pressure are given in problem 1.

Solution

a) As discussed in section 12.4, we need to apply the modified Rayleigh equation, Equation (12.20):

$$ln\left(\frac{W}{F}\right) = -\int_{x_{W,final}}^{x_F} \frac{dx_W}{x_D - x_W} = -Area$$

We first need to generate $1/(x_D - x_W)$ versus x_W data. We select a number of x_D values and plot the corresponding operating lines. For a fixed reflux ratio (R=1.75), all the operating lines are parallel with y-intercepts of $x_D/(1 + R)$. The family of operating lines is shown in the figure given below.

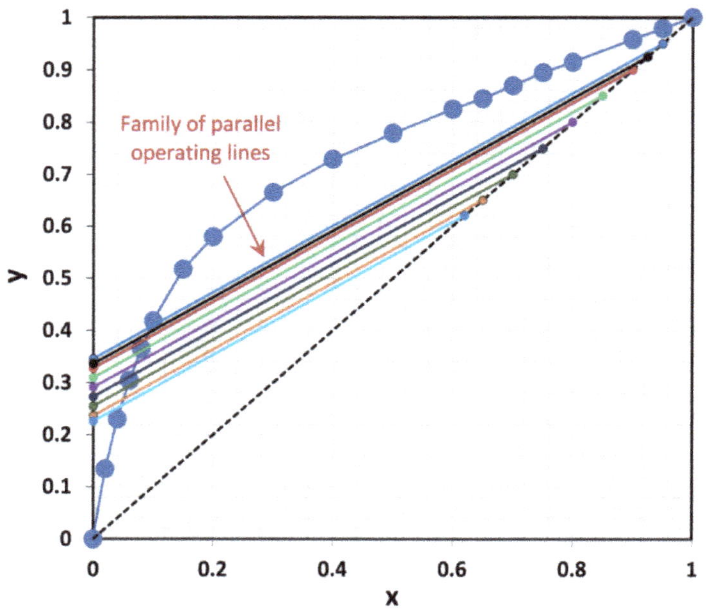

For any given operating line, the x_W value is determined by stepping off 3 stages (2 stages within the column and 1 still pot stage) as illustrated in the figure given below. The data of $1/(x_D - x_W)$ versus x_W is plotted and the area under the curve $1/(x_D - x_W)$ versus x_W is determined using the trapezoidal rule as shown in the figure given below.

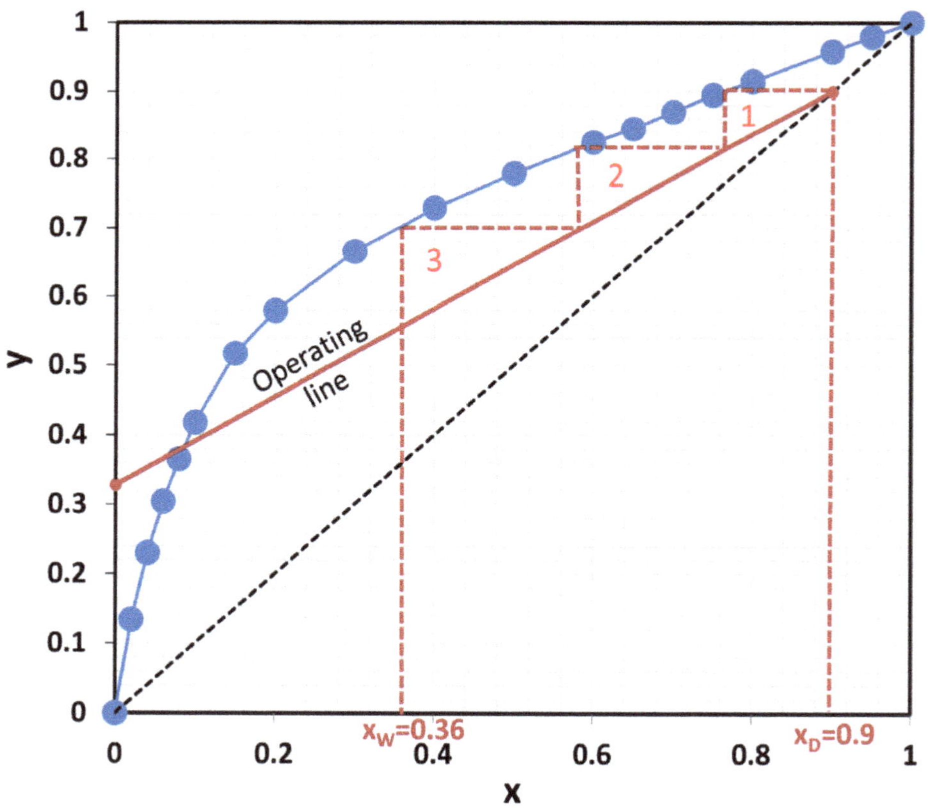

The total area under the curve is approximated by 4 trapezoids as shown in the figure below. The total area of 4 trapezoids is as follows:

$$Area = (0.13 - 0.1) \times \left(\frac{1.51 + 1.49}{2}\right) + (0.197 - 0.13) \times \left(\frac{1.49 + 1.53}{2}\right) + (0.36 - 0.197) \times \left(\frac{1.53 + 1.85}{2}\right) + (0.5 - 0.36) \times \left(\frac{1.85 + 2.35}{2}\right) = 0.716$$

Thus:

$$ln\left(\frac{W}{F}\right) = -0.716$$

$$\frac{W}{F} = 0.489$$

$$W = 0.489 \times 100 = 48.9 \text{ kmoles}$$

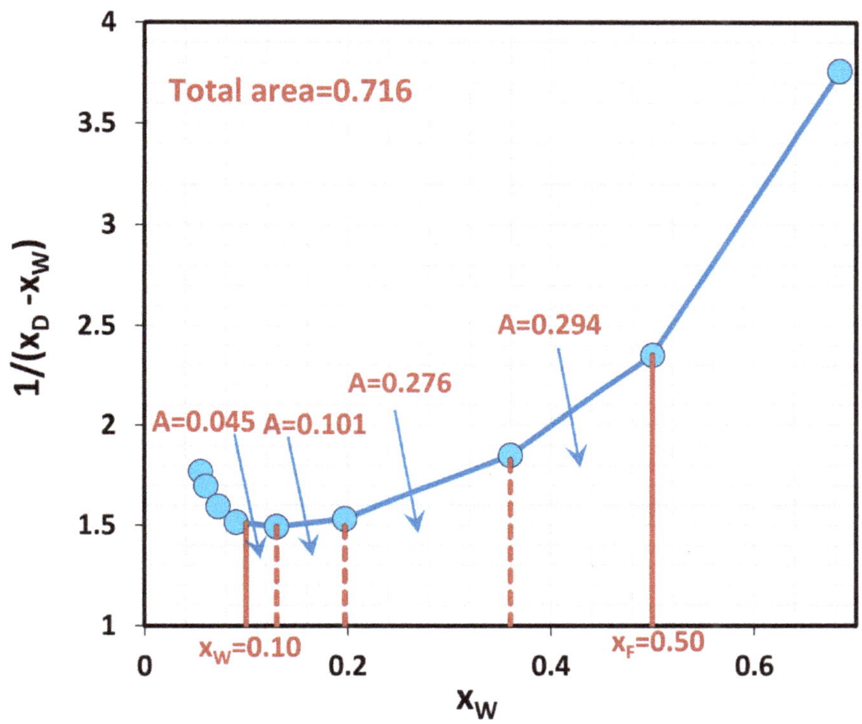

b) The amount of distillate collected is:

$$D = F - W = 100 - 48.9 = 51.1 \text{ kmoles}$$

c) The average composition of the distillate product is:

$$x_D = \frac{x_F F - x_W W}{D} = \frac{0.50 \times 100 - 0.10 \times 48.9}{51.1} = 0.883$$

Problem 5

A valuable volatile hydrocarbon (species A) is accidently contaminated with a non-volatile oil. The concentration of volatile hydrocarbon (species A) in the mixture is 75 mole%. We would like to recover species A (valuable hydrocarbon) from the mixture of species A and non-volatile oil. The mixture is batch distilled with water at atmospheric pressure. The mixture is immiscible in water. The vapor mixture (species A + steam) leaving the batch still is condensed using a condenser. The initial rate of production of pure A is 1 kmol/h.

a) Determine the distillation temperature at the beginning of the distillation process
b) What is the initial rate of steam generation?
c) What is the initial rate of heat addition to the process? Given that $\Delta H_W^{vap} = 40.8$ kJ/mol and $\Delta H_A^{vap} = 48$ kJ/mol
d) What is the temperature at the end of the distillation process when all of species A is recovered?

The Antoine equations for volatile hydrocarbon (species A) and water are as follows:

Species A:

$$log_{10}P^{Sat}(bar) = 4.34541 - \frac{1661.858}{T(K) - 74.048}$$

Water:

$$lnP^{Sat}(kPa) = 16.262 - \frac{3799.89}{T^oC + 226.35}$$

Solution

a) Distillation temperature at the beginning of the distillation process

The vapor pressure data (P_A^{Sat} and P_W^{Sat}) as a function of temperature are calculated from the Antoine equations. The total pressure $P = 101.325$ kPa. The mole fraction of volatile hydrocarbon (species A), that is, x_A, is 0.75 initially. The distillation temperature at the beginning of the distillation process is determined from the intersection of $P - x_A P_A^{Sat}$ and P_W^{Sat} plots as shown in the figure below. It is 98.8 °C. Note that this is much lower than the normal boiling point of pure species A which is approximately 184 °C.

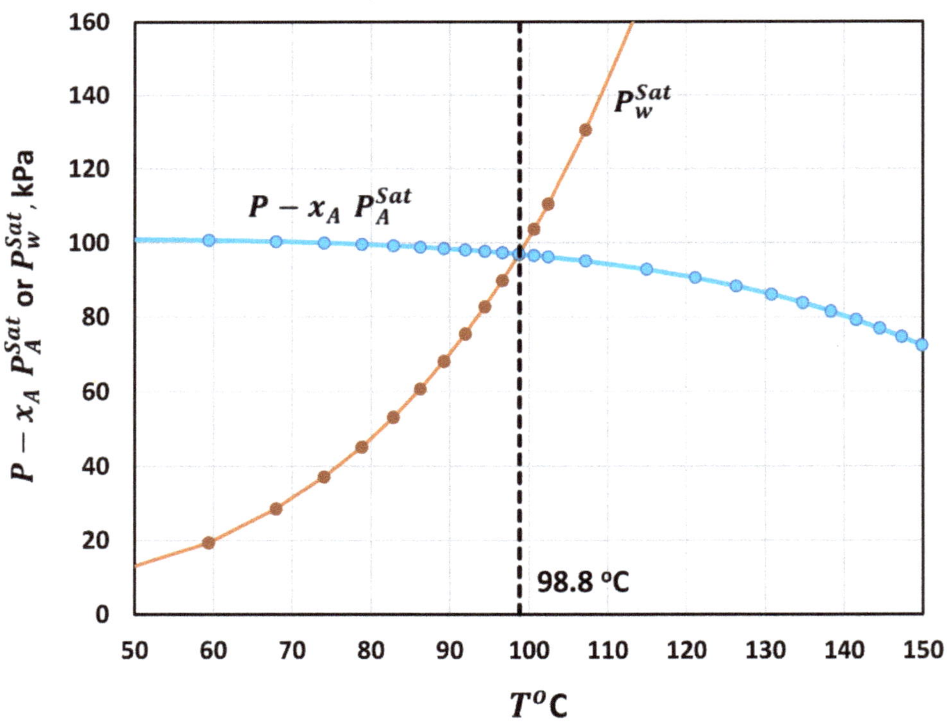

b) The initial rate of steam generation

$$n_W = \frac{P_W^{Sat}}{x_A P_A^{Sat}} n_A$$

$n_A = 1$ kmol/h, $x_A = 0.75$, P_A^{Sat} =5.846 kPa, $P_W^{Sat} = 97.07$ kPa

$$n_W = \frac{P_W^{Sat}}{x_A P_A^{Sat}} n_A = \frac{97.07}{0.75 \times 5.846} \times 1 = 22.139 \text{ kmol/h}$$

c) The initial rate of heat addition to the process

All the heat that is added goes into vaporization of species A and water. Thus, the rate of heat addition is:

$$\dot{Q} = n_W \Delta H_W^{vap} + n_A \Delta H_A^{vap}$$

$$\dot{Q} = 22.139 \times 40.8 \times 10^3 + 1 \times 48 \times 10^3 = 9.5127 \times 10^5 \text{ kJ/h} = 264.24 \text{ kW}$$

d) The temperature at the end of the distillation process when all of species A is recovered

The temperature at the end of the distillation process when all of species A is recovered is the normal boiling point of water, that is, 100 °C.

SHORT REVIEW QUESTIONS

1. In simple batch distillation, as the more volatile component is distilled off, the temperature in the still pot rises continuously. Is this true or false?
2. As batch distillation progresses, the instantaneous mole fraction of the more volatile component in the vapor leaving the still pot increases. Is this true or false?
3. Consider simple batch distillation with an initial charge $F = 125$ kmoles containing equimolar mixture of benzene and toluene. The time it takes to collect the distillate D in the amount of 55 kmoles is 5 h. Determine the relationship giving the amount of residue as a function of time. Assume that the liquid boilup rate in the still pot is constant.
4. Starting from one form of a Rayleigh equation, Equation (12.16), derive the other form of the Rayleigh equation, Equation (12.17).
5. What is the advantage of using liquid reflux in batch distillation?
6. Where is steam batch distillation used?
7. When the system consists of two immiscible liquid phases, each liquid phase exerts its own partial pressure at the prevailing temperature, independent of each other. Is this true or false?

PRACTICE PROBLEMS

Problem 1

A simple batch still is used to separate ethanol from water. The still is loaded with 100 kmoles of 23 mol% ethanol mixture. The distillation is carried out at 1 atm pressure. At the end of distillation, the still pot composition is 3 mol% ethanol.

Do the following:

a) Calculate the amount of distillate collected.
b) Calculate the amount of residue left in the still pot.
c) Calculate the average composition of distillate

The equilibrium data for ethanol-water system at 1 atm pressure are given below:

Composition, mole fraction of ethanol	
Liquid (x)	**Vapor (y)**
0	0
0.019	0.17
0.0721	0.3891
0.0966	0.4375
0.1238	0.4704
0.1661	0.5089
0.2377	0.5445
0.2608	0.558
0.3273	0.5826
0.3965	0.6122
0.5079	0.6564
0.5198	0.6599
0.5732	0.6841
0.6763	0.7385
0.7472	0.7815
0.8943	0.8943
1.0	1.0

Problem 2
50 kmoles of a binary mixture of benzene and toluene containing 40 mol% benzene is loaded in a batch still. The liquid boilup rate in the still pot (-dW/dt) is maintained at 4 kmol/h. After 5 hours of distillation, the composition of liquid left inside the pot is 19 mol% benzene.

a) Calculate the amount of liquid left inside the pot after 5 hours.
b) Calculate the amount of distillate collected after 5 hours.
c) Calculate the average composition of distillate after 5 hours.
d) Estimate the relative volatility using the Rayleigh equation.

Problem 3
An equimolar mixture of benzene and toluene is fed to a batch still and distilled at 1 atm pressure. The amount of mixture fed is 75 kmoles. The liquid boilup rate in the still pot (-dW/dt) is maintained at 7.5 kmol/h. The relative volatility of benzene with respect to toluene is nearly constant and has a value of approximately 2.40.

Do the following:

a) Plot instantaneous vapor composition as a function of time.
b) Plot still pot composition as a function of time.
c) Plot average distillate composition as a function of time.

Problem 4

The following equations are different forms of the Rayleigh equation valid for constant relative volatility (α) systems:

$$ln\left(\frac{W}{F}\right) = \frac{1}{(\alpha - 1)} ln\left[\frac{x_W(1 - x_F)}{x_F(1 - x_W)}\right] + ln\left[\frac{1 - x_F}{1 - x_W}\right]$$

$$\frac{n_A}{n_{A0}} = \left[\frac{n_B}{n_{B0}}\right]^{\alpha}$$

Starting from any one form of the Rayleigh equation, derive the other form. In other words, show that the above forms of the Rayleigh equation are equivalent.

Problem 5

100 kmol of ethanol-water mixture are boiled off from an initial amount of 175 kmol of ethanol-water mixture containing 55 mol% ethanol. The volatility of ethanol with respect to water is approximately 1.75. Determine the mole fraction of ethanol in the distillate product. What is the mole fraction of ethanol in the residual mixture?

Problem 6

A batch still is used to separate pentane from a mixture of pentane and hexane. The still is loaded with 125 kmol of a liquid containing 40 mol% pentane and 60 mol% hexane. The liquid boil-up rate (-dW/dt) in the still pot is 12 kmol/hr. The relative volatility of pentane with respect to hexane is 2.95. Determine how long would it take for the still pot pentane mole fraction (x_W) to decrease from 0.40 to 0.20. What would be the average distillate composition at that time?

Problem 7

Batch distillation with reflux is used to separate a mixture of ethanol and water. The distillation system consists of a still pot, a column with two equilibrium stages, and a total condenser. The feed to the still pot is 40 mol% ethanol mixture. The amount of initial feed is 100 kmoles. The reflux ratio is 0.75. The system is operated at 1 atm pressure. The final residue composition in the still pot is 6 mol% ethanol.

a) Calculate the amount of distillate collected.
b) Calculate the amount of residue left in the still pot.
c) Calculate the average composition of distillate

The equilibrium data for ethanol-water system at 1 atm pressure are given in problem 1.

Problem 8

A hydrocarbon mixture consists of 15 mol% volatile species A and 85 mol% of non-volatile components. We would like to recover species A from this hydrocarbon mixture. The mixture is batch distilled with water at atmospheric pressure. The mixture is immiscible in water.

a) Determine the distillation temperature at the beginning of the distillation process
b) What is the amount of species A produced per mole of steam generated at the beginning of the distillation process?
c) Determine the distillation temperature for the following values of x_A: 1, 0.9, 0.8, 0.7, 0.6, 0,5, 0.4, 0.2, 0. Plot the distillation temperature as a function of x_A.

The Antoine equations for volatile hydrocarbon (species A) and water are as follows:

Species A:

$$log_{10}P^{Sat}(\text{mmHg}) = 7.0151 - \frac{1246.33}{T^oC + 232.988}$$

Water:

$$lnP^{Sat}(kPa) = 16.262 - \frac{3799.89}{T^oC + 226.35}$$

Chapter 13

Multicomponent Distillation

13.1 LEARNING OUTCOMES

Upon completion of this chapter, students will be able to:

- Define key and non-key components
- Select key components from a multicomponent mixture
- Define Class 1 and Class 2 separations
- Derive Fenske equation for multicomponent distillation
- Derive Underwood equations for multicomponent distillation
- Apply FUG method to estimate the number of theoretical stages required for a given separation in multicomponent distillation
- Explain rigorous calculation procedure for the analysis and rating of distillation columns for multicomponent systems
- Develop MESH equations for a general equilibrium stage in multicomponent distillation
- Explain solution procedure for MESH equations of multicomponent distillation
- Explain sequencing of distillation columns for multicomponent distillation
- Develop possible sequences of distillation columns in multicomponent distillation
- Explain heuristics relevant to sequencing of distillation columns in multicomponent distillation

13.2 INTRODUCTION

In industrial distillation processes, the mixtures to be separated are often multi-component mixtures containing three or more components. The distillation of multicomponent mixtures is carried out using the same type of equipment: distillation columns, condensers and reboilers. However, there are significant differences between distillation of binary mixtures and multicomponent mixtures as far as the analysis and design of distillation columns is concerned.

In binary distillation, the mole fractions of only one component (more volatile component) need to be specified in the distillate and bottoms products. The mole fractions of the second less volatile component are automatically specified. Thus the compositions of the distillate and bottoms products are completely specified when designing binary distillation columns. Furthermore, only one distillation tower is sufficient in principle to separate a binary mixture (species A and B) into two pure product streams: pure A as distillate stream and pure B as bottoms product stream.

If we use a single distillation column for the separation of a multicomponent mixture, it is not possible to obtain more than one pure product (single-component) stream. Using a single distillation column, one can at best separate a multicomponent mixture of C species into one pure product stream and one mixture stream consisting of the remaining $C - 1$ components. In general, the product streams (distillate and bottoms) consist of mixtures. The distillate product is a mixture of lighter components and the bottoms product is a mixture of heavier components. In order to separate a multicomponent mixture of C species into pure components (that is, C pure product streams) we certainly need more than one distillation column. In fact, a total of $C - 1$ distillation columns are necessary to separate a C-component mixture into its pure components.

In the discussion to follow, we will first consider the separation of multicomponent mixtures using a single distillation column. The separation of multicomponent mixtures into pure component streams using a sequence of distillation columns is discussed next.

13.3 SEPARATION OF MULTICOMPONENT MIXTURES USING A SINGLE DISTILLATION COLUMN

There are approximate methods available for the design of distillation columns for multicomponent mixtures. For example, the approximate FUG (Fenske-Underwood-Gilliland) method of designing binary distillation columns discussed in Chapter 10 can be extended to multicomponent distillation. The extended FUG method allows the designer to determine the number of stages for a specified separation of so-called key components of a multicomponent mixture. Rigorous calculation programs are also available for rating the performance of existing (or specified) distillation columns in separation of multicomponent mixtures.

13.3.1 Key and Non-Key Components

In the design of a single distillation column for multicomponent separation, the distillate and the bottoms compositions cannot be specified completely. Only the mole fractions, or preferably fractional recoveries, of *two components* can and must be specified. Thus we have to *select* two components, referred to as *key components*, from a multicomponent mixture whose mole fractions or fractional recoveries would be specified in the distillate and bottoms products. The key components must be present in both distillate and bottoms products.

How to Select the Key Components

- Line up the components in order of their volatility, from highly volatile (lowest boiling point) to least volatile (highest boiling point)
- Select a component of high volatility (not necessarily the highest volatility) as a *light key* (*LK*). *LK* is the component that is desired to be kept out of the bottoms product, although not completely.
- Select a component of low volatility (not necessarily the lowest volatility) as a *heavy key* (*HK*). *HK* is the component that is desired to be kept out of the distillate product, although not completely.
- Both *LK* and *HK* components must be *distributed* meaning they must be present in both top (distillate) and bottoms product streams.
- All other components present in the feed are referred to as *non-key* (*NK*) components.
- The component lighter (more volatile) than the *LK*, if present, is called *light non-key component (LNK)* and the component heavier (less volatile) than the *HK*, if present, is called *heavy non-key component (HNK)*.
- The non-key components which are present in both top and bottoms products are called *distributed* components and the non-key components which are present only in one of the product streams (top or bottom) are called *non-distributed* components.

If the keys are listed next to each other in the list of components in order of volatility, the keys are called *adjacent keys*. The keys are called *split keys* if some other component or components lie between them in the order. The keys are usually chosen as adjacent keys as that leads to a sharp separation. When the keys are split keys, the separation is sloppy.

As an example, consider a multicomponent mixture of 5 components given in Table 13.1. The relative volatilities are calculated and listed in order with component 3 as the reference species. The light key is chosen as component 2 and the heavy key is chosen as component 3. The keys are “adjacent keys” as they are adjacent in the list of components in order of volatility. Component 1 is light non-key component and components 4 and 5 are heavy non-key components.

Table 13.1: A multicomponent system with adjacent keys

Component	Relative volatility (α)	Designation	Comments
1	3	LNK	The keys are "adjacent keys"
2	2.5	LK	
3	1	HK	
4	0.75	HNK	
5	0.50	HNK	

Table 13.2 describes a multicomponent mixture of 7 components. The relative volatilities are calculated and listed in order with component 5 as the reference species. The light key is chosen as component 2 and the heavy key is chosen as component 5. The keys are "split keys" as there are 2 components which lie between the key components.

Table 13.2: A multicomponent system with split keys

Component	Relative volatility (α)	Designation	Comments
1	2.5	LNK	The keys are "split keys"
2	2.0	LK	
3	1.5	NK	
4	1.2	NK	
5	1.0	HK	
6	0.75	HNK	
7	0.50	HNK	

13.3.2 Class 1 and Class 2 Separations

Multicomponent distillation can be classified into two groups: Class 1 separation and Class 2 separation. In Class 1 separation, all components in the feed appear in both distillate and bottoms products, that is, all components are distributed. This occurs when a narrow boiling range mixture is distilled. The degree of separation between the key components in Class 1 separation is generally sloppy. There is only *one pinch point* in class 1 separation at minimum reflux and the pinch condition occurs at the feed stage (see Figure 13.1). In Class 2 separations, one or more non-key components are non-distributed meaning these non-key components appear only in one of the product streams (distillate or bottoms). In such cases, we have two pinch points at minimum reflux– one above the feed stage and one below the feed stage (see Figure 13.2). Class 2 separations are more common in industrial applications. Note that the compositions of vapor and liquid phases hardly change from stage to stage near the pinch point. The regions of constant composition near the pinch point are referred to as *pinch-point zones* or *invariant zones*.

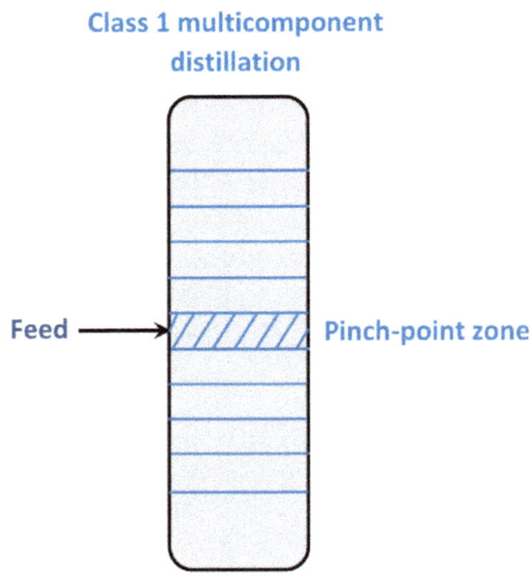

Figure 13.1 Multicomponent distillation- Class 1 separation. At minimum reflux, there is one pinch point zone at feed stage.

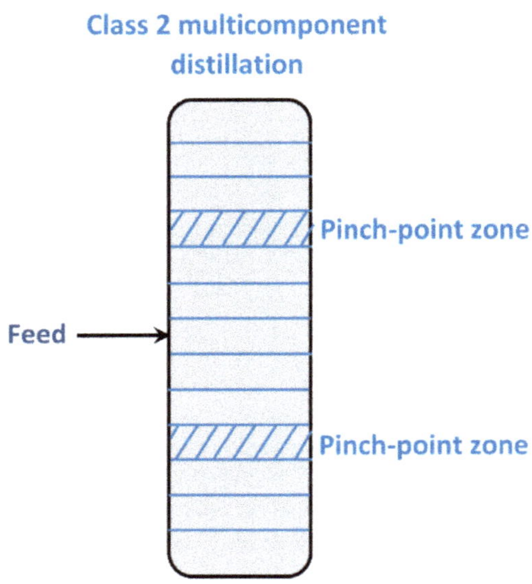

Figure 13.2 Multicomponent distillation- Class 2 separation. At minimum reflux, there are two pinch point zones, one above the feed stage and one below the feed stage.

13.3.3 Approximate Methods of Designing Multicomponent Distillation Columns

There are two approximate methods available for the design of multicomponent distillation columns. They are: pseudo-binary method and FUG method.

In the pseudo-binary method of designing multicomponent distillation columns, the multicomponent mixture is treated as a pseudo-binary mixture of two key components. Only the moles of *LK* and *HK* are considered in the calculations of the compositions. The McCabe-Thiele method is then used to design the distillation column. This method is a good approximation for Class 1 multicomponent separation when the amounts of non-key components present in the feed are small (less than 10%) and the mixtures are near-ideal in nature. When the amounts of the non-key components present in the mixture are substantial, this method is not accurate.

The FUG method of designing distillation columns for binary mixtures described in Chapter 10 can be readily extended to multicomponent distillation. To estimate the number of theoretical stages required for a given separation (specified in terms of the key components), we need to estimate N_{min} and R_{min}. The minimum number of stages N_{min} required for a given separation can be estimated from the Fenske equation adapted to multicomponent distillation. The minimum reflux ratio R_{min} can be estimated from the Underwood equations developed for multicomponent distillation.

Fenske Equation for Multicomponent Distillation

In Chapter 10, the following Fenske equation was derived for a binary system of species A and B:

$$N_{min} = \frac{ln[(x_A/x_B)_D/(x_A/x_B)_W]}{ln(\alpha_{AB})} \tag{13.1}$$

where subscript D refers to distillate product, subscript W refers to bottoms product, x is the mole fraction of a species, and α_{AB} is the volatility of species A relative to species B. The Fenske equation of binary system can be extended to any two components of a multicomponent system as (M.R. Fenske, Industrial and Engineering Chemistry, Vol. 24, 482-485, 1932):

$$N_{min} = \frac{ln\left[(x_i/x_j)_D/(x_i/x_j)_W\right]}{ln(\alpha_{ij})} \tag{13.2}$$

A more useful form of the Fenske equation is expressed in terms of the fractional recovery (FR) of components as follows:

$$N_{min} = \frac{ln\left[\{FR_i/(1-FR_i)\}_D/\{(1-FR_j)/FR_j\}_W\right]}{ln(\alpha_{ij})} \tag{13.3}$$

In order to calculate N_{min} from Equation (13.3), we just need the fractional recovery of species i in the distillate product and the fractional recovery of species j in the bottoms product, in addition to the relative volatility α_{ij}.

The following inverted form of the Fenske equation, Equation (13.3), is also useful:

$$FR_{i,D} = \frac{\alpha_{ij}^{N_{min}}}{\alpha_{ij}^{N_{min}}+\{FR_j/(1-FR_j)\}_W} \tag{13.4}$$

When applied to key and non-key components, Equations (13.2) to (13.4) can be re-written as:

$$N_{min} = \frac{ln[(x_{LK}/x_{HK})_D/(x_{LK}/x_{HK})_W]}{ln(\alpha_{LK,HK})} \tag{13.5}$$

$$N_{min} = \frac{ln[\{FR_{LK}/(1-FR_{LK})\}_D/\{(1-FR_{HK})/FR_{HK}\}_W]}{ln(\alpha_{LK,HK})} \tag{13.6}$$

$$FR_{NK,D} = \frac{\alpha_{NK,HK}^{N_{min}}}{\alpha_{NK,HK}^{N_{min}}+\{FR_{HK}/(1-FR_{HK})\}_W} \tag{13.7}$$

Equation (13.5) or (13.6) can be used to estimate the minimum number of stages at total reflux and Equation (13.7) can be used to determine the fractional recoveries of non-key components (fractional recoveries of the key components are already specified) in the distillate product at total reflux. Note that the fractional recovery of any component (key or non-key) in the bottoms product can be calculated from:

$$FR_{i,W} = 1 - FR_{i,D} \tag{13.8}$$

where

$$FR_{i,D} = x_{i,D}D/(z_iF) \tag{13.9}$$

$$FR_{i,W} = x_{i,W}W/(z_iF) \tag{13.10}$$

Here F is the feed flow rate, D is the distillate flow rate, W is the bottoms flow rate, and z_i is the mole fraction of species i in the feed.

Underwood Equations for Multicomponent Distillation

Underwood (A.J.V. Underwood, Chemical Engineering Progress, Vol. 44, 603-614, 1948) developed an approximate but fairly accurate method of determining the minimum reflux ratio R_{min} for multicomponent distillation. When one or more NK components are non-distributed (Class 2 separation), we have two pinch-point zones at minimum reflux, one above the feed stage and one below the feed stage, as already noted in section 13.3.2. In the pinch or invariant zone, composition and temperature are nearly constant. Consider the *rectifying section* of a distillation column operating at minimum reflux with a pinch-point zone above the feed stage (see Figure 13.3). A component mass balance on species i gives:

$$V_{min}y_{i,p} = x_{i,p}L_{min} + x_{i,D}D \tag{13.11}$$

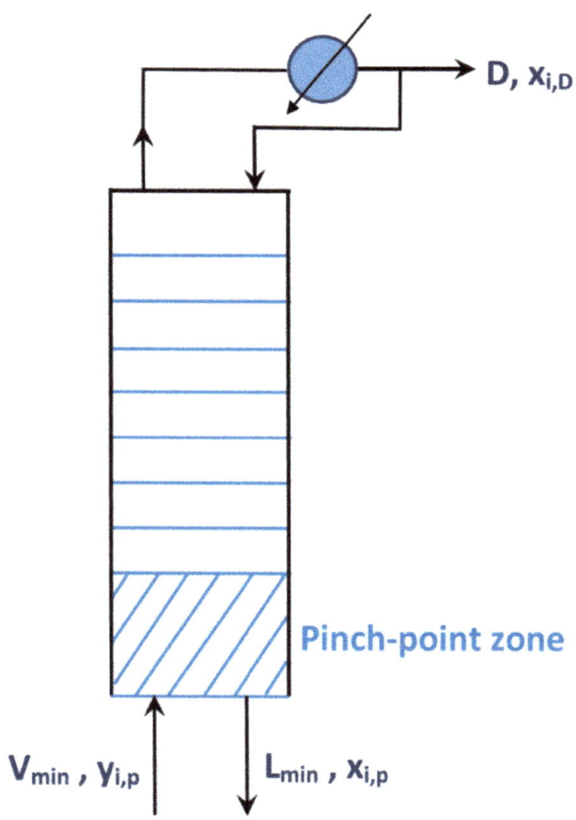

Figure 13.3 Rectifying section with pinch-point zone at minimum reflux.

The total mass balance gives:

$$V_{min} = D + L_{min} \tag{13.12}$$

As the liquid and vapor compositions are constant in the pinch-point zone, we can write the phase equilibrium relation as:

$$y_{i,p} = K_i x_{i,p} \tag{13.13}$$

Equation (13.13) could also be expressed in terms of the relative volatility α_{ir} as:

$$y_{i,p} = \alpha_{ir} K_r x_{i,p} \tag{13.14}$$

where α_{ir} is defined as:

$$\alpha_{ir} = K_i / K_r \tag{13.15}$$

From Equations (13.11) and (13.14), it follows that:

$$V_{min} y_{i,p} = \frac{\alpha_{ir} x_{i,D} D}{\left[\alpha_{ir} - \left(\frac{L_{min}}{K_r V_{min}}\right)\right]} \tag{13.16}$$

When summed over all components, Equation (13.16) gives:

$$V_{min} \sum y_{i,p} = V_{min} = \sum \frac{\alpha_{ir} x_{i,D} D}{\left[\alpha_{ir} - \left(\frac{L_{min}}{K_r V_{min}}\right)\right]} \tag{13.17}$$

Defining $\varphi_E = L_{min}/(K_r V_{min})$, Equation (13.17) can be re-written as:

$$V_{min} = \sum \frac{\alpha_{ir} x_{i,D} D}{[\alpha_{ir} - \varphi_E]} \tag{13.18}$$

From total mass balance, Equation (13.12), and Equation (13.18), we get:

$$R_{min} + 1 = \sum \frac{\alpha_{ir} x_{i,D}}{[\alpha_{ir} - \varphi_E]} \tag{13.19}$$

where $R_{min} = L_{min}/D$.

Let us now consider the *stripping section* of a distillation column operating at minimum reflux with a pinch-point zone below the feed stage (see Figure 13.4). A component mass balance on species i gives:

$$\bar{V}_{min} y_{i,p} = x_{i,p} \bar{L}_{min} - x_{i,W} W \tag{13.20}$$

As the liquid and vapor compositions are constant in the pinch-point zone, we can write the phase equilibrium relation as:

$$y_{i,p} = \alpha_{ir} K_r x_{i,p} \tag{13.21}$$

where $\alpha_{i,r}$ is defined in Equation (13.15). From Equations (13.20) and (13.21), it follows that:

$$-\bar{V}_{min} y_{i,p} = \frac{\alpha_{ir} x_{i,w} W}{\left[\alpha_{ir} - \left(\frac{\bar{L}_{min}}{K_r \bar{V}_{min}}\right)\right]} \tag{13.22}$$

When summed over all components, Equation (13.22) gives:

$$-\bar{V}_{min} \sum y_{i,p} = -\bar{V}_{min} = \sum \frac{\alpha_{ir} x_{i,w} W}{\left[\alpha_{ir} - \left(\frac{\bar{L}_{min}}{K_r \bar{V}_{min}}\right)\right]} \tag{13.23}$$

Defining $\varphi_S = \bar{L}_{min}/(K_r \bar{V}_{min})$, Equation (13.23) can be re-written as:

$$-\bar{V}_{min} = \sum \frac{\alpha_{ir} x_{i,w} W}{[\alpha_{ir} - \varphi_S]} \tag{13.24}$$

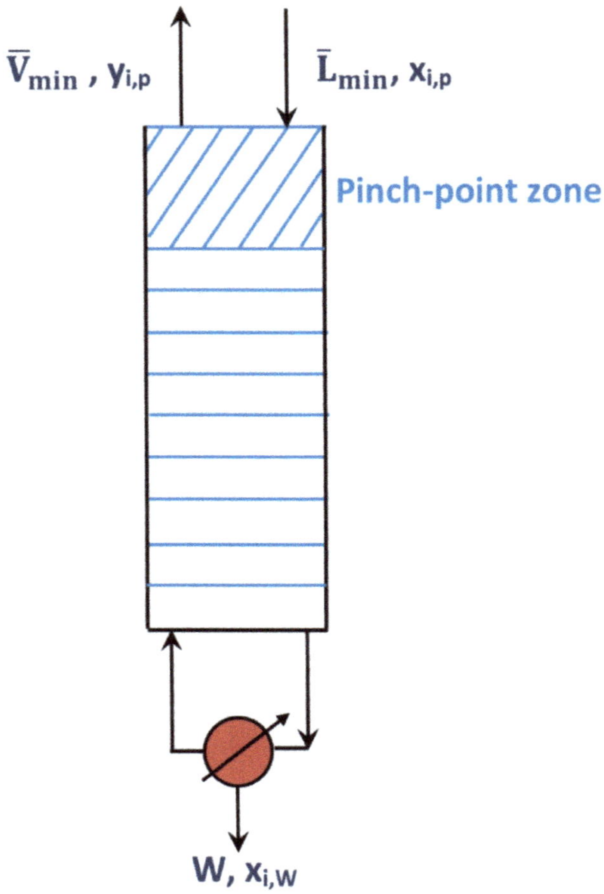

Figure 13.4 Stripping section with pinch-point zone at minimum reflux.

Underwood showed that if the relative volatilities are constant and CMO is valid in both enriching and stripping sections, there are common values of φ_E and φ_S which satisfy both enriching section equation (Equation (13.18) or (13.19)) and stripping section equation (Equation (13.24)). Thus we can take $\varphi_E = \varphi_S = \varphi$ and re-write Equations (13.18), (13.19) and (13.24) as follows:

$$V_{min} = \sum \frac{\alpha_{ir} x_{i,D} D}{[\alpha_{ir} - \varphi]} \tag{13.25}$$

$$R_{min} + 1 = \sum \frac{\alpha_{ir} x_{i,D}}{[\alpha_{ir} - \varphi]} \tag{13.26}$$

$$-\bar{V}_{min} = \sum \frac{\alpha_{ir} x_{i,w} W}{[\alpha_{ir} - \varphi]} \quad (13.27)$$

If we add Equations (13.25) and (13.27), we get:

$$V_{min} - \bar{V}_{min} = \sum \frac{\alpha_{ir} x_{i,D} D}{[\alpha_{ir} - \varphi]} + \frac{\alpha_{ir} x_{i,w} W}{[\alpha_{ir} - \varphi]} \quad (13.28)$$

From mass balance around the feed stage, it can be readily shown that:

$$V_{min} - \bar{V}_{min} = F(1 - q) \quad (13.29)$$

The overall component mass balance on the column gives:

$$F z_i = x_{i,D} D + x_{i,W} W \quad (13.30)$$

From Equations (13.28) to (13.29), it can be shown that:

$$1 - q = \sum \frac{\alpha_{ir} z_i}{\alpha_{ir} - \varphi} \quad (13.31)$$

In summary, the following two Underwood equations need to be solved to determine the minimum reflux ratio:

$$1 - q = \sum \frac{\alpha_{iHK} z_i}{\alpha_{iHK} - \varphi} \quad (13.32)$$

$$R_{min} + 1 = \sum \frac{\alpha_{iHK} x_{i,D}}{[\alpha_{iHK} - \varphi]} \quad (13.33)$$

Equation (13.32) is the same as Equation (13.31) with reference species as the heavy key. Equation (13.33) is the same as Equation (13.26) with reference species as the heavy key. Equation (13.32) is referred to as *First Underwood Equation* and Equation (13.33) is referred to as *Second Underwood Equation.* All components of feed are included in the summation of the first Underwood equation whereas only those components present in the distillate product are included in the summation of the second Underwood equation. The first Underwood equation, Equation (13.32), is a polynomial of degree C (number of components in the feed) with C different roots: one between adjacent pair of α-values and one between 0 and the smallest α. As an example, consider a 5 component system of Table 13.1. The five roots of Equation (13.32) in this case are as follows:

$$2.5 < \varphi_1 < 3\,;\; 1 < \varphi_2 < 2.5\,;\; 0.75 < \varphi_3 < 1\,;\; 0.5 < \varphi_4 < 0.75\,;\; 0 < \varphi_5 < 0.5 \quad (13.34)$$

Only one of the roots is valid for use in the second Underwood equation, Equation (13.33). The value of φ must be in the range

$$\alpha_{HK,HK} < \varphi < \alpha_{LK,HK}\,;\; 1 < \varphi < \alpha_{LK,HK} \quad (13.35)$$

The first Underwood equation is to be solved iteratively. Note that the equation is highly nonlinear with singularities and may pose convergence difficulties.

One possible approach to obtain the root of the first Underwood equation is to use the Newton-Raphson method. Equation (13.32) can be re-written as:

$$f(\varphi) = q - 1 + \sum \frac{\alpha_{iHK} z_i}{\alpha_{iHK} - \varphi} = 0 \tag{13.36}$$

The Newton-Raphson formula for estimating the improved value of the root (φ_{k+1}) over the previous value of the root (φ_k) is given as:

$$\varphi_{k+1} = \varphi_k - \frac{f(\varphi_k)}{f'(\varphi_k)} \tag{13.37}$$

where the derivative $f'(\varphi)$ is given as:

$$f'(\varphi) = \sum \frac{\alpha_{iHK} z_i}{(\alpha_{iHK} - \varphi)^2} \tag{13.38}$$

Equation (13.37) can be used to estimate φ by successive substitution. The range of φ is $1 < \varphi < \alpha_{LK,HK}$. The initial guess could be taken as $\varphi_o = (1 + \alpha_{LK,HK})/2$. The iteration process can be terminated when the deviation between two successive values is within the tolerance level, $[|(\varphi_{k+1} - \varphi_k)|/\varphi_k] \times 100 \leq \varepsilon$.

Once φ is known, the minimum reflux ratio R_{min} can be estimated from the second Underwood equation, Equation (13.33). No iteration is needed to calculate R_{min}.

Finally note that there are multiple solutions (multiple φ values) of the first Underwood equation within the range $\alpha_{HK,HK} < \varphi < \alpha_{LK,HK}$ when the keys are *split keys*, that is, there are one or more *NK* components present in the feed which have relative volatilities in the range $\alpha_{HK,HK} < \alpha_{NK,HK} < \alpha_{LK,HK}$. The correct value of φ is then obtained by solving the first and second Underwood equations simultaneously for φ and R_{min}.

Gilliland Correlation for Multicomponent Distillation

Once N_{min} and R_{min} have been estimated, we can use the Gilliland approach (E.R. Gilliland, Industrial and Engineering Chemistry, Vol. 32, 1220-1223, 1940) to estimate the number of equilibrium stages required for the actual reflux ratio R. Gilliland defined the following two variables:

$$X_G = \frac{R - R_{min}}{R + 1} \tag{13.39}$$

$$Y_G = \frac{N - N_{min}}{N + 1} \tag{13.40}$$

He found that a unique curve is obtained when Y_G is plotted as a function of X_G. The Y_G versus X_G curve can be described by the following equations:

$$Y_G = 1 - exp\left[\left(\frac{X_G - 1}{\sqrt{X_G}}\right)\left(\frac{1 + 54.4X_G}{11 + 117.2X_G}\right)\right] \tag{13.41}$$

$$Y_G = 0.75\left[1 - X_G^{0.5668}\right] \tag{13.42}$$

Equation (13.41) is the Gilliland-Molokanov equation (Y.K. Molokanov et al., Industrial and Engineering Chemistry, Vol. 12, 209-212, 1972) and Equation (13.42) is the Gilliland-Eduljee equation (H.E. Eduljee, Hydrocarbon Processing, Vol. 54, 120-122, 1975).

Kirkbride Equation for Multicomponent Distillation

Once the total number of equilibrium stages is estimated, the optimum location of the feed tray can be determined using the following empirical equation proposed by Kirkbride (C.G. Kirkbride, Petroleum Refiner, Vol. 23, 87-102, 1944):

$$\frac{n_F-1}{N-n_F}=\left[\left(\frac{z_{HK}}{z_{LK}}\right)\left(\frac{x_{LK,W}}{x_{HK,D}}\right)^2\frac{W}{D}\right]^{0.206} \tag{13.43}$$

where n_F is the feed stage number counting from top of the column and N is the total number of theoretical stages (including partial reboiler) required.

13.3.4 Rigorous Calculation Procedure for Rating the Performance of Multicomponent Distillation Columns

Rigorous or exact methods of analyzing and rating distillation columns are based on mass and energy balances on each tray/stage. Mass and energy balances along with equilibrium relationships and mole fraction summation relations together form the so-called MESH equations where "M" refers to mass balance, "E" refers to equilibrium relation, "S" refers to summation relation, and "H" refers to enthalpy/energy balance. The MESH equations are solved to determine temperatures, flow rates, and compositions at each stage.

Consider distillation of a multicomponent system with C components. The number of MESH equations for each stage of the distillation column are as follows: a) *M-equations*: there are C mass balance equations, one for each component; b) *E-equations*: there are C equilibrium relations, one for each component; c) *S-equations*: there are 2 summation relations; and d) *H-equations*: there is one enthalpy balance equation. Thus, there are $(3+2C)$ MESH equations for each stage.

MESH Equations for a Single Stage

Figure 13.5 shows the most general type of an equilibrium stage. The MESH equations for the general equilibrium stage shown in Figure 13.5 are described below.

M-equation (mole balance):

$$x_{i,n-1}L_{n-1}+y_{i,n+1}V_{n+1}+z_{i,n}F_n=x_{i,n}(L_n+U_n)+y_{i,n}(V_n+W_n) \tag{13.44}$$

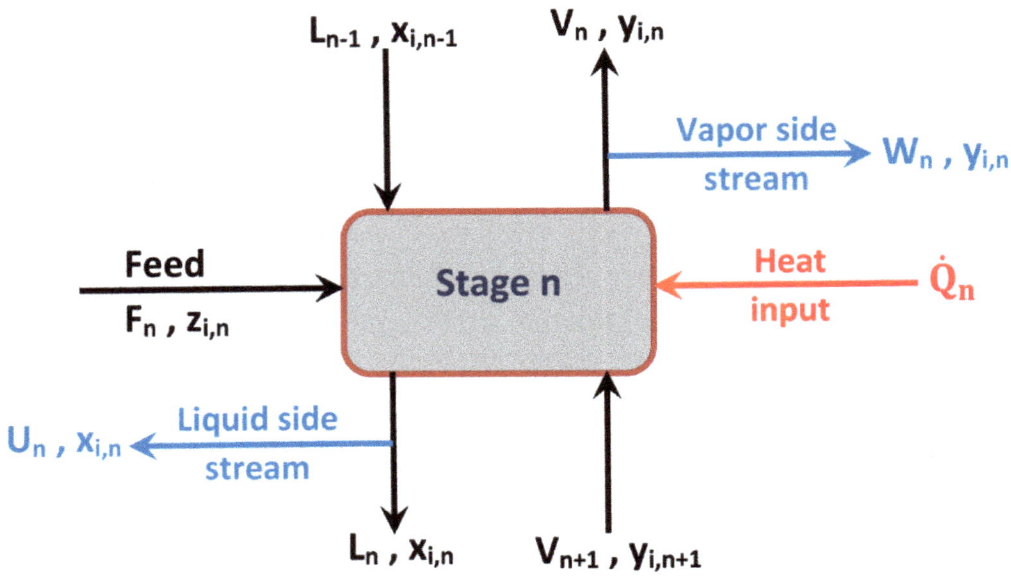

Figure 13.5 General equilibrium stage.

E-equation (equilibrium relation):

$$y_{i,n} = K_{i,n}x_{i,n} \tag{13.45}$$

where $K_{i,n}$ is the K-factor for species i at nth stage.

S-equations (summation relations):

$$\sum_{i=1}^{i=C} x_i = 1 \qquad 13.46)$$

$$\sum_{i=1}^{i=C} y_i = 1 \tag{13.47}$$

H-equation (enthalpy/energy balance):

$$\dot{Q}_n + h_{n-1}L_{n-1} + H_{n+1}V_{n+1} + h_{F,n}F_n = h_n(L_n + U_n) + H_n(V_n + W_n) \tag{13.48}$$

where h is molar enthalpy of liquid phase, H is molar enthalpy of vapor phase, and $\dot{Q}$ is the rate of heat input.

MESH Equations for the Whole Column

Figure 13.6 shows the most general type of a counter-current cascade of stages representing a distillation column. The cascade consists of N stages including partial condenser and reboiler. For each stage, there are $(3 + 2C)$ MESH equations. Thus, the total number of MESH equations for the whole column are: $N(3 + 2C)$

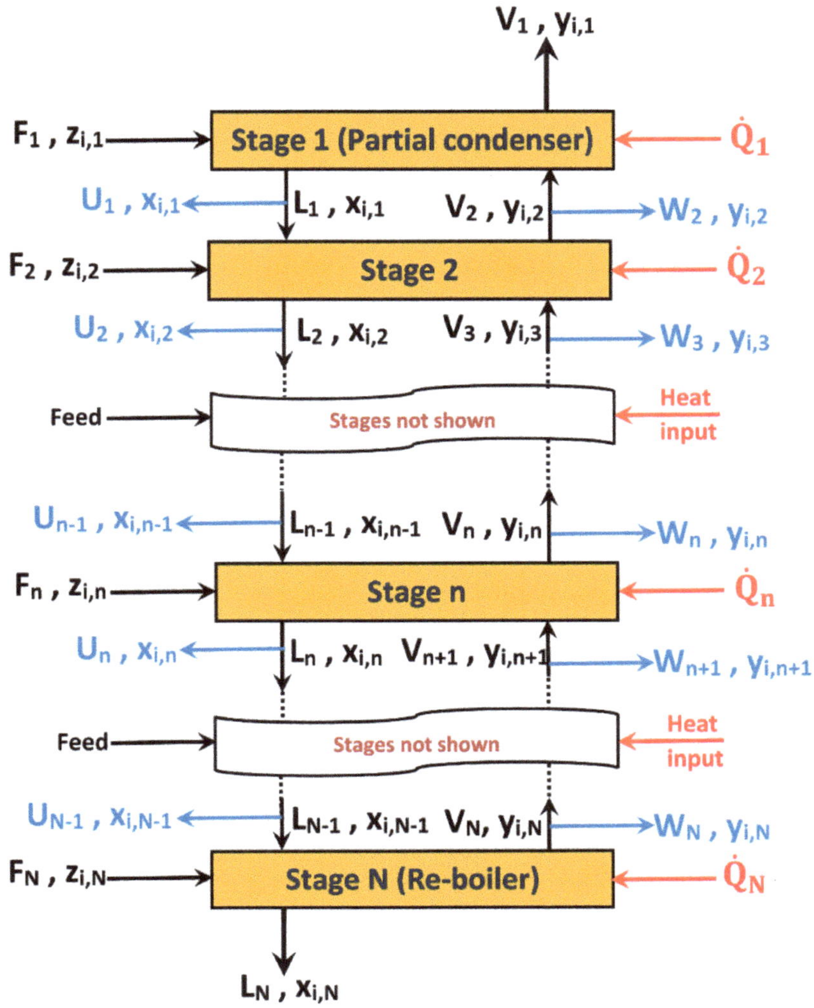

Figure 13.6 General counter-current cascade representing a distillation column.

Solution of MESH Equations

The total number of variables for the whole column are $N(10+3C)+1$. They are: a) $T_{F,n}$, $P_{F,n}$, F_n, P_n, $\dot{Q}_n$, L_n, V_n, T_n, W_n, U_n (total of 10 variables per stage); b) $x_{i,n}$, $y_{i,n}$, and $z_{i,n}$ (total of $3C$ variables); and c) total number of stages in the column including condenser and reboiler (1 variable). Thus we need to specify the following number of variables in order to solve the MESH equations:

$$\begin{pmatrix} Number\ of\ variables \\ to\ be\ specified \end{pmatrix} = [N(10+3C)+1] - [N(3+2C)] = N(C+7)+1 \qquad (13.49)$$

The following $N(C+7)+1$ variables are often specified: the number of equilibrium stages in the column (1 variable); feed composition at each stage (NC variables); and $T_{F,n}$, $P_{F,n}$, F_n, P_n, $\dot{Q}_n$, W_n, U_n at each stage (7 variables per stage, a total of $7N$ variables)

The following $N(3+2C)$ quantities are calculated using the MESH equations: liquid composition at each stage (NC variables); vapor composition at each stage (NC variables); liquid and vapor flow rates at each stage L_n and V_n (2 variable per stage, a total of $2N$ variables); and temperature T_n at each stage (a total of N variables).

The MESH equations are either solved stage-by-stage or simultaneously for the whole column. Details about different procedures used to solve the MESH equations can be found in the advanced level textbooks. Often an initial estimate of the number of stages required is made using the approximate methods such as the FUG method. The rigorous MESH equations are then solved to calculate compositions, flow rates, and temperatures. The estimates are revised and calculations are repeated until a satisfactory column design is achieved.

13.4 SEPARATION OF MULTICOMPONENT MIXTURES INTO PURE COMPONENTS USING A SEQUENCE OF DISTILLATION COLUMNS

Preceding sections have dealt with the analysis and design of single distillation columns for multicomponent separation. The feed stream is cut into two product streams on the basis of the key components. Most of the LK and LNK components go to the overhead distillate product stream and most of the HK and HNK components go to the bottoms product stream.

Industrial separation problems often deal with the separation of multicomponent mixture ($C > 2$) into more than two product streams. For example, one would like to separate a multicomponent mixture of C components into C pure product streams each consisting of only one component. As noted earlier, we need $(C - 1)$ distillation columns to separate a C- component mixture into C pure product streams.

There are many different ways in which multiple distillation columns can be sequenced for the separation of a multicomponent mixture. The choice of the sequence is based on capital and operating costs in addition to reliability and safety considerations. The total number of sequences possible to separate a multicomponent mixture of C components into C pure product (single component) streams using $(C - 1)$ distillation columns is given as follows (E.J. Henley, J.D. Seader, Equilibrium-Stage Separation Operations in Chemical Engineering, Wiley & Sons, New York, 1981):

$$S_C = \frac{[2(C-1)]!}{C!(C-1)!} \tag{13.50}$$

This formula is valid for simple sequences where : a) there is one feed stream per column; b) there are two product streams per columns; and c) the key components are *adjacent* in volatility.

Consider a multicomponent mixture of 3 species: A, B, and C. The order of relative volatility (with respect to a reference species) is as follows: $\alpha_A > \alpha_B > \alpha_C$ .We need 2 (= $C - 1$) distillation columns to separate this 3-component mixture into pure product streams. The columns can be sequenced in 2 ways ($S_C = 2$) as shown in Figures 13.7 and 13.8.

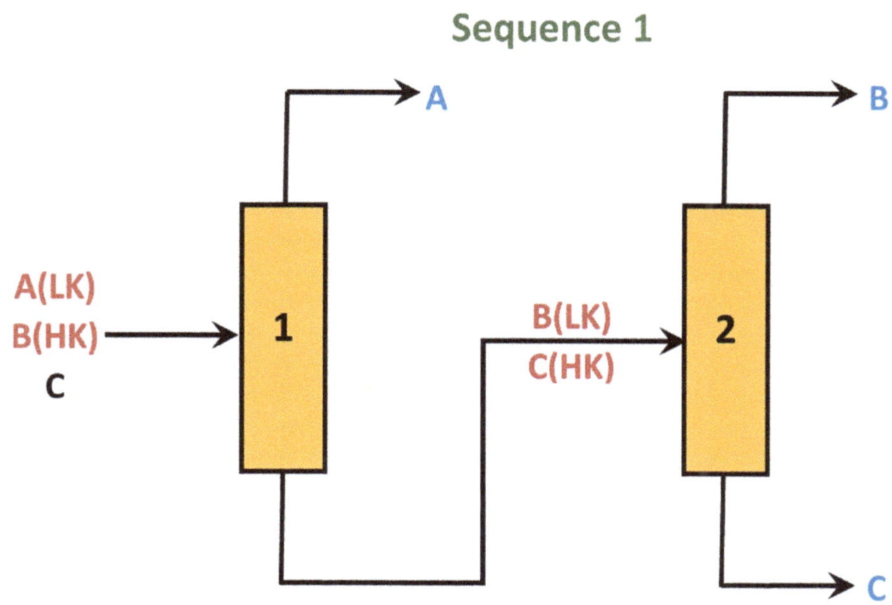

Figure 13.7 One possible sequence to separate 3 component mixture into pure products.

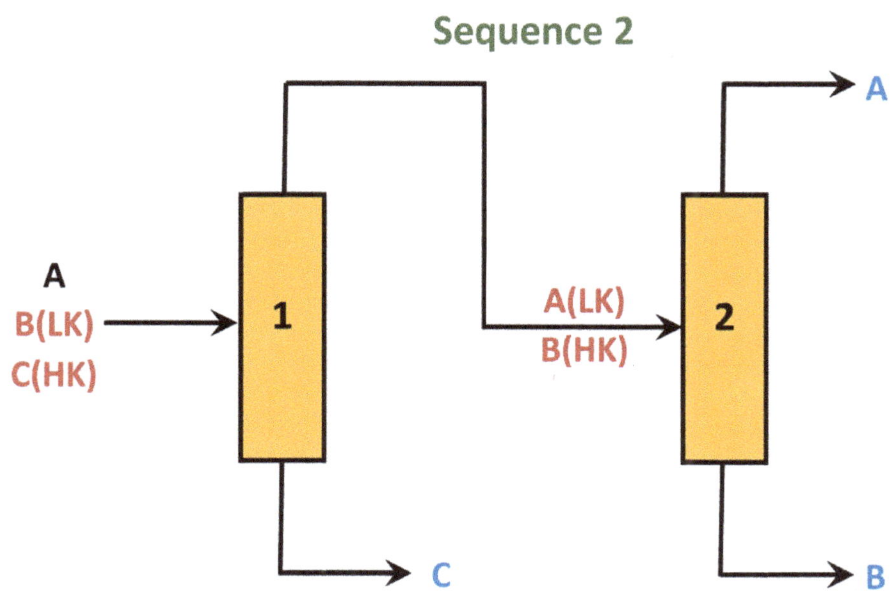

Figure 13.8 Another possible sequence to separate 3 component mixture into pure products.

Consider another example where the multicomponent mixture consists of 4 species: A, B, C and D. The order of relative volatility (with respect to a reference species) is as follows: $\alpha_A > \alpha_B > \alpha_C > \alpha_D$. We need 3 ($= C - 1$) distillation columns to separate this 4-component mixture into pure product streams. The columns can be sequenced in 5 ways ($S_C = 5$) as shown in Figures 13.9 to 13.13.

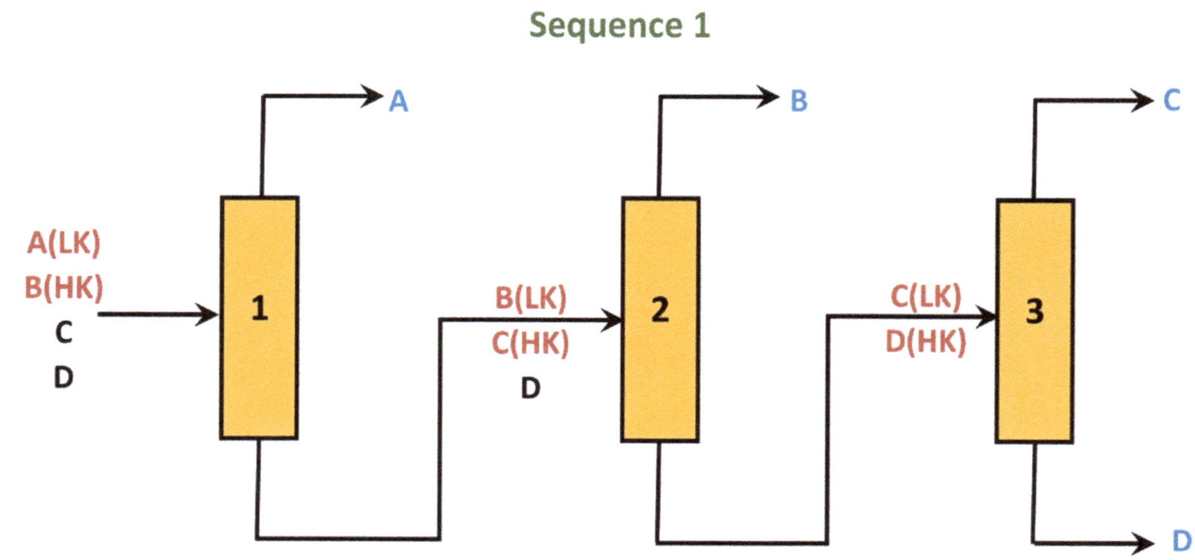

Figure 13.9 Sequence 1 of 5 possible sequences to separate 4 component mixture into pure products.

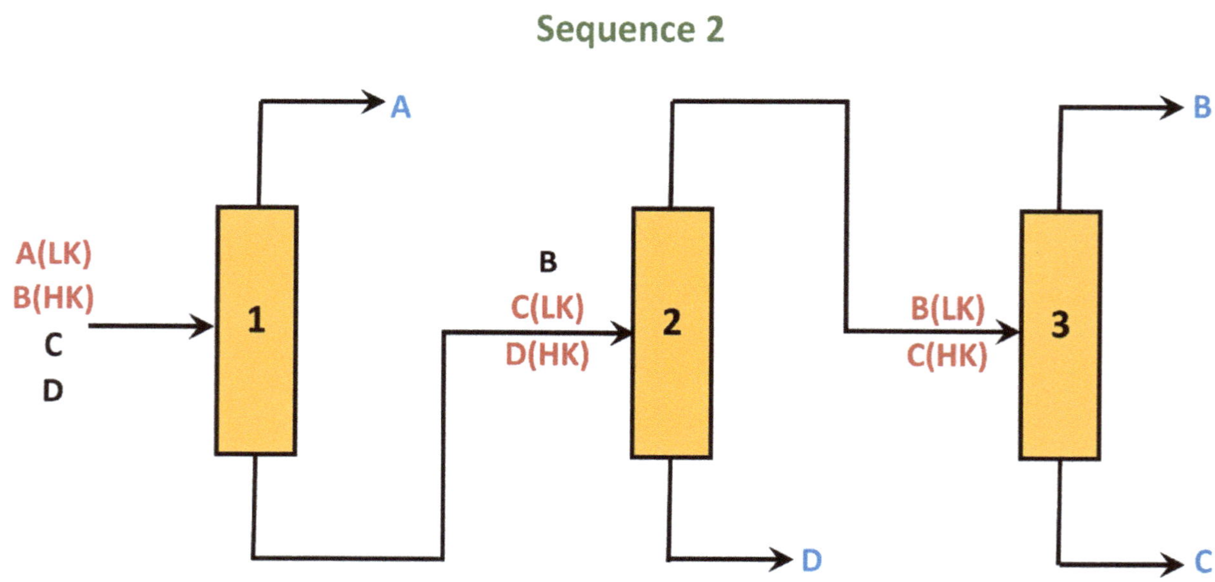

Figure 13.10 Sequence 2 of 5 possible sequences to separate 4 component mixture into pure products.

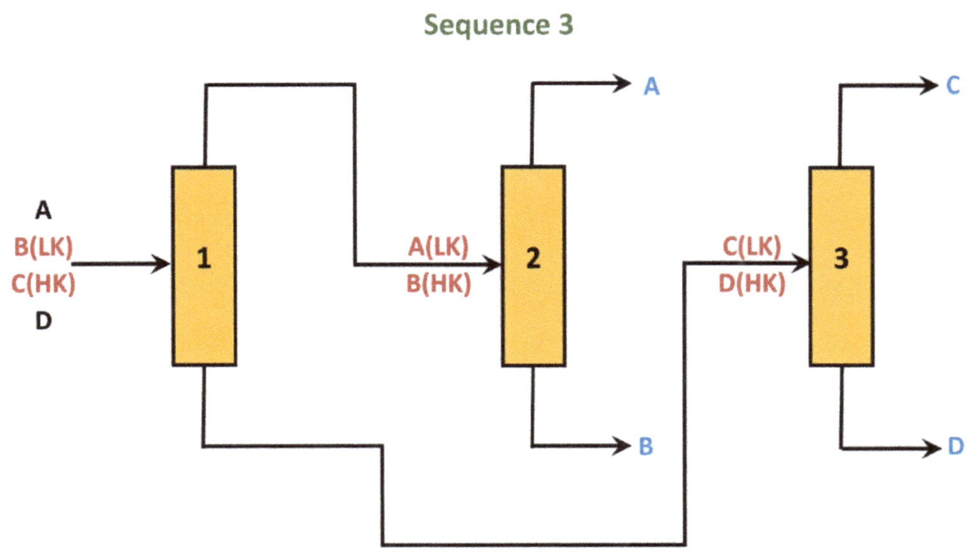

Figure 13.11 Sequence 3 of 5 possible sequences to separate 4 component mixture into pure products.

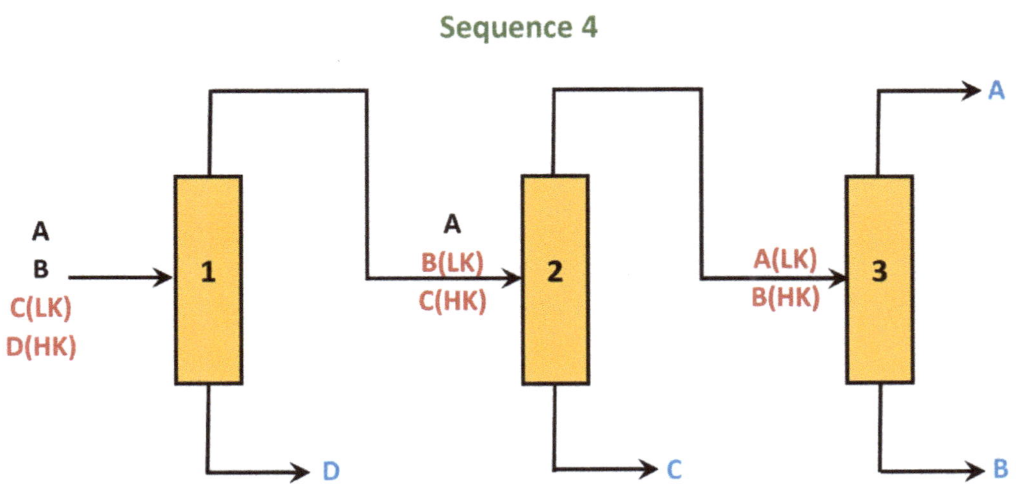

Figure 13.12 Sequence 4 of 5 possible sequences to separate 4 component mixture into pure products.

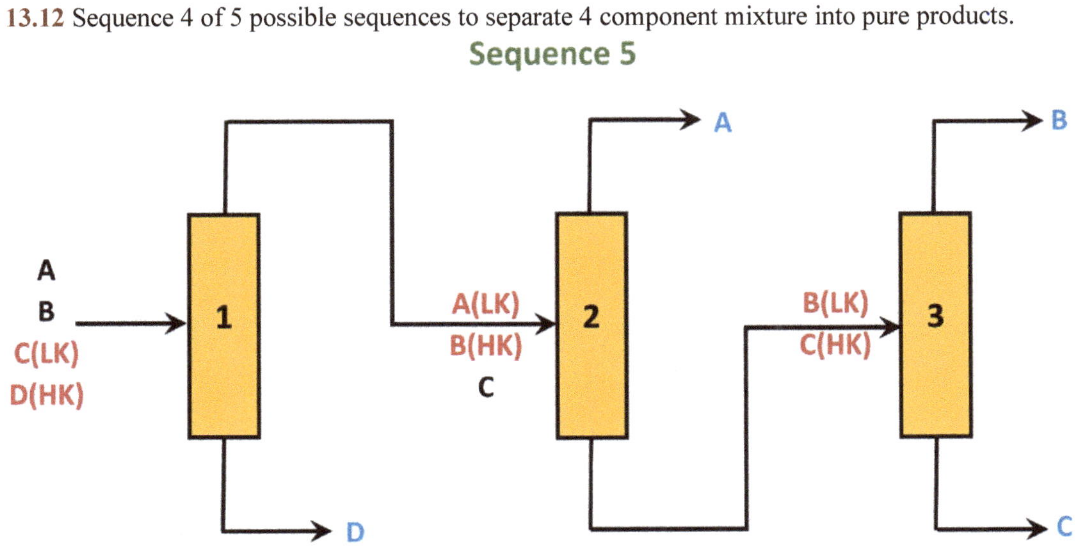

Figure 13.13 Sequence 5 of 5 possible sequences to separate 4 component mixture into pure products.

As a final example of column sequencing, consider a multicomponent mixture of 5 species (A,B,C,D, and E). The order of relative volatility (with respect to a reference species) is : $\alpha_A > \alpha_B > \alpha_C > \alpha_D > \alpha_E$. We need 4 distillation columns which can be sequenced in the following number of ways:

$$S_C = \frac{[2(5-1)]!}{5!(5-1)!} = 14 \tag{13.51}$$

The sequences can be grouped into 4 groups. Table 13.3 shows the group of 5 sequences starting with the same light and heavy keys (species A as light key and species B as heavy key). Table 13.4 shows another group of 5 sequences starting with the same light and heavy keys (species D as light key and species E as heavy key). Table 13.5 shows a group of 2 sequences starting with the same light and heavy keys (species B as light key and species C as heavy key). Table 13.6 shows another group of 2 sequences starting with the same light and heavy keys (species C as light key and species D as heavy key).

Table 13.3 Group of 5 sequences with species A as light key (LK) and species B as heavy key (HK).

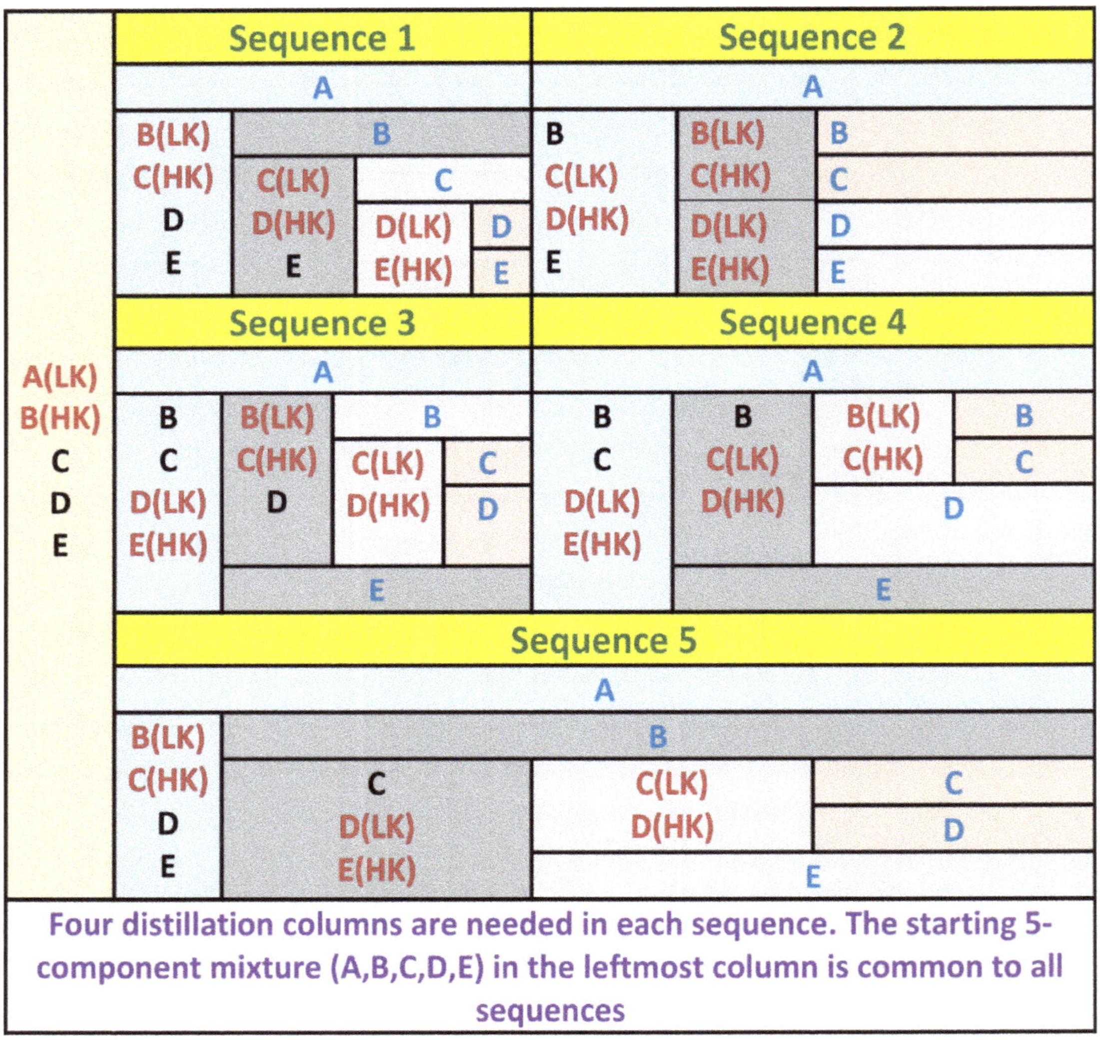

Table 13.4 Group of 5 sequences with species D as light key (LK) and species E as heavy key (HK).

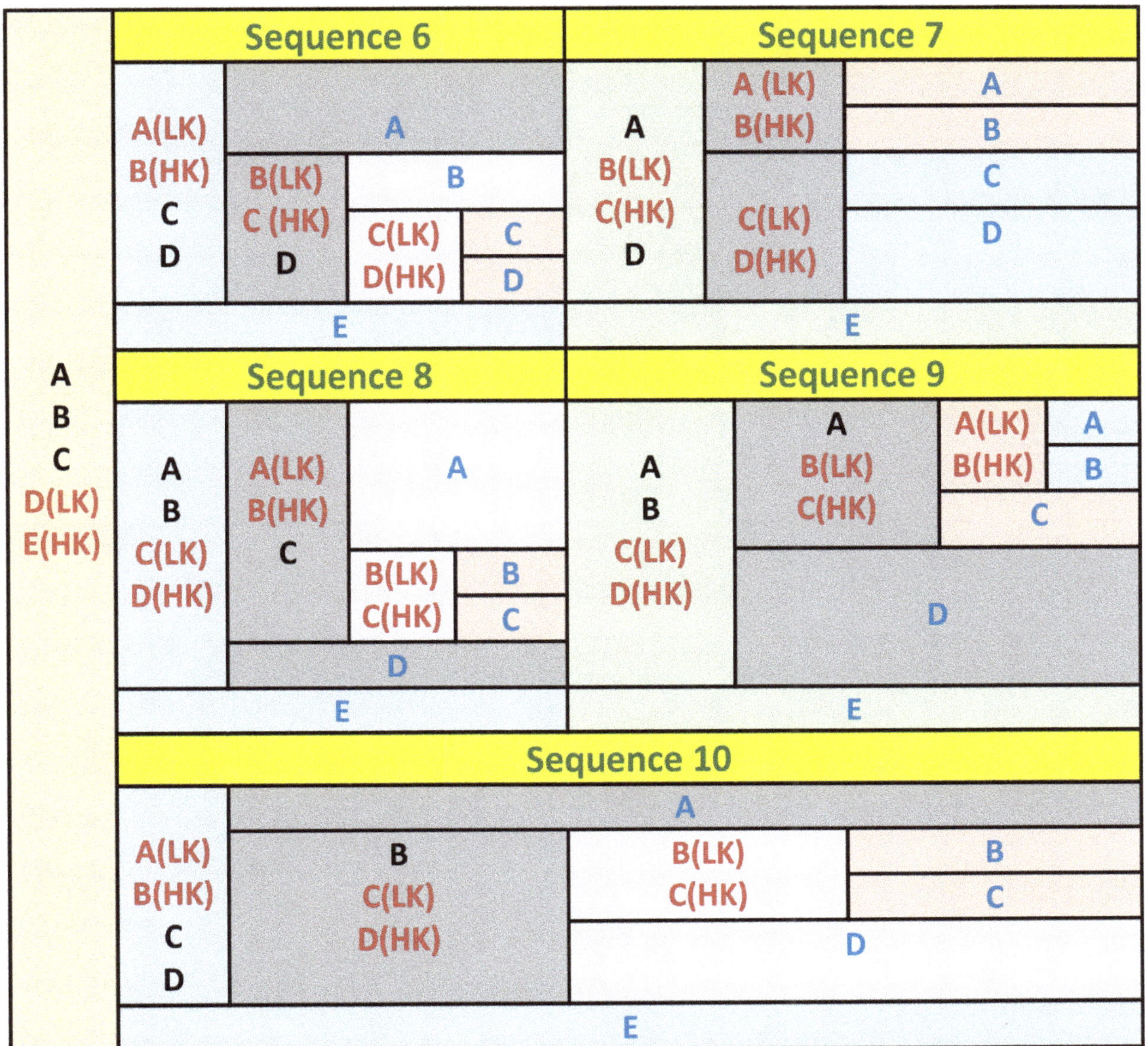

Four distillation columns are needed in each sequence. The starting 5-component mixture (A,B,C,D,E) in the leftmost column is common to all sequences

Table 13.5 Group of 2 sequences with species B as light key (LK) and species C as heavy key (HK).

	Sequence 11			Sequence 12		
A B(LK) C(HK) D E	A(LK) B(HK)	A		A(LK) B(HK)	A	
		B			B	
	C(LK) D(HK) E	C		C D(LK) E(HK)	C(LK) D(HK)	C
		D(LK) E(HK)	D			D
			E		E	
Four distillation columns are needed in each sequence. The starting 5-component mixture (A,B,C,D,E) in the leftmost column is common to both sequences						

Table 13.6 Group of 2 sequences with species C as light key (LK) and species D as heavy key (HK).

	Sequence 13			Sequence 14		
A B C(LK) D(HK) E	A(LK) B(HK) C	A		A B(LK) C(HK)	A(LK) B(HK)	A
		B(LK) C(HK)	B			B
			C		C	
	D(LK) E(HK)	D		D(LK) E(HK)	D	
		E			E	
Four distillation columns are needed in each sequence. The starting 5-component mixture (A,B,C,D,E) in the leftmost column is common to both sequences						

13.5 HEURISTICS APPROACH FOR THE DEVELOPMENT OF SEPARATION SEQUENCE

Heuristics are basically rules of thumb. Based on experience, a number of heuristics have been proposed for sequencing of distillation columns. Some of them are given below (P.C. Wankat, Separation Process Engineering, 3rd Edition, Prentice Hall, NJ, 2012; E.J. Henley, J.D. Seader, Equilibrium-Stage Separation Operations in Chemical Engineering, Wiley & Sons, NY, 1981):

- Remove corrosive or toxic component first. This would limit the spread of corrosive and or toxic component only to the first one or two columns.
- Sequence the splits in order of decreasing volatility.

- Remove the component present in excess first.
- Do the difficult separations such as close boiling point components at the end.
- Avoid distillation if $\alpha_{LK,HK}$ is below 1.05 or so.

NOTATION

C	number of components in a multicomponent mixture
D	molar flow rate of distillate liquid product
F	feed molar flow rate
F_n	feed molar flow rate in nth stage
FR_i	fractional recovery of species *i*, defined as molar flow rate of species *i* in the product (top or bottoms) to molar flow rate of species *i* in the feed
$FR_{i,D}$	fractional recovery of species *i* in the distillate product, defined as molar flow rate of species *i* in the distillate product to molar flow rate of species *i* in the feed, see Equation (13.9)
$FR_{i,W}$	fractional recovery of species *i* in the bottoms product, defined as molar flow rate of species *i* in the bottoms product to molar flow rate of species *i* in the feed, see Equation (13.10)
FR_j	fractional recovery of species *j*, defined as molar flow rate of species *j* in the product (top or bottoms) to molar flow rate of species *j* in the feed
FR_{LK}	fractional recovery of light key, defined as molar flow rate of light key in the product (top or bottoms) to molar flow rate of light key in the feed
FR_{HK}	fractional recovery of heavy key, defined as molar flow rate of heavy key in the product (top or bottoms) to molar flow rate of heavy key in the feed
$FR_{NK,D}$	fractional recovery of non-key, defined as molar flow rate of non-key in the distillate product to molar flow rate of non-key in the feed
FUG	Fenske-Underwood-Gilliland
h_{n-1}	molar enthalpy of liquid stream leaving (n-1)th stage
h_n	molar enthalpy of liquid stream leaving nth stage
$h_{F,n}$	molar enthalpy of feed stream entering nth stage
H_n	molar enthalpy of vapor stream leaving nth stage
H_{n+1}	molar enthalpy of vapor stream leaving (n+1)th stage
HK	heavy key component
HNK	heavy non-key component
K_i	distribution coefficient of species *i* at equilibrium, defined as y_i/x_i
$K_{i,n}$	distribution coefficient of species *i* at nth stage, defined in Equation (13.45)
K_r	distribution coefficient of reference species *r* at equilibrium, defined as y_r/x_r
LK	light key component
LNK	light non-key component
L_{min}	molar liquid flow rate in enriching section when pinch zone is above the feed stage
$\bar{L}_{min}$	molar liquid flow rate in stripping section when pinch zone is below the feed stage
L_{n-1}	molar flow rate of liquid stream leaving (n-1)th stage
L_n	molar flow rate of liquid stream leaving nth stage
$MESH$	refers to “Mass-Equilibrium-Summation-Enthalpy” relations
n	stage number
N	total number of theoretical stages (including partial reboiler)
n_F	feed stage number counting from top of the column
NK	non-key component
N_{min}	minimum number of theoretical stages (including partial reboiler) at total reflux
P_n	pressure of nth stage
$P_{F,n}$	pressure of feed stream at nth stage

q feed quality, defined in Chapter 9, see Equation (9.40)
$\dot{Q}_n$ heat input rate at nth stage
R actual reflux ratio
R_{min} minimum reflux ratio
S_C total number of sequences possible to separate a multicomponent mixture of C components into C pure product streams using $(C-1)$ distillation columns
T_n temperature of nth stage
$T_{F,n}$ temperature of feed stream at nth stage
U_n molar flow rate of liquid side stream leaving nth stage
V_n molar flow rate of vapor stream leaving any *n*th stage
V_{n+1} molar flow rate of vapor stream leaving any *n+1*th stage
V_{min} molar flow rate of vapor stream in enriching section when pinch zone is above the feed stage
$\bar{V}_{min}$ molar flow rate of vapor stream in stripping section when pinch zone is below the feed stage
W molar flow rate of bottoms liquid product of a column
W_n molar flow rate of vapor side stream leaving nth stage
x_A mole fraction of species A in liquid phase
x_B mole fraction of species B in liquid phase
x_i mole fraction of species *i* in liquid phase
$x_{i,D}$ mole fraction of species *i* in distillate liquid product of a column
$x_{i,n}$ mole fraction of species *i* in liquid stream leaving nth stage
$x_{i,n-1}$ mole fraction of species *i* in liquid stream leaving (n-1)th stage
$x_{i,p}$ mole fraction of species *i* in the liquid stream in the pinch-point zone
$x_{i,W}$ mole fraction of species *i* in the bottoms liquid product of a column
$x_{LK,W}$ mole fraction of *LK* component in the bottoms liquid product of a column
$x_{HK,D}$ mole fraction of *HK* component in the distillate liquid product of a column
X_G Gilliland parameter defined in Equation (13.39)
$y_{i,n}$ mole fraction of species *i* in vapor stream leaving nth stage
$y_{i,n+1}$ mole fraction of species *i* in vapor stream leaving (n+1)th stage
$y_{i,p}$ mole fraction of species *i* in the vapor stream in the pinch-point zone
Y_G Gilliland parameter defined in Equation (13.40)
z_{HK} mole fraction of *HK* component in the feed
z_i mole fraction of species *i* in the feed
$z_{i,n}$ mole fraction of species *i* in the feed entering nth stage
z_{LK} mole fraction of *LK* component in the feed

Greek Letters

α_{AB} relative volatility of species A with respect to species B
α_i relative volatility of species *i* with respect to some species
α_{ij} relative volatility of species *i* with respect to species *j*
α_{ir} relative volatility of species *i* with respect to reference species *r*
α_{iHK} relative volatility of species *i* with respect to *HK* component
$\alpha_{LK,HK}$ relative volatility of *LK* component with respect to *HK* component
$\alpha_{NK,HK}$ relative volatility of *NK* component with respect to *HK* component
ε tolerance level
φ parameter in Underwood equations, Equations (13.32) and (13.33)
φ_E Underwood parameter for enriching section, defined as $L_{min}/K_r V_{min}$, see Equation (13.18)
φ_o Initial guess value of Underwood parameter
φ_S Underwood parameter for stripping section, defined as $\bar{L}_{min}/K_r \bar{V}_{min}$, see Equation (13.24)

TUTORIAL PROBLEMS WITH SOLUTION

Problem 1

A conventional distillation column with total condenser and partial reboiler is being used to distill 1875 kmol/h of a saturated liquid feed. The feed consists of 4 components: propane (species 1, z_1=0.05), n-butane (species 2, z_2=0.30), n-pentane (species 3, z_3=0.50) and n-hexane (species 4, z_4=0.15). The column is operating at 1 atm pressure. The reflux ratio is 3.0.

The fractional recovery of n-butane (LK) in the distillate is desired to be 99% and the fractional recovery of n-pentane (HK) in the bottoms product is desired to be 98%. Calculate the following:

a) Composition and flow rate of distillate product.
b) Composition and flow rate of bottoms product.

You can assume that all of the propane (LNK) appears in the distillate product and that all of n-hexane (HNK) appears in the bottoms product.

Solution

Given information:

Component	Key or non-key	$z_{i,F}$	$FR_{i,D}$	$FR_{i,W}$
Propane (1)	LNK	0.05	1	0
n-butane (2)	LK	0.30	0.99	0.01
n-pentane (3)	HK	0.50	0.02	0.98
n-hexane (4)	HNK	0.15	0	1

A schematic diagram of the column is shown below.

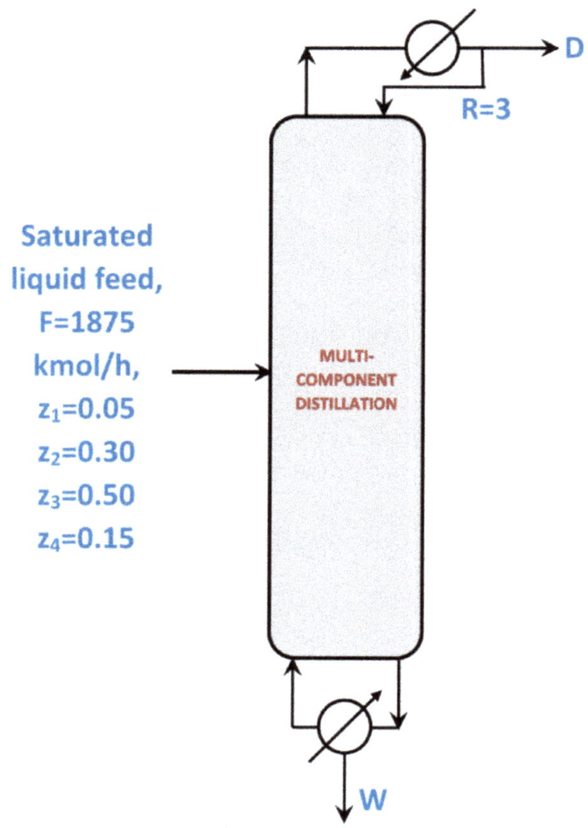

a) Composition and flow rate of distillate product:

We need to first calculate the component flow rates in the distillate product. They are as follows:

$$x_{i,D}D = FR_{i,D}(z_iF)$$

$$x_{1,D}D = FR_{1,D}(z_1F) = (1)(0.05 \times 1875) = 93.75\ kmol/h$$

$$x_{2,D}D = FR_{2,D}(z_2F) = (0.99)(0.30 \times 1875) = 556.875\ kmol/h$$

$$x_{3,D}D = FR_{3,D}(z_3F) = (0.02)(0.50 \times 1875) = 18.75\ kmol/h$$

$$x_{4,D}D = FR_{4,D}(z_4F) = (0)(0.15 \times 1875) = 0\ kmol/h$$

The total distillate flow rate D is:

$$D = \sum x_{i,D}D = 93.75 + 556.875 + 18.75 + 0 = 669.375\ kmol/h$$

The composition of the distillate product is:

$$x_{i,D} = \frac{x_{i,D}D}{D}$$

$$x_{1,D} = \frac{x_{1,D}D}{D} = \frac{93.75}{669.375} = 0.14$$

$$x_{2,D} = \frac{x_{2,D}D}{D} = \frac{556.875}{669.375} = 0.832$$

$$x_{3,D} = \frac{x_{3,D}D}{D} = \frac{18.75}{669.375} = 0.028$$

$$x_{4,D} = \frac{x_{4,D}D}{D} = \frac{0}{669.375} = 0$$

b) Composition and flow rate of bottoms product:

The component flow rates in the bottoms product are:

$$x_{i,W}W = FR_{i,W}(z_iF)$$

$$x_{1,W}W = FR_{1,W}(z_1F) = (0)(0.05 \times 1875) = 0\ kmol/h$$

$$x_{2,W}W = FR_{2,W}(z_2F) = (0.01)(0.30 \times 1875) = 5.625\ kmol/h$$

$$x_{3,W}W = FR_{3,W}(z_3F) = (0.98)(0.50 \times 1875) = 918.75\ kmol/h$$

$$x_{4,W}W = FR_{4,W}(z_4F) = (1)(0.15 \times 1875) = 281.25\ kmol/h$$

The total bottoms flow rate W is:

$$W = \sum x_{i,W}W = 0 + 5.625 + 918.75 + 281.25 = 1205.625\ kmol/h$$

The composition of the bottoms product is:

$$x_{i,W} = \frac{x_{i,W}W}{W}$$

$$x_{1,W} = \frac{x_{1,W}W}{W} = \frac{0}{1205.625} = 0$$

$$x_{2,W} = \frac{x_{2,W}W}{W} = \frac{5.625}{1205.625} = 0.0047$$

$$x_{3,W} = \frac{x_{3,W}W}{W} = \frac{918.75}{1205.625} = 0.762$$

$$x_{4,W} = \frac{x_{4,W}W}{W} = \frac{281.25}{1205.625} = 0.2333$$

Problem 2

A conventional distillation column with total condenser and partial reboiler is being used to separate a 4-component mixture. The feed composition and the relative volatilities of components are given in the table below.

Component	$z_{i,F}$	$\alpha_{i,HK}$
1(LNK)	0.08	2.3
2(LK)	0.44	1.8
3(HK)	0.42	1.0
4(HNK)	0.06	0.70

The fractional recovery of LK component in the distillate is 97%. The fractional recovery of HK component in the bottoms is 95%. Calculate the minimum number of equilibrium stages required? Specify the fractional recoveries of all components in the distillate and bottoms products at total reflux?

Solution: The figure below shows schematically the column operating at total reflux.

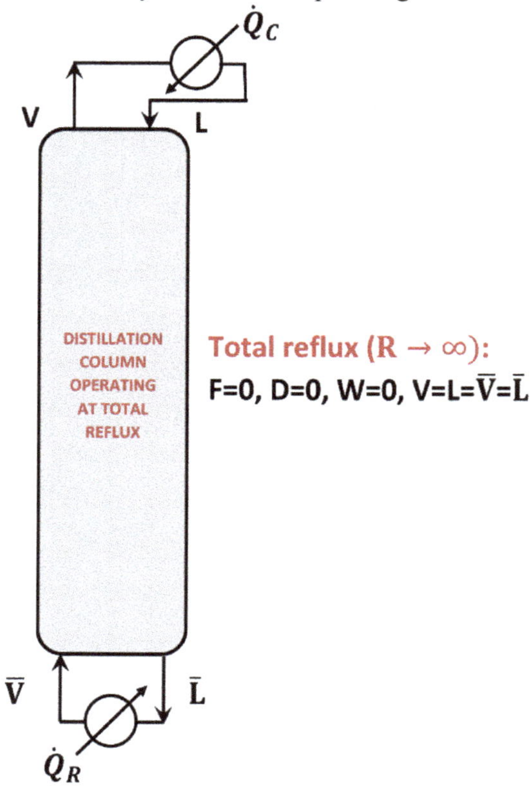

We apply Fenske equation, Equation (13.6), to calculate N_{min}

$$N_{min} = \frac{ln[\{FR_{LK}/(1-FR_{LK})\}_D/\{(1-FR_{HK})/FR_{HK}\}_W]}{ln(\alpha_{LK,HK})}$$

$$FR_{LK,D} = 0.97$$

$$FR_{HK,W} = 0.95$$

$$\alpha_{LK,HK} = 1.8$$

Thus,

$$N_{min} = \frac{ln[(0.97/(1-0.97))/((1-0.95)/0.95)]}{ln(1.8)} = \frac{\ln(614.333)}{\ln(1.8)} = 10.923$$

Using Equation (13.7), we can calculate the fractional recoveries of non-key components

$$FR_{NK,D} = \frac{\alpha_{NK,HK}^{N_{min}}}{\alpha_{NK,HK}^{N_{min}} + \{FR_{HK}/(1-FR_{HK})\}_W}$$

$$FR_{1,D} = \frac{\alpha_{1,HK}^{N_{min}}}{\alpha_{1,HK}^{N_{min}} + \{0.95/(1-0.95)\}} = \frac{(2.3)^{10.923}}{(2.3)^{10.923} + \{0.95/(1-0.95)\}} = 0.9979$$

$$FR_{4,D} = \frac{\alpha_{4,HK}^{N_{min}}}{\alpha_{4,HK}^{N_{min}} + \{0.95/(1-0.95)\}} = \frac{(0.7)^{10.923}}{(0.7)^{10.923} + \{0.95/(1-0.95)\}} = 0.00107$$

The fractional recoveries of different components in the bottoms product is simply:

$$FR_{i,W} = 1 - FR_{i,D}$$

The following table summarizes the fractional recoveries of all components in the distillate and bottoms products at total reflux.

Component	$FR_{i,D}$	$FR_{i,W}$
1(LNK)	0.9979	0.0021
2(LK)	0.97	0.03
3(HK)	0.05	0.95
4(HNK)	0.00107	0.99893

Problem 3

A saturated vapor feed ($q = 0$) is fed at a rate of 150 kmol/h to a conventional distillation column with total condenser and partial reboiler. The composition of the feed is 38 mol% benzene, 33 mol% toluene, and 29 mol% cumene. The feed is distilled so that 97% of the toluene is recovered in the distillate and 94% of the cumene is recovered in the bottoms. The relative volatilities can be assumed to be constant ($\alpha_{TC} = 4.76, \alpha_{BC} = 10.71, \alpha_{CC} = 1.0$ where subscript T refers to toluene, B refers to benzene, and C refers to cumene).

a) Calculate the number of equilibrium stages required at total reflux and fractional recovery of benzene in distillate
b) Calculate the minimum reflux ratio using the Underwood equations
c) Calculate the total number of equilibrium stages at $R = 1.3R_{min}$, and determine the optimum feed tray location.

Solution

Refer to the schematic diagram given below.

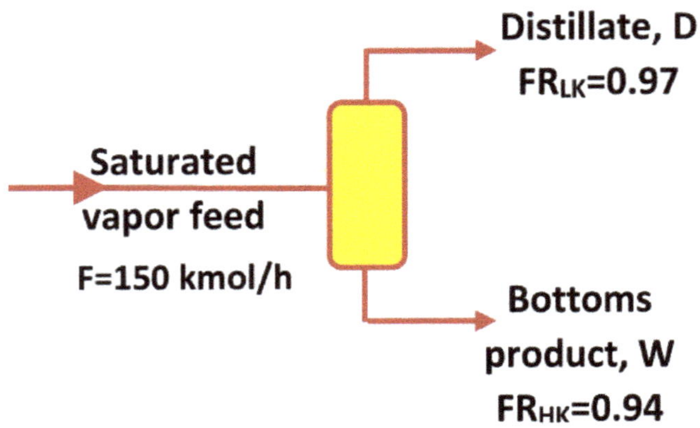

As the recoveries of toluene and cumene are specified, they are selected as the key components: more volatile toluene as light key (LK) and less volatile cumene as heavy key (HK). The information given is summarized in the table below.

Component	$z_{i,F}$	$\alpha_{i,HK}$	$FR_{i,D}$	$FR_{i,W}$
Benzene (LNK)	0.38	10.71	not given	not given
Toluene (LK)	0.33	4.76	0.97	0.03
Cumene (HK)	0.29	1.0	0.06	0.94

a) The number of equilibrium stages required at total reflux and fractional recovery of benzene in distillate are calculated using the different forms of the Fenske equation.

We apply Fenske equation, Equation (13.6), to calculate N_{min}

$$N_{min} = \frac{ln[\{FR_{LK}/(1-FR_{LK})\}_D/\{(1-FR_{HK})/FR_{HK}\}_W]}{ln(\alpha_{LK,HK})}$$

$$FR_{LK,D} = 0.97$$

$$FR_{HK,W} = 0.94$$

$$\alpha_{LK,HK} = 4.76$$

Thus,

$$N_{min} = \frac{ln[(0.97/(1-0.97))/((1-0.94)/0.94)]}{ln(4.76)} = 3.99$$

Using the other form of the Fenske equation, Equation (13.7), we can calculate the fractional recovery of light non-key (LNK) component benzene in the distillate as:

$$FR_{LNK,D} = \frac{\alpha_{LNK,HK}^{N_{min}}}{\alpha_{LNK,HK}^{N_{min}} + \{FR_{HK}/(1-FR_{HK})\}_W} = \frac{(10.71)^{3.99}}{(10.71)^{3.99} + (0.94/(1-0.94))} = 0.99878$$

b) Minimum reflux ratio using the Underwood equations

The First Underwood equation, Equation (13.32), can be written as:

$$F(1-q) = \sum \frac{\alpha_{iHK}(z_i F)}{\alpha_{iHK} - \varphi}$$

The only unknown in this equation is φ. Upon expansion of the summation term, we can write:

$$150(1-0) = \frac{(0.38 \times 150)(10.71)}{10.71 - \varphi} + \frac{(0.33 \times 150)(4.76)}{4.76 - \varphi} + \frac{(0.29 \times 150)(1)}{1 - \varphi}$$

$$150 = \frac{610.47}{10.71 - \varphi} + \frac{235.62}{4.76 - \varphi} + \frac{43.5}{1 - \varphi}$$

As discussed in section 13.3.3 (see Equation (13.35)), the solution of this equation lies within the range: $1 < \varphi < 4.76$. The solution can be obtained using the Newton-Raphson method. Another approach to solve this equation is to select several values of φ within the specified range, calculate the right hand side of the equation and plot $y = 150$ and $y = RHS$ on a y vs φ plot where RHS is the right hand side of the above equation. The intersection of the plots $y = 150$ and $y = RHS$ gives the solution. The solution is shown in the figure given below. We find $\varphi = 2.505$ satisfies the above equation.

Now we use the Second Underwood equation, Equation (13.33), to determine the minimum reflux ratio R_{min}:

$$D(R_{min} + 1) = \sum \frac{\alpha_{iHK}(x_{i,D} D)}{[\alpha_{iHK} - \varphi]}$$

No trial and error is needed to solve this equation for R_{min}. Before we solve this equation, we need to calculate the $(x_{i,D} D)$ terms and the total distillate flow rate D. The $(x_{i,D} D)$ terms are given as:

$$x_{i,D} D = (z_i F)(FR_{i,D})$$

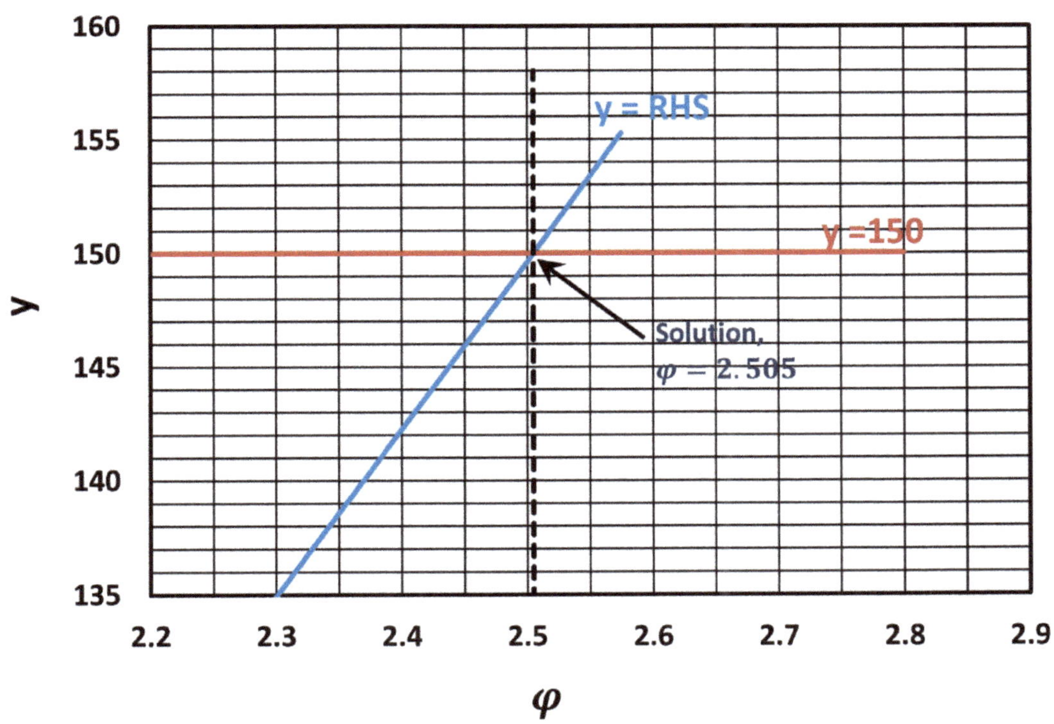

Thus,

$$x_{1,D}D = (z_1F)(FR_{1,D}) = (0.38 \times 150)(0.99878) = 56.9305\ kmol/h$$

$$x_{2,D}D = (z_2F)(FR_{2,D}) = (0.33 \times 150)(0.97) = 48.015\ kmol/h$$

$$x_{3,D}D = (z_3F)(FR_{3,D}) = (0.29 \times 150)(0.06) = 2.61\ kmol/h$$

$$D = 56.9305 + 48.015 + 2.61 = 107.56\ kmol/h$$

Now the Second Underwood equation can be solved for R_{min} as follows:

$$107.56 \times (R_{min} + 1) = \frac{10.71 \times 56.9305}{10.71 - 2.505} + \frac{4.76 \times 48.015}{4.76 - 2.505} + \frac{1 \times 2.61}{1 - 2.505} = 173.93$$

$$R_{min} = 0.617$$

c) The total number of equilibrium stages at $R = 1.3R_{min}$, and the optimum feed tray location

We can use the Gilliland approach to estimate the number of equilibrium stages at given reflux ratio.

$$R = 1.3 \times 0.617 = 0.8021$$

$$X_G = \frac{R - R_{min}}{R + 1} = \frac{0.8021 - 0.617}{0.8021 + 1} = 0.1027$$

Using the Gilliland-Molokanov equation, Equation (13.41), we get

$$Y_G = 1 - exp\left[\left(\frac{X_G - 1}{\sqrt{X_G}}\right)\left(\frac{1 + 54.4X_G}{11 + 117.2X_G}\right)\right] = 1 - exp[(0.28593)(-2.8)] = 0.551$$

$$Y_G = \frac{N - N_{min}}{N + 1} = 0.551$$

$$N = 10.114 \approx 10 \text{ stages}$$

Note that N includes partial reboiler.

The optimum feed tray location can be determined using the following Kirkbride equation:

$$\frac{n_F - 1}{N - n_F} = \left[\left(\frac{z_{HK}}{z_{LK}}\right)\left(\frac{x_{LK,W}}{x_{HK,D}}\right)^2 \frac{W}{D}\right]^{0.206}$$

where n_F is the feed stage number counting from top of the column and N is the total number of stages (including partial reboiler). $N = 10$, $z_{HK} = 0.29$, $z_{LK} = 0.33$, $D = 107.56$, and $W = F - D = 150 - 107.56 = 42.44\ kmol/h$. We need to calculate $x_{LK,W}$ and $x_{HK,D}$ as follows:

$$x_{LK,W}W = (1 - FR_{LK,D})(z_{LK}F) = (1 - 0.97)(0.33 \times 150) = 1.485$$

$$x_{LK,W} = 1.485/42.44 = 0.035$$

$$x_{HK,D}D = (1 - FR_{HK,W})(z_{HK}F) = (1 - 0.94)(0.29 \times 150) = 2.61$$

$$x_{HK,D} = 2.61/107.56 = 0.0243$$

The Kirkbride equation can now be solved for n_F :

$$\frac{n_F-1}{10-n_F}=\left[\left(\frac{0.29}{0.33}\right)\left(\frac{0.035}{0.0243}\right)^2\frac{42.44}{107.56}\right]^{0.206}=0.934$$

$$n_F=5.35\approx 5$$

Thus the optimum feed tray is the 5th tray (counting from top of the column). There are 4 trays above the feed tray and 4 trays below the feed tray (excluding the partial reboiler). The partial boiler is the 10th equilibrium stage.

Problem 4

A saturated liquid feed ($q=1.0$) is fed at a rate of 135 kmol/h to a conventional distillation column with total condenser and partial reboiler. The composition of the feed is 5 mol% n-pentane (C5), 39 mol% n-hexane (C6), 51 mol% n-heptane (C7), and 5 mol% n-octane (C8). The feed is distilled so that 97% of the n-hexane is recovered in the distillate. The amount of n-heptane recovered in the bottom product is 96%. The relative volatilities can be assumed to be constant ($\alpha_{C5,C7}=6.5, \alpha_{C6,C7}=2.5,\ \alpha_{C7,C7}=1.0, \alpha_{C8,C7}=0.40$).

a) Calculate the number of equilibrium stages required at total reflux and the fractional recoveries of C5 in distillate and C8 in bottoms product.
b) Calculate the minimum reflux ratio using the Underwood equations. Use graphical approach to obtain the solution of First Underwood Equation.
c) Calculate the total number of equilibrium stages at $R=1.35R_{min}$
d) Determine the optimum feed tray location.

Solution

Refer to the schematic diagram given below.

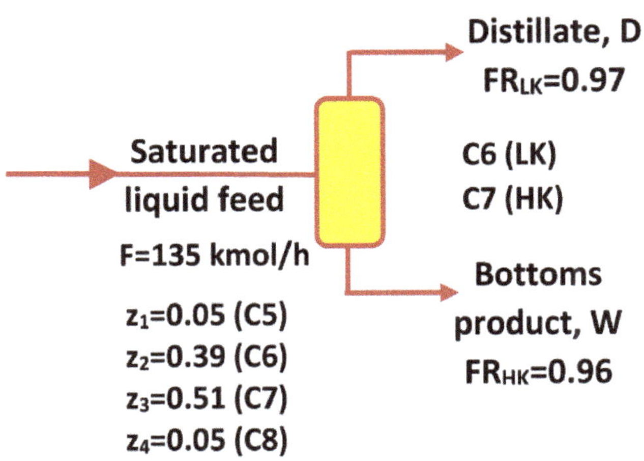

As the recoveries of n-hexane (C6) and n-heptane (C7) are specified, they are selected as the key components: more volatile n-hexane (C6) as light key (LK) and less volatile n-heptane (C7) as heavy key (HK). The information given is summarized in the table below.

Component	$z_{i,F}$	$\alpha_{i,HK}$	$FR_{i,D}$	$FR_{i,W}$
Species 1-C5(LNK)	0.05	6.5	not given	not given
Species 2- C6(LK)	0.39	2.5	0.97	0.03
Species 3 – C7 (HK)	0.51	1.0	0.04	0.96
Species 4–C8(HNK)	0.05	0.40	not given	not given

a) The number of equilibrium stages required at total reflux and fractional recoveries of C5 in distillate and C8 in bottoms product:

We apply Fenske equation, Equation (13.6), to calculate N_{min}

$$N_{min} = \frac{ln[\{FR_{LK}/(1-FR_{LK})\}_D/\{(1-FR_{HK})/FR_{HK}\}_W]}{ln(\alpha_{LK,HK})}$$

$$FR_{LK,D} = 0.97$$

$$FR_{HK,W} = 0.96$$

$$\alpha_{LK,HK} = 2.5$$

Thus,

$$N_{min} = \frac{ln[(0.97/(1-0.97))/((1-0.96)/0.96)]}{ln(2.5)} = 7.262$$

Using the other form of the Fenske equation, Equation (13.7), we can calculate the fractional recoveries of C5 (LNK) in distillate and C8 (HNK) in bottoms product as:

$$FR_{NK,D} = \frac{\alpha_{NK,HK}^{N_{min}}}{\alpha_{NK,HK}^{N_{min}}+\{FR_{HK}/(1-FR_{HK})\}_W}$$

$$FR_{LNK,D} = \frac{\alpha_{LNK,HK}^{N_{min}}}{\alpha_{LNK,HK}^{N_{min}} + \{FR_{HK}/(1 - FR_{HK})\}_W} = \frac{(6.5)^{7.262}}{(6.5)^{7.262} + (0.96/(1 - 0.96))} = 0.99997 \approx 1.0$$

$$FR_{HNK,D} = \frac{\alpha_{HNK,HK}^{N_{min}}}{\alpha_{HNK,HK}^{N_{min}}+\{FR_{HK}/(1-FR_{HK})\}_W} = \frac{(0.4)^{7.262}}{(0.4)^{7.262}+(0.96/(1-0.96))} = 5.37 \times 10^{-5} \approx 0$$

$$FR_{HNK,W} = 1 - FR_{HNK,D} = 1 - 5.37 \times 10^{-5} = 0.999946 \approx 1.0$$

b) Minimum reflux ratio using the Underwood equations:

The First Underwood equation, Equation (13.32), can be written as:

$$F(1-q) = \sum \frac{\alpha_{i,HK}(z_iF)}{\alpha_{i,HK}-\varphi}$$

The only unknown in this equation is φ. Upon expansion of the summation term, we can write:

$$135(1-1) = \frac{(0.05 \times 135)(6.5)}{6.5-\varphi} + \frac{(0.39 \times 135)(2.5)}{2.5-\varphi} + \frac{(0.51 \times 135)(1)}{1-\varphi} + \frac{(0.05 \times 135)(0.4)}{0.4-\varphi}$$

$$0 = \frac{43.875}{6.5-\varphi} + \frac{131.625}{2.5-\varphi} + \frac{68.85}{1-\varphi} + \frac{2.7}{0.4-\varphi}$$

As discussed in section 13.3.3 (see Equation (13.35)), the solution of this equation lies within the range: $1 < \varphi < 2.5$. The solution can be obtained using the Newton-Raphson method. Another approach to solve this equation is to select several values of φ within the specified range, calculate the right hand side of the equation and plot $y = 0$ and $y = RHS$ on a $y\ vs\ \varphi$ plot where RHS is the right hand side of the above equation. The intersection of the plots $y = 0$ and $y = RHS$ gives the solution. The solution is shown in the figure given below. We find $\varphi = 1.4994$ satisfies the above equation.

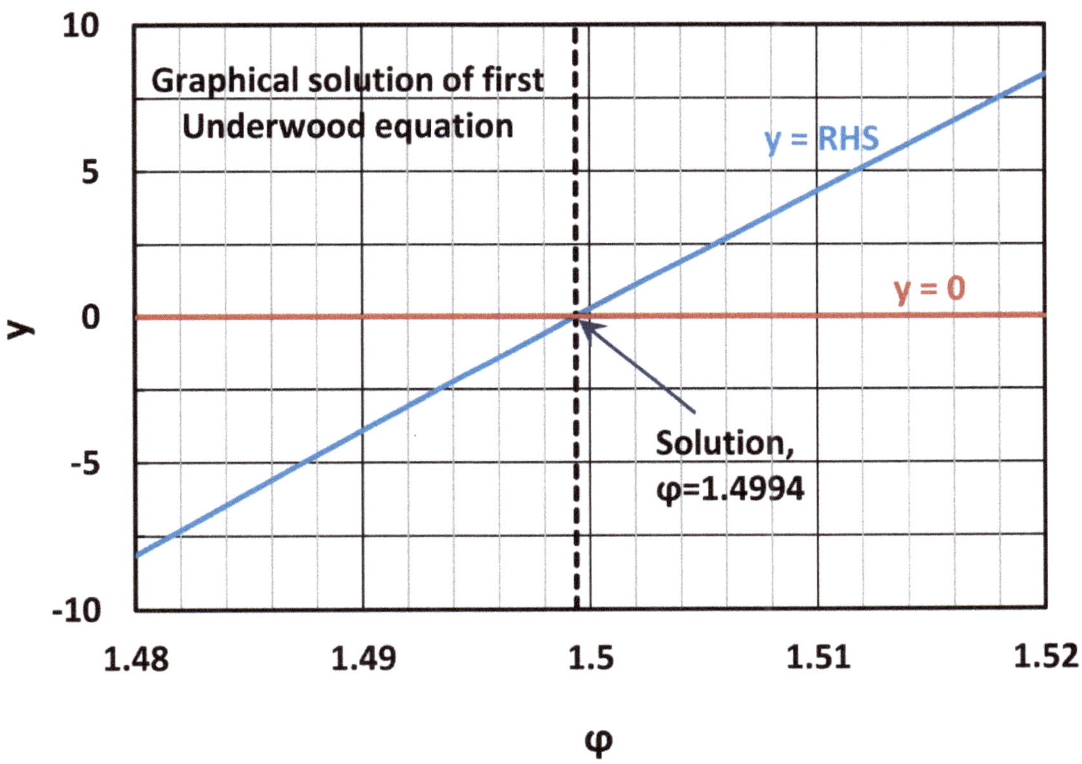

Now we use the Second Underwood equation, Equation (13.33), to determine the minimum reflux ratio R_{min}:

$$D(R_{min} + 1) = \sum \frac{\alpha_{iHK}(x_{i,D}D)}{[\alpha_{iHK} - \varphi]}$$

No trial and error is needed to solve this equation for R_{min}. Before we solve this equation, we need to calculate the $(x_{i,D}D)$ terms and the total distillate flow rate D. The $(x_{i,D}D)$ terms are given as:

$$x_{i,D}D = (z_i F)(FR_{i,D})$$

Thus,

$$x_{1,D}D = (z_1F)(FR_{1,D}) = (0.05 \times 135)(0.99997) = 6.75\ kmol/h$$

$$x_{2,D}D = (z_2F)(FR_{2,D}) = (0.39 \times 135)(0.97) = 51.07\ kmol/h$$

$$x_{3,D}D = (z_3F)(FR_{3,D}) = (0.51 \times 135)(0.04) = 2.754\ kmol/h$$

$$x_{4,D}D = (z_4F)(FR_{4,D}) = (0.05 \times 135)(5.37 \times 10^{-5}) = 0.000362\ kmol/h$$

$$D = 6.75 + 51.07 + 2.754 + 0.000362 = 60.574\ kmol/h$$

Now the Second Underwood equation can be solved for R_{min} as follows:

$$60.57 \times (R_{min} + 1) = \frac{6.5 \times 6.75}{6.5 - 1.4994} + \frac{2.5 \times 51.07}{2.5 - 1.4994} + \frac{1 \times 2.754}{1 - 1.4994} + \frac{0.40 \times 0.000362}{0.40 - 1.4994} = 130.858$$

$$R_{min} = 1.16$$

c) The total number of equilibrium stages at $R = 1.35R_{min}$, and the optimum feed tray location

We can use the Gilliland approach to estimate the number of equilibrium stages at given reflux ratio.

$$R = 1.35 \times 1.16 = 1.566$$

$$X_G = \frac{R - R_{min}}{R + 1} = \frac{1.566 - 1.16}{1.566 + 1} = 0.1582$$

Using the Gilliland-Molokanov equation, Equation (13.41), we get

$$Y_G = 1 - exp\left[\left(\frac{X_G - 1}{\sqrt{X_G}}\right)\left(\frac{1 + 54.4X_G}{11 + 117.2X_G}\right)\right] = 1 - exp[(-2.1164)(0.3252)] = 0.4975$$

$$Y_G = \frac{N - N_{min}}{N + 1} = 0.4975$$

$$N = 15.44 \approx 15 \text{ stages}$$

Note that N includes partial reboiler.

d) The optimum feed tray location can be determined using the following Kirkbride equation:

$$\frac{n_F - 1}{N - n_F} = \left[\left(\frac{z_{HK}}{z_{LK}}\right)\left(\frac{x_{LK,W}}{x_{HK,D}}\right)^2 \frac{W}{D}\right]^{0.206}$$

where n_F is the feed stage number counting from top of the column and N is the total number of stages (including partial reboiler). $N = 15$, $z_{HK} = 0.51$, $z_{LK} = 0.39$, $D = 60.574$, and $W = F - D = 135 - 60.574 = 74.426\ kmol/h$. We need to calculate $x_{LK,W}$ and $x_{HK,D}$ as follows:

$$x_{LK,W}W = (1 - FR_{LK,D})(z_{LK}F) = (1 - 0.97)(0.39 \times 135) = 1.5795$$

$$x_{LK,W} = 1.5795/74.426 = 0.02122$$

$$x_{HK,D}D = (1 - FR_{HK,W})(z_{HK}F) = (1 - 0.96)(0.51 \times 135) = 2.754$$

$$x_{HK,D} = 2.754/60.574 = 0.0455$$

The Kirkbride equation can now be solved for n_F :

$$\frac{n_F - 1}{15.44 - n_F} = \left[\left(\frac{0.51}{0.39}\right)\left(\frac{0.02122}{0.0455}\right)^2 \frac{74.426}{60.574}\right]^{0.206} = 0.8053$$

$$n_F = 7.44 \approx 7$$

Thus the optimum feed tray is the 7th tray (counting from top of the column). There are 6 trays above the feed tray and 7 trays below the feed tray (excluding the partial reboiler). The partial reboiler is the 15th equilibrium stage.

SHORT REVIEW QUESTIONS

1. How many distillation columns are required to separate a multicomponent mixture of C species into pure components?
2. For multicomponent separation using a single distillation column, the distillate and bottoms compositions cannot be specified completely. Is this true or false?
3. For multicomponent separation using a single distillation column, the mole fractions or fractional recoveries of only two components (key components) can be specified in the distillate and bottoms products. Is this true or false?
4. Define light key (LK) and heavy key (HK) components. How do you select these key components?
5. What is the difference between adjacent keys and split keys?
6. Both LK and HK components must be distributed. Is this true?
7. Define class 1 and class 2 separations in distillation.
8. How many pinch point zones are present in class 2 separation?
9. Describe pseudo-binary method of designing distillation columns for multicomponent systems.
10. Describe FUG method of designing distillation columns for multicomponent systems.
11. Define fractional recovery of any component i in the distillate product.
12. Define fractional recovery of any component i in the bottoms product.
13. The fractional recoveries of non-key components in the distillate product can be estimated from the Fenske equation provided that the fractional recoveries of the key components are specified in the distillate and bottoms products. Is this true or false?
14. When the keys are "adjacent keys", there exists only one solution of the first Underwood equation (Equation (13.32)) in the range $1 < \varphi < \alpha_{LK,HK}$. Is this true or false?
15. When the keys are "split keys", there are multiple solutions (multiple φ values) of the first Underwood equation (Equation (13.32)) in the range $1 < \varphi < \alpha_{LK,HK}$. Is this true or false?
16. What does MESH stand for?
17. For a multicomponent system of C species, how many MESH equations are there for each stage?
18. How many different sequences of distillation columns are possible to separate a multicomponent mixture of 7 components into 7 pure (single component) product streams?
19. Give two examples of heuristics proposed for sequencing of distillation columns?

PRACTICE PROBLEMS

Problem 1

A conventional distillation column with total condenser and partial reboiler is being used to separate a 5-component mixture. The feed composition and the relative volatilities of components are given in the table below.

Component	$z_{i,F}$	$\alpha_{i,HK}$
1(LNK)	0.07	2.5
2(LK)	0.39	2.0
3(HK)	0.41	1.0
4(HNK)	0.07	0.90
5(HNK)	0.06	0.70

The fractional recovery of LK component in the distillate is 98%. The fractional recovery of HK component in the bottoms is 97%. Calculate the minimum number of equilibrium stages required? Specify the fractional recoveries of all components in the distillate and bottoms products at total reflux?

Problem 2

We wish to distill 2000 kmol/h of a saturated liquid feed. The feed is 0.056 mole fraction propane,0.321 n-butane, 0.482 n-pentane, and the remainder n-hexane. The column operates at 101.3 kPa. The column has a total condenser and a partial reboiler. The reflux ratio (L/D) is 3.5, and the reflux is a saturated liquid. The optimum feed stage is to be used.

A fractional recovery of 99.4% n-butane is desired in the distillate and 99.7% of the n-pentane in the bottoms. The relative volatility of n-butane with respect to n-pentane is 2.45.

Calculate the following:

a) Flow rates of components in distillate product and total distillate flow rate.
b) Composition of distillate product (mole fractions of all components).
c) Flow rates of components in bottoms product and total bottoms flow rate.
d) Composition of bottoms product (mole fractions of all components).
e) Calculate the minimum number of equilibrium stages required.

You can take n-butane as LK and n-pentane as HK (the order of volatilities is: propane > n-butane > n-pentane > n-hexane). You can further assume that all the LNK collects in the distillate and all the HNK collects in the bottoms.

Problem 3

A saturated liquid feed ($q = 1.0$) is fed at a rate of 100 kmol/h to a conventional distillation column with total condenser and partial reboiler. The composition of the feed is 3 mol% n-pentane (C5), 41 mol% n-hexane (C6), 48 mol% n-heptane (C7), and 8 mol% n-octane (C8). The feed is distilled so that 95% of the n-hexane is recovered in the distillate. The amount of n-heptane recovered in the bottom product is 98%. The relative volatilities can be assumed to be constant ($\alpha_{C5,C7} = 6.5, \alpha_{C6,C7} = 2.5,\ \alpha_{C7,C7} = 1.0, \alpha_{C8,C7} = 0.40$).

a) Calculate the number of equilibrium stages required at total reflux and the fractional recoveries of C5 in distillate and C8 in bottoms product.
b) Calculate the minimum reflux ratio using the Underwood equations
c) Calculate the total number of equilibrium stages at $R = 1.4R_{min}$, and determine the optimum feed tray location.

Problem 4

A saturated vapor feed ($q = 0$) is fed at a rate of 200 kmol/h to a conventional distillation column with total condenser and partial reboiler. The composition of the feed is 25 mol% benzene, 45 mol% toluene, and 30 mol% cumene. The feed is distilled so that 98% of the toluene is recovered in the distillate and 96% of the cumene is recovered in the bottoms. The relative volatilities can be assumed to be constant ($\alpha_{TC} = 4.76, \alpha_{BC} = 10.71,\ \alpha_{CC} = 1.0$ where subscript T refers to toluene, B refers to benzene, and C refers to cumene).

a) Calculate the minimum number of equilibrium stages required at total reflux and fractional recovery of benzene in distillate.
b) Calculate the minimum reflux ratio using the Underwood equations. Use graphical approach to solve the first Underwood equation.
c) Calculate the total number of equilibrium stages at $R = 1.35R_{min}$
d) Determine the optimum feed tray location.

Problem 5

Using 4 distillation columns, show schematically the flow diagrams of 14 possible simple sequences to separate a 5-component mixture into pure components.

Chapter 14

Liquid-Liquid Extraction

14.1 LEARNING OUTCOMES

Upon completion of this chapter, students will be able to:

- Describe liquid-liquid extraction process and its practical applications
- Construct three-component equilateral and right-angled triangular phase diagrams
- Define plait point
- Analyze single-stage and multi-stage liquid-liquid extraction processes using triangular phase diagrams
- Estimate the number of equilibrium stages required for a given separation in multi-stage extraction processes dealing with immiscible solvent-diluent pair and partially miscible solvent-diluent pair
- Develop conjugate curve for the construction of tie lines on a triangular phase diagram

14.2 INTRODUCTION

As noted in Chapter 1, liquid-liquid extraction (LLX) involves the separation of solute from the feed liquid phase (typically aqueous phase) to the solvent liquid phase (typically organic solvent). The two phases are mixed intimately resulting in the formation of a dispersion of droplets of one liquid phase (solvent phase) in another (aqueous phase). For the solute (species A) to be transferred from feed to the solvent phase it should have high affinity for the solvent. Figure 14.1 shows the direction of mass transfer in a LLX process.

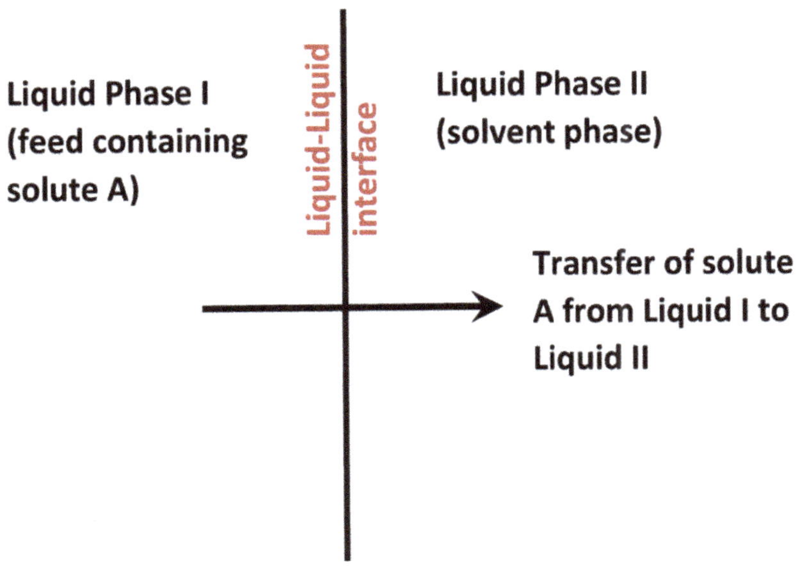

Figure 14.1 Direction of mass transfer of solute in a liquid-liquid extraction process.

Typically two pieces of equipment are involved in a single-stage LLX process, namely a mixer and a settler, as shown schematically in Figure 14.2. The two liquid streams, solvent (S) and feed (F), are mixed thoroughly in the mixing vessel using a mixer/stirrer and the mixture stream (M) is passed to a settling vessel for the separation of the two phases. During the intimate mixing process, the solute is transferred from the feed phase to the solvent phase. The solute-depleted feed phase (liquid phase I) leaving the process is referred to as *raffinate* (*R*) and the solute-rich solvent phase (liquid phase II) leaving the process is called *extract* (*E*).

LLX process is often utilized when separation of a solute by distillation is either very difficult or completely ineffective. For example, a close-boiling mixture is difficult to separate using distillation. If the mixture forms an azeotrope then ordinary distillation is completely ineffective. In some applications, the solute to be recovered is thermally fragile and therefore, distillation is not used. In all such applications where distillation is not suitable due to one reason or another, LLX is one of the main options to be considered. Some of the industrial applications where LLX is used to separate a component or components from a mixture are: separation of penicillin from fermentation broth using amyl or butyl acetate as a solvent; separation of aromatic hydrocarbons (benzene, toluene, xylene, etc.) from paraffins using sulfolane as a solvent; recovery of acetic acid from dilute aqueous solutions using a solvent such as ethyl or butyl acetate; and extraction of acetone from aqueous solutions using methyl isobutyl ketone (MIK) as a solvent.

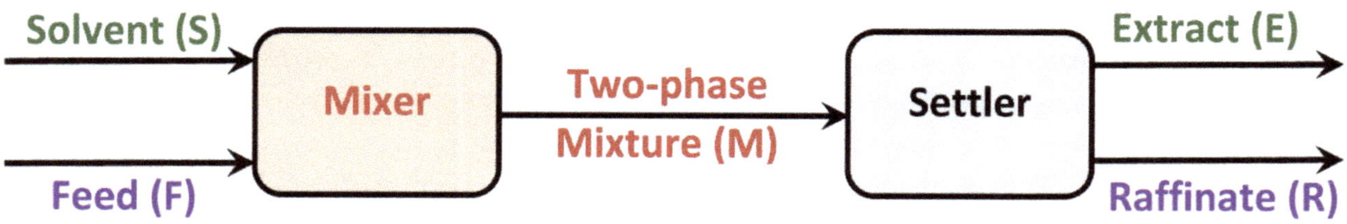

Figure 14.2 Single stage liquid-liquid extraction process.

An important point to note is that in LLX process, we have 3 species involved: solute (species A), diluent (species B) and solvent (species C). The diluent is the carrier fluid for the solute in the feed. The solute A is soluble in both solvent C and diluent B but its solubility is different in the two phases. The solvent and diluent are generally *partially miscible* in each other. Thus we need a *triangular phase diagram* for the analysis and design of a LLX process. In the special case of solvent and diluent being *completely immiscible* in each other, the analysis and design of a LLX process is no different from stripping covered in Chapters 3 and 5. No triangular phase diagrams are needed in the analysis and design of such systems where the solvent and diluent are completely immiscible in each other. Note that most solvent-diluent pairs can be treated as completely immiscible when the amount of solute present is small, that is, we have dilute solutions of solute. In concentrated systems where the amounts of solute present are large, the solvent-diluent pairs tend to become partially miscible. The presence of large amounts of solute makes the solvent and diluent chemically more similar and hence partially miscible.

14.3 TRIANGULAR PHASE DIAGRAMS

14.3.1 Partially Miscible Solvent – Diluent Pair

There are different ways of presenting triangular phase diagrams for partially miscible systems. First we will consider *equilateral triangle* (all sides equal) phase diagrams. Later on, *right-angled triangle* phase diagrams will be discussed.

Before we discuss an equilateral triangle phase diagram, it is important to know how to read the compositions from an equilateral triangle diagram. Figure 14.3 illustrates how to read the composition corresponding to any point located inside an equilateral triangle diagram. The triangle diagram involves 3 components: species A (solute), species B (diluent) and species C (solvent). The corners of the triangle represent the pure species. Any point located inside the triangle represents a 3-component mixture. To determine the mass fraction of a component A in the mixture, draw a line parallel to the edge opposite to corner A (100%) towards the x_A axis. For the mixture shown in Figure 14.3, $x_A = 40\%$. To determine x_B, draw a line parallel to the edge opposite to corner B (100%) towards the x_B axis. For the mixture shown in Figure 14.3, $x_B = 30\%$. In the same manner, x_C is found to be 30% for the mixture shown in Figure 14.3.

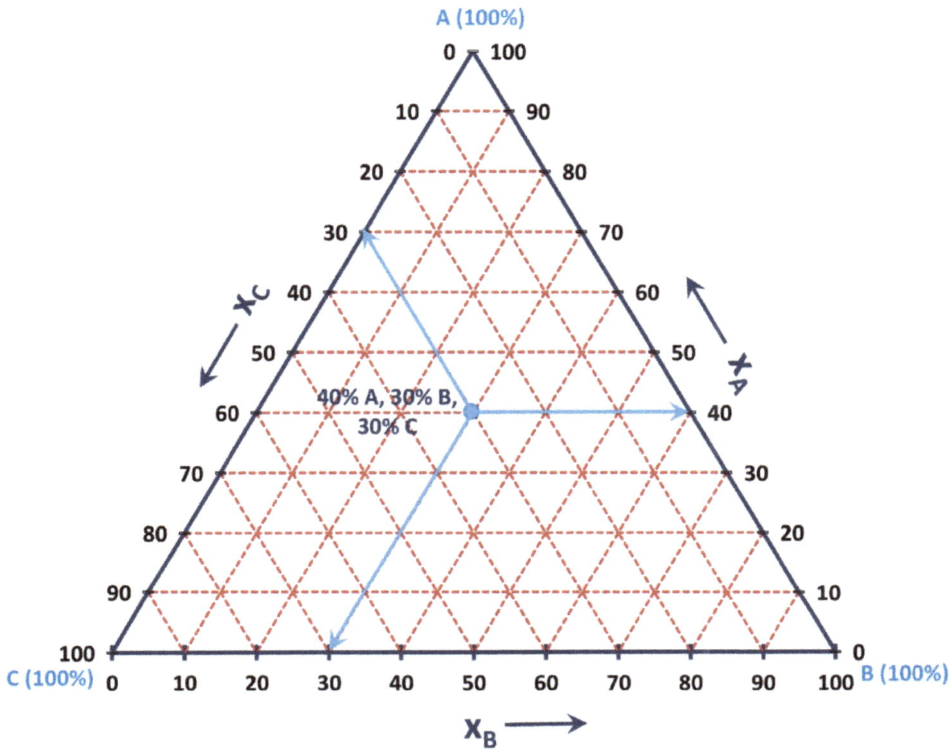

Figure 14.3 Reading the composition corresponding to a point located inside an equilateral triangle diagram.

Figure 14.4 shows a commonly observed equilateral triangle phase diagram at a fixed temperature. The solute A is completely miscible with diluent B and solvent C. The solvent C and diluent B are only partially miscible. The phase diagram consists of single-phase region and two-phase region (under the red envelope). In the single-phase region, the mixture of A,B, and C is a single-phase homogeneous solution. In the two-phase region, any point, say *M*, represents a two-phase mixture of *E* (extract phase rich in solvent C) and *R* (raffinate phase rich in diluent B) connected by an equilibrium tie-line. The tie lines connect the two phases (extract and raffinate) which are at equilibrium with each other. The *plait point* represents a condition where the 3-component mixture consists of two phases with identical compositions. The interface between the two phases vanishes just above the plait point resulting in a homogenous single-phase solution.

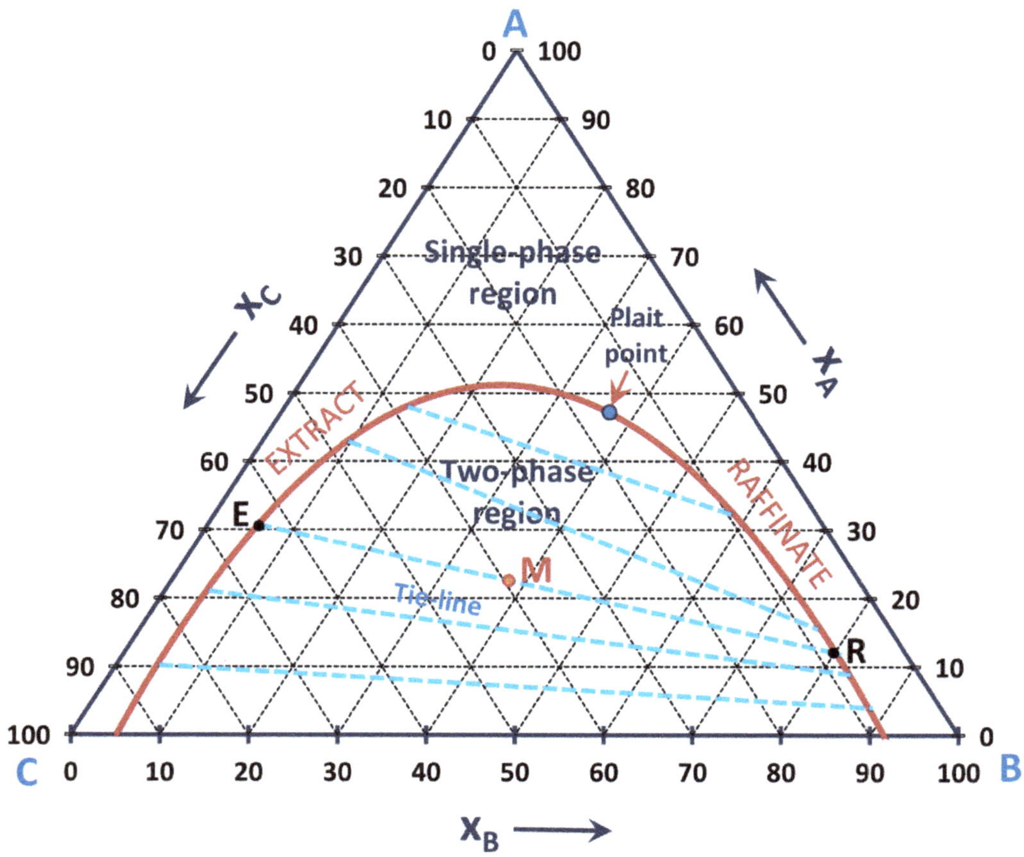

Figure 14.4 Equilateral triangle phase diagram where solute A is completely miscible with diluent B and solvent C, and the solvent C and diluent B are only partially miscible.

The phase diagram shown in Figure 14.4 represents *Type I* system where only one pair of species (B and C) is partially miscible. The other two pairs (A and C; A and B) are completely miscible. When two pairs of species are partially miscible, the system is called *Type II* system. For example, A & B and A & C pairs may be partially miscible and pair B & C is completely miscible. In this chapter, we will restrict ourselves to type I systems, shown in Figure 14.4, as these systems are the most commonly encountered in practical applications.

Although equilateral triangle diagrams are used extensively to represent the phase behavior of three-component systems, it is often more convenient to use right-angled triangular diagrams to analyze and design LLX process. Figure 14.5 shows the typical phase behavior of type I systems on a right-angled triangular diagram. We assume that species C is the solvent and species B is the diluent or carrier for solute species A. We need only two composition variable x_A and x_C to locate a point on the diagram. The hypotenuses represent the constant x_B lines. Under the equilibrium red curve (left side of the figure), we have the two-phase region. Above the equilibrium curve is the single-phase region. The tie lines connect the two-phases (raffinate and extract) at equilibrium.

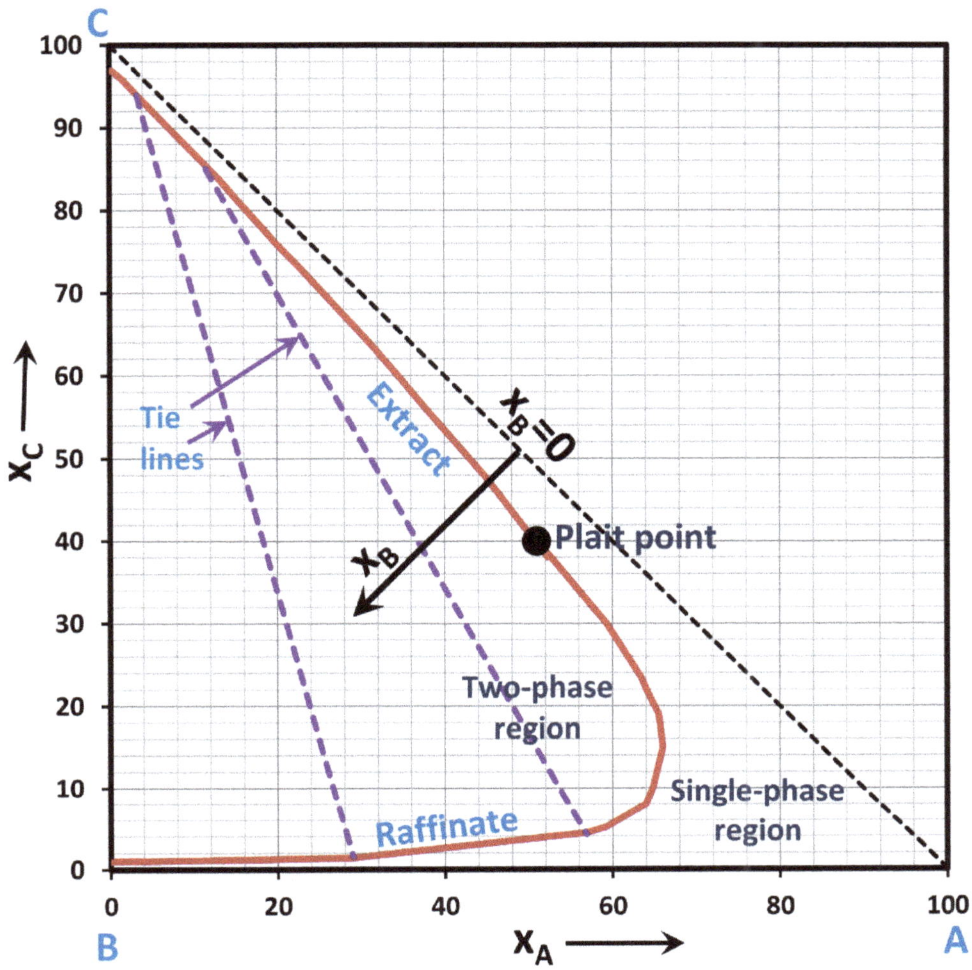

Figure 14.5 Right-angled triangle phase diagram where solute A is completely miscible with diluent or carrier B and solvent C, and the solvent C and diluent B are only partially miscible.

14.3.2 Completely Immiscible Solvent – Diluent Pair

When solvent-diluent pair is completely immiscible, the triangular phase diagram is simply a collection of tie-lines connecting $(A + B)$ mixtures in equilibrium with $(A + C)$ mixtures where A is solute, B is diluent, and C is solvent. The solute is assumed to be completely miscible in solvent and diluent. The entire triangle area in such systems is covered by a two-phase region. The triangular phase diagrams for a hypothetical system composed of immiscible solvent and diluent are shown in Figures 14.6 and 14.7.

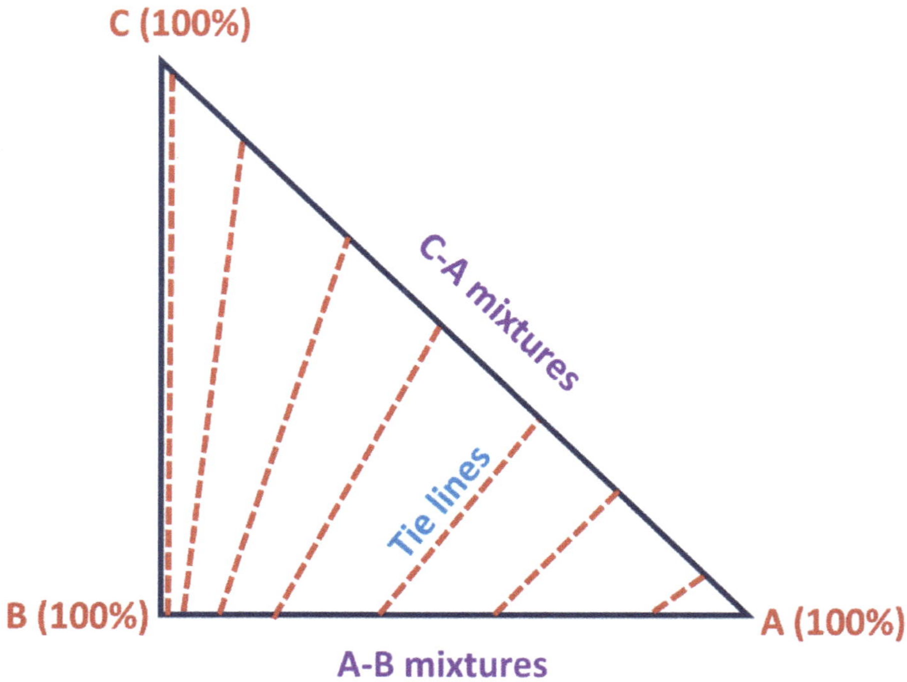

Figure 14.6 Right-angled triangle phase diagram of a hypothetical 3-component system where solute A is completely miscible with diluent B and solvent C, and the solvent C and diluent B are completely immiscible.

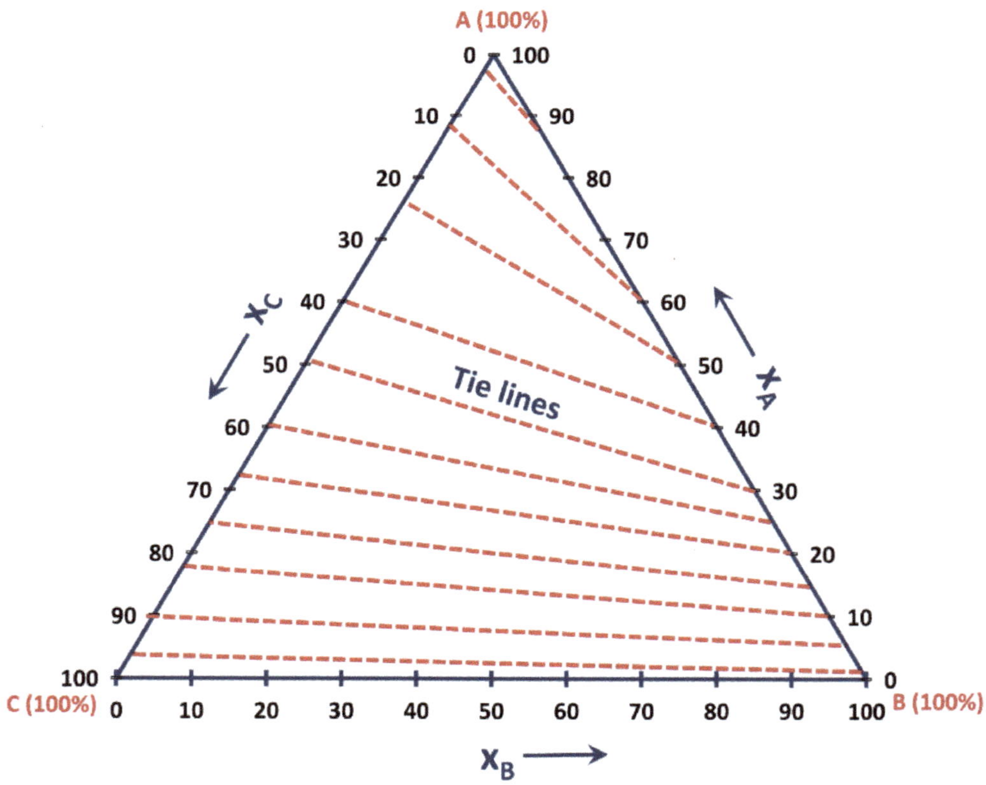

Figure 14.7 Equilateral triangle phase diagram of a hypothetical 3-component system where solute A is completely miscible with diluent B and solvent C, and the solvent C and diluent B are completely immiscible.

14.4 SINGLE-STAGE LIQUID-LIQUID EXTRACTION

14.4.1 Analysis of a Single-Stage LLX Process: Immiscible Solvent-Diluent Pair

Figure 14.8 shows a schematic diagram of a single-stage LLX process involving immiscible solvent-diluent pair. As the solvent and diluent are completely immiscible, the flow streams shown in Figure 14.8 consist of only two components (one of them is solute). Thus only the solute concentrations are specified.

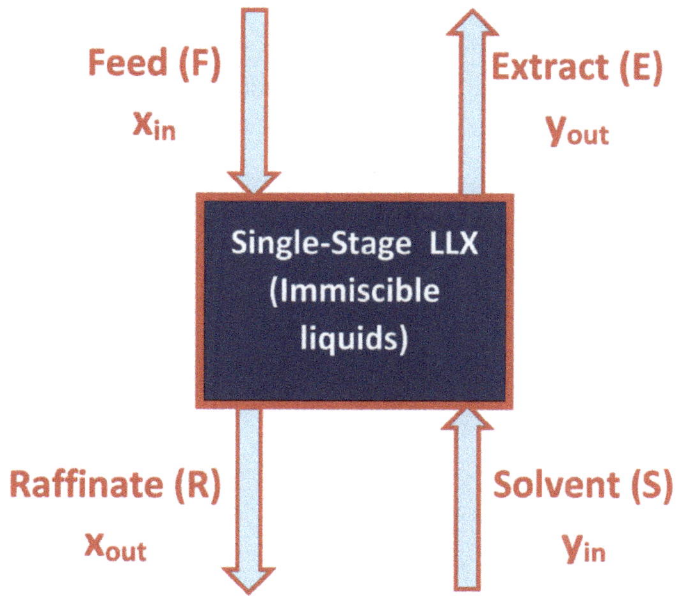

Figure 14.8 Single-stage LLX with immiscible solvent-diluent pair.

The overall material balance gives:

$$F + S = E + R \tag{14.1}$$

The solute material balance gives:

$$x_{in}F + y_{in}S = x_{out}R + y_{out}E \tag{14.2}$$

Assuming that the system is dilute (concentration of solute in any stream is small), we can regard:

$$F = R = L \tag{14.3}$$

$$S = E = V \tag{14.4}$$

where L refers to heavy liquid stream (feed or raffinate) and V refers to light liquid stream (solvent or extract).

Consequently, Equation (14.2) reduces to:

$$y_{out} = y_{in} + (-\tfrac{L}{V})(x_{out} - x_{in}) \tag{14.5}$$

This is the *operating line* of a single-stage LLX process using immiscible solvent-diluent pair. Using the condition of phase equilibrium, that is, the two streams leaving the stage are at equilibrium with each other, we can determine the unknown compositions of the extract and raffinate streams leaving the process. Figure 14.9 shows the graphical solution. Given the inlet streams compositions and their flow rates, and the equilibrium relationship, the intersection of the operating and equilibrium lines gives the unknown compositions (x_{out} and y_{out}).

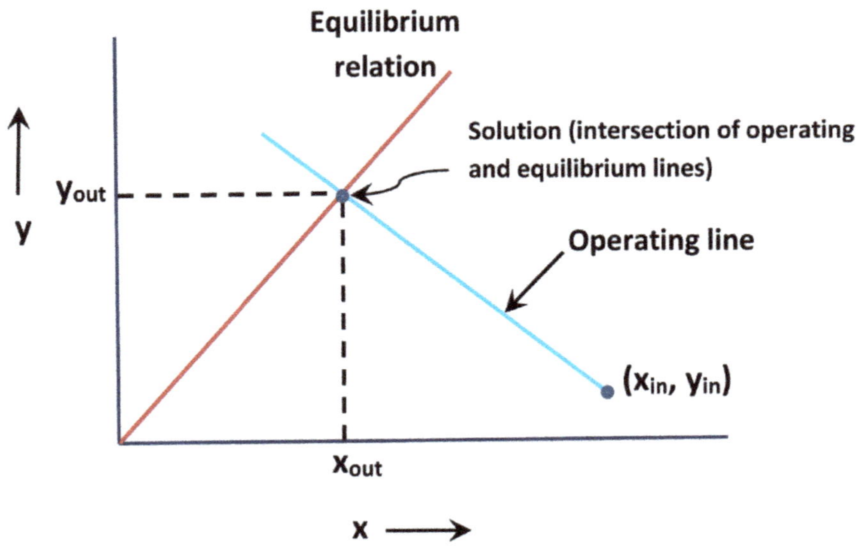

Figure 14.9 Graphical solution of single-stage LLX process with immiscible solvent-diluent pair.

The percent recovery of solute from a single-stage LLX process is given as:

$$\%\ Recovery\ of\ Solute = 100(1 - \frac{x_{out}}{x_{in}}) \tag{14.6}$$

An analytical solution of unknown compositions of the extract and raffinate streams can be obtained using the operating line (Equation (14.5)) and the equilibrium relation:

$$y = K_d x \tag{14.7}$$

where K_d is the distribution coefficient. For dilute solutions, K_d is constant at a given temperature. Since the two leaving streams (extract and raffinate) are at equilibrium with each other, Equation (14.7) can be applied and re-written as:

$$y_{out} = K_d x_{out} \tag{14.8}$$

From Equations (14.5) and (14.8), the following expression for x_{out} can be derived by eliminating y_{out}

$$\frac{x_{out}}{x_{in}} = \left(\frac{1}{1+E_o}\right)\left[1 + \frac{V}{L}\left(\frac{y_{in}}{x_{in}}\right)\right] \tag{14.9}$$

where E_o is the extraction factor, defined as:

$$E_o = \frac{slope\ of\ equilibrium\ line}{|slope\ of\ operating\ line|} = \frac{K_d}{L/V} \tag{14.10}$$

The extraction factor E_o is a measure of degree of separation of solute from the feed phase to the extract phase. To see the significance of E_o, consider a single-stage extraction with solute-free solvent entering the device. In this case, Equation (14.9) could be re-written as:

$$\frac{x_{out}}{x_{in}} = \left(\frac{1}{1+E_o}\right) \quad (14.11)$$

When $E_o \to 0, x_{out} = x_{in}$ and no separation of solute occurs. When $E_o \to \infty, x_{out} = 0$, that is, complete separation of solute occurs.

14.4.2 Analysis of a Single-Stage LLX Process: Partially Miscible Solvent-Diluent Pair

A single-stage LLX process with partially miscible solvent-diluent pair is shown schematically in Figure 14.10. The feed (F) and solvent (S) streams are mixed intimately in a mixing vessel to produce a two-phase mixture stream (M). The mixture stream is then allowed to settle in a settling tank into separate extract (E) and raffinate (R) streams. The flow streams now contain three components in general: solute A, diluent B and solvent C. However, we need to specify the concentrations of only two components (solute A and solvent C) in each stream. The concentration of the third component (diluent B) simply follows from mass balance.

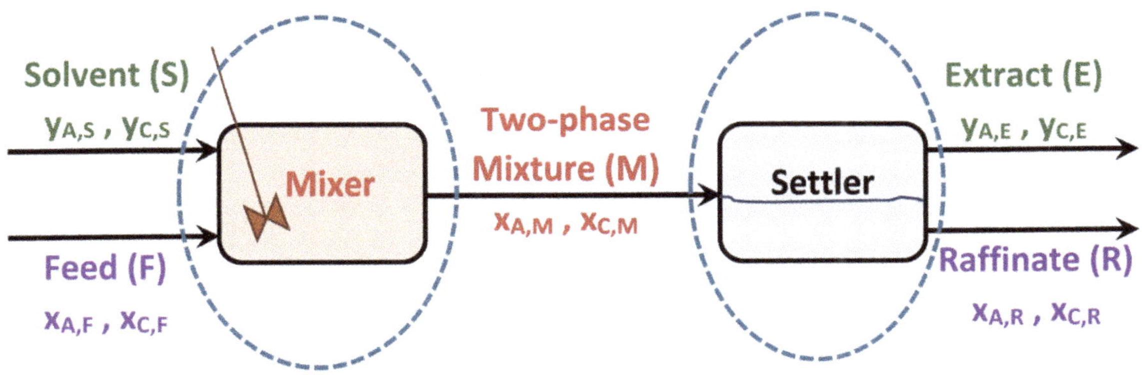

Figure 14.10 Single stage LLX process with partially miscible solvent-diluent pair.

Total material balance around the mixer (dashed blue envelope) gives:

$$S + F = M \quad (14.12)$$

Component material balances around the mixer give:

$$y_{A,S}S + x_{A,F}F = x_{A,M}M \quad (14.13)$$

$$y_{C,S}S + x_{C,F}F = x_{C,M}M \quad (14.14)$$

From Equations (14.12)-(14.14), it can be readily shown that:

$$x_{A,M} = \frac{y_{A,S}+x_{A,F}(F/S)}{1+(F/S)} \quad (14.15)$$

$$x_{C,M} = \frac{y_{C,S}+x_{C,F}(F/S)}{1+(F/S)} \quad (14.16)$$

Given the feed and solvent flow rates and compositions: $F, S, x_{A,F}, x_{C,F}, y_{A,S}, y_{C,S}$, we would like to calculate the extract and raffinate flow rates and compositions. The solution can be obtained graphically (see Figure 14.11) as follows:

- Locate the feed point $\boldsymbol{F}(x_{A,F}, x_{C,F})$ on the right-angled triangle phase diagram
- Locate the solvent point $\boldsymbol{S}(y_{A,S}, y_{C,S})$ on the right-angled triangle phase diagram
- Locate the mixture point $\boldsymbol{M}(x_{A,M}, x_{C,M})$ calculated from Equations (14.15) and (14.16) on the right-angled triangle phase diagram
- Connect the feed, solvent, and mixture points by a straight line as these three points are *collinear.*
- As the extract stream and raffinate stream are in *phase equilibrium* with each other, the extract and raffinate compositions can be determined graphically by drawing a *tie-line* passing through the mixture point as shown in Figure 14.11. Note that there is only one tie-line that passes through the points $\boldsymbol{E}, \boldsymbol{M}$ and $\boldsymbol{R}$; these three points are *collinear.* Read the compositions graphically: extract corresponding to point $\boldsymbol{E}(y_{A,E}, y_{C,E})$ and raffinate corresponding to point $\boldsymbol{R}(x_{A,R}, x_{C,R})$.
- Once the extract and raffinate compositions are determined, the extract and raffinate flow rates are determined using material balances.

Total material balance around the settler (dashed blue envelope, Figure 14.10) gives:

$$M = E + R \tag{14.17}$$

Solute material balance around the settler gives:

$$x_{A,M}M = y_{A,E}E + x_{A,R}R \tag{14.18}$$

From Equations (14.17) and (14.18), it follows that:

$$\frac{E}{M} = \frac{x_{A,M} - x_{A,R}}{y_{A,E} - x_{A,R}} \tag{14.19}$$

Equation (14.19) can be used to calculate the extract flow rate E. Note that M and all compositions are already known. The raffinate flow rate R can be calculated from Equation (14.17).

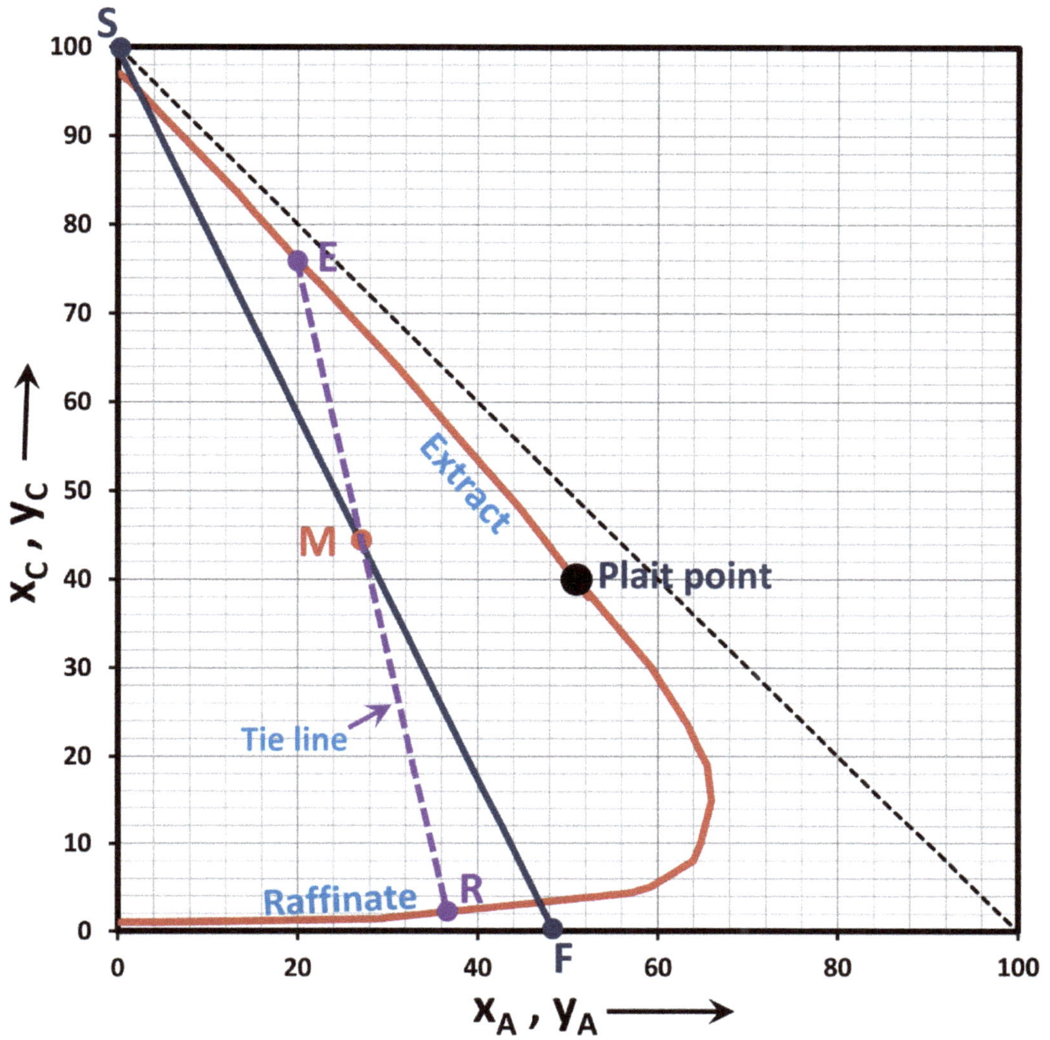

Figure 14.11 Graphical solution of single stage LLX process with partially miscible solvent-diluent pair.

14.5 MULTI-STAGE LIQUID-LIQUID EXTRACTION

The recovery of solute in a single-stage LLX process is often limited. In order to increase the percent recovery of solute from the feed stream, it may be necessary to use a multi-stage LLX process.

14.5.1 Analysis of a Multi-Stage LLX Process: Immiscible Solvent-Diluent Pair

Figure 14.12 shows a counter-current multi-stage LLX process flowsheet. The notation used for the analysis of a multi-stage LLX process with immiscible solvent-diluent pair is similar to that used for distillation. The lighter solvent phase (vapor in distillation) is specified by V and y (V is the flow rate and y is the concentration of solute in the light solvent phase). The heavier feed phase (liquid in distillation) is specified by L and x variable (L is the flow rate and x is the solute concentration in heavy feed phase).

The key calculation in the design of a multi-stage LLX process is the determination of the number of stages required to carry out the given separation. Thus, given the feed and solvent flow rates and compositions: L_a, V_b, x_a, x_b, y_b, we would like to determine the number of equilibrium stages required.

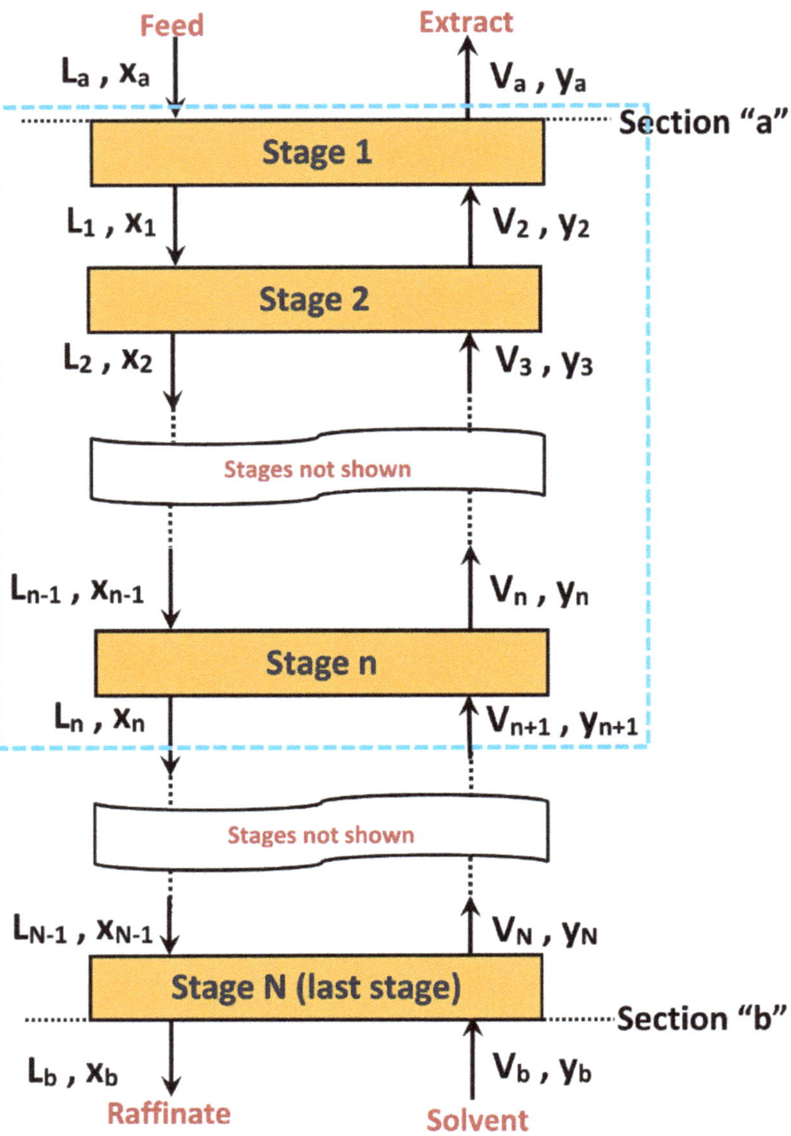

Figure 14.12 Counter-current multi-stage LLX with immiscible solvent-diluent pair.

The total material balance over the envelope indicated by blue dashed boundary gives:

$$L_a + V_{n+1} = L_n + V_a \tag{14.20}$$

The solute balance gives

$$x_a L_a + y_{n+1} V_{n+1} = x_n L_n + y_a V_a \tag{14.21}$$

Equation (14.21) can be re-written as:

$$y_{n+1} = \left(\frac{L_n}{V_{n+1}}\right) x_n + \frac{y_a V_a - x_a L_a}{V_{n+1}} \tag{14.22}$$

This is the equation of the operating line of counter-current cascade of Figure 14.12. For *dilute* systems, the flow rates of extract (solvent) stream and raffinate (feed) stream can be assumed to be constant and therefore, the slope L_n/V_{n+1} is constant leading to a linear operating line.

For dilute systems, one can write:

$$L_a = L_1 = \cdots = L_n = \cdots = L_b = L \tag{14.23}$$

$$V_a = V_1 = \cdots = V_n = \cdots = V_b = V \tag{14.24}$$

Consequently, Equation (14.22) reduces to:

$$y_{n+1} = y_a + (\tfrac{L}{V})(x_n - x_a) \tag{14.25}$$

This is the *operating line* of a multi-stage liquid-liquid extraction cascade assuming dilute systems. It a straight line of constant positive slope (L/V) passing through the terminal compositions (x_a, y_a) and (x_b, y_b). Given L, V, x_a, x_b, y_b, we can calculate y_a from solute balance, Equation (14.21). Once the terminal compositions are known, the operating line can be plotted.

Figure 14.13 shows the plots of the operating and equilibrium lines for a multi-stage LLX process with immiscible solvent-diluent pair. The figure also shows the driving force for mass transfer at any location within the extraction device. The equilibrium line for dilute solutions of solute is given as:

$$y = K_d x \tag{14.26}$$

where y is the concentration (usually weight fraction) of solute in the extract or solvent phase, x is the concentration of solute in the raffinate phase, and K_d is the distribution coefficient.

The number of stages in a multi-stage LLX extraction cascade can be determined graphically using the McCabe-Thiele diagram as shown in Figure 14.14.

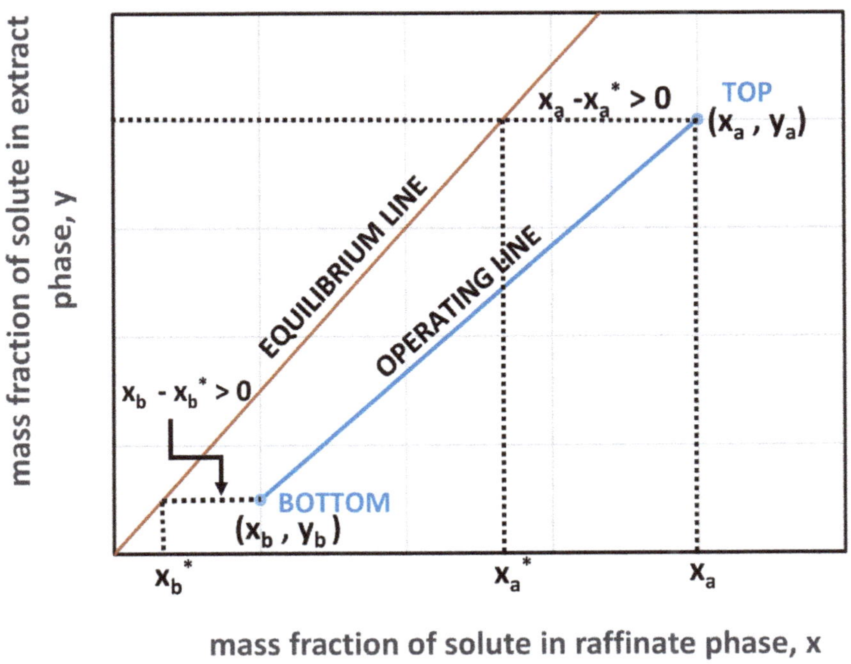

Figure 14.13 Plots of operating and equilibrium lines for multi-stage LLX of dilute systems.

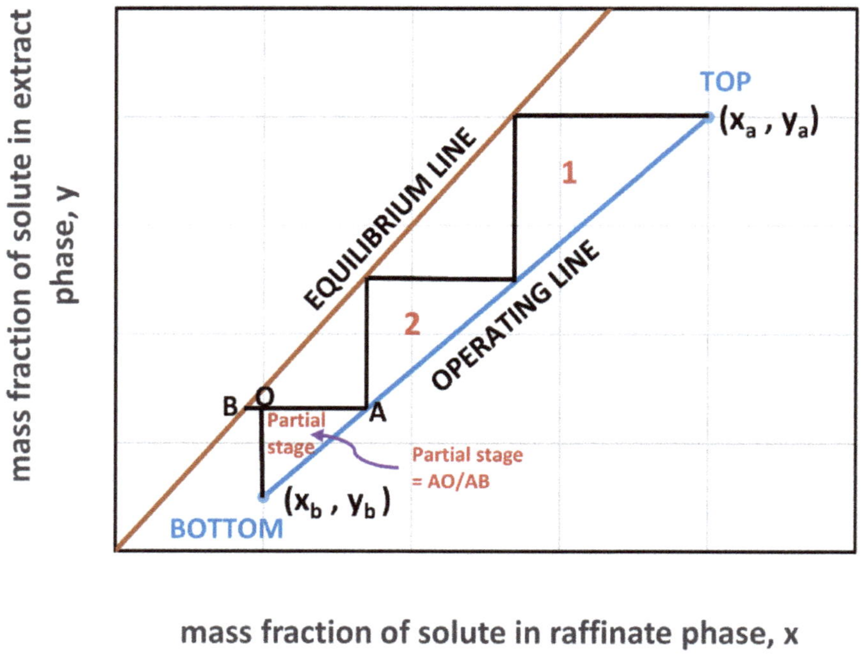

Figure 14.14 McCabe-Thiele diagram for the estimation of equilibrium stages in a multi-stage LLX process (dilute system with immiscible solvent-diluent pair).

The number of equilibrium stages can be determined analytically as well using the Kremser equations. The Kremser equations derived and discussed in Chapters 4 and 5 are applicable to a multi-stage LLX process with immiscible solvent-diluent pair.

The Kremser equations applicable to LLX are as follows:

$$N = \frac{\ln\left(\frac{x_a - x_a^*}{x_b - x_b^*}\right)}{\ln(E_o)} \quad when\ E_o \neq 1 \tag{14.27}$$

$$N = \frac{x_a - x_b}{x_b - x_b^*} = \frac{x_a - x_b}{x_a - x_a^*} \quad when\ E_o = 1 \tag{14.28}$$

where E_o is the extraction factor, defined as $E_o = K_d/(L/V)$, and the starred quantities are calculated using the equilibrium relation, Equation (14.26), as shown in Figure 14.15. Note that the extraction factor E_o is also equal to $(x_a - x_b)/(x_a^* - x_b^*)$.

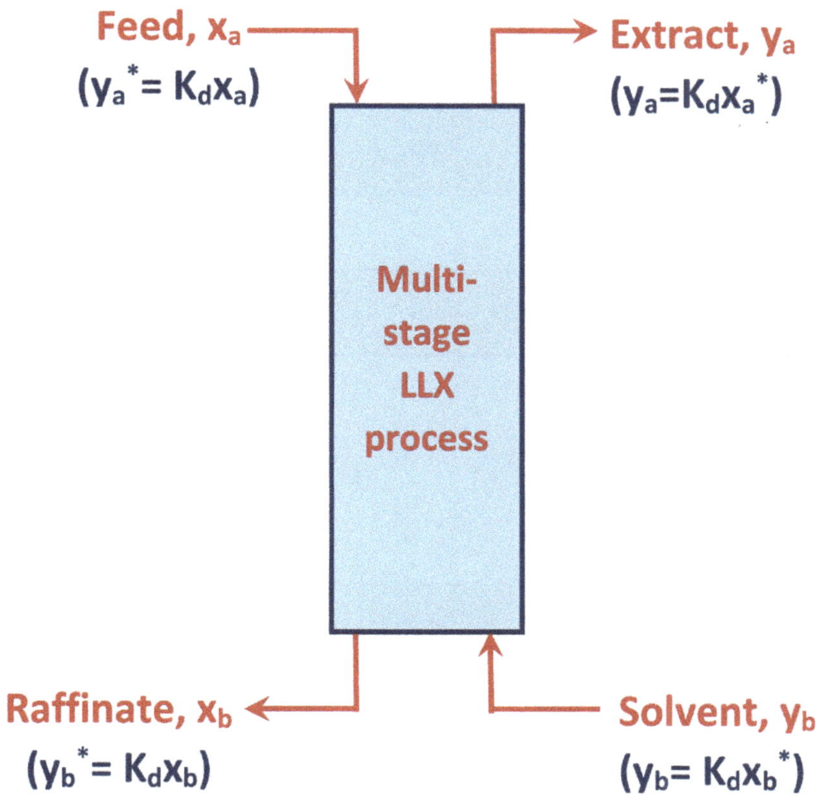

Figure 14.15 Multi-stage LLX tower indicating the quantities involved in the Kremser equations. Only x^* quantities are needed in the LLX analysis.

14.5.2 Analysis of a Multi-Stage LLX Process: Partially Miscible Solvent-Diluent Pair

In the analysis and design of a multi-stage LLX process involving partially miscible solvent-diluent pair, the triangular phase diagram is needed as all three components - solute A, diluent B and solvent C are generally present in each of the flow streams.

Consider the counter-current multi-stage liquid-liquid extraction process flowsheet shown in Figure 14.16. We would first like to calculate the extract and raffinate flow rates and compositions, given the feed and solvent flow rates and compositions: $L_o, V_{N+1}, x_{A,o}, x_{C,o}, y_{A,N+1}, y_{C,N+1}$. The concentration of solute in the raffinate $x_{A,N}$ is also specified.

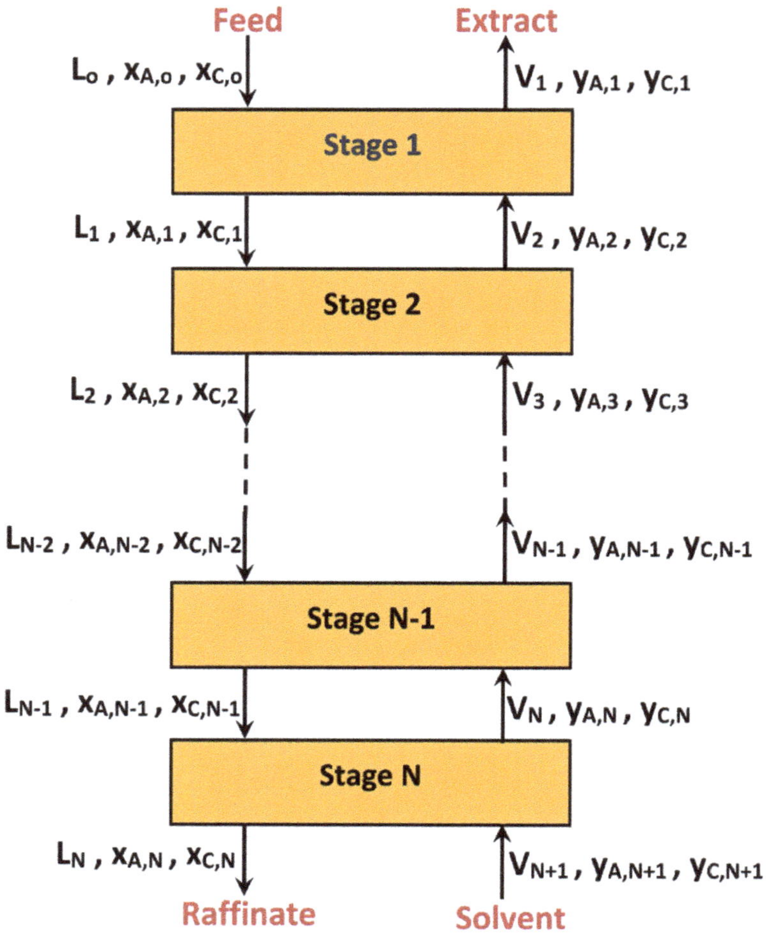

Figure 14.16 Counter-current multi-stage LLX process involving partially miscible solvent-diluent pair.

External Material Balances

As far as the external material balances are concerned, we need not know the internal stage by stage details (flow rates and compositions). Thus the process flowsheet of Figure 14.16 can be replaced by a blackbox as shown in Figure 14.17.

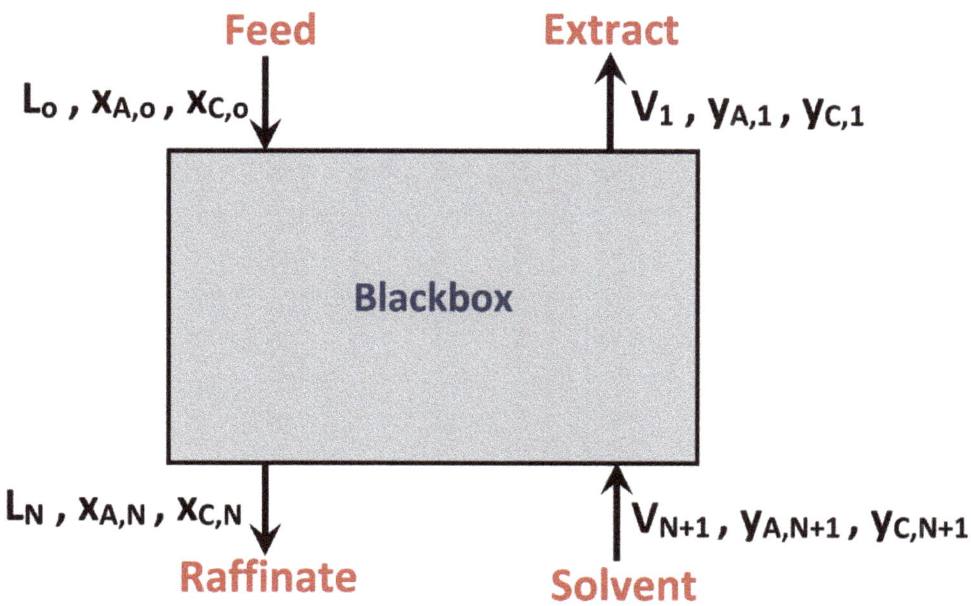

Figure 14.17 Counter-current multi-stage LLX process treated as a blackbox with inputs and outputs.

The feed and solvent flow rates and compositions are specified. We need to determine the unknown extract and raffinate flow rates and compositions (only $x_{A,N}$ in the raffinate is specified). In order to solve this problem, we create a fictitious mixture of the feed and solvent streams and split this fictitious mixture into the extract and raffinate streams, as shown in Figure 14.18.

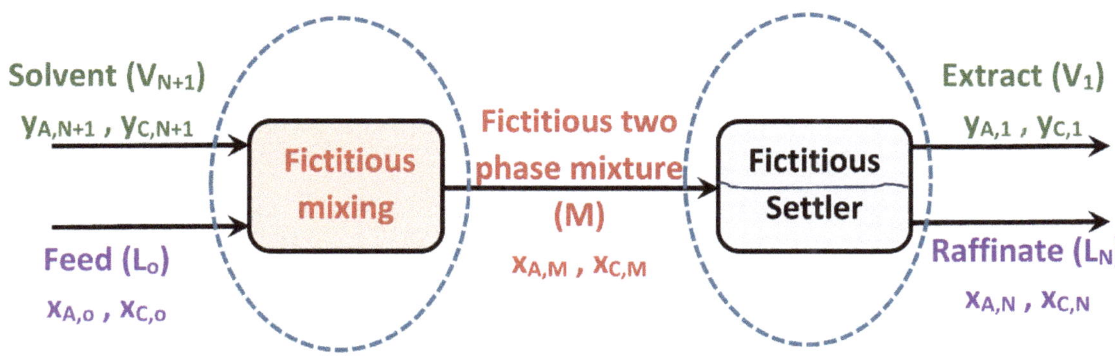

Figure 14.18 Counter-current multi-stage LLX process treated as an equivalent single-stage LLX process.

It is important to recognize the difference between the fictitious single-stage LLX process of Figure 14.18 and the actual single-stage LLX process. In the actual process, the extract and raffinate streams leaving the process are at equilibrium with each other whereas the extract and raffinate streams leaving the fictitious LLX process are not at equilibrium with each other.

Total material balance around the mixer of Figure 14.18 (dashed blue envelope) gives:

$$V_{N+1} + L_o = M \tag{14.29}$$

Component material balances around the mixer give:

$$y_{A,N+1}V_{N+1} + x_{A,o}L_o = x_{A,M}M \tag{14.30}$$

$$y_{C,N+1}V_{N+1} + x_{C,o}L_o = x_{A,M}M \tag{14.31}$$

From Equations (14.29)-(14.31), it can be readily shown that:

$$x_{A,M} = \frac{y_{A,N+1}+x_{A,o}(L_o/V_{N+1})}{1+(L_o/V_{N+1})} \tag{14.32}$$

$$x_{C,M} = \frac{y_{C,N+1}+x_{C,o}(L_o/V_{N+1})}{1+(L_o/V_{N+1})} \tag{14.33}$$

The unknown compositions of the extract and raffinate streams can now be determined graphically as explained below:

- Locate the feed point $\boldsymbol{L_o}(x_{A,o}, x_{C,o})$ on the right-angled triangle phase diagram
- Locate the solvent point $\boldsymbol{V_{N+1}}(y_{A,N+1}, y_{C,N+1})$ on the right-angled triangle phase diagram
- Locate the mixture point $\boldsymbol{M}(x_{A,M}, x_{C,M})$ calculated from Equations (14.32) and (14.33) on the right-angled triangle phase diagram
- Connect the feed, solvent, and mixture points by a straight line as these three points are *collinear*.
- Although the extract and raffinate streams are *not in phase equilibrium* with each other, the extract and raffinate compositions still fall on the equilibrium curve as they are equilibrium streams: extract stream emerging from stage 1 (see Figure 14.16) is in thermodynamic equilibrium with L_1 stream leaving stage 1; raffinate stream emerging from *Nth* stage (see Figure 14.16) is in thermodynamic equilibrium with V_N stream leaving stage N.
- The unknown compositions of the extract and raffinate streams can be determined graphically by drawing a straight line passing through the mixture point $\boldsymbol{M}(x_{A,M}, x_{C,M})$ and the point $\boldsymbol{L_N}(x_{A,N})$ on the equilibrium curve. The intersection of the line $L_N M$ with the equilibrium curve gives the unknown compositions as shown in Figure 14.19.
- Once the extract and raffinate compositions are determined, the extract and raffinate flow rates are determined using material balances.

Total material balance around the settler (dashed blue envelope, Figure 14.18) gives:

$$M = V_1 + L_N \tag{14.34}$$

Solute material balance around the settler gives:

$$x_{A,M}M = y_{A,1}V_1 + x_{A,N}L_N \tag{14.35}$$

From Equations (14.34) and (14.35), it follows that:

$$\frac{V_1}{M} = \frac{x_{A,M}-x_{A,N}}{y_{A,1}-x_{A,N}} \tag{14.36}$$

Equation (14.36) can be used to calculate the extract flow rate V_1. The raffinate flow rate L_N can be calculated from Equation (14.34).

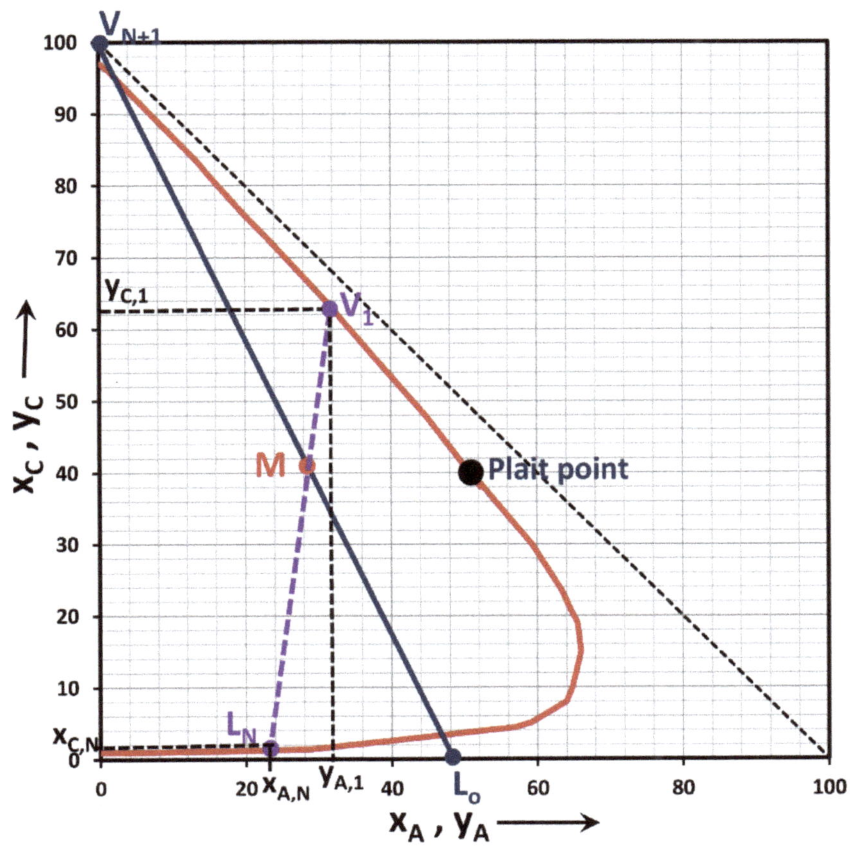

Figure 14.19 Graphical solution of multi-stage LLX process with partially miscible solvent-diluent pair.

Estimation of Number of Equilibrium Stages

Consider material balance on nth stage of a multistage LLX process (see Figure 14.20). The total material balance gives:

$$L_{n-1} - V_n = L_n - V_{n+1} \tag{14.37}$$

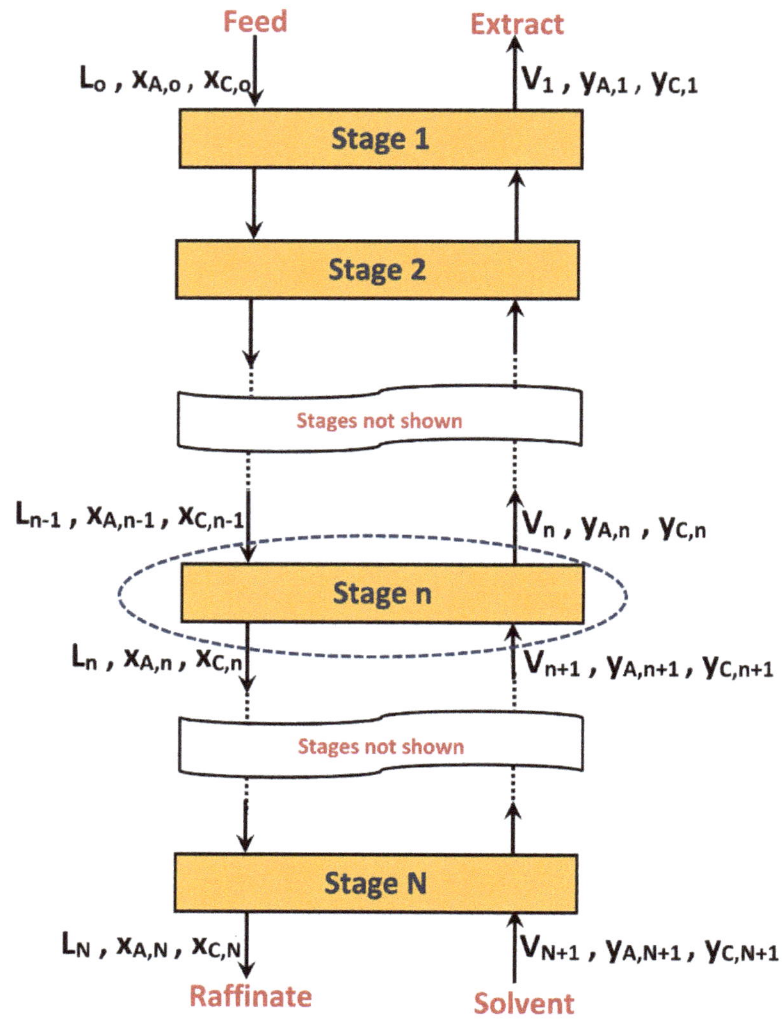

Figure 14.20 Material balance on nth stage of a multi-stage LLX process.

Equation (14.37) can be written for different values of n as:

$$L_o - V_1 = L_1 - V_2 \qquad (n = 1) \tag{14.38}$$

$$L_1 - V_2 = L_2 - V_3 \qquad (n = 2) \tag{14.39}$$

$$L_2 - V_3 = L_3 - V_4 \qquad (n = 3) \tag{14.40}$$

Clearly the net flow rate of material (total, including all components) in the downward direction is constant. Although the flow rate of L-stream or V-stream may change from one stage to another, the difference in the flow rates of passing L and V streams is constant at steady state.

The solute (species A) material balance gives:

$$x_{A,n-1}L_{n-1} - y_{A,n}V_n = x_{A,n}L_n - y_{A,n+1}V_{n+1} \tag{14.41}$$

Equation (14.41) can be written for different values of n as:

$$x_{A,o}L_o - y_{A,1}V_1 = x_{A,1}L_1 - y_{A,2}V_2 \qquad (n = 1) \qquad (14.42)$$

$$x_{A,1}L_1 - y_{A,2}V_2 = x_{A,2}L_2 - y_{A,3}V_3 \qquad (n = 2) \qquad (14.43)$$

Equations (14.41)-(14.43) indicate that the difference in solute flow rates between passing *L* and *V* streams is constant from one stage to another at steady state.

The solvent (species C) material balance gives:

$$x_{C,n-1}L_{n-1} - y_{C,n}V_n = x_{C,n}L_n - y_{C,n+1}V_{n+1} \qquad (14.44)$$

According to Equation (14.44), the difference in solvent flow rates between passing *L* and *V* streams is constant form one stage to another at steady state.

We now define the so-called *difference point* as follows:

$$x_{A,diff} = \frac{difference\ in\ solute\ flow\ rates\ of\ the\ passing\ streams}{difference\ in\ the\ total\ flow\ rates\ of\ the\ passing\ streams} \qquad (14.45)$$

$$x_{C,diff} = \frac{difference\ in\ solvent\ flow\ rates\ of\ the\ passing\ streams}{difference\ in\ the\ total\ flow\ rates\ of\ the\ passing\ streams} \qquad (14.46)$$

Thus the coordinates of a difference point are:

$$x_{A,diff} = \frac{x_{A,o}L_o - y_{A,1}V_1}{L_o - V_1} \qquad (14.47)$$

$$x_{C,diff} = \frac{x_{C,o}L_o - y_{C,1}V_1}{L_o - V_1} \qquad (14.48)$$

The difference point coordinates can be calculated from knowledge of feed stream (L_o) data and extract stream (V_1) data (V_1 information is obtained graphically, see Figure 14.19). The difference point could also be located graphically as the intersection of the operating lines $L_N V_{N+1}$ and $L_o V_1$.

Equation (14.47) upon re-arrangement gives:

$$\frac{L_o}{V_1} = \frac{x_{A,diff} - y_{A,1}}{x_{A,diff} - x_{A,o}} \qquad (14.49)$$

From Equation (14.48):

$$\frac{L_o}{V_1} = \frac{x_{C,diff} - y_{C,1}}{x_{C,diff} - x_{C,o}} \qquad (14.50)$$

Combining Equations (14.49) and (14.50):

$$\frac{L_o}{V_1} = \frac{x_{A,diff} - y_{A,1}}{x_{A,diff} - x_{A,o}} = \frac{x_{C,diff} - y_{C,1}}{x_{C,diff} - x_{C,o}} \qquad (14.51)$$

According to Equation (14.51), the following points are collinear: $(x_{A,diff}, x_{C,diff})$, $(y_{A,1}, y_{C,1})$ and $(x_{A,o}, x_{C,o})$. A line passing through these points represents the *operating line* for a multi-stage LLX process. A more general form of Equation (14.51) is as follows:

$$\frac{L_{n-1}}{V_n} = \frac{x_{A,diff} - y_{A,n}}{x_{A,diff} - x_{A,n-1}} = \frac{x_{C,diff} - y_{C,n}}{x_{C,diff} - x_{C,n-1}} \tag{14.52}$$

This relation indicates that we have a family of operating lines, all radiating from the difference point $P'(x_{A,diff}, x_{C,diff})$, as shown in Figure 14.21.

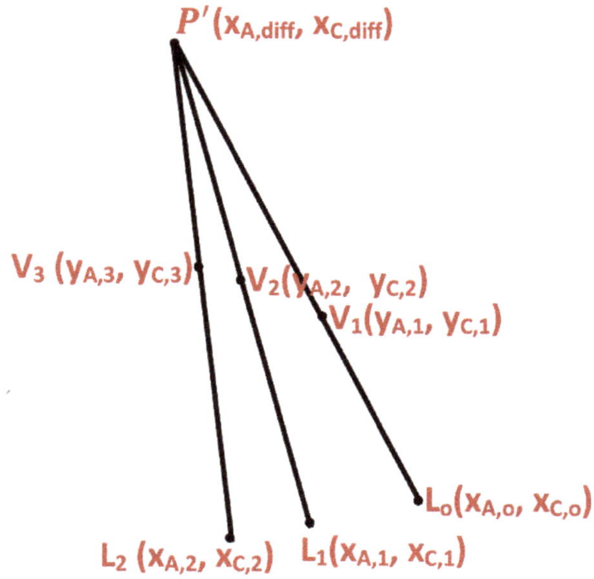

Figure 14.21 Family of operating lines radiating from the difference point P'.

Figure 14.22 illustrates the graphical construction of equilibrium stages on a triangular phase diagram for a multi-stage LLX extraction process. Each stage is represented by two operating lines and a tie-line (purple dashed lines in Figure 14.22). While the operating line connects the compositions of passing L and V streams at the same level, the tie-line connects the compositions of L and V streams leaving the stage at equilibrium with each other. The first operating line passes through the points L_o and V_1. The last operating line passes through the points L_N and V_{N+1}. All operating lines radiate from the difference point. For the separation shown in Figure 14.22, 3 equilibrium stages are required.

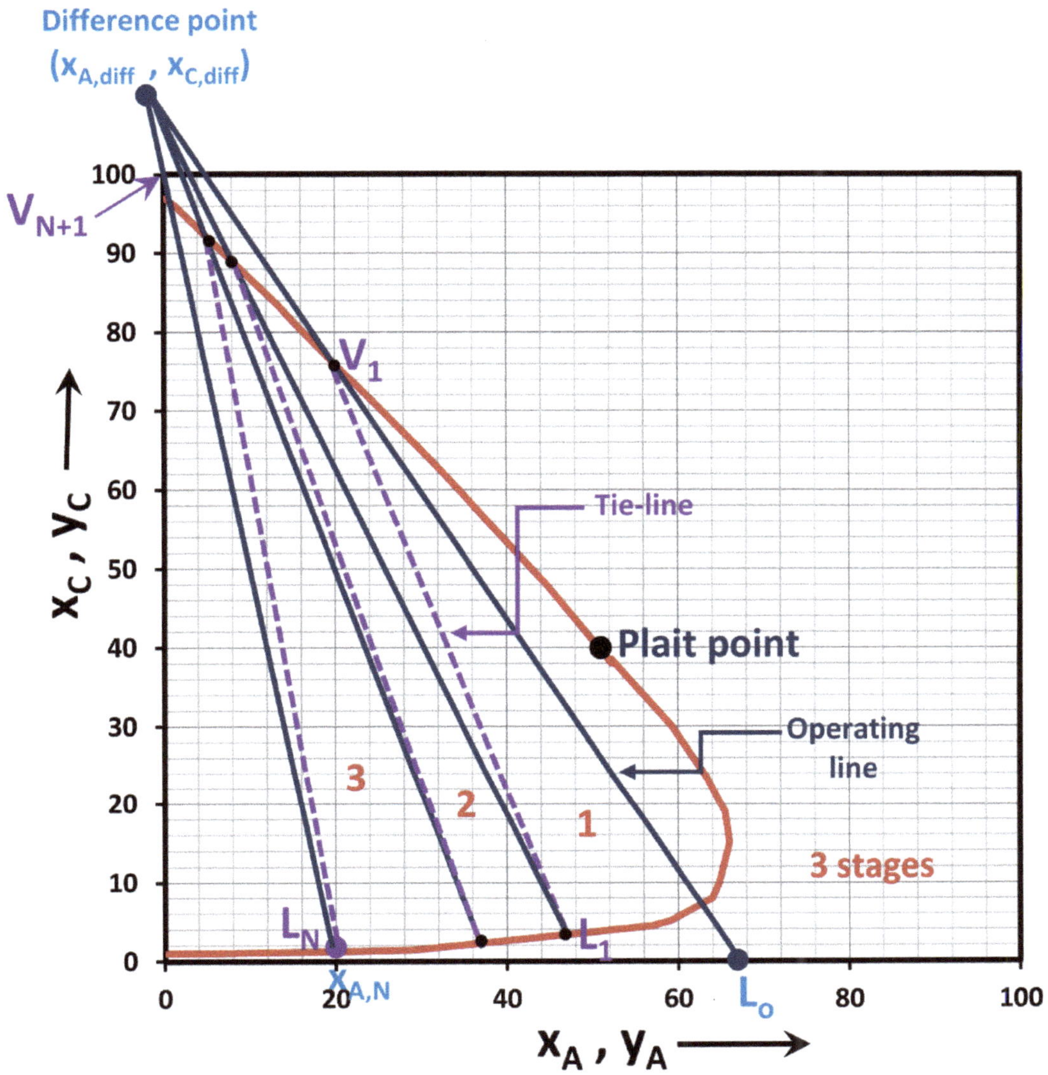

Figure 14.22 Construction of equilibrium stages on a triangular phase diagram for multi-stage LLX extraction process.

Conjugate Curve for the Construction of Tie-lines

As noted in Figure 14.22, we need to construct a number of tie-lines in the design of a multi-stage LLX process. The available experimental data on equilibrium tie-lines is usually limited to a few tie-lines. Based on the available experimental data on tie-lines, a so-called *conjugate curve* is first constructed. The conjugate curve is then used for the construction of more tie-lines. Figure 14.23 shows how to construct the conjugate curve from the given tie-lines. Once the conjugate curve is developed, it can be used to construct the tie-lines by reversing the process of construction of the conjugate curve.

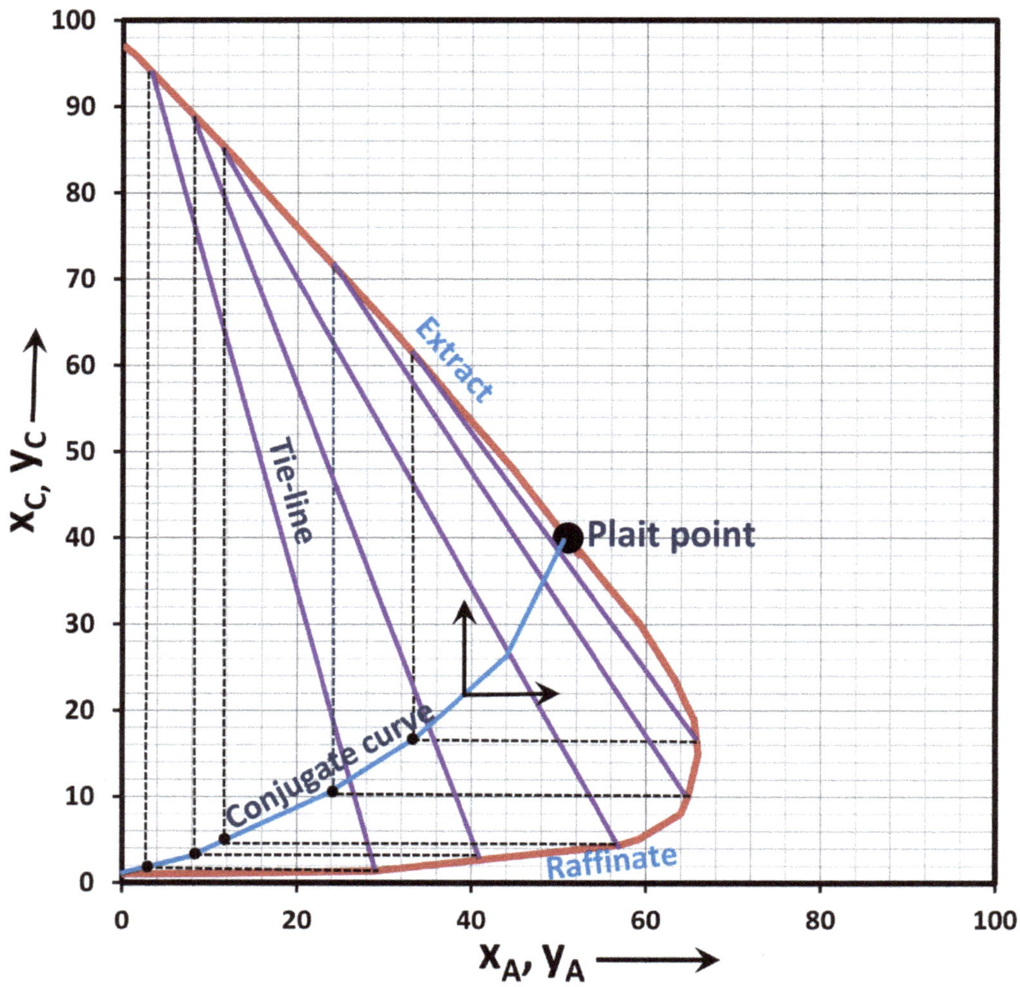

Figure 14.23 Construction of conjugate curve derived from experimental equilibrium data (tie-lines). The arrows on the conjugate curve indicate how to construct a tie-line. From any given point on the conjugate curve, draw horizontal and vertical lines up to the equilibrium curve. The intersections of the horizontal and vertical lines with the equilibrium curve give the two points necessary for the construction of a tie-line.

NOTATION

E	refers to extract phase, mass of extract phase or mass flow rate of extract stream
E_o	extraction factor, defined in Equation (14.10)
F	refers to feed phase, mass of feed or mass flow rate of feed stream
K_d	distribution coefficient of solute, defined as $K = y/x$ (also referred to as equilibrium constant), see Equation (14.7)
L	mass flow rate of raffinate stream in a cascade
L_a	mass flow rate of feed stream entering the first stage of a cascade
L_b	mass flow rate of raffinate stream leaving the last stage of a cascade
L_n	mass flow rate of raffinate stream leaving the nth stage of a cascade
L_{n-1}	mass flow rate of raffinate stream leaving (n-1)th stage of a cascade
L_N	mass flow rate of raffinate leaving the last Nth stage of a cascade (same as L_b)
L_o	mass flow rate of feed stream entering the first stage of a cascade
LLX	liquid-liquid extraction

M	refers to two-phase mixture, mass of mixture or mass flow rate of mixture stream
MIK	methyl isobutyl ketone
N	total number of equilibrium stages in a cascade
R	refers to raffinate phase, mass of raffinate or mass flow rate of raffinate stream
S	refers to solvent phase, mass of solvent phase or mass flow rate of solvent stream
V	mass flow rate of solvent stream in a cascade (solvent stream may include solute and diluent, in addition to solvent itself)
V_1	mass flow rate of solvent stream leaving the first stage of a cascade
V_a	mass flow rate of solvent stream leaving the first stage of a cascade (same as V_1)
V_b	mass flow rate of solvent stream entering the last Nth stage of a cascade
V_n	mass flow rate of solvent stream leaving the nth stage of a cascade
V_{n+1}	mass flow rate of solvent stream leaving the (n+1)th stage of a cascade
V_{N+1}	mass flow rate of solvent stream entering the last Nth stage of a cascade (same as V_b)
x	mass fraction of solute in the feed or raffinate stream
x_a	mass fraction of solute in the feed stream entering the cascade at the top
x_a^*	mass fraction of solute in the raffinate stream in equilibrium with the extract stream leaving the column at the top, see Figure 14.15
x_A	mass fraction of species *A* (species *A* is usually solute)
$x_{A,o}$	mass fraction of solute (species *A*) in the feed stream entering the cascade at the top
$x_{A,diff}$	difference point coordinate for solute (species *A*), defined in Equation (14.45)
$x_{A,F}$	mass fraction of solute (species *A*) in the feed stream
$x_{A,N}$	mass fraction of solute (species *A*) in raffinate stream leaving Nth stage
$x_{A,M}$	mass fraction of solute (species *A*) in two-phase mixture stream
$x_{A,R}$	mass fraction of solute (species *A*) in raffinate stream
x_b	mass fraction of solute in the raffinate stream leaving the cascade at the bottom
x_b^*	mass fraction of solute in the raffinate stream in equilibrium with the solvent stream entering the column at the botttom, see Figure 14.15
x_B	mass fraction of species *B* (species *B* is usually diluent, that is, carrier of solute)
x_C	mass fraction of species *C* (species *C* is usually solvent)
$x_{C,N}$	mass fraction of solvent (species *C*) in raffinate stream leaving Nth stage
$x_{C,M}$	mass fraction of solvent (species *C*) in two-phase mixture stream
$x_{C,o}$	mass fraction of solvent (species *C*) in the feed stream entering the cascade at the top
$x_{C,F}$	mass fraction of solvent (species *C*) in the feed stream
$x_{C,R}$	mass fraction of solvent (species *C*) in the raffinate stream
$x_{C,diff}$	difference point coordinate for solvent (species *C*), defined in Equation (14.46)
x_{in}	mass fraction of solute (species *A*) in the feed stream
x_n	mass fraction of solute in raffinate stream leaving the nth stage of the cascade
x_{out}	mass fraction of solute (species *A*) in the raffinate stream
y	mass fraction of solute in the solvent or extract
y_a	mass fraction of solute in the extract stream leaving the cascade at the top
y_a^*	mass fraction of solute in the extract stream in equilibrium with the raffinate stream leaving the column at the botttom, see Figure 14.15
$y_{A,1}$	mass fraction of solute (species *A*) in the extract stream leaving the column at the top, see Figure 14.16
$y_{A,E}$	mass fraction of solute (species *A*) in the extract stream
$y_{A,N+1}$	mass fraction of solute (species *A*) in the solvent stream entering Nth stage of a cascade at the bottom, see Figure 14.16
$y_{A,S}$	mass fraction of solute (species *A*) in the solvent stream
y_b	mass fraction of solute in the solvent stream entering the cascade at the bottom

y_b^* mass fraction of solute in the extract stream in equilibrium with the raffinate stream leaving the column at the botttom, see Figure 14.15

$y_{C,1}$ mass fraction of solvent (species *C*) in the extract stream leaving the first stage of a cascade at the top, see Figure 14.16

$y_{C,E}$ mass fraction of solvent (species *C*) in the extract stream

$y_{C,S}$ mass fraction of solvent (species *C*) in the solvent stream

$y_{C,N+1}$ mass fraction of solvent (species *C*) in the solvent stream entering Nth stage of a cascade at the bottom, see Figure 14.16

y_{in} mass fraction of solute (species *A*) in the solvent stream

y_{n+1} mass fraction of solute in solvent stream leaving the (n+1)th stage of a cascade

y_{out} mass fraction of solute (species *A*) in the extract stream

TUTORIAL PROBLEMS WITH SOLUTION

Problem 1

The equilibrium data in weight percent for the ternary system trimethylamine (TMA)-water-benzene are given below (TMA is the solute, water is solvent, and benzene is diluent or carrier of solute):

Water (solvent)	TMA (solute)	Benzene (diluent)
96.4	5	0.04
89.4	10	0.06
84	15	1
78	20	2
72	25	3
66.4	30	3.6
58	35	7
47	40	13
34	40	26
15	35	50
7	30	63
4	25	71
2	20	78
1	15	84
0	10	90
0	5	95

The tie line data for the system are given below:

TMA in water layer	TMA in benzene layer
39.5	31
21.5	14.5
13	9
8.3	6.8
4	3.5

a) Plot the triangular phase diagram using equilateral-triangle diagram
b) Plot the tie lines on the triangular phase diagram
c) Repeat (a) and (b) using right-angled triangle diagram

Solution

The solutions to parts (a) and (b) are shown in the first figure given below. The solution to part (c) is shown in the second figure given below.

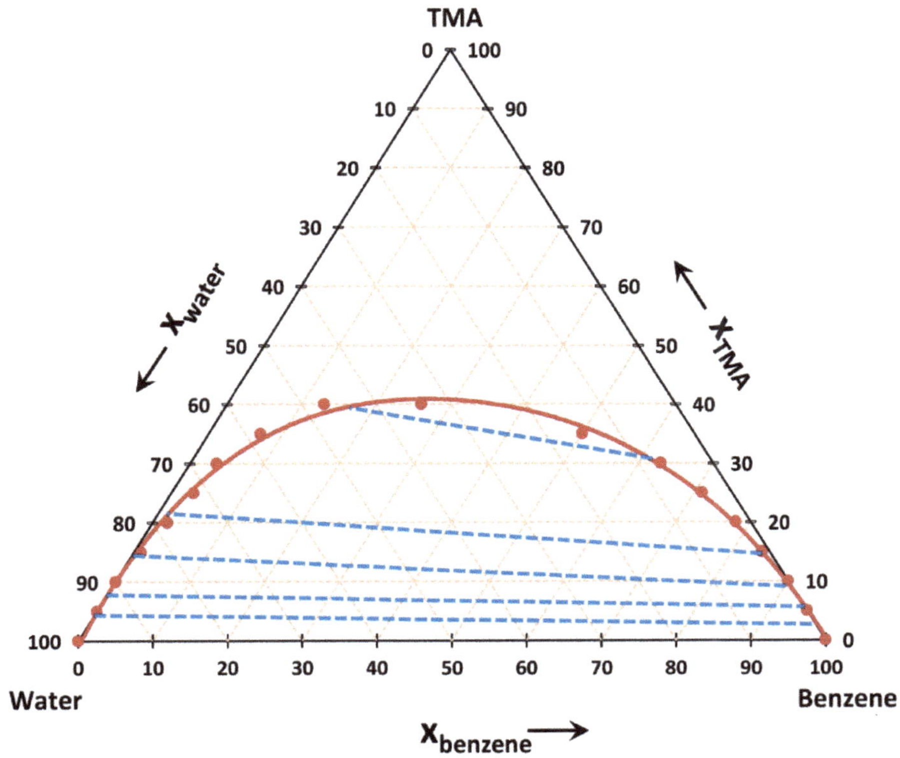

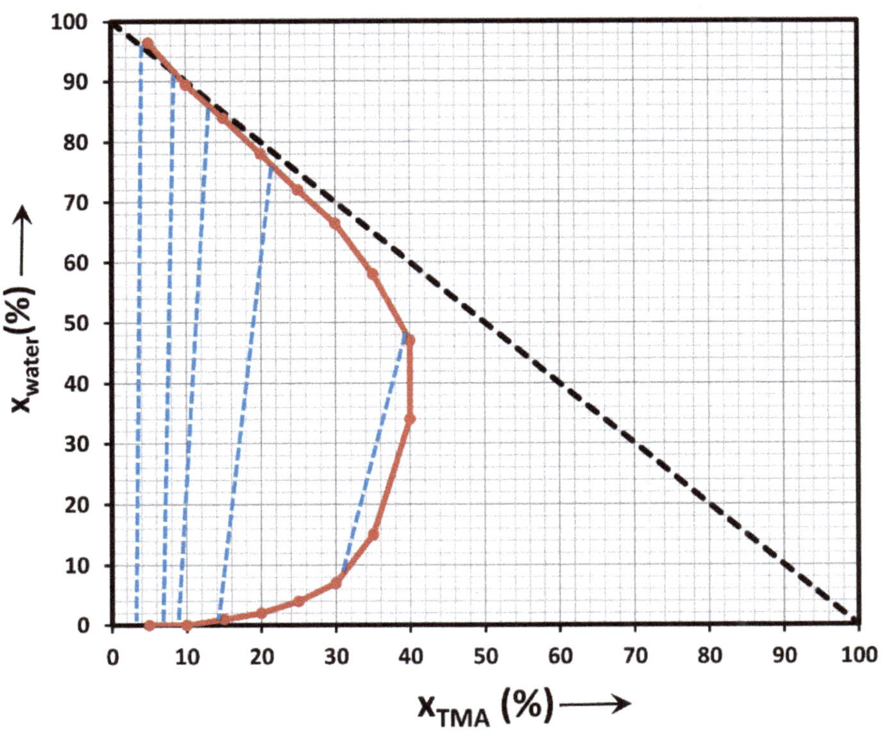

Problem 2

For the ternary system of tutorial problem 1 (trimethylamine (TMA)-water-benzene system), a single-stage extraction process is used in which 500 kg of a solution containing 30 wt% TMA in benzene is contacted with 500 kg of pure water.

Calculate the amounts and compositions of the water layer (extract) and benzene layer (raffinate). What percent of TMA is recovered?

Solution

The process is shown schematically below.

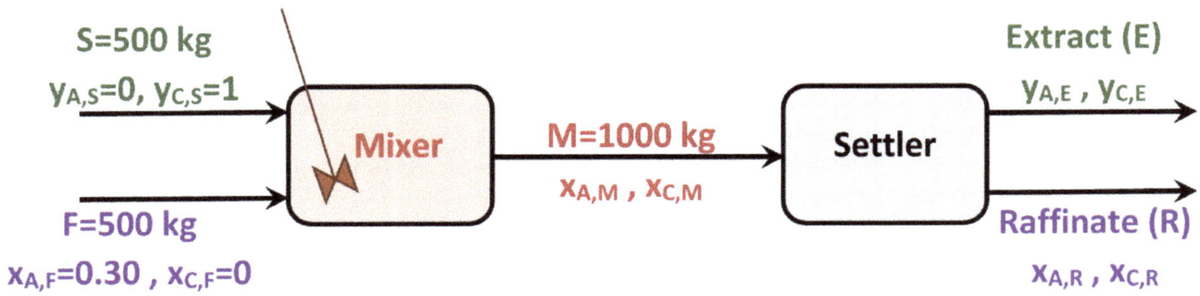

From Equations (14.15) and (14.16), we get the mixture composition:

$$x_{A,M} = \frac{y_{A,S}+x_{A,F}(F/S)}{1+(F/S)} = \frac{0+0.30\times(500/500)}{1+(500/500)} = 0.15$$

$$x_{C,M} = \frac{y_{C,S}+x_{C,F}(F/S)}{1+(F/S)} = \frac{1+0}{1+(500/500)} = 0.50$$

The solution is obtained graphically as shown below. The feed and solvent points are located and connected by a straight line (line SF) on a triangular diagram. The mixture point M falls on the SF line. A tie line passing through the mixture point M is drawn. The compositions of the extract and raffinate are indicated on the triangular diagram by the points E and R, respectively. The compositions are as follows:

Extract composition: $y_{A,E} = 0.175$ and $y_{C,E} = 0.80$

Raffinate composition: $x_{A,R} = 0.12$ and $x_{C,R} = 0.005$

From Equation (14.19), we get:

$$\frac{E}{M} = \frac{x_{A,M}-x_{A,R}}{y_{A,E}-x_{A,R}} = \frac{0.15-0.12}{0.175-0.12} = 0.545$$

Thus, $E = 545$ kg and $R = M - E = 455$ kg.

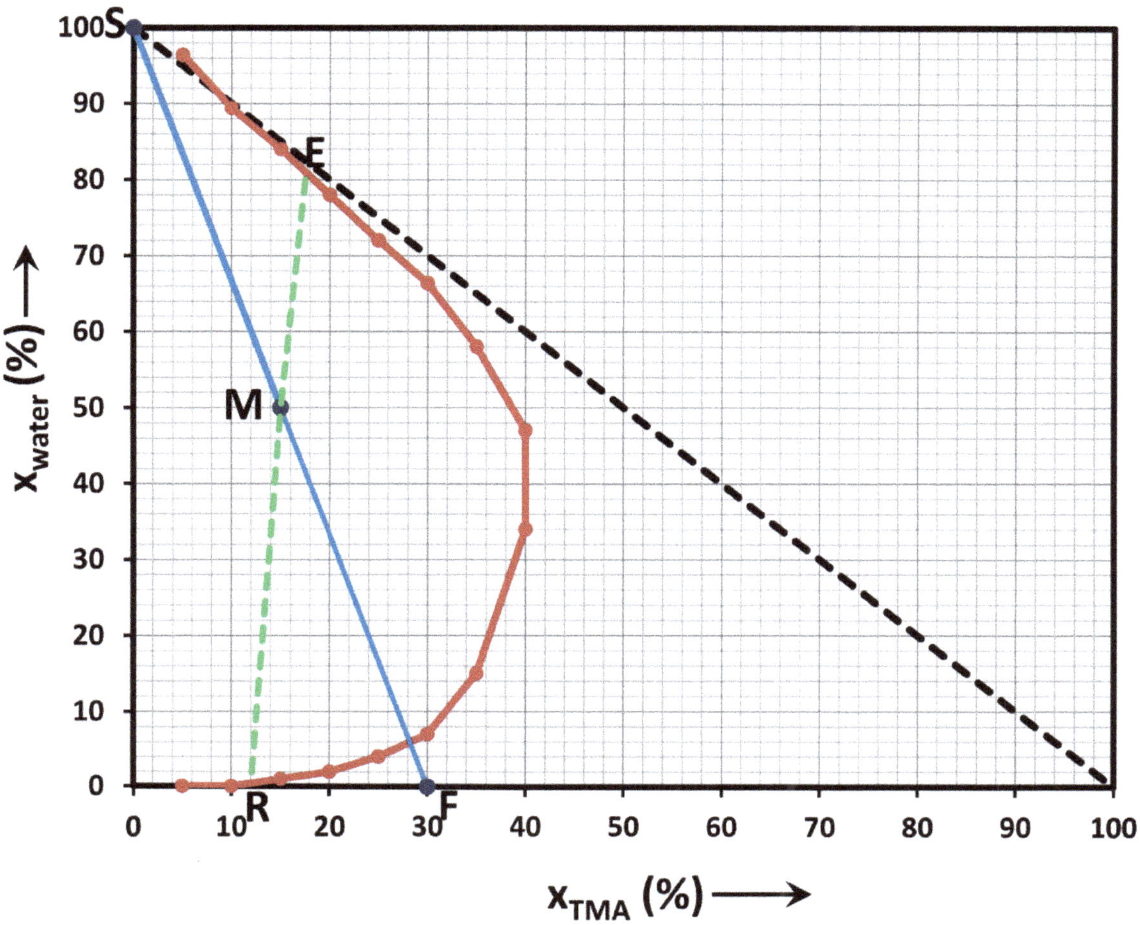

The percent recovery of solute TMA is as follows:

Recovery = $0.175 \times 545/(0.30 \times 500)$ =0.636, that is 63.6 %

Problem 3

Pure non-aqueous solvent (species C) is being used to extract a solute (species A) from aqueous feed containing 40 wt% A and 60 wt% water (species B) using countercurrent multistage extraction. The flow rate of entering solvent is 525 kg/h and the flow rate of feed solution is 185 kg/h. The desired exit solute concentration in the raffinate is 12 wt%.

Estimate the compositions and mass flow rates of extract and raffinate streams leaving the process.
The equilibrium data in terms of weight percent for the ternary system A-B-C are given below:

Composition of coexisting liquid phases in equilibrium					
Raffinate layer			Extract layer		
A%	B%	C%	A%	B%	C%
0.69	98.1	1.2	0.18	0.5	99.3
1.41	97.1	1.5	0.37	0.7	98.9
2.89	95.5	1.6	0.79	0.8	98.4
6.42	91.7	1.9	1.93	1.0	97.1
13.30	84.4	2.3	4.82	1.9	93.3
25.50	71.1	3.4	11.4	3.9	84.7

36.70	58.9	4.4	21.6	6.9	71.5
44.30	45.1	10.6	31.1	10.8	58.1
46.4	37.1	16.5	36.20	15.1	48.7

Solution

The procedure to solve this type of problem is discussed in section 14.5.2. A schematic diagram of the process is shown below.

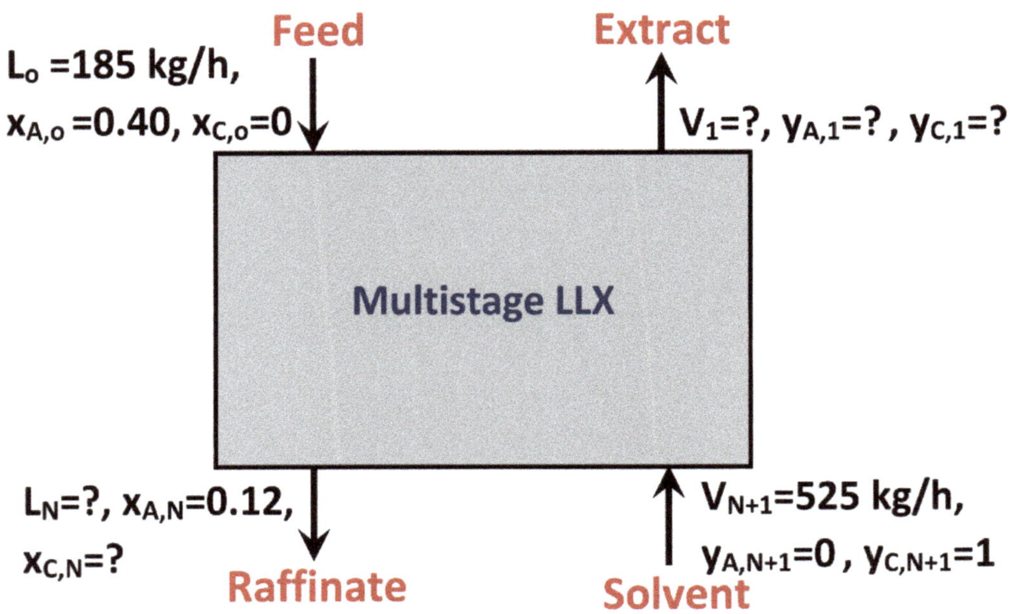

Using Equations (14.32) and (14.33), we obtain the mixture (M) composition as: $x_{A,M} = 0.1042$, $x_{C,M} = 0.7394$. The points V_{N+1}, L_o and M are located on the triangular diagram as shown below. The coordinates of these points are as follows: $V_{N+1}(y_{A,N+1} = 0, y_{C,N+1} = 1.0)$; $L_o(x_{A,o} = 0.40, x_{C,o} = 0)$; $M(x_{A,M} = 0.1042, x_{C,M} = 0.7394)$. Note that the points V_{N+1}, L_o and M are collinear. The point L_N is located next. This point falls on the equilibrium curve (raffinate portion) at $x_{A,N} = 0.12$. The intersection of the line $L_N M$ with the equilibrium curve (extract portion) locates the point V_1. From the triangular diagram, the compositions of extract and raffinate streams are as follows: Extract: $y_{A,1} = 0.101$; $y_{C,1} = 0.865$; Raffinate: $x_{A,N} = 0.12$; $x_{C,N} = 0.022$.

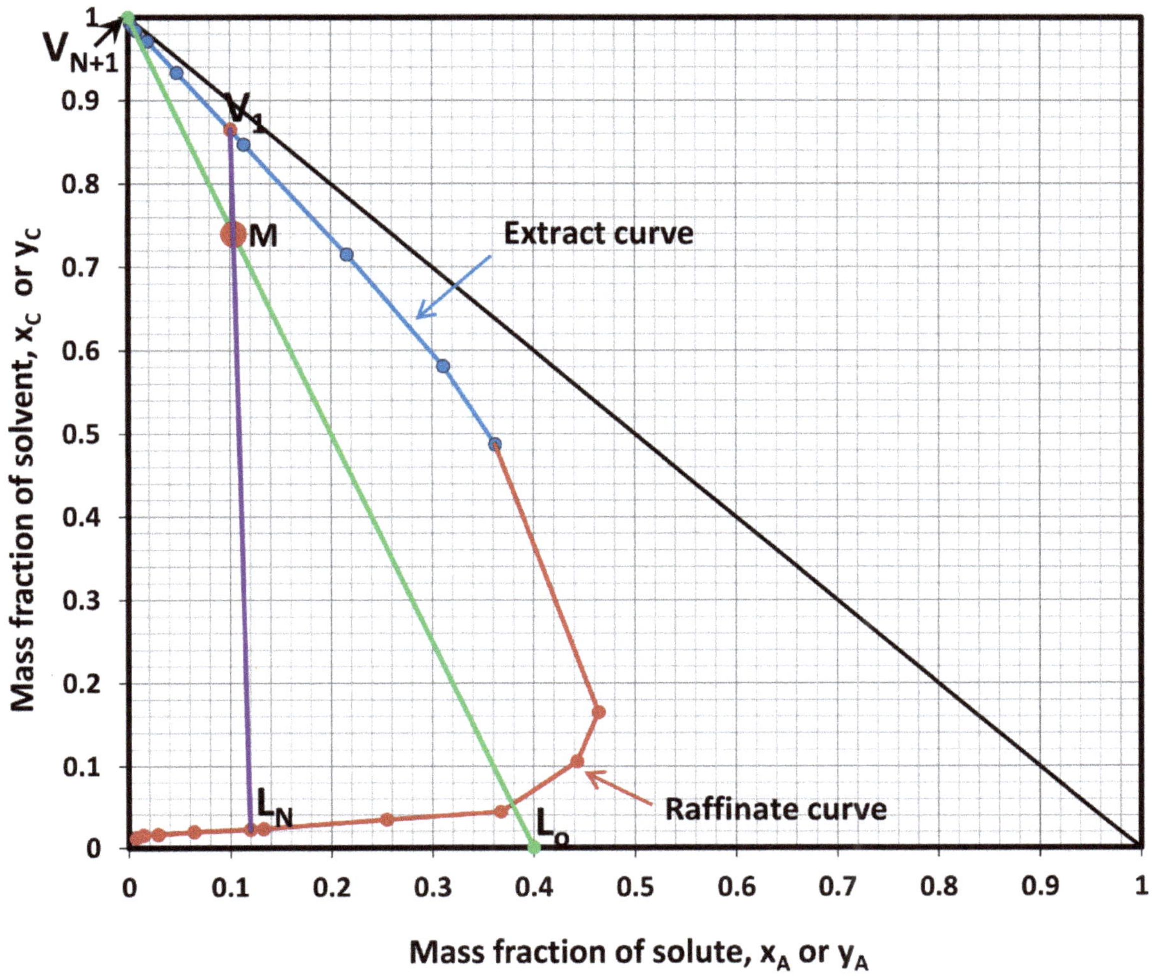

From Equation (14.36), we get:

$$\frac{V_1}{M} = \frac{x_{A,M} - x_{A,N}}{y_{A,1} - x_{A,N}} = \frac{0.1042 - 0.12}{0.101 - 0.12} = 0.8316$$

As $M = L_o + V_{N+1} = 185 + 525 = 710$ kg/h, it follows that: $V_1 = 590.44$ kg/h and $L_N = M - V_1 = 710 - 590.44 = 119.56$ kg/h.

Problem 4

Estimate the number of equilibrium stages required for the multistage liquid-liquid extraction process in problem 3.

Solution

In order to determine the number of stages, we need to determine the difference point. The coordinates of a difference point are given by Equations (14.47) and (14.48) as:

$$x_{A,diff} = \frac{x_{A,o}L_o - y_{A,1}V_1}{L_o - V_1} = \frac{0.40 \times 185 - 0.101 \times 590.44}{185 - 590.44} = -0.0354$$

$$x_{C,diff} = \frac{x_{C,o}L_o - y_{C,1}V_1}{L_o - V_1} = \frac{0 - 0.865 \times 590.44}{185 - 590.44} = 1.26$$

The difference point is located on the triangular diagram below. The stages are constructed graphically as discussed in section 14.5.2. All the operating lines radiate from the difference point.

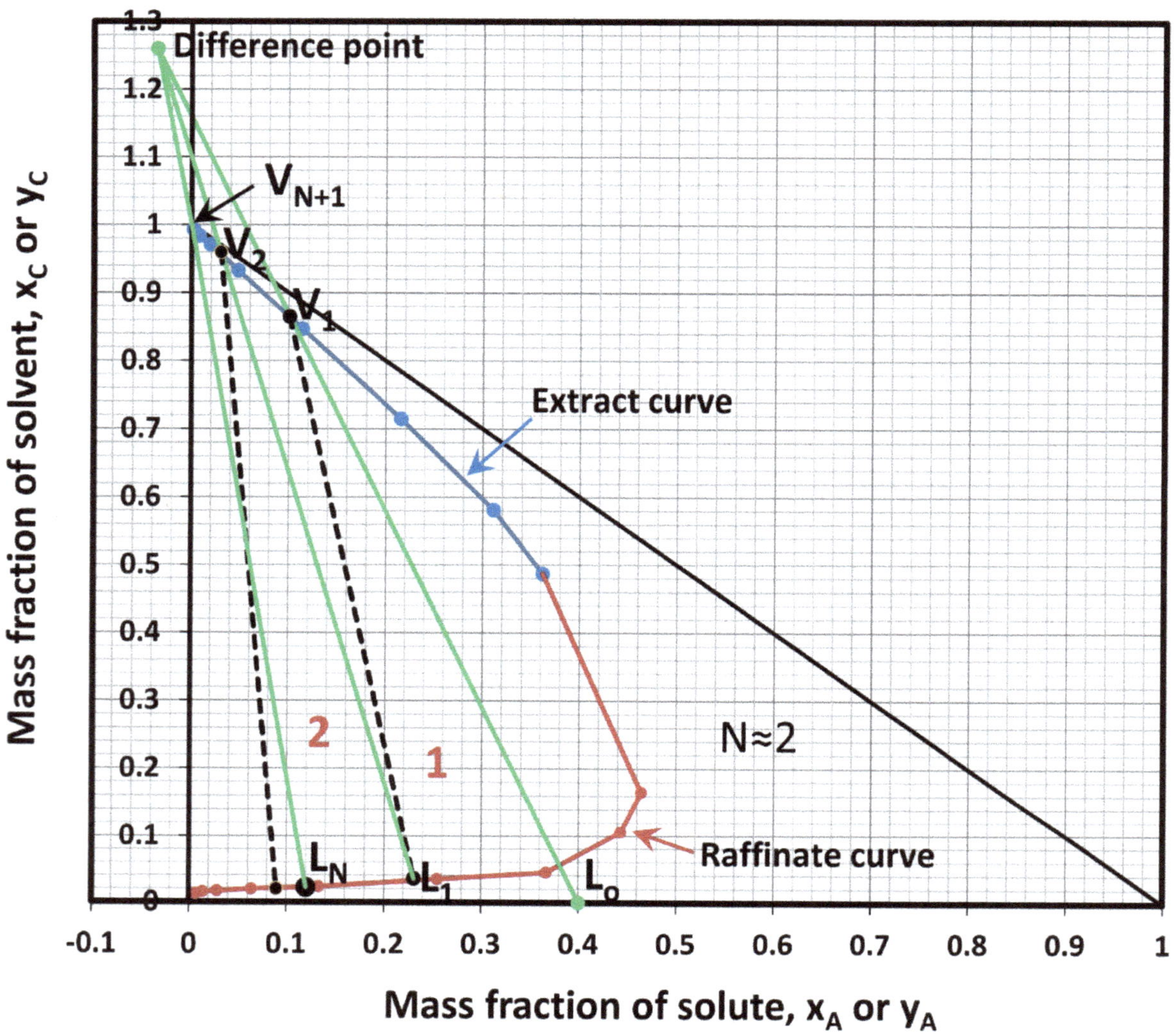

The first operating line connects the difference point to the feed point L_o and intersects the extract portion of the equilibrium curve at V_1. From V_1, a tie line is drawn. The intersection of this tie line with the raffinate equilibrium curve gives L_1. The next operating line is drawn by connecting L_1 with the difference point. The intersection of the second operating line with the extract equilibrium curve gives V_2. The tie line from V_2 goes slightly beyond the desired raffinate point L_N. Thus approximately 2 stages are required.

Interestingly, we can solve this problem using the McCabe-Thiele approach as well. We need the equilibrium curve of y_A versus x_A where y_A is the equilibrium mass fraction of solute in the extract phase and x_A is the equilibrium mass fraction of solute in raffinate phase. The y_A versus x_A equilibrium curve is obtained from the tie-line data. The figure given below shows the equilibrium curve for the problem under consideration. The next step is to plot the operating line. The operating line gives the solute concentrations in the passing streams. The operating line data are generated by drawing a number of operating lines arbitrarily from the difference point and reading y_A and x_A values as illustrated in the following figure. Note that the end points of the operating line are fixed as $(x_{A,o} = 0.40, y_{A,1} = 0.101)$ and $(x_{A,N} = 0.12, y_{A,N+1} = 0)$. Once the equilibrium curve and operating line are plotted, the stages are stepped off as shown in the McCabe-Thiele diagram above. The McCabe-Thiele approach gives the same number of stages as obtained from the triangular diagram.

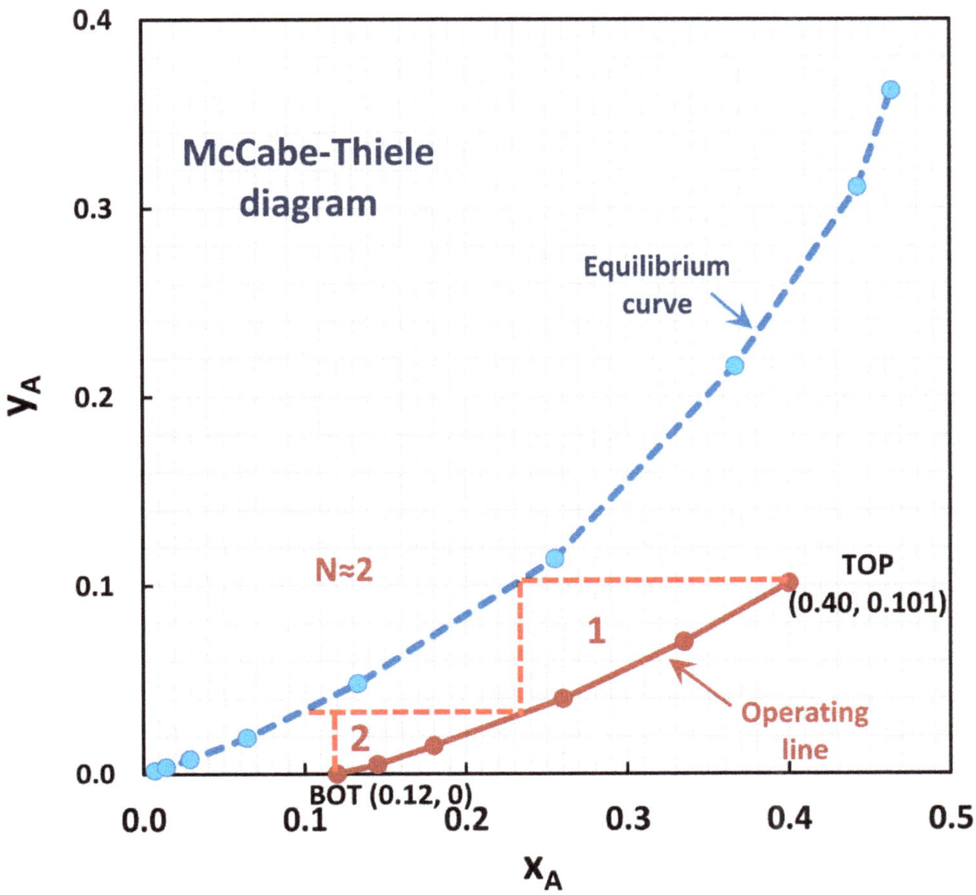

McCabe-Thiele diagram
Equilibrium curve
N≈2
1
2
TOP (0.40, 0.101)
Operating line
BOT (0.12, 0)
y_A
x_A
0.0
0.1
0.2
0.3
0.4
0.5

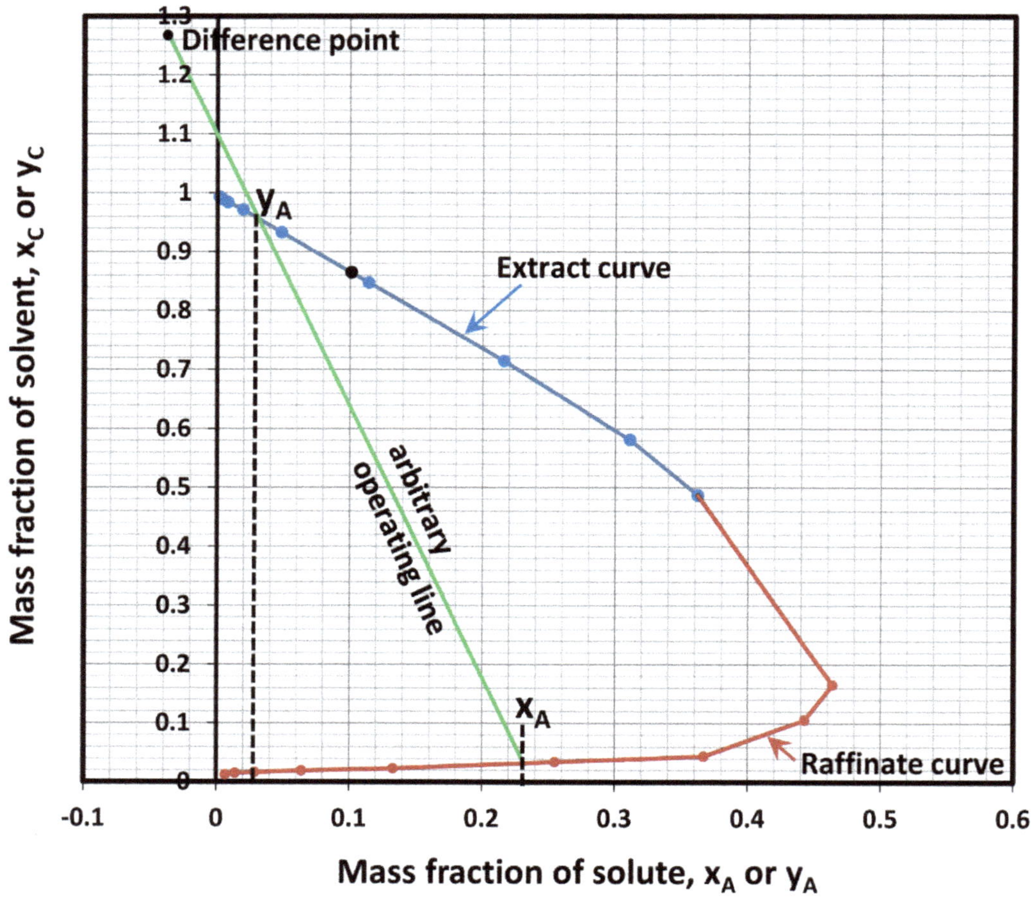

Problem 5

A single stage liquid-liquid extraction process is being used to recover antibiotic Penicillin F from dilute aqueous fermentation broth containing 0.6 wt% Penicillin F using pure solvent amyl acetate. The ratio of solvent mass flow rate to feed mass flow rate is 0.1 ($V/L = 0.10$). The equilibrium relation is $y = K_d x$ where K_d=85 (y and x are mass fractions of Penicillin F in extract and reffinate phases, respectively at equilibrium).

What are the compositions of the exit extract and raffinate streams?
What is the percent recovery of Penicillin F?
You can assume that solvent amyl acetate is *insoluble* in water.

Solution

The process is shown schematically below.

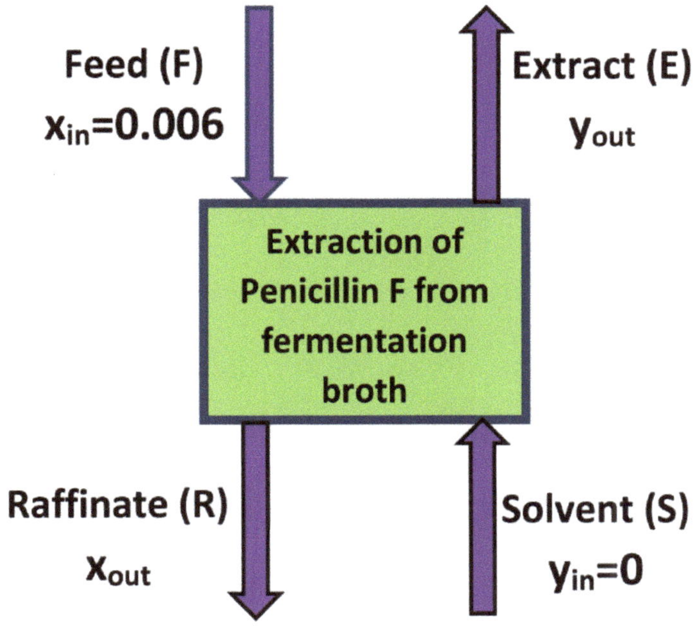

The extraction factor, defined in Equation (14.10), for the process is:

$$E_o = \frac{K_d}{L/V} = \frac{85}{10} = 8.5$$

Equation (14.9) gives

$$\frac{x_{out}}{x_{in}} = \left(\frac{1}{1+E_o}\right)\left[1 + \frac{V}{L}\left(\frac{y_{in}}{x_{in}}\right)\right] = \frac{1}{1+8.5} = 0.1053$$

$$x_{out} = 0.006 \times 0.1053 = 6.32 \times 10^{-4}$$

The extract composition is

$$y_{out} = 85 \times 6.32 \times 10^{-4} = 0.0537$$

The percent recovery of Penicillin F is:

$$\%\,Recovery\ of\ Penicillin = 100\left(1 - \frac{x_{out}}{x_{in}}\right) = 100(1 - 0.1053) = 89.5$$

Problem 6

A multistage liquid-liquid extraction process is being used to recover Penicillin F from dilute aqueous fermentation broth containing 0.6 wt% Penicillin F using pure solvent amyl acetate. The ratio of solvent mass flow rate to feed mass flow rate is 0.10 ($V/L = 0.10$). The equilibrium relation is $y = K_d x$ where K_d=85 (y and x are mass fractions of Penicillin F in extract and reffinate phases, respectively at equilibrium).

Estimate the number of equilibrium stages required for Penicillin recovery of 99 percent. You can use Kremser equation to estimate the number of stages. You can assume that solvent amyl acetate is *insoluble* in water.

Solution

The process is shown schematically below.

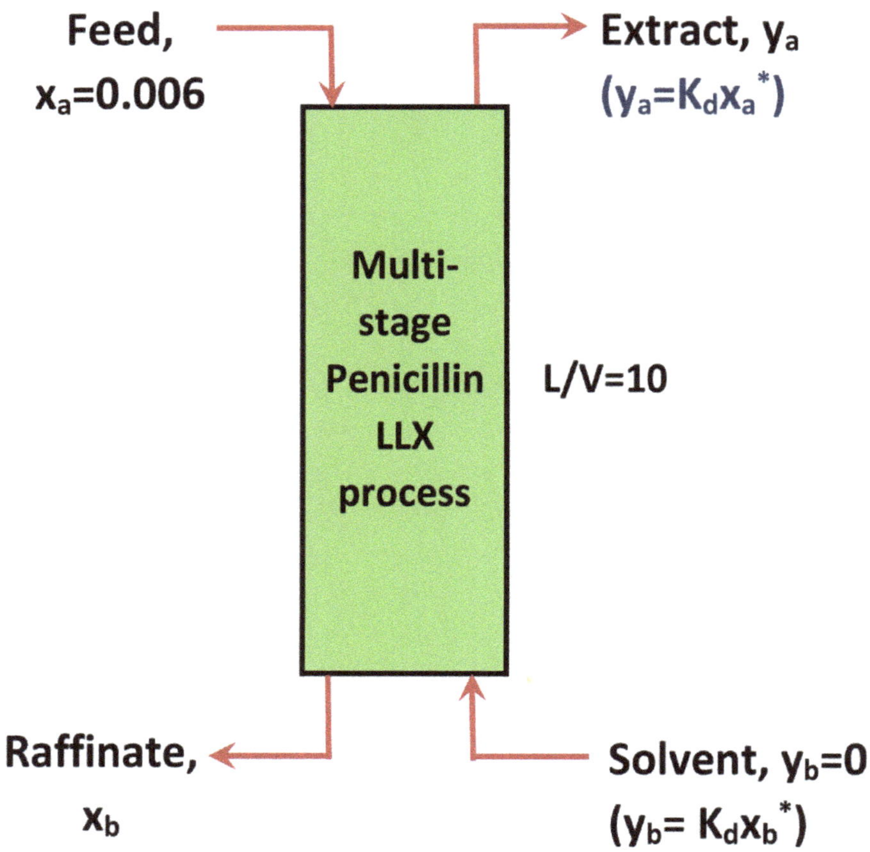

The recovery of Penicillin is 99%. Thus

$$x_b = \frac{1}{100} x_a = 0.6 \times 10^{-4}$$

The solute balance gives:

$$y_a = y_b - \left(\frac{L}{V}\right)(x_b - x_a) = -10(0.6 \times 10^{-4} - 0.006) = 0.0594$$

The extraction factor is:

$$E_o = \frac{K_d}{L/V} = \frac{85}{10} = 8.5$$

The following Kremser equation, Equation (14.27), can now be used to calculate the number of stages required:

$$N = \frac{\ln\left(\frac{x_a - x_a^*}{x_b - x_b^*}\right)}{\ln(E_o)} \; when \; E_o \neq 1$$

where $x_a^* = y_a/K_d = \frac{0.0594}{85} = 7 \times 10^{-4}$ and $x_b^* = y_b/K_d = 0$. The number of stages turn out to be:

$$N = \frac{\ln\left(\frac{0.006 - 7\times10^{-4}}{0.6\times10^{-4}}\right)}{\ln(8.5)} = 2.09$$

Thus approximately 2 stages are needed for 99% recovery of Penicillin under the given conditions.

SHORT REVIEW QUESTIONS

1. Define raffinate and extract phases.
2. Give two examples of industrial applications of liquid-liquid extraction.
3. In the analysis and design of liquid-liquid extraction systems, which type of extraction systems (immiscible diluent-solvent pair or miscible diluent-solvent pair) require triangular phase diagrams?
4. Define plait point
5. What is the difference between Type I and Type II ternary liquid-liquid systems?
6. Consider a single-stage liquid-liquid extraction process where the diluent and solvent are completely immiscible. However the solute is miscible in both diluent and solvent. In such an extraction process, how does the extraction efficiency vary with the ratio of feed to solvent flow rates?
7. The operating line of counter-current multi-stage liquid-liquid extraction cascade for immiscible solvent-diluent pair falls below the equilibrium line. Is this true or false?
8. The number of equilibrium stages required for multi-stage LLX process with immiscible solvent-diluent pair can be estimated using the Kremser equations provided that the operating line and equilibrium line are linear. Is this true or false?
9. Define the difference point utilized in the estimation of number of equilibrium stages in a multistage LLX process with partially miscible solvent-diluent pair.
10. The operating lines for a multistage LLX process with partially miscible solvent-diluent pair pass through three points: the difference point, a point on the extract portion of equilibrium curve, and a point on raffinate portion of the equilibrium curve. Is this true or false?

PRACTICE PROBLEMS

Problem 1

A mixture of two liquids A (solute) and B (diluent) are to be separated using solvent C. The equilibrium data in terms of weight percent for the ternary system A-B-C are given below:

Composition of coexisting liquid phases in equilibrium					
Extract layer			Raffinate layer		
A%	B%	C%	A%	B%	C%
0	0.18	99.82	0	99.89	0.11
5.21	0.32	94.47	5	94.82	0.18
10.79	0.49	88.72	10	89.79	0.21
16.20	0.63	83.17	15	84.78	0.24
22.23	0.79	76.93	20	79.69	0.31
29.01	1.11	69.82	25	74.58	0.42
37.48	1.72	60.80	30	69.42	0.58
43.28	2.33	54.39	35	64.22	0.78
49.44	3.05	47.51	40	58.69	1.36
54.92	4.28	40.80	45	52.76	2.24
59.19	7.24	33.57	50	46.28	3.72
61.79	13.83	24.38	55	38.69	6.31
60.58	25.66	13.76	60.58	25.66	13.76

a) Plot the triangular phase diagram using equilateral-triangle diagram. Show the Plait point.
b) Plot the tie lines on the triangular phase diagram
c) Repeat (a) and (b) using right-angled triangle diagram

Problem 2

For the ternary system of practice problem 1 given above, a single stage extraction process is used in which 350 kg of a feed solution containing 35 wt% solute A in diluent B is contacted with 350 kg of pure solvent C.

Calculate the amounts and compositions of the extract and raffinate layers. What is the percent recovery of solute A?

Problem 3

Pure solvent (species C) is being used to extract a solute (species A) from feed containing 40 wt% A and 60 wt% diluent (species B) using countercurrent multistage extraction. The flow rate of entering solvent is 525 kg/h and the flow rate of feed solution is 185 kg/h. The desired exit solute concentration in the raffinate is 5 wt%.

Estimate the compositions and mass flow rates of extract and raffinate streams leaving the process.

The equilibrium data in terms of weight percent for the ternary system A-B-C are given in practice problem 1 above.

Problem 4

A single stage liquid-liquid extraction process is being used to recover an antibiotic from dilute aqueous fermentation broth containing 0.75 wt% antibiotic using pure solvent. The ratio of solvent mass flow rate to feed mass flow rate is 0.25 ($V/L = 0.25$). The equilibrium relation is $y = K_d x$ where K_d=85 (y and x are mass fractions of antibiotic in extract and reffinate phases, respectively at equilibrium).

What are the compositions of the exit extract and raffinate streams?

What is the percent recovery of antibiotic?

You can assume that solvent is insoluble in water.

Problem 5

A single stage liquid-liquid extraction process is being used to recover solute A from dilute aqueous feed stream containing 1 wt% solute. The feed mass flow rate is 1000 kg/h. Pure solvent C is used to extract the solute.

If the concentration of solute in the raffinate is desired to be 0.1 wt%, estimate the mass flow rate of solvent. The equilibrium relation is $y = K_d x$ where K_d=0.75 (y and x are mass fractions of solute in extract and reffinate phases, respectively at equilibrium). You can assume that solvent is insoluble in water.

Problem 6

A multistage liquid-liquid extraction process is being used to remove solute from feed stream using pure solvent C. The feed mass flow rate is 50 kg/h and the solvent mass flow rate is 100 kg/h. The mass fraction of solute in the feed stream is 0.01.

Estimate the number of equilibrium stages required to reduce the concentration of solute in the reffinate stream leaving the process to 0.001.

The equilibrium relation is $y = K_d x$ where K_d=0.85 (y and x are mass fractions of solute in extract and reffinate phases, respectively at equilibrium).

You can use Kremser equation to estimate the number of stages.

Chapter 15

Solid-Liquid Leaching

15.1 LEARNING OUTCOMES

Upon completion of this chapter, students will be able to:

- Describe solid-liquid leaching process and its practical applications
- Construct phase diagrams relevant to solid-liquid leaching process
- Analyze single-stage leaching process
- Analyze multi-stage leaching process
- Estimate the number of equilibrium stages required in a multi-stage counter-current leaching process where complete dissolution of solute occurs in the solvent in the first stage itself
- Apply McCabe-Thiele graphical method and Kremser equations to estimate the number of equilibrium stages required in a multi-stage counter-current leaching process

15.2 INTRODUCTION

Leaching, also referred to as solid-liquid extraction, is a separation process whereby the solute present in the solid phase is transferred from the solid phase to the liquid phase. The solute present in the solid phase is completely soluble in the solvent whereas the solute-free solid (carrier or inert solid) is generally insoluble in the solvent. Thus there are three species involved in solid-liquid leaching process: solute species A, inert solid species B, and solvent species C. Figure 15.1 shows the direction of mass transfer in a solid-liquid leaching process.

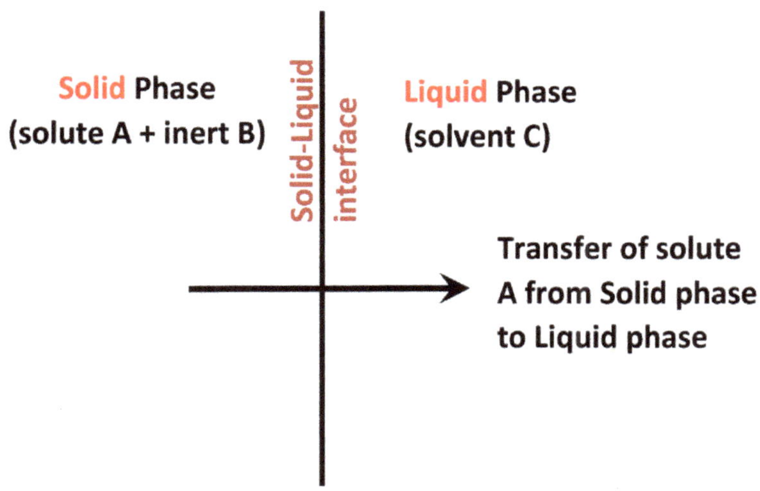

Figure 15.1 Direction of mass transfer of solute in a solid-liquid leaching process.

Typically two pieces of equipment are involved in a single-stage solid-liquid leaching process, namely a mixer and a settler, as shown schematically in Figure 15.2. The solid feed (solute A + inert solid B) in particulate form is intimately mixed with the solvent (species C) in a mixing vessel and the mixture (solids-liquid two-phase slurry) is transferred to a settling vessel for the separation of solute-rich solvent and solute-depleted solids slurry. In an *ideal* solid-liquid leaching process, it is generally assumed that the rate of leaching of solute is quite rapid and that the solute is infinitely soluble in the solvent. Consequently, the solute is completely removed from the carrier solids and transferred to the solvent during the mixing process. However, a significant amount of solute-rich solution is retained in the settled slurry of solute-free solid particles. The amount of solution retained in the settled slurry depends on the density and viscosity of the solution which vary with the solute content of the solution.

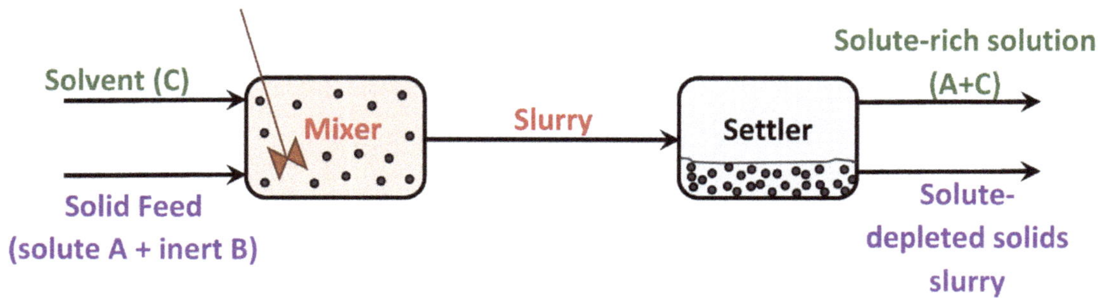

Figure 15.2 Single stage solid-liquid leaching process.

The practical applications of solid-liquid leaching process are many as noted in Chapter 1. For example, extraction of vegetable oil from oil seeds using some organic solvent, and removal of metals from metal ores using an appropriate solvent involve solid-liquid leaching.

15.3 PHASE DIAGRAMS FOR SOLID-LIQUID LEACHING

In the analysis and design of single-stage and multi-stage solid-liquid processes, knowledge of the phase diagrams is required. As noted earlier, three species are involved in solid-liquid leaching process: solute species A, inert solid species B, and solvent species C. There are two phases present in the settler at equilibrium, as shown schematically in Figure 15.3. The top phase, referred to as *overflow* solution, is solute-rich solution free of any inert solid B. Let y be the solute mass fraction (kg solute per kg solution) in the overflow solution. The bottom phase, referred to as *underflow*, is a slurry consisting of solute-depleted solids B and solution of A and C. Let x be the solute mass fraction (kg solute per kg solution) in the solution retained in settled slurry phase (underflow). At *equilibrium*, the overflow solution and the solution retained in the underflow slurry have the same compositions, that is, $y = x$ represents the equilibrium relationship connecting the compositions of solutions in the two phases (overflow and underflow).

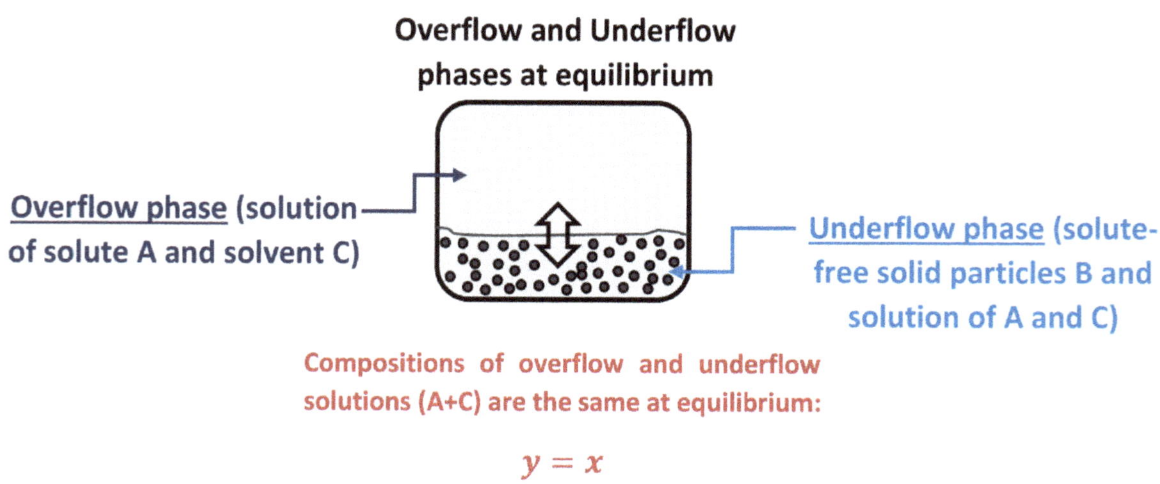

Figure 15.3 Phase equilibrium in solid-liquid leaching process.

Figure 15.4 shows the typical phase diagram relevant to solid-leaching process. The top curve shows the variation of N with x for the underflow slurry phase, where N is the ratio of mass of inert (solute-free) solids to mass of solution retained in the underflow slurry phase. The bottom curve shows the variation of N with y for the overflow phase, where N is the ratio of mass of inert (solute-free) solids to mass of solution present in the overflow phase. Thus the quantities describing the phase diagram are as follows:

$$N = \frac{mass\ of\ inert\ solid\ B}{mass\ of\ liquid\ solution\ of\ solute\ A\ and\ solvent\ C} \tag{15.1}$$

$$x = \frac{mass\ of\ solute\ A\ in\ underflow\ liquid\ solution}{mass\ of\ underflow\ liquid\ solution\ of\ A\ and\ C} \tag{15.2}$$

$$y = \frac{mass\ of\ solute\ A\ in\ overflow\ liquid\ solution}{mass\ of\ overflow\ liquid\ solution\ of\ A\ and\ C} \tag{15.3}$$

The tie-lines connecting the underflow and overflow phases at equilibrium are vertical lines as the concentration of solute in the solution is uniform at equilibrium ($y = x$), regardless of the phase (underflow or overflow). In the overflow phase, the value of N is zero as the overflow phase consists of just the solute-solvent solution free of any inert solids. In the underflow phase, the value of N decreases somewhat with the increase in the concentration of the solute as solutions tend to become more viscous and sticky with the increase in the solute concentration and hence more solution is retained in the underflow phase. In many applications, N vs. x is a horizontal line with negligible variation of N with solute concentration.

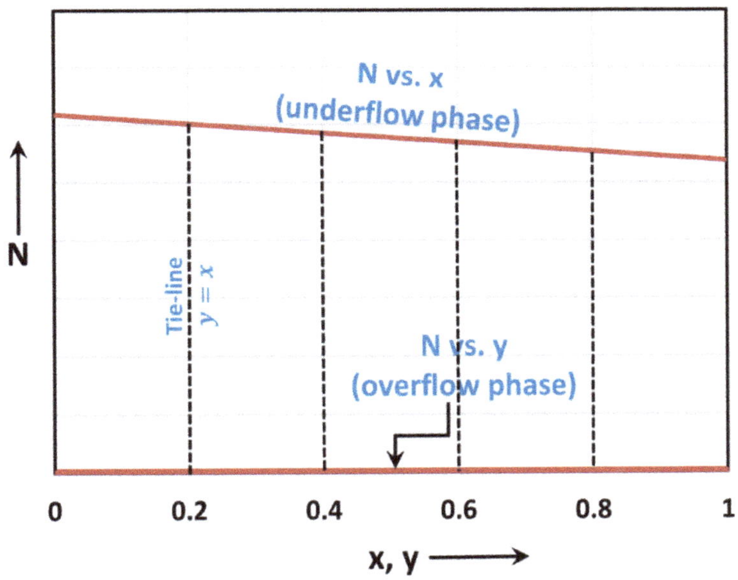

Figure 15.4 Phase diagram for solid-liquid leaching process.

15.4 ANALYSIS OF A SINGLE-STAGE SOLID-LIQUID LEACHING PROCESS

Figure 15.5 shows a schematic diagram of a single-stage solid-liquid leaching process. To analyze the process, we need to carry out material balances over the envelopes shown in Figure 15.5.

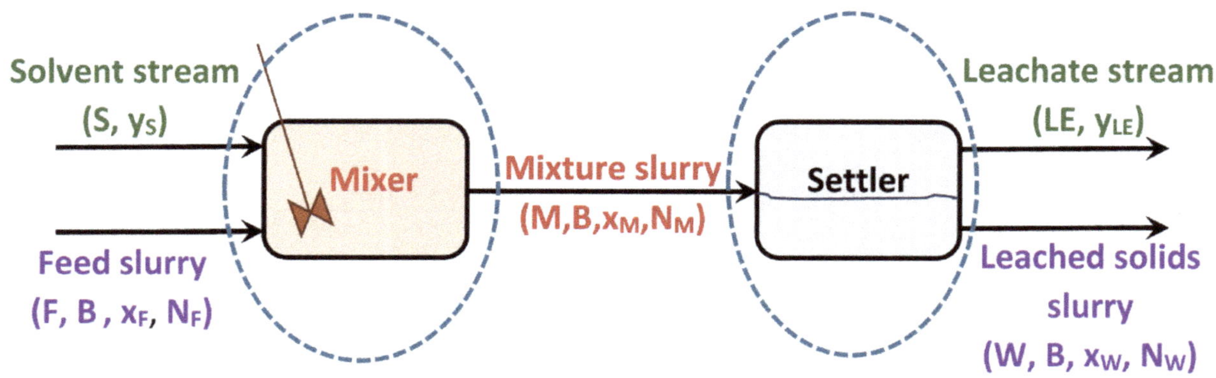

Figure 15.5 Material balance envelopes for the analysis of a single stage leaching process.

Total material balance on solid-free basis around the mixer (dashed blue envelope) gives:

$$F + S = M \tag{15.4}$$

where F is the mass flow rate of solution (solute + solvent) entering with the feed slurry, S is the mass flow rate of solvent (plus solute) stream entering the process and M is the mass flow rate of solid-free solution in the mixture slurry.

Solute material balance around the mixer gives:

$$x_F F + y_S S = x_M M \tag{15.5}$$

From Equations (15.4)-(15.5), it can be readily shown that:

$$x_M = \frac{y_S + x_F(F/S)}{1 + (F/S)} \tag{15.6}$$

Carrier material (inert-solid) balance around the mixer gives:

$$B = N_F F = N_M M \tag{15.7}$$

Given F, S, x_F, y_S, B, we would like to calculate LE, W, x_W, y_{LE}, that is, the amounts and compositions of the overflow leachate and underflow slurry streams leaving the stage.

The solution can be obtained graphically (see Figure 15.6) as follows:

- Locate the feed point $\boldsymbol{F}(x_F, N_F)$ on the phase diagram
- Locate the solvent point $\boldsymbol{S}(y_S, N_S = 0)$ on the phase diagram
- Locate the mixture point $\boldsymbol{M}(x_M, N_M)$ calculated from Equations (15.6) and (15.7) on the phase diagram
- Connect the feed, solvent, and mixture points by a straight line as these three points are *collinear*.
- As the overflow leachate stream and underflow slurry stream are in *phase equilibrium* with each other, the leachate and underflow solution compositions can be determined graphically by drawing a *tie-line* passing through the mixture point as shown in Figure 15.6. Note that there is only one tie-line that passes through the points $\boldsymbol{LE}, \boldsymbol{M}$ and $\boldsymbol{W}$; these three points are *collinear*. Read the compositions graphically: leachate corresponding to point $\boldsymbol{LE}(y_{LE}, N_{LE} = 0)$ and underflow solution corresponding to point $\boldsymbol{W}(x_W, N_W)$.

- Once the leachate and underflow solution compositions are determined, the leachate and underflow solution flow rates can be determined using material balances.

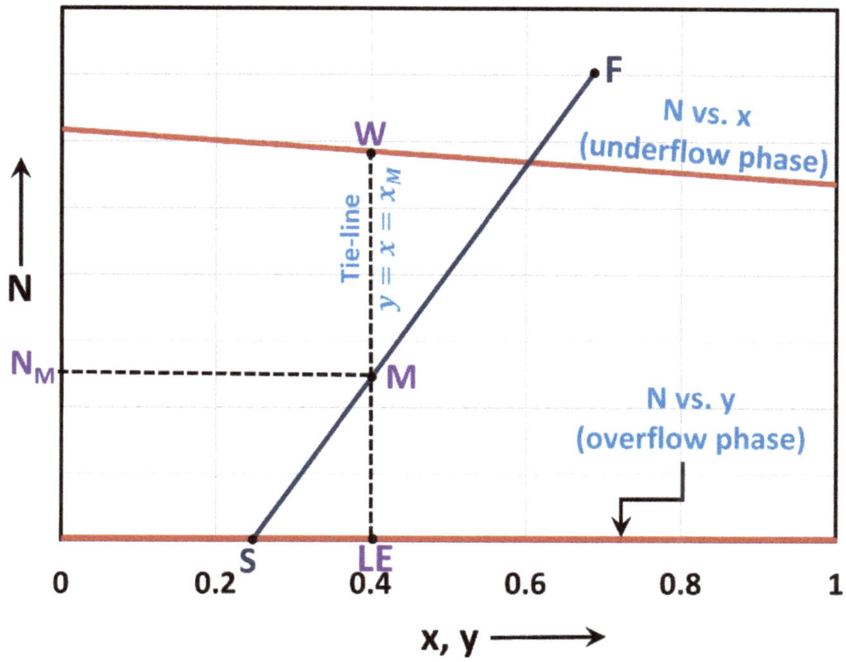

Figure 15.6 Graphical solution of single stage solid-liquid leaching process.

Total material balance on solid-free basis around the settler (dashed blue envelope) gives:

$$LE + W = M \tag{15.8}$$

where LE is the mass flow rate of overflow leachate stream leaving the process and W is the mass flow rate of solution in the underflow stream leaving the stage.

Carrier material (inert-solid) balance around the settler gives:

$$B = N_W W \tag{15.9}$$

Equation (15.9) can be used to calculate the flow rate W. The overflow leachate flow rate LE can then be calculated from Equation (15.8). Note that M and N_W are already known.

15.5 MULTI-STAGE SOLID-LIQUID LEACHING

The recovery of solute in a single-stage solid-liquid leaching process is often limited. In order to increase the percent recovery of solute from the feed stream, it is necessary to use a multi-stage leaching process.

15.5.1 Analysis of a Multi-Stage Solid-Liquid Leaching Process

Figure 15.7 shows a counter-current multi-stage solid-liquid leaching process. The notation used for the analysis of a multi-stage solid-liquid leaching process is as follows: the solvent stream (overflow phase) is specified by V and y variables where V is the mass flow rate of solvent stream and y is the concentration of solute in the solvent stream. The feed stream is specified by L and x variables where L is the mass flow rate of solution present in the underflow slurry and x is the solute concentration in the underflow solution.

We will assume that the rate of leaching is quite rapid and that the solute is infinitely soluble in the solvent so that all the solute present in the carrier solid is completely dissolved in the solvent in the first stage itself. In subsequent stages, the solute is simply washed out or recovered from the entrained solute-rich solution in the underflow slurry stream. It will be further assumed that the underflow and overflow streams leaving any stage are at equilibrium with each other. Note that at equilibrium, $y = x$.

In general, the solution flow rates in overflow V-stream and underflow slurry L-stream vary from one stage to another. For example, L_o is rarely equal to L_1. The entering underflow stream mass flow rate L_o is significantly different from L_1 as there is little or no solvent present in L_o.

We would first like to calculate the leached slurry (underflow) and exit leachate (overflow) mass flow rates (L_N, V_1) and (N_N, y_1), given the feed and solvent flow rates and compositions: $L_o, V_{N+1}, x_o, y_{N+1}$. The number of stages required for the given separation is determined later. Note that the concentration of solute in the exit leached slurry is usually set by the designer at some desired value of x_N.

Material balances are necessary in order to proceed with the analysis. The external material balances do not require internal stage by stage details (flow rates and compositions). Thus the process flowsheet of Figure 15.7 can be replaced by a blackbox as shown in Figure 15.8. Furthermore we create a fictitious mixture of the input feed and solvent streams and split this fictitious mixture into the exit leachate and leached slurry streams, as shown in Figure 15.9.

Total material balance around the mixer of Figure 15.9 (dashed blue envelope) gives:

$$V_{N+1} + L_o = M \tag{15.10}$$

Solute material balance around the mixer gives:

$$y_{N+1}V_{N+1} + x_o L_o = x_M M \tag{15.11}$$

From Equations (15.10)-(15.11), it follows that:

$$x_M = \frac{y_{N+1} + x_o(L_o/V_{N+1})}{1 + (L_o/V_{N+1})} \tag{15.12}$$

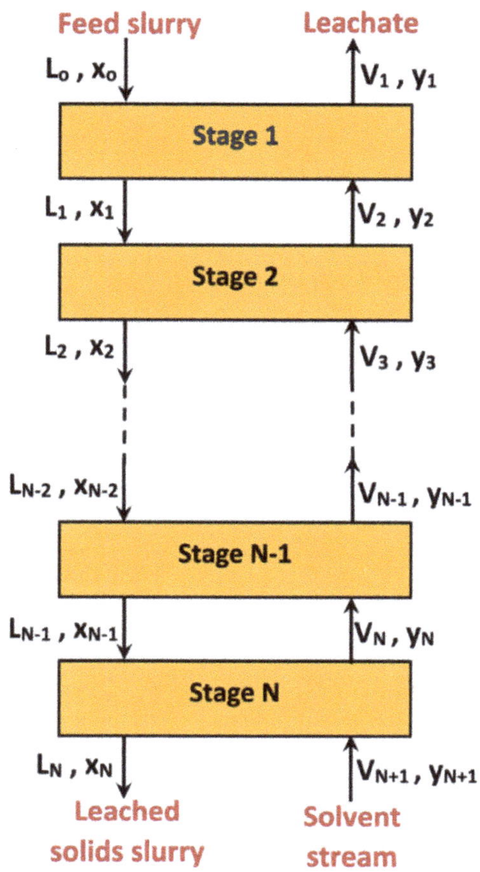

Figure 15.7 Counter-current multi-stage solid-liquid leaching process with variable flows.

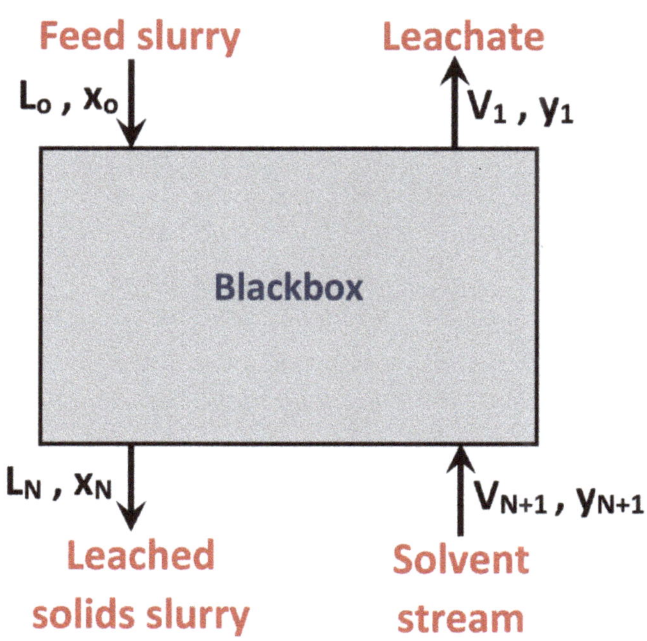

Figure 15.8 Counter-current multi-stage solid-liquid leaching process treated as a blackbox with inputs and outputs.

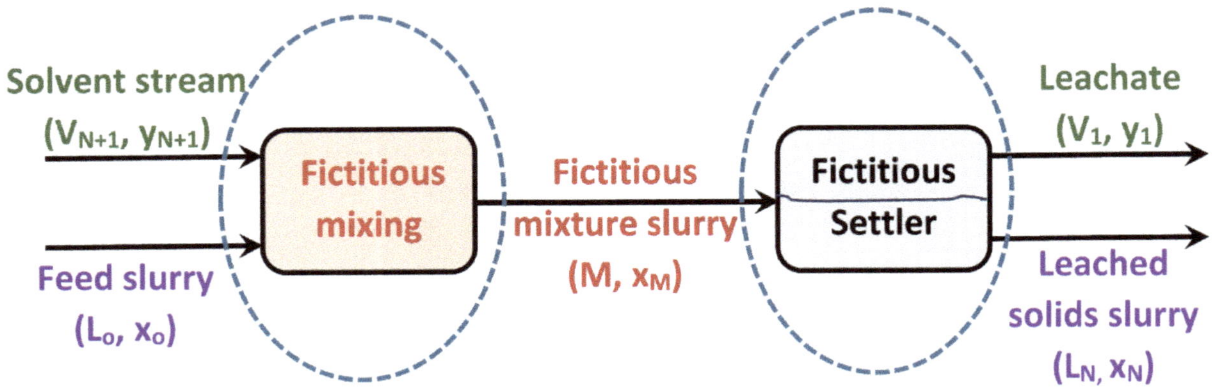

Figure 15.9 Counter-current multi-stage solid-liquid leaching process treated as an equivalent single-stage solid-liquid leaching process.

Carrier material (inert-solid) balance around the mixer gives:

$$B = N_M M \tag{15.13}$$

The unknown flow rates and compositions of the overflow leachate and underflow slurry streams can now be obtained graphically (see Figure 15.10) as explained below:

- Locate the feed point $\boldsymbol{L_o}(x_o, N_o)$ on the phase diagram
- Locate the solvent point $\boldsymbol{V_{N+1}}(y_{N+1}, N_{N+1} = 0)$ on the phase diagram
- Locate the mixture point $\boldsymbol{M}(x_M, N_M)$ calculated from Equations (15.12) and (15.13) on the phase diagram
- Connect the feed, solvent, and mixture points by a straight line as these three points are *collinear.*
- Although the exit leachate and leached slurry streams are *not in phase equilibrium* with each other, the leachate and leached solids slurry compositions still fall on the equilibrium curve as they are equilibrium streams: exit leachate stream emerging from stage 1 (see Figure 15.7) is in equilibrium with L_1 stream leaving stage 1; leached solids slurry stream emerging from *Nth* stage (see Figure 15.7) is in equilibrium with V_N stream leaving stage N.
- The unknown compositions of the leachate and leached slurry streams can be determined graphically by drawing a straight line passing through the mixture point $\boldsymbol{M}(x_M, N_M)$ and the point $\boldsymbol{L_N}(x_N)$ on the equilibrium curve. The intersection of the line $L_N M$ with the equilibrium curve gives the unknown (N_N, y_1), as shown in Figure 15.10.
- Once the leachate and leached slurry compositions are determined, the leachate and leached slurry flow rates (L_N, V_1) are determined using material balances.

Total material balance on solid-free basis around the settler (dashed blue envelope, Figure 15.9) gives:

$$V_1 + L_N = M \tag{15.14}$$

Carrier material (inert-solid) balance around the settler gives:

$$B = N_N L_N \tag{15.15}$$

Equation (15.15) can be used to calculate the flow rate L_N. The overflow leachate flow rate V_1 can then be calculated from Equation (15.14). Note that M and N_N are already known.

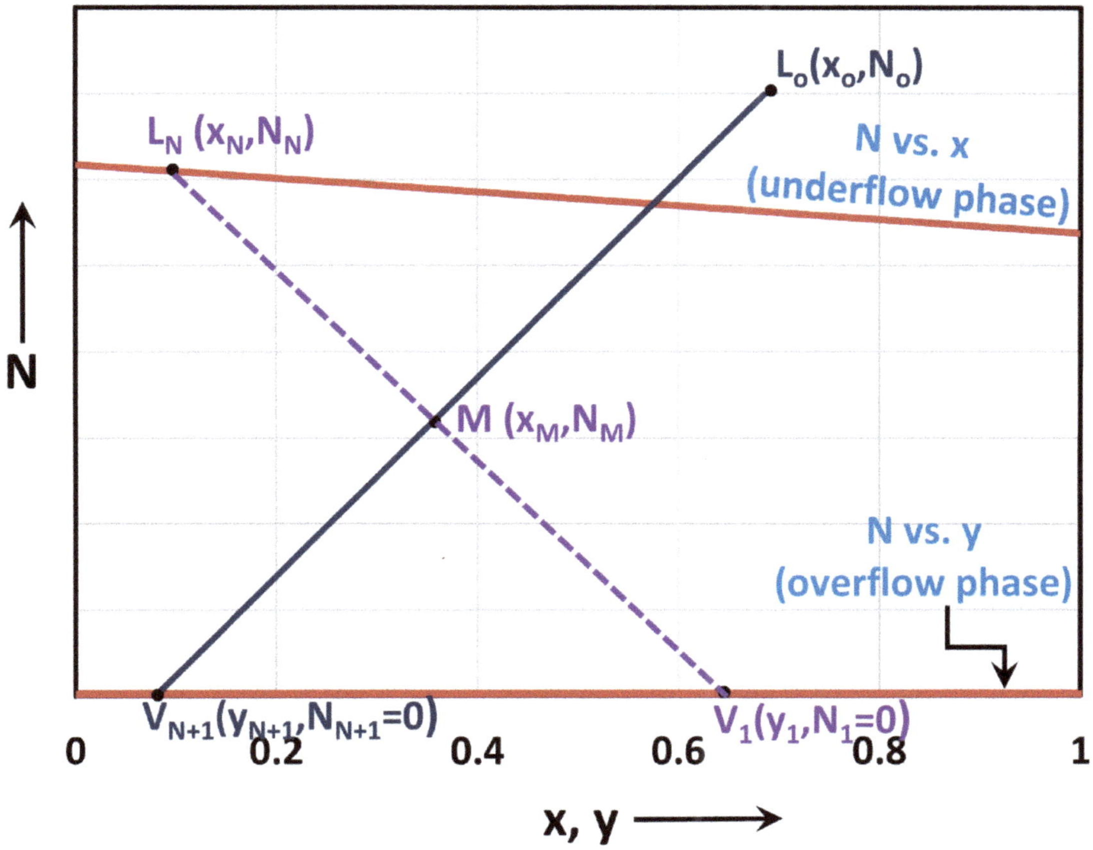

Figure 15.10 Graphical solution of multi-stage solid-liquid leaching process.

Estimation of Number of Equilibrium Stages

Consider material balance on nth stage of a multistage solid-liquid leaching process (see Figure 15.11). The total material balance on solid-free basis gives:

$$L_{n-1} - V_n = L_n - V_{n+1} \tag{15.16}$$

Equation (15.16) can be written for different values of n as:

$$L_o - V_1 = L_1 - V_2 \qquad (n = 1) \tag{15.17}$$

$$L_1 - V_2 = L_2 - V_3 \qquad (n = 2) \tag{15.18}$$

$$L_2 - V_3 = L_3 - V_4 \qquad (n = 3) \tag{15.19}$$

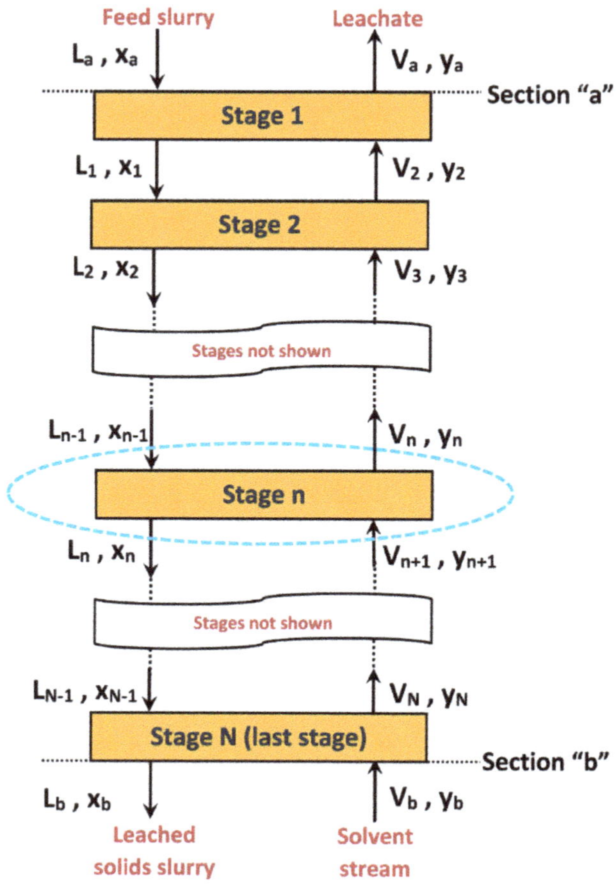

Figure 15.11 Material balance on nth stage of a multi-stage solid-liquid leaching process.

Clearly the net flow rate of material (solids-free basis) in the downward direction is constant. Although the flow rate of L-stream or V-stream may change from one stage to another, the difference in the flow rates of passing L and V streams is constant at steady state.

The solute material balance gives:

$$x_{n-1}L_{n-1} - y_n V_n = x_n L_n - y_{n+1}V_{n+1} \tag{15.20}$$

Equation (15.20) can be written for different values of n as:

$$x_o L_o - y_1 V_1 = x_1 L_1 - y_2 V_2 \qquad (n = 1) \tag{15.21}$$

$$x_1 L_1 - y_2 V_2 = x_2 L_2 - y_3 V_3 \qquad (n = 2) \tag{15.22}$$

Equations (15.20)-(15.22) indicate that the difference in solute flow rates between passing L and V streams is constant from one stage to another at steady state.

The inert-solid material balance gives:

$$N_{n-1}L_{n-1} - (0)V_n = N_n L_n - (0)V_{n+1} \tag{15.23}$$

Note that there are no solids present in the V-stream (overflow solvent stream).

Equation (15.23) can be written for different values of n as:

$$B = N_o L_o = N_1 L_1 \qquad (n = 1) \qquad (15.24)$$

$$N_1 L_1 = N_2 L_2 \qquad (n = 2) \qquad (15.25)$$

Thus the difference in inert-solid flow rates between passing L and V streams is constant, equal to B. We now define the so-called *difference point* as follows:

$$x_{diff} = \frac{difference\ in\ solute\ mass\ flow\ rates\ of\ the\ passing\ streams}{difference\ in\ the\ total\ mass\ flow\ rates\ of\ the\ passing\ streams} \qquad (15.26)$$

$$N_{diff} = \frac{difference\ in\ inert\text{-}solid\ mass\ flow\ rates\ of\ the\ passing\ streams}{difference\ in\ the\ total\ mass\ flow\ rates\ of\ the\ passing\ streams} \qquad (15.27)$$

Thus the coordinates of a difference point are:

$$x_{diff} = \frac{x_o L_o - y_1 V_1}{L_o - V_1} = \frac{x_N L_N - y_{N+1} V_{N+1}}{L_N - V_{N+1}} \qquad (15.28)$$

$$N_{diff} = \frac{N_o L_o}{L_o - V_1} = \frac{N_N L_N}{L_N - V_{N+1}} = \frac{B}{L_o - V_1} \qquad (15.29)$$

We are now in a position to determine the number of stages graphically for a multi-stage countercurrent leaching process. Figure 15.12 illustrates the graphical construction of equilibrium stages on a phase diagram. The construction process is as follows: a) Starting at point L_o, draw a straight line from L_o to the difference point to locate the point V_1. From the point V_1, draw a tie-line ($y = x$) to locate the point L_1. A line from L_1 to the difference point is drawn next to locate V_2 . A tie-line from V_2 locates L_2. This process of graphical construction of stages is continued until the end point L_N is reached. Each stage is represented by two operating lines and a tie-line (vertical dashed lines in Figure 15.12). While the operating line connects the compositions of passing L and V streams at the same level, the tie-line connects the compositions of L and V streams leaving the stage at equilibrium with each other. For the separation shown in Figure 15.12, 3 equilibrium stages are required. Note that the final stage may be a partial stage if the tie-line from point V_3 falls on the left of point L_N.

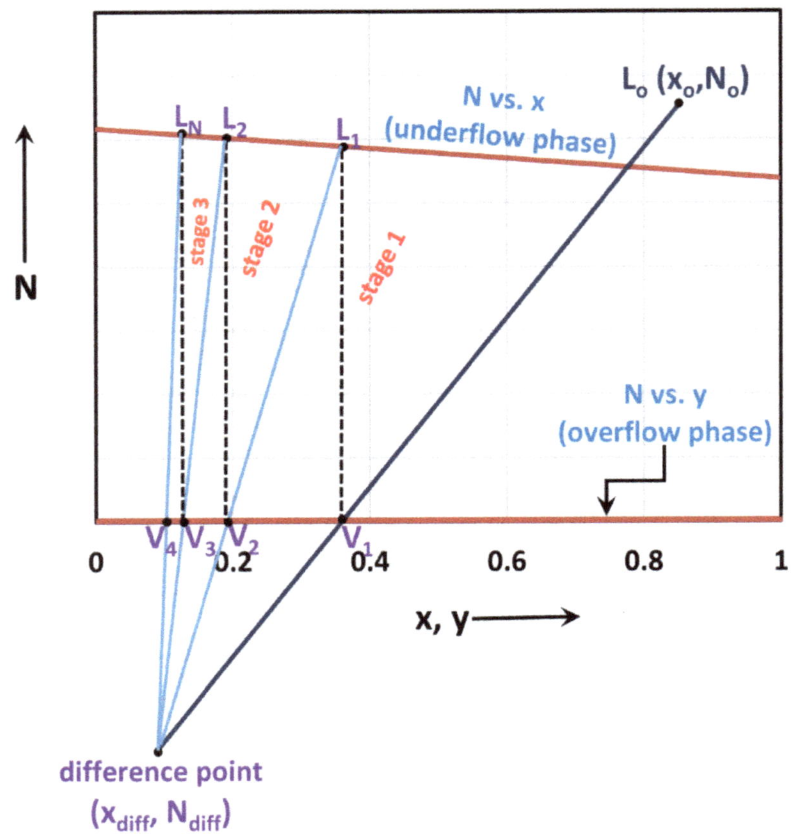

Figure 15.12 Construction of equilibrium stages for multi-stage countercurrent leaching process.

15.5.2 McCabe-Thiele Graphical Method of Estimating the Number of Equilibrium Stages for a Multi-Stage Countercurrent Leaching Process

The McCabe-Thiele graphical method can be applied to estimate the number of equilibrium stages for a multi-stage countercurrent leaching process. However, the underflow stream mass flow rate , represented by variable L, must be constant from one stage to another. When underflow stream flow rate is constant, the overflow stream mass flow rate, that is, V, is also constant. Thus the slope of the operating line, L/V, is constant and the operating line is a straight line.

For underflow stream mass flow rate to be constant, N must be constant and the plot of N versus x must be a horizontal line. Under such condition, the operating line for the multistage leaching process becomes a straight line of slope L/V. Note that N is the ratio of mass of inert (solute-free) solids to mass of solution retained in the underflow slurry phase.

With the exception of stage 1, the underflow stream mass flow rate is generally constant from one stage to another. The underflow flow rate L_1 leaving stage 1 is significantly different from the underflow feed flow rate L_a since L_a contains no solvent. Consequently, we have to treat the first stage separately from the rest of the cascade. The overall and solute mass balances (solid-free basis) over stage 1, see Figure 15.13, are as follows:

$$L_a + V_2 = L_1 + V_a \tag{15.30}$$

$$x_a L_a + y_2 V_2 = x_1 L_1 + y_a V_a \tag{15.31}$$

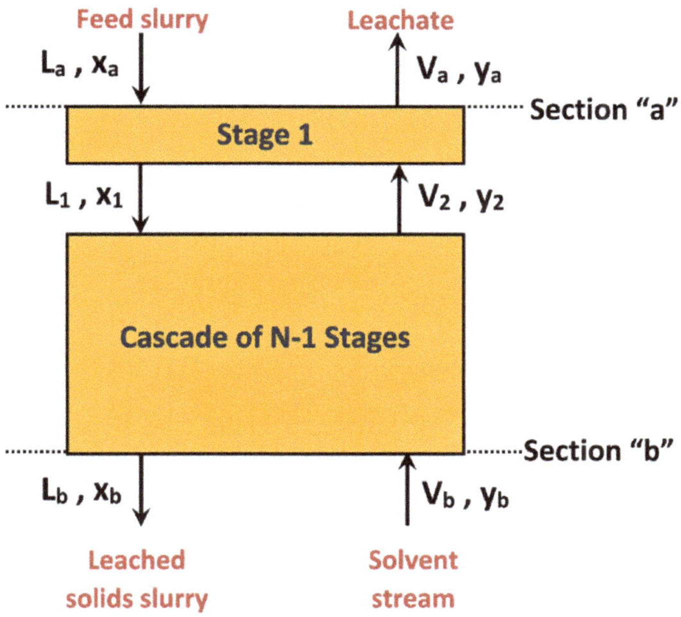

Figure 15.13 Material balance on stage 1 of a multi-stage solid-liquid leaching process.

Let us now develop an equation for the operating line of the multi-stage countercurrent leaching cascade excluding stage 1. The overall and solute mass balances (solid-free basis) over stages 2 to n, see Figure 15.14, are as follows:

$$L_1 + V_{n+1} = L_n + V_2 \tag{15.32}$$

$$x_1 L_1 + y_{n+1} V_{n+1} = x_n L_n + y_2 V_2 \tag{15.33}$$

The equation for the operating line can be obtained from the combination of Equations (15.32) and (15.33) as:

$$y_{n+1} = \left(\frac{L}{V}\right)(x_n - x_1) + y_2 \tag{15.34}$$

Note that we are assuming $L_1 = L_2 = \cdots = L_n = \cdots = L_b$ and $V_2 = V_3 = \cdots = V_n = \cdots = V_b$.

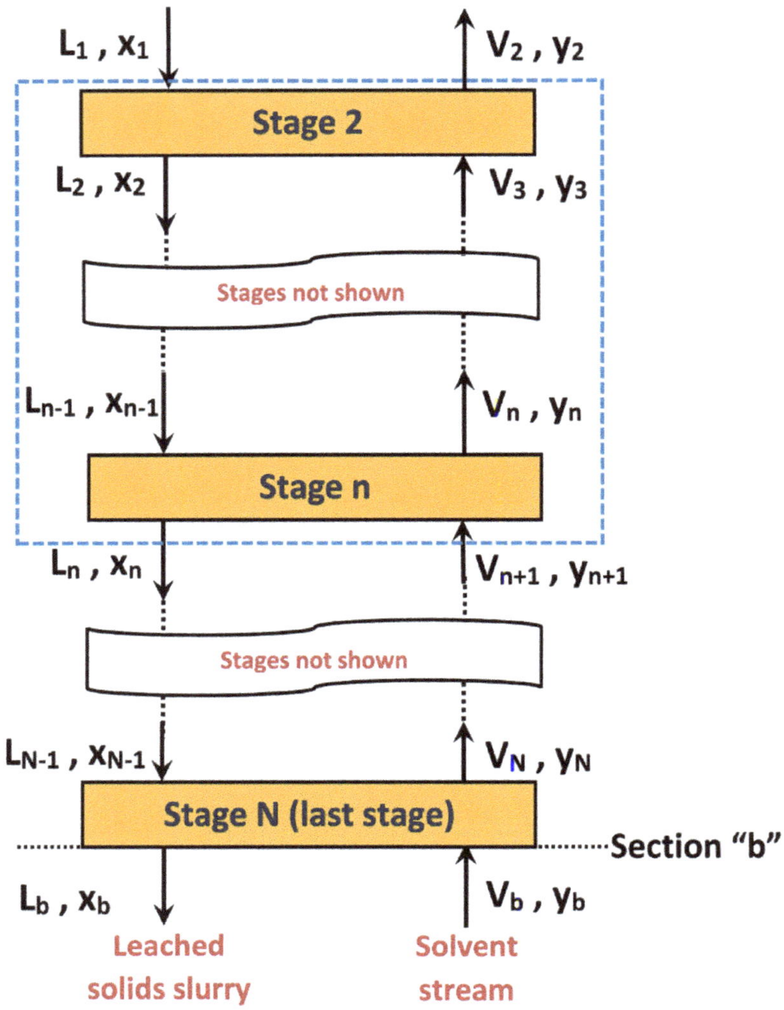

Figure 15.14 Material balance over stages 2 to n of a multi-stage solid-liquid leaching process.

The operating line relates the compositions of the passing underflow and overflow streams at the same level. It passes through the points (x_b, y_b) and (x_1, y_2). Enough information is given in the problem to calculate the points (x_b, y_b) and (x_1, y_2). A straight line connecting these points is the operating line. Once the operating line is drawn, one can then draw the equilibrium line. Note that the equilibrium line is simply $y = x$ for the leaching process.

The stepping off stages is shown in Figure 15.15. Stage 1 is also shown in the figure. Note that the terminal point (x_a, y_a) does not fall on the operating line as the underflow rate in stage 1 is not constant. The underflow rate changes from L_a to L_1 in stage 1.

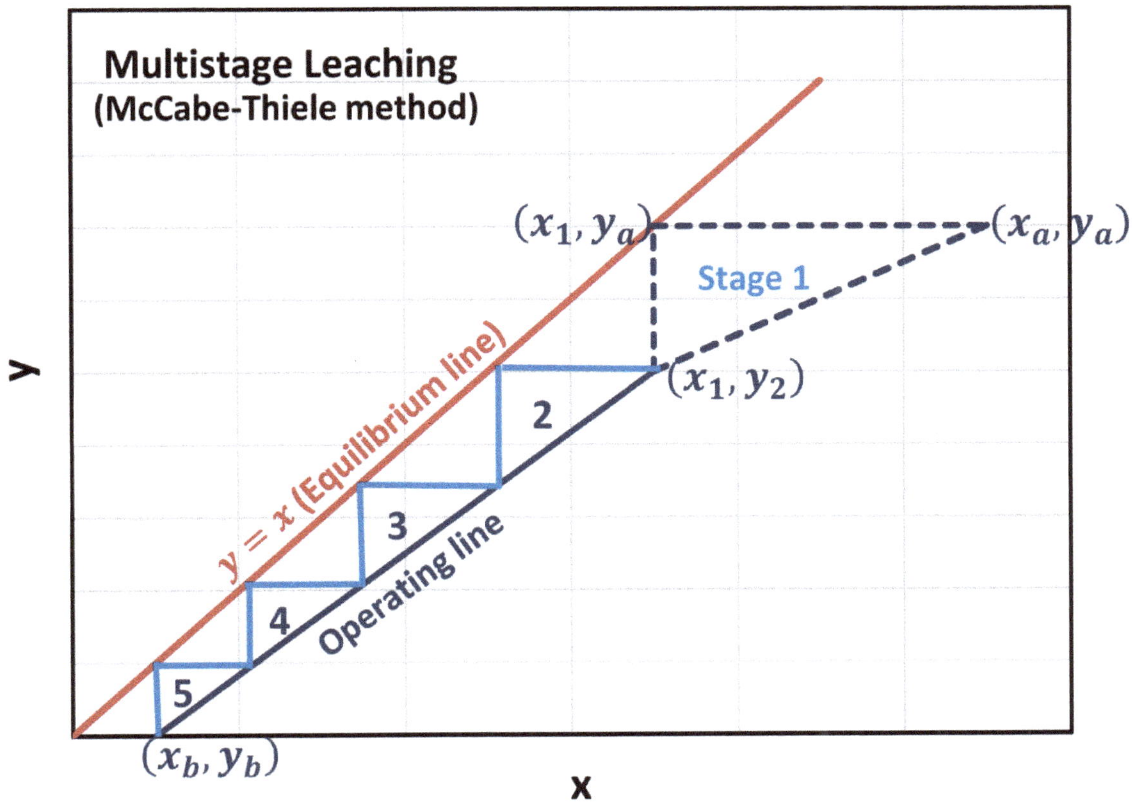

Figure 15.15 McCabe-Thiele graphical method of construction of equilibrium stages for a multi-stage solid-liquid leaching process

15.5.3 Application of Kremser Equation

The Kremser equation can be applied to estimate the number of equilibrium stages needed for a multistage countercurrent leaching process provided that the first stage is excluded. As the first stage is not part of the analysis, the Kremser equation gives $N-1$ stages when applied between the terminal points of the operating line: (x_b, y_b) and (x_1, y_2). Referring to Figure 15.16, the Kremser equation is given as:

$$N-1=\frac{\ln\left(\frac{x_1-x_1^*}{x_b-x_b^*}\right)}{\ln\left(\frac{x_1-x_b}{x_1^*-x_b^*}\right)} \tag{15.35}$$

$$N=1+\frac{\ln\left(\frac{x_1-x_1^*}{x_b-x_b^*}\right)}{\ln\left(\frac{x_1-x_b}{x_1^*-x_b^*}\right)} \tag{15.36}$$

Note that this Kremser equation assumes that the equilibrium line and operating line are not parallel to each other.

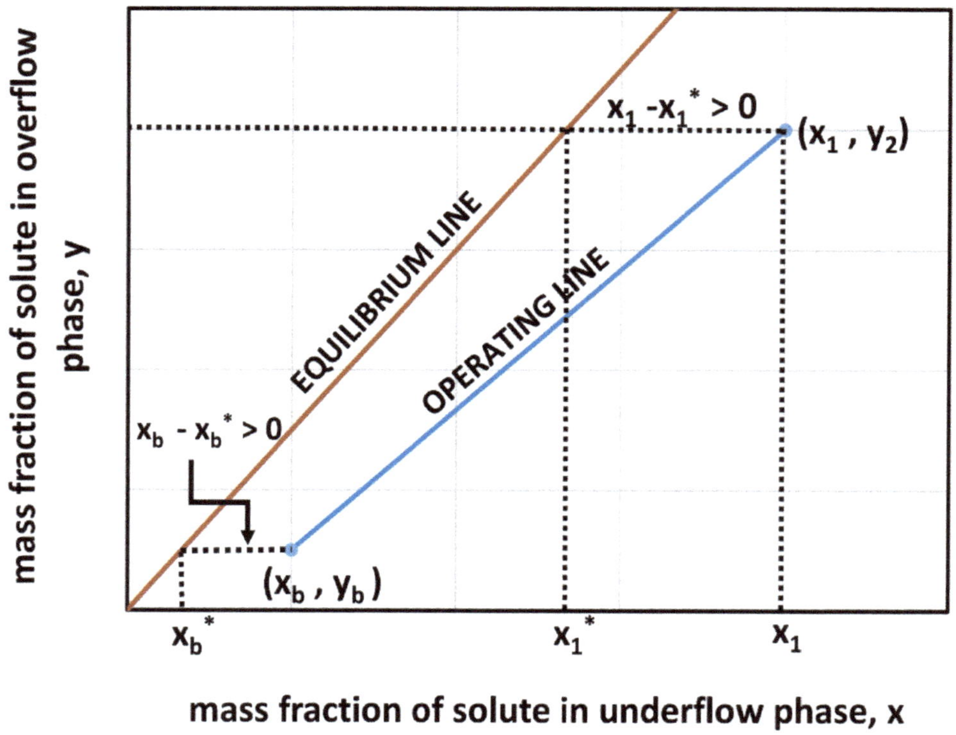

Figure 15.16 Equilibrium and operating lines for the application of Kremser equation

NOTATION

A	solute
B	refers to inert solid component or mass flow rate of inert solids (excluding solute and solvent)
C	solvent
F	refers to feed stream or mass flow rate of solution (solute + solvent) entering with the feed slurry
L	mass flow rate of solution (solute + solvent) present in the underflow solids slurry
L_a	mass flow rate of solution (solute + solvent) entering the cascade with feed slurry
L_b	mass flow rate of solution (solute + solvent) present in the underflow slurry leaving the cascade
L_o	mass flow rate of solution (solute + solvent) entering the cascade with feed slurry, see Figure 15.7
L_1	mass flow rate of solution (solute + solvent) leaving the first stage in the underflow slurry
L_n	mass flow rate of solution (solute + solvent) leaving the nth stage in the underflow slurry
L_{n-1}	mass flow rate of solution (solute + solvent) leaving the (n-1)th stage in the underflow slurry
L_N	mass flow rate of solution (solute + solvent) leaving the last Nth stage in the underflow slurry
LE	refers to overflow leachate phase or mass flow rate of overflow leachate stream
M	refers to mixture slurry or mass flow rate of mixture stream free of inert solids
N	either mass ratio parameter defined in Equation (15.1) or the total number of stages in a cascade
N_o	mass ratio parameter, defined in Equation (15.1), for feed slurry entering the cascade
N_F	mass ratio parameter, defined in Equation (15.1), for feed slurry
N_M	mass ratio parameter, defined in Equation (15.1), for the mixture slurry
N_N	mass ratio parameter, defined in Equation (15.1), for underflow slurry leaving the last Nth stage
N_W	mass ratio parameter, defined in Equation (15.1), for underflow slurry leaving the process
N_{diff}	difference point coordinate defined in Equation (15.27)
S	refers to solvent stream or mass flow rate of solvent stream (solute + solvent) entering the process
V	mass flow rate of overflow solvent stream (solute + solvent)

V_a mass flow rate of overflow solvent stream (solute + solvent) leaving the first stage of a cascade
V_b mass flow rate of overflow solvent stream (solute + solvent) entering the cascade
V_1 mass flow rate of overflow solvent stream (solute + solvent) leaving the first stage
V_n mass flow rate of overflow solvent stream (solute + solvent) leaving the nth stage
V_{n+1} mass flow rate of overflow solvent stream (solute + solvent) leaving the (n+1)th stage
V_{N+1} mass flow rate of overflow solvent stream (solute + solvent) entering the Nth stage
W refers to underflow slurry stream leaving the process or mass flow rate of solution (solute + solvent) leaving the process with the underflow slurry
x mass fraction of solute in solution (solute + solvent) retained in underflow slurry stream
x_a mass fraction of solute in solution (solute + solvent) present in the feed slurry stream entering a cascade
x_1^* mass fraction of solute in underflow solution (solute + solvent) in equilibrium with the leachate stream leaving stage 2
x_N mass fraction of solute in solution (solute + solvent) retained in underflow slurry stream leaving Nth stage, see Figure 15.7
x_o mass fraction of solute in solution (solute + solvent) present in the feed slurry stream, see Figure 15.7
x_{diff} difference point coordinate defined in Equation (15.26)
x_F mass fraction of solute in solution (solute + solvent) present in the feed slurry stream, see Figure 15.5
x_W mass fraction of solute in solution (solute + solvent) present in the underflow slurry stream leaving the process, see Figure 15.5
x_M mass fraction of solute in solution (solute + solvent) present in the mixture slurry stream, see Figure 15.5
x_n mass fraction of solute in solution (solute + solvent) present in the underflow slurry stream leaving nth stage
x_{n-1} mass fraction of solute in solution (solute + solvent) present in the underflow slurry stream leaving (n-1)th stage
x_b mass fraction of solute in solution (solute + solvent) present in the underflow slurry stream leaving a cascade
x_b^* mass fraction of solute in underflow solution (solute + solvent) in equilibrium with the overflow solvent stream entering the cascade at the bottom, see Figure 15.16
y mass fraction of solute in overflow solvent stream (solute + solvent)
y_a mass fraction of solute in overflow solvent stream (solute + solvent) leaving the first stage of a cascade
y_b mass fraction of solute in overflow solvent stream (solute + solvent) entering a
y_1 mass fraction of solute in overflow solvent stream (solute + solvent) leaving the first stage of a cascade
y_{LE} mass fraction of solute in overflow solvent stream (solute + solvent) leaving the process, see Figure 15.5
y_S mass fraction of solute in overflow solvent stream (solute + solvent) entering the process, see Figure 15.5
y_{N+1} mass fraction of solute in overflow solvent stream (solute + solvent) entering the Nth stage (last stage) of a cascade, see Figure 15.7
y_n mass fraction of solute in overflow solvent stream (solute + solvent) leaving nth stage of a cascade
y_{n+1} mass fraction of solute in overflow solvent stream (solute + solvent) leaving (n+1)th stage of a cascade

TUTORIAL PROBLEMS WITH SOLUTION

Problem 1

A single stage solid-liquid leaching process is being used to extract oil from flaked soybeans with a suitable solvent. 500 kg of soybeans is leached with 500 kg of pure solvent. The oil content of the soybeans is 19 wt%. At equilibrium, N is approximately 1.75 kg inert solid/kg solution retained for the underflow slurry and $N = 0$ for the overflow stream. The tie lines are vertical lines expressed as $y = x$.
Calculate the amounts and compositions of the exit leachate and leached solids slurry. What is the percent recovery of oil?

Solution

$$B = 0.81 \times 500 = 405\ kg \text{ (inert solids)}$$

$$F = 0.19 \times 500 = 95\ kg \text{ (Feed solution, 100\% solute as } x_F = 1)$$

$$S = 500\ kg \text{ (pure solvent, } y_S = 0)$$

$$B = N_F F;\ N_F = B/F = 405/95 = 4.263$$

$$F + S = M = 95 + 500 = 595\ kg$$

$$x_M = \frac{y_S + x_F(F/S)}{1 + (F/S)} = \frac{0 + 1(95/500)}{1 + (95/500)} = 0.16$$

$$B = N_M M;\ N_M = B/M = 405/595 = 0.681$$

We can now obtain the solution graphically as shown in the figure below.

- Locate the feed point $\boldsymbol{F}(x_F = 1, N_F = 4.263)$ on the phase diagram
- Locate the solvent point $\boldsymbol{S}(y_S = 0, N_S = 0)$ on the phase diagram
- Locate the mixture point $\boldsymbol{M}(x_M = 0.16, N_M = 0.681)$. $\boldsymbol{M}$ lies on the FS line

The vertical tie line drawn through the mixture point $\boldsymbol{M}$ gives the following results:

$$x_W = 0.16;\ N_W = 1.75;\ y_{LE} = 0.16;\ N_{LE} = 0$$

Thus,

$$W = B/N_W = 405/1.75 = 231.43\ kg$$

$$LE = M - W = 595 - 231.43 = 363.57\ kg$$

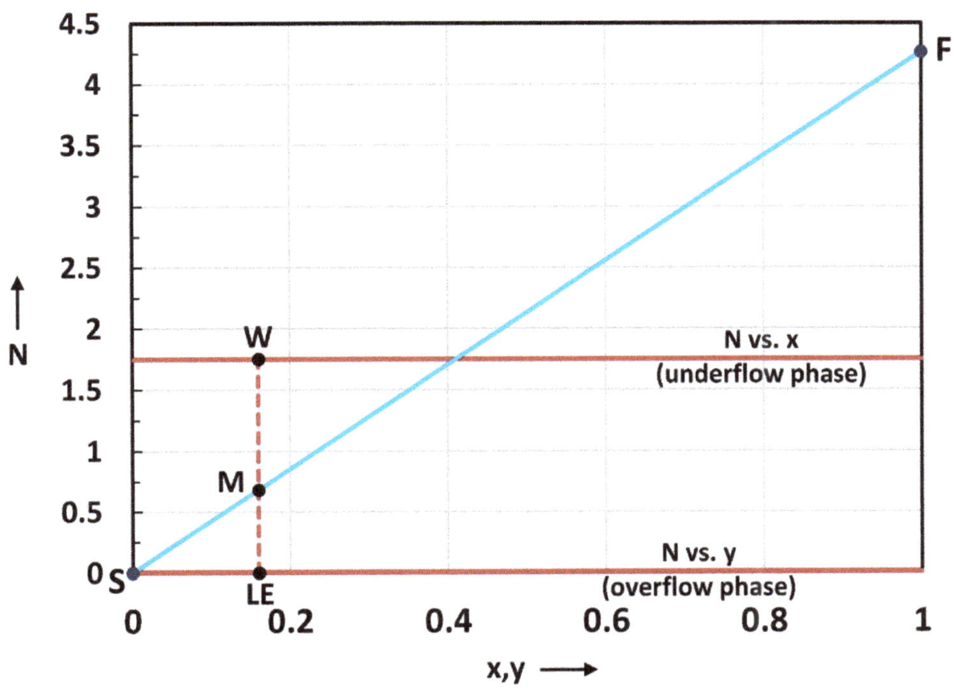

The percent recovery of oil is:

$$\%\text{Recovery} = \left(\frac{x_F F - x_W W}{x_F F}\right) \times 100 = \left(\frac{1 \times 95 - 0.16 \times 231.43}{1 \times 95}\right) \times 100 = 61.02$$

Problem 2

We wish to leach oil from meal by a countercurrent multistage leaching process using a pure solvent. The feed flow rate is 1500 kg/h of meal (inert solids +oil). The oil content of the meal is 17 wt%. The pure solvent flow rate entering the process is 900 kg/h. The desired mass fraction of oil in the underflow product solution is 0.05. Estimate the number of stages required. What are the flow rates and compositions of the streams leaving the process?

The equilibrium data are as follows:

x	0	0.1	0.2	0.3	0.4	0.5	0.6	0.7
N	2.000	1.980	1.942	1.887	1.818	1.751	1.681	1.613

Note that $N = 0$ for the overflow stream at equilibrium and the tie lines are vertical lines expressed as $y = x$.

Solution

The procedure to solve multistage leaching problem is discussed in section 15.5.1. We will first calculate the leached slurry (underflow) and exit leachate (overflow) mass flow rates (L_N, V_1) and the unknown (N_N, y_1), given the feed and solvent flow rates and compositions: $L_o, V_{N+1}, x_o, y_{N+1}$. The quantities specified are shown in the figure below.

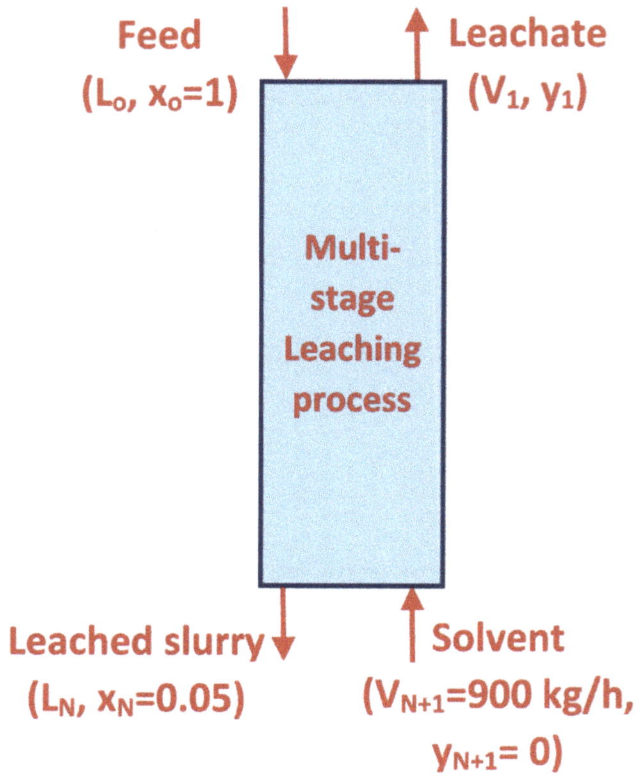

$$L_o = 0.17 \times 1500 = 255\ kg/h$$

$$B = 0.83 \times 1500 = 1245\ kg/h$$

$$B = N_o L_o;\ N_o = B/L_o = 1245/255 = 4.88$$

$$M = V_{N+1} + L_o = 900 + 255 = 1155\ kg/h$$

$$x_M = \frac{y_{N+1} + x_o(L_o/V_{N+1})}{1 + (L_o/V_{N+1})} = \frac{0 + 1(255/900)}{1 + (255/900)} = 0.221$$

$$B = N_M M;\ N_M = B/M = 1245/1155 = 1.0779$$

The unknowns (N_N, y_1), obtained graphically as shown in the figure below, are:

$$N_N = 1.99;\ y_1 = 0.423$$

Thus,

$$L_N = B/N_N = 1245/1.99 = 625.63\ kg/h$$

$$V_1 = M - L_N = 1155 - 625.63 = 529.37\ kg/h$$

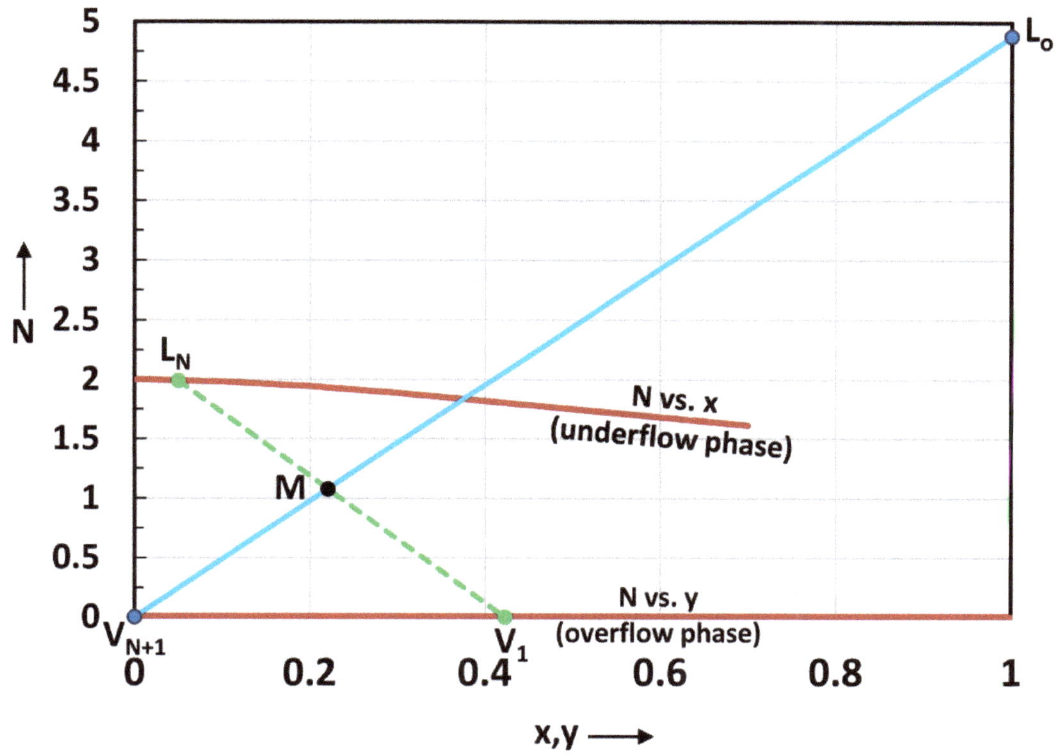

Let us now estimate the number of stages required. The difference point coordinates are:

$$x_{diff} = \frac{x_o L_o - y_1 V_1}{L_o - V_1} = \frac{(1 \times 255) - (0.423 \times 529.37)}{255 - 529.37} = -0.1133$$

$$N_{diff} = \frac{B}{L_o - V_1} = \frac{1245}{255 - 529.37} = -4.538$$

Given the difference point coordinates ($x_{diff} = -0.1133, N_{diff} = -4.538$) and feed point coordinates $L_o(x_o = 1, N_o = 4.88)$, we can estimate the number of stages graphically as shown in the figure below. Note that the point L_o is not shown as it falls outside the figure. Also the coordinates of the point L_N are ($x_N = 0.05, N_N = 1.99$).

The number of stages required is about 4.5.

The flow rates and compositions of the streams leaving the process are:

$$V_1 = 529.37\ kg/h,\ y_1 = 0.423$$
$$L_N = 625.63\ kg/h\ ,\ x_N = 0.05$$

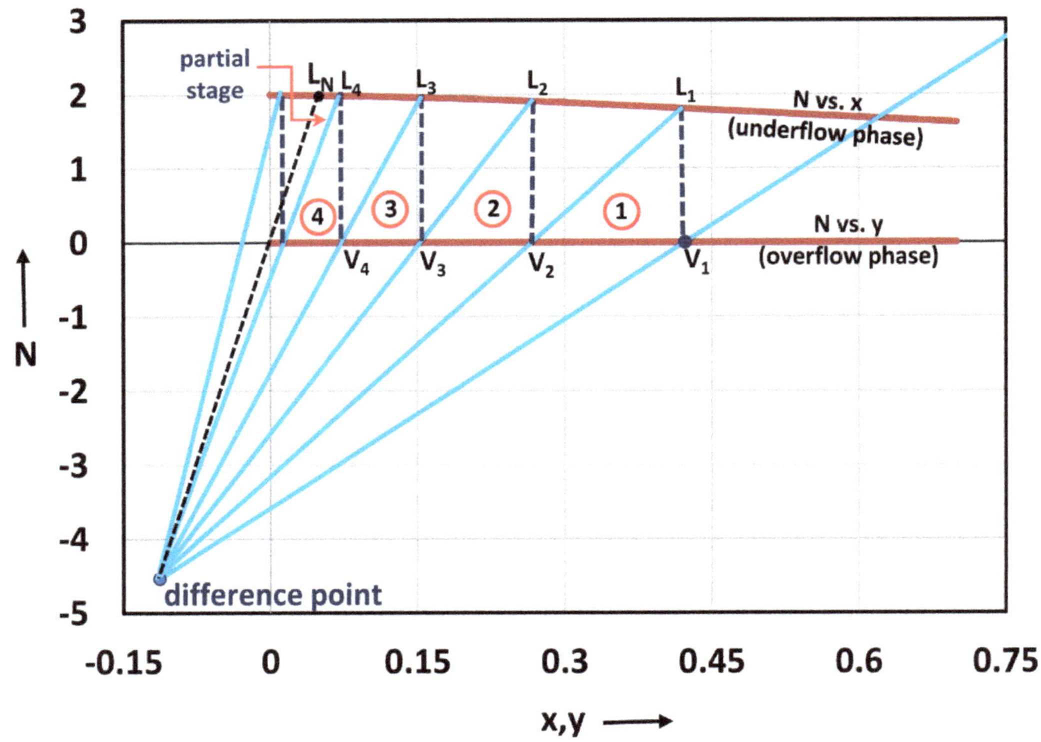

Problem 3

The feed to a multistage leaching process consists of 9,500 kg/h of solids (inert solids + solute in solid form). The solute content of the feed solids is 25% by weight. Pure solvent is used to extract the solute from the solids. The percent recovery of the solute desired is 85. The mass flow rate of pure solvent is 10,000 kg/h. The equilibrium relation between N and x is as follows:

$$N = 1/(1.15 + 0.30x)$$

Estimate the number of equilibrium stages needed for this leaching process. What are the flow rates and compositions of the streams leaving the process?

Solution

Refer to the diagram given below

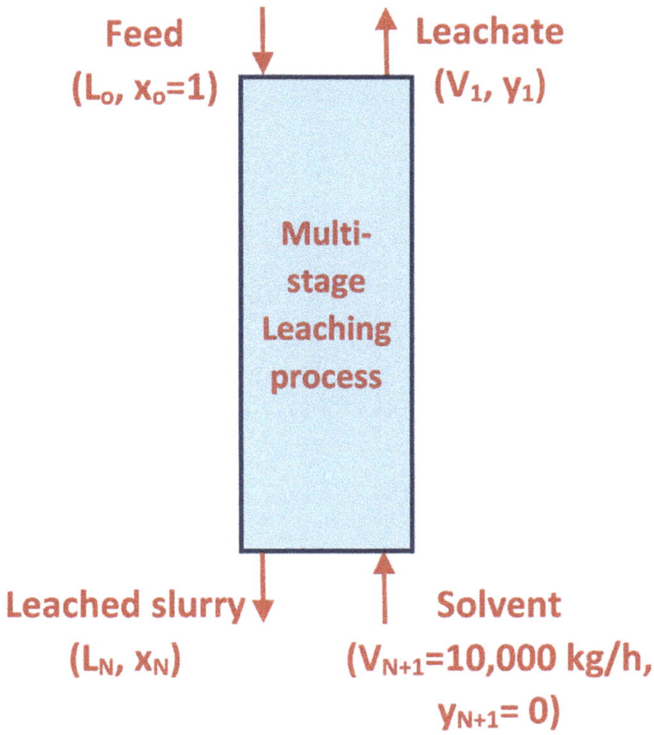

$$L_o = 0.25 \times 9500 = 2375\ kg/h$$

$$B = 0.75 \times 9500 = 7125\ kg/h$$

$$B = N_o L_o;\ N_o = B/L_o = 7125/2375 = 3$$

$$M = V_{N+1} + L_o = 10000 + 2375 = 12375\ kg/h$$

$$x_M = \frac{y_{N+1} + x_o(L_o/V_{N+1})}{1 + (L_o/V_{N+1})} = \frac{0 + 1(2375/10000)}{1 + (2375/10000)} = 0.192$$

$$B = N_M M;\ N_M = B/M = 7125/12375 = 0.57576$$

The concentration of solute in the exit underflow solution, x_N, can be determined from the given percent recovery as:

$$\%\text{Recovery} = \left(\frac{x_o L_o - x_N L_N}{x_o L_o}\right) \times 100 = 85$$

$$x_N L_N = (1 - 0.85)x_o L_o = 0.15 \times 1 \times 2375 = 356.25\ \text{kg/h}$$

As L_N is equal to B/N_N,

$$L_N = \frac{B}{N_N} = 7125(1.15 + 0.30x_N)$$

we get the following equation to solve for x_N:

$$7125x_N(1.15 + 0.30x_N) = 356.25$$

$$0.30x_N^2 + 1.15x_N - 0.05 = 0$$

The solution of this equation is: $x_N = 0.043$. We can now calculate N_N and determine y_1 graphically.

$$N_N = \frac{1}{1.15 + 0.30x_N} = \frac{1}{1.15 + 0.30 \times 0.043} = 0.8599$$

The unknown y_1 is obtained graphically as shown in the figure below:

$$y_1 = 0.494$$

Note that the feed point L_o of the line L_oV_{N+1} is not shown as it falls outside the graph. The mixture point M falls on the lines L_oV_{N+1} and L_NV_1 and y_1 corresponds to the point V_1 shown in the figure.

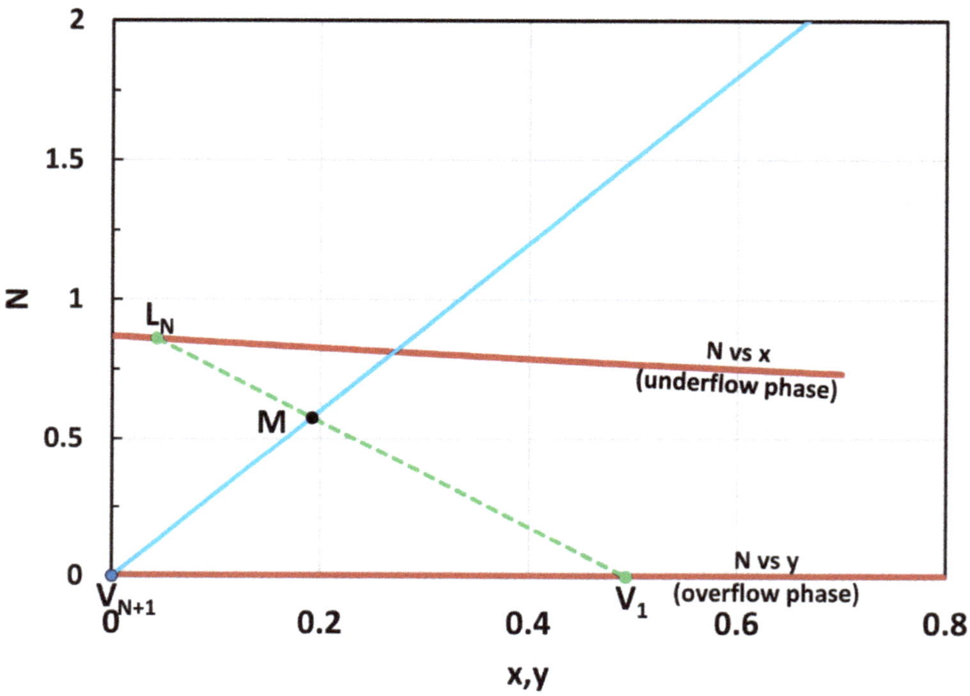

Let us now estimate the number of stages required. The difference point coordinates are:

$$x_{diff} = \frac{x_oL_o - y_1V_1}{L_o - V_1} = \frac{(1 \times 2375) - (0.494 \times 4089.15)}{2375 - 4089.15} = -0.2071$$

$$N_{diff} = \frac{B}{L_o - V_1} = \frac{7125}{2375 - 4089.15} = -4.1566$$

Given the difference point coordinates ($x_{diff} = -0.2071, N_{diff} = -4.1566$) and feed point coordinates $L_o(x_o = 1, N_o = 3)$, we can estimate the number of stages graphically as shown in the figure below. Note that the point L_o is not shown as it falls outside the figure. Also the coordinates of the point L_N are ($x_N = 0.043, N_N = 0.8599$).

The number of stages required is approximately 6.5.

We can calculate L_N and V_1 as shown below:

$$L_N = B/N_N = 7125/0.8599 = 8285.85\ kg/h$$

$$V_1 = M - L_N = 12375 - 8285.85 = 4089.15\ kg/h$$

Thus, the flow rates and compositions of the streams leaving the process are:

$$V_1 = 4089.15\ kg/h, y_1 = 0.494$$

$$L_N = 8285.85\ kg/h\ , x_N = 0.043$$

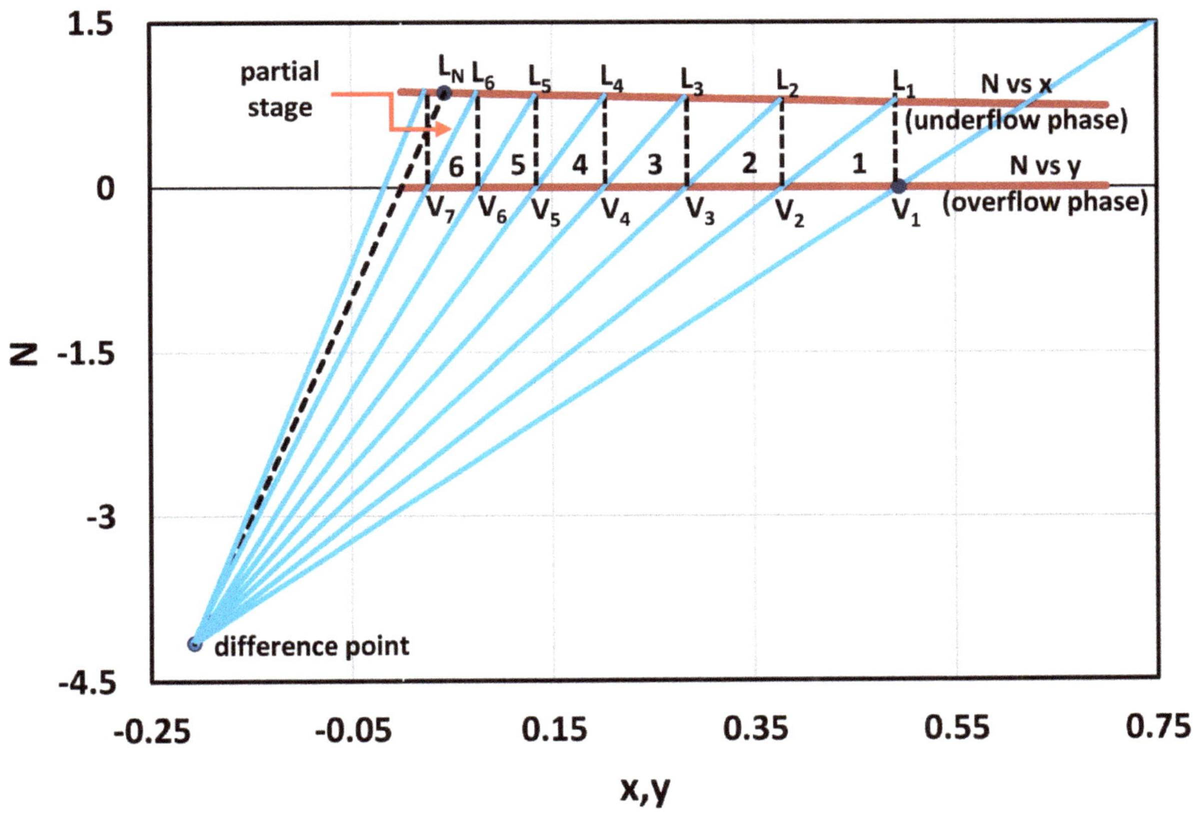

Problem 4

A multistage countercurrent leaching process is to be designed to extract coffee from ground coffee beans using hot water. The feed consists of ground coffee beans containing 39 percent by weight coffee and 61 percent by weight insoluble solids. The feed flow rate is 9,500 kg/h. The recovery of coffee desired is 97%. The insoluble solids retain 0.53 kg of coffee solution per kg of insoluble solids. The leachate leaving the process contains 49 percent by weight coffee. Estimate the number of equilibrium stages required for this process. Also determine the flow rates and compositions of all input and output streams.

Solution

Refer to the diagram given below.

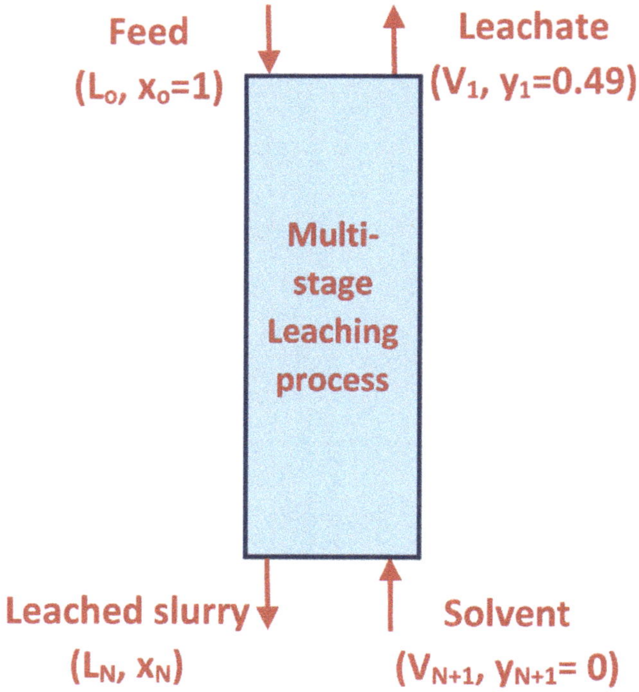

$$L_o = 0.39 \times 9500 = 3705\ kg/h$$

$$B = 0.61 \times 9500 = 5795\ kg/h$$

$$B = N_o L_o;\ N_o = B/L_o = 5795/3705 = 1.5641$$

The equilibrium relation is:

$$N = \text{constant} = \frac{1}{0.53} = 1.8868$$

The concentration of solute in the exit underflow solution, x_N, can be determined from the given percent recovery as:

$$\%\text{Recovery} = \left(\frac{x_o L_o - x_N L_N}{x_o L_o}\right) \times 100 = 97$$

$$x_N L_N = (1 - 0.97) x_o L_o = 0.13 \times 1 \times 3705 = 111.15 \text{ kg/h}$$

As L_N is equal to B/N_N,

$$L_N = \frac{B}{N_N} = \frac{5795}{1.8868} = 3071.34 \text{ kg/h}$$

Thus,

$$x_N = \frac{111.15}{3071.34} = 0.0362$$

From solute mass balance over the cascade:

$$x_oL_o + y_{N+1}V_{N+1} = y_1V_1 + x_NL_N$$

$$V_1 = \left(\frac{x_oL_o - x_NL_N}{y_1}\right) = \frac{3705 - 111.15}{0.49} = 7334.39 \text{ kg/h}$$

Note that $y_{N+1} = 0$. From overall mass balance over the cascade:

$$V_{N+1} = V_1 + L_N - L_o = 7334.39 + 3071.34 - 3705 = 6700.73 \text{ kg/h}$$

Let us now estimate the number of stages required. The difference point coordinates are:

$$x_{diff} = \frac{x_oL_o - y_1V_1}{L_o - V_1} = \frac{(1 \times 3705) - (0.49 \times 7334.39)}{3705 - 7334.39} = -0.0306$$

$$N_{diff} = \frac{B}{L_o - V_1} = \frac{5795}{3705 - 7334.39} = -1.5967$$

Given the difference point coordinates ($x_{diff} = -0.0306, N_{diff} = -1.5967$) and feed point coordinates $L_o(x_o = 1, N_o = 1.5641)$, we can estimate the number of stages graphically as shown in the figure below. Note that the coordinates of the point L_N are ($x_N = 0.0362, N_N = 1.8868$).

The number of stages required is approximately 3.6.

The flow rates and compositions of the input and output streams are summarized in the table below.

Flow stream	Mass flow rate (inert solids-free basis)	Mass fraction of solute (inert solids-free basis)
Input feed stream	$L_o = 3705\ kg/h$	$x_o = 1$
Output underflow (leached slurry) stream	$L_N = 3071.34$ kg/h	$x_N = 0.0362$
Input solvent stream	$V_{N+1} = 6700.73$ kg/h	$y_{N+1} = 0$
Output overflow (leachate) stream	$V_1 = 7334.39$ kg/h	$y_1 = 0.49$

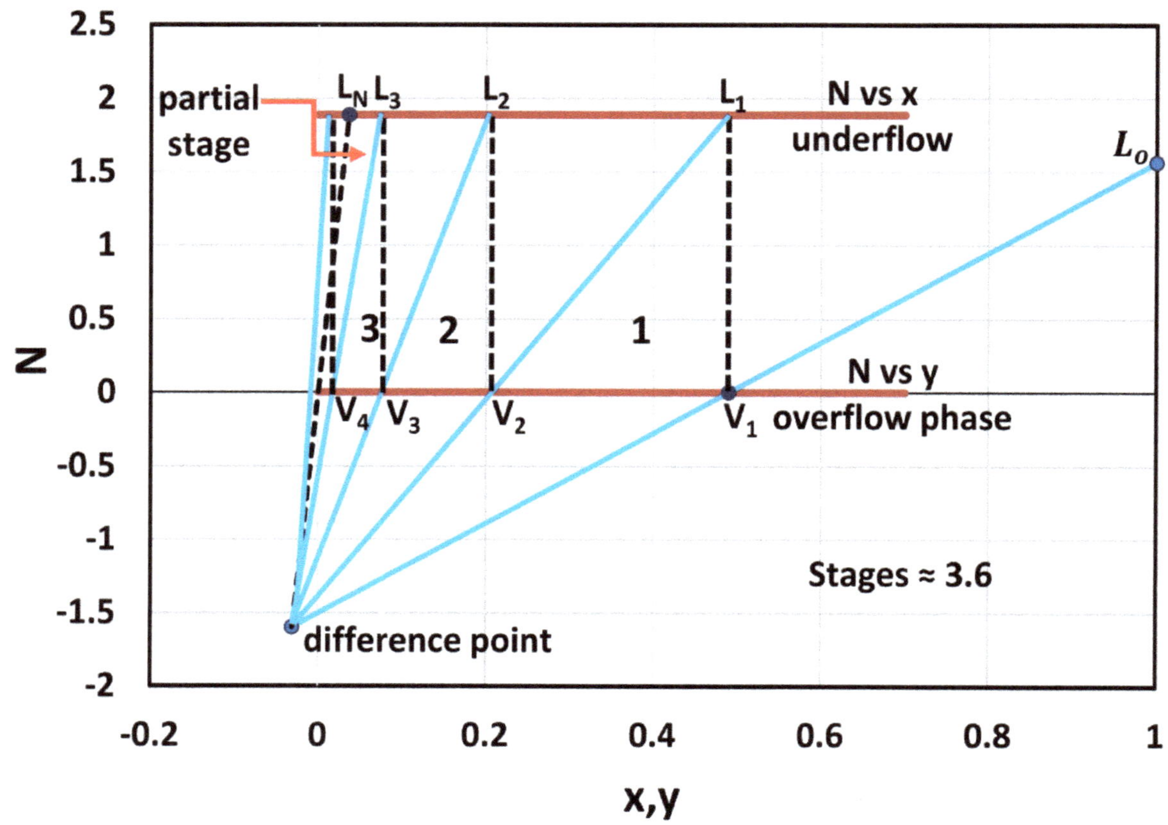

Problem 5

Estimate the number of equilibrium stages for the coffee leaching process described in problem 4 using the McCabe-Thiele graphical method.

Solution

The McCabe-Thiele graphical method is applicable as the underflow and overflow mass flow rates are constant in the cascade excluding the first stage. Note that N is constant ($N = 1.8868$).

Refer to the diagram given below where the first stage is shown separately from the rest of the cascade. Based on the information given in problem 4, the values of various flow rates and compositions are also shown in the diagram.

The operating line passes through the terminal points (x_1, y_2) and (x_b, y_b). From the given information, $x_b = 0.0362$ and $y_b = 0$. As the streams leaving stage 1 are at equilibrium, $x_1 = y_a = 0.49$. To determine y_2, we need to do solute balance over stage 1:

$$x_a L_a + y_2 V_2 = x_1 L_1 + y_a V_a$$

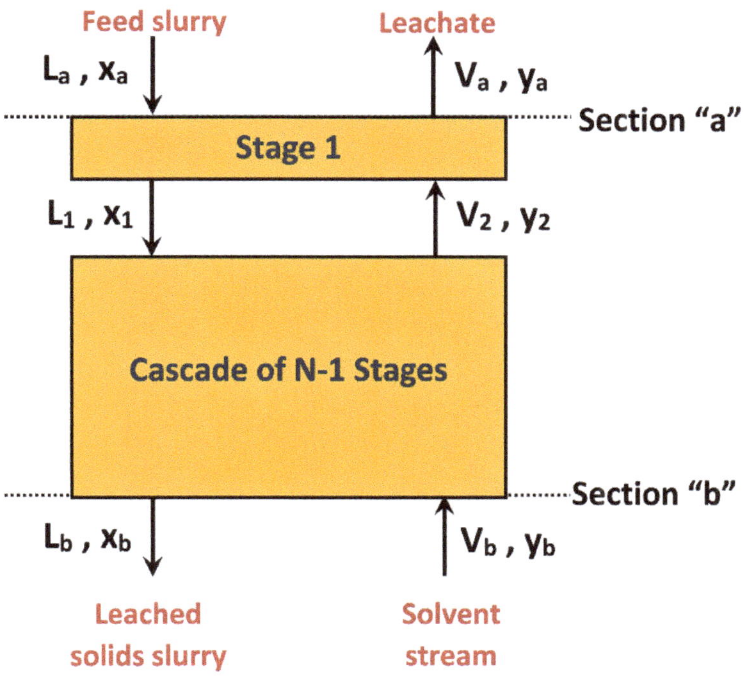

$$L_a = L_o = 3705\ kg/h\ ;\ x_a = x_o = 1$$

$$V_a = V_1 = 7334.39\ kg/h\ ;\ y_a = y_1 = 0.49$$

$$L_b = L_N = 3071.34\ kg/h\ ;\ x_b = x_N = 0.0362$$

$$V_b = V_{N+1} = 6700.73\ kg/h\ ;\ y_b = y_{N+1} = 0$$

Given $x_a = 1$, $L_a = 3705\ kg/h$, $V_2 = V_b = 6700.73\ kg/h$, $x_1 = y_a = 0.49$, $L_1 = L_b = 3071.34\ kg/h$, and $V_a = 7334.39\ kg/h$, we can calculate the unknown y_2. The solution for y_2 is: $y_2 = 0.208$.

Thus the operating line passes through the points (0.49,0.208) and (0.0362,0). The equilibrium line is $y = x$. Note that the operating line does not cover the first stage. The coordinates of the first stage are $(x_a, y_a) = (1, 0.49)$, $(x_1, y_2) = (0.49, 0.208)$, and $(x_1, y_a) = (0.49, 0.49)$. The stepping off stages is shown in the figure given below.

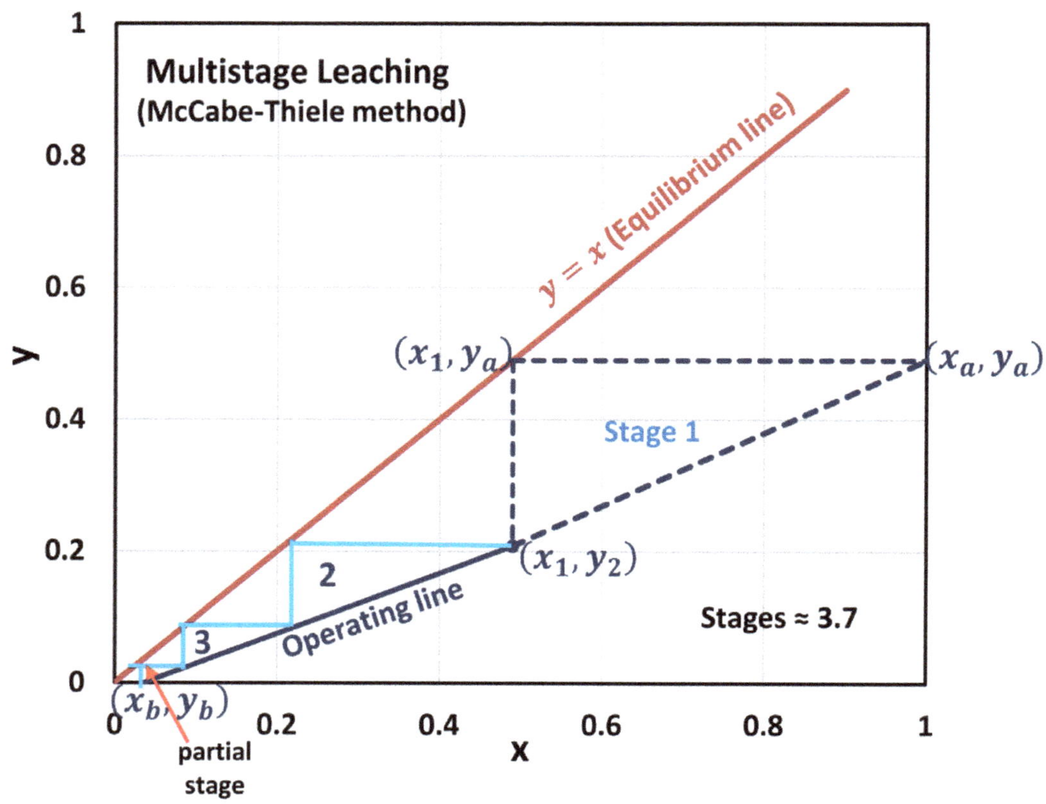

The McCabe-Thiele method gives almost the same value of the number of equilibrium stages required as that obtained in problem 4 using the N-x-y diagram.

Problem 6

Estimate the number of equilibrium stages for the coffee leaching process described in problem 4 using the Kremser equation.

Solution

As discussed in section 15.5.3, the Kremser equation can be applied when the underflow and overflow flow rates are constant. As the first stage is not part of the analysis, the Kremser equation gives $N-1$ stages when applied between the terminal points of the operating line: (x_b, y_b) and (x_1, y_2). The Kremser equation is given as:

$$N-1=\frac{\ln\left(\frac{x_1-x_1^*}{x_b-x_b^*}\right)}{\ln\left(\frac{x_1-x_b}{x_1^*-x_b^*}\right)}$$

Given that:

$x_1 = y_1 = 0.49$, $x_1^* = y_2 = 0.208$, $x_b = 0.0362$, and, $x_b^* = y_b = 0$

Thus,

$$N-1=\frac{\ln\left(\frac{0.49-0.208}{0.0362-0}\right)}{\ln\left(\frac{0.49-0.0362}{0.208-0}\right)}=2.63$$

$$N = 3.63$$

This value of number of stages required is almost the same as that obtained by N-x-y and McCabe-Thiele graphical methods.

SHORT REVIEW QUESTIONS

1. Describe briefly a solid-liquid leaching process with practical applications
2. Show schematically the direction of mass transfer of solute in a solid-liquid leaching process
3. How does N (ratio of mass of inert solids to mass of solution retained in the slurry) generally vary with the increase in solute concentration of the solution?
4. N is typically zero for the overflow phase True/False
5. The solution of solute and solvent generally tends to become more viscous and sticky with the increase in the solute concentration True/False
6. The tie-lines connecting the underflow and overflow phases at equilibrium are generally vertical lines on a N versus composition phase diagram. True/False
7. Draw a phase diagram (N versus composition) for solid-liquid leaching process.
8. The exit leachate and leached slurry are assumed to be at equilibrium with each other in a single-stage leaching process. True/False
9. The exit leachate and leached slurry are assumed to be at equilibrium with each other in a multi-stage leaching process. True/False
10. Define the difference point coordinates x_{diff} and N_{diff} in a multi-stage solid-liquid leaching process.
11. What are the assumptions made in the McCabe-Thiele analysis of multi-stage solid-liquid leaching process?
12. In order to apply the Kremser equation to calculate the number of equilibrium stages required in a multi-stage leaching process, the equilibrium relation and the operating line must be linear. True/False

PRACTICE PROBLEMS

Problem 1

A single stage solid-liquid leaching process is being used to extract oil from flaked oil beans with a suitable solvent. 750 kg of oil beans is leached with 650 kg of pure solvent. The oil content of the oil beans is 21 wt%. At equilibrium, N is approximately 2.1 kg inert solid/kg solution retained for the underflow slurry and $N = 0$ for the overflow stream. The tie lines are vertical lines expressed as $y = x$.

Calculate the amounts and compositions of the exit leachate and leached solids slurry.

Problem 2

We wish to leach oil from meal by a countercurrent multistage leaching process using a pure solvent. The feed flow rate is 1000 kg/h of meal (inert solids + oil). The oil content of the meal is 19 wt%. The pure solvent flow rate entering the process is 700 kg/h. The desired mass fraction of oil in the underflow product solution is 0.075. Estimate the number of stages required. What are the flow rates and compositions of the streams leaving the process?

The equilibrium data are as follows:

x	0	0.1	0.2	0.3	0.4	0.5	0.6	0.7
N	2.10	2.08	2.04	1.99	1.92	1.85	1.78	1.71

Note that $N = 0$ for the overflow stream at equilibrium and the tie lines are vertical lines expressed as $y = x$.

Problem 3

A multistage solid-liquid leaching process is being used to extract oil from flaked oil beans using hexane as a solvent. The feed flow rate is 1600 kg/h of meal (inert solids + oil). The oil content of the meal is 28 wt%. The pure solvent hexane enters the process at a mass flow rate of 800 kg/h. It is desired to recover 90% of the oil present in the feed stream. The equilibrium relationship is as follows:

$$N = \frac{1}{0.50 + 0.06x + 0.15x^2}$$

Estimate the number of equilibrium stages required. Determine the flow rates and compositions of the flow streams leaving the process.

Problem 4

A leaching battery consisting of a number of stages is to be designed for the extraction of sugar from sugar cane using pure hot water as the solvent. The feed containing 11wt% fiber (inert solids), 15wt% sugar, and 74wt% water, enters the process at a mass flow rate of 27,000 kg/h. How many stages are required if the desired recovery is 98% of the sugar present in the sugar cane. The sugar juice (leachate) produced in the process contains 11wt% sugar. At equilibrium, N is approximately 0.67 kg inert solids/kg solution retained for the underflow slurry and $N = 0$ for the overflow stream. Solve the problem using all three methods: N-x-y diagram, McCabe-Thiele diagram, and Kremser equation.

Problem 5

75 kg/h of waxed paper containing 23 wt% soluble wax and 77 wt% insoluble pulp (inert solids) is to be dewaxed in a multistage countercurrent leaching battery using kerosene as a solvent. The solvent stream entering the process contains 0.04 wt% wax and the remaining 99.96 wt% kerosene. The desired recovery is 99% of the wax present in the waxed paper feed. At equilibrium, N is approximately 0.42 kg inert solids/kg solution retained for the underflow slurry and $N = 0$ for the overflow stream. How many stages are required in this multistage leaching process? Solve the problem using all three methods: N-x-y diagram, McCabe-Thiele diagram, and Kremser equation.

Problem 6

Roasted copper ore consisting of 19 wt% $CuSO_4$ and 81 wt% inert solids is fed to a countercurrent leaching battery at a mass flow rate of 7,900 kg/h. Pure water is used as a solvent. The desired recovery of $CuSO_4$ is 97% of that present in the ore. At equilibrium, N is approximately 0.45 kg inert solids/kg solution retained for the underflow slurry and $N = 0$ for the overflow stream. If the leachate leaving the process contains 13 wt% $CuSO_4$, how many stages are required in this multistage leaching process? Solve the problem using all three methods: N-x-y diagram, McCabe-Thiele diagram, and Kremser equation.

Chapter 16

Analysis and Design of Continuous Contacting Packed Columns

16.1 LEARNING OUTCOMES

Upon completion of this chapter, students will be able to:

- Explain the difference between staged and packed columns used in mass transfer operations
- Explain the difference between random and structured packings
- Identify desired characteristics of packings
- Describe different types of packing materials used in packed columns
- Explain the importance of uniform liquid distribution
- Describe different types of liquid distributors
- Estimate the height of packed column using HETP method
- Estimate the height of packed column using HTU-NTU method
- Derive the relationship between HETP and HTU
- Describe flooding in packed columns
- Explain flooding curve
- Estimate column diameter using flooding curve or equation

16.2 INTRODUCTION

In this chapter, we will consider the analysis and design of *continuous* contacting devices, namely the packed columns. In packed columns, the heavier phase (liquid) flows downward in the column over the surface of some packing material and the lighter phase (gas or vapor) flows counter-currently in the upward direction. As the contact between the two phases is continuous, that is, maintained throughout the column, the packed columns are also called *continuous* contactors. The staged-columns studied in the previous chapters (for example in multi-stage absorption, stripping, and distillation) were *discontinuous* contactors as the contact between the two phases was not maintained throughout the column. In staged-columns, physically distinguishable discrete stages are present whereas in packed columns, no such physically distinguishable stages are present. Packed columns are also called rate-based mass exchange devices as mass transfer of solute from one phase to another occurs throughout the column. No phase equilibrium is established between the two phases in the column whereas in staged-columns, the two phases leaving any stage are at equilibrium or close to equilibrium with each other.

Packed towers are used for absorption, stripping, distillation, and liquid-liquid extraction. They are particularly attractive in applications where small diameter (< 2 ft) columns are needed. In small diameter columns, plates/trays are difficult to install and therefore, small diameter staged columns tend to be more expensive.

16.3 PACKINGS AND PACKING MATERIALS

The purpose of the packing in packed columns is to provide a large surface area for contact between the two phases. The liquid flows downward the column over the surface of the packing material in the form of thin films and the vapor flows upward in the free/void space available inside the packing. The packing should preferably have an open structure so as to provide low resistance to gas flow. The packing should also promote uniform distribution of liquid over the packing surface. Different materials are used for the generation of packings for columns. The popular materials are: a) ceramics – ceramics provide good wettability and are corrosion resistant. However, they are brittle in nature; b) metals – metals have good strength and good wettability; and c) plastics – plastics have good strength and are relatively inexpensive. However, they may have wettability issues. Packings come in many different styles. They can be broadly classified into two categories: *random* packing and *structured* packing.

16.3.1 Random Packings

In random packing, the packing consists of discrete particles. The particles are simply dumped into the column and the arrangement of particles in the column is random. Particles used for random packings come in different shapes and sizes. Figure 16.1 shows some of the popular shapes of particles. They are: Spheres, Raschig rings, Pall rings, and Saddles. Raschig rings are pieces of cylindrical tubes with diameter nearly the same as the length. Pall ring is a modification of Raschig ring. It has a more open structure to help maintain an even distribution of liquid and minimize plugging and fouling. Also it offers lower resistance to gas flow. Saddle-shaped particles are also popular for packing of columns. In some applications, spherical particles are used. Figure 16.2 shows random packings of Raschig rings and spherical particles.

Figure 16.1 Different shapes of particles used in random packings.

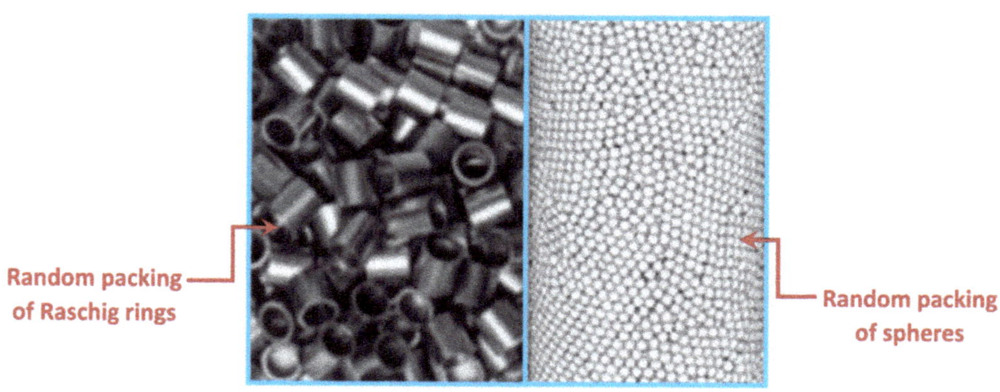

Figure 16.2 Random packings of Raschig rings and spheres.

16.3.2 Structured Packings

Structured packings are packings with regular geometry. They are usually made out of corrugated sheets of perforated metal, as shown in Figure 16.3. The layers of corrugated metal sheets are stacked together in a crisscrossing manner to create flow channels for the vapor or gas phase. The liquid spreads uniformly over the packing surface and flows downward as a continuous thin film held by surface tension.

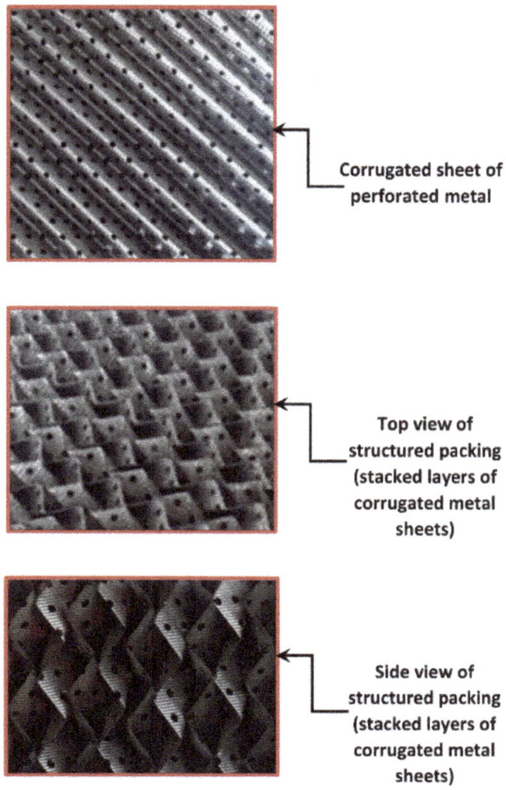

Figure 16.3 Intalox structured packing of corrugated sheets of porous metal.

16.3.3 Liquid Distribution

Even distribution of liquid over the entire cross-sectional area of the packed column is important. If the liquid is not distributed evenly over the entire cross-section of the packed column, the separation efficiency of the column is reduced significantly. Uneven distribution or channelling of liquid must be avoided. Ideally the entire packing surface area should be wetted by the liquid and the liquid should flow downwards in the form of thin films over the packing surface.

There are a number of different types of liquid distributors available commercially (see Figure 16.4). Some of the popular ones are: a) pan distributor, b) pipe-arm distributor, and c) spray nozzle distributor. Pan distributor consists of a pan with cylindrical gas risers and orifices at the bottom of the pan. Liquid is distributed through the orifices and gas passage is provided by gas risers. Pipe-arm distributor consists of a row of perforated pipes. The liquid is distributed evenly over the packed bed through perforated holes under the pipes. Spray nozzle distributor consist of a row of pipes with spray nozzles. There are numerous small holes on the nozzles to distribute the liquid.

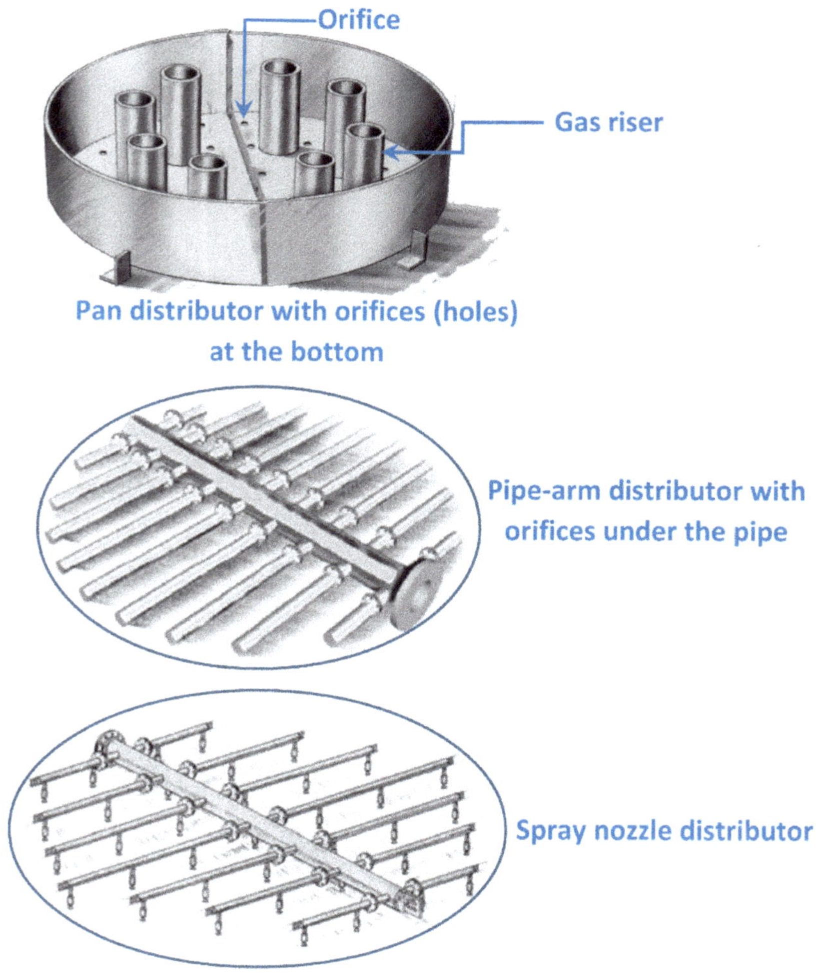

Figure 16.4 Examples of liquid distributors.

16.4 HEIGHT OF PACKED COLUMNS

16.4.1 HETP Method

Although packed column is a continuous contactor, it can still be treated like a staged-column using the concept of *HETP* (height equivalent to a theoretical plate). In the *HETP* method, it is imagined that the total height of the packed column can be divided into a number of layers or segments of packing of equal height, referred to as *HETP*, such that the gas and liquid phases leaving this imaginary segment are at equilibrium with each other. Thus *HETP* is the height of packing needed to obtain the same change in vapor (gas) composition as that obtained in one equilibrium stage. The concept of *HETP* is illustrated schematically in Figure 16.5. If at any point in the column, x is the solute concentration in the liquid phase and y is the solute concentration in the gas phase at the same point, then *HETP* is the height of the packing above this point where the gas concentration is y^* corresponding to equilibrium with the liquid concentration x .

The column height h can be expressed as:

$$h = (HETP) \times N_{theoretical} \tag{16.1}$$

where $N_{theoretical}$ is the number of equilibrium or theoretical stages estimated from the McCabe-Thiele graphical method.

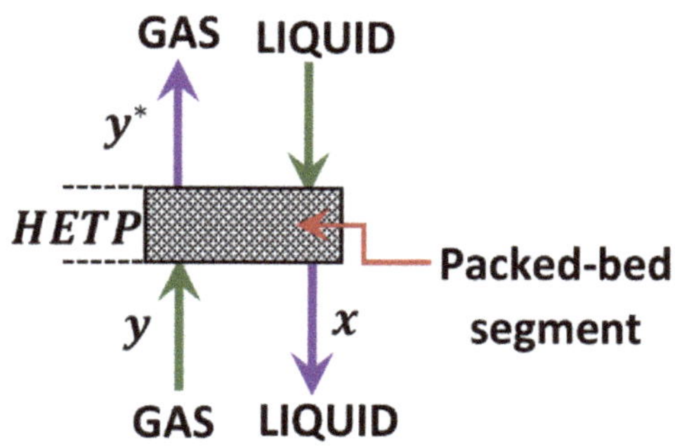

Figure 16.5 *HETP* of packing.

HETP can be readily determined experimentally for a given type of packing. *HETP* is related to the mass transfer area provided by the packing. Higher the surface contact area of the packing per unit volume of the column, more efficient is the packing with low value of *HETP*. Note that a low value of *HETP* implies a shorter column.

For randomly-packed columns, the surface area of the particles per unit volume of the column is inversely related to the equivalent diameter d_p of the particles. Consequently,

$$HETP \propto d_p \tag{16.2}$$

One rule of thumb for the estimation of *HETP* is as follows:

$$HETP(m) = 0.018\, d_p(mm) \tag{16.3}$$

For small diameter columns ($D < 2$ ft), $HETP \approx D$ regardless of the particle size. The *HETP* value is generally assumed to be 1 ft when column diameter is less than 1 ft.

For structured packings, the following rule of thumb may be used to estimate the *HETP:*

$$HETP(m) = \frac{100}{a} + 0.10 \tag{16.4}$$

where a is the surface area of the packing per unit volume.

Although the key parameter affecting the *HETP* is the surface area per unit volume a, *HETP* also depends on the physical properties to some extent. For example, Strigle (R.F. Strigle, Packed Tower Design and Applications, 2nd Edition, Gulf Pub. Co, 1994) proposed the following *HETP* correlation for Intalox corrugated sheet type structured packings:

$$ln(HETP) = n - 1.01 - 0.187 ln(\sigma) + 0.213 ln(\mu_L) \tag{16.5}$$

where *HETP* is in m, σ is the surface tension in N/m, and μ_L is the liquid viscosity in Pa.s. The parameter n is dependent on the type of Intalox structured packing (see Table 16.1).

Lockett (M.J. Lockett, Chemical Engineering Progress, Vol. 94, 1, 1998) proposed the following correlation for the estimation of *HETP* for high density corrugated sheet type structured packings:

$$HETP\ (m) = \frac{1.54g^{0.5}(\rho_L - \rho_V)^{0.5}\mu_r^{-0.06}}{a[1+0.78(\rho_V/\rho_L)^{0.25}exp(0.00058a)]^2} \quad (16.6)$$

where g is the acceleration due to gravity (9.8 m/s^2), $\mu_r = \mu_L(mPa.s)/0.978$, a is the specific area of packing in units of m^2/m^3, ρ is the density in kg/m^3.

Table 16.1 Values of n (Equation (16.5)) for different Intalox corrugated sheet type structured packings

Packing-type	Surface area per unit volume, m^2/m^3	n
1T, Intalox corrugated sheet structured packing	310	0.7683
2T, Intalox corrugated sheet structured packing	215	1.0128
3T, Intalox corrugated sheet structured packing	170	1.3868

16.4.2 HTU-NTU Method

In the HTU-NTU method of analysis and design of packed towers, it is imagined that the total height of the packed column can be divided into a number of layers of packing (referred to as number of transfer units, NTU) of equal height (referred to as height of transfer unit, HTU) such that the change in gas concentration over the HTU is equal to the average driving force for mass transfer over the whole column. The concept of *HTU* for gas absorption is illustrated schematically in Figure 16.6. If at any point in the column, x is the solute concentration in the liquid phase and y is the solute concentration in the gas phase at the same point, then *HTU* is the height of the packing above this point where the gas concentration is $y - (y - y^*)_{avg}$ where $(y - y^*)_{avg}$ is the average driving force for mass transfer of solute in the column from gas phase to liquid phase.

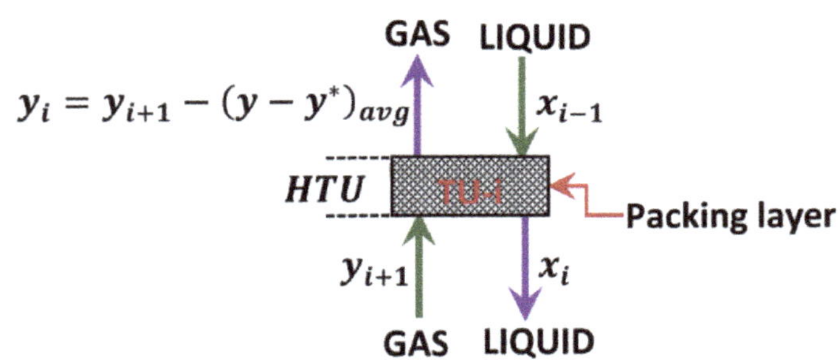

Figure 16.6 *HTU* of packing for gas absorption.

Figure 16.7 shows a schematic diagram of the packed column divided into N transfer units (NTU=N) of equal height, that is, HTU. Thus the column height h can be expressed as:

$$h = (HTU) \times (NTU) \quad (16.7)$$

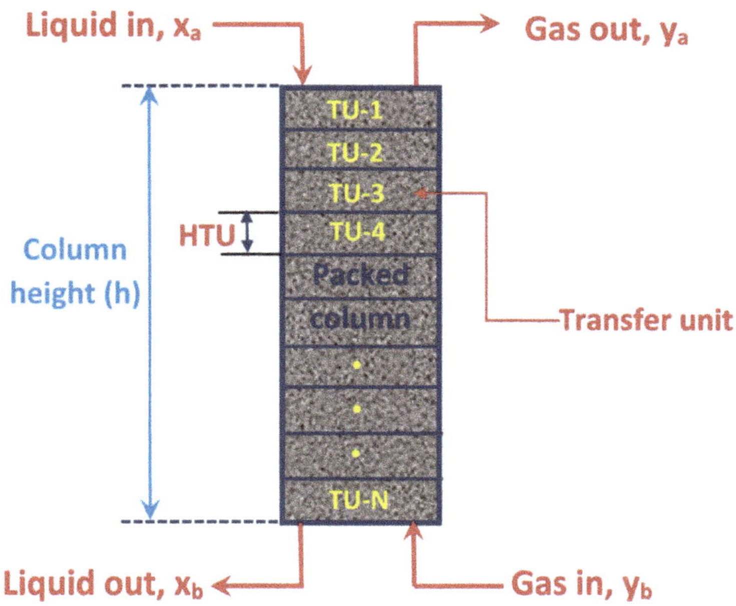

Figure 16.7 Packed column with imaginary transfer units.

The number of transfer units (NTU) depends on the required separation whereas the height of the transfer unit (HTU) depends on the efficiency of mass transfer. More efficient the mass transfer, smaller is the value of HTU.

In the discussion to follow, a detailed analysis is presented of counter-current *gas absorption* in a packed column where some solute, species A, present in the gas phase is being transferred to the liquid phase. The analysis of other mass transfer operations (stripping, binary distillation, liquid-liquid extraction, etc.) in packed columns can be treated in a similar manner.

Interphase Mass Transfer

In a packed column, interphase mass transfer of solute occurs from one phase to another by diffusion. Figure 16.8 shows the concentration profiles, in units of mole fraction, on both sides of the gas-liquid interface in gas absorption operation. The molar flux of solute N_A (moles/s.m^2) across the gas film on the left side of the gas-liquid interface can be expressed as:

$$N_A = k_y(y - y_i) \tag{16.8}$$

where k_y is the gas-phase mass transfer coefficient (moles/s.m^2). The mass transfer coefficient k_y is a measure of the ability of the system to transfer solute across the gas film. For the same driving force $(y - y_i)$, a higher k_y implies a higher rate of mass transfer of solute across the gas film, and hence, more efficient mass transfer.

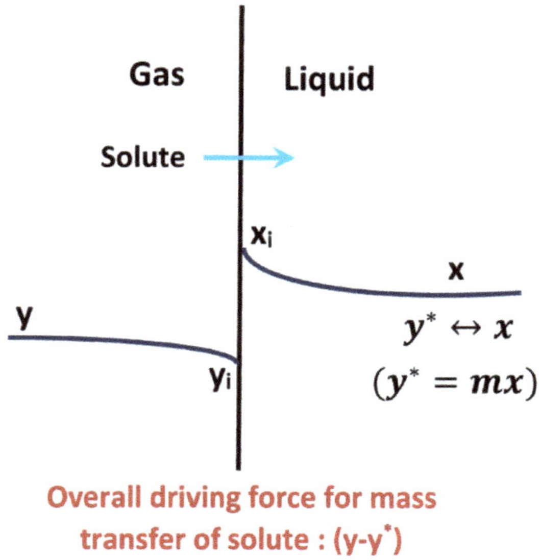

Figure 16.8 Interphase mass transfer of solute in gas absorption.

The molar flux of solute N_A (moles/s.m^2) across the liquid film on the right side of the gas-liquid interface can be expressed as:

$$N_A = k_x(x_i - x) \tag{16.9}$$

where k_x is the liquid-phase mass transfer coefficient (moles/s.m^2). The mass transfer coefficient k_x is a measure of the ability of the system to transfer solute across the liquid film. For the same driving force $(x_i - x)$, a higher k_x implies a higher rate of mass transfer of solute across the liquid film.

At steady state, there cannot be accumulation of solute at the interface and therefore,

$$N_A = k_y(y - y_i) = k_x(x_i - x) \tag{16.10}$$

The driving forces for mass transfer of solute included in the molar flux expressions, Equations (16.8) and (16.9), are the individual phase driving forces, $(y - y_i)$ for gas phase and $(x_i - x)$ for liquid phase. These driving forces contain the interface compositions (y_i, x_i) which are usually not known. In order to avoid the interface compositions, it is equally appropriate to express the molar flux of solute as:

$$N_A = K_y(y - y^*) \tag{16.11}$$

where K_y is the overall mass transfer coefficient based on gas phase, $(y - y^*)$ is the overall driving force for mass transfer of solute from gas phase to liquid phase, and y^* is the mole fraction of solute in the gas phase in equilibrium with the existing bulk liquid phase. Note that y^* is a fictitious quantity to be calculated from the equilibrium relation (Henry's law):

$$y^* = mx \tag{16.12}$$

where m is the equilibrium constant.

From Equations (16.10)-(16.12), it can be readily shown that:

$$\frac{1}{K_y} = \frac{1}{k_y} + \frac{m}{k_x} \tag{16.13}$$

If the solute is highly soluble in the liquid phase, m value is small, and the resistance to mass transfer in the liquid phase is small. Thus, the resistance to mass transfer of solute lies mainly in the gas phase. Consequently,

$$\frac{1}{K_y} \approx \frac{1}{k_y} \tag{16.14}$$

If the solute is only sparingly soluble in the liquid phase, m value is large, and the resistance to mass transfer of solute lies mainly in the liquid phase. Consequently,

$$\frac{1}{K_y} \approx \frac{m}{k_x} \tag{16.15}$$

Material Balance

Consider material balance of solute A over a thin section dz of the packed gas absorption column shown in Figure 16.9. Let the contact area of the packing per unit volume of packing be a. The interphase mass transfer rate of solute from gas phase to liquid phase over a section dz of the column is:

$$Rate\ of\ mass\ transfer = K_y a(y - y^*)Sdz \tag{16.16}$$

where S is the cross-sectional area of the column. The mass transfer rate of solute from gas phase to liquid phase over a section dz of the column is also given as:

$$Rate\ of\ mass\ transfer = Vy|_z - Vy|_{z+dz} \tag{16.17}$$

where V is the molar flow rate of gas. From Equations (16.16) and (16.17):

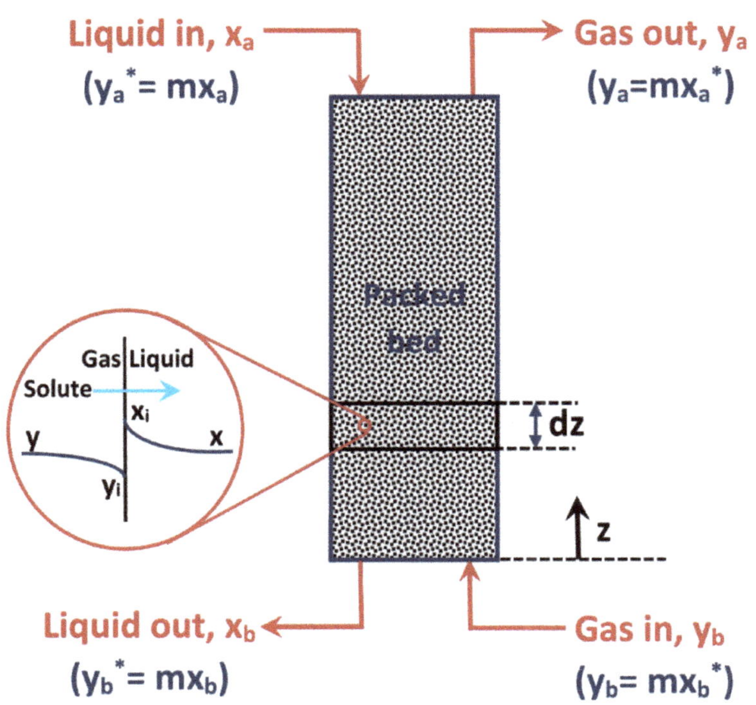

Figure 16.9 Packed gas absorption column.

$$Vy|_z - Vy|_{z+dz} = K_y a(y - y^*)Sdz \tag{16.18}$$

Thus,

$$-\frac{d(Vy)}{dz} = K_y a(y - y^*)S \tag{16.19}$$

Assuming the gas flow rate V to be constant, Equation (16.19) reduces to:

$$-V\frac{dy}{dz} = K_y a(y - y^*)S \tag{16.20}$$

Height of a Packed Column

Upon rearrangement and integration of the material balance equation, Equation (16.20), we obtain:

$$h = -\left(\frac{V/S}{K_y a}\right)\int_{y_b}^{y_a}\frac{dy}{y-y^*} \tag{16.21}$$

Upon inverting the limits of the integral:

$$h = \left(\frac{V/S}{K_y a}\right)\int_{y_a}^{y_b}\frac{dy}{y-y^*} \tag{16.22}$$

This equation could be recast as:

$$h = (HTU) \times (NTU) \tag{16.23}$$

where

$$HTU = \left(\frac{V/S}{K_y a}\right) \tag{16.24}$$

$$NTU = \int_{y_a}^{y_b}\frac{dy}{y-y^*} \tag{16.25}$$

The number of transfer units (*NTU*), defined in Equation (16.25), depends on the degree of desired separation $(y_b - y_a)$ and the average driving force for mass transfer $(y - y^*)_{avg}$. It is simply the ratio of $(y_b - y_a)$ to $(y - y^*)_{avg}$. The height of the transfer unit (HTU), given by Equation (16.24), is a measure of the efficiency of mass transfer. More efficient the mass transfer, smaller is the value of HTU. A larger mass transfer coefficient K_y implies a faster rate of interphase mass transfer of solute and hence a lower HTU. Likewise, a larger contact area of the packing (per unit volume of the packing) implies a faster rate of interphase mass transfer of solute and hence a lower HTU.

Given the mass transfer coefficient K_y , the packing contact area a per unit volume of the packing, the gas molar flow rate V, and the column diameter D $(S = \pi D^2/4)$, Equation (16.24) can be used to calculate the HTU. The estimation of *NTU* is described in the following section.

Estimation of NTU

The number of transfer units is given by the integral in Equation (16.25). In order to evaluate this integral, we need to develop an equation for the operating line of the packed column based on material balance. The solute material balance over the dashed envelope shown in Figure 16.10 gives:

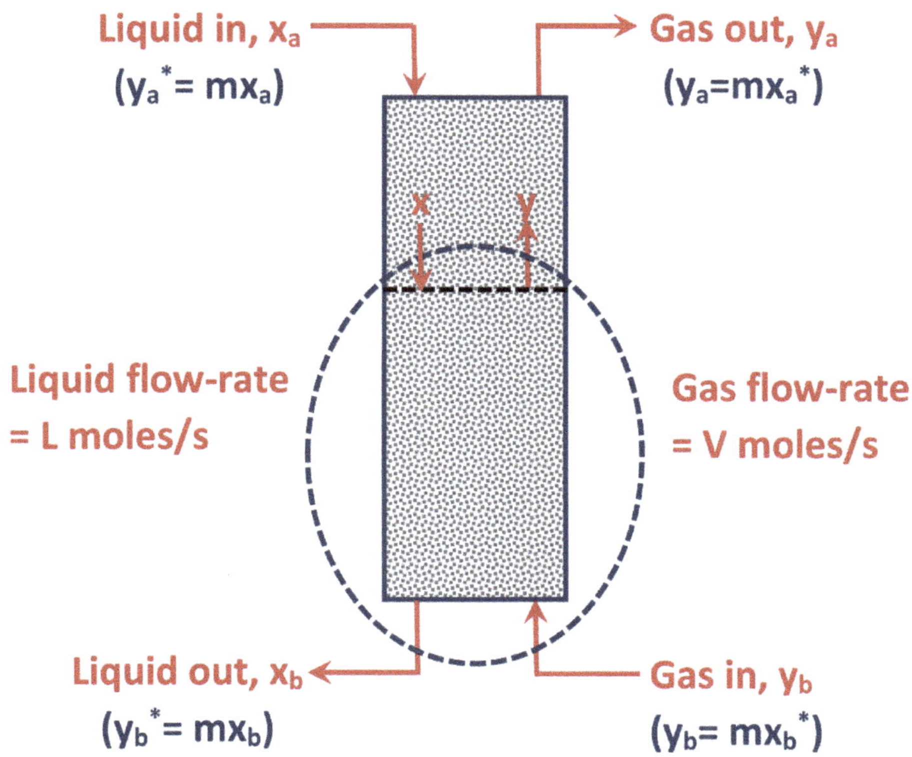

Figure 16.10 Material balance on a packed gas absorption column.

$$Lx + Vy_b = Lx_b + Vy \tag{16.26}$$

The liquid and gas flow rates are assumed to be constant. Upon rearrangement of Equation (16.26), we get:

$$y = y_b + (L/V)(x - x_b) \tag{16.27}$$

This is the equation for the *operating line* of the packed column. It relates the compositions of passing liquid and gas streams at any section of the column. The operating line has a slope of L/V and passes through the terminal compositions (x_a, y_a) and (x_b, y_b). Figure 16.11 shows the plots of the operating line, Equation (16.27), and the equilibrium relationship (assumed to be linear). The driving force for mass transfer of solute $(y - y^*)$ generally varies with y.

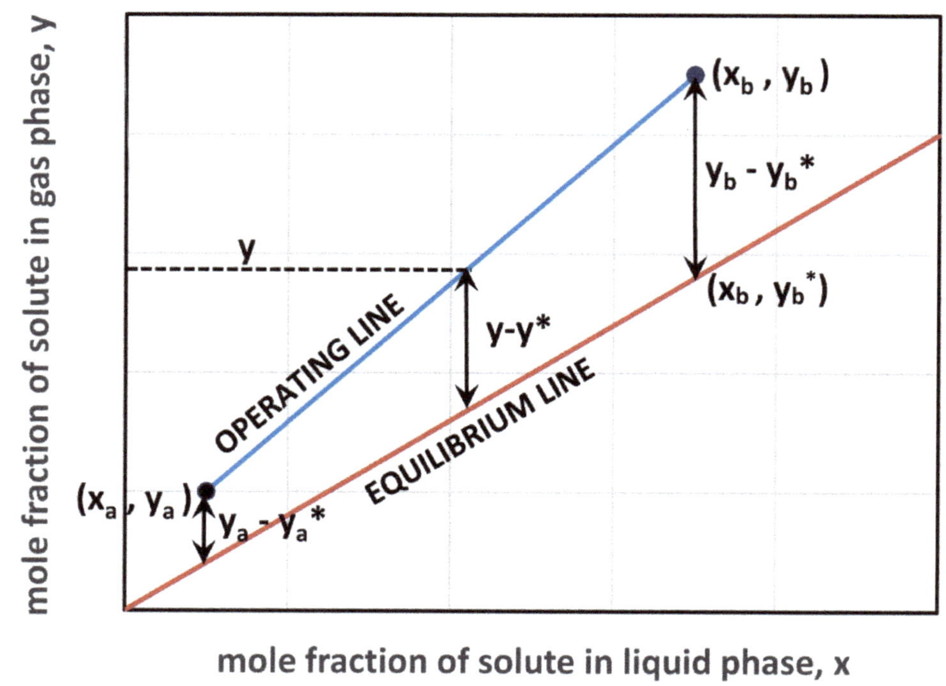

Figure 16.11 Plots of operating and equilibrium lines for counter-current absorption process.

For linear operating and equilibrium lines with different slopes, it can be readily shown that:

$$NTU = \int_{y_a}^{y_b} \frac{dy}{y-y^*} = \frac{y_b-y_a}{(y-y^*)_{log-mean}} \tag{16.28}$$

where

$$(y-y^*)_{log-mean} = \frac{(y_b-y_b^*)-(y_a-y_a^*)}{ln\left(\frac{y_b-y_b^*}{y_a-y_a^*}\right)} \tag{16.29}$$

In the special case when the slopes of the operating and equilibrium lines are the same, that is, the two lines are parallel and the driving force for mass transfer $(y-y^*)$ is constant,

$$NTU = \int_{y_a}^{y_b} \frac{dy}{y-y^*} = \frac{y_b-y_a}{(y_a-y_a^*)} = \frac{y_b-y_a}{(y_b-y_b^*)} \tag{16.30}$$

16.4.3 Relationship Between HETP and HTU

HETP and *HTU* are related to each other. Also *NTU* is related to the number of equilibrium or theoretical stages $N_{theoretical}$. These relationships are derived here for counter-current gas absorption in packed columns.

The absorption factor, *A*, is defined as:

$$A = \frac{L/V}{m} \tag{16.31}$$

where m is the Henry law constant (assuming equilibrium relationship is linear, that is, Henry law is applicable).

In the special case of $A = 1$, the number of theoretical or equilibrium stages given by the Kremser equation (see Chapter 4) is the same as *NTU* given by Equation (16.30). Thus,

$$N_{theoretical} = NTU = \frac{y_b - y_a}{(y_b - y_b^*)} = \frac{y_b - y_a}{(y_a - y_a^*)} \tag{16.32}$$

The total height h of a packed column can be expressed in two equivalent ways as:

$$h = HETP \times N_{theoretical} = HTU \times NTU \tag{16.33}$$

When $A = 1$, $N_{theoretical} = NTU$. Therefore,

$$HETP = HTU \tag{16.34}$$

In the general case when $A \neq 1$, the Kremser equation (see Chapter 4) gives the following number of theoretical (equilibrium) stages:

$$N_{theoretical} = \frac{ln\left(\frac{y_b - y_b^*}{y_a - y_a^*}\right)}{lnA} \tag{16.35}$$

Starting from Equation (16.28), it can be shown, after some algebra, that:

$$NTU = \frac{y_b - y_a}{(y - y^*)_{log-mean}} = \left(\frac{A}{A-1}\right) ln\left(\frac{y_b - y_b^*}{y_a - y_a^*}\right) \tag{16.36}$$

From Equations (16.35) and (16.36), it follows that:

$$NTU = \left(\frac{A}{A-1}\right)(lnA)N_{theoretical} \tag{16.37}$$

Upon substitution of *NTU* from Equation (16.37) into Equation (16.33), we get:

$$HETP = \left(\frac{A}{A-1}\right)(lnA)(HTU) \tag{16.38}$$

16.5 DIAMETER OF PACKED COLUMNS

The estimation of the column diameter is based on consideration of *flooding phenomenon* in packed columns. To discuss flooding, it is necessary to define superficial gas and liquid mass velocities. They are defined as follows:

$$G_V = \frac{\dot{m}_V}{A_{column}} = \frac{4\dot{m}_V}{\pi D^2} = \left(\frac{4M_V}{\pi D^2}\right)V \tag{16.39}$$

$$G_L = \frac{\dot{m}_L}{A_{column}} = \frac{4\dot{m}_L}{\pi D^2} = \left(\frac{4M_L}{\pi D^2}\right)L \tag{16.40}$$

where G is the superficial mass velocity, $\dot{m}$ is the mass flow rate, A is the cross-sectional area of the column, D is the column diameter, M is the molecular weight, V is the molar flow rate of gas, and L is the molar flow rate of liquid. For a given column and fluids, the mass velocity is directly proportional to the molar flow rate of the phase (gas or liquid). With the decrease in column diameter, the mass velocity increases keeping other factors constant (same fluid, same molar flow rate). In the discussion to follow, the mass velocity is often referred to as simply velocity.

16.5.1 Flooding in Packed Columns

Consider a counter-current gas absorption column of known diameter, packed with a packing of known type and size. The liquid flows in the downward direction in the column at some fixed molar flow rate L (fixed liquid velocity G_L to be more precise). The gas flows in the upward direction. What happens if we increase the gas molar flow rate (gas velocity G_V to be more precise)?

At low gas velocities, the downward flow of liquid through the packing is not affected by the upward flow of gas. The quantity of liquid retained in the packing, that is, liquid holdup of packing is constant unaffected by the changing gas flow rate or velocity. With further increase in gas velocity, the interphase friction between gas and liquid streams begins to hinder the downward flow of liquid. Consequently, more liquid is held up in the packing and liquid holdup of the packing increases. The gas velocity where liquid holdup of the packing begins to rise is called *loading point*. As the gas velocity is increased further, the accumulation of liquid within the packing increases. At some gas velocity, referred to as *flooding velocity*, the liquid is no longer able to flow down the packing and is blown out of the column with the rising gas. Obviously, one must operate the column at gas velocities below the flooding velocity.

16.5.2 Flooding Curve and its Interpretation

The flooding velocity is a function of liquid velocity, fluid properties, and packing characteristics. The flooding curve for random packings is shown in Figure 16.12. The flooding curve is a plot of $(G_V^2\psi)$ versus F_{LV}. The parameters ψ and F_{LV} are defined as follows:

$$\psi = \frac{F_p \phi \mu_L^{0.2}}{\rho_V \rho_L g_c} \tag{16.41}$$

$$F_{LV} = \frac{\dot{m}_L}{\dot{m}_V}\sqrt{\frac{\rho_V}{\rho_L}} = \frac{G_L}{G_V}\sqrt{\frac{\rho_V}{\rho_L}} \tag{16.42}$$

where F_p is the packing factor that has units of area/volume, that is, ft^{-1}, $\phi = \rho_{water}/\rho_L$, g_c= 32.2 lb$_m$(ft/s^2)/lb$_f$, μ_L is liquid viscosity in cP, ρ_V and ρ_L are densities in units of lb$_m$/ft^3, $\dot{m}$ is the mass flow rate in lb$_m$/s, and G is the mass velocity in units of lb$_m$/(s.ft^2). The packing factor F_p depends on the type and size of packing. Smaller packing sizes of the same packing family have larger values of packing factor and specific contact area. The values of the packing factors for various packings are given in Table 16.2. Also note that the ordinate of Figure 16.12, that is, $(G_V^2\psi)$, is not dimensionless and we must use the units of various quantities as already indicated here.

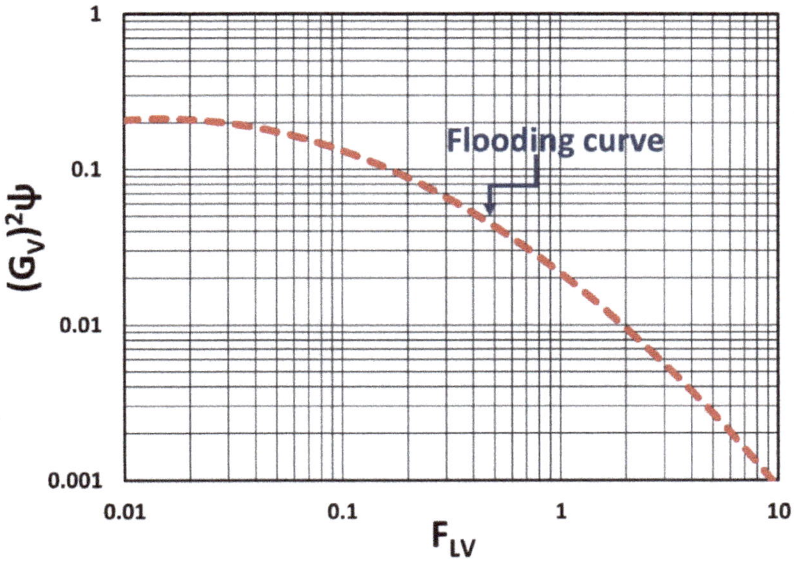

Figure 16.12 Flooding curve for random-packed columns, generated from Equation (16.43).

The flooding curve shown in Figure 16.12 is generated from the following equation proposed by Kessler and Wankat (D.P. Kessler, P.C. Wankat, Chemical Engineering, Vol. 95, 72-74, 1988):

$$log_{10}(G_V^2\,\psi) = -1.6678 - 1.085log_{10}(F_{LV}) - 0.29655[log_{10}(F_{LV})]^2 \quad (16.43)$$

We can interpret the flooding curve in different ways as discussed below:

- Given gas and liquid mass flow rates, fluid properties, type and size of packing, and column diameter, would there be flooding in the column under the given conditions? To answer this question, we need to calculate F_{LV} and $G_V^2\psi$. Then locate the point on Figure 16.12. If the point falls below the flooding curve, there occurs no flooding at the given conditions. If the point falls on or above the flooding curve, the column will be flooded at the given conditions.

- Given liquid mass flow rate, fluid properties, type and size of packing, and column diameter, what happens if the gas mass flow rate is increased? As the gas mass flow rate is increased, keeping other factors constant, we move in the direction indicated by the arrow in Figure 16.13. At some gas velocity, the flooding point is reached.

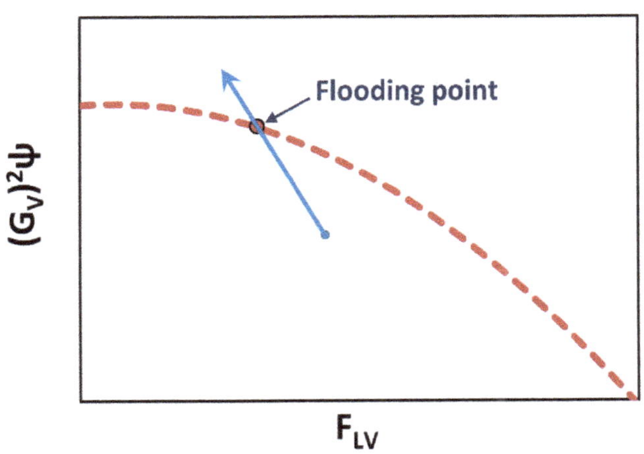

Figure 16.13 Flooding due to increase in gas velocity, keeping other factors constant.

- Both gas and liquid mass flow rates are increased in equal proportions, keeping other factors (fluid properties, type and size of packing, and column diameter) the same, would there be flooding in the column? As gas and liquid mass flow rates are increased in the same proportions, the flow parameter F_{LV} does not change. However, $G_V^2\psi$ increases due to an increase in gas velocity. Figure 16.14 shows the path followed. At some gas velocity (fixed F_{LV}) flooding occurs.

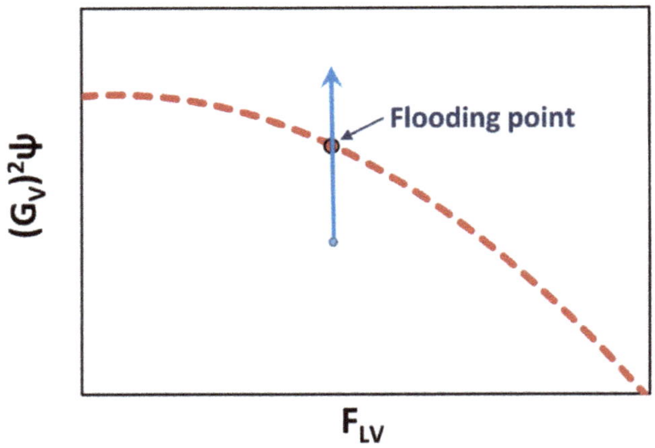

Figure 16.14 Flooding due to increase in gas and liquid mass flow rates in equal proportions, keeping other factors constant.

- Given gas mass flow rate, fluid properties, type and size of packing, and column diameter, what happens if the liquid mass flow rate is increased? As the liquid mass flow rate is increased, keeping other factors constant, we move in the direction indicated by the arrow in Figure 16.15. At some liquid velocity, the flooding point is reached.

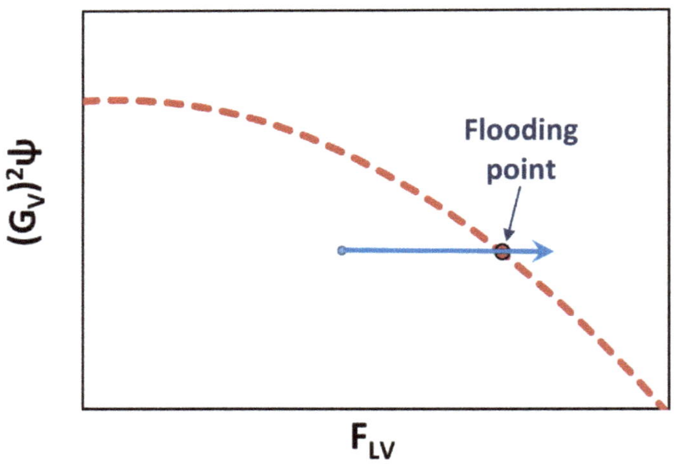

Figure 16.15 Flooding due to increase in liquid mass flow rate, keeping other factors constant.

- Given gas and liquid mass flow rates, fluid properties, type and size of packing, what happens if the column diameter is decreased? As the diameter of the column is decreased, keeping other factors constant, the flow parameter F_{LV} does not change. However, $G_V^2\psi$ increases due to an increase in the gas velocity. Figure 16.16 shows the path followed. At some column diameter (fixed F_{LV}) flooding occurs.

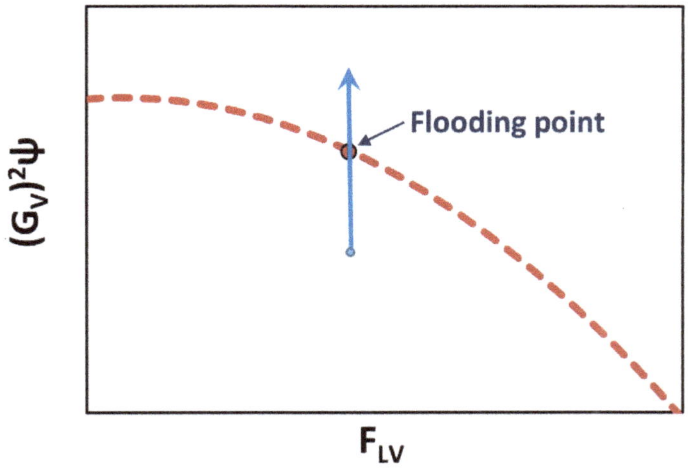

Figure 16.16 Flooding due to a decrease in column diameter, keeping gas and liquid mass flow rates and other factors constant.

Table 16.2 Packing factors for various packings

Packing -type	Nominal size (inches)	Packing factor, F_P (ft^{-1})
Raschig rings (metal, 1/16 in wall)	0.5	410
	1.0	137
	1.5	83
	2.0	57
Raschig rings (ceramic)	0.5	580
	1.0	155
	1.5	95
	2.0	65
Pall rings (metal)	0.625	70
	1.0	48
	1.5	28
	2.0	20
Pall rings (plastic)	0.625	97
	1.0	52
	1.5	32
	2.0	25
Intalox saddles (ceramic)	0.5	200
	1.0	98
	1.5	52
	2.0	40

Finally, it should be noted that the flooding curve shown in Figure 16.12 and the corresponding correlation, Equation (16.43), are valid for random packings. Flooding correlations have been proposed in the literature for structured packings as well. For example, Lockett (M.J. Lockett, Chemical Engineering Research and Design, Vol. 73, 379-384, 1995) proposed the following flooding correlation for high density structured packings:

$$U_V \rho_V^{0.5} = (\rho_L - \rho_V)^{0.5} \left[\frac{0.887 g^{0.25} (a)^{-0.25} \mu_r^{-0.03}}{1+\beta(\rho_V/\rho_L)^{0.25}(L/V)^{0.5}}\right]^2 \tag{16.44}$$

where β is given as:

$$\beta = 0.78exp(0.00058a) \tag{16.45}$$

In these equations, U_V is the flooding superficial gas velocity (m/s), g is the acceleration due to gravity ($9.8\ m/s^2$), $\mu_r = \mu_L(mPa.s)/0.978$, a is the specific area of packing in units of m^2/m^3, ρ is the density in kg/m^3.

Equation (16.44) can be modified as follows:

$$G_V^2\psi_S = \left[1 + \beta\sqrt{F_{LV}}\right]^{-4} \tag{16.46}$$

where ψ_S is given as:

$$\psi_S = \frac{1.615a\mu_r^{0.12}}{\rho_V(\rho_L-\rho_V)g} \tag{16.47}$$

Figure 16.17 shows the plots of flooding curves generated from Equation (16.46) for two different values of specific area of packing. With the increase in specific area (m^2/m^3) of packing, the flooding gas velocity decreases.

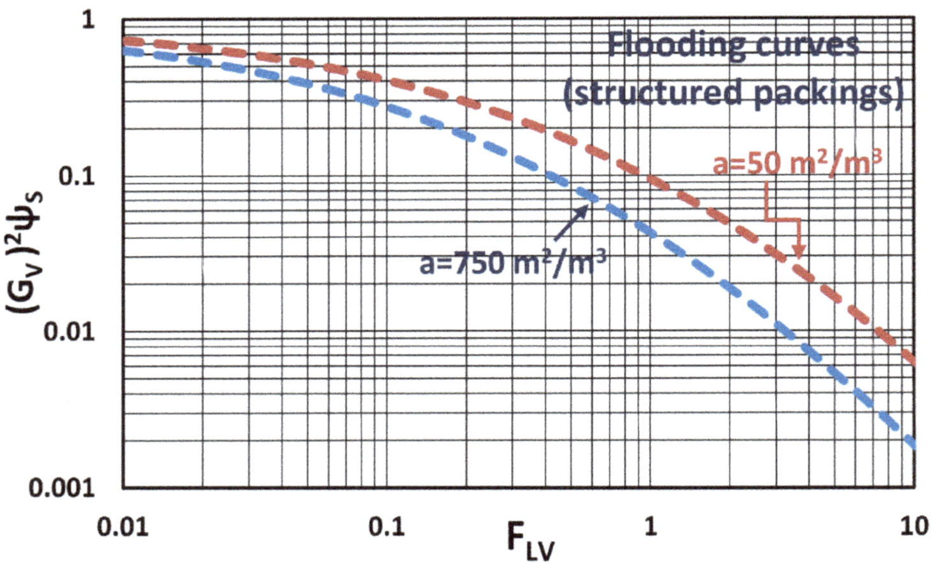

Figure 16.17 Flooding curves for structured packings, generated from Equation (16.46).

16.5.3 Estimation of Column Diameter using the Flooding Curve/Equation

The diameter of a packed column is selected such that the superficial mass velocity of gas G_V at which the column is operated is well below the flooding point. Thus:

$$(G_V)_{actual} = \alpha(G_V)_{flooding} \tag{16.48}$$

where $\alpha < 1$. For packed columns, the column is usually designed with α anywhere in the range: $0.50 < \alpha < 0.80$. More often than not, the packed columns operate at about 70% flooding ($\alpha = 0.70$). Once the value of α is selected, the column diameter is calculated from:

$$D = \sqrt{\frac{4\dot{m}_V}{\pi\alpha(G_V)_{flooding}}} \tag{16.49}$$

The flooding velocity of gas $(G_V)_{flooding}$ is determined either from flooding curves or flooding correlations, Equation (16.43) or (16.46).

NOTATION

a	surface area per unit volume of packing
A	absorption factor, defined in Equation (16.31)
D	column diameter
d_P	equivalent diameter of particles of a packing
F_{LV}	flow parameter, defined in Equation (16.42)
F_P	packing factor of a packing, see Table 16.2
g	acceleration due to gravity, 32.2 ft/s^2 or 9.8 m/s^2
g_c	conversion factor, 32.2 $lb_m.(ft/s^2)/lb_f$
G_L	superficial mass velocity of liquid
G_V	superficial mass velocity of gas or vapor
$(G_V)_{flooding}$	superficial mass velocity of gas or vapor corresponding to flooding in a column
h	height of a packed column
$HETP$	height equivalent to a theoretical plate
HTU	height of transfer unit
k_x	liquid phase mass transfer coefficient, see Equation (16.9)
k_y	gas phase mass transfer coefficient, see Equation (16.8)
K_x	overall mass transfer coefficient based on liquid phase
K_y	overall mass transfer coefficient based on gas phase, see Equation (16.11)
L	molar flow rate of liquid
m	Henry law constant of solute defined in Equation (16.12)
M_L	molecular weight of liquid phase
M_V	molecular weight of vapor or gas phase
$\dot{m}_L$	mass flow rate of liquid
$\dot{m}_V$	mass flow rate of vapor or gas
n	parameter in Equation (16.5)
NTU	number of transfer units
N_A	molar flux of solute
$N_{theoretical}$	total number of equilibrium or theoretical stages
S	cross-sectional area of a column
U_V	superficial flooding velocity of gas, see Equation (16.44)
V	molar flow rate of vapor or gas
x	mole fraction of solute in the liquid phase
x_a	mole fraction of solute in liquid stream entering the column at the top, see Figure 16.9
x_b	mole fraction of solute in liquid stream leaving the column at the bottom, see Figure 16.9
x_i	mole fraction of solute at the gas-liquid interface on the liquid side (see Figure 16.8) or mole fraction of solute in liquid stream leaving *i*th stage (see Figure 16.6)
x_{i-1}	mole fraction of solute in liquid stream entering *i*th stage (see Figure 16.6)
x_a^*	mole fraction of solute in the liquid phase in equilibrium with the gas stream leaving the column, see Figure 16.9
x_b^*	mole fraction of solute in the liquid phase in equilibrium with the gas stream entering the column, see Figure 16.9
y	mole fraction of solute in the gas phase

y_a mole fraction of solute in gas stream leaving the column at the top, see Figure 16.9
y_b mole fraction of solute in gas stream entering the column at the bottom, see Figure 16.9
y_i mole fraction of solute at the gas-liquid interface on the gas side (see Figure 16.8) or mole fraction of solute in gas stream leaving ith stage (see Figure 16.6)
y_{i+1} mole fraction of solute in gas stream entering ith stage (see Figure 16.6)
y^* mole fraction of solute in the gas phase if the gas phase was in equilibrium with the existing liquid phase
y_a^* mole fraction of solute in the gas phase in equilibrium with the liquid stream entering the column, see Figure 16.9
y_b^* mole fraction of solute in the gas phase in equilibrium with the liquid stream leaving the column, see Figure 16.9
z column height variable, see Figure 16.9

Greek Letters

β parameter defined in Equation (16.45)
μ_L liquid viscosity, in units of cP (Equation (16.41)) and mPa.s
μ_r relative viscosity given as $\mu_L(mPa.s)/0.978$, see Equation (16.44)
ρ density
ρ_L density of liquid
ρ_V density of gas or vapor
ρ_{water} density of water
σ surface tension
ϕ density ratio, ρ_{water}/ρ_L
ψ parameter defined in Equation (16.41)
ψ_S parameter defined in Equation (16.47)

TUTORIAL PROBLEMS WITH SOLUTION

Problem 1

A binary liquid mixture is to be distilled in a distillation column at 1 atm pressure. The feed is an equimolar of components. The more volatile component in the distillate product has a mole fraction of 0.98. The less volatile component in the bottoms product has a mole fraction of 0.95. The column is to be operated at a reflux ratio of 3. The number of theoretical (equilibrium) stages required is 10 (including the partial reboiler). The viscosity of feed liquid under column conditions is 0.5 cP. The average relative volatility (α) is 2.5.

Calculate the following:

a) overall efficiency using the O'Connell correlation and the actual number of trays if the column is a staged column.
b) the HETP and column height if the column is a packed tower consisting of 1.5-inch Raschig rings as packing particles. Use the rule of thumb, Equation (16.3), to estimate HETP.
c) the HETP and column height if the column is a packed tower consisting of 1T Intalox corrugated sheet structured packing. Use the rule of thumb, Equation (16.4), to estimate HETP.
d) the value of HETP for the structured packing of part (c) using the Strigle correlation, Equation (16.5). Take μ_L=0.5 cP and σ =5 mN/m.
e) the value of HETP for the structured packing of part (c) using the Lockett correlation, Equation (16.6). Take μ_L=0.5 cP, ρ_V =1.28 kg/m^3 and ρ_L =995 kg/m^3

Solution

a) O'Connell correlation (Equation (10.28), Chapter 10) gives the overall column efficiency as:

$$\eta_0 = 0.52782 - 0.27511\, log_{10}(\alpha\mu_L) + 0.044923\, [log_{10}(\alpha\mu_L)]^2$$

$$\eta_0 = 0.52782 - 0.27511\, log_{10}(2.5 \times 0.5) + 0.044923\, [log_{10}(2.5 \times 0.5)]^2 = 0.591$$

As $N_{ideal} = 9$, the actual number of trays required is $(9/0.591) = 15.23$ (excluding partial reboiler)

b) Using the rule of thumb, Equation (16.3):

$$HETP(m) = 0.018\, d_p(mm) = 0.018 \times 38.1 = 0.686\, m$$

$$h = (10 - 1) \times 0.686 = 6.17\, m$$

Note that $d_p = 38.1\, mm$ and that the partial reboiler is excluded from the number of given theoretical stages to calculate the tower height h.

c) Using the rule of thumb, Equation (16.4):

$$HETP(m) = \frac{100}{a} + 0.10 = \frac{100}{310} + 0.10 = 0.42\, m$$

$$h = (10 - 1) \times 0.42 = 3.78\, m$$

Note that $a = 310\ m^2/m^3$ for 1T Intalox corrugated sheet structured packing (see Table 16.1).

d) Strigle correlation, Equation (16.5), gives

$$ln(HETP) = n - 1.01 - 0.187 ln(\sigma) + 0.213 ln(\mu_L)$$

$$ln(HETP) = 0.7683 - 1.01 - 0.187 ln(5 \times 10^{-3}) + 0.213 ln(0.5 \times 10^{-3}) = -0.87$$

$$HETP = 0.419\, m$$

Note that $n = 0.7683$ is taken from Table 16.1

e) Lockett correlation, Equation (16.6), gives

$$HETP\ (m) = \frac{1.54 g^{0.5} (\rho_L - \rho_V)^{0.5} \mu_r^{-0.06}}{a[1 + 0.78(\rho_V/\, \rho_L)^{0.25} exp(0.00058 a)]^2}$$

$$HETP\ (m) = \frac{1.54(9.8)^{0.5}(995 - 1.28)^{0.5}(0.5/0.978)^{-0.06}}{310[1 + 0.78(1.28/\, 995)^{0.25} exp(0.00058 \times 310)]^2} = 0.37\, m$$

Note that $a = 310\ m^2/m^3$ for 1T Intalox corrugated sheet structured packing (see Table 16.1).

Problem 2

A packed tower is to be used to absorb species "A" from a gas stream using pure water as a solvent. The gas entering the tower contains 2.5 mol% A and 95 percent of it is to be absorbed. The tower is to operate at 25 °C and 1 atm, and the liquid and gas molar flow rates on a unit area basis are to be 150 kmol/h.ft^2 and 30 kmol/h.ft^2, respectively.

The equilibrium and overall mass transfer coefficient data are as follows:

$$y^* = 3x\ ;\ K_y a = 7.5\ kmol/h.ft^3$$

Determine the following:

a) the number of transfer units, NTUs
b) the height of the transfer unit, HTU
c) the height of the packed tower

You can assume that the gas and liquid molar flow rates are constant.

Solution

Refer to the schematic diagram of the packed tower shown below.

We first need to calculate x_b. From Equation (16.27):

$$y_a = y_b + (L/V)(x_a - x_b)$$

Upon rearrangement,

$$x_b = x_a + (y_b - y_a)/(L/V) = 0 + (0.025 - 0.00125)/(150/30) = 4.75 \times 10^{-4}$$

Note that y_a is 5% of y_b , that is, $y_a = 0.05 \times 0.025 = 0.00125$

Thus all the mole fractions at inlet and outlet of the tower are now known.

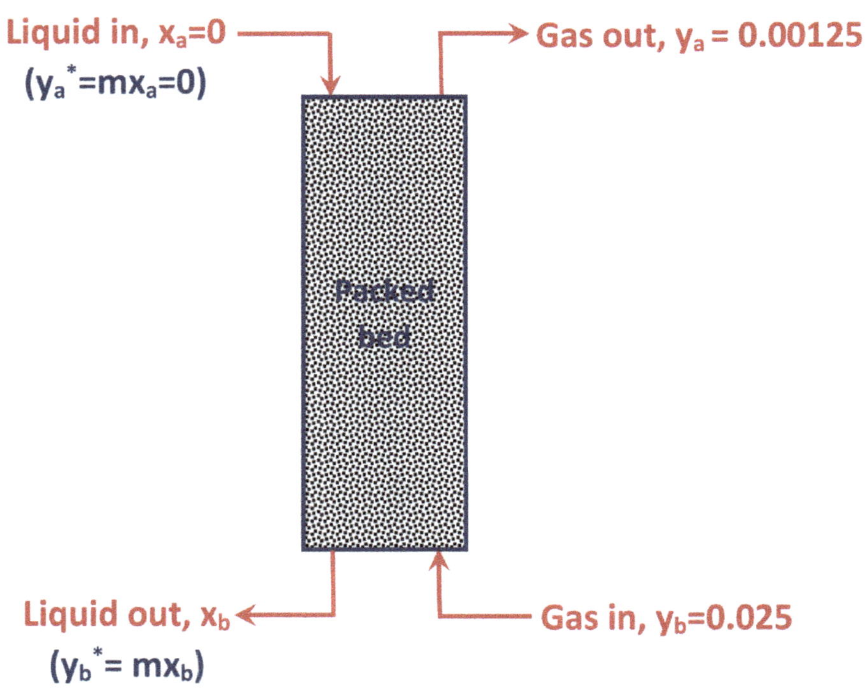

a) the number of transfer units, NTUs

For linear operating and equilibrium lines with different slopes, we can use Equation (16.28) to determine the NTUs:

$$NTU = \frac{y_b - y_a}{(y - y^*)_{log-mean}} = \frac{0.025 - 0.00125}{(y - y^*)_{log-mean}}$$

$$(y - y^*)_{log-mean} = \frac{(y_b - y_b^*) - (y_a - y_a^*)}{ln\left(\frac{y_b - y_b^*}{y_a - y_a^*}\right)} = \frac{(0.025 - 1.425 \times 10^{-3}) - (0.00125 - 0)}{ln\left(\frac{0.025 - 1.425 \times 10^{-3}}{0.00125 - 0}\right)}$$
$$= 7.6 \times 10^{-3}$$

$$NTU = \frac{0.025 - 0.00125}{7.6 \times 10^{-3}} = 3.125$$

Note that $y_a^* = 3x_a = 0$ and $y_b^* = 3x_b = 3(4.75 \times 10^{-4}) = 1.425 \times 10^{-3}$

b) the height of the transfer unit, HTU

From Equation (16.24):

$$HTU = \left(\frac{V/S}{K_y a}\right) = \frac{30}{7.5} = 4\ ft$$

c) the height of the packed tower

$$h = (HTU) \times (NTU) = 4 \times 3.125 = 12.5\ ft$$

Problem 3

A packed tower is to be used in the treatment of a gas stream that enters the tower at a volumetric flow rate of 9.81 ft^3/s. The tower is packed with 1-inch ceramic Raschig rings. The mass flow rates of liquid and gas streams are the same.

Estimate the diameter of the tower such that the gas mass velocity is 70% of the flooding limit. The properties of the fluids are:

$\rho_V = 0.062\ lb_m/ft^3$, $\rho_L = 62.4\ lb_m/ft^3$, $\mu_L = 1$ cP

For 1-inch ceramic Raschig rings, the packing factor $F_p = 155\ ft^{-1}$. Note that water is the liquid phase.

Solution

From Equation (16.42):

$$F_{LV} = \frac{\dot{m}_L}{\dot{m}_V}\sqrt{\frac{\rho_V}{\rho_L}} = 1 \times \sqrt{\frac{0.062}{62.4}} = 0.0315$$

Using Equation (16.43), we can estimate the flooding velocity:

$$log_{10}(G_V^2\ \psi) = -1.6678 - 1.085 log_{10}(F_{LV}) - 0.29655[log_{10}(F_{LV})]^2$$

$$log_{10}(G_V^2\,\psi) = -1.6678 - 1.085log_{10}(0.0315) - 0.29655[log_{10}(0.0315)]^2 = -0.684$$

$$G_V^2\,\psi = 0.207$$

From Equation (16.41):

$$\psi = \frac{F_p \phi \mu_L^{0.2}}{\rho_V \rho_L g_c} = \frac{155 \times 1 \times 1^{0.2}}{0.062 \times 62.4 \times 32.2} = 1.244$$

Thus, the flooding mass velocity G_V is:

$$G_{V,flooding} = \sqrt{\frac{0.207}{1.244}} = 0.408\,\frac{lb_m}{s.ft^2}$$

Using Equation (16.49),

$$D = \sqrt{\frac{4\dot{m}_V}{\pi\alpha(G_V)_{flooding}}} = \sqrt{\frac{4 \times 9.81 \times 0.062}{\pi \times 0.70 \times 0.408}} = 1.65\,ft$$

Problem 4

Ammonia is to be absorbed in a packed tower packed with 1T Intalox corrugated sheet structured packing using pure water as a solvent. The gas stream enters the tower at a mass flow rate of 600 kg/h. The mass flow rate of liquid is 1.75 times that of gas stream.

Estimate the diameter of the tower such that the gas mass velocity is 70% of the flooding limit.
The properties of the fluids are:

$\rho_V = 1$ kg /m^3 , $\rho_L = 1000$ kg /m^3 , $\mu_L = 0.001$ Pa.s

For 1T Intalox corrugated sheet structured packing, the surface area per unit volume is 310 m^2/m^3.

Solution

Refer to the schematic diagram of the packed tower given below.

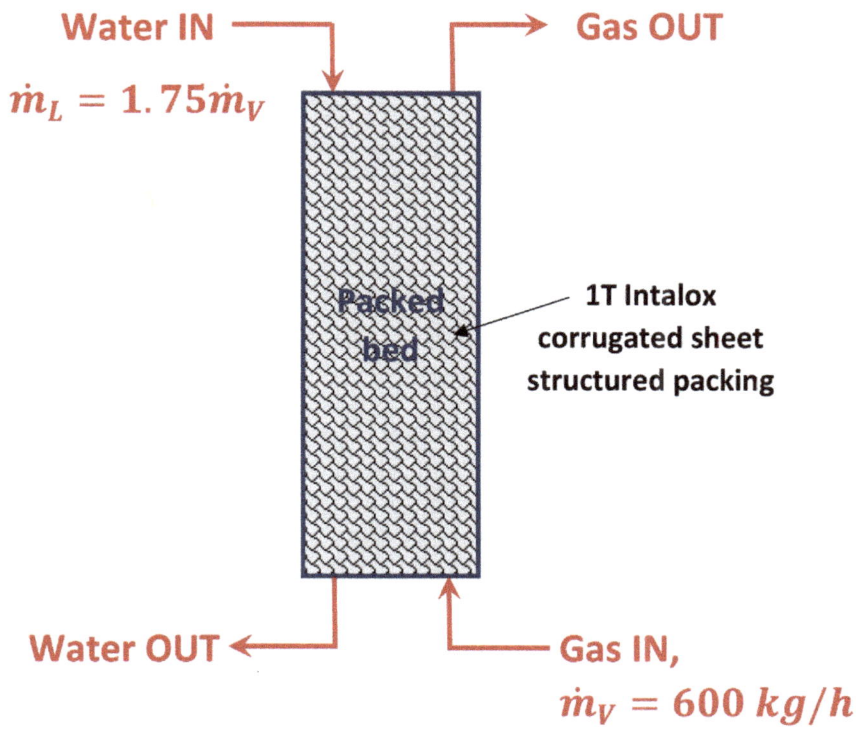

From Equation (16.42):

$$F_{LV} = \frac{\dot{m}_L}{\dot{m}_V}\sqrt{\frac{\rho_V}{\rho_L}} = 1.75 \times \sqrt{\frac{1}{1000}} = 0.055$$

Using Equation (16.46), we can estimate the flooding velocity:

$$G_V^2 \psi_S = \left[1 + \beta\sqrt{F_{LV}}\right]^{-4}$$

where β is given as:

$$\beta = 0.78exp(0.00058a) = 0.78\exp(0.00058 \times 310) = 0.934$$

Upon substitution of F_{LV} and β, we get:

$$G_V^2 \psi_S = \left[1 + (0.934)\sqrt{0.055}\right]^{-4} = 0.453$$

ψ_S is given by Equation (16.47) as:

$$\psi_S = \frac{1.615a\mu_r^{0.12}}{\rho_V(\rho_L - \rho_V)g} = \frac{1.615 \times 310 \times (1/0.978)^{0.12}}{1 \times (1000 - 1) \times 9.8} = 0.051$$

Thus,

$$G_{V,flooding} = \sqrt{\frac{0.453}{0.051}} = 2.98\ \frac{kg}{s.m^2}$$

Using Equation (16.49),

$$D = \sqrt{\frac{4\dot{m}_V}{\pi\alpha(G_V)_{flooding}}} = \sqrt{\frac{4 \times (600/3600)}{\pi \times 0.70 \times 2.98}} = 0.319\, m$$

Problem 5

Develop the equations for height of transfer unit (HTU) and number of transfer units (NTU) for liquid stripping in a packed column.

Solution

As described in the case of packed column gas absorption (see section 16.4.2), it is imagined here that the total height of the liquid stripping packed column can be divided into a number of layers of packing (referred to as number of transfer units, NTU) of equal height (referred to as height of transfer unit, HTU) such that the change in liquid concentration over the HTU is equal to the average driving force for mass transfer from liquid phase to gas phase over the whole column (see Figure 16.7).

The concept of *HTU* for liquid stripping is further illustrated schematically in the figure below. If at any point in the column, x is the solute concentration in the liquid phase and y is the solute concentration in the gas phase at the same point, then *HTU* is the height of the packing above this point where the liquid concentration is $x + (x - x^*)_{avg}$ where $(x - x^*)_{avg}$ is the average driving force for mass transfer of solute in the column from liquid phase to gas phase.

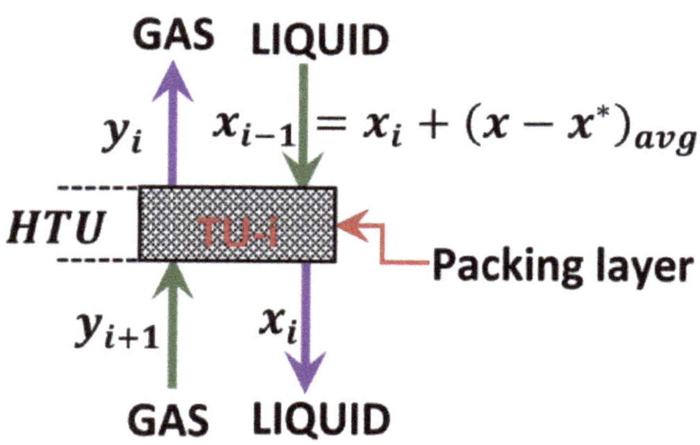

In a packed column, interphase mass transfer of solute occurs from one phase to another by diffusion. Figure given below shows the concentration profiles, in units of mole fraction, on both sides of the gas-liquid interface in liquid stripping operation. The molar flux of solute N_A (moles/s.m^2) across the liquid film on the left side of the gas-liquid interface can be expressed as:

$$N_A = k_x(x - x_i)$$

where k_x is the liquid-phase mass transfer coefficient (moles/s.m^2).

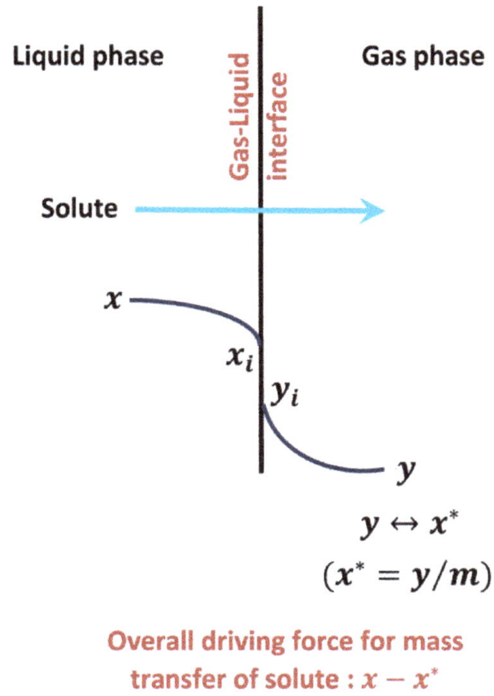

The molar flux of solute N_A (moles/s.m^2) across the gas film on the right side of the gas-liquid interface can be expressed as:

$$N_A = k_y(y_i - y)$$

where k_y is the gas-phase mass transfer coefficient (moles/s.m^2).

At steady state, there cannot be accumulation of solute at the interface and therefore,

$$N_A = k_x(x - x_i) = k_y(y_i - y)$$

In order to avoid the interface compositions, it is equally appropriate to express the molar flux of solute as:

$$N_A = K_x(x - x^*)$$

where K_x is the overall mass transfer coefficient based on liquid phase, $(x - x^*)$ is the overall driving force for mass transfer of solute from liquid phase to gas phase, and x^* is the mole fraction of solute in the liquid phase in equilibrium with the existing bulk gas phase. Note that x^* is a fictitious quantity to be calculated from the equilibrium relation (Henry's law):

$$y = mx^*$$

where m is the equilibrium constant. It can be readily shown that:

$$\frac{1}{K_x} = \frac{1}{k_x} + \frac{1}{mk_y}$$

$$\frac{1}{K_x a} = \frac{1}{k_x a} + \frac{1}{mk_y a}$$

We now consider material balance of solute over a thin section dz of the packed liquid stripping column shown in the figure below. Let the contact area of the packing per unit volume of packing be a. The interphase mass transfer rate of solute from liquid phase to gas phase over a section dz of the column is:

$$Rate\ of\ mass\ transfer = K_x a(x - x^*)Sdz$$

where S is the cross-sectional area of the column. The mass transfer rate of solute from liquid phase to gas phase over a section dz of the column is also given as:

$$Rate\ of\ mass\ transfer = Lx|_{z+dz} - Lx|_z$$

where L is the molar flow rate of liquid. Equating the two expressions, we get:

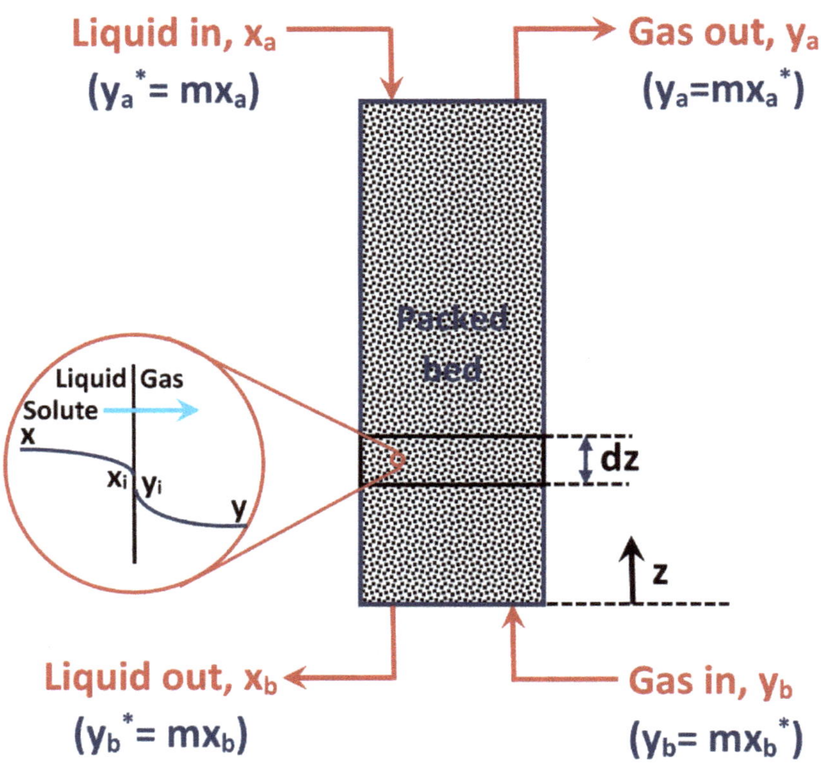

$$Lx|_{z+dz} - Lx|_z = K_x a(x - x^*)Sdz$$

Thus,

$$\frac{d(Lx)}{dz} = K_x a(x - x^*)S$$

Assuming the liquid flow rate L to be constant:

$$L\frac{dx}{dz} = K_x a(x - x^*)S$$

Upon rearrangement and integration,

$$h = \left(\frac{L/S}{K_x a}\right) \int_{x_b}^{x_a} \frac{dx}{x - x^*}$$

This equation could be recast as:

$$h = (HTU) \times (NTU)$$

where

$$HTU = \left(\frac{L/S}{K_x a}\right)$$

$$NTU = \int_{x_b}^{x_a} \frac{dx}{x - x^*}$$

The number of transfer units (*NTU*) depends on the degree of desired separation $(x_a - x_b)$ and the average driving force for mass transfer $(x - x^*)_{avg}$. It is simply the ratio of $(x_a - x_b)$ to $(x - x^*)_{avg}$. The height of the transfer unit (HTU) is a measure of the efficiency of mass transfer. More efficient the mass transfer, smaller is the value of HTU. A larger mass transfer coefficient K_x implies a faster rate of interphase mass transfer of solute and hence a lower HTU. Likewise, a larger contact area of the packing (per unit volume of the packing) implies a faster rate of interphase mass transfer of solute and hence a lower HTU. Given the overall mass transfer coefficient K_x , the packing contact area a per unit volume of the packing, the liquid molar flow rate L, and the column diameter D $(S = \pi D^2/4)$, *HTU* can be calculated.

The *NTU* can be calculated from the following equation:

$$NTU = \int_{x_b}^{x_a} \frac{dx}{x - x^*} = \frac{x_a - x_b}{(x - x^*)_{log-mean}}$$

where $(x - x^*)_{log-mean}$ is the log-mean driving force for mass transfer in the packed column, given as:

$$(x - x^*)_{log-mean} = \frac{(x_a - x_a^*) - (x_b - x_b^*)}{ln\left(\frac{x_a - x_a^*}{x_b - x_b^*}\right)}$$

Problem 6

A packed tower is used to remove a toxic chemical from wastewater. Pure air is used as the stripping gas. The wastewater entering the tower contains 5 mol% toxic chemical. It is desired to remove 98% of the toxic chemical. The equilibrium relation is $y = 0.85x$, where y and x are mole fractions of toxic chemical in air and water at equilibrium, respectively. The gas and liquid molar flow rates per unit cross-sectional area of the column are: 0.8 kmol/min.m^2 and 0.4 kmol/min.m^2, respectively. The mass transfer coefficients are:

$$k_x a = 6 \times 10^{-2}\ kmol/s.m^3\ ;\ k_y a = 4 \times 10^{-2}\ kmol/s.m^3$$

Determine the following:

a) the number of transfer units, NTUs
b) the height of the transfer unit, HTU
c) the height of the packed tower

You can assume that the gas and liquid molar flow rates are constant.

Solution

Refer to the diagram below

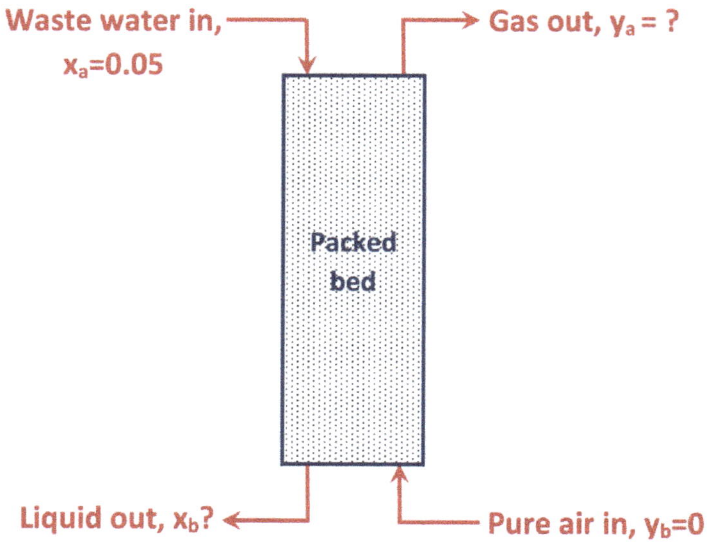

Let us first calculate the terminal compositions. As 98% of the toxic chemical is to be removed, $x_b = (1 - 0.98)x_a = 0.02 \times 0.05 = 0.001$. From the operating line equation,

$$y_a = y_b + (L/V)(x_a - x_b) = 0 + (0.4/0.8)(0.05 - 0.001) = 0.0245$$

a) the number of transfer units, NTUs

$$NTU = \int_{x_b}^{x_a} \frac{dx}{x - x^*} = \frac{x_a - x_b}{(x - x^*)_{log-mean}}$$

$$(x - x^*)_{log-mean} = \frac{(x_a - x_a^*) - (x_b - x_b^*)}{ln\left(\frac{x_a - x_a^*}{x_b - x_b^*}\right)} = \frac{(0.05 - 0.02882) - (0.001 - 0)}{ln\left(\frac{0.05 - 0.02882}{0.001 - 0}\right)} = 6.61 \times 10^{-3}$$

Note that: $x_a^* = y_a/m = 0.0245/0.85 = 0.02882$, $x_b^* = y_b/m = 0/0.85 = 0$. Thus, NTUs is

$$NTU = \frac{x_a - x_b}{(x - x^*)_{log-mean}} = \frac{0.05 - 0.001}{6.61 \times 10^{-3}} = 7.413$$

b) the height of the transfer unit, HTU

We first need to calculate the overall mass transfer coefficient $K_x a$ as:

$$\frac{1}{K_x a} = \frac{1}{k_x a} + \frac{1}{mk_y a} = \frac{1}{6 \times 10^{-2}} + \frac{1}{0.85 \times 4 \times 10^{-2}} = 46.07843$$

$$K_x a = 0.0217 \text{ kmol/s.m}^3$$

The HTU is given as:

$$HTU = \left(\frac{L/S}{K_x a}\right) = \frac{0.4}{0.0217 \times 60} = 0.3071 \text{ m}$$

c) the height of the packed tower

$$h = (HTU) \times (NTU) = 0.3071 \times 7.413 = 2.277 \text{ m}$$

SHORT REVIEW QUESTIONS

1. Packed towers are continuous contactors whereas staged towers are discontinuous contactors. true/false
2. Packed towers are preferred over staged towers when the tower diameter is small (< 2ft). true/false
3. Name three groups of materials used for packings in packed towers.
4. Explain the difference between random and structured packings.
5. Give examples of different shapes of particles used in random packings.
6. Explain the importance of liquid distribution in packed columns.
7. Describe three types of liquid distributors used in packed towers.
8. Define HETP
9. For random or structured packing, HETP is inversely related to the surface contact area of the packing per unit volume of packing. Higher the specific contact area, lower is the value of HETP and more efficient is the packing. true/false
10. A packing with high HETP value is more efficient than a packing with low value HETP. true/false
11. For random packing of discrete particles, HETP is directly proportional to the equivalent diameter of particles. true/false
12. Define HTU
13. HTU depends on the efficiency of mass transfer whereas NTU depends on the degree of desired separation. true/false
14. Write the expressions for molar flux of solute at the interface on both liquid and gas sides of the interface.
15. At steady state, the molar flux of solute at the interface on the gas side is equal to the molar flux on the liquid side of the interface. true/false
16. If the solute is highly soluble in the liquid phase, the resistance to mass transfer lies mainly in the gas phase. true/false
17. If the solute is sparingly soluble in the liquid phase, the resistance to mass transfer lies mainly in the liquid phase. true/false
18. When the absorption factor (A) is unity, the number of transfer units (NTU) is equal to the number of theoretical stages. true/false
19. When A=1, HETP=HTU. true/false
20. Describe flooding in packed columns.

21. The column diameter is selected on the basis of flooding in columns. The column diameter should be such that no flooding occurs in the column.
true/false
22. Flooding velocity of gas is a function liquid velocity, fluid properties, and packing characteristics.
true/false
23. Packed columns are often operated at about 70% flooding, that is the superficial mass velocity of gas is about 70% of the superficial mass velocity of gas where flooding occurs.
true/false

PRACTICE PROBLEMS

Problem 1

A packed tower is to be used to absorb SO_2 from air stream using pure water as a solvent. The gas entering the tower contains 2.0 mol% A and 85 percent of it is to be absorbed by water. The molar flow rates of gas and liquid streams are 0.05 kmol/s and 1.98 kmol/s, respectively. Under the tower conditions, the equilibrium relationship is as follows: $y = 35x$. The column height is 4 m and its diameter is 1.25 m.

Estimate the value of the mass transfer capacity coefficient $K_y a$. You can assume that the gas and liquid molar flow rates are constant.

Problem 2

A packed tower is to be used to absorb cyclohexane from airstream. A non-volatile oil is used to absorb cyclohexane. The air entering the tower contains 1.5 mol% cyclohexane and 90 percent of it is absorbed by oil. The oil entering the column is cyclohexane-free. The tower is operated at 30 °C and 1 atm. The gas and liquid molar flow rates on a unit area basis are 0.5 lb-mol/min.ft^2 and 0.17 lb-mol/min.ft^2, respectively. The equilibrium and mass transfer coefficient data are as follows:

$$y^* = 0.15x\ ;\ k_x a = 0.30\ lb - mol/h.ft^3\ ;\ k_y a = 14.0\ lb - mol/h.ft^3$$

Determine the following:

a) the number of transfer units, NTUs
b) the height of the transfer unit, HTU
c) the height of the packed tower

You can assume that the gas and liquid molar flow rates are constant.

Problem 3

A packed tower is to be designed to absorb ammonia from an air stream using pure water as a solvent. The gas enters the tower at a mass flow rate of 1500 lb_m/h. The mass flow rates of liquid is 3000 lb_m/h. The tower is packed with 1.5-inch metal Pall rings.

Estimate the diameter of the tower such that the gas mass velocity is 70% of the flooding limit.

The properties of the fluids are:

$$\rho_V = 0.062\ lb_m/ft^3\ ,\ \rho_L = 62.4\ lb_m/ft^3\ ,\ \mu_L = 1\ cP$$

For 1.5-inch metal Pall rings, the packing factor $F_p = 32\ ft^{-1}$.

Problem 4

Show that:

$$\frac{1}{K_y} = \frac{1}{k_y} + \frac{m}{k_x}$$

If m is small, which phase (gas or liquid) is controlling with respect to mass transfer? The controlling phase is one which offers the greatest resistance to mass transfer.

Problem 5

Show that:

$$\frac{1}{K_x} = \frac{1}{k_x} + \frac{1}{mk_y}$$

If m is large, which phase (gas or liquid) is controlling with respect to mass transfer? The controlling phase is one which offers the greatest resistance to mass transfer.

Problem 6

Show that for a packed gas absorption column,

$$NTU = \frac{y_b - y_a}{(y - y^*)_{log-mean}}$$

where

$$(y - y^*)_{log-mean} = \frac{(y_b - y_b^*) - (y_a - y_a^*)}{ln\left(\frac{y_b - y_b^*}{y_a - y_a^*}\right)}$$

What are the restrictions on this expression of NTU?

Problem 7

Show that for a liquid stripping packed column,

$$NTU = \frac{x_a - x_b}{(x - x^*)_{log-mean}}$$

where

$$(x - x^*)_{log-mean} = \frac{(x_a - x_a^*) - (x_b - x_b^*)}{ln\left(\frac{x_a - x_a^*}{x_b - x_b^*}\right)}$$

What are the restrictions on this expression of NTU?

Problem 8

A packed tower is used to remove a chemical from wastewater using pure air as the stripping gas. The wastewater entering the tower has a chemical mole fraction of 0.05. The chemical mole fraction in the exit water stream is 0.001. The equilibrium relation is $y = 1.5x$, where y and x are mole fractions of chemical in air and water at equilibrium, respectively. The gas and liquid molar flow rates per unit cross-sectional area of the column are: 0.9 kmol/min.m^2 and 0.6 kmol/min.m^2, respectively. The overall mass transfer coefficient based on the liquid phase is:

$$K_x a = 0.04\ kmol/s.m^3$$

Determine the following:

a) the number of transfer units, NTUs
b) the height of the transfer unit, HTU
c) the height of the packed tower

You can assume that the gas and liquid molar flow rates are constant.

Index

A

absorption factor, 50-51, 54-55, 57, 59-60, 64, 66-68, 79, 91, 95, 99, 102, 104, 109, 111-113, 119, 121, 506, 513, 525
absorption process, 5-6, 29, 45-51, 59-60, 63, 66-67, 72, 87-89, 91, 93-95, 110, 114, 184, 506
activity coefficient, 34, 143, 155
adjacent keys, 388-389, 421
Alexander Fleming, 14
antibiotic, 14, 458, 462
Antoine equation, 31, 38, 41-42, 140, 147-149, 159, 164, 334, 349-350
axial or longitudinal stress, 310
azeotrope, 22, 24, 26, 36-37, 39, 142, 163, 264-269, 426

B

batch distillation, 135, 363-365, 367-371, 373-375, 377-379, 381, 383, 385
batch distillation with reflux, 363, 368-371, 374, 379, 385
batch steam distillation, 363, 372, 374
bending moment, 315-318, 325
benzene-toluene system, 158, 205-206, 334, 336-337
binary distillation, 4-5, 219, 329, 387-388, 501
binary flash distillation, 183-184
boilup ratio, 232, 237, 246, 270, 280, 283
bubble point curve, 21-22, 154, 330, 333
bubble point pressure, 135, 143-145, 147, 160-161, 191, 205
bubble point temperature, 135, 144-145, 153, 164-165, 185-187, 198, 200, 205, 273, 329, 333, 335, 349-350, 373
bubble point temperature of feed, 187, 200, 205
bubble-cap trays, 302-303
BUBLT calculation, 147-148, 150-151, 165
BUBP calculation, 147

C

capacity factor, 308-309, 318
carbon dioxide capture and storage, 13

cascading of stages, 210
chemical absorption, 45
chemical design, 310
chemical potential, 18-19, 30, 42, 46, 55-57, 72, 79
circumferential or hoop stress, 310
Clapeyron equation, 139-140, 155-156, 163
Class 1 separation, 389-390
Class 2 separation, 389-390, 392, 421
Clausius-Clapeyron equation, 139-140, 155-156, 163
column diameter, 291, 300-301, 304, 306, 309, 324-325, 327, 495, 499, 504, 507, 509-513, 523, 526
column sequencing, 404
completely immiscible, 7, 53, 77, 426, 429-431, 461
composite trapezoid rule, 367
compressive stresses, 313-314, 326
concentrated systems, 45, 51-55, 68, 71, 77-79, 89, 426
condenser heat duty, 172, 176, 223, 339, 355
conjugate curve, 425, 447-448
constant boiling point, 264
continuous contacting devices, 495
continuous contactors, 495, 525
control volume, 167-172, 176-178, 189, 197, 221, 223, 256-257, 259-261, 269
corrugated sheets, 497
counter-current cascade, 17-18, 87, 89-90, 115, 398-399, 436
counter-current flow arrangement, 211, 213
counter-current multi-stage contacting, 17
counter-current multi-stage solid-liquid leaching, 467, 469-470
counter-current variable-temperature distillation cascade, 214
cylindrical vessel, 310-312, 325, 327

D

Dalton law, 19
dead load, 313
dehydration of natural gas, 11
DePriester charts, 141, 149, 194
derivation of Kremser equations, 96, 119
dew point curve, 21-22, 154, 330, 333
dew point of feed, 188, 245, 270
dew point pressure, 135, 143-144, 147, 150, 160-161, 205
dew point temperature, 135-136, 144, 153, 159, 185, 187, 197, 200, 205, 329, 333, 335, 350
DEWP calculation, 147, 150, 165
DEWT calculation, 148-149, 151, 159, 165
diameter of packed columns, 507
differential distillation, 363
dilute systems, 47, 52, 73, 77, 89-90, 92, 116, 436-437
direction of mass transfer, 1, 5-10, 45, 71, 425, 463, 493
discontinuous contactors, 495, 525
distillation cascade, 210-214, 283
distillation column with direct steam injection, 207, 262-264
distillation column with multiple feeds, 256
distillation column with partial reboiler, 250
distillation column with total reboiler, 251
distillation column with two feeds, 257-258
distributed, 388-389, 392, 421, 497
distribution coefficient, 19, 29, 50, 55, 76, 79, 91, 99, 121, 140, 155, 197, 407, 432, 437, 448

E

enriching column distillation, 252-253
enthalpy of fluids, 173
enthalpy of liquid solutions, 333
enthalpy of multicomponent mixtures, 175
enthalpy of vapor mixtures, 333
enthalpy-composition phase diagram, 329-331, 334
equilateral triangle, 426-428, 430
equilateral triangle phase diagram, 426-428, 430
equilibrium y-x curve, 21, 24, 26, 153, 228, 295
estimation of NTU, 504
external balances, 221
external energy balance, 172
external material balances, 259, 440, 468
extraction factor, 432-433, 438, 448, 459-460

F

feed line, 226-228, 235, 240, 246, 253, 257, 262, 272, 279, 284, 295, 345-346, 359
feed pinch, 207, 234, 284, 295, 325
feed pinch point, 234, 295, 325
feed quality factor, 245
Fenske equation, 291, 294, 296, 299, 322, 325, 387, 391, 412, 414, 418, 421
fermentation, 14, 426, 458-459, 462
first law of thermodynamics, 167
First Underwood equation, 395-396, 414, 417-418, 421, 423
flash distillation, 135, 152, 167, 183-187, 189, 191, 193-195, 197-199, 201, 203, 205-208
flash drum, 183-188, 190-192, 194, 197-198
flash temperature, 152, 186, 195
flooding, 291, 301, 305-309, 319, 324-325, 327, 495, 507-513, 517-519, 525-526
flooding curve, 495, 508-509, 511-512
flooding phenomenon, 507
flooding velocity, 291, 305-309, 324-325, 327, 508, 513, 517, 519, 526
flow parameter, 308-309, 318, 324, 510, 513
fractional recovery, 391-392, 407, 409, 412-414, 421-423

G

gas-phase mass transfer coefficient, 501, 521
general balance equation, 167
Gilliland correlation, 296, 396
Gilliland plot, 296
Gilliland-Eduljee equation, 296, 396
graphical solution, 49, 55, 58-60, 65, 68, 75, 79, 83-84, 185, 198, 432, 435, 443, 467, 471
greenhouse gases, 12

H

heat of mixing, 175
heavy key, 388-389, 395, 404-407, 414, 417, 421
heavy non-key component, 388, 407
height of column, 300, 318
height of packed columns, 498
HETP method, 495, 498
heuristics, 387, 406, 421
HTU-NTU method, 495, 500

I

ideal gas mixture, 19, 21, 56, 80, 140, 175-176, 333, 359
ideal liquid solution, 19, 26, 34, 42, 56, 80, 140-143, 153, 175-176
ideal solid-liquid leaching, 463
ideal solutions, 1, 26-27, 175, 180
ideal vapor mixture, 34, 42, 140, 142
internal balances, 223
interphase mass transfer, 501-504, 520, 522-523
invariant zones, 389
iterative procedure, 147-152

K

K-value, 19
key components, 387-392, 400, 413-414, 417, 421
Kirkbride equation, 291, 320, 323, 326, 397, 416-417, 420-421
Kremser equations, 87, 95-97, 115, 118-120, 122, 438-439, 461, 463

L

latent heat of vaporization, 139, 155, 163, 174, 177, 192, 198, 219, 223, 233, 245, 271, 273, 329-330, 334, 339, 350-351
law of conservation of energy, 167
law of conservation of mass, 167
leaching, 1, 3, 9-11, 13, 17-18, 39, 463-477, 479-481, 483-485, 487, 489-494
light key, 388-389, 404-407, 414, 417, 421
limits of operation, 186
liquid distribution, 495, 497, 525
liquid distributors, 495, 497-498, 525
liquid reflux, 172, 218, 247-248, 283, 383
liquid stripping towers, 117
liquid-phase mass transfer coefficient, 502, 520
loading point, 508
lower difference point, 342-343, 346-347, 350-351, 359

M

material of column construction, 310
maximum allowable stress, 311-312, 320
maximum boiling azeotrope, 26
maximum L/V ratio, 121
maximum-boiling azeotrope, 26, 264
McCabe-Thiele graphical method, 94, 101-102, 105, 107, 112-114, 118, 132, 207, 321, 329, 337, 339, 463, 474, 477, 490, 499
mechanical design, 291, 310, 317
MESH equations, 387, 397-400, 421
MESH equations for a single stage, 397
MESH equations for the whole column, 398
minimum boiling azeotrope, 26, 265-267
minimum gas flow rate, 115, 120
minimum L/V ratio, 98-99
minimum L/V requirement, 98
minimum liquid flow rate, 87, 98
minimum number of stages, 291, 318, 329, 391
minimum number of trays, 238
minimum shell thickness, 311-313, 317
minimum-boiling azeotrope, 26
modified Rayleigh equation, 369, 371, 379

molar enthalpy, 171, 175-176, 180, 189, 191, 193, 197-198, 221-222, 269-270, 349-350, 398, 407
molar flux, 501-502, 513, 520-521, 525
mole ratios, 52-53, 61-65, 68-69, 77, 84, 86, 89, 105-106, 123-124
multi-component flash distillation, 194
multi-stage absorption, 87-88, 90-94, 102, 105, 113-115, 495
multi-stage absorption towers, 93
multi-stage contacting, 16-17, 87-88
multi-stage distillation column, 207, 215-219
multi-stage gas absorption, 87, 92, 95, 115-117, 119
multi-stage liquid stripping, 115-121, 123, 125, 127, 129, 131, 133
multi-stage liquid-liquid extraction, 425, 435, 437, 439, 461
multi-stage solid-liquid leaching, 467, 469-472, 475-477, 493
multi-stage stripping column, 115-116, 131
multi-tray column, 16
multicomponent distillation, 219, 387-393, 395-397, 399, 401, 403, 405, 407, 409, 411, 413, 415, 417, 419, 421, 423
multicomponent mixtures, 175, 291, 387-388, 400
multiple feeds, 207, 256
Murphree tray efficiency, 291, 297, 319, 325

N

negative azeotrope, 264
negative deviation, 22, 24-26, 39-40
Newton-Raphson method, 195-197, 202, 204, 395, 415, 419
non-distillation processes, 3
non-distributed, 388-389, 392
non-ideal liquid solution, 26, 34, 142-143
non-ideal liquid solutions, 21-22
non-key component, 388, 407
non-optimum feed stage location, 232

O

O'Connell correlation, 299, 323, 326, 514-515
operating line of enriching section, 256
operating line of the stripping section, 257
optimum feed stage, 207, 226, 231-233, 347, 422
optimum feed stage location, 207, 232-233
optimum reflux, 207, 284
overall efficiency, 299-300, 319, 323, 326, 514
overall mass transfer coefficient, 502, 513, 516, 521, 523-524, 527
overflow leachate, 466-467, 470, 478

P

P-V-T phase diagram, 136-139
P-x-y phase diagram, 20, 22-25, 145, 153-154
packed columns, 495, 497-499, 501, 503, 505-509, 511-513, 515, 517, 519, 521, 523, 525-527
packed towers, 495, 500, 525
packing, 495-500, 503-504, 508-515, 517-518, 520, 522-523, 525-526
Pall rings, 496, 511, 526
partial condensation, 208-211, 213-214, 216, 283
partial pressure, 19-20, 26, 28-29, 33-34, 41-42, 50, 55, 68, 76, 79, 91, 99, 140-141, 372, 374, 383
partial vaporization, 152, 208-211, 213-214,

216, 283
partially miscible, 3, 7, 425-429, 433, 435, 439-440, 443, 461
penicillin, 14-16, 426, 458-461
phase equilibrium, 1, 18, 39, 42, 46, 48-49, 71-72, 75, 92, 96, 115-116, 139-140, 393, 432, 434, 442, 464, 466, 470, 495
physical absorption, 45
pinch condition, 87, 98, 113, 115, 120, 131-132, 234, 236, 348-349, 360, 389
pinch point, 98-99, 120-122, 234-235, 295, 325, 389-390, 421
pinch-point zones, 389, 392
plait point, 425, 427, 461-462
plate spacing, 300
Ponchon-Savarit method, 329-330, 337, 348-349, 352, 356, 358-362
positive azeotrope, 264
positive deviation, 22-24, 39-40
pressure-swing distillation, 207, 264, 270
pressure-swing distillation of azeotropes, 207, 264
principal stresses, 317
Production of penicillin, 14-15
pseudo-binary method, 390, 421
pseudo-equilibrium, 297-298
pump shaft work, 190

Q

q-line, 228, 236, 259, 272, 275, 279-280, 295, 320

R

Rachford-Rice flash equation, 183, 195, 205
raffinate, 7-8, 17, 426-428, 431-434, 436-437, 439, 441-442, 448-450, 452-454, 456, 458, 461-462
random packing, 496, 525
Raoult's law, 1, 19-22, 24, 26-28, 30-32, 34, 40, 135, 140, 147, 153, 158-160, 164, 202-203, 206, 285, 287, 320, 334
Raschig rings, 496, 511, 514, 517
rate-based mass exchange, 495
Rayleigh equation, 363, 365, 367-369, 371, 376, 378-379, 383-385
reboiler heat duties, 171-172, 352, 354, 360
reboiler heat duty, 172, 176, 223, 246, 355
rectifying section, 207, 217, 220, 226, 370, 392
reference state, 167, 173, 176-177, 180, 182, 191-193, 197
reference state of a material, 173, 191
reference states, 191-192, 333
Riedel equation, 174
right-angled triangle, 426, 429-430, 434, 442, 450, 462
right-angled triangle phase diagram, 429-430, 434, 442

S

Saddles, 496, 511
saturation condition, 136, 143
scrubbing, 45
Second Underwood equation, 395-396, 415-416, 419-420
separation of crude oil, 3
separation of multicomponent mixtures, 388, 400

sharp separation, 388
shell thickness, 291, 310-314, 317, 319, 325
side stream, 258-262, 270-271, 286, 408
side stream q-line, 259
side-stream draw-offs, 258
sieve trays, 122, 283, 291, 301, 303, 308
single-stage distillation processes, 207
single-stage gas absorption, 45-47, 49, 51, 53, 55, 57, 59, 61, 63, 65, 67, 69, 87
single-stage liquid stripping, 71-75, 77, 79, 81, 83, 85
single-stage liquid-liquid extraction, 8-9, 431, 461
single-stage solid-liquid leaching, 11, 463, 465, 467
Smoker equation, 231, 290
solid-liquid extraction, 9, 463
solvent-diluent pair, 425, 429, 431-433, 435-440, 443, 461
split keys, 388-389, 396, 421
Stage-by-stage calculations, 152
stripping column distillation, 254-255
stripping factor, 77-79, 81-85, 116, 118, 121, 124-125, 128, 130-131
structured packing, 496-497, 499-500, 514-515, 518, 525
subcooled liquid, 21, 138, 143-146, 152, 173, 191, 209, 240-242, 245-246, 254, 269, 273, 283, 330, 333
superheated vapor, 21, 136, 143-144, 146, 152, 173-174, 180, 182, 209-210, 240-241, 244-245, 272, 283, 330

T

T-x-y phase diagram, 145-146, 153-154, 217
tangent pinch, 207, 234-235, 284, 295, 325
tangent pinch point, 234-235, 295, 325
tensile stresses, 310, 314, 316, 325
thermal conditions of feed, 240
tie-lines, 330-331, 347-348, 352, 429, 447-448, 465, 493
total condenser, 228, 247-248, 284-285, 339, 351, 356, 360-362, 379, 385, 409, 411, 413, 417, 422-423
total longitudinal stress, 317
total number of stages, 99, 121, 221, 230, 280, 329, 346, 399, 416, 420, 478
total reboiler, 250-252, 284
trapezoid rule, 365-367
triangular phase diagram, 425-426, 429, 439, 446-447, 450, 462
Trouton rule, 163, 174
Type I system, 428
Type II system, 428

U

underflow, 464-468, 470, 474, 476, 478-481, 485, 488-490, 492-494
underflow slurry, 464-468, 470, 474, 478-480, 493-494
Underwood equation, 291, 295-296, 322, 395-396, 414-421, 423
upper difference point, 338-340, 346-349, 351, 353, 359

V

valve trays, 291, 303, 308
variable-pressure distillation cascade, 212
variable-temperature distillation cascade, 213-214
vegetable oil extraction, 13-14

W

Watson equation, 174
weeping phenomenon, 303, 305, 325
wind load, 291, 310, 314-318, 325

Y

y-x equilibrium curve, 21-22, 24, 37, 135, 153-154, 185, 228

www.ingramcontent.com/pod-product-compliance
Lightning Source LLC
Chambersburg PA
CBHW041302150926
51501CB00053B/1674
9781922617316